REGENERATIVE MEDICINE AND TISSUE ENGINEERING - CELLS AND BIOMATERIALS

Edited by **Daniel Eberli**

Regenerative Medicine and Tissue Engineering - Cells and Biomaterials
Edited by Daniel Eberli

Published by InTech
Janeza Trdine 9, 51000 Rijeka, Croatia

Publishing Process Manager Romina Krebel

Technical Editor Teodora Smiljanic

Cover Designer InTech Design Team

Additional hard copies can be obtained from orders@intechopen.com

Regenerative Medicine and Tissue Engineering - Cells and Biomaterials,
Edited by Daniel Eberli
p. cm.
ISBN 978-953-307-663-8

For EU product safety concerns:
IN TECH d.o.o., Prolaz Marije Krucifikse Kozulić 3, 51000 Rijeka, Croatia,
info@intechopen.com or visit our website at intechopen.com.

Meet the editor

Daniel Eberli MD. Ph.D. is a scientific physician working in the translational field of urologic tissue engineering. He has a medical degree from the Medical School in Zurich, Switzerland, and a Ph.D. in Molecular Medicine from Wake Forest University, Winston Salem, NC. He currently has a faculty position at the Department of Urology at the University Hospital Zurich, where he devotes half of his time to patient care. He is a lecturer at the Medical School of Zurich and the Swiss Federal Institute of Technology. Together with his research team, he is working on novel biomaterials for bladder reconstruction, improving autonomic innervation, cellular treatment of incontinence and tracking of stem cells.

Contents

Preface

Tissue Engineering is the first medical therapy where engineered tissues could potentially become fully integrated within the patient, thus offering a permanent cure for many diseases not curable today. The National Science Foundation (NSF) defined Tissue Engineering in 1988 as the "application of the principles and methods of engineering and life sciences toward fundamental understanding of structure-function relationships in normal and pathological mammalian tissues and the development of biological substitutes to restore, maintain or improve tissue function (Shalak and Fox, 1988). Two decades later, Tissue Engineering/Regenerative Medicine is still a growing and exciting field of research applying knowledge of biology, physiology and cell culture techniques to offer new treatment options for patients in need of replacement tissues.

The two prerequisites for the successful engineering of an organ are suitable cells and a biomaterial or extra celluar matrix component. A large variety of cells has been proposed for the use in tissue engineering, including pluripotent embryonic stem cells (ESC) with all their ethical controversies, adult stem cells found in most tissues, and committed precursor cells. While the plasticity of ESC offers the potential to grow an entire organ from a single cell source, the clear differentiation of these cells remains challenging. Currently, adult cells seem to have certain advantages regarding rapid clinical translation. Most biomaterials used in Tissue Engineering are based on acellular matrices or polyglycolic acid. Both materials must provide tissue support until the cells produce their own extracellular matrix. Ideally, they degrade thereafter without any toxic byproducts. Over the last years we started to understand the influence of the biomechanical environment allowing these cell-biomaterial composites to unfold their full functional potential. However, many fundamental questions regarding cells and biomaterials remain unanswered.

This book will be of interest to anyone interested in the application of Tissue Engineering. It offers a wide range of topics, including the use of stem cells and adult stem cells, their applications and the development of a tailored biomaterial, highlighting the importance of cell-biomaterial interaction. It offers insights into a wide variety of cells and biomaterials, explaining the groundwork required to open the avenue to the next generation biotechnology, which is Tissue Engineering.

Finally, I would like to express my appreciation to all authors who have contributed to this book.

Daniel Eberli
University Zurich
Switzerland

Part 1

Regenerative Medicine

1

Fundamental Technological Developments Required for Increased Availability of Tissue Engineering

Hideaki Kagami[1], Hideki Agata[1], Ryuji Kato[2], Fumiko Matsuoka[2] and Arinobu Tojo[1]
[1]*Tissue Engineering Research Group, Division of Molecular Therapy, The Institute of Medical Science, The University of Tokyo,*
[2]*Department of Biotechnology, Graduate School of Engineering, Nagoya University*
Japan

1. Introduction

Since the initial excitement surrounding successful clinical studies of skin tissue engineering more than 20 years ago (Gallico *et al.*, 1984), steady progress has been made in enhancing the capabilities of tissue engineering and regenerative medicine. Tissue engineering generally depends upon the use of cultured cells. Since living cells do not fall into any of the existing medical product categories, this has created a great challenge for both regulatory agencies and commercial entities. Although various treatment strategies have been developed, the fundamental technologies and infrastructure to support their widespread adoption are still limited.

In this chapter, attention was focused on fundamental technology development. Three major areas, i.e., introduction of serum-free culture media, development of cell storage technologies and methodological development for quality assurance of the products, are discussed with special reference to future development of tissue engineering.

2. Feasibility of tissue engineering using human bone marrow stromal cells cultivated in serum-free conditions

Tissue engineering is an interdisciplinary approach to regenerate tissue through integration of cell biology and biomaterial/biomedical sciences. The concept of tissue engineering is to regenerate target tissue by mimicking the developmental or regenerative process of that tissue. Thus, it can be considered an ideal therapeutic option for treating various tissue defects. Tissue engineering of skin, cartilage, and bone has already been shown both feasible and effective in several clinical studies, and its efficacy has attracted significant attention from both patients and doctors. However, there are several fundamental technologies which need to be improved before widespread practical use of tissue engineering in hospitals or clinics. In this chapter, the current status of cell culture media used for clinical tissue engineering and the need for the development of safe and reliable serum-free cell culture media will be discussed with special reference to bone tissue engineering.

Patients who lose healthy bone tissue as a result of inflammation or trauma need bone regenerative/reconstructive surgery in order to recover the function of the lost bone. To regenerate the lost bone tissue, autologous bone grafting is the current gold standard, though this technique is a great burden for patients because transplantable autologous bone must be harvested from a healthy site, which causes donor site morbidity and pain. Artificial bone substitutes have been developed as alternatives to autologous bone, though bone regeneration with them is inefficient because they lack osteo-inductive properties. Accordingly, tissue engineering of bone (bone tissue engineering) has attracted significant interest because it is considered less invasive than autologous bone grafting and more efficient than artificial bone substitutes. In fact, cell-based bone tissue engineering which utilizes cells, scaffolds, and bioactive molecules has been shown even more effective than artificial bone substitute in both basic and clinical studies.

For cell-based bone tissue engineering, various tissues derived cells are utilized since osteogenic cells can be harvested from bone marrow, periosteum, and adipose tissue, though recent studies indicate that bone marrow stromal cells (BMSCs, bone marrow derived multipotent mesenchymal stromal cells, or mesenchymal stem cells) are the most reliable cell source because of their superior osteogenic ability (Hayashi *et al.*, 2008). However, it is difficult to obtain adequate numbers of transplantable BMSCs from bone marrow aspirates, as they are rare in the bone marrow (less than 0.01% of marrow cells) (Montzka *et al.*, 2010). Therefore, *ex vivo* expansion of BMSCs is required to obtain a sufficient number of transplantable cells. Since BMSCs require several kinds of supportive factors for their growth, it is standard practice to use fetal bovine serum (FBS), while autologous human serum (HS) and pooled allogeneic HS have also been used. It has been suggested that FBS may not be favorable for clinical applications due to the possible risk of contamination (prions, viruses, zoonosis) or immunological reactions against xenogeneic serum antigens (Agata *et al.*, 2009). Although serious secondary effects of transplanted cells that were cultured in the presence of FBS have not been reported to date, a previous clinical study that utilized BMSCs cultivated in FBS-supplemented media for the treatment of osteogenesis imperfecta showed a 150-fold increase in antibody titer against FBS in the sera of one patient who received BMSCs infusions (Horwitz *et al.*, 2002). Theoretically, use of autologous HS could eliminate the risks of disease transmissions and immune reactions. However, it is not always possible to obtain a sufficient amount of autologous HS for *ex vivo* expansion of BMSCs. In fact, over 400 mL of peripheral blood is usually required to obtain 200 mL of autologous HS, which is only sufficient to support the growth of BMSCs for a few passages. Therefore, collection of a sufficient amount of autologous HS is a considerable burden for anaemic patients as well as for healthy female patients with a low body weight. Use of pooled allogeneic HS cannot overcome this problem because it has been shown that allogeneic HS does not fully support the growth of BMSCs (Kuznetsov *et al.*, 2000). Furthermore, even when a sufficient amount of autologous HS can be obtained from each patient, the constituents of individual HS could vary, which might lead to variations of cell culture outcome. Thus, it is desirable to develop efficient and safe serum-free culture media and eventually serum-independent cell expansion protocols for tissue engineering.

Recently, several companies have launched complete serum-free culture media that can support the growth of human mesenchymal stem cells without the addition of sera (Table 1). Although the number of studies that have investigated the potential of these serum-free media is still limited, it has been suggested that these serum-free media can support the growth of human somatic (postnatal/tissue) stem cells even more efficiently than conventional serum-based media (Lindroos *et al.*, 2009, Ishikawa *et al.*, 2009, Hartmann *et al.*, 2010). In support of this conclusion, our previous study of human BMSCs showed that the efficacy of cell growth was greater in StemPro® SFM (Invitrogen, Carlsbad, California, U.S.A.) than FBS-containing medium (Agata *et al.*, 2009). Similar findings have been reported with MesenCult®-XF (STEMCELL TECHNOLOGIES, Vancouver, BC, Canada), STK2® (DS Pharma Biomedical Co.,Ltd., Osaka, Japan), and the xenogenic-free (xeno-free) version of StemPro® SFM, all of which have been developed for xeno-free as well as serum-free cultivation of human somatic stem cells (Lindroos *et al.*, 2009, Ishikawa *et al.*, 2009, Hartmann *et al.*, 2010). These data indicate that currently available xeno-free, serum-free media may have the potential to replace conventional serum-based media in clinical tissue engineering, though further basic studies are required to ensure its safety and efficacy. To develop a protocol for bone tissue engineering with serum-free media, we now discuss current findings regarding the character of serum-free expanded cells.

Name of the medium	Company	Of note	*In vitro* osteogenic potential	*In vivo* osteogenic potential	Reference
StemPro SFM	Invitrogen (Carlsbad, California, U.S.A.)		Quantitative alkaline phosphatase (ALP) assay	Ectopic bone formation	Agata *et al.*, 2009,
StemPro SFM XenoFree	Invitrogen (Carlsbad, California, U.S.A.)	Xenofree	Quantitative ALP assay, ALP staining	?	Lindroos *et al*, 2009
MesenCult-XF	STEMCELL TECHNOLOGIES (Vancouver, BC, Canada)	Xenofree	Alizalin red staining	?	Hartmann *et al*, 2010
STK2	DS Pharma Biomedical Co.,Ltd. (Osaka, Japan)	Xenofree	Alizalin red staining	?	Ishikawa *et al.*, 2009

Table 1. List of currently available commercial serum-free media and the osteogenic ability of postnatal stem cells cultivated in each product

Since the type of expansion medium used in primary culture may affect the viability and type of cell population generated, it is important to compare the cell populations grown in serum-free and serum-containing medium. For this purpose, Lindroos *et al.* investigated cell surface marker expression by cells cultured in FBS- or HS-containing media and those cultured in serum-free media using human adipose stem cells. They reported that the expression profiles of examined cell surface antigens were not statistically different (Lindroos *et al.*, 2009). Our previous study investigated cell surface marker expression by human BMSCs cultured in serum-free medium. It also showed that the expression profiles of most of the examined antigens were comparable in both serum-free and serum-containing groups, though there were some differences in the expression of CD105 and CD146 (Agata *et al.*, 2009). Since the mean fluorescence intensity of the CD105 antigen was stronger in serum-free expanded BMSCs, it is possible that a larger population of CD105-positive cells was obtained by growth in serum-free medium. In contrast, the CD146-positive fraction was more evident in cells cultured in serum-based medium and only a limited number of cells were positive for CD146 in the serum-free group (Agata *et al.*, 2009). It is not clear whether serum-free conditions alter the expression of both of these surface markers or whether the conditions selectively support the growth of the $CD105^{positive}$

CD146^{dim} population. Nonetheless, cells grown in serum-free media do appear to be different from those grown in serum-containing media, and the information regarding BMSCs grown in serum-containing media may not be used as a reference. Therefore, the feasibility of bone tissue engineering with serum-free expanded BMSCs should be independently investigated from the beginning, though there have already been several clinical trials to show the safety and efficacy of bone tissue engineering with BMSCs grown in serum-containing media.

One of the most important things that should be assured for use in a clinical setting is that transplanted BMSCs do not form tumors in the recipient following transplantation. Since our previous study showed that transplanted BMSCs grown in serum-free medium did not form tumors in nude mice (Agata *et al.*, 2009), it might be possible that BMSCs expanded in serum-free medium are as safe as those expanded in serum-containing medium. However, further studies are required to confirm their safety because few studies have transplanted serum-free expanded somatic stem cells. Together with cell transplantation analyses, genomic and chromosomal stabilities must be analyzed, because these data can support the safety of serum-free expanded BMSCs.

In addition to confirming the safety of such transplants, assurance of the osteogenic differentiation ability of transplanted BMSCs is important in clinical bone tissue engineering. BMSCs grown in serum-containing media are known to differentiate into the osteogenic lineage when they are cultured in osteogenic induction medium (serum-containing media supplemented with dexamethasone, ascorbic acid, and β-glycerophosphate). However, it was still necessary to determine whether somatic stem cells grown in serum-free media would behave similarly in the presence of the same osteogenic components. To date, adipose stem cells, umbilical cord tissue-derived mesenchymal stem cells, and BMSCs those grown in serum-free media have been shown to differentiate into osteogenic cells in the conventional induction medium (Lindroos *et al.*, 2009, Ishikawa *et al.*, 2009, Hartmann *et al.*, 2010). However, it remains unknown whether conventional osteogenic induction medium is optimal for their differentiation, because some of the manufacturers recommend a specially formulated kit for osteogenic induction of serum-free expanded cells. Therefore, we explored osteogenic induction of BMSCs expanded in serum-free medium, using both a conventional osteogenic induction medium and the commercially supplied osteogenesis kit (Agata *et al.*, 2009). Results of alkaline phosphatase (ALP) assays showed that both treatments were able to induce osteogenic differentiation of serum-free expanded BMSCs, though the increase of ALP activity was more rapid with the osteogenesis kit (Fig. 1A). We also performed *in vivo* transplantation experiments to investigate possible differences in bone forming abilities between cells grown in the two media. As shown in Figure 1B - 1E, cells treated with both osteogenic medium and the osteogenesis kit were able to form bone *in vivo*, and there was no significant difference in the efficacy of bone formation (Fig. 1B, 1C: osteogenic medium; Fig. 1D, E: osteogenesis kit). These data indicate that bone tissue engineering with serum-free expanded BMSCs can be achieved with either the conventional osteogenic induction medium or the osteogenesis kit. However, these treatments may not be ideal for induction of osteogenic differentiation of serum-free expanded BMSCs, because both media (even the commercially supplied kit) contain some serum-derived components. Therefore, to enhance the safety of clinical bone tissue engineering, a completely serum-free osteogenic induction media should be developed.

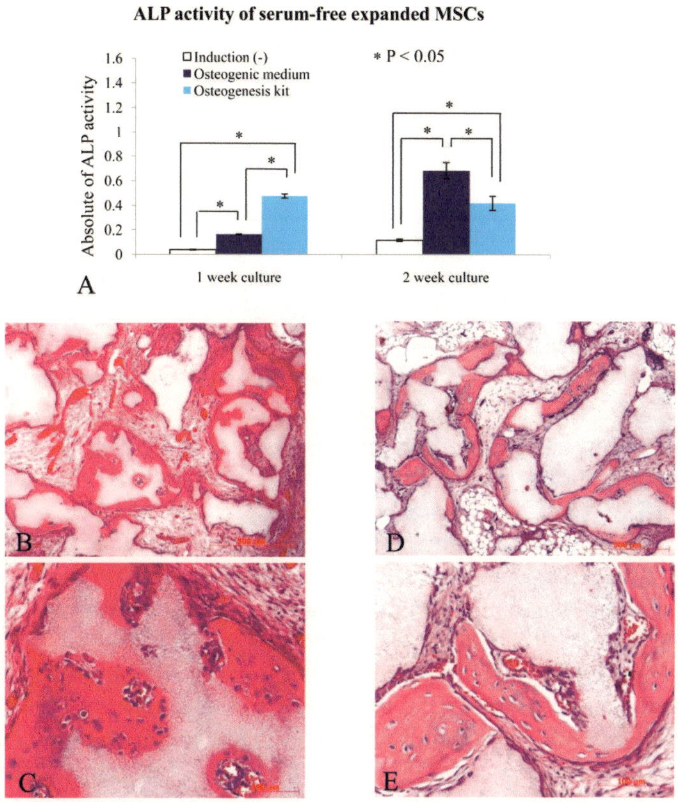

Fig. 1. Osteogenic abilities of serum-free expanded BMSCs after osteogenic induction with either osteogenic medium or osteogenesis kit (From Agata *et al.*, 2009 with permission)

Collectively, these data indicate that bone tissue engineering with BMSCs expanded in currently available commercial serum-free media is feasible, though further studies regarding the characteristics of the cells and the safety of serum-free expanded cells are required. In addition, further improvements in serum-free media are desirable because currently available xeno-free, serum-free media contain allogeneic human proteins, which may cause unknown disease-transmissions and immune reactions. Furthermore, related products for serum-free media such as cell culture dish coating materials, which are required for the efficient adhesion and proliferation of primary culture cells in serum-free culture system, should also be improved because no allogeneic-free materials are currently available.

3. Cell storage technologies

Cell storage technologies are essential for efficient, safe, and widespread use of tissue engineering. Storage technologies for cells and tissue-engineered products are required for their timely and efficient distribution. Furthermore, storage of stem cells (stem cell banking) is expected as a reservoir of stem cells for future use and also for public cell banking. Currently, cryopreservation is the most reliable and established technology to store tissues and cells. However, some novel technologies such as freeze dry technology have been investigated. In this chapter, we focus on the characteristic features of tissue-engineered products for cryopreservation and recent developments in storage technologies. Furthermore, potential future applications of stem cell banking are discussed.

3.1 Storage of tissue-engineered products

The storage of tissue-engineered products is an attractive target for technological development. Since tissue-engineered products usually consist of living cells, technical issues include the limited shelf life of the cells and the use of specialized conditions for transportation. Without storage, treatment with cells requires timely harvesting from the donor, which significantly affects the availability of tissue-engineered products. If the tissue-engineered products can be used as off-the-shelf products such as bioartificial bone substitutes, it would significantly enhance the adoption of this alternative.

3.1.1 Cryopreservation of tissue-engineered products

Currently, cryopreservation is the only available strategy for the storage of tissue-engineered products. However, tissue-engineered products usually consist of multiple layers of cells and, in most cases, the cells are seeded on scaffold made of biomaterials, which complicates the development of efficient freezing storage protocols (Pancrazio *et al.*, 2007). Furthermore, the scale-up of cryopreservation procedures from the cellular level to a macroscopic tissue scale introduces new problems related to heat and mass transfer phenomena in larger systems (Karlsson and Toner, 1996). Although it has been shown that frozen storage is feasible for some of the tissue-engineered products such as bone (Kofron *et al.*, 2003), it is more difficult than that for isolated cells and requires special considerations.

Water transport processes may cause difficulties for tissue-scale freezing. While cells at the surface layer would respond to freeze-induced osmotic changes much like cells in suspension, interior cells would dehydrate as a response to the increased intracellular tonicity in the dehydrate surface layers. Accordingly, interior cells dehydrate more slowly than surface cells, which may affect their survival (Karlsson and Toner, 1996). Heat transport limitation in larger tissue may also affect survival. Due to the macroscopic size of tissue-engineered products and its finite thermal conductivity, there may be large thermal gradients from the surface to the interior of the samples. The presence of a thermal gradient during cooling and warming phases makes it difficult to choose optimal temperature change protocols for both surface and interior cells. Moreover, osmotic effects (water movement from inside-unfrozen cells to outside-frozen cells) during cooling, reduces cell survival. Accordingly, it may not be possible to recover full viability throughout the tissue (Karlsson and Toner, 1996).

One of the key decisions in achieving successful freezing of tissue-engineered products is the choice of cryoprotectant. Cryoprotectants minimize damage caused by ice crystal formation and should induce an amorphous state, rather than ice crystals during the cooling and warming phases. Although the use of cryoprotectants is mandatory, currently available reagents are cytotoxic to some extent. Since tissue-engineered products are larger than isolated cells, longer incubation times with cryoprotectant are necessary which may result in a lower survival rate. On the other hand, short incubation times may not allow enough cryoprotectant to penetrate relatively thick tissue-engineered products and cause ice crystallization and cell death in internal layers. The pre-incubation time used for penetration of cryoprotectant should balance damage caused by toxicity and freezing/warming.

3.1.2 Technology development for the storage of tissue-engineered products

Two major approaches to cryopreservation are known, i.e., conventional freeze-thaw procedures and vitrification, which is defined as a glass-like solidification (Karlsson & Toner, 1996). While freeze-thaw procedures minimize the probability of intracellular ice formation, vitrification attempts to prevent ice formation throughout the entire sample during the cooling and warming process (Kuleshova *et al.*, 2007). Recently, the potential of vitrification has been tested for tissue-engineered constructs. Since tissue-engineered products consist of multicellular layers and often include biomaterials with varying coefficients of expansion compared with cells, cryopreservation using conventional freeze-thaw procedures with slow cooling rates has achieved limited success. Accordingly, vitrification could be an attractive alternative technology.

Vitrification has been investigated for tissue-engineered bone and blood vessels. Liu & McGrath (2003) explored the potential of vitrification for the cryopreservation of tissue-engineered bone constructs consisting of a hydroxyapatite scaffold-cell complex. Cell survival was 92.0% for suspended cells and 43.0% for attached cells. In terms of tissue-engineered blood vessel constructs, the effects of vitrification and conventional cryopreservation were compared (Elder *et al.*, 2005). Collagen-based vascular constructs were used as models in this study. Morphological changes associated with ice formation were visible within tissues preserved using traditional cryopreservation but not in tissue preserved using vitrification. The metabolic assay results indicated that vitrified tissue had viability similar to fresh controls. More recent study with tissue-engineered blood vessels using polyglycolic acid scaffold showed that ice formation in tissue-engineered blood vessels was negligible in the vitrified specimens but extensive (68.3 $\pm$ 4.5% of vessel area) in the extracellular matrix of frozen specimens. The vitrified tissue had a viability similar to fresh controls and the contractility results for vitrified samples were >82.7% of fresh controls but markedly reduced in the frozen samples (10.7% for fresh controls) (Dahl *et al.*, 2006). Vitrification is a feasible storage method for tissue-engineered blood vessel constructs, and their successful storage brings these constructs one step closer to clinical utility. Although it is a promising technique, higher concentration of cryoprotectant should be used, which could potentially damage the cells. Accordingly, it is still technically difficult and its utility for tissue-engineered products remains to be elucidated.

One of the most awaited technologies for the preservation of tissue-engineered product is long-term unfrozen storage (more specifically, dry storage) at ambient temperature. This approach allows storage without dependence on expensive freezers or liquid nitrogen, which require daily maintenance. This "off-the-shelf" availability and low cost should facilitate the usage of tissue-engineered products. Unfortunately, this is not yet a reality. However, many organisms can undergo a phenomenon called anhydrobiosis to survive in a completely dehydrated state for an extended time and resume activity upon rehydration (Crowe *et al.*, 2002). The sugar trehalose is found at high concentrations in many anhydrobiotic organisms. Thus, the addition of trehalose is considered a key factor in achieving freeze-dried storage. In mammalian cells, freeze-drying of platelets was reported using trehalose (Crowe *et al.*, 2005). Although recovered platelets were strongly attenuated, the survival rate exceeded 90%. Drying of nucleated cells is apparently more challenging. However, the addition of p29, a small a-crystallin stress protein, together with trehalose showed increased resistance to dryness in 293 cells (Ma *et al.*, 2005). Although "off-the-shelf" tissue-engineered products stored at ambient temperature are not available, the steady progress of research in this area may provide a functional protocol at some point in the future.

3.2 Cryopreservation of putative stem/progenitor cells

Storage of putative stem/progenitor cells is also an attractive research target in tissue engineering. Collection of cells for tissue engineering is less invasive than conventional tissue or organ transplantation. However, repetitive collection of cells can stress patients. If the stem/progenitor cells could be stored, it is possible that they could be repeatedly used for future therapy, reducing the burden and the cost to patients. The concept of banking stem/progenitor cells in storage is not new. Stem/progenitor cell populations decrease in size with age (D'Ippolito *et al.*, 1999; Zhou *et al.*, 2008). If the stem/progenitor cells were harvested at an early age and the cells could be stored, it is possible that those cells could be used later in the life of the donor for autologous transplantation. In this chapter, we also focus on cell banking.

3.2.1 Banking of somatic stem/progenitor cells

Currently, one of the most established and widely accepted stem cell bank systems is umbilical cord blood banking. However, the establishement of somatic stem cell bank other than cord blood (non-hematopoietic stem cells) is still underway. The nature of somatic stem/progenitor cells is much different from that of embryonic stem (ES) cells and induced pluripotent stem (iPS) cells. Somatic stem/progenitor cells possess limited ability to differentiate compared with ES cells and iPS cells. Furthermore, accumulating evidence suggests that somatic stem cells may lose their plasticity soon after cultivation is initiated (Sugiura *et al.*, 2004). Thus, expansion and storage of autologous somatic stem cells for personal future use could be impractical, particularly if the cells carry a genetic predisposition for the disease that is being treated. Improved understanding of somatic stem/progenitor cells is required for more varied applications.

3.2.2 Banking of ES cells and iPS cells

Recently, the establishment of "stem cell banks" has been reported in several countries. Those stem cell banks, mostly government-supported, aim to provide a resource for storing, characterising and supplying ethically collected, quality controlled stem cell lines for research and ultimately for treatment (London *et al.*, 2004; Nakamura, 2010). There is no doubt that those stem cell banks can facilitate research in this area and contribute to the availability of those rare resources such as human embryonic stem (ES) cells and somatic stem cells to researchers who do not have access. Recently, production of induced pluripotent stem (iPS) cell lines from human somatic cells has been reported (Nakagawa *et al.*, 2008). Theoretically, it is possible to generate personalized iPS cells for therapeutic use without ethical problems and also immunological rejection, though it may not be practical due to the time and cost required for the production and quality assurance. To overcome this problem, the establishment of HLA-haplotype banking for human iPS cells has been argued (Nakatsuji *et al.*, 2008). A similar idea was originally reported for human ES cell lines (Taylor *et al.*, 2005). HLA-haplotype banking may provide a more efficient and safe alternative. They estimated that only 50 iPS cell lines would be necessary to find a three-locus match for 90.7% of the Japanese population. At present, the safety of iPS cells for clinical use is the major concern. The development of safe and effective clinical applications with iPS cells would enhance their appeal.

4. Non-invasive image-based cell quality evaluation technology for cell therapy

In clinical tissue engineering, both safety and efficacy requirements must be satisfied. In the production of conventional chemical-based pharmaceutical drugs, the quality and efficacy of the product is guaranteed through adherence to strict regulations. However, cell-based materials, including tissue-engineered products, require different strategies to evaluate their quality.

Conventional cell quality evaluation technologies, such as RT-PCR or immuno-stainning, require cell-destruction processing such as cell-lysis or cell-fixations. If the quality of the cells can be evaluated without destruction process, both examined and non-examined cells can be used for therapies. Therefore, the development of such kind of non-invasive cell quality evaluation technologies has been awaited (Takagi, M. 2010).

4.1 Problems in cell therapy and their technological assessment

The safety and efficacy of cellular products, such as human cells, tissues, and cellular and tissue-based products or HCT/P, processed for cell therapy, are currently regulated by individual nations. However, the level and scope of regulation differs greatly among countries. Frequently referred references are Current Good Tissue Practice and 21 Code of Federal Regulations (CFR), Parts 1270 and 1271 of U.S. Food and Drug Administration (FDA) (FDA homepage, 2010).

Commonly, regulations attempt to limit the unique risks associated with cellular products. These regulations focus on the followings: (1) limiting the risk of transmission of communicable disease from donors to recipients, (2) establishing manufacturing practices that minimize the risk of contamination, and (3) requiring an appropriate demonstration of safety and effectiveness for cells and tissues that present greater risks due to their processing or their use. The regulations suggest that there are two major risks associated with HCT/Ps. One is the non-cellular contaminant risk, and the other is the risk from the cells themselves. Regulations have been established to assure sterility and/or the aseptic nature of the cellular product. However, there are still few criteria or regulations to control cell quality. Especially with stem cells, cell quality per se is not covered by licenses or regulations.

Cell quality issues in cell therapy include the health of the cells, and their ability to grow and differentiate as required without risk of tumorigencity. Historically, chromosomal tests have been conventionally used to assess tumorigenic risk. Animal implant tests are commonly used in safety tests for chemical-based pharmaceutical drugs. Biomarker tests, which assess specific tumorigenic marker genes/proteins by RT-PCR or flow cytometry are the most frequently used molecular biological techniques. However, as explained below, the availability of those conventional techniques do not help clinicians overcome the four major problems which inhibit the use of cell therapies. Those problems are as follows.

The most fundamental problem is guaranteeing the non-invasiveness of the engineered cells. Cells prepared for therapeutic use should also be high in viability, since artificial manipulation could trigger cellular abnormalities. Also, autologous cells from a patient are

commonly very limited in number due to the limitations of the source. For safety, fluorescent staining or gene transfer should also be avoided.
The second problem is a need for complete and exhaustive characterization of the engineered cells. In the case of conventional pharmaceuticals, the uniformity of products can be strictly controlled by regulation of production. However, with human cells, strict regulation of processing does not guarantee uniformity of the product since there are huge individual variations. Therefore, sampling does not always assure the quality of total cellular products.
The third problem is the time limitation inherent in sample characterization. A sample undergoing processing should be assessed repeatedly, until the day of therapy to assure the highest level of safety. However, currently utilized assays such as sterilization test take a few weeks to show the safety properties of cell-products.
The fourth and final problem is instability of the cell quality throughout the culture process. Accordingly, the cell quality of the cell-based products should be assessed just before operation, though conventional evaluation techniques require a few days or even weeks for the assessment of cell-products.
Here, we introduce image-based cell quality prediction technique which might be able to overcome all the problems listed above. Image-based cell quality prediction enables non-invasive, complete, on-time, and predictive evaluation of cells.
Image cytometry is an exciting new area in cell research (Kim, JS. *et al.*, 2010) and could support cell quality evaluation of tissue-engineered products. Given the advances in hardware and software, there are commercially available analysis systems for image cytometry, such as the IN Cell Analyzer (GE Healthcare, Chalfont St Giles, Buckinghamshire, United Kingdom) (GE Healthcare home page, 2011). Image cytometry provides exhaustive high content information characterizing intact/fixed cells. It can also provide very detailed data describing the localization or expression of molecules and organelles in cells. However, current image cytometry analysis is based on fluorescence. Although there are many "less-damaging" technologies for fluorescent labelling of cells, they still give some changes to the cells;, thus, incompatible to assess patient cells in clinical tissue engineering.
In contrast, there are developing technologies that evaluate/estimate cellular activities by "non-invasive" measurement technologies (Takagi, M. 2010). Takagi reviewed these non-invasive cell imaging technologies, and indicated that cellular activities can be estimated by cellular morphologies. Recent non-invasive technologies have also been applied to the cell quality evaluation of three dimensionally cultured tissue engineering products (Kino-oka *et al.*, 2008).
Here, we review some important aspects of basic image analysis strategy, especially designed for “cell quality prediction system” in tissue engineering and cell therapy, and show examples of image-based cell quality prediction technology.

4.2 Basics of cell image analysis for cell quality prediction

For image-based cell quality prediction, there are four major steps (Fig. 2): (1) image data collection, (2) image processing, (3) experimental data collection, and (4) data analysis. By adding a fifth step, (5) Prediction, conventional correlation analysis (in-sample analysis) is extended to achieve predictive performance (out-of sample analysis) for the evaluation of new samples.

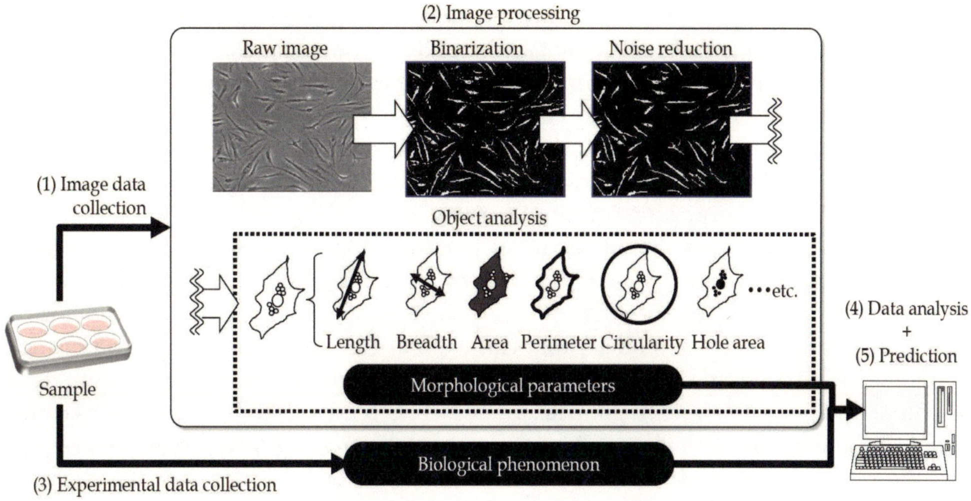

Fig. 2. Work flow of non-invasive image-based cell quality prediction

4.2.1 Image data collection

To assess cellular quality by digital imagery, a large number of detailed images should be collected. There are reports of various types of cellular images applied to non-invasive image analysis, such as phase-contrast microscopy, differential interference contrast microscopy, light-field microscopy, phase-shifting scanning confocal laser microscopy, etc.
In our laboratory, phase contrast microscopic images are obtained by BioStaition CT (Nikon Corporation, Tokyo, Japan), a fully-automatic cell monitoring system.

4.2.2 Image processing

After collecting data, raw images can be processed by image processing software, such as ImageJ, CellProfiler, Scion Image, Metamorph, CellClassifier, etc. The image processing scheme for cell image analyses combines (1) binarization, (2) noise reduction, and (3) object analysis.
There is no gold standard threshold for image analysis. Based on the type of images, different processing filters and orders should be examined to gain the best processing result.
For the analysis, object analysis is the most informative and essential process, that is the conversion of cellular images into numerical morphological parameters. The objects in the images that correspond to cells are individually labelled and their morphological parameters are measured. Although there are some variations in morphological parameters calculated by software, the basic parameters are length, breadth, area, perimeter, centroid, inner/outer radius, and area of holes of cells.

4.2.3 Experimental data collection

To find out the correlation between the cellular morphology and cell quality, cellular images should be compared with the data from biological experimentals. Therefore, the samples are evaluated by conventional biological techniques after the image acquisition.

4.2.4 Data analysis

Choosing the best approach to data analysis depends on the goal. Bioinformatics, the computational and statistical science that analyzes data from molecular biology, offers many effective solutions for the analysis of image data and biological experimental data. If the aim is evaluation of cell quality, regression analysis is an effective solution. Regression analysis could be easily introduced in laboratories by commercially available statistical software, such as SPSS. Free statistical programming platform "R" and BioConductor (R Project homepage, 2011) provide appropriate applications.

The basic strategy for image-based cell quality prediction is to select the proper regression type and construct a model by using a dataset containing both morphological data and experimental data. In the regression analysis, the algorithm searches for the best mathematical function ("the model") that can explain the correlation between a change in morphological data (input data) and biological data (output data). Hence, once a model that links input and output data with good accuracy has been established, future results could be predicted using image data alone. This is analogous to arriving at a prognosis using gene expression from microarray studies. SPSS (IBM, New York, U.S.A.) and the R platform provide multiple optional functions for parameter selection.

4.2.5 Prediction

Prediction is the final process for practical image-based cell quality assessment. The prediction process is similar to the process of data analysis. Briefly, after the completion of regression analysis, the resultant model can function as a "prediction model". However, compared to the analytic process that uses all data (known as "in-sample analysis"), partitioned data are separately used for "model construction" and "model test" (known as "out-of-sample analysis") for evaluating the generalization performance. In practice, the total data are commonly partitioned into training data and test data. Using the regression model constructed only with the training data, the error rate between the actual output values of test data and the predicted values calculated from the input values of test data are compared. When the average error rate for all test data is small, the constructed model could be considered a "reliable model" to use with new data. In such a data model validation process, leave-one-out or several-fold cross validation is commonly used. Since this final process requires the partitioning of total data, the total data pool should be large and varied. When the quantity of data is too small, the constructed model will have limited prediction value with new data. Also, when data variations are very biased to certain conditions, such as with only one lot of cells, the constructed model may perform well for the cell lot examined, but would not work with new cell lots.

4.3 A prediction system based on cell image analysis which supports clinical tissue engineering

Here, we introduce practical examples of non-invasive cell quality prediction to support clinical tissue engineering.

4.3.1 Prediction of cell yield

In clinical tissue engineering, cell yield on the day of surgery is a critical factor. Seeding cell density or cell number is commonly critically defined in most cell therapy protocols to assure a certain therapeutic effect. Therefore, the surgery schedule is projected based on the experienced observation of cellular behaviour in the culture flasks. However, such scheduling is very tentative and there are cases in which the cell yield is insufficient on the scheduled day of surgery.

To overcome such problems in clinical tissue engineering, we tried to predict the future cell yield (14 days later) from early cellular images (images one to three days into the culture period). For image data, we collected total 270 phase contrast microscopic images (4x) of cultured primary dermal fibroblasts, obtained from 10 healthy volunteers (3 males, 7 females, 29 - 72 years old). Informed consent was obtained according to a protocol approved by the ethics committee of Nagoya University Hospital. Biological data (cell growth rate) within 14 days was obtained by manual cell count. All cells were prepared using passage three or four from the primary expansion in modified Eagle's medium (DMEM) containing 10% FBS at 37°C in the presence of 5% CO_2. A schematic image is provided in Fig. 3.

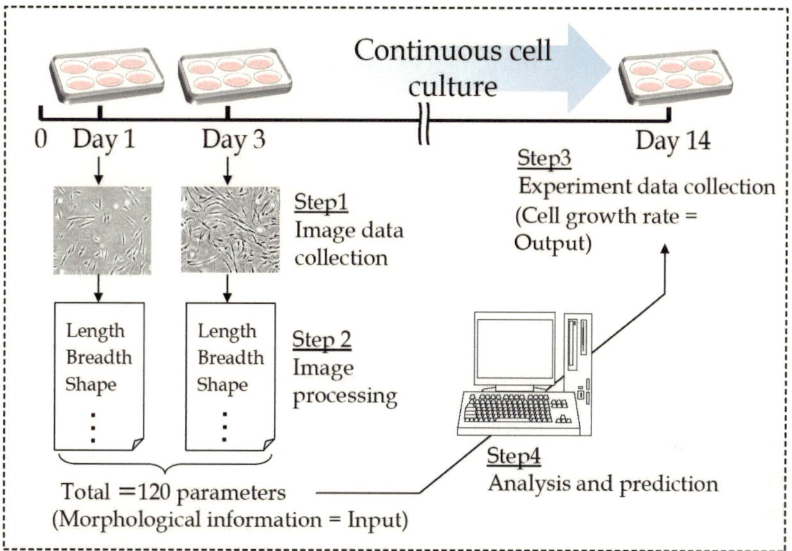

Fig. 3. Schematic diagram of cell yield prediction

The cell attachment rates and growth curves differed greatly among individuals (data not shown). However, by multiple-regression analysis (MRA) with parameter selection, the predicted growth rates had very small error rates compared to the actual growth rates (Fig. 4). Among 120 parameters extracted from the image data, three parameters, such as (P1) change rate of the variation of elliptical rate (1st day to 3rd day), (P2) Size of the inner radius on the third day, and (P3) cell number on the first day, were found to be the best combination of cell culture parameters to predict future cell yield (average squared error = 0.14) (Fig. 4C). It is interesting that other parameters, intentionally selected by cell culture experts badly correlated with the cell yield (Fig. 5). By comparing all the data, we arrived at three conclusions: (1) morphological cell information is informative for cell growth prediction, (2) objectively selected parameters are more effective in cell growth prediction, and (3) multiple combinational parameters work better than a single parameter in cell growth prediction.

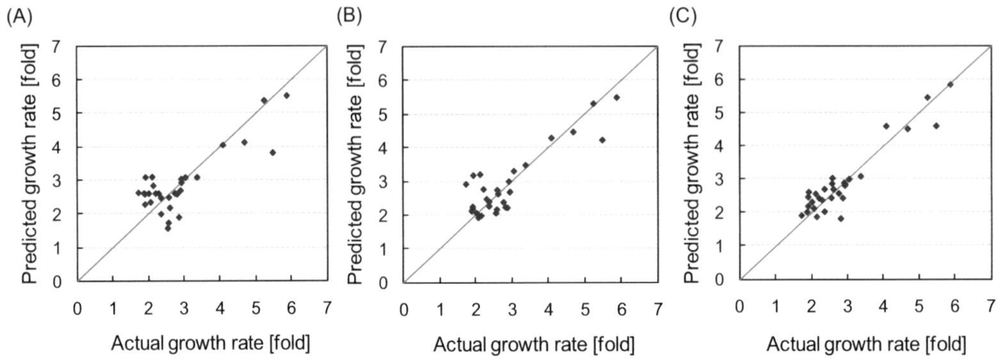

Fig. 4. Accuracy of cell yield prediction models. (A) one-parameter prediction model (P1 only), (B) two-parameter prediction model (P1 and P2), (C) three-parameter prediction model (P1, P2, P3). P1, change rate of the variation of elliptical rate (first day to third day); P2, size of the inner radius on the third day; P3, cell number on first day

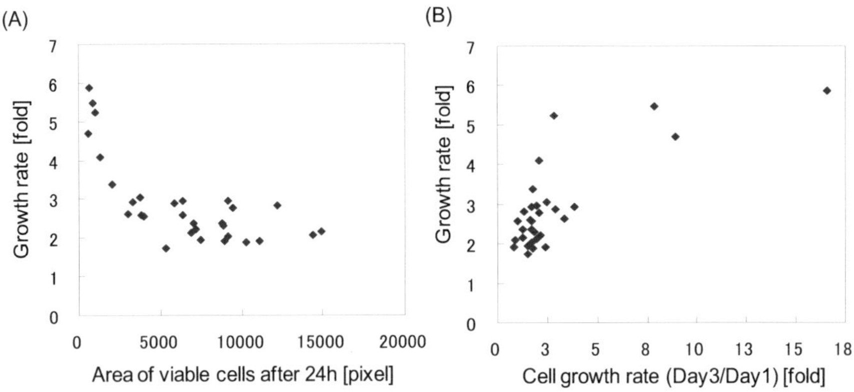

Fig. 5. Correlation plot with the image-derived parameter selected by cell culture experts. (A) Correlation with area of viable cells after 24 h, (B) Correlation with total cell growth rate (day three/day one).

4.3.2 Predicting the level of osteogenic differentiation

In stem cell therapy for bone regeneration, the extent of differentiation critically affects the *in vivo* bone formation after implantation. However, good early markers that predict the future differentiation level of cells have been unavailable. If the differentiation level of cells could be predicted in advance, scheduling of surgery could be set to optimize therapeutic outcome.

As a prediction model of cell differentiation ability, we attempted to predict the alkaline phophatase (ALP) activity of human mesenchymal stem cells (hMSCs). For image data, we collected 1,170 phase contrast microscopic images (4x) of cultured commercial hMSCs (three lots; 20 years old male, 22 year old male, and 19 year old male). Cells at the same passage number were cultureed in the differentiation medium (Agata *et al.*. 2009) at 37°C in the presence of 5% CO_2. For biological phenomenon data, we collected the ALP activity of all the samples using the method described previously (Agata *et al.*, 2009). We compared the two groups of induction positives and negatives. The protocol is described in Fig. 6.

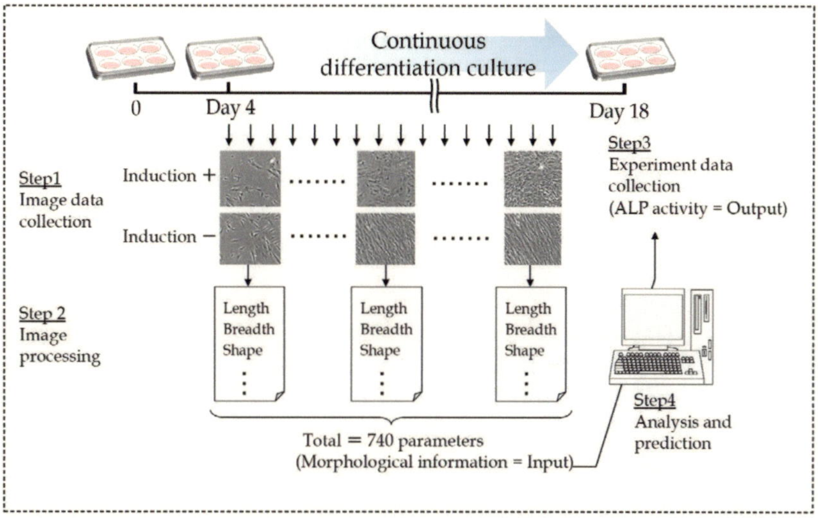

Fig. 6. Schematic diagram of osteogenic differentiation prediction

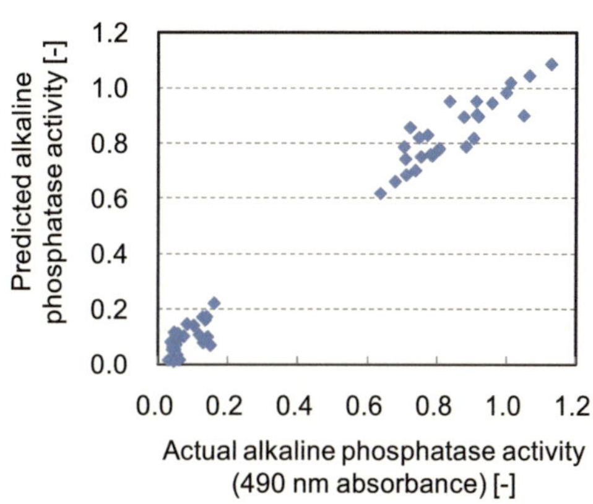

Fig. 7. Prediction accuracy of osteogenic differentiation model

By using ten morphological parameters selected by MRA, we found that ALP activity could be predicted with high accuracy (Fig. 7). Although there were slight differences in differentiation ability among the examined MSCs, the regression model could be extended to respond to any MSCs. In other words, patient differences could be compensated for by such a model-based analysis.

5. Conclusions

Despite rapid progress in treatment technologies using tissue engineering and regenerative medicine, those novel therapies have yet to be extended to standard medical practice and only limited numbers of patients have thus far benefited from these less invasive and potentially efficient therapies. The development of the technologies described here should facilitate general medical acceptance of tissue engineering. Increased availability of tissue engineering should contribute to the quality of life of patients with a wide range of diseases.

6. Acknowledgments

This work was supported in part by a grant for "Research on Human Genome and Tissue Engineering" from the Ministry of Health, Labor and Welfare of Japan, and Grant for Industrial Technology Research (Financial support to young researchers, 09C46036a) contracted with New Energy and Industrial Technology Development Organization (NEDO).

7. References

Agata et al. (2009). Feasibility and efficacy of bone tissue engineering using human bone marrow stromal cells cultivated in serum-free conditions. *Biochem Biophys Res Commun*, Vol. 382, No. 2, (March 2009), pp. 353-358, ISSN 0006-291X

Berger et al. (2006). Cell culture medium composition and translational adult bone marrow-derived stem cell research, *Stem Cells*, Vol. 24, No. 12, (December 2006), pp. 2888–2890, ISSN 1066-5099

Dahl et al. (2006). Feasibility of vitrification as a storage method for tissue-engineered blood vessels. *Tissue Eng*, Vol. 12, No. 2, (February 2006), pp. 291-300, ISSN 1937-3341

D'Ippolito et al. (1999). Age-related osteogenic potential of mesenchymal stromal stem cells from human vertebral bone marrow. *J Bone Miner Res*, Vol. 14, No. 7, (July 1999), pp. 1115-1122, ISSN 0884-0431

Elder et al. (2005). Enhanced tissue strength in cryopreserved, collagen-based blood vessel constructs. *Transplant Proc*, Vo. 37, No. 10, (December 2005), pp. 4625-4629, ISSN 0041-1345

FDA (12.07.2010) *Tissue & Tissue Products*, In: FDA homepage, 25.03.2011, Available from: http://www.fda.gov/BiologicsBloodVaccines/TissueTissueProducts/default.htm

Gallico et al. (1984). Permanent coverage of large burn wounds with autologous cultured human epithelium. *N Engl J Med*, Vol. 311, No. 7, (August 1984), pp. 448-451, ISSN 0028-4793

GE Healthcare (March 2011) *High Content Analysis Reference Database*, In: GE Healthcare homepage, 26.03.2011, Available from: http://www.biacore.com/high-content-analysis/applications/Publications_Database/index.html

Hartmann et al. (2010). Umbilical cord tissue-derived mesenchymal stem cells grow best under GMP-compliant culture conditions and maintain their phenotypic and functional properties. *J Immunol Methods*, Vol. 363, No. 1, (December 2010), pp. 80-89, ISSN 0022-1759

Hayashi et al. (2008). Comparison of osteogenic ability of rat mesenchymal stem cells from bone marrow, periosteum, and adipose tissue. *Calcif Tissue Int.* Vol. 82, No.3, (March 2008), pp. 238-247, ISSN 0171-967X

Held et al. (2010) CellCognition: time-resolved phenotype annotation in high-throughput live cell imaging, *Nature methods*, Vol. 7, (July 2010), pp. 747-754, ISSN 1548-7091

Horwitz et al. (2002). Isolated allogeneic bone marrow-derived mesenchymal cells engraft and stimulate growth in children with osteogenesis imperfecta: Implications for cell therapy of bone. *Proc Natl Acad Sci U S A*, Vol. 99, No. 13, (June 2002), pp. 8932-8937, ISSN 0027-8424

Ishikawa et al. (2009). Effectivity of the novel serum-free medium STK2 for proliferating human mesenchymal stem cells (articles in Japanese). *Yakugaku Zasshi*, Vol 129, No. 3, (March 2009), pp. 381-384, ISSN: 0031-6903

Jung S et al. (2010). Identification of growth and attachment factors for the serum-free isolation and expansion of human mesenchymal stromal cells. *Cytotherapy*, Vol.12, No. 5, (September 2010), pp. 637-657, ISSN 1465-3249

Karlsson and Toner (1996). Long-term storage of tissues by cryopreservation : critical issues. *Biomaterials*, Vol. 17, No. 3, (February 1996), pp. 243-256, ISSN 0142-9612

Kofron et al. (2003). Cryopreservation of tissue engineered constructs for bone. *J Orthop Res*, Vol. 21, No. 6, (November 2003), pp. 1005-1010, ISSN 0736-0266

Kim et al. (2010) Ongoing development of image cytometers. *Bioanalysis*, Vol. 2, No. 10, (October 2010), pp. 1755-1765, ISSN 1757-6180

Kino-Oka et al. (2008) Characterization of spatial growth and distribution of chondrocyte cells embedded in collagen gels through a stereoscopic cell imaging system. *Biotechnol Bioeng.*, Vol. 99, No. 5, pp. 1230-1240, ISSN 0006-3592

Kuznetsov et al. (2000). Effect of serum on human bone marrow stromal cells: ex vivo expansion and in vivo bone formation. *Transplantation*, Vol. 70, No. 12, (December 2000), pp 1780-1787, ISSN 0041-1337

Lindroos et al. (2009). Serum-free, xeno-free culture media maintain the proliferation rate and multipotentiality of adipose stem cells in vitro. *Cytotherapy*, Vol. 11, No. 7, (November 2009), pp 958-972, ISSN 1465-3249

Liu and McGrath (2003). Vitrification solutions for the cryopreservation of tissue-engineered bone. *Cell Preservation Technology*, Vol. 2, No. 2, (June 2004), pp. 133-143, ISSN 1538-344X

London SM (2004). First stem cell bank in the world is opened in UK. BMJ. (May 2004),Vol. 328, No. 7451, (May 2004), pp. 1277, ISSN 0959-8138

Lu et al. (2006). Defined culture conditions of human embryonic stem cells. *Proc Natl Acad Sci U S A*, Vol. 103, No. 15, (April 2006), pp. 5688-5693, ISSN 0027-8424

Misselwitz et al. (2010) Enhanced cell classifier a multi-classification tool for microscopy images. *BMC Bioinformatics*, Vol. 11, No. 30, (January 2010), pp.1471-2105, ISSN 1471-2105

Montzka et al. (2010). Expansion of human bone marrow-derived mesenchymal stromal cells: serum-reduced medium is better than conventional medium. *Cytotherapy*, Vol. 12, No. 5, (September 2010), pp 587-592, ISSN 1465-3249

Nakagawa et al. (2008). Generation of induced pluripotent stem cells without Myc from mouse and human fibroblasts. *Nat Biotechnol*, Vol. 26, No. 1, (Janurary 2008), pp.101-106, ISSN 1087-0156

Nakamura Y (2010). Bio-resource of human and animal-derived cell materials. Exp Anim, Vol. 59, No. 1, (March 2010), pp. 1-7, ISSN 1341-1357

Pancrazio et al. (2007). Enabling tools for tissue engineering. *Biosensors Bioelectronics*, Vol.22, No. 12, (June 2007), pp. 2803-2811, ISSN 0956-5663

Stute et al. (2004). Autologous serum for isolation and expansion of human mesenchymal stem cells for clinical use. *Exp Hematol*, Vol. 32, No. 12, (December 2004), pp. 1212-1225, ISSN 0301-472X

Sugiura et al. (2004). Osteogenic potential of rat mesenchymal stem cells after several passages. *Biochem Biophys Res Commun*, Vo. 316, No. 1. (March 2004), pp. 233-239, ISSN 0006-291X

Takagi M. (2010) Noninvasive quality estimation of adherent mammalian cells for transplantation, *Biotechnology and Bioprocess Engineering*, Vol. 15, No. 1, (February 2010), pp. 54-60, ISSN 1226-8372

The R project (2011) The R project for statistical analysis, 25.06.2011, Available from: http://www.r-project.org/

Thirumala et al. (2010). Evaluation of methylcellulose and dimethyl sulfoxide as the cryoprotectants in a serum-free freezing media for cryopreservation of adipose-derived adult stem cells. *Stem Cells Dev*, Vol. 19, No. 4, (April 2010), pp. 513-522, ISSN 1547-3287

Zhou et al. (2008). Age-related intrinsic changes in human bone-marrow-derived mesenchymal stem cells and their differentiation to osteoblasts. *Aging Cell*, Vol. 7, No. 3, (June 2008), pp. 335-343, ISSN 1474-9718

Part 2

Cells for Regenerative Medicine

2

Bone and Cartilage from Stem Cells: Growth Optimalization and Stabilization of Cell Phenotypes

Jan O. Gordeladze et al.*
Department of Medical Biochemistry, Institute of Basic Medical Science, University of Oslo, Norwegian Center for Stem Cell Research, The National Hospital, Oslo Norway

1. Introduction

Engineered cells replacing tissues should mimic the three-dimensional (3D) structure and reflect the different cell phenotypes exhibited by the lost or damaged tissue (Raimondi 2006; Keung, Healy et al. 2010). The engineered cells should also demonstrate a certain plasticity, i.e. an ability to adapt to the environment, in which they are deposited, reflecting the minute differences in features necessary to rebuild a functional tissue, which is able to renew itself over time (Grad and Salzmann 2009; Ohishi, Chiusaroli et al. 2009; Tare, Kanczler et al. 2010).

Osteoblastic cells in bone need to be able to involve themselves in a remodelling cycle with osteoclasts (Hanada, Hanada et al. 2010; Trouvin and Goeb 2010), which may be recruited from surrounding bone structures, and/or may be furnished as preosteoclasts within the population of osteoblasts. Furthermore, the osteoblasts should be able to undergo a distinct alteration in terms of life-span defined characteristics (Lian and Stein 2003; Lian, Stein et al. 2006; Gordeladze, Reseland et al. 2009), including the transition to osteocytes. Finally, the newly formed engineered tissue does not survive unless it develops a vascular network (Matsumoto, Kuroda et al. 2008; Grellier, Bordenave et al. 2009) furnishing the bone tissue with oxygen, growth factors and nutrients.

Chondrocytes in engineered cartilage should be able to produce an extracellular matrix reflecting the composition, water-binding capacity and mechanical characteristics of true

* Janne E Reseland[3], Tommy A Karlsen[1,2], Rune B Jakobsen[1,2], Lars Engebretsen[4,5], Ståle P Lyngstadaas[3], Christian Jorgensen[6,7], Hans Yssel[6] and Jan E Brinchmann[1,2]
[1]*Department of Medical Biochemistry, Institute of Basic Medical Science, University of Oslo, Norway,*
[2]*Norwegian Center for Stem Cell research, The National Hospital, Oslo, Norway,*
[3]*Department of Biomaterials, Faculty of Dentistry, University of Oslo, Norway,*
[4]*Orthopaedic Center, Ullevål University Hospital, Oslo, Norway,*
[5]*Faculty of medicine, University of Oslo, Norway,*
[6]*INSERM U844, Montpellier France,*
[7]*Unité Immuno-Rhumatologie Thérapeutique, Centre Hospitalier Universitaire (CHU) Lapeyronie, Montpellier, France*

hyaline cartilage (Knecht, Vanwanseele et al. 2006; Heinegard 2009; Bertrand, Cromme et al. 2010; Goldring and Goldring 2010). This type of cartilage exhibits certain features, such as hypoxic conditions and chondrocytes demonstrating gradients of gene transcript levels (cell phenotype plasticity) between the juxta-luminal and bone-lining surfaces of a joint (Grimshaw and Mason 2001; Lu, Subramony et al. 2010; Oh, Kim et al. 2010).

The features listed above have been extensively described and confirmed in the literature, however, a joint approach to produce well adapting engineered osteoblasts and chondrocytes has hitherto not been the subject of review articles or book chapters. The present outline encompasses a combined literature review of phenomena to take into consideration when engineering such cells from stem cells (SCs): sources of SCs to use (Logeart-Avramoglou, Anagnostou et al. 2005; van Osch, Brittberg et al. 2009), genes or microRNAs to manipulate (Goldring, Tsuchimochi et al. 2006; Betz 2008; Grundberg, Brandstrom et al. 2008; Duggal, Fronsdal et al. 2009; Gordeladze, Djouad et al. 2009; Lin 2009; Granchi, Ochoa et al. 2010; Sun 2010; Herlofsen, Kuchler et al. 2011), selection of gene and microRNA delivery systems (Saraf and Mikos 2006; Phillips, Gersbach et al. 2007), choice of humoral growth factors to facilitate SC differentiation (Shahdadfar, Fronsdal et al. 2005; Boeuf and Richter 2010; van der Kraan, Davidson et al. 2010), selection of appropriate scaffolds to support "asymmetric" SC differentiation (Vinatier, Bouffi et al. 2009; Oh, Kim et al. 2010; Seidi, Ramalingam et al. 2011), combination of stem cell niches and/or co-cultures to ensure organ mimicry reflecting proper cell-cell interactions (Grad and Salzmann 2009; Grellier, Bordenave et al. 2009; Boeuf and Richter 2010; Tare, Kanczler et al. 2010), mechano-stimulation of cells (Kelly and Jacobs 2010; Nowlan, Sharpe et al. 2010), and three-dimensional (3D) organ printing (Williams 2009; Visconti, Kasyanov et al. 2010).

Furthermore, this review also discusses how to stabilize osteoblasts and chondrocytes obtained by differentiation of SCs, i.e. how we can make the subject cells resilient to the detrimental effects of inflammatory cytokines and T-cells (Gordeladze, Reseland et al. 2009; Gruber 2010; Hanada, Hanada et al. 2010; Pacifici 2010), and exosomes shredded from immune cells (Valadi, Ekstrom et al. 2007; Camussi, Deregibus et al. 2010; Zomer, Vendrig et al. 2010).

2. Sources of stem cells for osteoblast and chondrocyte differentiation

Osteoblast differentiation

Bone marrow, which is the natural repository of osteoblasts, is widely used as source for bone engineering. Under appropriate conditions, bone-derived stem cells (bone mesenchymal stem cells = bone MSCs) can differentiate into osteoblasts, chondrocytes, adipocytes, and stromal cells (Javazon, Beggs et al. 2004; Otto and Rao 2004; Logeart-Avramoglou, Anagnostou et al. 2005). The differentiating potency of bone MSCs was enhanced when embedded in diffusion chambers or organ capsules, however, there is now a plethora of scaffolds securing the development of "proper" osteoblasts to produce bone for tissue replacement purposes. The advantage of using bone MSCs is related to the large number of obtainable osteoblasts, their high number of passages before the differentiating potential is lost, and their ability to be stored frozen for a long period. And the default pathway of bone MSCs is the osteogenic pathway (Logeart-Avramoglou, Anagnostou et al. 2005).

During the past 10-12 years, many other stem cell sources with osteogenic potential have been isolated, including blood, adipose tissue, lung, synovium, skeletal muscle and tooth pulp (for review, (Barry and Murphy 2004; Logeart-Avramoglou, Anagnostou et al. 2005; Gordeladze, Reseland et al. 2009; Bodle, Hanson et al. 2011; Levi and Longaker 2011; Witkowska-Zimny and Walenko 2011)). However, it seems that adipose stem cells (ASCs), provided that they are similar to MSCs in terms of surface receptor molecule profile (STRO-1, CD34, CD45, CD117 negative; CD44, CD49 CD29, CD90, CD105, CD106 positive) (Logeart-Avramoglou, Anagnostou et al. 2005; Niemeyer, Krause et al. 2006), may serve as a good source for bone engineering (Bodle, Hanson et al. 2011; Levi and Longaker 2011; Monaco, Bionaz et al. 2011). Irrespective of whether the source encompasses MSCs or ASCs, it seems that "proper" osteoblasts may be obtained if incubation conditions (i.e. the choice of growth factor source) and appropriate scaffolds are employed (Logeart-Avramoglou, Anagnostou et al. 2005; Kanczler and Oreffo 2008; Kwan, Slater et al. 2008; Gordeladze, Reseland et al. 2009; Tiainen, Lyngstadaas et al. 2010; Rahaman, Day et al. 2011; Sabetrasekh, Tiainen et al. 2011).

Chondrocyte differentiation

As for bone engineering, cartilage engineering relies firmly on the use of MSCs and ASCs (van Osch, Brittberg et al. 2009; Boeuf and Richter 2010; O'Sullivan, D'Arcy et al. 2011; Witkowska-Zimny and Walenko 2011). Some other sources, i.e. ectodermal cells like skin and hair follicles, as well as perinatal tissue and umbilical cord blood (Kuhn and Tuan 2010). One article refers to the use of synovial membrane stem cells (SMSCs) and compares their potency for chondrocyte differentiation with MSCs and ASCs (Segawa, Muneta et al. 2009). The criteria for selection of cell source may vary, but the authors focus on the necessity to analyse chondrocytes differentiated from these stem cells and choose the better source depending on how close they resemble the gene expression profile of mature chondrocytes isolated from hyaline cartilage (Segawa, Muneta et al. 2009; Vinatier, Bouffi et al. 2009; Vinatier, Mrugala et al. 2009).

Differentiating MSCs and ASCs produce all the components constituting ECM and represent the cells of choice for engineering articular cartilage. However, adult chondrocytes isolated from various sources like articular cartilage, nasal septum, ribs or ear cartilage (Kafienah, Jakob et al. 2002; Isogai, Kusuhara et al. 2006) produce *de novo* cartilage displaying the characteristics of its original tissue (Isogai, Kusuhara et al. 2006). It is therefore more appropriate to use hyaline cartilage as the preferred source of chondrocytes, and a comparison between different sources of hyaline chondrocytes (nasal, costal, and articular) has shown the superiority of costal and nasal chondrocytes in terms of quantity of cartilage formed after subcutaneous transplantation (Isogai, Kusuhara et al. 2006). One major limit related to the use of chondrocytes, is their instability in monolayer culture resulting in loss of phenotype (i.e. loss of collagen II, aggrecan and superficial zone protein = SZP) (Darling and Athanasiou 2005). Loss of the chondrocytic phenotype is accompanied by a phenotypic shift towards fibroblast like cells, which is characterized by an enhanced expression of collagen I (Schnabel, Marlovits et al. 2002). This dedifferentiation process is reversible, and dedifferentiated chondrocytes arranged in a three-dimensional (3D) lattice may retrieve their differentiated phenotype (Domm, Schunke et al. 2002; Malda, van Blitterswijk et al. 2003). This is especially true for dedifferentiated chondrocytes, having been "reversed" some 7-10 days before assuming the fibroblast phenotype (Brinchmann et al., unpublished observations).

The use of chondrocytes from osteoarthritic (OA) cartilage has also been contemplated. However, OA chondrocytes are subject to metabolic alterations leading to a low response to inductive environmental factors (Fukui, Purple et al. 2001; Sandell and Aigner 2001). Although chondrocytes derived from OA patients seem to be less appropriate for articular cartilage repair, it has been reported that OA chondrocytes may resume a normal chondrocytic phenotype upon 3D-cultivation *in vitro* (Tallheden, Bengtsson et al. 2005).

3. Genes and microRNAs as determinants of bone and cartilage quality

Characteristics of transcriptomes

When tissue replacement with the aid of tissue engineering is the ultimate therapeutic goal, it is vital to understand the differentiation process from precursor cells in terms of gene expression. Hence, it is necessary to identify a transcriptome, which is reflecting the "true" osteoblast and chondrocyte phenotypes pertaining to the function and localization of such cells in the skeleton. Many excellent articles have addressed this task over the past 10 years, of which only some are mentioned here (Kulterer, Friedl et al. 2007; Grundberg, Brandstrom et al. 2008; Duggal, Fronsdal et al. 2009; Morsczeck, Schmalz et al. 2009; Sundelacruz and Kaplan 2009; Bernstein, Sticht et al. 2010; Granchi, Ochoa et al. 2010; Piek, Sleumer et al. 2010; Sun, Mauerhan et al. 2010; van der Zande, Walboomers et al. 2010; Herlofsen, Kuchler et al. 2011). Two transcriptomes featuring putative gene markers of osteoblasts and chondrocytes (two each), respectively, will be described in detail.

Grundberg et al. (Grundberg, Brandstrom et al. 2008) used human trabecular osteoblasts stimulated with BMP-2 and dexamethasone for 24 hours. The article refers to genes specific for trabecular bone cells (osteoblasts) and genes up-regulated after 2 and 24 hours incubation with BMP-2 and dexamethasone, as well as genes altered upon 24 hours incubation with dexamethasone alone. Top similarity pathways of cell phenotype modulation (as assessed by Ingenuity®) were the IGF-1-, Leptin-, BMP-2- and Wnt-pathways, indicating a good correlation with the bulk of literature on osteoblast differentiation (Gordeladze, Drevon et al. 2002; Komori 2006; Marie 2008; Gordeladze, Reseland et al. 2009). The second paper on osteoblastogenesis (Granchi, Ochoa et al. 2010) was based on incubation of human MSCs in differentiating medium with dexamethasone for 24 hours, and thereafter in mineralizing medium with dexamethasone for 7 days. The main groups for the classifications of up-regulated genes were characterized by: angiogenesis, apoptosis, bone development, cell communication, cell cycle, embryonal development, TGFβ-signalling, and Wnt-signalling. The cumulative gene lists from these reports constituted the osteoblast transcriptome (188 genes) used as osteoblast reference to evaluate osteoblast differentiation (see paragraph 11. Bone and cartilage engineering revisited).

As for the chondrocytic differentiation, it is referred to the papers of Bernstein et al. (Bernstein, Sticht et al. 2010) and Herlofsen et al. (Herlofsen, Kuchler et al. 2011). Bernstein and co-workers used chondrocytes from articular cartilage and MSCs in an intricate array of manipulations. The cells were incubated in a differentiating medium containing TGFβ3 or control medium without growth factor. Transcriptomes obtained from the cells were categorized in gene transcripts up-regulated, down-regulated or "unsteady". The following results were obtained: Genes were classified as belonging to groups designated TGFβ-related, Wnt-pathway, glycans, actin metabolism, integrins/motility, bone development, muscle development, neuronal development, sperm development, and lipid metabolism. Comparisons with published gene ontology (GEO) datasets revealed that 1) MSC differentiation towards the chondrogenic lineage resembled MSC differentiation in mouse embryo limb buds (endochondral differentiation), and 2) an increasing confluence of proliferating MSCs will resemble the pellet situation in a timely delayed manner (transition from proliferation to differentiation). The paper by Herlofsen et al. describes MSCs differentiated to chondrocytes in alginate beads for 21 days in a standard differentiating medium with BMP-2. The following gene transcripts COL1A1, COL2A1, COL10A1, SOX5, SOX6, SOX9, ACAN, COMP, VCAN, MMP13, ALPL, RUNX2, and SOX8 were analysed by Q-PCR for verification of differentiation. However, a similarity search, where the transcripts for COL2A1 (up-regulated upon differentiation) and CXCL12 (down-regulated upon differentiation) was compared with the time-course of other genes (1072 up-regulated, 898 down-regulated). From these exercises, a joint list was compiled, consisting of 261 genes. This cumulative list of genes was used as chondrocyte reference to evaluate chondrocyte differentiation (see paragraph 11. Bone and cartilage engineering revisited).

These are just some examples of transcriptomes characterizing cell phenotypes subsequent to exposure to differentiating conditions *in vitro*. Since these experiments have been conducted *ex vivo*, it is reasonable to anticipate that the ultimate transcriptome can only be revealed if factors and/or conditions like cell sources, growth factors (adapted incubation media), scaffolds or organ printing, mechano-stimulation, gene and/or microRNA manipulations, or gene and microRNA delivery systems are all taken into account when a final "tissue engineering" process or algorithm is selected.

Spectrum of microRNAs expressed

MicroRNAs are short (20-24 nt) non-coding single-stranded RNA molecules that play an important role in regulating cellular gene expression. They function at the post-transcriptional level, by binding mRNA molecules (Gordeladze, Djouad et al. 2009; Lin 2009; Beezhold, Castranova et al. 2010). MicroRNAs have been found to play important roles in mediating fundamental biological processes like proliferation and differentiation in a variety of cells within defined tissues types. Recent reports have suggested that microRNAs may play a significant role in bone and cartilage development (Gordeladze, Djouad et al. 2009; Lin 2009; Sun 2010; Karlsen, Shahdadfar et al. 2011).

MicroRNAs suppress target gene translation by binding to the 3'-untranslated region (3'-UTR) of mRNAs, thus repressing translation and/or enhancing mRNA degradation. This requires that the 3'-UTR contains at least one specific 6-7 nt sequence which exhibits at least partial complementarity to a so-called "seed site," located within the 5'-region of the microRNA molecule (Lin 2009; Beezhold, Castranova et al. 2010).

Despite mounting evidence that miRNAs play a significant role in embryonic development and other biological processes, the function of only a handful of miRNAs has been determined thus far. And of these miRNAs, only a small subset has been implicated in cartilage and/or bone development. These are mir-140 (targeting HDAC4), mir-199a and mir-26a (both targeting SMAD1), mir-126 (targeting VCAM1 and HOXA9), mir-125b (targeting ERBB2), mir-145, and mir-146 (Lin 2009). In a recent review article, Gordeladze et. al. summarized reports on microRNA species like mir-29b (targeting HDAC4, TGFβ3, ACVR2A, CTNNBIP1, and DUSP), mir-125b (target not specified), mir-133b and mir-135a (both targeting SMAD5 and RUNX2), and mir-199b (target not specified) somehow being involved in osteochondral development and skeletogenesis (Gordeladze, Djouad et al. 2009). Other microRNAs involved in TF-interactions belong to the microRNA family of mir-23a-27a-24-2 (targeting APC2, RUNX2 and SATB2) (Hassan, Gordon et al. 2010).

Other studies involving differentiation of human mesenchymal stem cells into osteocytes and chondrocytes implicated a different subset of miRNAs. Mir-638 and mir-663 were found to be up-regulated in chondrocytes, while mir-24, let-7a, let-7b, let-7c, mir-138, and mir-320 were associated with osteocyte maturation (Lakshmipathy, Love et al. 2007). We have also found (Gordeladze et al., unpublished observations) that the microRNA species 638 and 663 are up-regulated in chondrocytes as early as 3 days of differentiation from MSCs, but these microRNAs are also heavily up-regulated in Th-17 cells differentiated from CD4+ naive T-cells after 5 days (Yssel et al., unpublished observations). Mir-638 and mir-663 appear to have the following targets in common (JUN, FOSB, SP3, and MYC, all of which are important for osteoblastogenesis). Finally, a recent survey of the literature revealed that several microRNA species (mir-335-5p, mir-27, and mir-29) directly target molecules involved in the Wnt-signalling pathway (Kapinas, Kessler et al. 2009; Kapinas, Kessler et al. 2010; Wang and Xu 2010; Zhang, Tu et al. 2011).

MicroRNAs directly targeting specific gene markers of bone and cartilage structural ECM molecules have not been indisputably identified, however, many microRNAs have been shown to affect the steady state levels of such molecules, though probably indirectly. These microRNAs are mir-34a, mir-675, mir-21, mir-146a (COL2A1) (Yamasaki, Nakasa et al. 2009; Abouheif, Nakasa et al. 2010; Dudek, Lafont et al. 2010; Kongcharoensombat, Nakasa et al. 2010), mir-140 (COL2A1, ACAN) (Miyaki, Nakasa et al. 2009), and mir-29b (also directly targeting COL1A1, COL5A3, AND COL4A2, as evidenced by the use of 3'-UTR reporter assays) (Li, Hassan et al. 2009).

Interestingly, the expression of microRNAs in osteoblasts and chondrocytes seems to be reciprocal, in the sense that the microRNA species highly expressed in chondrocytes are virtually absent in osteoblasts (and vice versa). A series of articles on microRNA expression profiles in osteoblasts grown in a variety of scaffold composite material have recently been published (Annalisa, Furio et al. 2008; Palmieri, Pezzetti et al. 2008; Palmieri, Pezzetti et al. 2008; Palmieri, Pezzetti et al. 2008), however, the microRNA profiles are not overtly compatible with single microRNA studies where proof-of-microRNA-binding has been shown.

The possible role of microRNAs in disease processes like rheumatoid arthritis (RA) and osteoarthritis (OA) has been addressed, and a significant up-regulation of mir-155 and miR-146a in synovial fibroblasts (RASFs) and synovial fluid derived from patients with RA have been documented (Stanczyk, Pedrioli et al. 2008; Duroux-Richard, Presumey et al. 2011). These findings unite the concepts of cell-specific microRNA signatures and microRNA-exchange between cells in the form of exosomes (Valadi, Ekstrom et al. 2007; Zomer, Vendrig et al. 2010). Lastly, predicted polymorphisms in binding sequences for mir-146 in the promoter region of the FGF2 have been found (Lei, Papasian et al. 2011), implicating microRNAs even more closely with development and treatment of disease states.

4. Gene and microRNA manipulations and selection of delivery systems

Osteoblasts and bone engineering

Biologists have identified several bioactive factors being able to induce or support bone generation, including BMPs, TGFβ, IGF-1, FGFs, LIM mineralizing protein-1 (LMP-1), VEGF and caALK2 (activin-receptor like kinase-2, mediating BMP-signalling) (for review, see (Betz 2008)). Delivery systems frequently used are viral vectors, adenoviral vectors, retroviral and lentiviral vectors, adeno-associated vectors (AAV), and non-viral vectors (mostly plasmids). Standard transfer procedures comprise electroporation, lipofection and gene-activated matrices (GAM) (Betz 2008). Other osteogenic factors of interest for gene manipulation are RUNX2, SMADs, DLX3, DLX5, AP-1, and SP7 (osterix) (Marie 2008; Gordeladze, Djouad et al. 2009).

Gene therapy may be based on single genes, however, more successful attempts have been made by using combinations of genes, such as BMP + VEGF, BMP2 + RUNX2, VEGF + RANKL, BMP-2 + IGF-1, and BMP-2 + SMAD8 (Gersbach, Phillips et al. 2007). Another strategy is to deliver the above mentioned osteoinductive growth factors or hormones in a scaffold material to render a "kick-start" in terms of osteoblast function, cell organization and bone building (Fischer, Kolk et al. 2011). Other applications of gene therapy for osteogenesis, such as for periodontal and craniofacial regeneration, have been described elsewhere (Scheller and Krebsbach 2009; Rios, Lin et al. 2011).

Chondrocytes and cartilage engineering

The concepts of gene therapy for cartilage repair have been thoroughly reviewed by Steinert et al. (Steinert, Noth et al. 2008). Approaches mentioned are stimulation of chondrogenic differentiation (using TGFβs, BMPs, WNT, SMADs, SOX9, Brachyury), stimulation of cartilage matrix synthesis and/or cell proliferation (TGFβs, BMPs, IGF-1, PDGF, type 2 Collagen minigene, COMP, GlcAT-1), inhibition of osteogenesis/hypertrophy growth factors (Noggin, Chordin, PTHrP, SMAD6,7), the use of anti-inflammatory agents (IL-1 blockage, TNFα-inhibition, MMP-inhibition), senescence inhibition, and inhibition of apoptosis (Saraf and Mikos 2006; Steinert, Noth et al. 2008). The delivery systems for chondrogenic genes show many common features to the ones described for enhancing osteoblastogenesis (see above) (Saraf and Mikos 2006; Betz 2008).

5. Choice of humoral factors for differentiation purposes

The differentiation of progenitor cells to osteoblasts or chondrocytes *in vitro* has been conducted in media containing differentiating factors like Calcitriol, Dexamethasone, BMP2, IGF-1, PDGF, EGF, FGFs, TGFβs, HGF, PTH/PTHrP, (Logeart-Avramoglou, Anagnostou et al. 2005; Steinert, Noth et al. 2008; Tilg, Moschen et al. 2008; Gordeladze, Reseland et al. 2009; Boeuf and Richter 2010; Levi and Longaker 2011; Witkowska-Zimny and Walenko 2011). The choice of such factors, either as a single remedy, or in combinations, most certainly will affect cell phenotype acquisition in different ways (Kulterer, Friedl et al. 2007; Grundberg, Brandstrom et al. 2008; Duggal, Fronsdal et al. 2009; Sundelacruz and Kaplan 2009; Bernstein, Sticht et al. 2010; Granchi, Ochoa et al. 2010; Piek, Sleumer et al. 2010; Herlofsen, Kuchler et al. 2011). Thus, the outcome of the differentiation process is not easy to predict.

In vitro differentiation normally requires fetal bovine serum (FBS), however, FBS rises a concern over infections, possible immunological reactions to xenogenic peptides and inorganic compounds (like non-human sialic acid) (Hattori, Nogami et al. 2008). Hence, the use of serum-free incubation media is warranted. It has been shown that MSCs grown in serum free-media will acquire both osteoblast and chondrocyte phenotypes when exposed to EGF and bFGF, stimulating the ERK-pathway (Solmesky, Lefler et al. 2010), and similar results have been obtained by others (Gigout, Buschmann et al. 2009; Felka, Schafer et al. 2010). Waese et al. report on a one-step successful generation of chondrocytes in a serum-free monolayer system (with the addition of TGFβ3 or BMP-4) (Waese and Stanford 2011), and several articles underscore the importance of serum source for optimal differentiation and inhibition of senescence in engineered chondrocytes (Shahdadfar, Fronsdal et al. 2005; Dahl, Duggal et al. 2008; Duggal and Brinchmann 2011).
Hence, the combination of factors inducing optimal differentiation and the selection of serum-free media to produce good chondrocytes and osteoblasts for tissue engineering purposes represents a major task to elucidate.

6. Properties of scaffold materials in bone and cartilage engineering

Trauma (including bone fractures and cartilage destruction), cancer metastases, rheumatoid arthritis and osteoarthritis represent a therapeutic challenge, which previously has been approached by implanting autologus tissues (Gordeladze, Reseland et al. 2009; Torroni 2009; Giannoudis and Dinopoulos 2010; Khan, Johnson et al. 2010; Lu, Subramony et al. 2010; Takeda, Nakagawa et al. 2011). The modern approach of using scaffolds as artificial cell- and tissue-supporting material is promising and has been extensively reviewed by Sundelacruz and Kaplan (Sundelacruz and Kaplan 2009). Basicly, che choice of scaffold biomaterial and biocompatibility is vital for support of cell proliferation, differentiation, and suitability for implantation *in vivo*. Secondly, the geometry and architecture is important determinants of support of 3D tissue growth, control of morphology of the growing tissue, support of cell proliferation, and favourisation of cell differentiation into particular lineages. Thirdly, the porosity of the scaffold is important for the support of cell differentiation, recruitment, aggregation, and vascularisation. Furthermore, the mechanical properties, degradation rate, and biochemical stimuli are determinant of the scaffold's ability to permit new tissue ingrowth, allow remodelling of the ECM formed, match the healing rate of the new tissue, and stimulate progenitor cells to assume a functional and stable cell phenotype. The following scaffolds have been tested in different ostechondral tissue engineering settings: PET, PLDL, PLA, PGA, PLGA, HA, TCP, and silk fibroin (porosity and pore size), HA, TCP, various synthetic polymers and co-polymers, polymer-ceramic composites (pore inter-connectivity), natural synthetic polymers, including collagen, silk, PLGA, and PCL (degradation), natural synthetic polymers, bioactive glasses and ceramic material (mechanical strength), and finally PLGA, CaP, TCP, chitosan, HA, collagen, and silk fibroin (incorporation of biochemical signalling).
In order to arrive at the very best system for tissue engineering, large experimental permutations of the above mentioned factors including cell sources, humoral factors and gene therapeutic approaches, should be performed to obtain the better cell phenotype for osteochondral tissue replacement.
However, some recent articles featuring the use of polymeric scaffold structures in osteochondral engineering deserve citation here. Hydrogels incorporating agarose, alginate,

collagen, hyaluronic acid polymer and gelatine have been successfully applied to support stem cell differentiation and 3D-structuring (Vinatier, Bouffi et al. 2009; Vinatier, Mrugala et al. 2009; Hunt and Grover 2010). MSCs embedded in fibrin hydrogel showed superior differentiation to osteoblasts compared to cells grown in monolayers, however, they did not assume a preferred phenotype after 28 days of incubation. Tiainen and co-workers have reported on an ultra-porous titanium oxide (TiO_2) scaffold with high compressive strength (above 2.5 MPa at 80-90% porosities) (Tiainen, Lyngstadaas et al. 2010), satisfying criteria for mechano-stimulation and pore size favouring cell differentiation, recruitment, aggregation and vascularisation. Another report on TiO_2 confirms its applicability in producing a proper bone replacement material (Sabetrasekh, Tiainen et al. 2011). Rahman et al. used bioactive glass, which, despite its brittleness, showed physical characteristics favouring neo-vascularisation being necessary for the perpetuation of engineered bone, when implanted *in vivo* (Rahaman, Day et al. 2011). And finally, it should be mentioned that osteoblast-like cells cultured in a bone-mimicking material made of poly-L-lactide + carbon nanotubes + micro-hydroxyapatite differentiated well into proper, bone-forming osteoblasts, as ascertained by genetic profiling (van der Zande, Walboomers et al. 2010).

Sabetrasekh and co-workers showed that Hydroxylprolyl-methyl Cellulose Hydrogel (Histocare™) (Sabetrasekh 2011) supported the differentiation of MSCs and preosteoblasts and cell clusters forming an artificial tissue favouring cell-cell interactions. Duggal et al. (Duggal, Fronsdal et al. 2009) showed that MSCs exposed to high-guluronic tripeptide arginine-glycine-aspartic acid (RDG) alginate scaffolds, facilitating binding to integrin, differentiated well into chondrocytes in the absence of any growth factors. Integrins are extracellular receptors conveying mechano-stimulation to the interior of the cell (Liu, Calderwood et al. 2000; Weyts, Li et al. 2002; Kapur, Baylink et al. 2003; Gordeladze, Reseland et al. 2009), and the use of RDG alginate scaffolds makes the addition of growth factors less critical for chondrocyte phenotype acquisition (as shown by transcriptome analyses). Hyalouronan (HYAFF-11®) scaffolds have been shown to produce useful cells for osteochondral tissue replacement, provided that MSCs were applied instead of ASCs in the presence of TGFβ1 (Loken, Jakobsen et al. 2008; Jakobsen, Shahdadfar et al. 2010). Finally, it should be mentioned that scaffold material (e.g. polycaprolactone, PLGA/Hap/, Collagen/Hap, agarose/gelatine hydrogel, polyacryl-amide hydrogel, PLGA nanofiber, agarose gel, polyacryl-amide gel, poly(2-hydroxyetylmethylmethacrylate) micro-porous gel, and silk fibroin) made with a gradient in pore size (Sundelacruz and Kaplan 2009; Oh, Kim et al. 2010; Seidi, Ramalingam et al. 2011) is especially well suitable for interface (i.e. ligament-to-bone, tendon-to-bone and cartilage-to-bone) tissue engineering.

The concept of scaffolds/biomaterial presently extends to include biopolymers, self assembled systems, nanoparticles, carbon nanotubes and quantum dots (Williams 2009). This definition also includes micro-structured surfaces (Kolind, Dolatshahi-Pirouz et al. 2010), shown to inhibit cell proliferation and favour differentiation, as well as UV-bioimprinting of single cell surfaces (Muys, Alkaisi et al. 2006), favouring propagation of surface-cell-cell interactions, ensuring proper development of a defined cell phenotype in a 3D-structure. Application of the scaffolds principle to create a functional 3D-tissue structure can also be refined to what is called organ printing. Organ printing is a process which is scaffold free or involving hydrogels, and is defined as layer-by-layer additive robotic bio-fabrication of 3D-functional living macro-tissues and organ constructs using tissue spheroids as building blocks. These spheroids are subject to tissue fusion, constituting the final 3D-structure of living tissue (Mironov, Visconti et al. 2009). The principles consists of three steps, including a) the production of homo-cellular aggregates, b) building hetero-cellular aggregates, and finally c) the assembly of organ-mimetics containing a 3D-vascular bed (Mironov, Visconti et al. 2009; Visconti, Kasyanov et al. 2010).

Organ print design of tissues may solve some of the problems encountered in osteochondral tissue engineering, namely vascularisation of bone tissue and gradient expression of genes from the luminal space to the bone interface of chondrocytes in hyaline cartilage, due to a lack of blood-born delivery of nutritional substances and oxygen (Salim, Nacamuli et al. 2004; Gibson, Milner et al. 2008). Transient changes in oxygen tension inhibit osteogenic differentiation, as demonstrated by reduced transcription of gene classes related to angiogenesis, family of matrix proteins, HIF-1α, as well as RUNX2, osteocalcin, and COL1A1 (Salim, Nacamuli et al. 2004). As for chondrocytes, it has been shown that high O_2 tension makes them shift from producing normal articular isoforms of collagen (types II, IX, and XI) to collagen types I, III, and V (Gibson, Milner et al. 2008). High O_2 levels also suppress the expression of SOX9, necessary for chondrocytic differentiation and Aggrecan synthesis (Murphy and Polak 2004). Interestingly, the microRNA species mir-210 has been shown to be enhanced by HIF-1α, thus improving tissue tolerance to low O_2 levels (Huang, Le et al. 2010). This is consistent with the fact that mir-210 is down-regulated in dedifferentiated human articular chondrocytes assuming a more fibroblast/stem cell like phenotype (Karlsen, Shahdadfar et al. 2011).

Hence, there exists a plethora of scaffold materials to be considered, when optimal osteochondral tissue engineering conditions are to be defined.

7. Selection of stem cell niches and/or cell co-cultures

MSCs can be obtained from various tissues (Aicher, Buhring et al. 2010). Today the main source for isolation of MSCs in mammals is the bone marrow. However, bone marrow and other sources including placenta and adipose tissue contain MSCs displaying heterogeneous cell populations. Only a restricted number of appropriate stem cell markers have been explored so far, and it seems that the expression profile of CD-molecules differ on MSCs isolated from bone marrow, trabecular bone, dental pulp, articular cartilage, synovial membrane, adipose tissue, perivascular sites, term placenta, amnionic fluid, umbilical cord and pancreas (Bartholomew, Sturgeon et al. 2002; Dean and Bishop 2003; Le Blanc, Tammik et al. 2003; Niemeyer, Krause et al. 2006; Drosse, Volkmer et al. 2008; Gordeladze, Reseland et al. 2009; Aicher, Buhring et al. 2010). Knowledge of the phenotypical characteristics and the functional consequences of such subsets of MSCs might allow the development of improved regimens for regenerative medicine. MSCs, which express the specific cell adhesion molecule CD146, also known as MCAM, are well suited for bone repair. MSCs expressing CD56, CD146 and/or CD271 seem to be adaptable for the regeneration of bone, cartilage and intervertebral disks (Ohishi, Chiusaroli et al. 2009; Aicher, Buhring et al. 2010).

Using two or more MSC niches may thus prove beneficial for the generation of bone tissue (Matsumoto, Kuroda et al. 2008; Grellier, Bordenave et al. 2009). CD34-positive, VEGF-secreting endothelial/skeletal progenitor cells have been shown to enhance the vascularisation and speed up fracture healing (Matsumoto, Kuroda et al. 2008). Such progenitor cells are normally recruited to the bone-forming site by the CXR4/SDF-1 pathway (Otsuru, Tamai et al. 2008). Grellier and co-workers have reviewed the literature as to the use of co-cultures of osteogenic and endothelial cells (Grellier, Bordenave et al. 2009). They describe combinations of osteogenic cells and endothelial cells like osteoblasts, osteoprogenitor cells, umbilical vein endothelial cells, endothelial progenitor cells, saphenous vein endothelial cells, outgrowth endothelial cells, and dermal vascular endothelial cells cultured in 2D- or 3D-structures of various scaffold materials (Villars, Bordenave et al. 2000; Wenger, Stahl et al. 2004; Kaigler, Krebsbach et al. 2005; Stahl, Wu et al. 2005; Kaigler, Krebsbach et al. 2006; Clarkin, Emery et al. 2008; Guillotin, Bareille et al. 2008; Grellier, Ferreira-Tojais et al. 2009; Grellier, Granja et al. 2009) where the endothelial cells form a tubular structure surrounded by ECM-producing and mineralizing osteoblasts (Grellier, Bordenave et al. 2009).

In conclusion, co-cultures of niches of MSCs and/or vascularisation of appropriate scaffolds (e.g. scaffolds supporting "asymmetric" differentiation of tissue-generating cells) might secure a better functional and long-lasting engineered tissue.

8. Mechano-stimulation of progenitor cells during differentiation

Mechano-biology is a relatively new research field, where most of the insight related to osteochondral tissue engineering comes from embryonic skeletal development (Nowlan, Sharpe et al. 2010). However, the "mechanostat" principle was launched several decades ago by Frost and colleagues (Frost 2003; Skerry 2006; Mulvihill and Prendergast 2008). Genetic lesions or immobilization (surgical or drug-induced) lead to muscle less limbs, reduced muscle fibre size/number, or non-contractile muscles, and to underdeveloped joints and bones, mostly due to a lack of mechano-stimulation (Gomez, David et al. 2007; Kahn, Shwartz et al. 2009; Nowlan, Bourdon et al. 2010; Nowlan, Sharpe et al. 2010).

Several humoral factors, growth factors and receptors/ECM-protein/anchoring proteins share important signalling pathways, thus eventually leading to osteochondral differentiation of progenitor cells, for review, see (Gordeladze, Reseland et al. 2009; Kelly and Jacobs 2010; Potier, Noailly et al. 2010). Osteochondral progenitor cells may be subjected to shear stress (by fluid flow), compressive load (scaffold compression, hydrostatic pressure), or stretching (uni-, bi-, or equi-axial) leading to both proliferation and differentiation (Potier, Noailly et al. 2010). Several mechano-modulatory regimens (featuring detailed molecular mechanisms, type of mechano-stimulation, mechanical load applied, static or intermittent load, frequencies, as well as time frame during osteoprogenitor cell differentiation) using both 2D- and 3D-incubation systems, have extensively been described elsewhere (Angele, Schumann et al. 2004; Huang, Hagar et al. 2004; Woods, Wang et al. 2005; Campbell, Lee et al. 2006; Miyanishi, Trindade et al. 2006; Sumanasinghe, Bernacki et al. 2006; Mauck, Byers et al. 2007; McMahon, Campbell et al. 2008; McMahon, Reid et al. 2008; Thorpe, Buckley et al. 2008; Wagner, Lindsey et al. 2008; Arnsdorf, Tummala et al. 2009; Arnsdorf, Tummala et al. 2009; Gordeladze, Reseland et al. 2009; Haudenschild, Hsieh et al. 2009; Li, Kupcsik et al. 2010). However, the permutation of various factors enlisted above, yielding the optimal osteochondral cells for further studies *in vivo*, is difficult to envisage.

Cell shape, determined by the RhoA-Rho kinase = ROCK (influencing the actin cytoskeleton), has received much attention as a controller of cell development (McBeath, Pirone et al. 2004; Arnsdorf, Tummala et al. 2009; Kelly and Jacobs 2010). This has renewed the interest in scaffold material made by nanotechnology, which is able to deliver 2D- and 3D-surfaces mimicking the ultimate surface pattern of osteoblasts and chondrocytes encountered in live tissues (Muys, Alkaisi et al. 2006; Kolind, Dolatshahi-Pirouz et al. 2010).

9. Stabilization of the osteoblast and chondrocyte cell phenotypes

In order to succeed replacing tissues like bone and cartilage, it is vital that the differentiated cells, whether pre-embedded in scaffolds or not, do not develop tumours or alter phenotype within a short period after implantation. The preferable phenotype should not lose acquired features or assume new ones. However, it has been speculated that engineered osteoblasts may be subject to premature senescence, acquire "drag-over" adipocyte characteristics, lose their ECM-synthesizing and mineralizing ability, while also enhancing osteoclast-mediated resorption yielding negative bone mass through multiple remodelling cycles. Furthermore, engineered hyaline cartilage chondrocytes may possibly shift their collagen-synthesizing and non-collagenous ECM producing profile towards hypertrophic and mineralizing chondrocytes. And chondrocytes may also recruit, activate and over-stimulate osteoclasts to resorb adjoining bone structures. Finally, it should be mentioned that engineered cartilage to replace hyaline articular cartilage also will be subject to remodelling, e.g. via the IL-1 induced Syndecan4-ERK-MMP3-ADAMT5 cleavage of Aggrecan, which is up-regulated in osteoarthritic joints (Bertrand, Cromme et al. 2010). It should also be mentioned that immune cells (e.g Th-17 cells) secrete interleukins known to differentiate and activate osteoclasts from monocytes (Weitzmann and Pacifici 2007; Adamopoulos and Bowman 2008; Tilg, Moschen et al. 2008; Hanada, Hanada et al. 2010; Pacifici 2010), and that chondrocytes exposed to exosome-like structures or certain microRNA antago- or pre-mirs (e.g. antagomir-222), are detrimental to the chondrocyte phenotype (Gordeladze et al., unpublished observations).

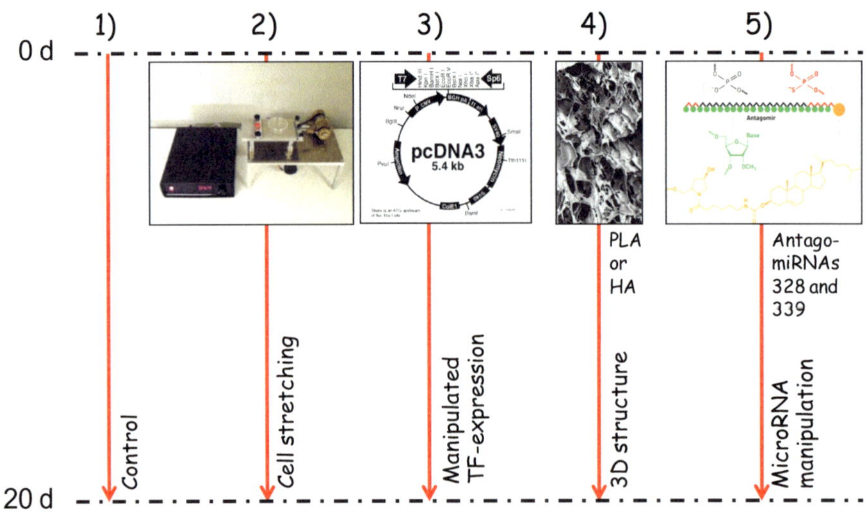

Fig. 1A. Osteoblast differentiating scheme. Human mesenchymal stem cells (hMSCs) were incubated for 20 days in standard differentiating medium (containing dexamethasone), subjected to mechanical loading, transfected with the pcDNA3-Runx2 containing plasmid, grown in a 3D-lattice (PLA- or HA-based scaffolds), or transfected with antagomiRNAs corresponding to mir-328 and mir-339

It is therefore suggested that gene manipulations (at least temporal transcription control) should be considered as part of a strategy to create and stabilize *in vivo* engineered bone or cartilage for tissue replacement. Potentially, one should consider the transient manipulations of microRNAs, since these short RNA-molecules are known to interfere with a plethora of cell specific transcriptions factors (Gordeladze, Djouad et al. 2009). MicroRNAs are also targeting epigenetic factors (Roach and Aigner 2007; Dahl, Duggal et al. 2008; Haberland, Montgomery et al. 2009; Lee, Jung et al. 2011; McGee-Lawrence and Westendorf 2011) like HDACs involved in the differentiation of stem cells and stabilization of various cell phenotypes (Li, Xie et al. 2009; Li, Hassan et al. 2009; Lee, Jung et al. 2011).

10. Bone and cartilage engineering revisited

Permutation of factors influencing cell phenotypes

There are numerous reports in the literature featuring the results of manipulations of single or a few variables known to affect the result of cell engineering based on stem cell or progenitor cell differentiation towards osteoblasts or chondrocytes to be implanted to heal osteochondral tissue lesions. These factors relate to cell source(s), application of growth factors, the use of gene therapy, application of mechano-stimulation and the selection of scaffold material (Isogai, Kusuhara et al. 2006; Gordeladze, Reseland et al. 2009; Aicher, Buhring et al. 2010; Granchi, Ochoa et al. 2010).

To find the combination of factors rendering engineered cells functional enough to assume a "proper" phenotype, generating tissues not deviating from their original counterparts with given characteristics, represents a painstaking task. It seems insurmountable, since the number of permutations necessary to explore all possible additive or synergistic interactions are numerous. It is therefore probably a good approach to define a set of measurable end-point characteristics for osteochondral tissues to evaluate the experimental steps taken, when going from bench to patient. Osteochondral tissues represent certain geometrical and mechanical properties (Knecht, Vanwanseele et al. 2006; Gordeladze, Reseland et al. 2009), as well as gene expression profiles (Grundberg, Brandstrom et al. 2008; Duggal, Fronsdal et al. 2009; Granchi, Ochoa et al. 2010; Herlofsen, Kuchler et al. 2011), which may guide the selection of major combinations of treatments, as envisaged by the permutation process. To shed light on this exercise, some bioinformatics exercises have been conducted, and some selected experiments have been described.

Permutations encompassing mechano-stimulation, 3D-growth, and manipulations of genes and microRNAs

MSCs were differentiated in standard media towards osteoblasts or chondrocytes, by subjecting them to cyclical mechano-stimulation (uni-axial stretch), gene manipulations, growth in 3D-lattices, and finally to manipulations of microRNA levels. The following test battery was used: Q-PCR analyses of osteoblast and chondrocyte "specific" transcription factors (TFs) and marker genes (e.g. RUNX2, OSTERIX = SP7, VDR, RANK-L, OPG, SOX9, GLI3, FOXO3A, WNT5A, ALPL, COL1A1, OSTEOCALCIN, OSTEOPONTIN, COL2A1, COL10A1, AGGRECAN); Q-PCR of mir-326, mir-339, mir-24, and mir-149; immunohistochemistry of COL2a1 and AGGRECAN; cell staining using Alizarin Red S and Alcian Blue; mineralization (radiology and histology) in SCID mice (*m. tibialis*); GAG/DNA-ratio, clinical score for micropellets and alginate beads; osteoclast resorption assay (using PBMCs + RANK-L/CSF-U on dentine slices). Some of the results obtained with MSCs differentiated towards osteoblasts are referred to in Figures 1 and 2, panels A and B.

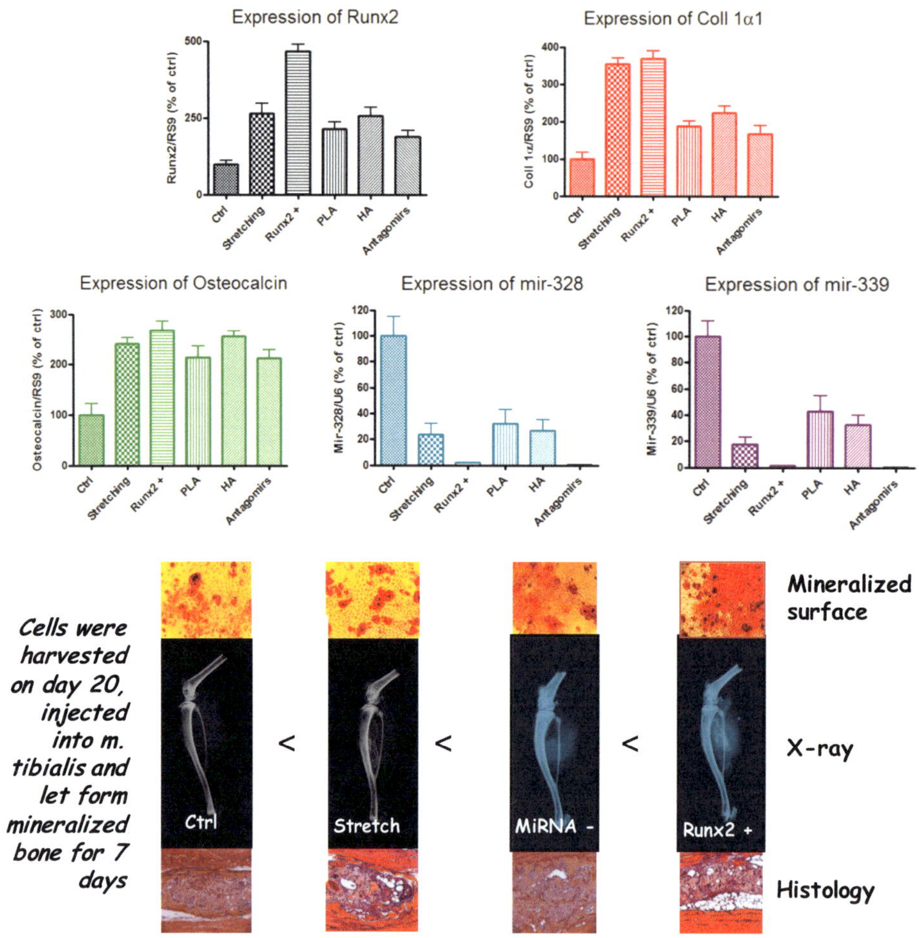

Fig. 1B. Selected results of the experiment described in Figure 1A. Mechanical loading was performed in monolayers using uni-axial cell diameter alteration by 1000 µE (1E = 1 micro-strain = 1/1,000,000 alteration of the cell's diameter) for 30 min every other day. The antago-miRNAs were transfected (by lipofection) into cells in monolayers every 5 days. Expression of genes like Runx2, Collagen-1, Osteocalcin and the microRNAs 328 and 339, were performed using Q-PCR. Furhtermore, Alizarin Red staining (indicating mineralized surface) was performed at day 20, and cells being deposited (for an additional 7 days) in the tibial muscle of SCID mice were X-rayed, harvested and subjectd to histological analyses

Figure 1A indicates the manipulation of MSCs grown in: 1) osteoblast differentiating medium 2) in mono-layers, 3) exposed to mechano-stimulation, 4) subject to up/down-regulation of TFs, and 5) grown in PLA- or HA-scaffolds (cylinders), or 5) transfected with pre- or antago-microRNAs. Figure 1B features some of the results of these single manipulations, indicating that RUNX2 over-expression is superior in terms of osteoblast differentiation, however, mechano-stimulation, and suppression of mir-328 and mir-339 also give promising osteoblasts for *in vivo* implantation.

Figure 2A describes the manipulation of MSCs grown in chondrocyte differentiating medium 1) in mono-layers, 2) exposed to mechano-stimulation, 3) subject to up/down-regulation of TFs, 4) grown in alginate beads or micropellets, or 5) transfected with pre- or antago-microRNAs. Figure 2B summarizes selected results of these single manipulations, indicating that suppression of RUNX2 is no better than incubation in micropellets or alginate beads, or transfecting the cells with premir-24 and premir-149. All in all, manipulating the microRNA species seem to give superior results.

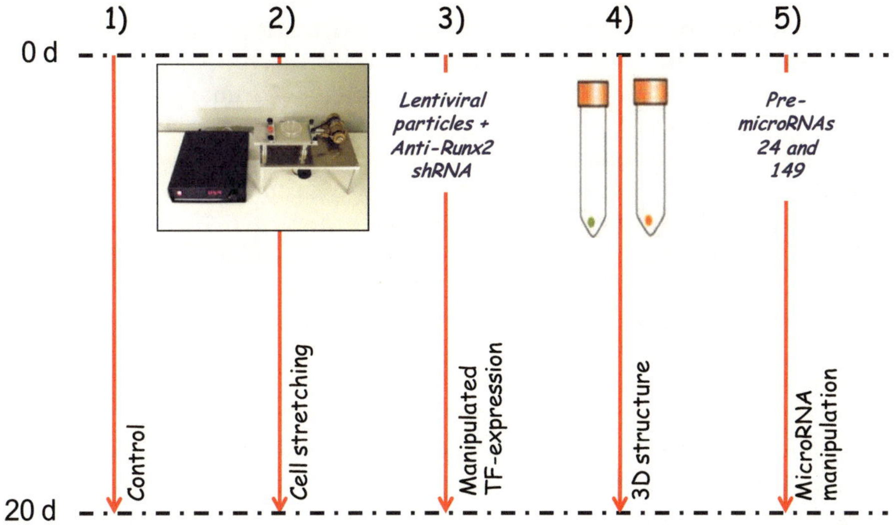

Fig. 2A. Chondrocyte differentiation scheme. Human mesenchymal stem cells (hMSCs) were incubated for 20 days in standard differentiating medium (containing TGFβ3), subjected to mechanical loading (1000 μE), infected with anti-Runx2-shRNA (contained within a lentiviral construct), grown in a 3D-lattice (micropellet or alginate), or transfected with premiRNAs corresponding to mir-24 and mir-149

From the above experiments, a 20 day differentiating scheme was envisaged, where gene- and microRNA-manipulated MSCs, grown in standard differentiating media, were mechano-stimulated for 10 days and thereafter incubated for another 10 days in a 3D-structure (HA-scaffold for osteoblast, and alginate beads for chondrocytes). These incubation schemes are shown in Figure 3A, while results of the experiments are summarized in Figure 3B. By combining the different manipulations, it was shown that osteoblast and chondrocyte "specific" markers were enhanced some 3-4 fold over control MSCs differentiated in mono-layers compared to 2-3 fold for single condition manipulations. To assess the influence of inflammation (using incubation media containing interleukins and TNFα) on osteoblast or chondrocyte phenotype stability and osteoclast activation, cells were exposed to IL-1β, IL-6, IL-17 and TNFα for 14 days. Osteoclasts differentiated from PBMCs for 7 days were then co-culture with the osteoblasts or chondrocytes, and resorption pit surface was assessed. It became quite clear that inflammatory cytokines were detrimental to the ostechondral cell phenotypes and microRNA profile, and they also enhanced their ability to stimulate bone resorption through activation of osteoclasts. From these experiments, it seems that one might chose transient microRNA manipulations in combination with either cell stretching or growth in scaffold/hydrogel, if a permanent gene manipulation (e.g. alteration of RUNX2- and possibly also SOX9-expression) may render the cells less prone to negative influence encountered within their new environment.

Bioinformatics networking using micro-arrays of translated RNAs and non-translated microRNAs

To elucidate the concept of permutations of variables pertaining to the differentiation of stem cells (SCs) to become preferred osteoblasts or chondrocytes for tissue replacement, we will present an interesting exercise with transcriptomes, microRNA micro-arrays and a literature survey. Based on osteoblast derived transcriptomes (Grundberg, Brandstrom et al. 2008; Granchi, Ochoa et al. 2010), featuring gene transcripts from cells in human trabecular hip bone explants, and differentiating human mesenchymal stem cells (MSCs) undergoing differentiation and mineralization phases, respectively, a combined transcriptome of 188 genes were constructed. This transcriptome was run against two microRNA micro-arrays obtained from a) human MSCs differentiated to osteoblasts within a hydroxyapatite (HA) scaffold for 28 days, and from b) human MSCs differentiated to osteoblasts in monolayers for 3 days only, using a bioinformatics program designated Mir@nt@n (Le Bechec, Portales-Casamar et al. 2011). Furthermore, a transcriptome of genes pertinent to the chondocyte phenotype, consisting of 261 genes, was compiled by Brinchmann et al. (Duggal, Fronsdal et al. 2009; Herlofsen, Kuchler et al. 2011). These MSCs, grown in PRONOVA-LVG alginate for 21 days, represented genes displaying the same time course over 21 days as did COL2A1 or CXCL12. The present transcriptome was run against two microRNA micro-arrays obtained from a) human chondrocytes embedded in hyaline cartilage and dedifferentiated for 28 days, and b) human MSCs differentiated to chondrocytes in micropellets for 3 days.
The bioinformatics procedure featuring some comprehensive examples is given in Figure 4A. Twelve genes involved in WNT- and NOTCH-mediated signalling (according to KEGG's pathways) and a set of fourteen transcription factors (TFs) known to be important for osteoblastogenesis (Komori 2006; Marie 2008; Gordeladze, Djouad et al. 2009) were loaded into Mir@nt@n and two small networks emerged. All the microRNAs 16, 22, 24, 93, 125b, 141, 149, 200a and 206 have been shown to be down-regulated in osteoblastic cells (Gordeladze, Djouad et al. 2009; Lin 2009), which would be consistent with an up-regulation of TFs (SATB2, ETS1, and RNF11), and WNT (signalling molecule binding to FRIZZLED-LRP5/6) (Gordeladze, Reseland et al. 2009). However, NOTCH3 (known to inhibit osteoblastogenesis through interactions with the canonical WNT-pathway and Runx2) (Gordeladze, Reseland et al. 2009) would also be up-regulated. Interestingly, ETS1 seems to be involved in a regulatory network involving NOTCH3, RNF11, and six microRNA species, where mir-206 is reciprocally interacting with ETS1. Mir-206 is marginally down-regulated in osteoblasts, however, significant over-expression of this microRNA species in mice leads to bone loss (Inose, Ochi et al. 2009). Finally DKK2 (an inhibitor of LPR5/6) would be up-regulated, and the present prediction cannot be given a straight-forward, simple interpretation.

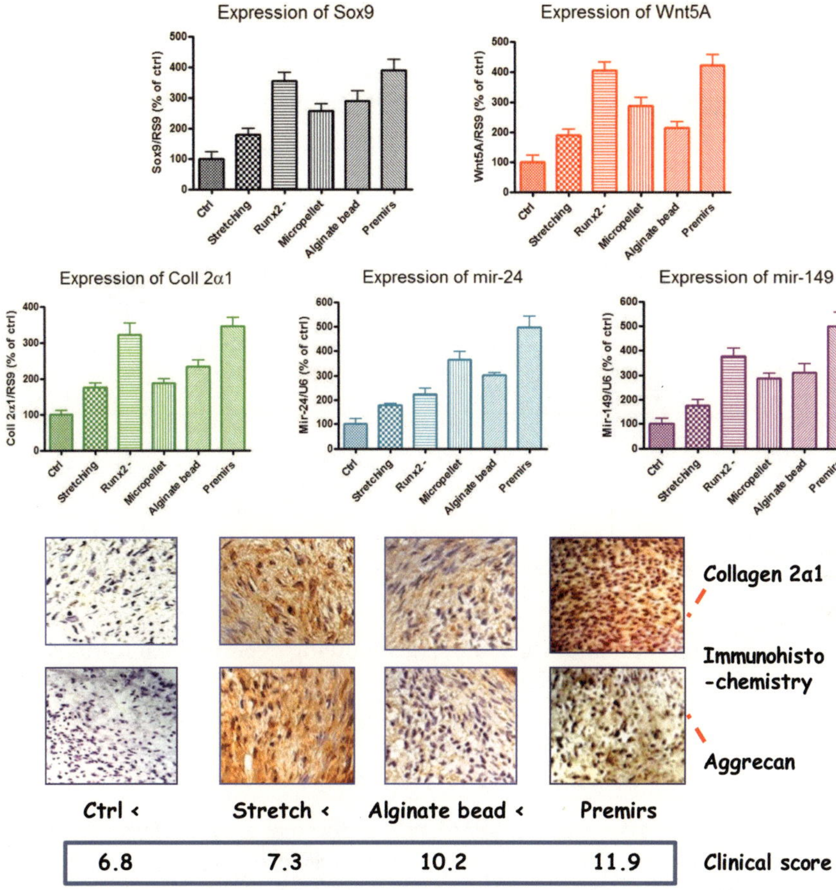

Fig. 2B. Selected results of the experiment described in Figure 2A. Mechanical loading of MSCs was performed in monolayers using uni-axial cell diameter alteration (1000 µE) for 30 min every other day. The pre-miRNAs were transfected into cells in monolayers every 5 days. Expression of genes like Sox9, Wnt5A, Collagen-2, and the microRNAs 24 and 149, were performed using Q-PCR. Furhtermore, immunohistochemistry of Collagen 2 and Aggrecan was performed, and clinical scoring (featuring GAG/DNA-ratio, immuno-histochemistry, and distance between cells in micropellets and alginate beads) were also measured

Out of seventeen putative interactions (reciprocal or not) between microRNAs known to be down-regulated in osteoblasts, eight are compatible with a direct inhibitory effect on translation, yielding 47% consistency according to the concept of microRNA-TF interactions (Zhou, Ferguson et al. 2007; Aguda, Kim et al. 2008; Hobert 2008; Do and Scholer 2009). The conclusion to be drawn from this example is that the list of marker genes and microRNAs describing the differentiation of MSCs to osteoblasts is too slim to warrant its use as a predictor of the acquisition of a proper osteoblast phenotype to be employed in bone replacement therapy. But, after all, the marker genes and microRNAs were all just picked from various, independent articles on osteoblast differentiation and from two KEGG's pathways charts.

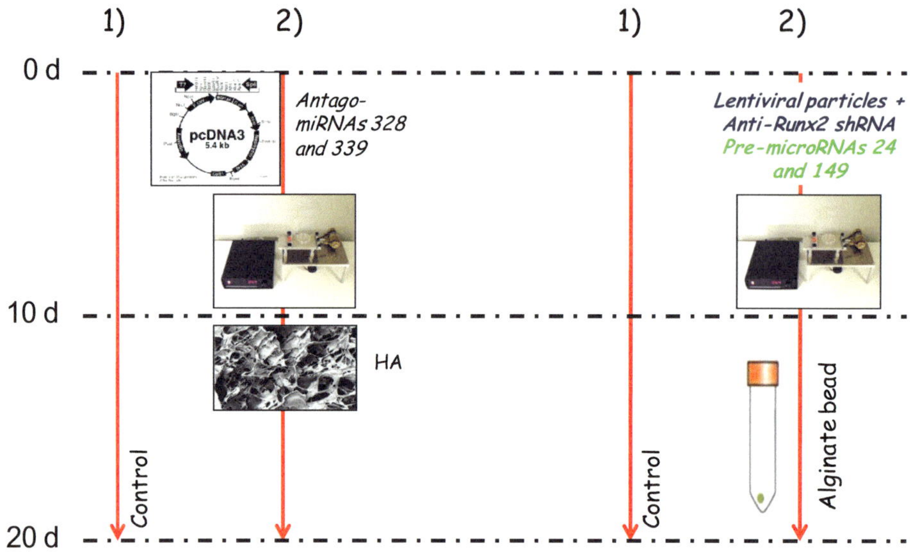

Fig. 3A. Multipurpose differentiation scheme. The following experimental settings were selected: Human MSCs were incubated in standard differentiating medium containing dexamethasone or TGFβ3 alone, respectively, or manipulated for 10 days interfering with gene expression, microRNA levels, and cell shape, and thereafter incubated in a 3D-stucture (HA-scaffold or alginate hydrogel, respectively) for another 10 days

More interestingly, Figure 4B describes the results of the use of the Mir@nt@n networking algorhitm, where the applied lists of target genes are based on cells derived from healthy human bone and cartilage, and the microRNA species are retrieved from micro-arrays obtained from the tabulated experiments. The four experimental conditions summarized here (involving on average some 30 microRNA species and 225 genes per experiment) clearly indicate that cell manipulations performed in a 3D-structure, and over a prolonged time frame of 28 days, yield a preferred osteoblast or chondrocyte phenotype, since the per cent compatibility demonstrated by microRNA - target gene interactions were 76% versus 19% (osteoblasts) and 88% versus 16% (chondrocytes), respectively.

Some examples of expected regulation patterns are given underneath: MSCs differentiated into osteoblasts in a HA scaffold show an up-regulated level of mir-143. In parallel, transcripts of putative target genes like DUSP2 (inactivates the MAPK pathway used by TNFs and TGFβs), BMP1 (involved in chondrogenesis), ID1 (belongs to the TGFβ pathway involved in chondrogenesis), TNFAIP6 (TNFα-induced protein 6), and FBN1 (sequestering TGFβ within ECM) were up-regulated. The first five interactions are expected, the last one is not (giving a per cent compatibility of 83%). Furthermore, the mir-29 family (MIR-29a,b,c) of microRNAs was down-regulated, and putative target transcripts of genes like COL1A1, COL4A1, COL4A5, COL5A1, COL21A1, BMP1, ID1, TNFAIP6, and FBN1 should be up-regulated. According to their alleged function, the interaction of the 29-family of microRNAs is expected when the collagens are concerned (Eyre 2002; Almarza and Athanasiou 2004; Goldring, Tsuchimochi et al. 2006; Davies, Chang et al. 2007; Shahdadfar,

Loken et al. 2008; Heinegard 2009; Van Agtmael and Bruckner-Tuderman 2010), but does not comply with the expected down-regulation of BMP1, ID1, and TNFAIP6. Cumulative compatibility score is now down to 73%. According to the literature (Li, Hassan et al. 2009), mir-29b does *de facto* bind to the 3'-UTR of the COL1A1, COL5A3 and COL4A2. Furthermore, the mir-29 family of microRNAs has also been shown to be involved in the regulation of Wnt-signalling through a positive feed-back loop (Kapinas, Kessler et al. 2010) and via suppression of SPARC (osteonectin) (Kapinas, Kessler et al. 2009). Scrutinizing the effect of mir-376c (being down-regulated) reveals that putative targets are DLX1 and RUNX2 (important TFs ensuring osteoblastogenesis), SPP1 (involved in bio-mineral tissue development and ossification) and PAFAH1B1 (involved in cell cycle adaptation to differentiation). Now, the cumulative compatibility score is 80%. The final cumulative score for this experiment (MSCs to osteoblasts in HA scaffold for 28 days) converged towards 76%. As expected, the 3 days incubation of MSCs seeded in culture flasks in osteogenic medium yielded a compatibility score of only 18%, where some of the microRNAs being modulated in the 28 days experiments with MSCs seeded into HA scaffolds did not appear as significantly altered (e.g. mir-376c).

Characteristics obtained with combined differentiation strategies (based on rank scores)	Osteo Control	Osteo Combined	Chondro Control	Chondro Combined
Effect on osteoblast differentiation (Q-PCR of TFs and marker genes)	100	326		
Effect on chondrocyte differentiation (Q-PCR of TFs and marker genes)			100	365
Effect on osteoblast differentiation (Mineralization/ALP-positive surface, *in vivo* mineralization and histology)	100	287		
Effect on chondrocyte differentiation (Alcian blue surface, GAG/DNA-ratio, immunohistochemistry, histological score)			100	345
Effect on engineered osteoblasts to "reverse" detrimental biological effect of inflammatory cytokines	100	388		
Effect on engineered chondrocytes to "reverse" detrimental biological effect of inflammatory cytokines			100	276
Effect on engineered osteoblasts to resist "increase" in pertinent microRNAs (i.e. MiR-328, -339)	100	445		
Effect on engineered chondrocytes to resist "loss" of pertinent microRNAs (i.e.MiR-24, -149)			100	412

Fig. 3B. Selected results obtained in the experiment described in Figure 3A showing synergism of the single manipulations used in combination. The following parameters were analysed: effect on cell differentiation, as estimated by Q-PCR of transcription factors (TFs) and marker genes, or as mineralized/ALP positive surfaces, *in vivo* mineralization in SCID mice and *de novo* bone tissue production (histology), and proteoglycan positive surface (Alcian blue colouration). Furthermore, the impact on cell phenotype stability upon exposure to cytokines (IL-1β, IL-6, IL-17, and TNFα) in terms of osteoclast activation and microRNA stability, was determined

Chondrocytes embedded within hyaline cartilage were dedifferentiated within their native matrix for 28 days, and a micro-RNA micro-array was obtained. Running these micro-RNA species using the Mir@nt@n algorithm together with the transcriptome of 261 genes gave a compatibility score of 88% (see Figure 4B). The following microRNAs and putative target gene transcripts should be mentioned: Mir-143 was up-regulated upon dedifferentiation, and putatively interacts with gene transcripts like SMO (involved in hedgehog = Hh activation of Gli1/2/3-mediated chondrogenesis) (Bale 2002; Takebe, Harris et al. 2011), COL1A1 (serves as bone matrix protein), WNT10B (involved in osteoblast differentiation), ADAMTSL1 (exhibits metalloproteinase activity), and HAS3 (synthesizes hyaluran). The mir-143 mediated suppression of all the above listed genes are expected when chondrocytes are dedifferentiated. Mir-140-3p was down-regulated and coupled to the modulation of gene transcripts like KLF4 (transcription factor activated by the Wnt-pathway) (Saulnier, Puglisi et al. 2011) , FOXQ1 (serves as a down-stream mediator of TGFβ1 signalling) (Feuerborn, Srivastava et al. 2011), CITED4 (serves as a co-activator of CEP/p300, TFAP2, and SMAD4 transcription factors involved in stem cell differentiation) (Braganca, Swingler et al. 2002), and PTCH4 (receptor activated by Hh, thus stimulating the SMO-GLI pathway of gene transcription) (Takebe, Harris et al. 2011).

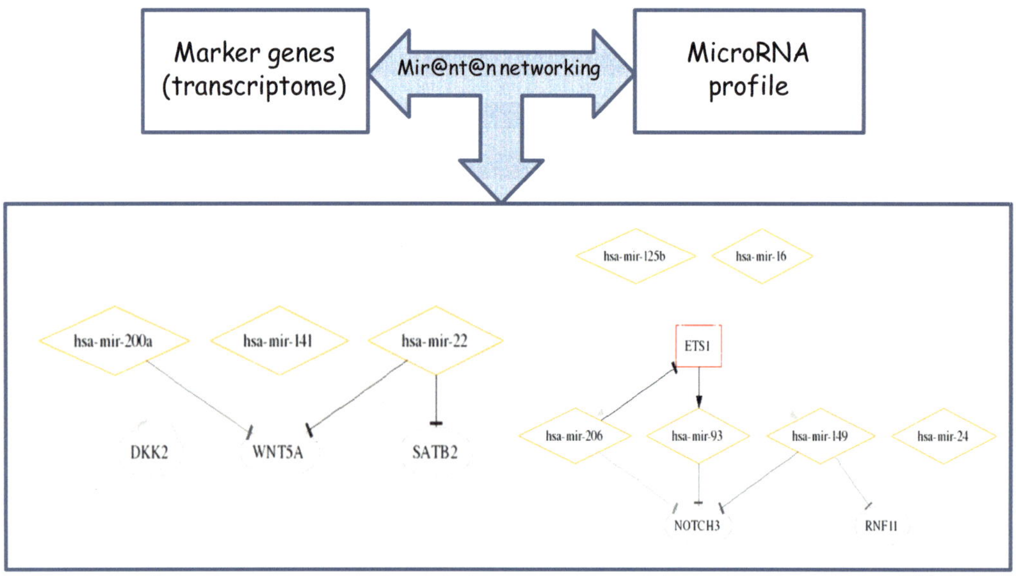

Fig. 4A. Bioinformatics-based marker gene and microRNA networking. A list consisting of marker genes taken from the canonical WNT- and the NOTCH-pathways (see KEGG's pathways), as well as transcription factors (TFs) and microRNAs demonstrated to be involved in differentiation of osteoblasts (Gordeladze, Djouad et al. 2009; Hassan, Gordon et al. 2010; Kapinas, Kessler et al. 2010) was loaded into the Mir@nt@n algorithm, searching for interaction networks. Within the complicated network obtained, two types of interactions emerged: 1) microRNAs target several gene transcripts (putatively binding to the 3'-UTR region of the subject mRNAs) (left-hand chart), and 2) microRNAs may be involved in regulatory loops with TFs (right-hand chart)

A fall in mir-140-3p is compatible with an up-regulation of the four above mentioned gene transcripts and loss of chondrocyte phenotype. So far, the compatibility score is 100%. Another microRNA up-regulated in dedifferentiated chondrocytes is mir-382, which putatively targets the transcripts of the following genes: PLCG2 (PLCγ2 activates NF-κB, AP-1, and NFATc1 induced gene expression important for osteoblastogenesis) (Chen, Wang et al. 2008; Marie 2008; Gordeladze, Reseland et al. 2009), DKK2 (serves as an inhibitor of the Wnt-signalling pathway), and RUNX3 (cooperates with RUNX2 to induce chondrogenesis through Hh synthesis) (Komori 2005). Mir-15b proved to be up-regulated upon dedifferentiation of the mature, hyaline cartilage-embedded chondrocytes, and putatively targets the following gene transcripts: FGF2 (involved in chondrogenesis) (Goldring, Tsuchimochi et al. 2006), CCND1, LRP6, FZD4 (Katoh 2007). All genes targeted by mir-382 and mir-15b are associated with the osteochondral phenotype and therefore, the compatibility score remains at 100%. Lastly, mir-21 and mir-495 were up-regulated upon dedifferentiation of the hyaline cartilage chondrocytes. Putative targets were SOX5 (transcription factor favouring chondrogenesis), MEF2C (necessary factor for collagen X transcription, and interacting with Dlx5/6 to enhance RUNX2 expression) (Solomon, Berube et al. 2008), and SOX6 (transcription factor favouring chondrogenesis, FGF7 (involved in chondrogenesis), CDH13 (predisposing factor along with TGFβ3, PTHR1, and PRG1 in ossification of ligaments of the spine) (Furushima, Shimo-Onoda et al. 2002), GLI3 (early transcription factor appearing during chondrogenesis) (Bale 2002; Takebe, Harris et al. 2011), respectively. This completes the random selection of microRNAs, however, at this point the compatibility score was still a staggering 100%. Subsequent to the analysis of all putative interactions, the score fell to 88%. The same exercise performed on microRNA-arrays from MSCs differentiated in micropellets for 3 days revealed a compatibility score of some 16% only, despite a similar number of microRNAs and gene transcripts significantly modulated compared to controls.

	Experiment conducted	Mir@nt@n networking		
		MicroRNA species	Gene transcripts targeted	Per cent (%) compatibility with expected modulation
Osteoblast: 188 genes	MSCs: differentiation in HA scaffold (3D) for 28 days	34	89 (14%)	76
	MSCs: differentiation in monolayers (2D) for 3 days	17	53 (28%)	18
Chondrocytes: 261 genes	Chondrcytes: Embedded in hyaline cartilage, dedifferentiated in 3D for 28 days	52	136 (52%)	88
	MSCs: differentiation in micropellets (3D) for 3 days	54	142 (54%)	16

Fig. 4B. Computation of compatibility score between osteoblast and chondrocyte transcriptomes and microRNA profiles. Gene transcript (mRNA) and microRNA networks were generated using osteoblast mRNA fingerprints from separate experiments (published in the literature) and microRNA-arrays from own experiments (see chapter text). Percentage of predicted mRNA-microRNA interactions in accordance with expected up-and down-regulation of gene expression in osteoblasts and chondrocytes were calculated. The higher the percentage, the better the differentiation process obtained, and (theoretically) the higher probability of success when using engineered osteochondral cells for tissue replacement

In conclusion, the more *in vivo* like incubation conditions, the more tissue-adapted osteoblasts and chondrocytes will be obtained when performing *in vitro* cell engineering. This exercise does not take into considerations all possible favourable factors (like stem cell source, differentiation media, optimal scaffolds, mechano-stimulation, gene-manipulations including phenotype protection by microRNAs etc.), but it is reasonable to believe that a permutation of selected conditions will aid in arriving at osteoblasts and chondrocytes highly suitable for long-lasting tissue replacements. Finally, it should be emphasized that one must improve on the selection of genes (and microRNAs) to constitute the preferred profile of proper osteoblasts and chondrocytes for successful tissue replacements.

Of special interest are the observations that the use of microRNA manipulations seems to protect the engineered osteoblasts and chondrocytes from losing their phenotypic characteristics in an environment where inflammation still is active, as well as protecting them from over-activating osteoclasts within the space (i.e. knee joint) where they might be replacing damaged tissue.

11. Summary and future perspectives

This chapter summarizes the concept of single factor permutations in order to arrive at the optimal scheme for generating osteochondral cells for tissue replacement. To be considered is the use of trimmed osteoblast or chondrocyte transcriptomes (between 200 and 400 transcripts) obtained from clean cell populations residing within healthy bone and cartilage, along with a defined number of microRNA species (not more than 20-30) as markers and guidance for the use of a set of manipulations eventually leading to functional and stable cell phenotypes.

One scheme may consist of the following materials and factors: MSCs or ASCs exposed to a growth factor in a serum-free differentiating medium, mechano-stimulation (adapted to optimalize differentiation of osteoblasts or chondrocytes), preferably within scaffolds (designed to display a porosity gradient), transient adjustments of the levels of certain microRNA species (down-regulated in differentiating osteoblasts, up-regulated in differentiating chondrocytes).

If the disease necessitating tissue replacement can be handled/treated successfully, manipulations of the engineered cells to withstand phenotype alterations, may not be necessary. However, in the case of osteochondral replacement in joints being subject to inflammation, it may be necessary to protect the engineered cells from changing their function (e.g. stimulating osteoclastogenesis) or showing an accelerated development of senescence, by permanently modulating expression of selected genes or microRNAs.

12. Acknowledgements

This project was supported by EU FP6 integrated project "Genostem cells engineering for connective tissue disorders", Norwegian Center for Stem Cell research, The National Hospital, Oslo, Norway, and The Research Council of Norway.

13. References

Abouheif, M. M., T. Nakasa, et al. (2010). Silencing microRNA-34a inhibits chondrocyte apoptosis in a rat osteoarthritis model in vitro. *Rheumatology* 49(11): 2054-2060.

Adamopoulos, I. E. and E. P. Bowman (2008). Immune regulation of bone loss by Th17 cells. *Arthritis research & therapy* 10(5): 225.

Aguda, B. D., Y. Kim, et al. (2008). MicroRNA regulation of a cancer network: consequences of the feedback loops involving miR-17-92, E2F, and Myc. *Proc Natl Acad Sci U S A* 105(50): 19678-19683.

Aicher, W. K., H. J. Buhring, et al. (2010). Regeneration of cartilage and bone by defined subsets of mesenchymal stromal cells-Potential and pitfalls. *Advanced drug delivery reviews.*

Almarza, A. J. and K. A. Athanasiou (2004). Design characteristics for the tissue engineering of cartilaginous tissues. *Annals of biomedical engineering* 32(1): 2-17.

Angele, P., D. Schumann, et al. (2004). Cyclic, mechanical compression enhances chondrogenesis of mesenchymal progenitor cells in tissue engineering scaffolds. *Biorheology* 41(3-4): 335-346.

Annalisa, P., P. Furio, et al. (2008). Anorganic bovine bone and a silicate-based synthetic bone activate different microRNAs. *J Oral Sci* 50(3): 301-307.

Arnsdorf, E. J., P. Tummala, et al. (2009). Non-canonical Wnt signaling and N-cadherin related beta-catenin signaling play a role in mechanically induced osteogenic cell fate. *PloS one* 4(4): e5388.

Arnsdorf, E. J., P. Tummala, et al. (2009). Mechanically induced osteogenic differentiation--the role of RhoA, ROCKII and cytoskeletal dynamics. *Journal of cell science* 122(Pt 4): 546-553.

Bale, A. E. (2002). Hedgehog signaling and human disease. *Annual review of genomics and human genetics* 3: 47-65.

Barry, F. P. and J. M. Murphy (2004). Mesenchymal stem cells: clinical applications and biological characterization. *The international journal of biochemistry & cell biology* 36(4): 568-584.

Bartholomew, A., C. Sturgeon, et al. (2002). Mesenchymal stem cells suppress lymphocyte proliferation in vitro and prolong skin graft survival in vivo. *Exp Hematol* 30(1): 42-48.

Beezhold, K. J., V. Castranova, et al. (2010). Microprocessor of microRNAs: regulation and potential for therapeutic intervention. *Molecular cancer* 9: 134.

Bernstein, P., C. Sticht, et al. (2010). Expression pattern differences between osteoarthritic chondrocytes and mesenchymal stem cells during chondrogenic differentiation. *Osteoarthritis and cartilage / OARS, Osteoarthritis Research Society* 18(12): 1596-1607.

Bertrand, J., C. Cromme, et al. (2010). Molecular mechanisms of cartilage remodelling in osteoarthritis. *The international journal of biochemistry & cell biology* 42(10): 1594-1601.

Betz, V. M., Betz, O.B., Harris, M.B., Vrahas, M.S., Evans, C.H. (2008). Bone tissue engineering and repair by gene therapy. *Frontiers in Bioscience* 13: 833-841.

Bodle, J. C., A. D. Hanson, et al. (2011). Adipose-Derived Stem Cells in Functional Bone Tissue Engineering: Lessons from Bone Mechanobiology. *Tissue engineering. Part B, Reviews.*

Boeuf, S. and W. Richter (2010). Chondrogenesis of mesenchymal stem cells: role of tissue source and inducing factors. *Stem cell research & therapy* 1(4): 31.

Braganca, J., T. Swingler, et al. (2002). Human CREB-binding protein/p300-interacting transactivator with ED-rich tail (CITED) 4, a new member of the CITED family, functions as a co-activator for transcription factor AP-2. *The Journal of biological chemistry* 277(10): 8559-8565.

Campbell, J. J., D. A. Lee, et al. (2006). Dynamic compressive strain influences chondrogenic gene expression in human mesenchymal stem cells. *Biorheology* 43(3-4): 455-470.

Camussi, G., M. C. Deregibus, et al. (2010). Exosomes/microvesicles as a mechanism of cell-to-cell communication. *Kidney international* 78(9): 838-848.

Chen, Y., X. Wang, et al. (2008). Phospholipase Cgamma2 mediates RANKL-stimulated lymph node organogenesis and osteoclastogenesis. *The Journal of biological chemistry* 283(43): 29593-29601.

Clarkin, C. E., R. J. Emery, et al. (2008). Evaluation of VEGF-mediated signaling in primary human cells reveals a paracrine action for VEGF in osteoblast-mediated crosstalk to endothelial cells. *Journal of cellular physiology* 214(2): 537-544.

Dahl, J. A., S. Duggal, et al. (2008). Genetic and epigenetic instability of human bone marrow mesenchymal stem cells expanded in autologous serum or fetal bovine serum. *The International journal of developmental biology* 52(8): 1033-1042.

Darling, E. M. and K. A. Athanasiou (2005). Rapid phenotypic changes in passaged articular chondrocyte subpopulations. *Journal of orthopaedic research : official publication of the Orthopaedic Research Society* 23(2): 425-432.

Davies, S. R., L. W. Chang, et al. (2007). Computational identification and functional validation of regulatory motifs in cartilage-expressed genes. *Genome research* 17(10): 1438-1447.

Dean, R. M. and M. R. Bishop (2003). Graft-versus-host disease: emerging concepts in prevention and therapy. *Curr Hematol Rep* 2(4): 287-294.

Do, J. T. and H. R. Scholer (2009). Regulatory circuits underlying pluripotency and reprogramming. *Trends Pharmacol Sci* 30(6): 296-302.

Domm, C., M. Schunke, et al. (2002). Redifferentiation of dedifferentiated bovine articular chondrocytes in alginate culture under low oxygen tension. *Osteoarthritis and cartilage / OARS, Osteoarthritis Research Society* 10(1): 13-22.

Drosse, I., E. Volkmer, et al. (2008). Tissue engineering for bone defect healing: an update on a multi-component approach. *Injury* 39 Suppl 2: S9-20.

Dudek, K. A., J. E. Lafont, et al. (2010). Type II collagen expression is regulated by tissue-specific miR-675 in human articular chondrocytes. *The Journal of biological chemistry* 285(32): 24381-24387.

Duggal, S. and J. E. Brinchmann (2011). Importance of serum source for the in vitro replicative senescence of human bone marrow derived mesenchymal stem cells. *Journal of cellular physiology*.

Duggal, S., K. B. Fronsdal, et al. (2009). Phenotype and gene expression of human mesenchymal stem cells in alginate scaffolds. *Tissue engineering. Part A* 15(7): 1763-1773.

Duroux-Richard, I., J. Presumey, et al. (2011). MicroRNAs as new player in rheumatoid arthritis. *Joint, bone, spine : revue du rhumatisme* 78(1): 17-22.

Eyre, D. (2002). Collagen of articular cartilage. *Arthritis research* 4(1): 30-35.

Felka, T., R. Schafer, et al. (2010). Animal serum-free expansion and differentiation of human mesenchymal stromal cells. *Cytotherapy* 12(2): 143-153.

Feuerborn, A., P. K. Srivastava, et al. (2011). The Forkhead factor FoxQ1 influences epithelial differentiation. *Journal of cellular physiology* 226(3): 710-719.

Fischer, J., A. Kolk, et al. (2011). Future of local bone regeneration - Protein versus gene therapy. *Journal of cranio-maxillo-facial surgery: official publication of the European Association for Cranio-Maxillo-Facial Surgery* 39(1): 54-64.

Frost, H. M. (2003). Bone's mechanostat: a 2003 update. *Anat Rec A Discov Mol Cell Evol Biol* 275(2): 1081-1101.

Fukui, N., C. R. Purple, et al. (2001). Cell biology of osteoarthritis: the chondrocyte's response to injury. *Current rheumatology reports* 3(6): 496-505.

Furushima, K., K. Shimo-Onoda, et al. (2002). Large-scale screening for candidate genes of ossification of the posterior longitudinal ligament of the spine. *Journal of bone and mineral research : the official journal of the American Society for Bone and Mineral Research* 17(1): 128-137.

Gersbach, C. A., J. E. Phillips, et al. (2007). Genetic engineering for skeletal regenerative medicine. *Annual review of biomedical engineering* 9: 87-119.

Giannoudis, P. V. and H. T. Dinopoulos (2010). Autologous bone graft: when shall we add growth factors? *The Orthopedic clinics of North America* 41(1): 85-94; table of contents.

Gibson, J. S., P. I. Milner, et al. (2008). Oxygen and reactive oxygen species in articular cartilage: modulators of ionic homeostasis. *Pflugers Archiv: European journal of physiology* 455(4): 563-573.

Gigout, A., M. D. Buschmann, et al. (2009). Chondrocytes cultured in stirred suspension with serum-free medium containing pluronic-68 aggregate and proliferate while maintaining their differentiated phenotype. *Tissue engineering. Part A* 15(8): 2237-2248.

Goldring, M. B. and S. R. Goldring (2010). Articular cartilage and subchondral bone in the pathogenesis of osteoarthritis. *Annals of the New York Academy of Sciences* 1192: 230-237.

Goldring, M. B., K. Tsuchimochi, et al. (2006). The control of chondrogenesis. *J Cell Biochem* 97(1): 33-44.

Gomez, C., V. David, et al. (2007). Absence of mechanical loading in utero influences bone mass and architecture but not innervation in Myod-Myf5-deficient mice. *Journal of anatomy* 210(3): 259-271.

Gordeladze, J. O., F. Djouad, et al. (2009). Concerted stimuli regulating osteo-chondral differentiation from stem cells: phenotype acquisition regulated by microRNAs. *Acta pharmacologica Sinica* 30(10): 1369-1384.

Gordeladze, J. O., C. A. Drevon, et al. (2002). Leptin stimulates human osteoblastic cell proliferation, de novo collagen synthesis, and mineralization: Impact on differentiation markers, apoptosis, and osteoclastic signaling. *J Cell Biochem* 85(4): 825-836.

Gordeladze, J. O., J. E. Reseland, et al. (2009). From stem cells to bone: phenotype acquisition, stabilization, and tissue engineering in animal models. *ILAR journal / National Research Council, Institute of Laboratory Animal Resources* 51(1): 42-61.

Grad, S. and G. M. Salzmann (2009). [Chondrocytes - one cell type, different subpopulations : characteristics and behavior of different types of chondrocytes and implications for tissue engineering applications]. *Der Orthopade* 38(11): 1038-1044.

Granchi, D., G. Ochoa, et al. (2010). Gene expression patterns related to osteogenic differentiation of bone marrow-derived mesenchymal stem cells during ex vivo expansion. *Tissue engineering. Part C, Methods* 16(3): 511-524.

Grellier, M., L. Bordenave, et al. (2009). Cell-to-cell communication between osteogenic and endothelial lineages: implications for tissue engineering. *Trends in biotechnology* 27(10): 562-571.

Grellier, M., N. Ferreira-Tojais, et al. (2009). Role of vascular endothelial growth factor in the communication between human osteoprogenitors and endothelial cells. *Journal of cellular biochemistry* 106(3): 390-398.

Grellier, M., P. L. Granja, et al. (2009). The effect of the co-immobilization of human osteoprogenitors and endothelial cells within alginate microspheres on mineralization in a bone defect. *Biomaterials* 30(19): 3271-3278.

Grimshaw, M. J. and R. M. Mason (2001). Modulation of bovine articular chondrocyte gene expression in vitro by oxygen tension. *Osteoarthritis and cartilage / OARS, Osteoarthritis Research Society* 9(4): 357-364.

Gruber, R. (2010). Cell biology of osteoimmunology. *Wiener medizinische Wochenschrift* 160(17-18): 438-445.

Grundberg, E., H. Brandstrom, et al. (2008). Systematic assessment of the human osteoblast transcriptome in resting and induced primary cells. *Physiological genomics* 33(3): 301-311.

Guillotin, B., R. Bareille, et al. (2008). Interaction between human umbilical vein endothelial cells and human osteoprogenitors triggers pleiotropic effect that may support osteoblastic function. *Bone* 42(6): 1080-1091.

Haberland, M., R. L. Montgomery, et al. (2009). The many roles of histone deacetylases in development and physiology: implications for disease and therapy. *Nature reviews. Genetics* 10(1): 32-42.

Hanada, R., T. Hanada, et al. (2010). Physiology and pathophysiology of the RANKL/RANK system. *Biological chemistry* 391(12): 1365-1370.

Hassan, M. Q., J. A. Gordon, et al. (2010). A network connecting Runx2, SATB2, and the miR-23a~27a~24-2 cluster regulates the osteoblast differentiation program. *Proceedings of the National Academy of Sciences of the United States of America* 107(46): 19879-19884.

Hattori, H., Y. Nogami, et al. (2008). Expansion and characterization of adipose tissue-derived stromal cells cultured with low serum medium. *Journal of biomedical materials research. Part B, Applied biomaterials* 87(1): 229-236.

Haudenschild, A. K., A. H. Hsieh, et al. (2009). Pressure and distortion regulate human mesenchymal stem cell gene expression. *Annals of biomedical engineering* 37(3): 492-502.

Heinegard, D. (2009). Proteoglycans and more--from molecules to biology. *International journal of experimental pathology* 90(6): 575-586.

Herlofsen, S. R., A. M. Kuchler, et al. (2011). Chondrogenic differentiation of human bone marrow-derived mesenchymal stem cells in self-gelling alginate discs reveals novel chondrogenic signature gene clusters. *Tissue engineering. Part A* 17(7-8): 1003-1013.

Hobert, O. (2008). Gene regulation by transcription factors and microRNAs. *Science* 319(5871): 1785-1786.

Huang, C. Y., K. L. Hagar, et al. (2004). Effects of cyclic compressive loading on chondrogenesis of rabbit bone-marrow derived mesenchymal stem cells. *Stem cells* 22(3): 313-323.

Huang, X., Q. T. Le, et al. (2010). MiR-210--micromanager of the hypoxia pathway. *Trends in molecular medicine* 16(5): 230-237.

Hunt, N. C. and L. M. Grover (2010). Cell encapsulation using biopolymer gels for regenerative medicine. *Biotechnology letters* 32(6): 733-742.

Inose, H., H. Ochi, et al. (2009). A microRNA regulatory mechanism of osteoblast differentiation. *Proceedings of the National Academy of Sciences of the United States of America* 106(49): 20794-20799.

Isogai, N., H. Kusuhara, et al. (2006). Comparison of different chondrocytes for use in tissue engineering of cartilage model structures. *Tissue engineering* 12(4): 691-703.

Jakobsen, R. B., A. Shahdadfar, et al. (2010). Chondrogenesis in a hyaluronic acid scaffold: comparison between chondrocytes and MSC from bone marrow and adipose tissue. *Knee surgery, sports traumatology, arthroscopy : official journal of the ESSKA* 18(10): 1407-1416.

Javazon, E. H., K. J. Beggs, et al. (2004). Mesenchymal stem cells: paradoxes of passaging. *Experimental hematology* 32(5): 414-425.

Kafienah, W., M. Jakob, et al. (2002). Three-dimensional tissue engineering of hyaline cartilage: comparison of adult nasal and articular chondrocytes. *Tissue engineering* 8(5): 817-826.

Kahn, J., Y. Shwartz, et al. (2009). Muscle contraction is necessary to maintain joint progenitor cell fate. *Developmental cell* 16(5): 734-743.

Kaigler, D., P. H. Krebsbach, et al. (2006). Transplanted endothelial cells enhance orthotopic bone regeneration. *Journal of dental research* 85(7): 633-637.

Kaigler, D., P. H. Krebsbach, et al. (2005). Endothelial cell modulation of bone marrow stromal cell osteogenic potential. *The FASEB journal: official publication of the Federation of American Societies for Experimental Biology* 19(6): 665-667.

Kanczler, J. M. and R. O. Oreffo (2008). Osteogenesis and angiogenesis: the potential for engineering bone. *European cells & materials* 15: 100-114.

Kapinas, K., C. Kessler, et al. (2010). miR-29 modulates Wnt signaling in human osteoblasts through a positive feedback loop. *The Journal of biological chemistry* 285(33): 25221-25231.

Kapinas, K., C. B. Kessler, et al. (2009). miR-29 suppression of osteonectin in osteoblasts: regulation during differentiation and by canonical Wnt signaling. *Journal of cellular biochemistry* 108(1): 216-224.

Kapur, S., D. J. Baylink, et al. (2003). Fluid flow shear stress stimulates human osteoblast proliferation and differentiation through multiple interacting and competing signal transduction pathways. *Bone* 32(3): 241-251.

Karlsen, T. A., A. Shahdadfar, et al. (2011). Human primary articular chondrocytes, chondroblasts-like cells, and dedifferentiated chondrocytes: differences in gene, microRNA, and protein expression and phenotype. *Tissue engineering. Part C, Methods* 17(2): 219-227.

Katoh, M. (2007). WNT signaling pathway and stem cell signaling network. *Clinical cancer research : an official journal of the American Association for Cancer Research* 13(14): 4042-4045.

Kelly, D. J. and C. R. Jacobs (2010). The role of mechanical signals in regulating chondrogenesis and osteogenesis of mesenchymal stem cells. *Birth defects research. Part C, Embryo today: reviews* 90(1): 75-85.

Keung, A. J., K. E. Healy, et al. (2010). Biophysics and dynamics of natural and engineered stem cell microenvironments. *Wiley interdisciplinary reviews. Systems biology and medicine* 2(1): 49-64.

Khan, W. S., D. S. Johnson, et al. (2010). The potential of stem cells in the treatment of knee cartilage defects. *The Knee* 17(6): 369-374.

Knecht, S., B. Vanwanseele, et al. (2006). A review on the mechanical quality of articular cartilage - implications for the diagnosis of osteoarthritis. *Clinical biomechanics* 21(10): 999-1012.

Kolind, K., A. Dolatshahi-Pirouz, et al. (2010). A combinatorial screening of human fibroblast responses on micro-structured surfaces. *Biomaterials* 31(35): 9182-9191.

Komori, T. (2005). [Functions of BMPs, Runx2, and osterix in the development of bone and cartilage]. *Nippon rinsho. Japanese journal of clinical medicine* 63(9): 1671-1677.

Komori, T. (2006). Regulation of osteoblast differentiation by transcription factors. *J Cell Biochem* 99(5): 1233-1239.

Kongcharoensombat, W., T. Nakasa, et al. (2010). The effect of microRNA-21 on proliferation and matrix synthesis of chondrocytes embedded in atelocollagen gel. *Knee surgery, sports traumatology, arthroscopy: official journal of the ESSKA* 18(12): 1679-1684.

Kuhn, N. Z. and R. S. Tuan (2010). Regulation of stemness and stem cell niche of mesenchymal stem cells: implications in tumorigenesis and metastasis. *Journal of cellular physiology* 222(2): 268-277.

Kulterer, B., G. Friedl, et al. (2007). Gene expression profiling of human mesenchymal stem cells derived from bone marrow during expansion and osteoblast differentiation. *BMC genomics* 8: 70.

Kwan, M. D., B. J. Slater, et al. (2008). Cell-based therapies for skeletal regenerative medicine. *Human molecular genetics* 17(R1): R93-98.

Lakshmipathy, U., B. Love, et al. (2007). MicroRNA expression pattern of undifferentiated and differentiated human embryonic stem cells. *Stem Cells Dev* 16(6): 1003-1016.

Le Bechec, A., E. Portales-Casamar, et al. (2011). MIR@NT@N: a framework integrating transcription factors, microRNAs and their targets to identify sub-network motifs in a meta-regulation network model. *BMC bioinformatics* 12: 67.

Le Blanc, K., C. Tammik, et al. (2003). HLA expression and immunologic properties of differentiated and undifferentiated mesenchymal stem cells. *Exp Hematol* 31(10): 890-896.

Lee, S., J. W. Jung, et al. (2011). Histone deacetylase regulates high mobility group A2-targeting microRNAs in human cord blood-derived multipotent stem cell aging. *Cellular and molecular life sciences : CMLS* 68(2): 325-336.

Lei, S. F., C. J. Papasian, et al. (2011). Polymorphisms in predicted miRNA binding sites and osteoporosis. *Journal of bone and mineral research : the official journal of the American Society for Bone and Mineral Research* 26(1): 72-78.

Levi, B. and M. T. Longaker (2011). Concise review: adipose-derived stromal cells for skeletal regenerative medicine. *Stem cells* 29(4): 576-582.

Li, H., H. Xie, et al. (2009). A novel microRNA targeting HDAC5 regulates osteoblast differentiation in mice and contributes to primary osteoporosis in humans. *The Journal of clinical investigation* 119(12): 3666-3677.

Li, Z., M. Q. Hassan, et al. (2009). Biological functions of miR-29b contribute to positive regulation of osteoblast differentiation. *The Journal of biological chemistry* 284(23): 15676-15684.

Li, Z., L. Kupcsik, et al. (2010). Mechanical load modulates chondrogenesis of human mesenchymal stem cells through the TGF-beta pathway. *Journal of cellular and molecular medicine* 14(6A): 1338-1346.

Lian, J. B. and G. S. Stein (2003). Runx2/Cbfa1: a multifunctional regulator of bone formation.*Curr Pharm Des* 9(32): 2677-2685.

Lian, J. B., G. S. Stein, et al. (2006). Networks and hubs for the transcriptional control of osteoblastogenesis. *Reviews in endocrine & metabolic disorders* 7(1-2): 1-16.

Lin, C. J., Liu, C.-J. (2009). MicroRNAs in Skeletogenesis. *Frontiers in Bioscience* 14: 2757-2764.

Liu, S., D. A. Calderwood, et al. (2000). Integrin cytoplasmic domain-binding proteins. *J Cell Sci* 113 (Pt 20): 3563-3571.

Logeart-Avramoglou, D., F. Anagnostou, et al. (2005). Engineering bone: challenges and obstacles. *Journal of cellular and molecular medicine* 9(1): 72-84.

Loken, S., R. B. Jakobsen, et al. (2008). Bone marrow mesenchymal stem cells in a hyaluronan scaffold for treatment of an osteochondral defect in a rabbit model. *Knee surgery, sports traumatology, arthroscopy : official journal of the ESSKA* 16(10): 896-903.

Lu, H. H., S. D. Subramony, et al. (2010). Tissue engineering strategies for the regeneration of orthopedic interfaces. *Annals of biomedical engineering* 38(6): 2142-2154.

Malda, J., C. A. van Blitterswijk, et al. (2003). Expansion of bovine chondrocytes on microcarriers enhances redifferentiation. *Tissue engineering* 9(5): 939-948.

Marie, P. J. (2008). Transcription factors controlling osteoblastogenesis. *Arch Biochem Biophys* 473(2): 98-105.

Matsumoto, T., R. Kuroda, et al. (2008). Circulating endothelial/skeletal progenitor cells for bone regeneration and healing. *Bone* 43(3): 434-439.

Mauck, R. L., B. A. Byers, et al. (2007). Regulation of cartilaginous ECM gene transcription by chondrocytes and MSCs in 3D culture in response to dynamic loading. *Biomechanics and modeling in mechanobiology* 6(1-2): 113-125.

McBeath, R., D. M. Pirone, et al. (2004). Cell shape, cytoskeletal tension, and RhoA regulate stem cell lineage commitment. *Developmental cell* 6(4): 483-495.

McGee-Lawrence, M. E. and J. J. Westendorf (2011). Histone deacetylases in skeletal development and bone mass maintenance. *Gene* 474(1-2): 1-11.

McMahon, L. A., V. A. Campbell, et al. (2008). Involvement of stretch-activated ion channels in strain-regulated glycosaminoglycan synthesis in mesenchymal stem cell-seeded 3D scaffolds. *Journal of biomechanics* 41(9): 2055-2059.

McMahon, L. A., A. J. Reid, et al. (2008). Regulatory effects of mechanical strain on the chondrogenic differentiation of MSCs in a collagen-GAG scaffold: experimental and computational analysis. *Annals of biomedical engineering* 36(2): 185-194.

Mironov, V., R. P. Visconti, et al. (2009). Organ printing: tissue spheroids as building blocks. *Biomaterials* 30(12): 2164-2174.

Miyaki, S., T. Nakasa, et al. (2009). MicroRNA-140 is expressed in differentiated human articular chondrocytes and modulates interleukin-1 responses. *Arthritis and rheumatism* 60(9): 2723-2730.

Miyanishi, K., M. C. Trindade, et al. (2006). Dose- and time-dependent effects of cyclic hydrostatic pressure on transforming growth factor-beta3-induced chondrogenesis by adult human mesenchymal stem cells in vitro. *Tissue engineering* 12(8): 2253-2262.

Monaco, E., M. Bionaz, et al. (2011). Strategies for regeneration of the bone using porcine adult adipose-derived mesenchymal stem cells. *Theriogenology* 75(8): 1381-1399.

Morsczeck, C., G. Schmalz, et al. (2009). Gene expression profiles of dental follicle cells before and after osteogenic differentiation in vitro. *Clinical oral investigations* 13(4): 383-391.

Mulvihill, B. M. and P. J. Prendergast (2008). An algorithm for bone mechanoresponsiveness: implementation to study the effect of patient-specific cell mechanosensitivity on trabecular bone loss. *Comput Methods Biomech Biomed Engin* 11(5): 443-451.

Murphy, C. L. and J. M. Polak (2004). Control of human articular chondrocyte differentiation by reduced oxygen tension. *Journal of cellular physiology* 199(3): 451-459.

Muys, J. J., M. M. Alkaisi, et al. (2006). Cellular replication and atomic force microscope imaging using a UV-Bioimprint technique. *Nanomedicine: nanotechnology, biology, and medicine* 2(3): 169-176.

Niemeyer, P., U. Krause, et al. (2006). Mesenchymal stem cell-based HLA-independent cell therapy for tissue engineering of bone and cartilage. *Curr Stem Cell Res Ther* 1(1): 21-27.

Nowlan, N. C., C. Bourdon, et al. (2010). Developing bones are differentially affected by compromised skeletal muscle formation. *Bone* 46(5): 1275-1285.

Nowlan, N. C., J. Sharpe, et al. (2010). Mechanobiology of embryonic skeletal development: Insights from animal models. *Birth defects research. Part C, Embryo today : reviews* 90(3): 203-213.

O'Sullivan, J., S. D'Arcy, et al. (2011). Mesenchymal chondroprogenitor cell origin and therapeutic potential. *Stem cell research & therapy* 2(1): 8.

Oh, S. H., T. H. Kim, et al. (2010). Investigation of pore size effect on chondrogenic differentiation of adipose stem cells using a pore size gradient scaffold. *Biomacromolecules* 11(8): 1948-1955.

Ohishi, M., R. Chiusaroli, et al. (2009). Osteoprotegerin abrogated cortical porosity and bone marrow fibrosis in a mouse model of constitutive activation of the PTH/PTHrP receptor. *The American journal of pathology* 174(6): 2160-2171.

Otsuru, S., K. Tamai, et al. (2008). Circulating bone marrow-derived osteoblast progenitor cells are recruited to the bone-forming site by the CXCR4/stromal cell-derived factor-1 pathway. *Stem cells* 26(1): 223-234.

Otto, W. R. and J. Rao (2004). Tomorrow's skeleton staff: mesenchymal stem cells and the repair of bone and cartilage. *Cell proliferation* 37(1): 97-110.

Pacifici, R. (2010). The immune system and bone. *Archives of biochemistry and biophysics* 503(1): 41-53.

Palmieri, A., F. Pezzetti, et al. (2008). Titanium acts on osteoblast translational process. The *Journal of oral implantology* 34(4): 190-195.

Palmieri, A., F. Pezzetti, et al. (2008). Medpor regulates osteoblast's microRNAs. *Bio-medical materials and engineering* 18(2): 91-97.

Palmieri, A., F. Pezzetti, et al. (2008). PerioGlas regulates osteoblast RNA interfering. *Journal of prosthodontics : official journal of the American College of Prosthodontists* 17(7): 522-526.

Phillips, J. E., C. A. Gersbach, et al. (2007). Virus-based gene therapy strategies for bone regeneration. *Biomaterials* 28(2): 211-229.

Piek, E., L. S. Sleumer, et al. (2010). Osteo-transcriptomics of human mesenchymal stem cells: accelerated gene expression and osteoblast differentiation induced by vitamin D reveals c-MYC as an enhancer of BMP2-induced osteogenesis. *Bone* 46(3): 613-627.

Potier, E., J. Noailly, et al. (2010). Directing bone marrow-derived stromal cell function with mechanics. *Journal of biomechanics* 43(5): 807-817.

Rahaman, M. N., D. E. Day, et al. (2011). Bioactive glass in tissue engineering. *Acta biomaterialia.*

Raimondi, M. T. (2006)."Engineered tissue as a model to study cell and tissue function from a biophysical perspective. *Current drug discovery technologies* 3(4): 245-268.

Rios, H. F., Z. Lin, et al. (2011). Cell- and Gene-Based Therapeutic Strategies for Periodontal Regenerative Medicine. *Journal of periodontology.*

Roach, H. I. and T. Aigner (2007). DNA methylation in osteoarthritic chondrocytes: a new molecular target. *Osteoarthritis and cartilage / OARS, Osteoarthritis Research Society* 15(2): 128-137.

Sabetrasekh, R., Reseland, J.E., Busch, C., Lyngstadaas, S.P. (2011). Hydroxyproly-methyl Cellulose hydrogel (HISTOCARETM) Supports 3D Expansion of Mesenchymal Cells for Tissue Engineering. *Cells & Materials.*

Sabetrasekh, R., H. Tiainen, et al. (2011). A novel ultra-porous titanium dioxide ceramic with excellent biocompatibility. *Journal of biomaterials applications* 25(6): 559-580.

Salim, A., R. P. Nacamuli, et al. (2004). Transient changes in oxygen tension inhibit osteogenic differentiation and Runx2 expression in osteoblasts. *The Journal of biological chemistry* 279(38): 40007-40016.

Sandell, L. J. and T. Aigner (2001). Articular cartilage and changes in arthritis. An introduction: cell biology of osteoarthritis. *Arthritis research* 3(2): 107-113.

Saraf, A. and A. G. Mikos (2006). Gene delivery strategies for cartilage tissue engineering. *Advanced drug delivery reviews* 58(4): 592-603.

Saulnier, N., M. A. Puglisi, et al. (2011). Gene profiling of bone marrow- and adipose tissue-derived stromal cells: a key role of Kruppel-like factor 4 in cell fate regulation. *Cytotherapy* 13(3): 329-340.

Scheller, E. L. and P. H. Krebsbach (2009). Gene therapy: design and prospects for craniofacial regeneration. *Journal of dental research* 88(7): 585-596.

Schnabel, M., S. Marlovits, et al. (2002). Dedifferentiation-associated changes in morphology and gene expression in primary human articular chondrocytes in cell culture. *Osteoarthritis and cartilage / OARS, Osteoarthritis Research Society* 10(1): 62-70.

Segawa, Y., T. Muneta, et al. (2009). Mesenchymal stem cells derived from synovium, meniscus, anterior cruciate ligament, and articular chondrocytes share similar gene expression profiles. *Journal of orthopaedic research : official publication of the Orthopaedic Research Society* 27(4): 435-441.

Seidi, A., M. Ramalingam, et al. (2011). Gradient biomaterials for soft-to-hard interface tissue engineering. *Acta biomaterialia* 7(4): 1441-1451.

Shahdadfar, A., K. Fronsdal, et al. (2005). In vitro expansion of human mesenchymal stem cells: choice of serum is a determinant of cell proliferation, differentiation, gene expression, and transcriptome stability. *Stem cells* 23(9): 1357-1366.

Shahdadfar, A., S. Loken, et al. (2008). Persistence of collagen type II synthesis and secretion in rapidly proliferating human articular chondrocytes in vitro. *Tissue engineering. Part A* 14(12): 1999-2007.

Skerry, T. M. (2006). One mechanostat or many? Modifications of the site-specific response of bone to mechanical loading by nature and nurture. *J Musculoskelet Neuronal Interact* 6(2): 122-127.

Solmesky, L., S. Lefler, et al. (2010). Serum free cultured bone marrow mesenchymal stem cells as a platform to characterize the effects of specific molecules. *PloS one* 5(9).

Solomon, L. A., N. G. Berube, et al. (2008). Transcriptional regulators of chondrocyte hypertrophy. *Birth defects research. Part C, Embryo today : reviews* 84(2): 123-130.

Stahl, A., X. Wu, et al. (2005). Endothelial progenitor cell sprouting in spheroid cultures is resistant to inhibition by osteoblasts: a model for bone replacement grafts. *FEBS letters* 579(24): 5338-5342.

Stanczyk, J., D. M. Pedrioli, et al. (2008). Altered expression of MicroRNA in synovial fibroblasts and synovial tissue in rheumatoid arthritis. *Arthritis and rheumatism* 58(4): 1001-1009.

Steinert, A. F., U. Noth, et al. (2008). Concepts in gene therapy for cartilage repair. *Injury* 39 Suppl 1: S97-113.

Sumanasinghe, R. D., S. H. Bernacki, et al. (2006). Osteogenic differentiation of human mesenchymal stem cells in collagen matrices: effect of uniaxial cyclic tensile strain on bone morphogenetic protein (BMP-2) mRNA expression. *Tissue engineering* 12(12): 3459-3465.

Sun, X., Fu, X., Han, W., Zhao, Y., Liu, H. (2010). Can controlled cellular reprogramming be achieved using microRNAs? *Ageing Research Reviews* 9: 475-483.

Sun, Y., D. R. Mauerhan, et al. (2010). Analysis of meniscal degeneration and meniscal gene expression. *BMC musculoskeletal disorders* 11: 19.

Sundelacruz, S. and D. L. Kaplan (2009). Stem cell- and scaffold-based tissue engineering approaches to osteochondral regenerative medicine. *Seminars in cell & developmental biology* 20(6): 646-655.

Takebe, N., P. J. Harris, et al. (2011). Targeting cancer stem cells by inhibiting Wnt, Notch, and Hedgehog pathways. *Nature reviews. Clinical oncology* 8(2): 97-106.

Takeda, H., T. Nakagawa, et al. (2011). Prevention and management of knee osteoarthritis and knee cartilage injury in sports. *British journal of sports medicine* 45(4): 304-309.

Tallheden, T., C. Bengtsson, et al. (2005). Proliferation and differentiation potential of chondrocytes from osteoarthritic patients. *Arthritis research & therapy* 7(3): R560-568.

Tare, R. S., J. Kanczler, et al. (2010). Skeletal stem cells and bone regeneration: translational strategies from bench to clinic. *Proceedings of the Institution of Mechanical Engineers. Part H, Journal of engineering in medicine* 224(12): 1455-1470.

Thorpe, S. D., C. T. Buckley, et al. (2008). Dynamic compression can inhibit chondrogenesis of mesenchymal stem cells. *Biochemical and biophysical research communications* 377(2): 458-462.

Tiainen, H., S. P. Lyngstadaas, et al. (2010). Ultra-porous titanium oxide scaffold with high compressive strength. *Journal of materials science. Materials in medicine* 21(10): 2783-2792.

Tilg, H., A. R. Moschen, et al. (2008). Gut, inflammation and osteoporosis: basic and clinical concepts. *Gut* 57(5): 684-694.

Torroni, A. (2009). Engineered bone grafts and bone flaps for maxillofacial defects: state of the art. *Journal of oral and maxillofacial surgery : official journal of the American Association of Oral and Maxillofacial Surgeons* 67(5): 1121-1127.

Trouvin, A. P. and V. Goeb (2010). Receptor activator of nuclear factor-kappaB ligand and osteoprotegerin: maintaining the balance to prevent bone loss. *Clinical interventions in aging* 5: 345-354.

Valadi, H., K. Ekstrom, et al. (2007). Exosome-mediated transfer of mRNAs and microRNAs is a novel mechanism of genetic exchange between cells. *Nature cell biology* 9(6): 654-659.

Van Agtmael, T. and L. Bruckner-Tuderman (2010). Basement membranes and human disease. *Cell and tissue research* 339(1): 167-188.

van der Kraan, P. M., E. N. Davidson, et al. (2010). Bone morphogenetic proteins and articular cartilage: To serve and protect or a wolf in sheep clothing's? *Osteoarthritis and cartilage / OARS, Osteoarthritis Research Society* 18(6): 735-741.

van der Zande, M., X. F. Walboomers, et al. (2010). Genetic profiling of osteoblast-like cells cultured on a novel bone reconstructive material, consisting of poly-L-lactide, carbon nanotubes and microhydroxyapatite, in the presence of bone morphogenetic protein-2. *Acta biomaterialia* 6(11): 4352-4360.

van Osch, G. J., M. Brittberg, et al. (2009). Cartilage repair: past and future--lessons for regenerative medicine. *Journal of cellular and molecular medicine* 13(5): 792-810.

Villars, F., L. Bordenave, et al. (2000). Effect of human endothelial cells on human bone marrow stromal cell phenotype: role of VEGF? *Journal of cellular biochemistry* 79(4): 672-685.

Vinatier, C., C. Bouffi, et al. (2009). Cartilage tissue engineering: towards a biomaterial-assisted mesenchymal stem cell therapy. *Current stem cell research & therapy* 4(4): 318-329.

Vinatier, C., D. Mrugala, et al. (2009). Cartilage engineering: a crucial combination of cells, biomaterials and biofactors. *Trends Biotechnol* 27(5): 307-314.

Visconti, R. P., V. Kasyanov, et al. (2010). Towards organ printing: engineering an intra-organ branched vascular tree. *Expert opinion on biological therapy* 10(3): 409-420.

Waese, E. Y. and W. L. Stanford (2011). One-step generation of murine embryonic stem cell-derived mesoderm progenitors and chondrocytes in a serum-free monolayer differentiation system. *Stem cell research* 6(1): 34-49.

Wagner, D. R., D. P. Lindsey, et al. (2008). Hydrostatic pressure enhances chondrogenic differentiation of human bone marrow stromal cells in osteochondrogenic medium. *Annals of biomedical engineering* 36(5): 813-820.

Wang, T. and Z. Xu (2010). miR-27 promotes osteoblast differentiation by modulating Wnt signaling. *Biochemical and biophysical research communications* 402(2): 186-189.

Weitzmann, M. N. and R. Pacifici (2007). T cells: unexpected players in the bone loss induced by estrogen deficiency and in basal bone homeostasis. *Annals of the New York Academy of Sciences* 1116: 360-375.

Wenger, A., A. Stahl, et al. (2004). Modulation of in vitro angiogenesis in a three-dimensional spheroidal coculture model for bone tissue engineering. *Tissue engineering* 10(9-10): 1536-1547.

Weyts, F. A., Y. S. Li, et al. (2002). ERK activation and alpha v beta 3 integrin signaling through Shc recruitment in response to mechanical stimulation in human osteoblasts. *J Cell Biochem* 87(1): 85-92.

Williams, D. F. (2009). On the nature of biomaterials. *Biomaterials* 30(30): 5897-5909.

Witkowska-Zimny, M. and K. Walenko (2011). Stem cells from adipose tissue. *Cellular & molecular biology letters* 16(2): 236-257.

Woods, A., G. Wang, et al. (2005). RhoA/ROCK signaling regulates Sox9 expression and actin organization during chondrogenesis. *The Journal of biological chemistry* 280(12): 11626-11634.

Yamasaki, K., T. Nakasa, et al. (2009). Expression of MicroRNA-146a in osteoarthritis cartilage. *Arthritis and rheumatism* 60(4): 1035-1041.

Zhang, J., Q. Tu, et al. (2011). Effects of miR-335-5p in modulating osteogenic differentiation by specifically down-regulating Wnt antagonist DKK1. *Journal of bone and mineral research : the official journal of the American Society for Bone and Mineral Research.*

Zhou, Y., J. Ferguson, et al. (2007). Inter- and intra-combinatorial regulation by transcription factors and microRNAs. *BMC Genomics* 8: 396.

Zomer, A., T. Vendrig, et al. (2010). Exosomes: Fit to deliver small RNA. *Communicative & integrative biology* 3(5): 447-450.

3

Mesenchymal Stem Cell-Based Bone Engineering for Bone Regeneration

Mohamadreza Baghaban Eslaminejad and Faezeh Faghihi
Department of Stem Cell and Developmental Biology, Royan Institute for Stem Cell Biology and Technology, ACECR, Tehran Iran

1. Introduction

Many people worldwide suffer from bone defects due to trauma or disease. About 5-10% of football injury-related bone fractures, in addition to bone tumor resection and massive traumatic bone loss result in critical-sized bone defects that cannot regenerate autonomously (Cattermole et al., 1996; Low et al., 2004; Reuss et al., 2007 as cited in Porter et al., 2009). Usually small bone defects heal spontaneously but large defects cannot regenerate without intervention. There are several options for the reconstruction of large bone defects that include bone grafts (autograft, allograft and xenograft) as well as bone constructs created by bone tissue engineering principles.

Among these different strategies, bone grafts from healthy donors have been used as therapies for decades, but nowadays they are used less due to inherent limitations associated with their application. For example, autografts are usually obtained from the iliac crest (Porter et al., 2009) and have been used in clinics for a long time. However some disadvantages, including supply limitation, risk of donor site morbidity, pain, paresthesia, hematoma and inflammation, as well as the need for prolonged rehabilitation (For review see Heary et al., 2002; Krelow et al., 2007; Maddela et al., 2010; Nakajima et al., 2007) limit their applications. Allografts, on the other hand, are the other option that may be chosen for bone defect regeneration. This type of bone graft can be derived from viable or sterilized non-viable (cadaver) human sources. Orthopedic allografts exhibit certain drawbacks by transmitting donor pathogens to the recipient body and triggering host immune responses (For review see Hou et al., 2005 and Nishida et al., 2008). Xenografts obtained from non-human sources may be considered as the other alternative for reconstructing bone defects, but this is a last resort which may be taken because it is not an approved option in transplantation, owing to the obvious risk of viral and disease transmission, infection, toxicity and immunogenicity as well as rejection by the recipient's body (For review see Laurencin et al., 2008; Yang et al., 2007).

Concerns associated with applications of bone graft for critical size bone defects have challenged scientists to search for suitable options. Attempts to replace appropriate substitutes for bone graft have resulted in opening a new window in modern regenerative biomedicine and the emergence of bone constructs elaborated by tissue engineering principles. The term tissue engineering is defined as the application of the principles and methods of engineering and life science toward fundamental understanding of structure-function relationship in normal and pathological mammalian tissues and the development of biological substitutes for the repair or regeneration of tissue or organ function (Persidis 1999; Chapekar et al., 2000). To develop biologic substitute tissue engineering uses three building blocks including scaffold, cells and growth factors. Each of these elements alone can promote tissue regeneration but constructs fabricated with the use of component combination would be more effective. The objective of the present chapter is therefore to describe bone construct fabricated with tissue engineering principles for bone regeneration. Among cellular candidates, mesenchymal stem cells (MSCs) possess some characteristics that make them more appropriate for bone tissue engineering. Hence, the focus of the current chapter is MSC-based bone constructs.

2. Mesenchymal stem cells as cellular candidates for bone engineering

Bone constructs typically consist of three elements: scaffolds, growth factors and cells. Several cell types can potentially be used as cellular material for elaborating a bone construct. In this section, commonly used cells with potential applications in the field of bone engineering will be summarized, followed by special focus on MSCs.

2.1 Osteoblastic cells

Osteoblastic cells may seem to be more appropriate for bone engineering because they are resident cells in natural bone and can be utilized at autogenic settings. In bone engineering strategies, osteoblastic cells are usually obtained from biopsies taken autologously from patient´s bone. Despite of this obvious advantage, there are some concerns with the use of osteoblasts in the process of bone engineering. Osteoblasts are present in limited numbers in bone biopsies since bone is indeed a tissue rich in ECM rather than cells. Furthermore, tissue engineering strategies need considerable cell numbers, however osteoblast proliferation is slow (Bruder et al., 1999; Heath et al., 2000). For these reasons stem cells having enormous proliferative capacity are preferable.

2.2 Embryonic stem cells

Embryonic stem cells (ESCs) are pluripotent cells derived from a blastocyst inner cell mass. Murine ESCs were described in 1981 (Evans et al., 1981) and human ESCs introduced in 1998 (Thomson et al., 1998). Their indefinite self-renewal potential, differentiation capacity to all three germ layers, and osteoblastic cells in particular (Arpornmaekilong et al., 2009; Hwang et al., 2009a; Warotayanont et al., 2009 as cited in Seong et al., 2010) have persuaded scientists that ESCs could be appropriate for clinical applications. ESCs have the characteristics of self-renewal as long as they are exposed to a feeder cell layer or leukemia inhibitory factor (LIF). Differentiation initiates upon removal of the feeder cell layer or LIF, resulting in the formation of three dimensional cell aggregates known as embryoid bodies (EBs). These EBs can regionally differentiate into derivatives of three germ layers: mesoderm, ectoderm and endoderm (Itskovitz-Eldor et al., 2000 as cited in Seong et al., 2010). Generally, for osteogenic induction of ESCs, EBs or single cells from EBs are replated and induced by beta-glycerophosphate, ascorbic acid and vitamin D3 (Buttery et al., 2001; Kawaguchi et al., 2005; Woll et al., 2006). Thus ESCs can be a potential stem cell source to fabricate bone-like tissue constructs in the field of tissue engineering; however, immunologic incompatibility and the possibility of teratoma formation in transplantations as well as certain ethical concerns make scientists hesitant to use them as cellular materials with which to fabricate bone construct for bone regeneration (Undale, 2009).

2.3 Induced pluripotent stem cells

Because of the above-mentioned concerns regarding ESCs, scientists have tried to establish ESC-like stem cells, known as induced pluripotent stem cells (iPSCs) from somatic cells by plasmid or adenovirus-based transduction (Takahashi and Yamanaka, 2006, Park et al., 2008, Kang et al., 2009 as cited in Feng et al., 2010). Actually, iPSCs are patient-specific ESCs without ethical concerns that do not trigger an immune response (Feng et al., 2010). The differentiation potential of these cells into various cell lineages, such as neural cells, cardiomyocytes and hematopoetic cells as well as osteoblasts have been confirmed (Hanna et al., 2007; Wernig et al., 2007). Despite this prominent capability, there is an important issue which needs to be considered before their clinical applications. The method by which iPSCs are generated, i.e. through plasmid or adenovirus-based transduction is a main concern.

2.4 Mesenchymal stem cells

Among stem cells, mesenchymal stem cells (MSCs) seem to be more suitable for bone engineering compared to ESCs, iPSCs or osteoblastic cells due to several characteristics that they possess. First, the osteogenic differentiation potential of MSCs is the first differentiation capacity reported at the time MSCs were discovered. Today, it is one of the most obvious characteristics of MSCs which is maintained for an extended time. Secondly, autologous MSCs are easily accessible from patient's multiple tissues, including bone marrow aspirates. Additionally, because of MSCs ability to modulate immune responses, the use of allogeneic MSCs may be feasible without a substantial risk of immune rejection (Undale et al., 2009).

2.4.1 Characteristics of MSCs

MSCs are defined as non-hematopoietic cells derived from bone marrow as well as other mesenchmal tissues. These cells possess two important capacities: the potential to self-renew for a relatively long time and the ability to differentiate along multiple cell lineages including bone, cartilage and adipose cells. MSCs express many surface antigens including STRO-1, CD105, SH3, CD29, CD44, CD71, CD90, CD106 and CD124 (for review see Pittenger et al., 1999; Colter et al., 2001). Besides bone marrow, multiple tissues have been reported to contain MSCs including adipose tissue (Dragoo et al., 2003), trabecular bone (Noth et al., 2002), periosteum (Fukumoto et al., 2003), synovial membrane (Wickham et al., 2003), skeletal muscle (Jankowsk et al., 2002), as well as teeth (Miura et al., 2003).

Cohnheim, a German pathologist, initially suggested the presence of MSCs when he attempted to study wound healing in rabbits. By intravenous injection of non-soluble aniline stain, Cohnheim could detect some stained cells at the site of the wound which had been experimentally created in the animal's distal limb. He concluded that the stained fibroblastic cells were derived from bone marrow and transferred to the wound site via the circulatory system (Prockop, 1997; Ross, et al., 1970). Since then, experimental studies on bone marrow transplantation have confirmed the osteogenic and chondrogenic differentiation capacities of bone marrow cells, but no one could clearly indicate the exact responsible cell types (Friedenstein et al., 1966; Petrakova et al., 1963). Finally, Friedenstein et al. have determined that the osteo/chondrogenic differentiation potential of bone marrow is due to the existence of a fibroblastic population referred to as colony forming unit-fibroblasts (Friedenstein et al., 1973). Thus far, these fibroblast-like cells have been referred to as marrow stromal cells; marrow progenitor cells (MPCs) and marrow stromal fibroblasts (MSFs), as well as mesenchymal stem cells (MSCs). MSC is the more frequently used nomination particularly in recently published investigations.

MSCs occur in low quantity in bone marrow aspirate and constitute approximately 0.001%-0.01% of the entire bone marrow cells. In spite of their limited numbers, MSCs can easily be expanded through standard culture techniques. The expansion of these cells is strongly dependent on the bovine serum content of the culture media. The cells assume spindly-shaped morphology upon cultivation. MSCs primary culture has been reported to be heterogeneous and contains multiple colonies with various differentiation capacities. Pittenger et al., in 1999, have shown that nearly one third of these colonies have osteogenic, adipogenic and chondrogenic differentiation potentials and the other two thirds exhibit either bipotent or unipotent capacity to differentiate into osteogenic/chondrogenic and adipogenic lineages, respectively (Pittenger et al., 1999). In addition to differentiating into bone, cartilage and adipose cells, MSCs have been reported to possess differentiation capacity along non-mesenchymal cell lineages such as neurons, keratinocytes, liver, intestine and kidney epithelial cells (for review see Sugaya 2003; Chapel et al. 2003). This property is referred to as MSCs plasticity or transdifferentiation.

2.4.2 Osteogenic differentiation potential of MSCs

Osteogenic differentiation of MSCs is a complex process in which various environmental factors are involved. Dexamethasone, ascorbic acid and beta glycerol phosphates are the most commonly used chemicals that promote in vitro osteogenic differentiation of MSCs. Among these compounds, dexamethasone plays a pivotal role such that in its absence, no differentiation occurs in human MSC culture (Porter et al., 2003). Ascorbic acid, on the other hand, has been found to be an important but not necessary component of osteogenic medium. The addition of ascorbic acid into osteogenic medium results in enrichment of the deposited matrix with collagen (Choi et al., 2008). Beta glycerol phosphate, as a phosphate enriched organic compound, plays some role in matrix mineralization (Coelho et al., 2000). Besides these three routinely used compounds, hormones and growth factors have osteogenic effects on MSCs differentiation. These include 1, 25-di-hydroxyvitaminD3 (vitD3) (Rickard et al., 1995), estrogen (Holzer et al., 2002), leptin and parathyroid hormone (Holzer et al., 2002), prostaglandin E2 (Scutt et al., 1995), sonic hedgehog (Spinella-Jaegles et al., 2001), IGF-1 (Koch et al., 2005), BMP-2, 4, 6, 7 (Gori et al., 1999; Diefenderfer et al., 2003; Gruber et al., 2004), FGF (Jaiswal et al., 2000), as well as bio- and lithium chloride (de Boer et al., 2004; Eslaminejad et al., 2008).

Each osteogenic factor inserts its effect through a distinct signaling pathway. Some of these pathways, such as the dexamethasone pathway, are unknown whereas others are recognized to some extent. For example, BMPs mediate their osteogenic effects through the BMP signaling pathway, via activation of smad transcription factors (Massague et al., 2000) and parathyroid hormone induces the expression of osteogenic genes via the G-protein coupled receptor signaling pathway (Carpio et al., 2001).

Induction of the above-mentioned signaling pathways ends consequently in expression of some specific signaling proteins and specific osteoblastic transcription factors. Core binding factor alpha 1 (Cbfa1), also referred to as Runx2, is the most important transcription factor involved in osteogenesis (Yamaguchi et al., 2000). This transcription factor induces the expression of different bone-related genes including osteocalcin, osteopontin, bone sialoprotein and the parathyroid hormone receptor (Ducy et al., 2000). Furthermore, osterix is the other transcription factor reported to be expressed upon commitment of cells toward osteogenic differentiation (Tu et al., 2006).

2.4.3 Self-renewal in MSCs

Self-renewal and differentiation potential are hallmarks for stem cells. The ability of a cell to produce similar replicates in a defined period of time via mitosis is called self-renewal. In this type of mitosis, genetic properties and karyotype of daughter cells remain intact, the same as the mother cell. There are two models that explain stem cell self-renewal property: symmetric and asymmetric cell division. In symmetric cell division, the stem cell divides into two daughter cells, similar to the mother cell, which have the capacity to differentiate under appropriate conditions (Potten and Loffler, 1990). In asymmetric cell division, each stem cell divides into one stem cell and one progenitor cell. The stem cell continues to divide and replenish the stem cell pool, while the progenitor cell differentiates under appropriate conditions (Sherely et al., 1995). Therefore, through this type of cell division, the number of stem cells remains intact.

While identification the underlying molecular mechanisms of MSCs self-renewal would be beneficial to stop replicative senescence in MSCs, this issue remains largely unknown. According to the literature, various cytokines and growth factors such as LIF (Jiang et al., 2002), FGFs (Zaragosi et al., 2006) and Wnt proteins (Kleber et al., 2004) have, however, some roles in keeping MSCs in a stemness state.

2.4.4 Immunomodulatory properties of MSCs

The immunomodulatory property of MSCs is one of the most considerable issues in the field of regenerative medicine. Numerous evidences exist on the suppressive effects of MSCs on immune cell activity. Co-culturing MSCs with T-lymphocytes results in the inhibition of T cell proliferation (Di Nicola et al., 2002). This effect is believed to be mediated by secretion of soluble factors that include interleukin 10, prostaglandins and hepatocyte growth factor as well as TGF-beta or via direct cell to cell contact (for review see Bassi et al., 2011). Besides T-lymphocytes, MSCs insert their immunomodulatory effect by inhibition of B cell proliferation and antibody secretion (Rasmusson et al., 2007). Reportedly, they also suppress differentiation, maturation and activation of dendritic cells (Nauta et al., 2006 as cited in Bassi et al., 2011). Moreover, MSCs can suppress natural killer (NK) cell proliferation and change their phenotype, cytokine secretion and cytotoxic properties (Sotiropoulou et al., 2006). MSCs express an intermediate level of MHC class I, very low levels of MHC class II and do not express costimulatory molecules (Klyushnenkova et al., 2005).

Although the hypo-immunogenic properties of MSCs have roused medical interest to establish allogenic MSC banks for clinical application in the field of regenerative medicine, there remains one important question that must be taken into consideration: Do MSCs keep their immunomodulatory properties even after differentiation? Liu et al. have used MSCs from a rabbit model to answer this question (Liu et al., 2006). Their results have shown that osteogenic differentiated MSCs did not express MHC class II in vitro; however, transplantation of the differentiated cells makes these cells lose their immunomodulatory properties in vivo.

3. MSC-based bone constructs

As mentioned, bone construct fabricated using scaffolds, cells and growth factors would be appropriate substitutes for bone grafts. Cells were previously described. In this part of the chapter, scaffolds and growth factors will briefly be considered and followed by some important issues in the field of bone engineering, including cell seeding onto scaffold, commonly-used bioreactors in bone engineering and the issue of the construct vascularity.

3.1 Scaffolds

Bone, as a dynamic and supportive tissue, is a nanocomposite structure. It is in fact a complex of well-organized inorganic–organic nanomaterials including hydroxyapatite [HA, Ca10 $(Po_4)_6$ $(OH)_2$)] and collagen fibrils arranged in a masterly order (Maddela et al., 2010). Within this biomaterial composite, osteocytes lie in cavities referred to as lacunae. In bone tissue engineering, material scientists attempt to make biocompatible and biodegradable scaffolds with appropriate porosity, mechanical strength and hydrophobicity comparable to native bone materials (Lee et al., 2007 as cited in Maddela et al., 2010). Biocompatibility is the most important characteristic of scaffolds, which means that the scaffold material must not be toxic or promote inflammation upon transplantation. All scaffold degradation products should also be biocompatible. The scaffold must also be biodegradable meaning that it gradually degrades at the implantation site in harmony with bone regeneration in order to provide spaces for natural bone growth. Scaffold porosity is another characteristic that merits consideration in designing scaffolds. Incorporation of open and interconnective pores into the scaffold structure is essential in terms of cell growth and distribution, facilitation of vessel formation and diffusion of nutrient and waste products. The size of pores is recommended to be around 200-900 micrometers. If they are too small cells may adhere to the scaffold's superficial part. Large pore size, on the other hand, compromises scaffold strength. Since bone construct should be implanted at the defective bone site where it must temporary support the fracture, therefore bone scaffolds must possess mechanical properties comparable to native bone. The scaffold's surface property, in terms of chemical and topographical features, is a parameter that influences scaffold interaction with cells. Scaffolds must possess appropriate surfaces upon which cells can adhere, proliferate and differentiate.

Materials used to fabricate bone scaffolds include natural and synthetic bioceramics, natural and synthetic polymers and composites of polymers with bioceramics.

3.1.1 Bioceramics

There are two categories of bioceramics: naturally occurring (i.e., coral hydroxyappatite) and synthesized (i.e., synthetic hydroxyapatite and tricalcium phosphate). These materials are widely used in medical trials, orthopedics (for review see Block et al., 2000; Heini et al., 2001; Katti et al., 2004; Taksali et al., 2004 as cited in Habraken et al., 2010) and plastic surgery (Gladstone et al., 1995) due to their high mechanical strength and body response (Habraken et al., 2010). In particular, hydroxyapatite-based calcium phosphate compounds and bioactive glass are a focus of interest for bone engineering because they possess osteoconductive properties (LeGeros et al., 2002; Kenny et al., 2003). There are, however, several disadvantages that limit the application of bioceramics in the tissue engineering field. These materials possess low biodegradability and they bind strongly to growth factors such as bFGF and TGF-β3, releasing them at very slow rates unsuitable for clinical applications (Habraken et al., 2010). Moreover, ceramics possess low tensile strength and are very brittle.

3.1.2 Polymers

As a whole, based on origin, polymers can be divided into two categories: natural and synthetic (Lee et al., 2007). The most widely used natural polymers in bone tissue engineering include collagen, alginate and chitosan. Collagen can be easily solubilized in physiological fluid (Lee et al., 2007; Maddela et al., 2010). Beside this advantage, immunogenecity, hard processing and the possibility of pathogen transmission are the most prominent disadvantages of natural polymers. Among natural polymers, collagen type I is very interesting since it is the pivotal organic component of bone matrix that is actively secreted by osteoblast cells (Lee et al., 2007). Although this protein in the form of a gel, nanofiber, porous scaffold and film is biocompatible, it cannot tolerate mechanical forces and consequently undergoes degradation at the implant site. Thus, natural polymers are rarely used alone.

Synthetic biodegradable polymers include polyglycolic acid (PGA), polylactic acid (PLA), polylactic-co-glycolic acid (PLGA), polycaprolacton (PCL), polypropylene fumarate (PPF), polycaprolacton fumarat and polycaprolacton diacrylate. Synthetic polymers are appropriate options as bone engineering scaffolds because they can be fabricated in large quantities, the risk of infection and toxicity is less compared to natural polymers. In addition, they possess less immunogenicity. Other advantages of synthetic polymers are their mechanical, physical, and chemical properties, elastic module, acidity and hydrophilicy which can easily be adjusted to match bone tissue. The disadvantages of synthetic polymers however include their very low strength that worsens with the introduction of pores to form tissue engineering scaffolds. Furthermore, most of these materials are considered non-osteoconductive (Behravesh et al., 1999; Middletone et al., 2000).

3.1.3 Composite scaffolds

The composite strategy has emerged with the purpose of seeking better scaffolds for bone tissue engineering. In this strategy, the objective is to combine bioceramic material with polymers to make use of the advantages of each biomaterial. In this manner, many polymer-based scaffolds that consist of bioactive bioceramics such as hydroxyapatite/PLLA have been produced (Zandi et al., 2010; Eslaminejad et al., 2007). In composite scaffolds, bioceramics serve to make the scaffold osteoconductive and provide reinforcement whereas the polymer serves to solve the problem of poor degradability, low tensile strength and brittleness of ceramics (Wang 2005).

3.2 Growth factors

Bone normal development involves a variety of hormones, cytokines and growth factors. These biomolecules regulate osteoprogenitor proliferation, migration and differentiation in a controlled manner (for review sees Kain et al., 2005, Schmidmaier et al., 2006). The following growth factors have been reported to play some role in bone differentiation; hence they can be used in bone engineering.

3.2.1 Bone morphogenetic protein

Urist et al. in 1965 have reported that decalcified bone can induce the formation of ectopic bone; therefore they concluded that osteoinductive molecules could be present in the bone matrix and direct the differentiation of precursor cells into bone cells. Later these molecules were called bone morphogenetic proteins (BMPs) and are members of the TGF-beta superfamily. BMP-2, 4 and 7 are the main members of this category whit a considerable effect on the induction of bone formation (Kirker et al., 2000; Yoon et al., 2004).

3.2.2 Insulin-like growth factor

Insulin-like growth factor (IGF) induces proliferation and chemotactic migration of many types of cells. This growth factor has an important role in bone metabolism (Matsuda et al., 1992) especially at the time of fracture healing (Chen et al., 2006).

3.2.3 Fibroblast growth factor

The fibroblast growth factor (FGF) family has a conflicting role in osteogenesis and bone healing. While some authors have reported a positive effect of FGF-2 on bone healing (Kawaguchi et al., 1994; Kato et al., 1998), others have indicated that bFGF induces osteoclast formation (Nakagawa et al., 1999).

3.2.4 Vascular endothelial growth factor

This growth factor exerts its effects through induction of vascularization which in turn plays an important role in bone growth and development (Lee et al., 2007). Vascular endothelial growth factor (VEGF) also supports the survival and activity of bone-forming cells (Hsiong et al., 2000) as well as migration and differentiation of primary human osteoblasts (Mayr-Wohlfart et al., 2002; Orlandini et al., 2006).

3.2.5 Growth factor delivery in bone tissue engineering

In order to have an efficient bone healing procedure, selection of suitable vehicles and types of bioactive molecules are crucial. There are two types of immobilization methods for binding of bioactive molecules to the carrier: non-covalent (physical entrapment, surface adsorption, affinity binding or ionic complexation) or covalent (chemical conjugation).

In bone engineering either polymeric or bioceramic carriers can be used to deliver growth factors to the site of the tissue defect. PLA and PGA are two well- known polymers of alpha-hydroxyl esters and copolymers of these monomers (PLGA) are good vehicles to carry osteoinductive factors (for review see Behravesh et al., 1999; Saito et al., 2001a, as cited in Lee et al., 2007). Calcium phosphate cement (CPC), bioactive glasses, HA and beta-TCP are non-polymeric, inorganic materials that have also been processed to deliver growth factors such as TGF, beta1 and BMPs (for review see Hedberg et al., 2005; Laffargue et al., 1999; Ripamonti et al., 1992).

The design of a delivery system, in the form of three-dimensional matrices, injectable gels, and micro/nano particulates determines release kinetics and stability of growth factors in constructs (Lee et al., 2007). Furthermore, physical structure and degradation time of polymers are important parameters to determine the release behavior of growth factors (Holland et al., 2006, as cited in Lee et al., 2007).

3.3 Cell seeding onto scaffold surfaces

Bone construct is a term used to denote an engineered material that consists of different components, including scaffold/growth factors, scaffold/cells and scaffold/cells/growth factors. In the previous section incorporation of growth factor into the delivery vehicle (scaffold) was summarized. In the following sections, seeding of cells on scaffold surfaces - a crucial procedure in bone construct fabrication is discussed. Indeed, good cell-to-cell contact and acceptable dissemination of the cells within a scaffold are the results of an appropriate seeding procedure. These parameters have extensive impact on uniform well-developed bone tissue regeneration and mineralization after construct transplantation (Holy et al., 2000; Ishaug- Riley et al., 1998).

3.3.1 Static cell seeding method

Although the addition of the cell suspension onto a scaffold (static cell seeding system) is the most simple and frequently used method (Scaglione et al., 2008; Li et al., 2001), it is not a good approach for a well-elaborated bone construct with high cell density and homogenous cell distribution. This method is associated with a low seeding efficiency of approximately 10-25% (Roh et al., 2007) and a low rate of cell penetration inside the scaffold (Mo et al., 2004; Ravi et al., 2009; Ma, 2008). Even penetrated cells are unable to establish attachment with scaffold surfaces.

3.3.2 Dynamic cell seeding

Dynamic cell seeding process is a type of cell seeding usually performed with bioreactor systems. It can be categorized into two different systems: rotational seeding based on the use of hydrostatic forces (Hsu et al., 2005; Nasseri et al., 2003) or vacuum seeding using pressure differentials (van Wachem et al., 1990; Williams et al., 2004). Cell seeding yields by these techniques range from 38 to 90% and 60 to 90%, respectively, compared to 10-25% for static seeding. Although the long seeding time in the rotational system and its bad impact on cell morphology limits its application, no adverse effects have thus far been reported for the vacuum seeding method (Gustavo et al., 2010).

3.3.3 Magnetic cell seeding method

Magnetic cell seeding is another technique proposed to efficiently incorporate cells inside scaffold porosity. This method involves the use of a magnetic force to attract magnetic nanoparticles attached to desired cells (Gustavo et al., 2010). There are two main approaches in this method. The first one involves the application of superparamagnetic monosized polymers such as dynabeads with the ability to bind specifically to a desired cell or protein and subsequent seeding of the cells in the scaffold by producing a temporary magnetic field. The efficiency of this method has been reported to be as much as 99% (Perea et al., 2006; Tiwari et al., 2003, as cited in Gustavo et al., 2010). The second approach is to first label desired cells by cationic liposomes that contain superparamagnetic iron oxide particles, then seed them into a scaffold by the administration of a transient magnetic force. The efficiency of this protocol has been reported to be as high as 90% (Ito et al., 2004; Shimizu et al., 2007 as cited in Gustavo et al., 2010). Rapid graft production and reproducible results are the main advantages of using this method in tissue engineering; albeit cell viability, cell morphology, as well as the fate of magnetic particles in the body and their adverse effects on other tissues need further evaluation (Gustavo et al., 2010).

3.4 Bioreactors

Bioreactors, as containers or vessels, have long been used to hold microorganisms and different types of eukaryotic cells for the purpose of harnessing their natural biochemical processes in the production of drug and recombinant proteins. The main reason to use bioreactors in industry and medicine is the large scale production of desired products at an optimal level of gas, nutrients, temperature and tolerable amount of waste products for an extended period of time (Haasper et al., 2008). Bioreactors have also found some application in the bone tissue engineering field due to valuable advantages they offer toward producing optimal bone construct. Bioreactor systems help bone tissue engineering in several ways. The bioreactor cell culture process is automated. In this manner, the risks of cellular/ microbial contamination, labor intensity and laboratory costs during cell expansion and differentiation are minimized (Andrew et al., 2011). Furthermore, automated cell cultures have the advantages of improved cell seeding efficiency (Wendt et al., 2003) as well as cell proliferation (Grayson et al., 2008). Bioreactors can be used to improve nutrition of the cells located in the deep areas of the constructs during the culture period. With the use of an appropriate bioreactor system, different stress protocols (such as shear) can be applied on bone construct. There are three main commonly used bioreactor systems in bone tissue engineering: spinner flask, rotating wall and perfusion system.

3.4.1 Spinner flask

This bioreactor is a commonly used system in bone tissue engineering (for review see Ichinohe et al., 2008; Sikavitsas et al., 2002). It is composed of a vessel containing culture medium in which the tissue construct is suspended through a wire. A rotating magnetic bar causes the culture medium to agitate (Andrew et al., 2011). This system is more efficient in terms of bone differentiation compared with static and rotating wall culture systems. In spite of this positive point, the spinner flask possesses a deficiency regarding construct nutrition (Sikavitsas et al., 2002). One way to improve nutrient and waste product transport in and out of the construct is to increase the amount of rotation in the magnetic bar but this may lead to increased turbulent flow which in turn increases shear stress on the construct. It should be mentioned that although shear stress produced in the bioreactor has a positive influence on bone differentiation and mineralization (Bancroft et al., 2002, 2003; Bilodeau et al., 2006), exceeded stress may cause cellular damage. For this reason, there must always be a balance between nutrient transport efficiency and shear stress within the spinner flask.

3.4.2 Rotating wall bioreactors

This type of bioreactor is composed of two cylinders. The smaller cylinder is a stationary one inside the system and provides for gas exchange, while the outer one rotates. The cell-seeded scaffold moves freely in the medium between these two cylinders (Andrew et al., 2011; Sikavitsas et al., 2002). It is believed that this system is not a very good culture system in bone tissue engineering due to the haphazard movement of the scaffold in the system, collision with the wall of the bioreactor and the low level of shear stress provided by this regime (Andrew et al., 2011).

3.4.3 Perfusion bioreactors

This system is a widely-used bioreactor in bone tissue engineering (for review see Bancroft et al., 2003; Gomes et al., 2003; Grayson et al., 2008; Sikavitsas et al., 2005). This bioreactor is composed of a perfusion cartridge connected to a pump and a medium reservoir through a tubing circuit. The perfusion cartridge surrounds the scaffold tightly, thus the medium cannot flow around the scaffold so it perfuse directly through the scaffold (Andrew et al., 2011). There are various types of this bioreactor. Those commonly used in bone tissue engineering include the flow perfusion culture bioreactor (for review see Bancroft et al., 2002, Gomes et al., 2003; Grayson et al., 2008; Sikavitsas et al., 2005), radial channel perfusion system (Grayson et al., 2008) and direct perfusion bioreactors (for review see Janssen et al., 2006a, 2006b, 2010). Use of the perfusion bioreactor improves nutrient and waste product transport within the construct. Cell loading efficiency is also improved in these systems since they are designed in such a way that the flow direction is repeatedly changed. Moreover, shear stress produced in this bioreactor enhances osteogenic differentiation of the loaded cells.

3.5 Vascularization

Bone construct usually encounters nutritional limitation due to a lack of microvasculature which occurs at two stages, during in vitro culture and the early days after implantation. As previously described, in vitro nutritional limitations can somewhat be improved with the application of an appropriate bioreactor system. To overcome nutritional limitation during the early days of implantation several strategies have been developed with the objective of inducing angiogenesis inside the construct.

3.5.1 Angiogenic growth factors

One strategy is to use growth factors known to promote angiogenesis when fabricating bone construct. According to the literature, angiogenic growth factors include VEGF (Mayr-Wohlfart et al., 2002; Deckers et al., 2000), FGF (Saadeh et al., 2000), BMP-2, endothelin-1 (ET-1) (Von Schroeder et al., 2003; Bouletreau et al., 2002a), PDGF-BB (Bouletreau et al., 2002b), IGF and TGF-beta (Bouletreau et al., 2002b; Saadeh et al., 1999). VEGF has a direct effect on angiogenesis whereas the others exert their effect indirectly through regulation of VEGF secretion. Although VEGF plays a pivotal role in angiogenesis, unfortunately high doses of this recombinant protein are necessary to reach an optimal level of angiogenesis (Barralet et al., 2009).

3.5.2 Design and architecture of the scaffold

Design and architecture of a bone scaffold facilitates blood vessel formation in the fabricated construct. For example, nano/micro fiber combined scaffolds have been shown to promote endothelial cell migration and organization into capillary-like structures within the scaffold (Santos et al., 2008 as cited in Marina et al., 2010). The inclusion of a network with vasculature geometry in a biocompatible polymer using microfabrication techniques has been reported as an alternative way to create vessel-like structures in scaffolds. It should be mentioned that such methods are commonly used in developing a vasculature tree in soft organs rather than hard tissues such as bone (Marina et al., 2010).

3.5.3 Co-culture system

Since endothelial cells are able to establish microcapillary-like structures, the use of them in bone construct can be considered an alternative strategy to promote vessel formation inside the construct. For this purpose endothelial progenitor cells (EPCs) from bone marrow may be preferred. There are two types of EPCs, early and late (Veleva et al., 2008). Early cells appear within 4-7 days in culture and exhibit some endothelial as well as monocytic properties with restricted proliferative capacity. In contrast, late cells are those that appear 2-3 weeks after culture initiation and keep their expansion potential for a long time. According to investigations, while early cells contribute in neovascularization indirectly through cytokines and matrix metalloproteinase-8 (MMP-8) secretion, late cells contribute to vessel formation as building blocks as well as through MMP-2 secretion (Yoon et al., 2005, as cited in Marina et al., 2010).

3.5.4 Microsurgery strategies

Microsurgery techniques offer another approach to create blood vessel inside bone construct. Two of the most popular microsurgery methods for creating vascularized bone construct include flap fabrication and the creation of an arteriovenous loop (Kneser et al., 2006). In flap fabrication, bone construct including the scaffold, cells and growth factors pre-transplants in a rich vascular bed (i.e., muscle) from which some blood vessels grow into the construct (Scheufler et al., 2008; Polykandriotis et al., 2007). Donor-site morbidity and two surgical interventions are two disadvantages of this method (Ren et al., 2008). In the arteriovenous loop method, vascularization of the porous scaffold is performed by implantation of an arterovenous loop around the construct (Kneser et al., 2006).

4. Bone regeneration promoted by MSC-based bone constructs

As mentioned earlier, large bone defects need clinical intervention for regeneration. For this purpose bone construct fabricated using tissue engineering principles is considered as a promising choice. Different types of bone constructs will be described followed by some examples in which MSC-based bone constructs have been used to regenerate bone defects in either animal models or humans.

4.1 Types of bone construct based on constituting components

Past investigations regarding bone regeneration have used the following constructs which differed in terms of their constituting elements: scaffold, cells, scaffold/growth factor, scaffold/cells, scaffold/cells/growth factor and scaffold/DNA. Scaffold alone is more suitable for small bone defects and usually comprised of tricalcium phosphate ceramics. Cell constructs involve the administration of MSCs in an injectable form with or without genetic manipulation. Scaffold/growth factor constructs are constructs in which growth factors are linked chemically or physically to a carrier or scaffold. Scaffold/MSCs constructs are the most frequently used one in tissue engineering and many studies have been designed with this model. Sometimes the scaffold/MSC construct contains transfected cells and this indeed is a combination of gene therapy and tissue engineering. Scaffold/cells/growth factor construct, on the other hand are rarely used. An excellent example of such construct is Gronthos´ study (see below). Finally, the scaffold/DNA construct involves loading a scaffold by a DNA construct, usually in the form of plasmid DNA (Bonadio et al., 1999).

4.2 MSC-based bone construct transplantation in animal models

Before clinical application, a newly developed bone construct should be tested in animal models in term of its functionality. By pre-clinical studies scientists understand whether regeneration occurs due to the presence of MSCs in bone constructs or the osteoconductive/osteoinductive properties of the scaffold/growth factor. Among different preclinical models, large bone defect models are more commonly used since they are the type of defects problematic in the field of orthopedics as well as maxillofacial surgery (for review sees Perka et al., 2000). Various worldwide animal and clinical studies have been undertaken to cure this type of bone disease. Following are some examples of animal model studies that have used MSC-based constructs.

4.2.1 MSCs alone

An example of the application of MSCs in an animal bone defect model is a study conducted by Tsuda et al. who used MSCs transfected with BMP-2 to improve an osteoporetic disorder of aged rats (Tsuda et al., 2005).

4.2.2 Scaffold/MSCs

The application of bone constructs that consist of scaffold/MSCs is a common strategy to heal large bone defects, also referred to as critical sized defects in animal models. In most studies bioceramics are the biomaterial of choice. Kadiyala et al. have used autologous bone marrow-derived MSCs seeded onto ceramic scaffolds to regenerate an 8-mm experimentally-created defect in the rat femora. After eight weeks, bone formation was detected in the defective area (Kadiyala et al., 1997b). Similar animal studies have also been reported for canine and sheep models using scaffold/MSCs constructs (For review see Arinzeh et al., 2003; Kon et al., 2000). Application of varying ceramics in different ratios would be very crucial for fabricating a good construct. For example, Buruder et al. have fabricated MSC-based scaffolds comprised of 35% beta tricalcium phosphate and 65% hydroxyapatite which was implanted in a 21-mm experimentally-created segmental defect in the canine femur. They observed good integration of the construct with the host bone (Bruder et al., 1998). Using ceramic biomaterials we have also conducted some investigations. According to our studies, the use of rat MSCs in conjunction with natural scaffolds (i.e., Bio-Oss or human deproteinized and decellularized bone tissue) could enhance bone regeneration in rat calvarial defects more than platelet-rich plasma treatment (Khojasteh et al., 2009). In another study, we have compared the bone regeneration capacity of HA/TCP/MSCs with Bio-Oss/MSC constructs. Our results have shown the enhanced bone regeneration potential of HA/TCP/MSCs construct in a canine full-thickness alveolar defect model (Jafarian et al., 2008) compared with Bio-Oss/MSCs constructs.

Other research groups have studied the regenerative role of MSCs in combination with polymeric scaffolds in animal bone defects. For example, in a study by Holy et al., 1.2 cm bone defects in a rabbit femur were regenerated by using MSCs-loaded PLGA scaffolds. Their results have shown significant bone regeneration in MSC-based PLGA scaffolds compared with PLGA alone (Holy et al., 2003).

The composite scaffold/MSCs construct has also been tested in animal bone defects. For example, Diao et al. have manufactured a construct comprised of umbilical cord blood MSCs seeded on composite scaffold consisting of hydroxyapatite, PLLA and collagen. The composite, then, was implanted subcutaneously into SCID mice. Twelve weeks later, well-promoted bone formation was observed in histological sections (Diao et al., 2009).

MSCs derived from sources other than bone marrow have been successfully applied to regenerate bone defects in combination with scaffolds. For example, the construct made of porous cylindrical PLA scaffolds and autologus adipose-derived stem cells have been reported to successfully regenerate skull defects in New Zealand white rabbits (Bella et al., 2008). Similarly, Jang et al. have noted the efficiency of construct comprised of beta TCPs and canine umbilical cord blood MSCs in canine cortical defect regeneration (Jang et al., 2008).

4.3 MSC- based bone construct in clinical trials

The ultimate objective of elaborating bone construct using principles of tissue engineering is to find an appropriate substitute for autologous bone graft which is considered the golden standard for regeneration of bone defects. The excellent regenerating effects of autologous bone graft may be related to its osteoconductive, osteoinductive and osteogenic capacities. Constructs fabricated by engineering principles using all three main building blocks have potentially all components necessary for real representation of autologous bone graft.

In bone construct, MSCs can be present either in an undifferentiated or differentiated state. Using MSCs as undifferentiated cells may have the disadvantage of their unwanted differentiation into non-bone cells where they are supposed to generate osteogenic cells. On the other hand, transplantation of MSCs as fully differentiated cells would be an alternative way to deliver cells into a bone defect. This requires long-term culture of the cells which is undesirable in a cell therapy strategy. Both strategies have been used in clinical trials.

At the moment, multiple clinical trials have been accomplished on human problematic bone lesions. Several others are ongoing and registered at http://clinicaltrials.gov, the official clinical trials website. In this regard, Royan Institute has registered a number of clinical trials regarding nonunion fractures, delayed union, bone cyst and distraction osteogenesis, using MSC-based bone constructs. In the following section some accomplished trials will be noted which are categorized according to the composition of the construct used.

4.3.1 MSCs alone

One well known example of the use of MSCs alone to regenerate osteogenic defects is in the treatment of osteogenesis imperfecta (OI), a heterogenous group of inherited disorders. Horwitz et al. have reported a trial in which allogenic bone marrow cells (from HLA-identical or single- antigen-mismatched siblings) have been intravenously infused into children with severe OI. According to their findings, there have been signs of improvement after transplantation as they observed a reduced number of osteocytes, linearly organized osteoblasts, lamellar bone formation as well as mineralization in the trabecular bones of affected children. Although this study demonstrated osteoblastic differentiation of MSCs, the lack of reliable controls and the absence of a long follow up period were the weak points of this clinical study (Horwitz et al., 1999). To fulfill this trial, Horwitz continued his study in another project in 2001 in which seven children with OI were selected. Five received cell therapy while two were in the control group. Six months after transplantation, growth acceleration was observed in the treated group in comparison with control children (Horwitz et al., 2001).

Another example of MSCs injection is with tibial achondroplasia and pseudoarthrosis in which distraction osteogenesis is necessary. To accomplish this, in vitro osteogenic differentiated marrow MSCs accompanied by platelet-rich plasma were injected into the distracted callus. After a period of time, healing was observed in treated patients (Kitoh et al., 2004).

4.3.2 Scaffold/MSCs/growth factor

Constructs consisting of three components have rarely been utilized to regenerate human bone defects. An example is the work by Gronthos who has tried to regenerate 7-cm length mandibular defect generated due to tumor resection in a 56 year old patient. In this study, he designed a hollow titanium mesh, filled it with hydroxapatite, recombinant human BMP-7 and MSCs, and transplanted it into the patient's lathismus dorsi muscle in order to encourage ectopic bone formation and blood vessel ingrowth. Seven weeks later, vascularized bone construct was removed and implanted into the patient´s mandible. Four weeks post-transplantation, the mandible was functional and the patient could chew food (Gronthos, 2004).

4.3.3 Scaffold/MSCs

Scaffold/MSCs constructs are commonly used constructs in most clinical trials. There are several examples of such approaches in the literature. One such example is the work by Quarto et al. who have designed bone marrow MSC-loaded hydroxyapatite scaffolds to regenerate 7-cm length human bone defects. Two months after transplantation, a large callus formed in the defect site with integration of the construct into host bone (Quarto et al., 2001). In a similar study, Morishita et al. have published successful bone tissue engineering approaches in the treatment of bone tumors by the implantation of authologous MSCs-loaded hydroxyapatite scaffold (Morishita et al., 2006). Similarly Marcacci et al. have prepared constructs using autologous MSCs and hydroxyapatite scaffold, and have transplanted them into long bone defects in humans. Their trial was prominent since they had a prolonged follow up of 6-7 years. Using this method, they could detect good integration of the loaded scaffolds with surrounding bone, new vascular ingrowth and new bone formation inside the scaffold. Recovery of limb function was also reported and maintained after the 6-7 year follow up (Marcacci et al., 2007).

5. Conclusion

Bone constructs elaborated with tissue engineering principles are a promising substitute for autologous bone graft and have long been considered the golden standard for repair of large bone defects. Autologus bone graft owes its excellent repair effects to three crucial properties of osteoinduction, osteoconduction and osteogenesis. The building blocks that are used to fabricate bone construct impart three key properties as autologous bone graft (cells impart osteogenic property, scaffold cause osteoconductive and osteoinductive capacity and growth factors give osteoinductive potential). However, before tissue-engineered bone construct are to be routinely used in the clinic setting instead of bone grafts, several issues must be addressed. Although application of MSCs as cellular material facilitates the construct fabrication, there is still some issue with MSC preparation. MSC propagation is largely dependent on fetal bovine serum. Furthermore, natural bone is a composite of nano hydroxyapatite particles with collagen nanofibers which impart the tissue's unique properties. Unfortunately developing a scaffold with similar properties is still challenging. Finally, perhaps the main challenge in the field of bone tissue engineering is formation of blood vessels inside the fabricated constructs. Several strategies including the addition of angiogenic growth factors and cells to the construct, angiogenic design and architecture of bone scaffold, and microsurgery techniques have been proposed to promote angiogenesis inside the constructs. However, there is no reliable, reproducible and practical strategy in this developing field.

6. References

Andrew, B., Yeatts, J., Fisher, P. (2011). Bone tissue engineering bioreactors: Dynamic culture and the influence of shear stress. *Bone*, Vol.48,pp.(171–181).

Anita, H., Undale, M.B., Westendorf, J.J., Yaszemski, M.J., Sundeep Khosla, M.D. (2009). Mesenchymal Stem Cells for Bone Repair and Metabolic Bone Diseases, *Mayo Clinical Proceeding*. Vol. 84, NO.10, pp.(893-902).

Bancroft, G.N., Sikavitsast, V.I., van den Dolder, J., Sheffield, T.L., Ambrose, C.G., Jansen, J.A.(2002). Fluid flow increases mineralized matrix deposition in 3D perfusion culture of marrow stromal osteloblasts in a dose-dependent manner. *Proc Natl Acad Sci U SA,* Vol.99,pp.(12600–5).

Bassi, E.J., Aita, C.A.M., Câmara, N.O.S. (2011). Immune regulatory properties of multipotent mesenchymal stromal cells: Where do we stand?. *World Journal of Stem Cells,*Vol. 3, N.1, pp.(1 -8).

Behraves, E., Yasko, A.W., Engel, P.S., Mikos, A.G. (1999). Synyhetic biodegradable polymers for orthopedic applications. *Clinical Orthopedic related Research,* Vol. 367, Pp. (S118-129).

Bella, C.D., Farlie, P. & Penington, A.J.(2008). Bone regeneration in a rabbit critical-sized skull defect using autologous adipose-derived cells. *Tissue Engineering A,* Vol.14, pp.(483–90).

Bilodeau, K., Mantovani, D. (2006). Bioreactors for tissue engineering: focus on mechanical constraints. A comparative review. Tissue Engineering, Vol.12,pp.(2367–83).

Bouletreau, P. J., Warren, S. M., Spector, J. A., Peled, Z. M., Gerrets, R. P., Greenwald, J. A. M., Longaker, T. (2002a). *Plastic Reconstruction Surgery,*Vol.109,pp.(2384-).

Bonadio, J., Smiley, E., Patil, P., Goldstein, S. (1999). Localized, direct plasmid gene delivery in vivo: prolonged therapy results in reproducible tissue regeneration. *Nature Medicine,* Vol.5,pp.(753-9).

Bouletreau, P. J., Warren, S. M., Spector, J. A., Steinbrech, D. S., Mehrara, B. J., Longaker, M. T. (2002b). *Plastic Reconstruction Surgery.* Vol.110, pp.(139-148).

Bruder, S.P., Jaiswal, N., Haynesworth, S.E. (1997). Growth kinetics, Self-renewal, and the osteogenic potential of purified human mesenchymal stem cells during extensive subcultivation and following cryopreservation. *Journal of Cell Biochemistry,* Vol. 64, pp.(278_294).

Bruder, S.P., Kraus, K.H., Goldberg, V.M., Kadiyala, S. (1998). The effect of implants loaded with autologous mesenchymal stem cells on the healing of canine segmental bone defects. *Journal of Bone and Joint Surgery,* Vol.80, pp.(985-96).

Bruder, S.P., Fox, B.S. (1999). Tissue engineering of bone: cell based strategies. *Clinical Orthopedie,* Vol. 367, pp.(S68-83).

Buttery, L.D., Bourne, S., Xynos, J.D., Wood, H., Hughes, F.J., Hughes, S.P., Episkopou, V., & Polak, J.M.(2001). Differentiation of osteoblasts and in vitro bone formation from murine embryonic stem cells. *Tissue Engineering,*Vol. 7, pp. (89–99)

Caplan, A.I. (1994). The mesengenic process. *Clinical Plastic Surgery,* Vol. 21, pp.(429-435).

Carpio, L., Gladu, J., Goltzman, D., Rabbani, S.A. (2001). Induction of osteoblast differentiation indexes by PTHrP in MG-63 cells involves multiple signaling pathways. *American Journal of Physiology Endocrinology and Metabolism*, Vol. 281, pp. (489-99).

Chapekar, M.S. (2000). Tissue engineering: challenges and opportunities: *J Biomed Mater Res*, Vol. 53, pp.(617-620)

Chen,F.M., Zhao,Y.M., Wu,H., Deng, Z.H., Wang, Q.T., Zhou, W., Liu, Q., Dong, G.Y., Li, K., Wu, Z.F., Jin, Y. (2006). Enhancement of periodontal tissue regeneration by locally controlled delivery of insulin-like growth factor-I from dextran-co-gelatin microspheres, *J. Control. Release*, Vol.114, pp.(209–222).

Choi, K.M., Seo, Y.K., Yoon, H.H., Song, K.Y., Kwon, S.Y., Lee, H.S., Park, J.K.(2008). Effect of ascorbic acid on bone marrow-derived mesenchymal stem cell proliferation and differentiation. *Journal of Bioscience and Bioengineering*, Vol. 105, pp. (586-94).

Coelho, M.J, Fernandes, M.H.(2000). Human bone cell cultures in biocompatibility testing. Part II: effect of ascorbic acid, beta-glycerophosphate and dexamethasone on osteoblastic differentiation. *Biomaterial*, Vol. 21, pp.(1095-102).

Colter, D.C., Sekiya, I., & Prockop, D.J.(2001). Identification of a subpopulation of rapidly self-renewing and multipotential adult stem cells in colonies of human marrow stromal cells. *Proc.Natl Acad. Sci. USA*. Vol. 98, pp.(7841–5)

Deckers, M.L., Karperien, M., van der Bent, C.,Yamashita, T., Papapoulos, S.E., Lowik, C.W.G.M. (2000). *Endocrinology*, Vol. 141,pp.(1667-1674).

de Boer, J., Siddappa, R., Gaspar, C., van Apeldoorn, A., Fodde, R., van Blitterswijk, C. (2004). Wnt signaling inhibits osteogenic differentiation of human mesenchymal stem cells. *Bone*, Vol. 34, pp. (818-26).

Diao, Y., Ma, Q., Cui, F. & Zhong, Y.(2009). Human umbilical cord mesenchymal stem cells: osteogenesis in vivo as seed cells for bone tissue engineering. *Journal of Biomedical Materials Research* , Vol.91,pp.(123–131).

Diefenderfer, D.L., Osyczka, A.M., Reilly, G.C., Leboy PS. (2003). BMP responsiveness in human mesenchymal stem cells. *Connective Tissue Research*, Vol. 44, Suppl. 1,pp. (305-11).

Di Nicola, M., Carlo-Stella, C., Magni, M., Milanesi, M., Longoni, P.D., Matteucci, P., Grisanti, S., Gianni, A.M. (2002). Human bone marrow stromal cells suppress T-lymphocyte proliferation induced by cellular or nonspecific mitogenic stimuli. *Blood*,Vol.99, pp.(3838-3843).

Diptiman, C., Kumar S., & Ponnazhagan, S. (2010). Therapeutic Potential of Adult Bone Marrow-Derived Mesenchymal Stem Cells in Diseases of the Skeleton. *Journal of Cellular Biochemistry*, Vol.111, pp.(249–257).

Dominici, M., Le Blanc, K., Mueller, I., Slaper-Cortenbach, I., Marini, F., Krause, D., Deans, R., Keating, A., Prockop, D.j., Horwitz, E.(2006). Minimal criteria for defining multipotent mesenchymal stromal cells. The International Society for Cellular Therapy position statement.*Cytotherapy*, Vol. 8, pp.(307-315).

Dragoo, J.L., Samini, B., Zhu, M., Hame, S.L., Thomas, B.J., Lieberman, J.R., Hedrick, M.H., Benhaim, P. (2003).Tissue-engineered cartilage and bone using stem cells from human infrapatellar fat pads. *Journal of Bone Joint Surgery*, Vol. 85, pp.(740-747).

Ducy, P., Schinke, T., Karsenty, G.(2000b). The osteoblast: a sophisticated fibroblast under central surveillance. *Science,* Vol.289, pp. (1501-4).

Ehrhart, N.P., Hong, L., Morgan, A.L., Eurell, J.A., Jamison, R.D. (2005). Effect of transforming growth factor-beta1 on bone regeneration in critical-sized bone defects after irradiation of host tissues. *American Journal ofVeterinary Research,* Vol.66,pp.(1039–1045).

Eslaminejad, M.B., Mirzadeh, H., Nickmahzar, A., Mohamadi, Y. (2007). Bone differentiation of marrow- derived mesenchymal stem cells using beta- tricalcium phosphate-alginate- gelatin hybrid scaffolds. *Journal of Tissue Engineering & Regenerative Medicine,* Vol.1,No.6, pp.(417- 424).

Eslaminejad, M.B., Talkhabi, M., Zainali, B. (2008) Effect of Lithium chloride on proliferation and bone differentiation of rat marrow-derived mesenchymal stem cells in culture. *Iranian Journal of Basic Medical Sciences,* Vol. 3, pp.(143-151).

Evans, M.J., & Kaufman, M.H. (1981). Establishment in culture of pluripotential cells from mouse embryos. *Nature,*Vol. 292, pp. (154–6).

Feng, L., Bronson, S., & Niyibizi, C. (2010). Derivation of Murine Induced Pluripotent Stem Cells (iPS) and assessment of Their Differentiation Toward Osteogenic Lineage. *Journal of Cellular Biochemistry.* Vol. 109. pp. (643–652)

Friedenstein, A.J., Piatetzky-Shapiro, II., Petrakova, K.V. (1966). Osteogenesis in transplants of bone marrow cells. *Journal of Embryology and Experimental Morphology,* Vol. 16, pp. (381-90).

Friedenstein, A.J. (1973). Detrimined and inducible osteogenic precursure cells. *Hard Tissue Growth, Repair and Remineralization.*pp:(169-185).

Friedenstein, A. J, Deriglasova, U.F., Kulagina, N.N., Panasuk, A.F., Rudakowa, S.F., Luria, E.A., & Ruadkow, I.A. (1974). Precursors for fibroblasts in different populations of hematopoietic cells as detected by the in vitro colony assay method. *Experimental Hematology.* Vol. 2,pp.(83–92).

Fukumoto, T., Sperling, J.W., Sanyal, A., Fitzsimmons, J.S., Reiholz, G.G., Conover, C.A., O'Driscoll, S.W. (2003). Combined effects of insulin-like growth factor1- and transforming growth factor-beta1 on periosteal mesenchymal stem cells during chondrogenesis in vitro . *Osteoarthritis Cartilage,* Vol. 11, pp.(55-64).

Galmiche, M.C., Koteliansky, V.E., Briere, J., Herve, P., Charbord, P.(1993) .Stromal cells from human long-term marrow cultures are mesenchymal cells that differentiate following a vascular smooth muscle differentiation pathway. *Blood,* Vol. 1, No. 82, pp. (66-76).

Gladstone, M.W., McDermott, D., Cooke. (1995). Implants for cranioplasty, Otolaryngol, *Clin. North Am,* Vol.28, No. 2, pp.(381–400).

Gomes, M.E., Sikavitsas, V.I., Behravesh, E., Reis, R.L., Mikos, A.G. (2003). Effect of flow perfusion on the osteogenic differentiation of bone marrow stromal cells cultured on starchbased three-dimensional scaffolds. *Journal of Biomedical Material Research* A, Vol.67A, pp.(87–95).

Gori, F., Thomas, T., Hicok, K.C., Spelsberg, T.C., Riggs, B.L. (1999). Differentiation of human marrow stromal precursor cells: bone morphogenetic protein-2 increases OSF2/CBFA1, enhances osteoblast commitment, and inhibits late adipocyte maturation. *Journal of Bone and Mineral Research,* Vol. 14, pp.(1522-35).

Grayson, W.L., Bhumiratana, S., Cannizzaro, C., Chao, P.H.G., Lennon, D.P., Caplan, A.I.(2008). Effects of initial seeding density and fluid perfusion rate on formation of tissueengineered bone. *Tissue Engineering A,* Vol.14,pp.(1809–20).

Gronthos, S. (2004). Reconstruction of human mandible by tissue engineering. *Lancet,* Vol.3,pp.(735-736).

Gruber, R., Kandler, B., Fuerst, G., Fischer, M.B., Watzek, G. (2004). Porcine sinus mucosa holds cells that respond to bone morphogenetic protein (BMP)-6 and BMP-7 with increased osteogenic differentiation in vitro. *Clinical Oral Implants Research,* Vol. 15, pp. (575-80).

Villalona, G.A., Udelsman, V.B., Duncan, D.R., McGillicuddy, E., Sawh-Martinez, R.F., Hibino, N., Painter, C., Mirensky, T., Erickson, B., Shinoka, T., and Breuer, C.K.(2010). Cell-Seeding Techniques in Vascular Tissue Engineering, *Tissue engineering:part B,* Vol.16,No. 3, pp.(341-350).

Haasper, C., Zeichen, J., Meister, R., Krettek, C., Jagodzinski, M.(2008). Tissue engineering of osteochondral constructs in vitro using bioreactors. *Injury,* 39S1,pp.(S66-S76).

Habraken, W.J.E.M., Wolke, J.G.C., Jansen, J.A. (2007). Ceramic composites as matrices and scaffolds for drug delivery in tissue engineering, Advanced drug delivery reviews, Vol. 59, pp.(234-248).

Hanna, J., Wernig, M., Markoulaki, S., Sun, C.W., Meissner, A., Cassady, J.P., Beard, C., Brambrink, T., Wu, L.C., Townes, T.M., Jaenisch, R.(2007). Treatment of sickle cell anemia mouse model with iPS cells generated from autologous skin. *Science,* Vol. 318, pp.(1920–1923).

Heary, R.F., Schlenk, R.P., Sacchieri, T.A., Barone, D., Brotea, C. (2002). Persistent iliac crest donor site pain: independent outcome assessment. *Neurosurgery,* Vol.50, pp.(510–516).

Heath, C.A. (2000). Cells for tissue engineering. *Trends in Biotechnology.* Vol. 18, pp. (17-19).

Horwitz, E.M., Prockop, D.J., Fitzpatrick, L.A., Koo, W.W., Gordon, P.L., Neel, M., Sussman, M., Orchard, P., Marx, J.C., Pyeritz, R.E., Brenner, M.K. (1999). Transplantability and therapeutic effects of bone marrow-derived mesenchymal cells in children with osteogenesis imperfecta. *Nature Medicine,* Vol.5, pp.(309–313).

Horwitz, E.M., Prockop, D.J., Gordon, P.L., Koo, W.W., Fitzpatrick, L.A., Neel, M.D., McCarville, M.E., Orchard, P.J., Pyeritz, R.E., Brenner, M.K.(2001). Clinical responses to bone marrow transplantation in children with severe osteogenesis imperfecta. *Blood,* Vol.97,pp.(1227–1231).

Holy, C.E., Fialkov, J.A., Davies, J.E., Shoichet, M.S. (2003). Use of a biomimetic strategy to engineer bone. *Journal of Biomedical Materials Research A,* Vol.65,pp.(447-53).

Holzer, G., Einhorn, T.A., Majeska, R.J. (2002).Estrogen regulation of growth and alkaline phosphatase expression by cultured human bone marrow stromal cells. *Journal of Orthopedic Research,* Vol. 20, pp. (281-8).

Hou, C.H., Yang, R.S., Hou, S.M. (2005). Hospital-based allogenic bonebank—10-year experience. *Journal of Hospital Infections,* Vol. 59, pp.(41–45).

Hsu, S.H., Tsai, I.J., Lin, D.J., Chen, D.C. (2005).The effect of dynamic culture conditions on endothelial cell seeding and retention on small diameter polyurethane vascular grafts. *Med Eng Phys,* Vol.27, pp. (267-272).

Huss, R., Hong, D.S., McSweeney, P.A., Hoy, C.A., Deeg, H.J. (1995). Differentiation of canine bone marrow cells with hematopoietic characteristic from an adherent stromal cell precursors. *Proc Natl Acad Sci USA, Vol.* 92, pp.(748-752).

Ichinohe, N., Takamoto, T., Tabata, Y. (2008). Proliferation, osteogenic differentiation, and distribution of rat bone marrow stromal cells in nonwoven fabrics by different culture methods. *Tissue Engineering A* , Vol.14, pp.(107–16).

Isenmann, S., Cakouros, D., Zannettino, A., Shi, S., Gronthos, S. (2007). hTERT transcription is repressed by Cbfa1 in human mesenchymal stem cell populations. *Journal of Bone and Mineral Research,* Vol.22, pp. (897-906).

Ishaug-Riley, S.L., Crane, G.M., Gurlek, A., Miller, M.J., Yasko, A.W., Yaszemski, M.J., Mikos, A.G. (1997). Ectopic bone formation by marrow stromal osteoblast transplantation using poly(DL-lactic-co-glycolic acid) foams implanted into the rat mesentery. *Journal of Biomedical Materials Research,* Vol. 36, pp.(1-8).

Jafarian, M., Eslaminejad, M.B., Khojasteh, A., Mashhadi Abbas, F, Dehghan, M.M., Hassanizadeh, R., Houshmand, B. (2008). Marrow- derived mesenchymal stem cells- directed bone regeneration in dog mandible: a comparison between biphasic calcium phosphate and natural bone mineral. *Oral Surg Oral Med Oral Pathol Oral Radiol Endod,* Vol.105,No. 5, pp.(14-24).

Jaiswal, R.K., Jaiswal, N., Bruder, S.P., Mbalaviele, G., Marshak, D.R., Pittenger, M.F. (2000). Adult human mesenchymal stem cell differentiation to the osteogenic or adipogenic lineage is regulated by mitogen-activated protein kinase. *Journal of Biological Chemistry,* Vol. 275, pp. (9645- 52).

Jang, B.J., Byeon, Y.E., Lim, J.H., Ryu, H.H., Kim, W.H., Koyama, Y., Kikuchi, M., Kang, K.S., & Kweon, O.K. (2008). Implantation of canine umbilical cord blood-derived mesenchymal stem cells mixed with beta-tricalcium phosphate enhances osteogenesis in bone defect model dogs. *Journal of Veterinology Science,* Vol.9, pp.(387–93).

Jankowsk, R.J., Deasy, B.M., Huard, J. (2002). Muscle-derived stem cells. *Gene therapy,* Vol. 9, pp. (642-647).

Janssen, F.W., Oostra, J., van Oorschot, A., van Blitterswijk, C.A. (2006a). A perfusion bioreactor system capable of producing clinically relevant volumes of tissue-engineered bone: in vivo bone formation showing proof of concept. *Biomaterials*,Vol.27,pp.(315–23).

Janssen, F.W., Hofland, I., van Oorschot, A., Oostra, J., Peters, H., van Blitterswijk, C.A.(2006). Online measurement of oxygen consumption by goat bone marrow stromal cells in a combined cell-seeding and proliferation perfusion bioreactor. *Journal of Biomedical Material Research* A , Vol.79A,pp.(338–48).

Janssen, F.W., van Dijkhuizen-Radersma, R., Van Oorschot, A., Oostra, J., de Bruijn, J.D., Van Blitterswijk, C.A. (2010). Human tissue-engineered bone produced in clinically relevant amounts using a semi-automated perfusion bioreactor system: a preliminary study. *Journal of Tissue Engineering and Regenerative Medicine,* Vol.10, No.4,pp.(12–24).

Jiang, Y., Jahagirdar, B.N., Reinhardt, R.L., Schwartz, R.E., Dirk, Keene, C., Ortiz-Gonzalez, X.R., Reyes, M., Lenvik, T., Lund, T., Blackstad, M., Du, J., Aldrich, S., Lisberg, A., Low, W.C., Largaespada, D.A., Verfaillie, C.M. (2002). Pluripotency of mesenchymal stem cells derived from adult marrow. *Nature,* Vol. 418, pp.(41-49).

Kadiyala, S., Young, R.G., Thiede, M.A., Bruder, S.P. (1997a).Culture expanded canine mesenchymalstem cells possess osteochon drogenic potential in vivo and in vitro. *Cell Transplant,* Vol. 6, pp.(125-134).

Kadiyala, S., Jaiswal, N., Bruder, S.P. (1997b). Culture-expanded, bone marrowderived mesenchymal stem cells regenerate a critical-sized segmental bone defect. *Tissue Eng*ineering, Vol.3,No.2, pp.(173-185).

Kain, M.S., Einhorn, T.A. (2005). Recombinant human bone morphogenetic proteins in the treatment of fractures, *Foot Ankle Clinics,* Vol.10, pp.(639–650).

Kato, T., Kawaguchi, H., Hanada, K., Aoyama, I., Hiyama, Y., Nakamura, T., Kuzutani, K., M. Tamura, T., Kurokawa, K. Nakamura. (1998).Single local injection of recombinant fibroblast growth factor-2 stimulates healing of segmental bone defects in rabbits. *Journal of Orthopedical Research,* Vol.16, pp.(654–659).

Kawaguchi, H., Kurokawa, T., Hanada, K., Hiyama, Y., Tamura, M., Ogata, E., Matsumoto, T. (1994). Stimulation of fracture repair by recombinant human basic fibroblast growth factor in normal and streptozotocin diabetic rats. *Endocrinology,* Vol.135,pp.(774–781).

Kawaguchi, J., Mee, P.J., & Smith, A.G.(2005). Osteogenic and chondrogenic differentiation of embryonic stem cells in response to specific growth factors. *Bone,* Vol. 36, pp. (758–69).

Kenny, S.M., Buggy, M. (2003). Bone cements and fillers: a review. *Journal of Material Science and Material Medicine,* Vol.14, pp.(923–938).

Khojasteh, A., Eslaminejad, M.B., Nazarian, H.(2008). Mesenchymal stem cells enhance bone regeneration in rat calvarial critical size defect more than platelete rich plasma. *Oral Surg, Oral Med, Oral Pathol, Oral Radiol, Endodont,* Vol.106, pp.(356-362).

Kimelman, N., Pelled, G., Gazit, Z., Gazit, D. (2006). Applications of gene therapy and adult stem cells in bone bioengineering. *Regenerative Medicine,* Vol. 1, pp.(549-561).

Kirker-Head, C.A. (2000). Potential applications and delivery strategies for bone morphogenetic proteins, Adv. *Drug Deliery. Reviews,* Vol. 43,pp.(65–92).

Kitoh, H., Kitakoji, T., Tsuchiya, H., Mitsuyama, H., Nakamura, H., Katoh, M., Ishiguro, N. (2004). Transplantation of marrow-derived mesenchymal stem cells and platelet-rich plasma during distraction osteogenesis--a preliminary result of three cases. *Bone,* Vol.35, pp.(892-8).

Kleber, M., Sommer, L. (2004). Wnt signaling and the regulation of stem cell function. *Current Opinions in Cell Biology,* Vol.16, pp.(681-687).

Klyushnenkova, E., Mosca, J.D., Zernetkina, V., Majumdar, M.K., Beggs, K.J., Simonetti, D.W., Deans, R.J., McIntosh, K.R. (2005). T cell responses to allogeneic human mesenchymal stem cells: immunogenicity, tolerance, and suppression. *Journal of Biomedical Sciences,*Vol.12, pp.(47-57).

Kneser, U., Schaefer, D.J., Polykandriotis, E., Horch, V.J. (2006). Tissue engineering of bone : the surgeon's point of view. *Journal of Cellular and Molecular Medicine,* Vol.10,pp.(7-19).

Koch, H., Jadlowiec, J.A., Campbell, P.G. (2005). Insulin-like growth factor-I induces early osteoblast gene expression in human mesenchymal stem cells. *Stem Cells and Development,* Vol. 14, pp. (621-31).

Kretlow, J.D., Mikos, A.G. (2007).Mineralization of synthetic polymer scaffolds for bone tissue engineering. *Tissue Engineering*, Vol. 13, pp.(927–938).

Laffargue, P., Hildebrand, H.F., Rtaimate, M., Frayssinet, P., Amoureux, J.P., Marchandise, X. (1999). Evaluation of human recombinant bone morphogeneticprotein-2-loaded tricalcium phosphate implants in rabbits' bone defects.*Bone,* Vol. 25, pp.(55S–58S).

Lee S.H., Shin, H. (2007).Matrices and scaffolds for delivery of bioactive molecules in bone and,cartilage tissue engineering. *Advanced Drug Delivery Reviews*, Vol. 59, pp.(339–359).

LeGeros, R.Z. (2002). Properties of osteoconductive biomaterials: calcium phosphates, *Clin. Orthop. Relat. Res*, Vol.395, pp.(81–98).

Li, Y., Ma, T., Kniss, D.A., Lasky, L.C., Yang, S.T. (2001). Effects of filtration seeding on cell density, spatial distribution, and proliferation in nonwoven fibrous matrices. *Biotechnol Prog*, Vol.17, pp.(935–44).

Li, R.,Stewart, D.J., von Schroeder, H.P., Mackinnon, E.S., Schemitsch, E.H. (2009). Effect of cell-based VEGF gene therapy on healing a segmental bone defect. *Journal of Orthopedic Research*. Vol. 27, pp.(8-14).

Liu, Y., Titus, L., Barghouthi, M., Viggeswarapu, M., Hair, G., Boden, S.D. (2004). Glucocorticoid regulation of human BMP-6 transcription. *Bone,* Vol. 35, pp. (673-81).

Liu, H., Kemeny, D.M., Heng, B.C., Ouyang, H.W., Melendez, A.J., Cao, T. (2006). The immunogenicity and immunomodulatory function of osteogenic cells differentiated from mesenchymal stem cells. *Journal of Immunology*, Vol. 176, pp.(2864-71).

Marcacci, M., Kon, E., Moukhachev, V. (2007). Stem cells associated with macroporous bioceramics for long bone repair: 6- to 7-year outcome of a pilot clinical study. *Tissue Engineering,* Vol.13, No.6, pp.(947-955).

Marina, I., Santos, Rui, L. Reis. (2010). Vascularization in Bone Tissue Engineering: Physiology, Current Strategies, Major Hurdles and Future Challenges. *Macromolecular bioscience,* Vol.10,pp.(12–27).

Massagué, J., Wotton, D. (2000). Transcriptional control by the TGF-beta/Smad signaling system. *The EMBO Journal,* Vol. 19, pp. (1745-54).

Matsuda, N., Lin, W.L., Kumar, N.M., Cho, M.I., Genco, R.J. (1992).Mitogenic, chemotactic, and synthetic responses of rat periodontal ligament fibroblastic cells to polypeptide growth factors in vitro, *J. Periodontol*., Vol.63,pp.(515–525).

Mayr-Wohlfart, J., Waltenberger, H., Hausser, S., Kessler, K. P., Gunther, C., Dehio, Brenner R. E. (2002). Vascular endothelial growth factor stimulate chemotactic migration of primary human osteoblasts. *Bone,* Vol.30,pp.(472-477).

Meyers, V.E, Zayzafoon, M., Douglas, J.T., McDonald, J.M. (2005). RhoA and cytoskeletal disruption mediate reduced osteoblastogenesis and enhanced adipogenesis of human mesenchymal stem cells in modeled microgravity. *Journal of Bone and Mineral Research,* Vol. 20, pp.(1858-66).

Middleton, J.C., Tipton, A.J. (2000). Synthetic biodegradable polymers as orthopedic devices. *Biomaterials*, Vol 23, pp. (2335-2346).

Miura, M., Grothos, S., Zhao, M., Lu, B., Fisher, L.W., Robery, P.G., Shi, S. (2003). Stem cells from human exfoliated decidusus teeth. *Proceed nation acad Sci USA*, Vol. 100, pp.(5807-5812).

Morishita, T., Honoki, K., Ohgushi, H., Kotobuki, N., Matsushima, A., Takakura, Y. (2006). Tissue engineering approach to the treatment of bone tumors: three cases of cultured bone grafts derived from patients' mesenchymal stem cells.*Artifitial Organs,* Vol.30,pp.(115-8).

Nakagawa,N., Yasuda, H., Yano, K., Mochizuki, S., Kobayashi, N., Fujimoto,H., Shima, N., Morinaga, T., Chikazu, D., Kawaguchi, H., Higashio,K.(1999). Basic fibroblast growth factor induces osteoclast formation by reciprocally regulating the production of osteoclast differentiation factor and osteoclastogenesis inhibitory factor in mouse osteoblastic cells. *Biochemical and Biophysical Research Communications,* Vol.265, pp.(158–163).

Nakajima, T., Iizuka, H., Tsutsumi, S., Kayakabe. M., Takagishi, K. (2007). Evaluation of posterolateral spinal fusion using mesenchymal stem cells: differences with or without osteogenic differentiation. *Spine,* Vol.32, pp.(2432–2436).

Nasseri, B.A., Pomerantseva, I., Kaazempur-Mofrad, M.R.Sutherland, F.W., Perry, T., Ochoa, E. (2003). Dynamic rotational seeding and cell culture system for vascular tube formation. *Tissue Engineering,*Vol.9, pp.(291-299).

Nishida, J., Shimamura, T. (2008). Methods of reconstruction for bone defect after tumor excision: a review of alternatives. *Med Sci Monit,* Vol. 14,pp. (107–113).

Noth, U., Osyczka, A.M., Tuli, R., Hickok, N.J., Danielson, K.G., Tuan, R.S. (2002). Multilineage mesenchymal differentiation of human trabecular bone-derived cells. *Journal of Orthopedics Research,* Vol. 20, pp.(1060-1069).

Perka, C., Schultz, O., Spitzer, R.S., Lindenhayn, K., Burmester, G.R., Sittinger. M.(2000). Segmental bone repair by tissue-engineered periosteal cell transplants with bioresorbable fleece and fibrin scaffolds in rabbits. *Biomaterials,* Vol.21,pp.(1145-53).

Persidis, A.(1999). Tissue engineering. *Nat Biotechnol,* Vol. 17, pp.(508-510)

Petrakova, K.V., Tolmacheva, A.A., Fridenstein, A.J. (1963).Bone formation occuring in bone marrow transplantation in diffusion chambers, *Biull Eksp Biol Med,* Vol. 56, pp. (87-91).

Pittenger, M.F., Mackay, A.M., Beck, S.C., Jaiswal, R.K., Douglas, J.D.(1999). Multilineage potential of adult human mesenchymal stem cells. *Science,* Vol. 284, pp.(143–7).

Porter, J.R., Ruckh, T.T., & Popat, K.C. (2009). Bone Tissue Engineering: A Review in Bone Biomimetics and Drug Delivery Strategies. *American Institute of Chemical Engineers Biotechnol. Prog,* Vol. 25, pp.(1539–1560).

Potten, C.S., Loffler, M. (1990). Stem cells: attributes, cycles, spirals, pitfalls and uncertainties. Lessons for and from the crypt. *Development,* Vol. 110,pp. (1001-1020).

Prockop, D.J.(1997). Marrow stromal cells as stem cells for nonhematopoietic tissues. *Science.* Vol. 276, pp.(71-74).

Quarto, R., Mastrogiacomo, M., Cancedda, R., Kutepov, S.M., Mukhachev, V., Lavroukov, A., Kon, E., Marcacci, M. (2001). Repair of large bone defects with the use of autologous bone marrow stromal cells. *National English Journal of Medicine,*Vol.1, pp.(385-386).

Rickard, D.J., Kazhdan, I., Leboy, P.S. (1995). Importance of 1,25-dihydroxyvitamin D3 and the nonadherent cells of marrow for osteoblast differentiation from rat marrow stromal cells. *Bone,* Vol. 16, pp.(671-8).

Rasmusson, I., Le Blanc, K., Sundberg, B., Ringdén, O. (2007). Mesenchymalstem cells stimulate antibody secretion in human B cells. *Scand J Immunol,* Vol. 65, pp.(336-343).

Ripamonti, U., Ma, S.S., van den Heever, B., Reddi, A.H. (1992). Osteogenin, a bone morphogenetic protein, adsorbed on porous hydroxyapatite substrata, induces rapid bone differentiation in calvarial defects of adult primates. *Plastic Reconstruction Surgery,* Vol.90, pp.(382–393).

Roh, J.D., Nelson, G.N., Udelsman, B.V., Brennan, M.P., Lockhart, B., Fong, P.M., Lopez-Soler, R.I., Saltzman, W.M.,and Breuer, C.K. (2007).Centrifugal seeding increases seeding efficiency and cellular distribution of bone marrow stromal cells in porous biodegradable scaffolds. *Tissue Engineering,* Vol.13,pp.(2743-2749).

Ross, R., Everett, N.B., Tyler, R. (1970). Wound healing and collagen formation. The origin of the wound fibroblast studied in parabiosis. *Journal of Cell Biology,* Vol. 44, pp. (645-654).

Saadeh, P.B., Mehrara, B.J., Steinbrech, D.S., Spector, J.A., Greenwald, J.A., Chin, G.S., Ueno, H., Gittes, G.K., Longaker, M.T. (2000). Mechanisms of fibroblast growth factor-2 modulation of vascular endothelial growth factor expression by osteoblastic cells. *Endocrinology,* Vol.141, pp. (2075-2083).

Scaglione, S.,Wendt, D., Miggino, S., Papadimitropoulos, A., Fato, M., Quarto, R. (2008). Effects of fluid flow and calcium phosphate coating on human bone marrow stromal cells cultured in a defined 2D model system. *J Biomed Mater Res A,*Vol.86, pp.(411–9).

Schmidmaier, G., Lucke, M., Schwabe, P., Raschke, M., Haas, N.P., Wildemann, B. (2006). Collective review: bioactive implants coated with poly(D,Llactide) and growth factors IGF-I, TGF-beta1, or BMP-2 for stimulation of fracture healing, *Journal Long-Term Effects in Medical Implants,* Vol.16, pp.(61–69).

Scutt, A., Bertram, P. (1995). Bone marrow cells are targets for the anabolic actions of prostaglandin E2 on bone: induction of a transition from nonadherent to adherent osteoblast precursors. *Journal of Bone and Mineral Research,* Vol. 10, pp.(474-87).

Seong, J.M., Kim, B.C., Park, J.H., Kwon, I.K., Mantalaris, A., & Hwang Y.S. (2010). Stem cells in bone tissue engineering. *Biomedical Materials,*Vol. 5,pp. (1-15)

Sherley, J.L., Stadler, P.B., Johnson, D.R. (1995). Expnession of the wild-type P53 antioncogene induces guanine nucleotide-dependent stem cell division kinetics. *Proc.Natl.Acad.sci.USA,* Vol. 2, pp.(136-140).

Sikavitsas, V.I., Bancroft, G.N., Mikos, A.G. (2002). Formation of three-dimensional cell/ polymer constructs for bone tissue engineering in a spinner flask and a rotatingwall vessel bioreactor. *Journal of Biomedical Material Research,* Vol.62,pp.(136–48).

Sotiropoulou, P.A., Perez, S.A., Gritzapis, A.D., Baxevanis, C.N., Papamichail, M. (2006). Interactions between human mesenchymal stem cells and natural killer cells. *Stem Cells,* Vol. 24, pp.(74-85).

Spinella-Jaegle, S., Rawadi, G., Kawai, S., Gallea, S., Faucheu, C., Mollat, P., Courtois, B., Bergaud, B., Ramez, V., Blanchet, A.M., Adelmant, G., Baron, R., Roman-Roman, S. (2001).Sonic hedgehog increases the commitment of pluripotent mesenchymal cells into the osteoblastic lineage and abolishes adipocytic differentiation. *Journal of Cell Science,* Vol. 114, pp. (2085-94).

Swetha, M., Sahithi, K., Moorthi, A., Srinivasan, N., Ramasamy, K., Selvamurgan, N. (2010)Biocomposites containing natural polymers and hydroxyapatite for bone tissue engineering. *International journal of biological macromolecules,* Vol. 47,pp.(1-4).

Thomson, J.A., Itskovitz-Eldor, J., Shapiro, S.S., Waknitz, M.A., Swiergiel, J.J., Marshall, V.S., Jones, J.M. (1998). Embryonic stem cell lines derived from human blastocysts. *Sciences.* Vol. 282, pp.(1145-1147).

Tsuda, H., Wada, T., Yamashita, T., Hamada, H.(2005). Enhanced osteoinduction by mesenchymal stem cells transfected with a fiber-mutant adenoviral BMP2 gene. *J Gene Med,* Vol. 7,pp.(1322–1334).

Tu, Q., Valverde, P., Chen, J. (2006). Osterix enhances proliferation and osteogenic potential of bone marrow stromal cells. *Biochemical and Biophysic Research Communication.* Vo. 341, pp. (1257-1265)

Undale, A.H., Westendorf, J.J., Yaszemski, M.J., Khosla, S. (2009). Mesenchymal Stem Cells for Bone Repair and Metabolic Bone Diseases. *Mayo Clinical Proceedings,* Vol. 84, No. 10, pp. (893-902).

Urist,M.R.(1965). Bone: formation by autoinduction, *Science,* Vol. 150, pp.(893–899).

van Wachem, P.B., Stronck, J.W., Koers-Zuideveld, R., Dijk, F., and Wildevuur, C.R. (1990).Vacuum cell seeding: a new method for the fast application of an evenly distributed cell layer on porous vascular grafts. *Biomaterials,* Vol.11,pp.(602-606).

Veleva, A. N., Heath, D.E., Cooper, S.L., Patterson, C. (2008). Selective endothelial cell attachment to peptide-modified terpolymers. *Biomaterials,* Vol.29,pp.(3656- 3661).

von Schroeder, H. P., Veillette, C. J., Payandeh, J., Qureshi, A., Heersche, J. N. (2003). Endothelin-1 promotes osteoprogenitor proliferation and differentiation in fetal rat calvarial cell cultures. *Bone,*Vol. 33,pp.(673-684).

Wang, M. (2006). Composite scaffolds for bone tissue engineering. *American journal of Biochemistry and Biothecnology,* Vol. 2, pp. (80-84)

Wendt, D., Marsano, A., Jakob, M., Heberer, M., Martin, I.(2003). Oscillating perfusion of cell suspensions through three-dimensional scaffolds enhances cell seeding efficiency and uniformity. *Biotechnol Bioeng, Vol.*84,pp.(205–14).

Wernig, M., Meissner, A., Foreman, R., Brambrink, T., Ku, M., Hochedlinger, K., Bernstein, B.E., Jaenisch, R.(2007). In vitro reprogramming of fibroblasts into a pluripotent ES-cell-like state. *Nature,* Vol. 448. Pp. (318–324).

Wickham, M.Q., Erickson, G.R., Gimble, J.M., Vail, T.P., Guilak, F. (2003). Multipotent stromal cells derived from infrapatellar fat pad of knee. *Clinical Orthopedics,* pp. (196-212).

Williams, C., and Wick, T.M. (2004). Perfusion bioreactor for small diameter tissue-engineered arteries. *Tissue Engineering,* Vol.10,pp.(930-941).

Woll, N.L, Heaney, J.D., & Bronson, S.K.(2006). Osteogenic nodule formation from single embryonic stem cell-derived progenitors. *Stem Cell Development,* Vol. 15, pp.(865–79).

Yamaguchi, A., Komori, T., Suda, T. (2000). Regulation of osteoblast differentiation mediated by bone morphogenetic proteins, hedgehogs, and Cbfa1. *Endocrine Reviews,* Vol. 21, pp. (393-411).

Yang, Y.G., Sykes, M. (2007). Xenotransplantation: current status and a perspective on the future. *Nature Review Immunology,* Vol. 7, pp.(519–531).

Yoon, B.S., Lyons, K.M. (2004). Multiple functions of BMPs in chondrogenesis. *Journal of Cellular Biochemistry*, Vol.93,pp.(93–103).

Young, R.G., Butler, D.L., Weber, W., Caplan, A.I., Gordon, S.L., Fink, D.J. (1998). Use of mesenchymal stem cells in a collagen matrix for Achilles tendon repair. *Journal of Orthopedics Research,* Vol. 16, pp. (406-413).

Zandi, M., Mirzadeh, H., Urch, H., Eslaminejad, M.B., Bagheri, F., Mivechi, H. (2010). Biocompatibility evaluation of nano- rod hydroxyapatite/ gelatin coated with nano- Hap as a novel scaffold using mesenchymal stem cell. *Journal of Biomedical Materials Research* A, Vol.92, pp.(1244- 55).

Zaragosi, L.E., Ailhaud, G., Dani, C. (2006). Autocrine FGF2 signaling is critical for self-renewal of human multipotent adipose-derived stem cells. *Stem Cells,* Vol.24, pp.(2412-2419).

4

From Multipotent Cells to Fully Differentiated Connective Tissue Cells for Regenerative Medicine: Emerging Applications of Mesenchymal Stem Cells

Ali Mobasheri[1], Stephen M. Richardson[2], Judith A. Hoyland[2], Constanze Buhrmann[3] and Mehdi Shakibaei[3]

[1]Musculoskeletal Research Group, Division of Veterinary Medicine, School of Veterinary Medicine and Science, Faculty of Medicine and Health Science, The University of Nottingham, Sutton Bonington Campus, Leicestershire,
[2]Developmental Biomedicine Research Group, School of Biomedicine, Faculty of Medical and Human Sciences, The University of Manchester, Stopford Building, Oxford Road, Manchester,
[3]Musculoskeletal Research Group, Institute of Anatomy, Ludwig-Maximilian-University Munich, Munich
[1,2]United Kingdom
[3]Germany

1. Introduction

Mesenchymal stem cells (MSCs) are a heterogeneous subset of stromal cells that can be isolated from many adult tissues (Uccelli et al., 2008). Adult MSCs can be isolated from bone marrow, marrow aspirates, skeletal muscle, adipose tissue (Guilak et al., 2010), synovium and many other connective tissues (Barry, 2003). Due to their culture-dish adherence, they can be expanded in culture while maintaining their multipotency (Caplan, 2007). Their multipotency is an important property that allows them to differentiate into cells of the mesodermal lineage, giving rise to a range of specialized connective tissue cells including adipocytes, osteoblasts, chondrocytes and tenocytes as well as cells of other embryonic lineages (Figure 1) (Uccelli et al., 2008). MSCs have been isolated from humans and a variety of animal species including rodents, rabbits, dogs and horses. MSCs show considerable promise for use in repairing and rebuilding damaged or diseased mesenchymal tissues (Caplan, 2007). After *in vivo* administration, MSCs can induce peripheral tolerance and migrate to injured tissues where they have the capacity to exert immunosuppressive properties (Uccelli et al., 2007) and inhibit the release of pro-inflammatory cytokines and promote the survival of existing cells and the repair of damaged tissue (Uccelli et al., 2008). They are being clinically explored as a new therapeutic for treating a variety of immune-mediated diseases (Parekkadan and Milwid, 2010). Thus, they have potential applications in tissue engineering and regenerative medicine and may represent an attractive option for bone, cartilage, tendon and ligament regeneration.

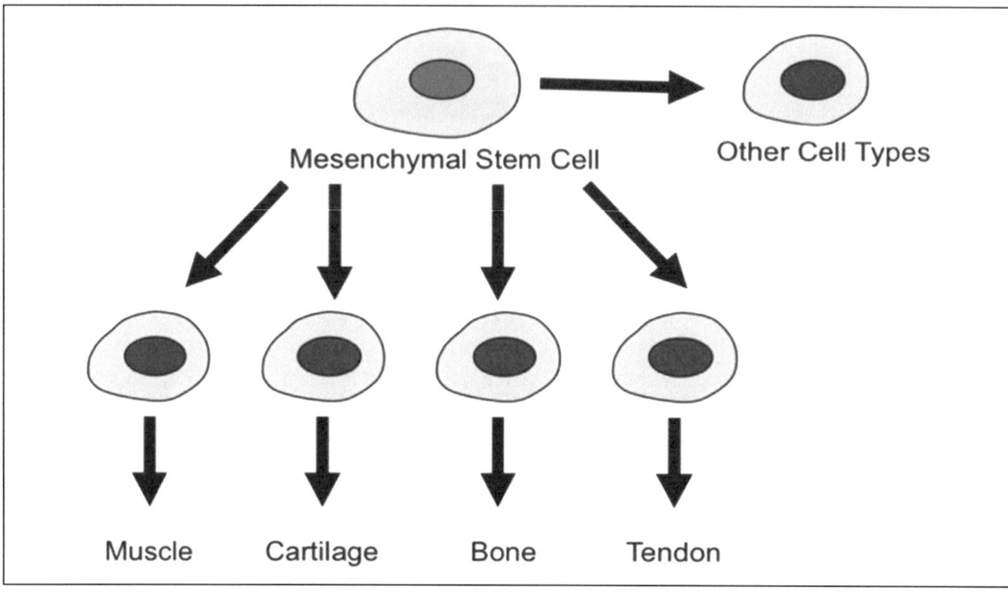

Fig. 1. Multipotency of mesenchymal stem cells (MSCs). MSCs can give rise to myocytes, chondrocytes, osteoblasts, tenocytes and a variety of other connective tissue cells

This chapter focuses on the potential of MSCs as an alternative to cells derived from patient tissues in autologous transplantation and tissue engineering. The prospects of using MSCs in regenerative medicine strategies are discussed and the advantages and disadvantages of these cells in articular cartilage tissue engineering are summarized. There is currently little consensus on how MSCs are best isolated, identified and characterized. This is partly due to the paucity of standardized specific cell surface markers. Although MSCs cells have been isolated and expanded in culture, their use for therapeutic strategies requires protocols and technologies that have not yet been clinically trialled and perfected (Caplan, 2005). Also, there are no established guidelines from governmental and intergovernmental agencies for their use in clinical applications. Even if these guidelines existed, more research is needed to help understand their basic biology since the therapeutic effects afforded by MSC transplantation are likely to be short-lived and related to dynamic, paracrine interactions between MSCs and host cells (Parekkadan and Milwid, 2010). MSCs possess a fibroblastic morphology but the published literature suggests that there is no well-defined phenotype for these cells. More work is needed to characterise the phenotype of these cells. This chapter will also focus on the conceptual and practical difficulties associated with differentiating and preconditioning MSCs for subsequent survival in a physiologically harsh extracellular matrix, an environment that will be highly hypoxic, acidic, and nutrient deprived. Interestingly, published data suggests that MSCs better differentiate into chondrocytes in low oxygen environments.

2. Increasing life expectancy and the growing burden of musculoskeletal and joint diseases

For the past 160 years human life expectancy has consistently increased by a quarter of a year every year (Oeppen and Vaupel, 2002). It is predicted that life expectancy will continue to increase by 2.5 years each decade, meaning that the western world's average life expectancy should reach and exceed 100 within the next 50 years (Oeppen and Vaupel, 2002). The increase in life expectancy has been mainly due to the significant advances in medicine and healthcare. However, the increased longevity of humans has greatly expanded the elderly population in the western world and has resulted in the increased prevalence of a range of arthritic, rheumatic and musculoskeletal disorders, which are placing an ever-greater socioeconomic burden on health systems around the world as the population ages. According to the World Health Organization (WHO)[1], orthopaedic, rheumatic and musculoskeletal conditions comprise over 150 diseases and syndromes, which are usually progressive and associated with pain and disability. They can broadly be categorized as joint diseases, physical disability, spinal disorders, and conditions resulting from trauma. These conditions are leading causes of morbidity, giving rise to enormous healthcare expenditures and loss of productivity. Knowledge of the key determinants of disability in musculoskeletal conditions is critical for reducing their burden on the world's growing population (Oeppen and Vaupel, 2002; Weigl et al., 2008).

The United Nations, the WHO and 37 other countries have proclaimed the years 2000-2010 as the Bone and Joint Decade[2] (McGowan, 2003; Woolf and Pfleger, 2003). This global initiative is intended to improve the lives of people with musculoskeletal disorders, such as arthritis, and to advance understanding and treatment of musculoskeletal disorders through prevention, education and research. The 10-year global initiative launched by the UN urges governments around the world to start taking action to draw attention to the growing pervasiveness and impact of musculoskeletal diseases and to reduce the social and financial burdens to society. Support for this global initiative will raise awareness of musculoskeletal health, stimulate research and improve people's quality of life.

Musculoskeletal diseases are one of the major causes of disability around the world and have been a significant reason for the development of the Bone and Joint Decade (Brooks, 2002; McGowan, 2003; Woolf and Pfleger, 2003). Rheumatoid arthritis (RA), osteoarthritis (OA), gout and back pain are important causes of disability-adjusted-life years in both the developed and developing world (Brooks, 2006). The Arthritis Foundation[3] in the United States plays a key role in co-ordinating efforts during the Bone and Joint Decade as a supporter. Its aims are to:

- Raise awareness and educate the world on the increasing societal impact of musculoskeletal injuries and disorders
- Empower patients to participate in decisions about their care and treatment
- Increase global funding for prevention activities and treatment research
- Continually seek and promote cost-effective prevention and treatment of musculoskeletal injuries and disorders

The major consequence of all forms of arthritis is joint dysfunction, disability, chronic pain, and significant morbidity. Aside from analgesics, there are currently no effective pharmacotherapies capable of restoring the structure and function of damaged synovial tissues in any form of arthritis. Consequently, there is an acute need for developing new tissue engineering and regenerative medicine strategies for arthritis.

---

[1] http://www.who.int/en/

[2] http://www.arthritis.org/bone-joint-decade.php

[3] http://www.arthritis.org/

3. The inflammatory component of musculoskeletal and joint diseases

Many rheumatic diseases and arthritic conditions are essentially inflammatory disorders. The term "arthritis" characterizes a group of conditions involving damage to synovial joints. Arthritis literally means an inflammation (*itis*) of the joints (*arthr*). It involves pain, redness, heat, swelling and other harmful effects of inflammation within the joint. The most common form OA (also known as osteoarthrosis or degenerative joint disease) can result from trauma to the joint, infection of the joint, or simply as a consequence of ageing. Other forms of arthritis include psoriatic arthritis, and autoimmune diseases in which the body's immune system attacks itself such as RA. Inflammation plays slightly different roles in OA and RA. Inflammation is primarily the result of OA but in RA it is the root cause. The major consequence of arthritis is pain and disability. Pain is a constant and daily feature in well-established forms of the disease. Arthritis pain occurs due to inflammation that occurs around the joint, damage to the joint from disease, daily wear and tear of joint, muscular strains caused by movement against stiff, painful joints and fatigue. Disability in patients with arthritis is a consequence of degeneration in the joint and surrounding tissues and is further enhanced by inflammation-induced pain.

4. Rheumatoid Arthritis (RA), Osteoarthritis (OA) and Lower Back Pain (LBP)

The number of rheumatoid arthritis (RA) and osteoarthritis (OA) patients steadily rises as the elderly population grows in Western Europe, North America, and the rest of the developing world. RA, OA, and lower back pain (LBP) are important causes of disability adjusted-life years in both the developed and developing world (Brooks, 2006). Back and knee pain are common in the community and are likely to increase with the aging population (Brooks, 2006). Until recently OA was viewed as a "degenerative" or "wear-and tear" disease and held little interest for most clinicians. It is now accepted that the age related degeneration of articular cartilage as part of the clinical syndrome of OA is one of the most common causes of pain and disability in middle-aged and older people (Brooks, 2006). RA, OA and LBP pain are among the major causes of disability and morbidity within the adult population. OA is the most common form of joint disease, with the majority of the population over 65 years of age demonstrating radiographic evidence of OA in at least one joint. Likewise, around two- thirds of the adult population suffer from LBP at some point in their lifetime. LBP is also a major cause of disability and suffering globally (Grabias and Mankin, 1979; Jackson, 2004). LBP OA and RA are important causes of disability and are increasingly among the ageing population. LBP is also common among younger athletes and those older individuals who enjoy sports such as golf (Baker and Patel, 2005; McHardy and Pollard, 2005; Reed and Wadsworth, 2010).

The current inadequacy of treatments for these conditions, combined with their increasing prevalence exacerbates the burden on healthcare systems. Therefore, researchers and clinicians are striving for novel, innovative treatment options and the emergence of the fields of tissue engineering and regenerative medicine offer hope that long-term tissue repair may be possible. However, one of the current limiting factors for treatment of joint diseases is a source of cells. While a range of cell sources have been proposed, adult mesenchymal stem cells (MSCs) offer the greatest potential for clinical application.

Many studies have explored the prevalence of arthritis (OA and RA), related musculoskeletal conditions, LBP and intervertebral disc degeneration among elderly population. Chronic LBP is a major complaint of elderly patients and restricts most of their activities (with a higher prevalence in women). Interestingly LBP increases with age in men to reach the same frequency as women in those aged 90 years or older. Many studies have shown that a major cause of LBP is intervertebral disc degeneration, which is, primarily, an ageing related phenomenon. Degeneration of the intervertebral disc impacts not only on the disc, but also on the surrounding tissues such as the muscles and ligaments and affects the spine's ability to cope with the physiologically normal loads it experiences during a daily routine. As well as accelerating degeneration these changes also cause pain and reduce the mobility of the spine. The drug treatments currently available for LBP are inadequate and restricted to symptomatic pain control in mild cases and complicated and invasive surgical intervention (i.e. discectomy or spinal fusion) in severe cases. The increasing prevalence of LBP among the ageing population will become a major socioeconomic issue as it exacerbates the burden on healthcare systems throughout the developed world. Therefore new treatments are needed which restore full disc function and normalise disc cell biology.

5. Regenerative medicine for RA and OA

Bone and cartilage defects are common features of joint diseases, such as RA and OA (Noel et al., 2002). They have significant social and economic impact on the aging population. Despite progress in orthopaedic surgery, bone and cartilage repair is a major challenge as large defects will not spontaneously heal (Noel et al., 2002). Regenerative medicine is an emerging field that seeks to repair or replace injured tissues and organs through natural or bioengineered means. Recent research on stromal MSCs has provided a new and exciting opportunity for bone and cartilage tissue engineering. We have learned a great deal about the isolation, cultivation, and characterization of MSCs in recent years. A huge amount of research effort is now focused on their differentiation and models that exploit their regenerative potential. MSCs have generated a great deal of public, scientific, and media interest because of their potential use in regenerative medicine and tissue engineering (Tolar et al., 2010).

6. Structure and function of articular cartilage

Cartilage is a flexible and mechanically compliant connective tissue found at the end of long bones in articulating joints and in the intervertebral disc. It is sub-classified into three different types: elastic cartilage, hyaline cartilage and fibrocartilage, which differ in the relative amounts of its three principal components, namely collagen fibres, ground substance (proteoglycans) and elastic fibres. Articular or hyaline cartilage is a load-bearing tissue with unique biological characteristics (Figure 2). The biochemical properties of cartilage depend on the structural design of the tissue, the molecular composition of the extracellular matrix (ECM) that makes up the bulk of the tissue volume and the interactions between its resident cells and the ECM (Buckwalter and Mankin, 1998). Chondrocytes are the only cells found within the cartilage ECM. They are architects of cartilage (Muir, 1995), building the macromolecular framework of its ECM from three distinct classes of macromolecules: collagens (predominantly type II collagens), proteoglycans (mainly aggrecan), and a variety of non-collagenous proteins. Collagens type II, IX, and XI form a fibrillar meshwork that gives cartilage tensile stiffness and strength (Buckwalter and Mankin, 1998; Eyre, 2004; Kuettner et al., 1991), whereas collagen type VI forms part of the matrix immediately surrounding the chondrocytes, enabling them to attach to the macromolecular framework of the ECM and act as a transducer of biomechanical and biochemical signals in the articular cartilage (Guilak et al., 2006; Roughley and Lee, 1994). Large aggregating proteoglycans (aggrecan) are embedded in the ECM and give cartilage its stiffness to compression, its resilience and contribute to its long-term durability (Dudhia, 2005; Kiani et al., 2002; Luo et al., 2000; Roughley and Lee, 1994). Figure 3 illustrates the major molecular components of the cartilage ECM.

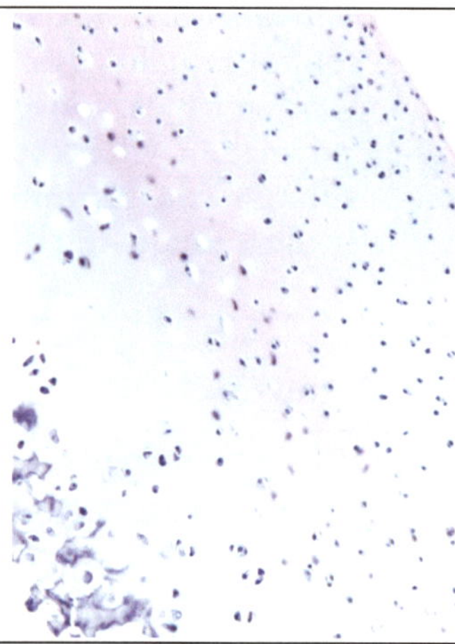

Fig. 2. A section of porcine articular cartilage showing the superficial, middle, deep and calcified zones

ECM proteins in cartilage are of great significance for the regulation of the cell behaviour, proliferation, differentiation and morphogenesis (Enomoto-Iwamoto et al., 1997; Gonzalez et al., 1993; Hewitt et al., 1982; Kosher and Church, 1975; Kosher et al., 1973; Ramachandrula et al., 1992; Ruoslahti and Reed, 1994; Sommarin et al., 1989; von der Mark et al., 1977). Small proteoglycans, including decorin, biglycan, and fibromodulin are further embedded in the ECM. Decorin and fibromodulin both interact with the type II collagen fibrils in the matrix and play a role in fibrillogenesis and interfibril interactions. Biglycan is mainly found in the immediate surrounding of the chondrocytes, where it may interact with collagen type VI (Buckwalter and Mankin, 1998; Roughley and Lee, 1994). Modulation of the ECM proteins is regulated by an interaction of a diversity of growth factors with chondrocytes (Hunziker et al., 1994; Isgaard, 1992; Jenniskens et al., 2006; Sah et al., 1994; Trippel et al., 1989). In fact, it has been reported recently, that IGF-I and TGF-β stimulate the chondrocyte surface expression of integrins, and that this event is accompanied by increasing adhesion of chondrocytes to matrix proteins (Loeser, 1997). Other non-collagenous proteins in articular cartilage such as cartilage oligomeric matrix protein (COMP) are less well studied and may have a value as a biomarker of turnover and degeneration of cartilage (Di Cesare et al., 1996), while tenascin and fibronectin influence interactions between the chondrocytes and the ECM (Buckwalter and Mankin, 1998; Burton-Wurster et al., 1997). The ECM surrounds chondrocytes and protects them from biomechanical stress arousing during normal joint motion, determines the types and concentrations of molecules that reach the cells and helps to maintain the chondrocyte phenotype.

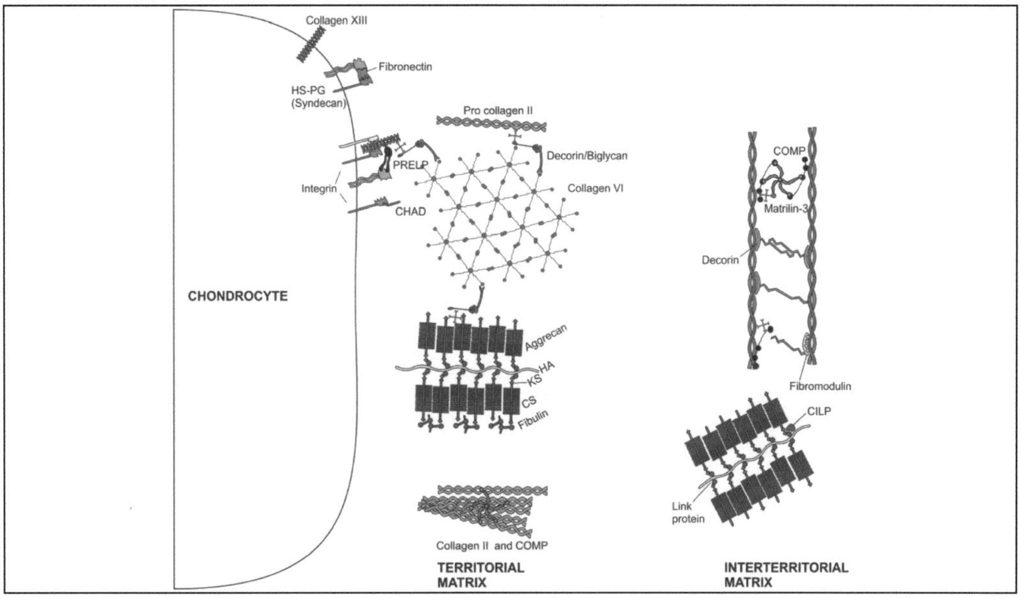

Fig. 3. Molecular composition of the extracellular matrix (ECM) of articular cartilage

Throughout life, cartilage undergoes continuous internal remodelling as chondrocytes replace matrix macromolecules lost through degradation. Evidence indicates that ECM turnover depends on the ability of chondrocytes to detect alterations in the macromolecular composition and organization of the matrix, including the presence of degraded macromolecules, and to respond by synthesizing appropriate types and amounts of new ECM components. It is known that mechanical loading of cartilage creates mechanical, electrical, and physicochemical signals that help to direct the synthesizing and degrading activity of chondrocytes (Mobasheri et al., 2002). In addition, the ECM acts as a signal transducer for chondrocytes (Millward-Sadler and Salter, 2004). A prolonged and severe decrease in the use of the joint leads to alterations in the composition of the ECM and eventually to a loss of tissue structure and its specific biomechanical properties, whereas normal physical strain stimulates the synthesizing activity of chondrocytes and possibly the internal tissue remodelling (Buckwalter and Lane, 1997; Maffulli and King, 1992).

Although articular cartilage can tolerate a tremendous amount of intensive and repetitive physical stress, it manifests a striking inability to heal even the most minor injury (Buckwalter, 2003; Buckwalter and Lane, 1997; Martin et al., 2004; Newman, 1998). This makes joints particularly sensitive to degenerative processes (Solursh, 1991). Furthermore, aging leads to alterations in the ECM composition and alters the activity of the chondrocytes, including their ability to respond to a variety of stimuli such as growth factors (Eckstein et al., 2001; Hudelmaier et al., 2001; Ralphs and Benjamin, 1994). All these alterations increase the probability of cartilage degeneration (Buckwalter, 2003; Poole, 1999; Sarzi-Puttini et al., 2005; Setton et al., 1999) and emphasize the importance of interaction of chondrocytes with their surrounding ECM since this interaction regulates their growth, differentiation, and survival in normal and pathophysiological conditions such as OA (Shakibaei et al., 1999).

7. Identification and application of Adult Stem Cells

There are two types of stem cells: embryonic stem cells and non-embryonic "somatic" or "adult" stem cells. Embryonic stem cells are found in the blastocyst whereas adult stem cells are found in adult tissues. Adult stem cells maintain the normal turnover of organs with a high intrinsic regenerative capacity. These include blood, skin and intestinal epithelium. Adult stem cells can be found in children, adolescents as well as adults and are generally unipotent or multipotent. Pluripotent adult stem cells are very rare and are generally found in small numbers. However, they can be found in a number of tissues including umbilical cord blood. The best studied adult stem cells are multipotent and are generally referred to by their tissue origin (i.e. haematopoietic stem cells that differentiate into erythrocytes, white blood cells, platelets, etc. and bone marrow stromal cells also known as MSCs which have the capacity to differentiate into connective tissue cells (Pittenger et al., 1999) (see Figure 4).

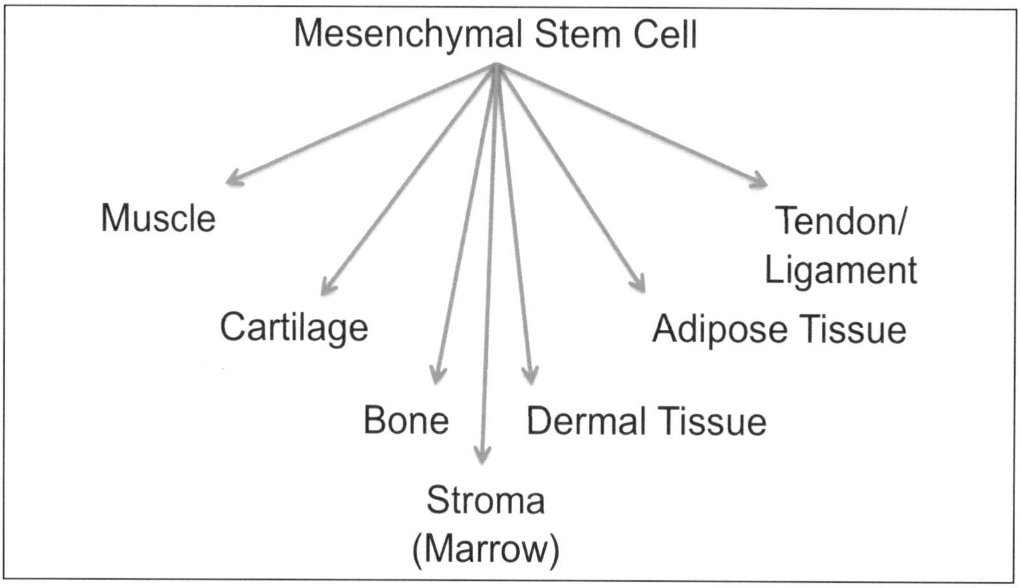

Fig. 4. Differentiation potential of human mesenchymal stem cells

These connective tissues include bone (Arinzeh, 2005; Hong et al., 2006; Noel et al., 2002), fat (Barry and Murphy, 2004; Helder et al., 2007), cartilage (Barry and Murphy, 2004; Caplan, 2007; Noel et al., 2002), intervertebral disc (Richardson et al., 2007; Trubiani et al., 2005; Trubiani et al., 2006), ligament (Sonoyama et al., 2006; Trubiani et al., 2005; Trubiani et al., 2006) and muscle (Barry and Murphy, 2004). Evidence suggests that these cells are also capable of differentiation along myogenic and neurogenic lineages, although these pathways are not normally utilised to demonstrate multipotentiality of isolated MSCs. Adult mesenchymal stem cells were originally isolated from bone marrow in 1999 by Pittinger and co-workers, who demonstrated their multilineage differentiation potential (Pittenger et al., 1999). Subsequent studies have identified the presence of stem cells in a number of adult tissues, including adipose, muscle, dermis, periosteum, synovial membrane, synovial fluid and articular cartilage. Thus far MSC-like progenitor cells have been isolated from bone marrow (Grigoriadis et al., 1988), periosteum, trabecular bone, adipose tissue, synovium, skeletal muscle and deciduous teeth (Barry and Murphy, 2004; Sonoyama et al., 2006). In fact they were known as marrow stromal cells long before they were known as MSCs (Grigoriadis et al., 1988). The International Society for Cellular Therapies has recently proposed a definiton for MSCs (Dominici et al., 2006). While there are no definitive markers of MSCs a range of cell surface markers are routinely used as markers of MSCs. These include immunopositivity for STRO-1, CD73, CD105, CD106 CD145 and CD166, combined with negative immunoreactivity for CD11b, CD31, CD34, CD45 and CD117. These markers can also be used to identify a more homogeneous population of cells than previous methods utilising either density-gradient centrifugation, or even simple plastic adherence. The general heterogeneity of bone marrow cell populations can lead to variable results; however MSCs are generally regarded to be capable of differentiation along the chondrogenic, osteogenic and adipogenic pathways. Work from our laboratories suggests that MSCs are capable of differentiation to nucleus pulposus (NP) cells of the IVD, chondrocytes and osteoblasts (Csaki et al., 2009; Csaki et al., 2007; Mobasheri et al., 2009; Richardson et al., 2007). However, since definite markers of NP cells have only recently been described (Minogue et al., 2010), a range of chondrocyte markers, with which they share a large phenotypic similarity, are routinely used (see Figure 5 and Table 1 for details of some of the currently investigated MSC markers).

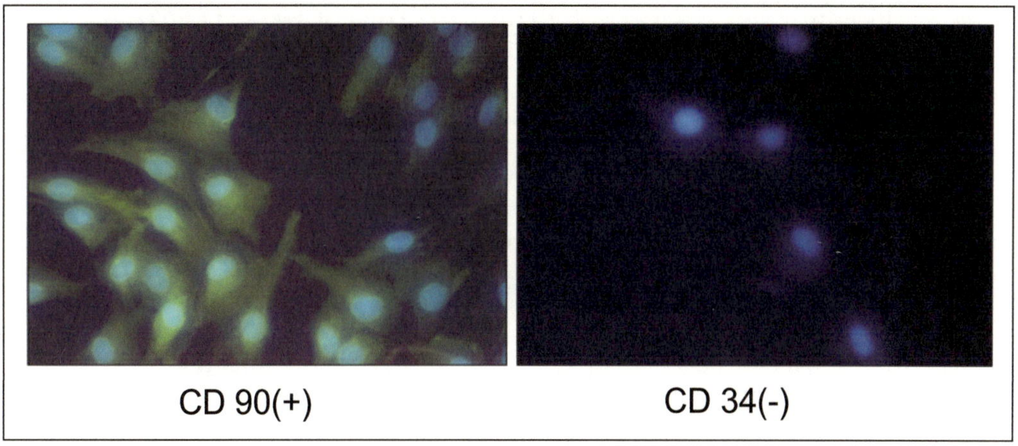

Fig. 5. Positive expression of CD 90 and lack of expression of CD 34 in canine mesenchymal stem cells

8. Proliferation of research on Adult Stem Cells

Since they were first identified by Pittinger and co-workers (Pittenger et al., 1999) research on adult stem cells has proliferated at a staggering pace. Work using adult stem cells does not involve many of the ethical challenges associated with using embryonic stem cells. Another reason for the rapid expansion in this area is the availability of a range of adult tissues from humans and animals. MSCs also posses neuroprotective (Uccelli et al., 2011) and cardioregenerative (Flynn and O'Brien, 2011) properties as well as the potential for musculoskeletal regeneration (O'Sullivan et al., 2011).

Species	Source	Cell surface markers	
		Positive	**Negative**
Human	Adipose tissue	CD13, CD29, CD44, CD73 and CD90	
Canine	Bone marrow	CD90 and MHC-I	CD34, CD45 and MHC-II
Human	Bone marrow Umbilical cord blood Adipose tissue	CD29, CD44, CD73, CD90, CD105, CD106 and HLA-I	CD14, CD34, CD45, CD133, CD144 and HLA-II
Equine	Bone marrow	CD90, fibronectin, perlecan and collagen type IV	---
Equine	Umbilical cord matrix	CD54, CD90, CD105, CD146, Oct4, SSEA-4, c-Kit, SSEA-3 and TRA-1-60	CD34, CD45 and CD133
Canine	Bone marrow	CD90 and CD105	CD34 and CD45
Equine	Umbilical cord blood	Oct4, TRA-1-60, TRA-1-81 and SSEA-1	___
Equine	Adipose tissue	Oct4, TRA-1-60 and TRA-1-81	SSEA-1 and SSEA-4
Ovine	Bone marrow	CD44, vimentin and CD105	CD34 and CD45
Human	Adipose tissue	CD44, CD90, CD105, CD146, CD166 and STRO-1	CD14, CD31 and CD45
Human	Bone marrow	CD44, CD90, CD105 and CD166	CD45 and CD117
Human	Bone marrow	CD29, CD44, CD73, CD90, CD105 and CD146	CD14, CD34 and CD45
Human	Infrapatellar fat pad	CD13, CD29, CD44, CD90, CD105 and 3G5	CD34 CD56, LNGFR and STRO-1
Human	Infrapatellar fat pad	CD13, CD29, CD44, CD90 and CD105	CD35, CD56, LNGFR and STRO-1
Human	Infrapatellar fat pad	CD13, CD29, CD44, CD90 and CD105	CD34, CD56, LNGFR and STRO-1
Ovine	Bone marrow	CD29, CD44 and CD166	CD31 and CD45
Human	Bone marrow	CD44^c, CD54, CD90 and STRO-1	CD14 and CD45
Human	Placental chorionic villi	CD49a/VLA-1, CD146, CD106, 3G5, α-SMA and STRO-1	X63 and vWF
Canine	Adipose tissue	CD29, CD44 and CD90	CD13, CD31, CD34, CD45, CD73, CD105 and CD117
Canine	Umbilical cord blood	CD29, CD33, CD44, CD105, CD184 and Oct4	CD4, CD8a, CD10, CD14, CD20, CD24, CD31, CD34, CD38, CD41a, CD41/61, CD62p, CD73, CD90, CD133 and HLA-DR
Equine	Bone marrow	Sox-2, Oct4 and Nanog	CD34
Equine	Adipose tissue	CD90 and CD44	CD13
Canine	Adipose tissue	CD44, CD90, CD140a and CD117	CD34 and CD45
Canine	Adipose tissue	CD90 and CD105	CD34 and CD45

Table 1. Cell surface markers of MSCs from various species and tissue sources.
Data from Penny et al., (in press)

9. Musculoskeletal tissue regeneration

Connective tissues, such as bone, cartilage, tendon, ligament and the IVD all suffer from both traumatic and age-related degenerative injuries. While bone repairs relatively well, the avascular and hypocellular nature of cartilage and in particular the IVD means these tissues suffer from a very limited self-repair capacity. During OA and IVD degeneration (one of the main causes of LBP) there is an imbalance between synthesis and degradation of the extra-cellular matrix which leads to an overall loss of tissue. This tissue destruction over time leads to pain and reduced mobility. While the increased repair capacity of bone prevents this, a number of metabolic and degenerative disorders, such as osteoporosis, osteonecrosis and Paget's disease, can cause bone loss or abnormal bone turnover. Traumatic injury and loss of bone following surgery for other diseases such as cancer can also lead to bone lesions which are too large for the body's self-regeneration capacity to cope with. Current clinical interventions for large bone defects, such as bone grafting, suffer from problems with donor site morbidity and limited material for large scale autologous grafts and infection or immune rejection with allogeneic grafts. The main histological, ECM and phenotypic markers of osteoblasts, chondrocytes and adipocytes differentiated from mesenchymal stem cells are shown in Table 2.

	Osteoblasts	Chondrocytes	Adipocytes
Histological Stains	Alizarin Red	Toluidine Blue	Oil Red O
	von Kossa	Alcian Blue	
		Safranin O	
		Masson trichrome stain C	
ECM Markers	Type I collagen	Type II collagen	Type I collagen
	Beta 1 integrin	Aggrecan	Beta 1 integrin
Other Markers	Alkaline phosphatase (ALP)	Sox9	Adiponectin
	Osteonectin (OSTN)		Peroxisome proliferator-activated receptor-gamma (PPAR-gamma)
	Osteocalcin (OCN)		
	Runt-related transcription factor-2 (Runx2) / Core binding factor alpha 1		Fatty Acid Binding Protein 4 (FABP4)
	Osteopontin (OPN)		Lipoprotein Lipase (LPL)
	Bone Sialoprotein (BSP)		Leptin
	Vitamin D receptor (VDR) (14)		
	Homeobox gene MSX2 (14)		

Table 2. The main histological, ECM and phenotypic markers of osteoblasts, chondrocytes and adipocytes differentiated from mesenchymal stem cells. Data from Penny et al., (in press)

These limitations in current clinical treatments for bone defects and disorders have led to the development of tissue engineering strategies using both synthetic and natural scaffold implants. Materials such as tricalcium phosphates (Guo et al., 2004a; Guo et al., 2004b; Jiang et al., 2007; Shao et al., 2006; Solchaga et al., 1999), calcium carbonates (Kreklau et al., 1999), hydroxyapatite (Chajra et al., 2008; Reddi, 2000; Yoshikawa and Myoui, 2005) and Bioglass (Helen and Gough, 2008; Wilda and Gough, 2006) have all been utilised alongside growth factors to promote new growth, or act as supports for implanted cells to regenerate new tissue. These materials all aim to mimic the specialised microenvironment or bone and cartilage and are growing in their clinical application.

Articular cartilage and the IVD share similarities in cellular phenotype and extra-cellular matrix composition. Consequently similar approaches have been adopted for tissue engineering cartilage and IVD (Kalson et al., 2008; Richardson et al., 2007). As with bone repair, there are major limitations to current clinical treatments for OA and IVD degeneration. Scaffold-free autologous chondrocyte implantation (ACI) is currently used clinically to treat small cartilage lesions, such as those caused by traumatic injury. For this treatment a biopsy is taken from a non-load bearing region of cartilage, the chondrocytes enzymatically extracted, expanded in monolayer culture, then reimplanted into the lesion and covered with a periosteal flap (Brittberg et al., 1994). The therapy has been used to treat over 12,000 patients worldwide and offers cartilage repair and reduced pain equivalent to existing surgical cartilage repair therapies (Peterson et al., 2002). However, while this treatment appears to work well for small cartilage lesions it has limited practical application for larger osteoarthritic lesions.

Cell implantation into degenerate IVDs has been demonstrated in animal models to both inhibit degeneration and regenerate tissue (Ganey et al., 2003). A small-scale study in human patients comparing standard discectomy for disc prolapse with discectomy plus autologous disc cell implantation demonstrated improvements in pain reduction, preservation of disc height and prevention of adjacent disc segment degeneration in patients who underwent the cell implantation therapy (Meisel et al., 2006; Meisel et al., 2007). Clinical follow-up data is limited and there is no biochemical or biomechanical data, but this study demonstrates the potential for cell-based therapies for IVD tissue regeneration. However, studies suggest that isolation of cells from degenerate IVD could accelerate degeneration, while isolation of cells from non-degenerate levels may induce degenerative changes and needle puncture is a common method of inducting degeneration in animal models of disease (Kim et al., 2005; Korecki et al., 2008; Masuda et al., 2006; Zhang et al., 2009). Additionally, as IVD degeneration, like OA in cartilage, affects the phenotype of resident cells the identification of a suitable cell source for novel tissue engineering and regenerative medicine strategies is one of the key determinants of its success.

Cells used in tissue engineering must survive within the graft site and produce a suitable and functional matrix, which mimics, or even improves on the original tissue. Autologous chondrocyte implantation (ACI) has been shown to be effective in the treatment of osteoarthritic cartilage lesions, producing a type II collagen and proteoglycan-rich matrix which restores function to the joint (Gillogly et al., 1998). Similarly, implantation of chondrocytes within a 3-dimensional carrier system such as a collagen gel has also been shown to produce a cartilaginous matrix, which could be of clinical use (Gavenis et al., 2006; Sakai et al., 2006).

10. Microenvironmental challenges for MSC-based tissue regeneration: The 3-D environment and oxygenation

One of the major challenges facing MSC-based tissue engineering is mimicking and recreating the microenvironment of the tissue that is being regenerated. MSCs destined for tissue engineering must be able to survive in a physiologically harsh extracellular matrix, an environment that will be highly hypoxic, acidic, and nutrient deprived. In addition, this environment may be exposed to dynamic compressive forces or rich in pro-inflammatory mediators that could induce catabolic and apoptotic responses in MSCs. Culturing the cells in a high-density 3-D environment is a good starting point and clearly has a number of benefits. High-density culture (see Figure 6) is a 3-dimensional system that has been in the literature since the early 1970's. High-density suspension cultures of chondrocytes were initially used to study cartilage matrix protein synthesis by mammalian chondrocytes (Kuettner et al., 1982a; Kuettner et al., 1982b), specifically mucoprotein (Nevo and Dorfman, 1972) and proteoglycan (Handley and Lowther, 1976) biosynthesis. In an important study published in 1977 the method was refined by von der Mark and von der Mark who used tissue culture plastic dishes on an agar base to monitor chondrogenesis of stage-24 chick limb mesodermal cells *in vitro* by analyzing the onset of type II collagen synthesis (von der Mark and von der Mark, 1977). This work demonstrated that high-density cultures may be used for the analysis of cell specific types of collagen and provide a useful model for detailing the specific events in the differentiation of mesenchymal cells *in vitro*. Several subsequent studies used chick limb bud mesenchymal cells and chick sternal chondrocytes cultured in high-density to study chondrogenesis and proteoglycan synthesis (Lohmander et al., 1979) and fibronectin expression (Hassell et al., 1979). We have used this culture system as a model for chondrogenic, osteogenic and tenogenic differentiation of MSCs and redifferentiation of dedifferentiated chondrocytes (Csaki et al., 2009; Csaki et al., 2007; Seifarth et al., 2009; Shakibaei et al., 2006) (see Figure 7). The high-density model exhibits a number of characteristics that make it particularly suitable for studied on chondrogenesis. In the first 24 hours of the high-density chondrocyte culture, cells form prechondrogenic areas composed of densely packed cells with intercellular interactions (gap junctions); these are surrounded by a perichondrium of flat fibroblast-like cells (Shakibaei et al., 1993) resembling the situation during the early stage of chondrogenesis *in vivo*. The extensive cell-cell interactions during the first step of chondrogenesis are crucial for cell aggregation (Denker et al., 1999; Loty et al., 2000). The high-density culture system allows investigations during chondrogenesis starting from early blastema/mesenchymal stem cell condensations until cartilage maturation. This chondrogenic development is accompanied by enlargement of cartilage nodules, which is not due to cell proliferation but rather is due to appositional growth, which includes the transition of perichondral cells into chondrocytes. These conclusions are based on the almost complete absence of mitotic figures in cartilage and perichondrium and on the continuous increase in nodule size during cultivation (Denker et al., 1999).

In addition to 3-D environment, the presence of member(s) of the transforming growth factor (TGF-β) family and low oxygen tension have been reported to promote the *in vitro* differentiation of MSCs (Ronziere et al., 2010). Appropriate levels of oxygenation can be powerful determinants of MSC differentiation. For example, low oxygen tension is a more potent promoter of chondrogenic differentiation than dynamic compression (Meyer et al., 2010). Enhanced chondrogenic differentiation of human bone marrow-derived MSCs has been observed in a low oxygen environment in micropellet cultures (Khan et al., 2010; Markway et al., 2010). Clearly oxygen tension is an important determinant of MSC differentiation; hypoxia, cell aggregation and TGF-β delivery are crucial for achieving complete chondrogenesis (Mobasheri et al., 2009; Richardson et al., 2007). Although MSCs may possess a superior potential to generate a functional repair tissue in low oxygen tensions, in the context of *in vitro* cartilage tissue engineering, cells maintained in normoxic conditions in the presence of TGF-β generate the most mechanically functional tissue

(Buckley et al., 2010). Physiologically low oxygen tension during monolayer expansion of MSCs has been shown to be advantageous in order to improve cartilage tissue engineering in a sheep model (Zscharnack et al., 2009). The information available in the literature leads to the conclusion that MSCs in culture function optimally in an atmosphere of reduced oxygen that more closely approximates documented *in vivo* oxygen tension (Lennon et al., 2001). These observations apply to cells derived from OA patients (Khan et al., 2007). Clearly fine adjustments to these conditions in future studies could produce an engineered tissue with the desirable mechanical and biological properties.

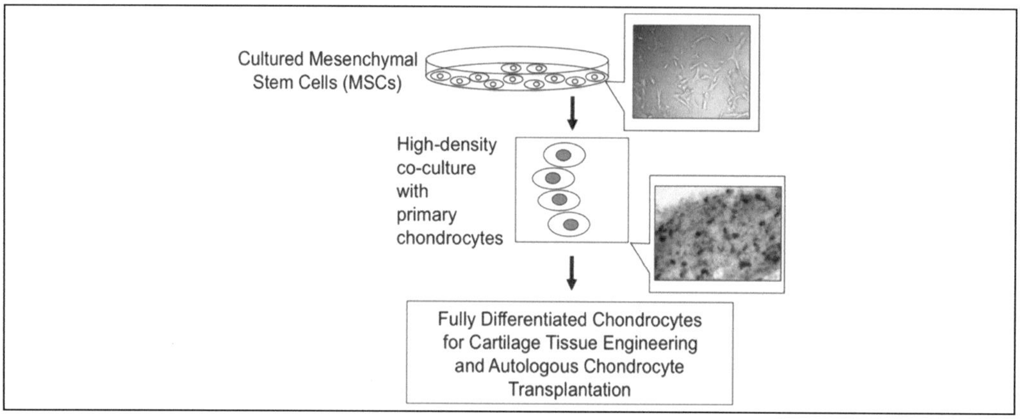

Fig. 6. 3-Dimensional high-density culture of mesenchymal stem cells (MSCs) and primary chondrocytes. The MSCs in this high-density model are stimulated to undergo chondrogenesis by the co-cultured primary chondrocytes

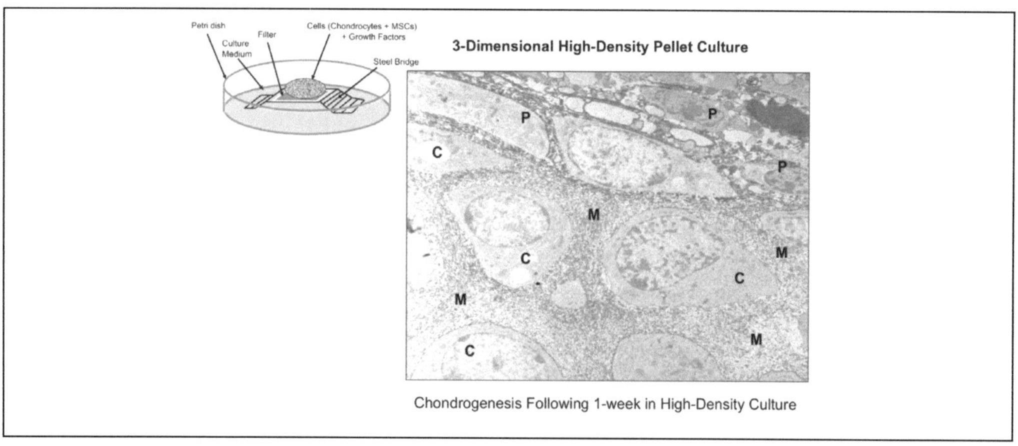

Fig. 7. 3-Dimensional high-density culture. In this model chondrocytes (C) are maintained in a pellet at the air-liquid interface, which will allow them to maintain their phenotype and produce an extracellular matrix (M). Cells at the periphery form a perichondrium-like layer of cells (P)

One proposed method to ensure MSC survival and function following implantation into the degenerate tissue is to pre-condition, or pre-differentiate the cells. A recent study by Wuertz and co-workers (Wuertz et al., 2008) demonstrated that when cultured in an intervertebral disc-like environment containing low glucose, low pH and high osmolarity, MSCs demonstrated a lower proliferation rate and lower expression of matrix genes compared to standard conditions, highlighting the importance of the microenvironmental niche on MSC biology. In particular, low pH has been shown to inhibit expression of aggrecan by MSCs (Wuertz et al., 2009a), while the presence of catabolic cytokines IL-1β and TNF- α, which are increased in both disc degeneration and OA, inhibited chondrogenesis of MSCs (Wehling et al., 2009).

11. High density cultures of MSCs: "3Rs" models for engineering biomimetic models of articular cartilage

A number of animal models of arthritis have been developed to study arthritic disease pathogenesis and evaluate the efficacy of candidate anti-inflammatory and anti-arthritic drugs for clinical development (Aigner et al., 2010; Ameye and Young, 2006; Brandt, 1991, 1994; Goldring, 1999; Koch and Betts, 2007; van den Berg, 2001). These animal models may involve injection of inflammatory agents into the joint, surgical creation of joint instability or surgical replication of joint trauma. Such models not only cause a considerable amount of pain and suffering but also none of them have a proven track record of predictability in human disease. Consequently, there is an acute need for developing novel and alternative *in vitro* models that mimic the biology of cartilage (Freed et al., 2006; Oreffo and Triffitt, 1999; Reddi, 1998, 2000; Sittinger et al., 2004). Many studies have used articular cartilage and synovial tissues to establish *in vitro* models of the joint. In our laboratories we have developed explant models of articular cartilage inflammation, co-cultures of primary synoviocytes and chondrocytes as well as 3-dimensional high-density cultures of chondrocytes and MSCs (Buhrmann et al., 2010; Csaki et al., 2009; Csaki et al., 2007; Mobasheri et al., 2009; Richardson et al., 2010). We have also determined the potential of phytochemicals as stimulators of chondrogenic differentiation. These culture models of MSCs, fully differentiated cells and co-cultures have potential of as "proof of principle" ethically acceptable "3Rs" models for engineering biomimetic models of articular cartilage *in vitro* for the purpose of replacing and reducing the use of animals in arthritis research. This approach could help reduce the number of laboratory animals used in arthritis research and may provide a realistic alternative to experimental animals.

12. Conclusions

One of the major conceptual difficulties in designing or applying clinically based MSC therapies to cartilage and intervertebral disc degeneration is the complexity and heterogeneity of disease among patients. New information obtained from *in vitro* studies and animal models are useful. However, the information gained from some animal models cannot be interpolated directly for translational studies in human patients with cartilage or disc degeneration. For example, stem cell therapy may not be applied across the entire spectrum of cartilage and intervertebral disc degeneration; the therapy may well work in cases that involve very early morphological and degenerative changes but success may not be guaranteed in cases where the damage and degeneration is extensive. It is difficult to conceive how MSCs and tissue engineering may be clinically useful in extreme cases where the cartilage has completely worn away or the disc has collapsed completely. Therefore, the variability among the human patient population complicates the concept of stem cell based therapy and introduces a very large element of risk, which may inhibit or impede the translational potential of regenerative medicine using MSCs.

Several clinical trials have already begun to evaluate the use and efficacy of mesenchymal stem cells and biomaterials in arthritic diseases of joints and the spine. At the present time too many gaps remain, both clinically and scientifically, before such approaches can be used safely and rationally. Interest in tissue engineering of cartilage and intervertebral disc is increasing and cartilage tissue engineering has been investigated for over 20 years. However, none of the approaches available for cartilage for example have been able to achieve the consistency, effectiveness and reliability required for clinical applications. Tissue engineering of a mechanically resilient disc that meets the structural and functional criteria for effective functional integration into a defect site in the host is a difficult endeavour. One of the fundamental weaknesses of all the tissue engineered models available to date is that none of them possess the normal zonal organization of cells that is seen *in vivo* (i.e. annulus fibrosus and nucleus pulposus regions) and the local composition of extracellular matrix in each zone. This structural organisation is a pre-requisite for normal disc function and the success of any future clinical applications.

For the foreseeable future the molecular determinants of MSC differentiation into chondrogenic and osteogenic lineages continue to be elucidated using existing 2D models. However, new research using 3D models is more likely to shed light on fundamental processes such as paracrine-mediated differentiation, and mechanisms responsible for the recruitment of endogenous progenitor populations in response to chemical, biological, and physical cues. Furthermore, three-dimensional and high-density culture models have huge potential as "3Rs" models for replacing and reducing the use of animals in arthritis research. There are numerous animal models for OA and other types of arthritis. However, none of them have a proven track record of predictability for the equivalent disease in humans and the majority of them are plagued with problems. Three-dimensional culture conditions, hypoxia and the use of specially supplemented culture media will ultimately need to be assessed in pre-clinical animal models that closely mimic the human disease. However, 3D culture models utilizing tissue engineered articular cartilage and IVD should help us carry out better and more focused research as well as reducing the number and variety of experimental animals used before moving onto larger and more expensive pre-clinical animal models.

Bone marrow and adipose tissue derived MSCs possess the potential for chondrogenic, osteogenic and tenogenic differentiation in the presence of appropriate cellular stimuli. Stable induction can be achieved with a combination of endocrine and environmental factors (i.e. co-culture of MSCs and primary cells), or through culturing MSCs with spent media from primary chondrocyte, osteoblast or tendon cell cultures. Therefore, these results point to new ways in which MSCs might be manipulated in culture for future therapeutic strategies for cartilage, bone and tendon tissue engineering.

Regenerative medicine relies on the optimistic view that stem cells, allogeneic tissue transplantation, patient derived adult stem cells and biomaterials may be used for repairing and regenerating tissues and organs in the future. Although we should maintain this optimistic view, it would be prudent to consider the numerous hurdles and complicating factors that need to be overcome as research progresses in this exciting and rapidly expanding field.

13. Acknowledgments

The authors thank their collaborators Professor Anthony J. Freemont (University of Manchester) and Professor Pablo Martin-Vasallo (University of La Laguna, Tenerife). This study was made possible by grants from The Wellcome Trust, the National Centre for the Replacement, Refinement and Reduction of Animals in Research (NC3Rs) (grant number: Mobasheri.A.28102007), the Biotechnology and Biological Sciences Research Council (BBSRC) (grants BBSRC/S/M/2006/ 13141 and BB/G018030/1) and the Engineering and Physical Sciences Research Council (EPSRC), UK. We would like to thank the members of our laboratories for their support and collaboration.

14. References

Aigner, T., Cook, J.L., Gerwin, N., Glasson, S.S., Laverty, S., Little, C.B., McIlwraith, W., Kraus, V.B., 2010. Histopathology atlas of animal model systems - overview of guiding principles. *Osteoarthritis Cartilage* 18 Suppl 3, S2-6.

Ameye, L.G., Young, M.F., 2006. Animal models of osteoarthritis: lessons learned while seeking the "Holy Grail". *Curr Opin Rheumatol* 18, 537-547.

Arinzeh, T.L., 2005. Mesenchymal stem cells for bone repair: preclinical studies and potential orthopedic applications. *Foot Ankle Clin* 10, 651-665, viii.

Baker, R.J., Patel, D., 2005. Lower back pain in the athlete: common conditions and treatment. *Prim Care* 32, 201-229.

Barry, F.P., 2003. Mesenchymal stem cell therapy in joint disease. *Novartis Found Symp* 249, 86-96; discussion 96-102, 170-104, 239-141.

Barry, F.P., Murphy, J.M., 2004. Mesenchymal stem cells: clinical applications and biological characterization. *Int J Biochem Cell Biol* 36, 568-584.

Brandt, K.D., 1991. Transection of the anterior cruciate ligament in the dog: a model of osteoarthritis. *Semin Arthritis Rheum* 21, 22-32.

Brandt, K.D., 1994. Insights into the natural history of osteoarthritis provided by the cruciate-deficient dog. An animal model of osteoarthritis. *Ann N Y Acad Sci* 732, 199-205.

Brittberg, M., Lindahl, A., Nilsson, A., Ohlsson, C., Isaksson, O., Peterson, L., 1994. Treatment of deep cartilage defects in the knee with autologous chondrocyte transplantation. *N Engl J Med* 331, 889-895.

Brooks, P.M., 2002. Impact of osteoarthritis on individuals and society: how much disability? Social consequences and health economic implications. *Curr Opin Rheumatol* 14, 573-577.

Brooks, P.M., 2006. The burden of musculoskeletal disease--a global perspective. *Clin Rheumatol* 25, 778-781.

Buckley, C.T., Vinardell, T., Kelly, D.J., 2010. Oxygen tension differentially regulates the functional properties of cartilaginous tissues engineered from infrapatellar fat pad derived MSCs and articular chondrocytes. *Osteoarthritis Cartilage* 18, 1345-1354.

Buckwalter, J.A., 2003. Sports, joint injury, and posttraumatic osteoarthritis. *J Orthop Sports Phys Ther* 33, 578-588.

Buckwalter, J.A., Lane, N.E., 1997. Athletics and osteoarthritis. *Am J Sports Med* 25, 873-881.

Buckwalter, J.A., Mankin, H.J., 1998. Articular cartilage: tissue design and chondrocyte-matrix interactions. *Instr Course Lect* 47, 477-486.

Buhrmann, C., Mobasheri, A., Matis, U., Shakibaei, M., 2010. Curcumin mediated suppression of nuclear factor-kappaB promotes chondrogenic differentiation of mesenchymal stem cells in a high-density co-culture microenvironment. *Arthritis Res Ther* 12, R127.

Burton-Wurster, N., Lust, G., Macleod, J.N., 1997. Cartilage fibronectin isoforms: in search of functions for a special population of matrix glycoproteins. *Matrix Biol* 15, 441-454.

Caplan, A.I., 2005. Review: mesenchymal stem cells: cell-based reconstructive therapy in orthopedics. *Tissue Eng* 11, 1198-1211.

Caplan, A.I., 2007. Adult mesenchymal stem cells for tissue engineering versus regenerative medicine. *J Cell Physiol* 213, 341-347.

Chajra, H., Rousseau, C.F., Cortial, D., Ronziere, M.C., Herbage, D., Mallein-Gerin, F., Freyria, A.M., 2008. Collagen-based biomaterials and cartilage engineering. Application to osteochondral defects. *Biomed Mater Eng* 18, S33-45.

Csaki, C., Matis, U., Mobasheri, A., Shakibaei, M., 2009. Co-culture of canine mesenchymal stem cells with primary bone-derived osteoblasts promotes osteogenic differentiation. *Histochem Cell Biol* 131, 251-266.

Csaki, C., Matis, U., Mobasheri, A., Ye, H., Shakibaei, M., 2007. Chondrogenesis, osteogenesis and adipogenesis of canine mesenchymal stem cells: a biochemical, morphological and ultrastructural study. *Histochem Cell Biol* 128, 507-520.

Denker, A.E., Haas, A.R., Nicoll, S.B., Tuan, R.S., 1999. Chondrogenic differentiation of murine C3H10T1/2 multipotential mesenchymal cells: I. Stimulation by bone morphogenetic protein-2 in high-density micromass cultures. *Differentiation* 64, 67-76.

Di Cesare, P.E., Carlson, C.S., Stolerman, E.S., Hauser, N., Tulli, H., Paulsson, M., 1996. Increased degradation and altered tissue distribution of cartilage oligomeric matrix protein in human rheumatoid and osteoarthritic cartilage. J Orthop Res 14, 946-955.

Dominici, M., Le Blanc, K., Mueller, I., Slaper-Cortenbach, I., Marini, F., Krause, D., Deans, R., Keating, A., Prockop, D., Horwitz, E., 2006. Minimal criteria for defining multipotent mesenchymal stromal cells. The International Society for Cellular Therapy position statement. *Cytotherapy* 8, 315-317.

Dudhia, J., 2005. Aggrecan, aging and assembly in articular cartilage. *Cell Mol Life Sci* 62, 2241-2256.

Eckstein, F., Reiser, M., Englmeier, K.H., Putz, R., 2001. In vivo morphometry and functional analysis of human articular cartilage with quantitative magnetic resonance imaging--from image to data, from data to theory. *Anat Embryo*l (Berl) 203, 147-173.

Enomoto-Iwamoto, M., Iwamoto, M., Nakashima, K., Mukudai, Y., Boettiger, D., Pacifici, M., Kurisu, K., Suzuki, F., 1997. Involvement of alpha5beta1 integrin in matrix interactions and proliferation of chondrocytes. *J Bone Miner Res* 12, 1124-1132.

Eyre, D.R., 2004. Collagens and cartilage matrix homeostasis. *Clin Orthop Relat Res,* S118-122.

Flynn, A., O'Brien, T., 2011. Stem cell therapy for cardiac disease. *Expert Opin Biol Ther* 11, 177-187.

Freed, L.E., Guilak, F., Guo, X.E., Gray, M.L., Tranquillo, R., Holmes, J.W., Radisic, M., Sefton, M.V., Kaplan, D., Vunjak-Novakovic, G., 2006. Advanced tools for tissue engineering: scaffolds, bioreactors, and signaling. *Tissue Eng* 12, 3285-3305.

Ganey, T., Libera, J., Moos, V., Alasevic, O., Fritsch, K.G., Meisel, H.J., Hutton, W.C., 2003. Disc chondrocyte transplantation in a canine model: a treatment for degenerated or damaged intervertebral disc. *Spine* (Phila Pa 1976) 28, 2609-2620.

Gavenis, K., Schmidt-Rohlfing, B., Mueller-Rath, R., Andereya, S., Schneider, U., 2006. In vitro comparison of six different matrix systems for the cultivation of human chondrocytes. *In Vitro Cell Dev Biol Anim* 42, 159-167.

Gillogly, S.D., Voight, M., Blackburn, T., 1998. Treatment of articular cartilage defects of the knee with autologous chondrocyte implantation. *J Orthop Sports Phys Ther* 28, 241-251.

Goldring, M.B., 1999. The role of cytokines as inflammatory mediators in osteoarthritis: lessons from animal models. *Connect Tissue Res* 40, 1-11.

Gonzalez, F.A., Seth, A., Raden, D.L., Bowman, D.S., Fay, F.S., Davis, R.J., 1993. Serum-induced translocation of mitogen-activated protein kinase to the cell surface ruffling membrane and the nucleus. *J Cell Biol* 122, 1089-1101.

Grabias, S.L., Mankin, H.J., 1979. Pain in the lower back. *Bull Rheum Dis* 30, 1040-1045.

Grigoriadis, A.E., Heersche, J.N., Aubin, J.E., 1988. Differentiation of muscle, fat, cartilage, and bone from progenitor cells present in a bone-derived clonal cell population: effect of dexamethasone. *J Cell Biol* 106, 2139-2151.

Guilak, F., Alexopoulos, L.G., Upton, M.L., Youn, I., Choi, J.B., Cao, L., Setton, L.A., Haider, M.A., 2006. The pericellular matrix as a transducer of biomechanical and biochemical signals in articular cartilage. *Ann N Y Acad Sci* 1068, 498-512.

Guilak, F., Estes, B.T., Diekman, B.O., Moutos, F.T., Gimble, J.M., 2010. 2010 Nicolas Andry Award: Multipotent adult stem cells from adipose tissue for musculoskeletal tissue engineering. *Clin Orthop Relat Res* 468, 2530-2540.

Guo, X., Wang, C., Duan, C., Descamps, M., Zhao, Q., Dong, L., Lu, S., Anselme, K., Lu, J., Song, Y.Q., 2004a. Repair of osteochondral defects with autologous chondrocytes seeded onto bioceramic scaffold in sheep. *Tissue Eng* 10, 1830-1840.

Guo, X., Wang, C., Zhang, Y., Xia, R., Hu, M., Duan, C., Zhao, Q., Dong, L., Lu, J., Qing Song, Y., 2004b. Repair of large articular cartilage defects with implants of autologous mesenchymal stem cells seeded into beta-tricalcium phosphate in a sheep model. *Tissue Eng* 10, 1818-1829.

Handley, C.J., Lowther, D.A., 1976. Inhibition of proteoglycan biosynthesis by hyaluronic acid in chondrocytes in cell culture. *Biochim Biophys Acta* 444, 69-74.

Hassell, J.R., Pennypacker, J.P., Kleinman, H.K., Pratt, R.M., Yamada, K.M., 1979. Enhanced cellular fibronectin accumulation in chondrocytes treated with vitamin A. *Cell* 17, 821-826.

Helder, M.N., Knippenberg, M., Klein-Nulend, J., Wuisman, P.I., 2007. Stem cells from adipose tissue allow challenging new concepts for regenerative medicine. T*issue Eng* 13, 1799-1808.

Helen, W., Gough, J.E., 2008. Cell viability, proliferation and extracellular matrix production of human annulus fibrosus cells cultured within PDLLA/Bioglass composite foam scaffolds in vitro. *Acta Biomater* 4, 230-243.

Hewitt, A.T., Varner, H.H., Silver, M.H., Martin, G.R., 1982. The role of chondronectin and cartilage proteoglycan in the attachment of chondrocytes to collagen. *Prog Clin Biol Res* 110 Pt B, 25-33.

Hong, L., Colpan, A., Peptan, I.A., 2006. Modulations of 17-beta estradiol on osteogenic and adipogenic differentiations of human mesenchymal stem cells. *Tissue Eng* 12, 2747-2753.

Hudelmaier, M., Glaser, C., Hohe, J., Englmeier, K.H., Reiser, M., Putz, R., Eckstein, F., 2001. Age-related changes in the morphology and deformational behavior of knee joint cartilage. *Arthritis Rheum* 44, 2556-2561.

Hunziker, E.B., Wagner, J., Zapf, J., 1994. Differential effects of insulin-like growth factor I and growth hormone on developmental stages of rat growth plate chondrocytes in vivo. *J Clin Invest* 93, 1078-1086.

Isgaard, J., 1992. Expression and regulation of IGF-I in cartilage and skeletal muscle. *Growth Regul* 2, 16-22.

Jackson, K.C., 2nd, 2004. Pharmacotherapy in lower back pain. *Drugs Today* (Barc) 40, 765-772.

Jenniskens, Y.M., Koevoet, W., de Bart, A.C., Weinans, H., Jahr, H., Verhaar, J.A., DeGroot, J., van Osch, G.J., 2006. Biochemical and functional modulation of the cartilage collagen network by IGF1, TGFbeta2 and FGF2. *Osteoarthritis Cartilage* 14, 1136-1146.

Jiang, C.C., Chiang, H., Liao, C.J., Lin, Y.J., Kuo, T.F., Shieh, C.S., Huang, Y.Y., Tuan, R.S., 2007. Repair of porcine articular cartilage defect with a biphasic osteochondral composite. *J Orthop Res* 25, 1277-1290.

Kalson, N.S., Richardson, S., Hoyland, J.A., 2008. Strategies for regeneration of the intervertebral disc. *Regen Med* 3, 717-729.

Khan, W.S., Adesida, A.B., Hardingham, T.E., 2007. Hypoxic conditions increase hypoxia-inducible transcription factor 2alpha and enhance chondrogenesis in stem cells from the infrapatellar fat pad of osteoarthritis patients. *Arthritis Res Ther* 9, R55.

Khan, W.S., Adesida, A.B., Tew, S.R., Lowe, E.T., Hardingham, T.E., 2010. Bone marrow-derived mesenchymal stem cells express the pericyte marker 3G5 in culture and show enhanced chondrogenesis in hypoxic conditions. *J Orthop Res* 28, 834-840.

Kiani, C., Chen, L., Wu, Y.J., Yee, A.J., Yang, B.B., 2002. Structure and function of aggrecan. *Cell Res* 12, 19-32.

Kim, K.S., Yoon, S.T., Li, J., Park, J.S., Hutton, W.C., 2005. Disc degeneration in the rabbit: a biochemical and radiological comparison between four disc injury models. *Spine* (Phila Pa 1976) 30, 33-37.

Koch, T.G., Betts, D.H., 2007. Stem cell therapy for joint problems using the horse as a clinically relevant animal model. *Expert Opin Biol Ther* 7, 1621-1626.

Korecki, C.L., Costi, J.J., Iatridis, J.C., 2008. Needle puncture injury affects intervertebral disc mechanics and biology in an organ culture model. *Spine* (Phila Pa 1976) 33, 235-241.

Kosher, R.A., Church, R.L., 1975. Stimulation of in vitro somite chondrogenesis by procollagen and collagen. *Nature* 258, 327-330.

Kosher, R.A., Lash, J.W., Minor, R.R., 1973. Environmental enhancement of in vitro chondrogenesis. IV. Stimulation of somite chondrogenesis by exogenous chondromucoprotein. *Dev Biol* 35, 210-220.

Kreklau, B., Sittinger, M., Mensing, M.B., Voigt, C., Berger, G., Burmester, G.R., Rahmanzadeh, R., Gross, U., 1999. Tissue engineering of biphasic joint cartilage transplants. *Biomaterials* 20, 1743-1749.

Kuettner, K.E., Aydelotte, M.B., Thonar, E.J., 1991. Articular cartilage matrix and structure: a minireview. *J Rheumatol Suppl* 27, 46-48.

Kuettner, K.E., Memoli, V.A., Pauli, B.U., Wrobel, N.C., Thonar, E.J., Daniel, J.C., 1982a. Synthesis of cartilage matrix by mammalian chondrocytes in vitro. II. Maintenance of collagen and proteoglycan phenotype. *J Cell Biol* 93, 751-757.

Kuettner, K.E., Pauli, B.U., Gall, G., Memoli, V.A., Schenk, R.K., 1982b. Synthesis of cartilage matrix by mammalian chondrocytes in vitro. I. Isolation, culture characteristics, and morphology. *J Cell Biol* 93, 743-750.

Lennon, D.P., Edmison, J.M., Caplan, A.I., 2001. Cultivation of rat marrow-derived mesenchymal stem cells in reduced oxygen tension: effects on in vitro and in vivo osteochondrogenesis. *J Cell Physiol* 187, 345-355.

Loeser, R.F., 1997. Growth factor regulation of chondrocyte integrins. Differential effects of insulin-like growth factor 1 and transforming growth factor beta on alpha 1 beta 1 integrin expression and chondrocyte adhesion to type VI collagen. *Arthritis Rheum* 40, 270-276.

Lohmander, L.S., Hascall, V.C., Caplan, A.I., 1979. Effects of 4-methyl umbelliferyl-beta-D-xylopyranoside on chondrogenesis and proteoglycan synthesis in chick limb bud mesenchymal cell cultures. *J Biol Chem* 254, 10551-10561.

Loty, S., Foll, C., Forest, N., Sautier, J.M., 2000. Association of enhanced expression of gap junctions with in vitro chondrogenic differentiation of rat nasal septal cartilage-released cells following their dedifferentiation and redifferentiation. *Arch Oral Biol* 45, 843-856.

Luo, W., Guo, C., Zheng, J., Chen, T.L., Wang, P.Y., Vertel, B.M., Tanzer, M.L., 2000. Aggrecan from start to finish. *J Bone Miner Metab* 18, 51-56.

Maffulli, N., King, J.B., 1992. Effects of physical activity on some components of the skeletal system. *Sports Med* 13, 393-407.

Markway, B.D., Tan, G.K., Brooke, G., Hudson, J.E., Cooper-White, J.J., Doran, M.R., 2010. Enhanced chondrogenic differentiation of human bone marrow-derived mesenchymal stem cells in low oxygen environment micropellet cultures. *Cell Transplant* 19, 29-42.

Martin, J.A., Brown, T., Heiner, A., Buckwalter, J.A., 2004. Post-traumatic osteoarthritis: the role of accelerated chondrocyte senescence. *Biorheology* 41, 479-491.

Masuda, K., Imai, Y., Okuma, M., Muehleman, C., Nakagawa, K., Akeda, K., Thonar, E., Andersson, G., An, H.S., 2006. Osteogenic protein-1 injection into a degenerated disc induces the restoration of disc height and structural changes in the rabbit anular puncture model. *Spine* (Phila Pa 1976) 31, 742-754.

McGowan, J.A., 2003. Perspectives on the future of bone and joint diseases. *J Rheumatol Suppl* 67, 62-64.

McHardy, A., Pollard, H., 2005. Lower back pain in golfers: a review of the literature. *J Chiropr Med* 4, 135-143.

Meisel, H.J., Ganey, T., Hutton, W.C., Libera, J., Minkus, Y., Alasevic, O., 2006. Clinical experience in cell-based therapeutics: intervention and outcome. *Eur Spine J* 15 Suppl 3, S397-405.

Meisel, H.J., Siodla, V., Ganey, T., Minkus, Y., Hutton, W.C., Alasevic, O.J., 2007. Clinical experience in cell-based therapeutics: disc chondrocyte transplantation A treatment for degenerated or damaged intervertebral disc. *Biomol Eng* 24, 5-21.

Meyer, E.G., Buckley, C.T., Thorpe, S.D., Kelly, D.J., 2010. Low oxygen tension is a more potent promoter of chondrogenic differentiation than dynamic compression. *J Biomech* 43, 2516-2523.

Millward-Sadler, S.J., Salter, D.M., 2004. Integrin-dependent signal cascades in chondrocyte mechanotransduction. *Ann Biomed Eng* 32, 435-446.

Minogue, B.M., Richardson, S.M., Zeef, L.A., Freemont, A.J., Hoyland, J.A., 2010. Characterization of the human nucleus pulposus cell phenotype and evaluation of novel marker gene expression to define adult stem cell differentiation. *Arthritis Rheum* 62, 3695-3705.

Mobasheri, A., Carter, S.D., Martin-Vasallo, P., Shakibaei, M., 2002. Integrins and stretch activated ion channels; putative components of functional cell surface mechanoreceptors in articular chondrocytes. *Cell Biol Int* 26, 1-18.

Mobasheri, A., Csaki, C., Clutterbuck, A.L., Rahmanzadeh, M., Shakibaei, M., 2009. Mesenchymal stem cells in connective tissue engineering and regenerative medicine: applications in cartilage repair and osteoarthritis therapy. *Histol Histopathol* 24, 347-366.

Muir, H., 1995. The chondrocyte, architect of cartilage. Biomechanics, structure, function and molecular biology of cartilage matrix macromolecules. *Bioessays* 17, 1039-1048.

Nevo, Z., Dorfman, A., 1972. Stimulation of chondromucoprotein synthesis in chondrocytes by extracellular chondromucoprotein. *Proc Natl Acad Sci U S A* 69, 2069-2072.

Newman, A.P., 1998. Articular cartilage repair. *Am J Sports Med* 26, 309-324.

Noel, D., Djouad, F., Jorgense, C., 2002. Regenerative medicine through mesenchymal stem cells for bone and cartilage repair. *Curr Opin Investig Drugs* 3, 1000-1004.

O'Sullivan, J., D'Arcy, S., Barry, F.P., Murphy, J.M., Coleman, C.M., 2011. Mesenchymal chondroprogenitor cell origin and therapeutic potential. *Stem Cell Res Ther* 2, 8.

Oeppen, J., Vaupel, J.W., 2002. Demography. Broken limits to life expectancy. *Science* 296, 1029-1031.

Oreffo, R.O., Triffitt, J.T., 1999. Future potentials for using osteogenic stem cells and biomaterials in orthopedics. *Bone* 25, 5S-9S.

Parekkadan, B., Milwid, J.M., 2010. Mesenchymal stem cells as therapeutics. *Annu Rev Biomed Eng* 12, 87-117.

Penny, J., Harris, P., Shakesheff, K., Mobasheri, A., 2011. The biology of equine mesenchymal stem cells: phenotypic characterization, cell surface markers and multilineage differentiation. *Front Biosci* (in press)

Peterson, L., Brittberg, M., Kiviranta, I., Akerlund, E.L., Lindahl, A., 2002. Autologous chondrocyte transplantation. Biomechanics and long-term durability. *Am J Sports Med* 30, 2-12.

Pittenger, M.F., Mackay, A.M., Beck, S.C., Jaiswal, R.K., Douglas, R., Mosca, J.D., Moorman, M.A., Simonetti, D.W., Craig, S., Marshak, D.R., 1999. Multilineage potential of adult human mesenchymal stem cells. *Science* 284, 143-147.

Poole, A.R., 1999. An introduction to the pathophysiology of osteoarthritis. *Front Biosci* 4, D662-670.

Ralphs, J.R., Benjamin, M., 1994. The joint capsule: structure, composition, ageing and disease. *J Anat* 184 (Pt 3), 503-509.

Ramachandrula, A., Tiku, K., Tiku, M.L., 1992. Tripeptide RGD-dependent adhesion of articular chondrocytes to synovial fibroblasts. *J Cell Sci* 101 (Pt 4), 859-871.

Reddi, A.H., 1998. Role of morphogenetic proteins in skeletal tissue engineering and regeneration. *Nat Biotechnol* 16, 247-252.

Reddi, A.H., 2000. Morphogenesis and tissue engineering of bone and cartilage: inductive signals, stem cells, and biomimetic biomaterials. *Tissue Eng* 6, 351-359.

Reed, J.J., Wadsworth, L.T., 2010. Lower back pain in golf: a review. *Curr Sports Med Rep* 9, 57-59.

Richardson, S.M., Hoyland, J.A., Mobasheri, R., Csaki, C., Shakibaei, M., Mobasheri, A., 2010. Mesenchymal stem cells in regenerative medicine: opportunities and challenges for articular cartilage and intervertebral disc tissue engineering. *J Cell Physiol* 222, 23-32.

Richardson, S.M., Mobasheri, A., Freemont, A.J., Hoyland, J.A., 2007. Intervertebral disc biology, degeneration and novel tissue engineering and regenerative medicine therapies. *Histol Histopathol* 22, 1033-1041.

Ronziere, M.C., Perrier, E., Mallein-Gerin, F., Freyria, A.M., 2010. Chondrogenic potential of bone marrow- and adipose tissue-derived adult human mesenchymal stem cells. *Biomed Mater Eng* 20, 145-158.

Roughley, P.J., Lee, E.R., 1994. Cartilage proteoglycans: structure and potential functions. *Microsc Res Tech* 28, 385-397.

Ruoslahti, E., Reed, J.C., 1994. Anchorage dependence, integrins, and apoptosis. *Cell* 77, 477-478.

Sah, R.L., Chen, A.C., Grodzinsky, A.J., Trippel, S.B., 1994. Differential effects of bFGF and IGF-I on matrix metabolism in calf and adult bovine cartilage explants. *Arch Biochem Biophys* 308, 137-147.

Sakai, D., Mochida, J., Iwashina, T., Watanabe, T., Suyama, K., Ando, K., Hotta, T., 2006. Atelocollagen for culture of human nucleus pulposus cells forming nucleus pulposus-like tissue in vitro: influence on the proliferation and proteoglycan production of HNPSV-1 cells. *Biomaterials* 27, 346-353.

Sarzi-Puttini, P., Cimmino, M.A., Scarpa, R., Caporali, R., Parazzini, F., Zaninelli, A., Atzeni, F., Canesi, B., 2005. Osteoarthritis: an overview of the disease and its treatment strategies. *Semin Arthritis Rheum* 35, 1-10.

Seifarth, C., Csaki, C., Shakibaei, M., 2009. Anabolic actions of IGF-I and TGF-beta1 on Interleukin-1beta-treated human articular chondrocytes: evaluation in two and three dimensional cultures. *Histol Histopathol* 24, 1245-1262.

Setton, L.A., Elliott, D.M., Mow, V.C., 1999. Altered mechanics of cartilage with osteoarthritis: human osteoarthritis and an experimental model of joint degeneration. *Osteoarthritis Cartilage* 7, 2-14.

Shakibaei, M., John, T., De Souza, P., Rahmanzadeh, R., Merker, H.J., 1999. Signal transduction by beta1 integrin receptors in human chondrocytes in vitro: collaboration with the insulin-like growth factor-I receptor. *Biochem J* 342 Pt 3, 615-623.

Shakibaei, M., Schroter-Kermani, C., Merker, H.J., 1993. Matrix changes during long-term cultivation of cartilage (organoid or high-density cultures). *Histol Histopathol* 8, 463-470.

Shakibaei, M., Seifarth, C., John, T., Rahmanzadeh, M., Mobasheri, A., 2006. Igf-I extends the chondrogenic potential of human articular chondrocytes in vitro: molecular association between Sox9 and Erk1/2. *Biochem Pharmacol* 72, 1382-1395.

Shao, X., Goh, J.C., Hutmacher, D.W., Lee, E.H., Zigang, G., 2006. Repair of large articular osteochondral defects using hybrid scaffolds and bone marrow-derived mesenchymal stem cells in a rabbit model. *Tissue Eng* 12, 1539-1551.

Sittinger, M., Hutmacher, D.W., Risbud, M.V., 2004. Current strategies for cell delivery in cartilage and bone regeneration. *Curr Opin Biotechnol* 15, 411-418.

Solchaga, L.A., Dennis, J.E., Goldberg, V.M., Caplan, A.I., 1999. Hyaluronic acid-based polymers as cell carriers for tissue-engineered repair of bone and cartilage. *J Orthop Res* 17, 205-213.

Solursh, M., 1991. Formation of cartilage tissue in vitro. *J Cell Biochem* 45, 258-260.

Sommarin, Y., Larsson, T., Heinegard, D., 1989. Chondrocyte-matrix interactions. Attachment to proteins isolated from cartilage. *Exp Cell Res* 184, 181-192.

Sonoyama, W., Liu, Y., Fang, D., Yamaza, T., Seo, B.M., Zhang, C., Liu, H., Gronthos, S., Wang, C.Y., Shi, S., Wang, S., 2006. Mesenchymal stem cell-mediated functional tooth regeneration in swine. *PLoS ONE* 1, e79.

Tolar, J., Le Blanc, K., Keating, A., Blazar, B.R., 2010. Concise review: hitting the right spot with mesenchymal stromal cells. *Stem Cells* 28, 1446-1455.

Trippel, S.B., Corvol, M.T., Dumontier, M.F., Rappaport, R., Hung, H.H., Mankin, H.J., 1989. Effect of somatomedin-C/insulin-like growth factor I and growth hormone on cultured growth plate and articular chondrocytes. *Pediatr Res* 25, 76-82.

Trubiani, O., Di Primio, R., Traini, T., Pizzicannella, J., Scarano, A., Piattelli, A., Caputi, S., 2005. Morphological and cytofluorimetric analysis of adult mesenchymal stem cells expanded ex vivo from periodontal ligament. *Int J Immunopathol Pharmacol* 18, 213-221.

Trubiani, O., Orsini, G., Caputi, S., Piatelli, A., 2006. Adult mesenchymal stem cells in dental research: a new approach for tissue engineering. *Int J Immunopathol Pharmacol* 19, 451-460.

Uccelli, A., Benvenuto, F., Laroni, A., Giunti, D., 2011. Neuroprotective features of mesenchymal stem cells. *Best Pract Res Clin Haematol* 24, 59-64.

Uccelli, A., Moretta, L., Pistoia, V., 2008. Mesenchymal stem cells in health and disease. *Nat Rev Immunol* 8, 726-736.

Uccelli, A., Pistoia, V., Moretta, L., 2007. Mesenchymal stem cells: a new strategy for immunosuppression? *Trends Immunol* 28, 219-226.

van den Berg, W.B., 2001. Lessons from animal models of osteoarthritis. *Curr Opin Rheumatol* 13, 452-456.

von der Mark, K., Gauss, V., von der Mark, H., Muller, P., 1977. Relationship between cell shape and type of collagen synthesised as chondrocytes lose their cartilage phenotype in culture. *Nature* 267, 531-532.

von der Mark, K., von der Mark, H., 1977. Immunological and biochemical studies of collagen type transition during in vitro chrondrogenesis of chick limb mesodermal cells. *J Cell Biol* 73, 736-747.

Weigl, M., Cieza, A., Cantista, P., Reinhardt, J.D., Stucki, G., 2008. Determinants of disability in chronic musculoskeletal health conditions: a literature review. *Eur J Phys Rehabil Med* 44, 67-79.

Wilda, H., Gough, J.E., 2006. In vitro studies of annulus fibrosus disc cell attachment, differentiation and matrix production on PDLLA/45S5 Bioglass composite films. *Biomaterials* 27, 5220-5229.

Woolf, A.D., Pfleger, B., 2003. Burden of major musculoskeletal conditions. *Bull World Health Organ* 81, 646-656.

Yoshikawa, H., Myoui, A., 2005. Bone tissue engineering with porous hydroxyapatite ceramics. *J Artif Organs* 8, 131-136.

Zhang, H., La Marca, F., Hollister, S.J., Goldstein, S.A., Lin, C.Y., 2009. Developing consistently reproducible intervertebral disc degeneration at rat caudal spine by using needle puncture. *J Neurosurg Spine* 10, 522-530.

Zscharnack, M., Poesel, C., Galle, J., Bader, A., 2009. Low oxygen expansion improves subsequent chondrogenesis of ovine bone-marrow-derived mesenchymal stem cells in collagen type I hydrogel. *Cells Tissues Organs* 190, 81-93.

5

Skeletal Regeneration by Mesenchymal Stem Cells: What Else?

José A. Andrades[1,2] et al.*
[1]Laboratory of Bioengineering and Tissue Regeneration (LABRET-UMA), Department of Cell Biology, Genetics and Physiology, Faculty of Sciences, University of Málaga, Málaga,
[2]Networking Biomedical Research Center in Bioengineering, Biomaterials and Nanomedicine (CIBER-BBN)
Spain

1. Introduction

1.1 The definition of MSCs

Bone marrow (BM) was the first tissue described as a source of plastic-adherent, fibroblast-like cells that develops colony-forming unit fibroblastic (CFU-F) when seeded in tissue culture plates (Friedenstein et al., 1982; 1987). These cells, originally designated stromal cells, elicited much attention, and the main goal of thousands of studies conducted using these cells was to find an ultimate pure cell population that could be further utilized for regenerative purposes. In these studies, cells were isolated using several methods and were given names such as mesenchymal stem cells (MSCs), mesenchymal progenitors, stromal stem cells, among others. Lately, a committee of the International Society for Cytotherapy suggested the name "multipotent mesenchymal stromal cells" (Dominici et al., 2006). However, most scientists have been referring to them simply as "MSCs".

The precise definition of these cells remains a matter of debate. Nevertheless, to date MSCs are widely defined as a plastic-adherent cell population that, under closely controlled conditions, can be directed to differentiate *in vitro* into cells of osteogenic, chondrogenic, adipogenic, myogenic, tenogenic, or hematopoietic-supportive stromal lineages (Pittenger et al., 1999; Javazon et al., 2004; Alonso et al., 2008; Prockop, 2009) (Fig. 1).

As part of their stem cell nature, MSCs proliferate and give rise to daughter cells that have the same pattern of gene expression and phenotype and, therefore, maintain the "stemness"

---

* Silvia Claros[1,2], Pedro Jiménez-Palomo[1,2], José Mª López-Puerta[3,2], Plácido Zamora-Navas[4,2], Enrique Guerado[5,2], Manuel Monleón[6,2], María C. Araque[6,2] and José Becerra[1,2]
[1]*Laboratory of Bioengineering and Tissue Regeneration (LABRET-UMA), Department of Cell Biology, Genetics and Physiology. Faculty of Sciences, University of Málaga, Málaga,*
[2]*Networking Biomedical Research Center in Bioengineering, Biomaterials and Nanomedicine (CIBER-BBN),*
[3]*Department of Orthopaedic Surgery and Traumatology, Universitary Hospital Virgen del Rocío, Sevilla,*
[4]*Department of Orthopaedic Surgery and Traumatology, Universitary Hospital Virgen de la Victoria, Málaga,*
[5]*Department of Orthopaedic Surgery and Traumatology, Hospital Costa del Sol, Marbella,*
[6]*Center of Biomaterials and Tissue Engineering, Universidad Politécnica de Valencia,*
Center of Research Príncipe Felipe, Valencia
Spain

of the original cells. Self-renewal and differentiation potential are two criteria that define MSCs as real stem cells; however, these characteristics have only been proved after *in vitro* manipulation, in bulk and at single-cell level, and there is no clear description of the characteristics displayed by unmanipulated MSCs *in vivo* (Lee et al., 2010).

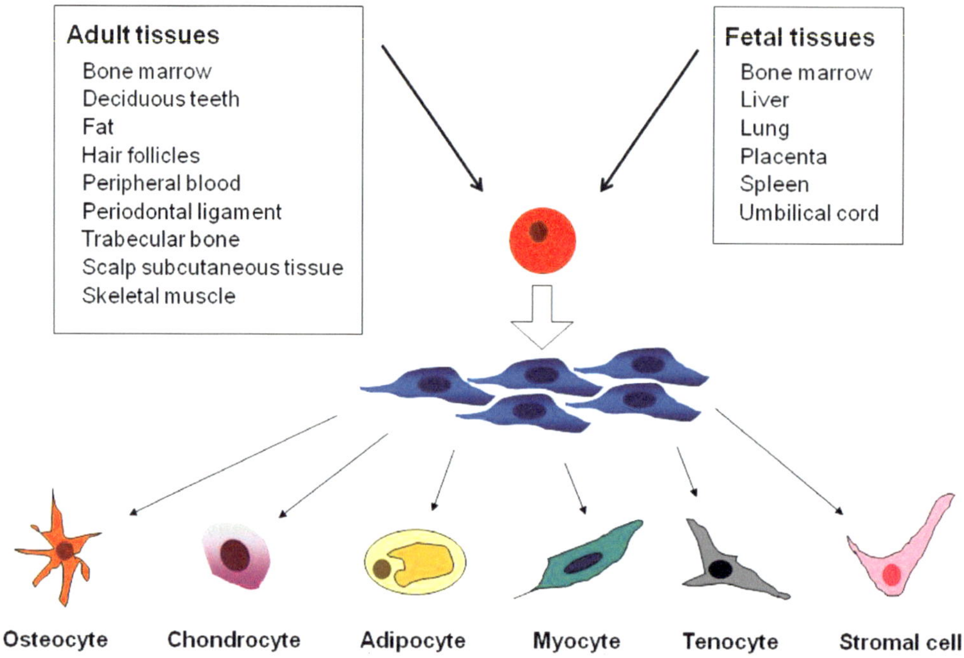

Fig. 1. Source of MSCs and multipotent differentiation capacity. MSCs can be isolated from bone marrow, deciduous teeth, fat, hair follicles, peripheral blood, periodontal ligament, trabecular bone, scalp subcutaneous tissue, and skeletal muscle in adult tissues, and from bone marrow, liver, lung, placenta, spleen and umbilical cord in fetal tissues. MSCs can generate multiple mesoderm-type cell lineages, such as osteocytes, chondrocytes, adipocytes, myocytes, tenocytes, and stromal cells

In contrast to other stem cells such as hematopoietic stem cells (HCSs), which are identified by the expression of the CD34 surface marker, MSCs lack a unique marker. The CD105 surface antigen (endoglin) has been recently used to isolate human MSCs (hMSCs) from BM and such an approach enabled the characterization of freshly isolated hMSCs before culture. A distinct expression of certain surface antigens such as CD31 and CD45 was demonstrated in freshly isolated hMSCs and the expression of these molecules was lower in culture-expanded hMSCs (Aslan et al., 2006). These data suggest, again, the alterations that hMSCs may undergo during culture (Boquest et al., 2005).

In several studies, cultured MSCs have been characterized either by using cell surface antigens and/or by examining the cells' differentiation potential. Lately, the Mesenchymal and Tissue Stem Cell Committee of the International Society for Cellular Therapy proposed minimal criteria to define human MSCs: (1) MSCs must be plastic-adherent when maintained in standard culture conditions and form CFU-Fs in primary cultures, (2) MSCs must express CD73, CD90, CD105, and lack expression of CD11b, CD14, CD19, CD34, CD45,

CD79alpha, and HLA-DR surface molecules, and (3) MSCs must differentiate to osteoblasts, adipocytes, and chondroblasts *in vitro* (Dominici et al., 2006; Claros et al., 2008).

1.2 The stem cell nature of MSCs

Stem cells are defined by their ability to self-renew and by their potential to undergo differentiation into functional cells under the right conditions. The ongoing public discussion regarding whether MSCs are strictly stem cells requires a revision of the definition of stem cells, as MSCs apply to a wide cluster of non-hematopoietic stem-like cells isolated from mesenchymal tissues such as BM, adipose, amniotic fluid, and blood vessels. The central question would be whether they might be differentiated into cells of other than a mesenchymal nature. Researchers have reported that MSCs from BM and other tissues can be differentiated into epithelial, endothelial, and neural cells (Spees et al., 2003; Greco & Rameshwar, 2007; Yue et al., 2008). As stated above, there is a consensus on specific MSC markers, but a unique marker of "stemness" and multipotentiality has not yet been defined, since culture-expanded MSCs may lose some of these markers and acquire others, which are non-specific, but cells retain their multipotentiality (Jones & McGonable, 2008). The molecular signature and *in vivo* distribution status of MSCs remain unknown and, as such, subject to investigation, even though *ex vivo*-expanded MSCs have been widely used in numerous studies (Prockop, 2007; Kubo et al., 2009; Pricola et al., 2009).

In local models, direct injection of hMSCs into the brain tissue of rats resulted in the cells' long-term engraftment and subsequent migration along pathways similar to those used by neural stem cells (Azizi et al., 1998). The results of these studies demonstrate the multilineage differentitation potential of BM-derived adult MSCs and aid in defining them as suitable candidates for the regeneration of several mesenchymal tissues. These data suggest that stem cells require a specific tissue environment to develop their intrinsic potency (Weissman, 2000). A stem cell niche is defined as a complex, multifactorial local microenvironment required for the maintenance of the stem cell biology. The stem cell niche consists of stem cells, non-stem cells, an extracellular matrix (ECM) and molecular signals. Inside the niche, the stem cell can divide asymmetrically giving rise to both new stem cells and proliferating progenitor cells. These proliferating cells give rise to a cell population that undergoes differentiation (Becerra et al., 2011). Recent scientific advances have lead to a substantial increase in the amount of information regarding stem cell niche data. Some of the best-characterized stem cell niche models are *Drosophila* germarium or testis, vertebrate hair follicle, intestinal crypts, BM, and brain subventricular/subgranular zones (Fuchs et al., 2004; Mitsiadis et al., 2007). *In vitro* culture conditions of single stem cells form intestinal crypts can give rise to organoids which may behave as self-organizing structures in the absence of other non-epithelial cellular niche components (Sato et al., 2009).

1.3 MSCs and tissue engineering

The chronic shortage of donor organs and tissues for donor organs and tissues for transplantation has provided the impetus for intense research in the field of tissue engineering (TE). Unlike pharmacology and physiotherapies they are mainly palliative, TE and cellular therapy seek to augment, replace, or reconstruct damage of diseased tissues (Chai & Leong, 2007). Tissue engineering is an emerging field that offers outstanding opportunities for regenerative medicine. The most common concept underlying TE is to combine a scaffold or matrix, living cells and/or biologically active molecules to form a "TE construct" to promote the repair and regeneration of tissues. The scaffold supports cell colonization, migration, growth and differentiation, and often guides the development of the required tissue or acts as a drug delivering vehicle. Hence, TE can be defined as a discipline that seeks to create or to induce the formation of a specific tissue in a specific location through the selection and manipulation of cells, scaffolds, biologic stimuli (Muschler & Midura, 2002), and vascular support (angiogenesis and/or vasculogenesis), on which the TE paradigm is based on (Fig. 2).

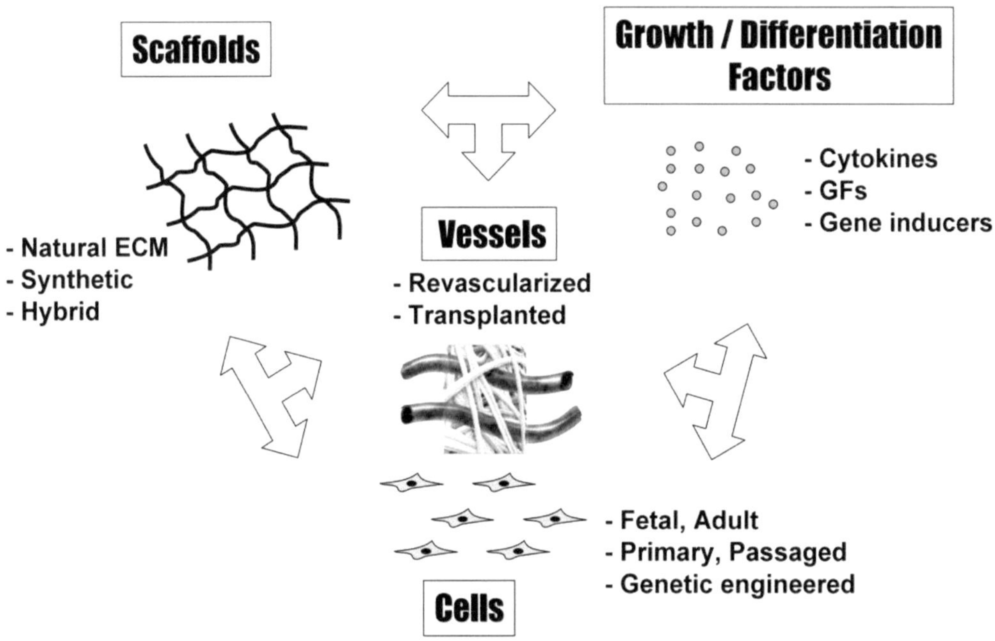

Fig. 2. Graphical illustrations showing the principles underlying tissue engineering. *From the same authors, published by Cellular and Molecular Biology, vol. 54 (1), 40-51. Copyright 2008 C.M.B. Edition*

The right knowledge of these interactions will create exciting new opportunities that might be useful in a broad array of clinical applications. As a logic consequence, today, a great number of multi-disciplinary groups with different backgrounds (Fig. 3) focus on various problems associated with TE, including cell isolation, characterization, and manipulation of cell proliferation/differentiation for stem cell therapy, design and elaboration of appropriate biomaterials as well as for development of bioreactors to enlarge tissue/organ engineering as a strategy to be applied in regenerative medicine.

MSCs have recently received much attention for their therapeutic potential in regenerative medicine, due to their capacity to secrete soluble factors that have beneficial effects (Caplan & Dennis, 2006; Cuenca-López et al., 2008; Caplan, 2009). Several studies have demonstrated that these MSC trophic factors may enhance regeneration ability of injured tissues, inhibit apoptosis, limit pathologic fibrotic remodelling, stimulate proliferation and differentiation of endogenous stem-like progenitors, decrease inflammatory oxidative stress and modulate immune reactions (Meirelles et al., 2009; Lee, 2010; Rodrigues et al., 2010).

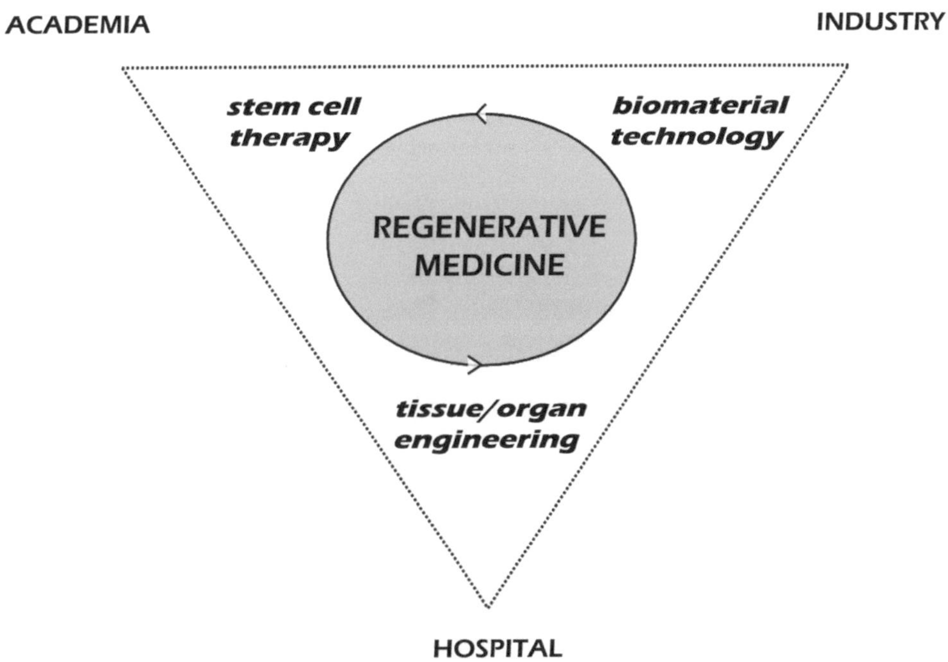

Fig. 3. Graphical illustration showing the multidisciplinary and complexity of interactions within the context of regenerative medicine. *From the same authors, published by Cellular and Molecular Biology, vol. 54 (1), 40-51. Copyright 2008 C.M.B. Edition*

Moreover, it has been described that growth factors (GFs) can act by an autocrine, paracrine, endocrine, juxtacrine, ECM mediated or intracrine process (Fig. 4) (Nimni, 1997). Autocrine action is the secretion of GFs for which the cell possesses receptors (1). Some experiments have suggested that this interaction may even occur within the cell, a process called intracrine interaction (6). Paracrine action is defined as the release of soluble GFs which diffuse into the extracellular space and act upon adjacent or closely located cells (2). In the case of endocrine action, GFs are carried in the bloodstream and may act on distant sites much like a hormone (3). Juxtacrine stimulation is when one cell has surface bound GFs which interact with an adjacent cell containing receptors for the GF (4). Furthermore, some ECM molecules can bind GFs, modulating their activities and regulating how they interact with cells (5).

Protein chain reaction, flow cytometry, enzyme-linked immunosorbent assay and Western blotting have shown that the secretion of MSCs include many major GF families and multitude of chemokine, including transforming growth factor beta-1 (TGF-β1), fibroblast growth factor (FGF), hepatocyte growth factor (HGF), insulin-like growth factor (IGF), platelet-derived growth factor (PDGF), vascular endothelial growth factor (VEGF), brain-derived neurotrophic factor (BDNF), nerve growth factor beta (NGF-β), Wnts, interleukin-1 (IL-1), IL-1β, IL-3, IL-6, IL-7, IL-11, among others (Gnecchi et al., 2008; Caplan, 2009; Zisa et al., 2009).

TGF-β1 is a multifunctional GF with a broad range of biological activities in various cell types in many different tissues. In general, TGF-β1 is known to influence cells from the chondro- and osteogenic lineage, promoting initial stages of mesenchymal condensation, regulating cell proliferation and cell differentiation, and stimulating production of ECM

(Andrades et al., 2003; van der Kraan et al., 2009; Janssens et al., 2005). Other of the trophic benefits produced by TGF-β1 is the immunomodulatory effect. Several authors have reported that MSCs suppress natural killer (NK) cell proliferation, alloreactive T-lymphocyte proliferation and activation, cytokine production, and cytotoxicity against HLA-class 1 expressing targets via TGF-β1 and prostaglandin E2 secretion (Groh et al., 2005; Sotiropoulou et al., 2006).

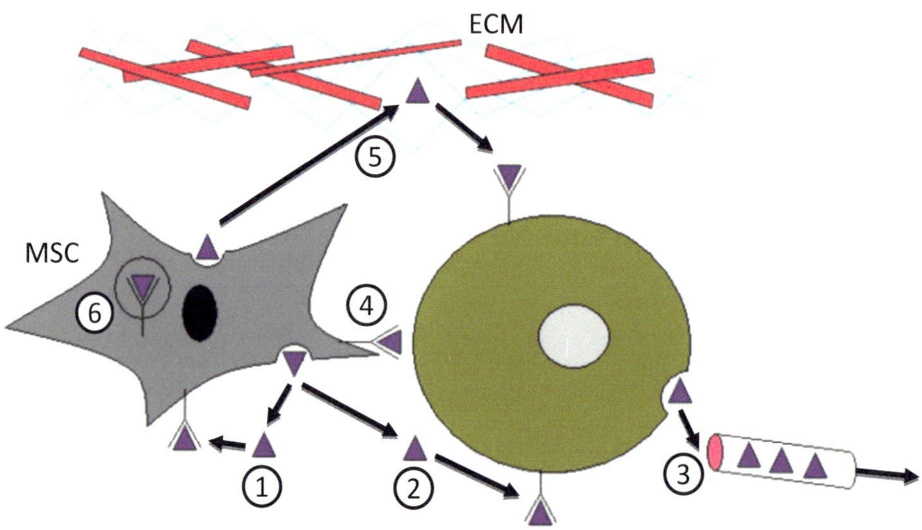

Fig. 4. Chemical signaling mechanisms of growth factors. GFs (triangles) produced by the cells can act within itself or in adjacent or remote cells to modulate their activities by reacting with specific receptors. (1) Autocrine; (2) paracrine; (3) endocrine; (4) juxtacrine; (5) extracellular matrix mediated; (6) intracrine. *Modified from Nimni ME. Polypeptide growth factors: targeted delivery systems. Biomaterials. 1997; 18(18):1201-1225*

FGFs are a family of GFs that act in an autocrine and paracrine way as a mitogen on many cell types (Chiou et al., 2006; Makino et al., 2010). They also regulate events in normal embryonic development, angiogenesis, wound repair, and cell differentiation. FGFs are involved in endothelial cell proliferation and the physical organization of endothelial cells into tube-like structures. They thus promote angiogenesis, the growth of new blood vessels from the pre-existing vasculature (Cao et al., 2003; Behr et al., 2010). Moreover, FGFs are important players in wound repair, stimulating the proliferation of fibroblasts that give rise to granulation tissue, which fills up a wound cavity early in the wound healing process (Schmid et al., 2009). Respect to cell differentiation, it is known that FGFs not only maintain MSC proliferation, they also retain osteogenic, adipogenic and chondrogenic differentiation potentials throughout many mitotic divisions (Tsutsumi et al., 2001; Kawazoe et al., 2008; Rodrigues et al., 2010).

The anti-fibrotic effects of MSCs have been demonstrated in different animal models, but the molecular mechanisms are not yet fully understood. Recently, HGF has been shown to be involved in the prevention of fibrosis (Li et al., 2009). The results reveal that, in a situation of tissue injury, MSCs become proliferative and secrete HGF, which in turn mediates anti-fibrotic and immunomodulatory effects. Additionally, the secretion of HGF and IGF-I is essential for activation of cardiac stem cells, which may contribute to endogenous repair mechanisms (Linke et al., 2005). On the other hand, IGFs play an important role in stem cell maintenance within the niche (Bendall et al., 2007), and bone metabolism. During osteogenesis, bone cells secrete amounts of IGFs that are stored at the highest concentration of all GFs in the bone matrix (Wang et al., 2011). Principally, IGFs have an anti-apoptotic effect on (pre)osteoblast and enhance bone matrix synthesis (Meinel et al., 2003; Niu & Rosen, 2005).

PDGF is considered one of the key regulators of general tissue repair. The main functions of PDGF are to stimulate cell replication of healing capable stem cells, as well as proliferation of endothelial cells. This will cause budding of new capillaries into the wound (angiogenesis), a fundamental part of all wound healing. In addition, PDGF also seems to promote the migration of perivascular healing capable cells into a wound and to modulate the effects of other GFs (Hollinger et al., 2008).

VEGF is best known for inducing angiogenesis (Gerstenfeld et al., 2003). Further, VEGF induce vascular changes: increased vascular support network and permeabilization of capillaries (Neufeld et al., 1999; Hansen-Algenstaedt et al., 2006), which could be responsible for the stimulation of the recruitment of MSCs (Fiedler et al., 2005). In addition, VEGF secretion is also increased during MSC osteogenesis (Wang et al., 2010), thus accounting for the infiltration of blood vessels during bone development (Shum et al., 2003).While VEGF has been reported to decrease the synthesis of osteocalcin (a marker of osteoblast maturation) in osteogenic MSCs (Villars et al., 2000), it has also been observed to increase mineralization (Mayer et al., 2005) and to be regulated by proteins involved in osteoblast differentiation (Deckers et al., 2000), suggesting that it usually acts as and enhancer of both angiogenesis and osteogenesis.

The neuroprotective effects of MSCs are principally attributed to production of neurotrophic factors, such as BDNF and NGF-β, that support neuronal cell survival, induce endogenous cell proliferation and promote nerve fiber regeneration at sites of injury (Li et al., 2002; Mahmood et al., 2004; Alexanian et al., 2010). Additionally, MSCs also express various neuro-regulatory proteins, adhesion molecules and receptors that likely contribute to the MSC-induced effects on neuronal cell survival and neurite formation (Crigler et al., 2006).

Another family of proteins critical in MSC biology is the Wnt family of signalling molecules (Ling et al., 2009). Wnts are highly conserved proteins that are essential to limb development and musculoskeletal morphogenesis in vertebrates (Yang, 2003). At least Wnts 2, 4, 5a, 11 and 16 are expressed by MSCs along with several Wnt receptors (Frizzleds 2-6) (Etheridge et al., 2004). Wnts are very active in determining MSC patterns of proliferation and differentiation (Boland et al., 2004; Baksh & Tuan, 2007), and they also play a role in cell adhesion and migration through their indirect interactions with the cadherin pathway.

Therefore, MSCs themselves secrete a large number of regulatory proteins. *In vivo,* these signals work together to regulate the regenerative abilities of MSCs. By learning more about the mechanisms of action and interactions of each member of this microenvironment, it will be possible to develop the full potential of MSCs for future therapeutic applications.

2. Skeletal tissue regeneration by MSCs

2.1 Therapy for bone fractures and arthrodesis

The treatment aim of bone fractures and of many orthopaedic deformities is the fusion between two or more bone ends or bone structures. Therapy for diaphyseal, metaphyseal or paraarticular fractures of long or short bones, either simple or complex, is always addressed to achieve permanent solid bone continuity between fractured bone fragments. The same comes true for orthopaedic corrections such as bone osteotomies, scoliosis or whatever situation where a bone discontinuity, either traumatic or therapeutic mediated, exists (Guerado, 2005; Guerado et al., 2005). Nowadays surgical treatment of these conditions is the choice for both young and old patients, as conservative treatment requires prolonged neighbor joints rest which results in articular stiffness and unacceptable deformities. Further, in many instances, non-surgical treatment will be unsuccessful in dealing with many low or high energy fractures, and also with orthopaedic corrections as non-union or malunion will occur.

Current surgical treatment of fractures or orthopaedic deformities aimed to promote osteogenesis for bony fusion is still far away from being undoubtedly successful. Surgical site infection, implant failure, and many other complications will facilitate failure of achieving the main aim of surgical treatment. In many instances, even carrying out appropriately surgical technique, failure of what had been planned will be the final result.

Although many techniques have been developed for non union treatment, complexity of them is, as far as biomechanics and biology are concerned, very difficult for accomplishing a successful outcome (Guerado, 2005; Guerado et al., 2005). Nonetheless, diaphyseal non-unions and even paraarticular ones, although in some cases present difficult challenges to bony union, can be managed, at least, under favorable biomechanical situations; further, its biology can be managed with autograft. Conversely, since in spinal fusion the size of any vertebra and the distance to each other make the biomechanical and biological situation very difficult for addressing a successful treatment of fusion or non-union, current and forthcoming new "biological" therapies have important perspectives. In relation to bone therapy, spinal fusion is currently the most significant surgical technique expecting for progress in biological treatments.

Surgical technique of spinal arthrodesis combines a hardware system for mechanical stabilization and a biological substance for bone formation enhancement. Both surgical steps have the aim of creating an optimal biomechanical situation together with a biological environment which promotes definite spinal bony fusion (Guerado, 2005; Guerado et al., 2005).

2.1.1 Biomechanics

Biomechanics is indeed a very important issue in spinal surgery. As a general concept, spinal arthrodesis is just the treatment of multiple foci of fractures once neat decortication of host bone has been carried-out; consequently, correct application of any of the four main principles of osteosynthesis (compression, neutralization, buttressing, and tension band, together with accessory bridging) play a major role depending on the sort and position of graft used; further, not only biology but also biomechanical features of host and donor bones are decisive. Using the same graft, either cortical or cancellous, different placements may need different nature of osteosynthesis, and may make the own graft behaves in different ways. Therefore, only when correct stabilization has been performed, as in any fracture treatment, biology behaves in the wished way. Likewise provided biomechanics is correct, the same graft behaves in different ways depending on the biomechanical forces acting on it (Guerado et al., 2005). So far, there has been a long list of bone graft substitutes, and strategies for bone promotion (Hecht et al., 1999; Bostrom & Seigerman, 2005; Huang et al., 2006; Ma et al., 2007; Thalgott et al., 2009). However only cells synthesize osteoid, and none of the nowadays known bone substitutes, apart from cells transplantation, are real "bone substitutes", since they do not have osteoprogenitor cells.

2.1.2 Biology

Grafting enhances bone fusion, and that results in permanent stability. Otherwise hardware fails with time. The three main properties a therapy for optimum fusion enhacement must have are osteogenesis, osteoconduction, and osteoinduction, together with avoidance of histoincompatibility, and, in the clinical setting, as said above, a favorable biomechanical situation (Boatright & Boden, 2002). Osteogenesis consists basically on cells differentiation into osteogenic linage, being bone cells the only agent with osteogenetic power. Osteoinduction is also very important; as mentioned before, molecules commonly called GFs induce MSCs to amplification (increase of cell population in number) and subsequently to differentiation into osteoprogenitor cells, specially the TGFs superfamily which also includes the well-known BMPs (Li et al., 2002). Osteoconduction is also the complement to osteogenesis and osteoinduction; a tridimensional scaffold structure of bone or bone substitute is an indispensable scaffolding for cell population growth. An osteoconductive microstructure of 200-400 microns is nowadays very easy to be manufactured. However osteoinductive GFs doses to be added to it are unknown; further osteogenetic properties of "bone substitutes" are lacking for osteoconductive materials, as osteogenesis is an exclusive property of mature osteoprogenitor cells.

Since bone synthesis is exclusively made by bone cells, a large population of osteoprogenitor cells is necessary for successful fracture or spinal fusion. The larger the area to be fused the more number of osteoprogenitor cells are needed. Because MSCs population is very scarce in elderly people, this age group has a higher risk for non-union (Caplan, 1991). Nowadays it is known that differentiated cells do not reproduce themselves, being undifferentiated cells, particularly MSCs, responsible for cell amplification and subsequent differentiation into osteogenic linage; moreover mature osteoblasts can recapitulate back to former less differentiated MSCs in order to achieve amplification for increasing the final population of differentiated osteoblasts (McCulloch & Till, 2005; Daley & Scadden, 2008; Jones & Wagers, 2008; Kuang & Rudnicki, 2008; Morrison & Sprading, 2008). During the entire process, the new-forming blood vessels that grow inside the callus act as a source of new MSCs; in fact, recent publications have given strong evidences of a perivascular origin for the MSCs (Crisan et al., 2008; Da Silva Meirelles et al., 2008; Koob et al., 2010); therefore neat decortication of host bone until bleeding, becomes of overwhelming importance. On the other hand fracture focus stability prevents new vessels and incipient callus from breakage.

Currently, the combination of biomechanical stabilization ensuring appropriate environment for bone growth, osteoprogenitor bone cells for osteoid synthesis, vascularization of fusion site for cells and metabolite transport, signaling molecules (TGFs, particularly BMPs) to encourage cells to amplification and differentiation, and scaffold for tridimensional cells growth has been denominated the "diamond concept" (Giannoudis et al., 2007; 2008), as to the ideal situation for fusion to take place. Hence clinical situations lacking of any of these variables are in less than optimal condition for successful spinal fusion. This concept is indeed extended to any fracture of the skeleton or joint arthrodesis (Giannoudis et al., 2007; 2008).

MSCs are, therefore, the centre of the osteogenesis concept, as to its amplification and commitment to osteoprogenitor line signifies that biomechanics is already achieved into its optimum, and that the appropriate environment has been created. Finding a natural source of viable MSCs for therapy is the star point of the diamond concept.

2.1.3 Autograft as a source of osteoprogenitor cells. The relation between biomechanics and biology in spinal surgery

Autograft is the standard biological agent to be added in fracture or arthrodesis focus for bone repair enhancement (Figs. 5-8). Autograft is the appropriate biological source of cells whenever bone healing is required in any part of the skeleton, as its biological properties are considered to be optimum: autograft has a good volume effect, and also potent osteogenic, osteoinductive, and osteoconductive properties (Boatright & Boden, 2002; Niu et al., 2009). Autograft is also the standard for spinal arthrodesis as the natural source of osteoprogenitor cells (Fig. 6). It can be used in spinal surgery either as cancellous chips or as a tricortical alive implant in order to improve the immediate strength of the construct. So far, no allogenic or synthetic bone substitutes have achieved its osteogenicity, although careful autograft implantation for spinal surgery must include shortening harvesting-to-implant interim, because these properties diminish with time (Sandhu et al., 1999).

Nevertheless, morbidity in donor site after autograft collection is very high, and fractures of iliac bone, particularly after tricortical graft has been harvested, with ensuing non-union are not rare, making the need for new bone substitutes (Arrington et al., 1999; Banwart et al., 1995; Delawi et al., 2010; Epstein, 2008; Glassman et al., 2010; Slosar et al., 2007). Limited availability together with longer operative time are also the most important burdens in conventional autograft harvesting, and new less invasive aspiration techniques of cancellous bone from femoral and tibial metaphysis by a "Reamer Irrigator Aspirator" (RIA) apparently provide larger graft amounts than conventional iliac crest harvesting (Belthur et al., 2008; Kobbe et al., 2008a; 2008b) and with higher concentration of GFs (Porter et al., 2008; Schmidmaier et al., 2006); further, RIA would provoke less postoperative pain and shorter length of stay (Belthur et al., 2008). However studies on RIA for spinal surgery are lacking.

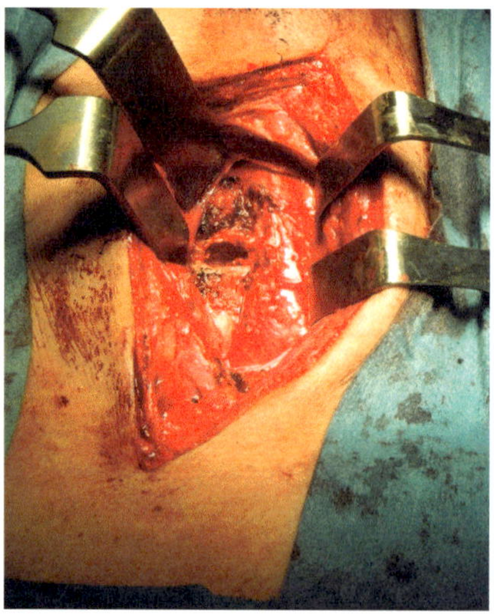

Fig. 5. Cervical spine. C_6-C_7 interbody disc has been removed for interbody fusion by a tricortical autograft harvested from iliac crest

This basic knowledge on autograft has been achieved after animal experiments, making clinical inference to human very uncertain. Cancellous bone graft has greater cellular activity than cortical, whereas cortical graft is much more dense and stronger (Day et al., 2000). Spongy structure of cancellous bone allows osteoprogenitor cells to have better vascularization, developing superior osteoconduction and osteogenesis. Anyhow although histological incorporation with prompt vascularization and MSCs invasion stars at the second day of implantation, mechanical properties of cancellous bone are compromised during remodelation phase, taking a few months to become structurally stronger (Pape et al., 2010). Cortical autograft having less biological properties than cancellous bone also suffers biomechanical weakening up to the 75% (19) at 6 to 24 weeks postimplantation, returning to normal strength some 48 weeks thereafter (Enneking et al., 1975).

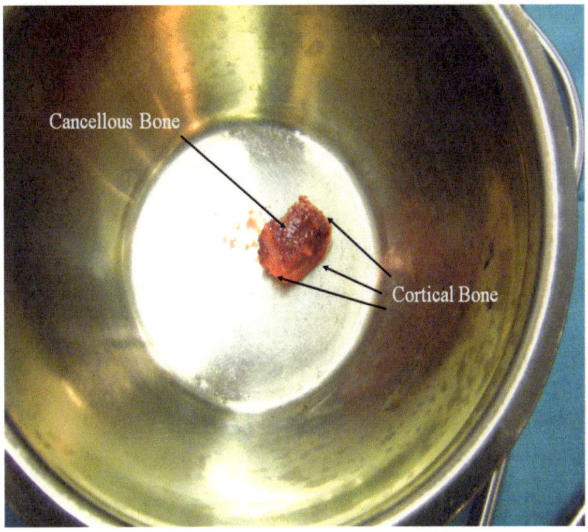

Fig. 6. Harvested tricortical autograft is prepared for implantation. Cortical layers provide strength for interbody compression. Cancellous bone provides biological support for osteogenesis enhacement and faster fusion

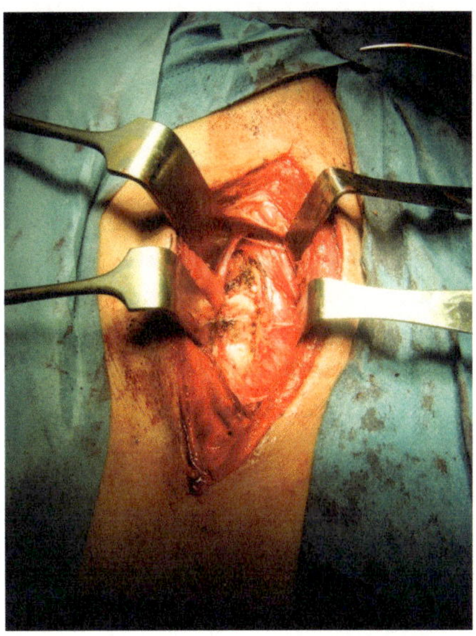

Fig. 7. Tricortical graft in place

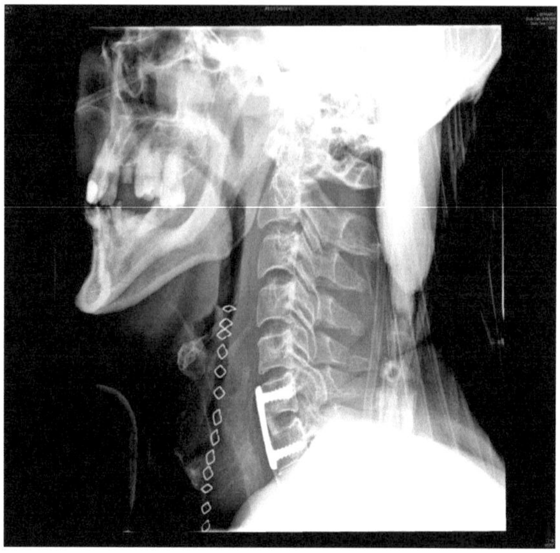

Fig. 8. X-ray control shows good position of graft and osteosynthesis stabilization by a plate. The cortical layers of the graft are positioned anteriorly facing the plate, superiorly facing the above vertebral body, and inferiorly facing the below vertebral body. The posterior face has not cortical bone

Consequently, according to biological and biomechanical properties of autograft, spinal posterior artrodesis is better enhanced by cancellous autograft whereas tricortical bone under interbody compression is more suitable for buttressing in anterior interbody fusion, together with a posterior transpedicular screw system in order to provide a tension band principle; alternatively an interbody plating together with the interbody tricortical autograft provides buttressing leaving the tricortical graft healing without supporting compression; this last possibility may be used in any anterior surgical approach (Figs. 5-8); however in clinical setting it is usually preferred, by some surgeons, a 360° fusion, as surgeons feel more confident in case of implant failure. Eventually, during the course of thoracotomy, removed ribs for anterolateral thoracic or thoracolumbar approaches can substitute the use of tricortical grafts, suppressing the morbidity provoked by iliac crest graft harvesting. In this case the graft can provide limited buttressing, as ribs are not as strong as tricortical iliac crest grafts; full buttressing and neutralization must be supported by the hardware; anterior interbody plating for this purpose appears to be ideal, and ultimately posterior pedicular instrumentation either alone or, ideally, within a 360° fusion concept. In conclusion, failure in the correct combination of graft type with proper instrumentation principles will result in poorer outcome, although the right graft had been used or the appropriate osteosynthesis applied.

2.1.4 Therapy with laboratory treated MSCs

Many studies have been made during the last years in experimental MSCs application for therapy in animals. Nevertheless biomechanics and biology of animal used are quite different from humans (Heineck et al., 2010); moreover clinical situations are different from one patient to another. Stem cells are currently being studied for use in numerous clinical applications, ranging from neurodegenerative diseases to cardiac insufficiency. The use of MSCs in spinal surgery is also compelling, especially with the increasing age of the general population. In spinal surgery, the use of MSCs is focused in intervertebral disc repair and regeneration and in spinal arthrodesis procedures. Although the routine use of cellular therapies by spine surgeons to improve outcome after a variety of surgical procedures is rapidly approaching with uncontrolled enthusiasm (Helm & Gazit, 2005), knowledge on its real effectiveness is far from being identified.

Experimental studies on the effect of low intensity pulsed ultrasound on rabbit posterolateral intertransverse processes spinal fusion with MSC-derived osteogenic cells and bioceramic composite have shown that this combination promote clinical fusion. The mechanism was likely to be mediated through better osteointegration between the host bone and implanted materials and enhanced endochondral ossification at the fusion site (Hui et al., 2011). However it is known that osteogenesis in spinal rabbit is achieved just by stripping the periostium. Also MSCs that had been cultured with osteogenic differentiation medium may induce the formation of new bone in experimental posterolateral intertransverse process spinal fusion in rabbits. Nonetheless suitability level of osteogenic differentiation of MSCs as well as the most appropriate carrier for MSCs is unkown (Nakajima et al., 2007).

New approaches and carrier have been introduced in the armamentarium for MSCs transplantation. It has been shown that MSCs and platelet lysate seeded in a fibrin or collagen scaffold can improve the new bone formation around an uncemented hip prosthesis stem in a sheep model. In vitro expanded MSCs suspended in platelet lysate and either mixed with collagen or fibrin gel as delivery vehicle inserted inside the femoral canal, in a press-fit femoral stem model provides higher bone-prosthesis contact (Dozza et al., 2011). But again biology is subjected to biomechanics and human and animal hip biomechanics are not similar.

MSCs transplantation will be a very important therapeutic principle, however deep knowledge not only on cells in cultures but also in biomechanical situations resembling clinical setting is needed. We know now that MSCs behave in different ways depending on the biomechanical situation. In fact this principle is worldwide known since more than 100 years by any orthopaedic surgeon as Wolf's law (Wolf, 2010).

2.2 Cartilage regeneration

Degenerative disease of articular cartilage (AC), generically known as osteoarthritis (OA), is an irreversible evolution process towards terminal articular failure. Due to its high prevalence on population and its socioeconomic impact, this condition is of great concern, and this way more resources and effort are dedicated to the research on its development.

OA is the result of several mechanical changes and biological events that destabilize the balance between normal degradation and synthesis of AC, ECM and subchondral bone. Knowledge on OA physiopathology has improved considerably, evolving from a purely mechanic approach to a molecular and inflammatory view, taking into consideration that cartilage, synovium and subchondral bone contribute to the joint restructuring.

Mature AC is a tissue free of blood vessels and nerves, located at the diarthrodial joints of the skeleton, making easier its gliding and the lubrication between the articular surfaces; it absorbs traumatisms and distributes load on the adjacent bone. It is made up of one ECM with proteoglycan, rich in type II collagen, and about only 5% of the tissue volume occupied by cells. These cells, the chondrocytes, are spherical, and can be found embedded inside lacunae filled with pericellular matrix and have no contact with the distant neighbour cells. Although human cartilage can reach up to 7-8 mm thickness, its supply with nutrients and oxygen is limited by diffusion, which is, however, facilitated by a cyclic compressive loading; this provides a pumping mechanism during joint movements. There are several reasons that contribute to a lessened capacity of response from chondrocytes before an injury regarding other parenchymal cells. As it has no vascularisation, the typical provisional matrices made from fibrin deposits do not exist, nor the inflammatory cells afflux occurs, there is no delivery to the environment of cell mediators through degranulation, whereas in other tissues, arrival to the focus of such mediators with mitogenic, fibrogenic and phagocytosis-inducing capacities are responsible for the repair. Although most of the actions promoted by the mediators of inflammation can enhance the injury magnitude (such as the proteolythic capacity, free radicals production, cell death or proliferation), the best guarantee for the *restitutio ad integrum* is its early delivery at the injured focus.

Chondrocytes are well differentiated cells with a limited proliferative and migratory capacity. Before physiological stimuli, it synthesizes its own ECM at a very constant and slow rate. A macroscopic structural injury involves higher demand of matrix synthesis, resulting in an insufficient cell response to achieve the tissue regeneration. However, the regenerative capacity of the organs lies on the persistence of cell progenitors in some areas of the adult tissues. Mesenchymal non-differentiated cells found in the AC prove that cartilage, as other tissues, has the capacity to regenerate its own structure. This suggests the existence of a lapse of time in which the capacity or repair is complete; this period would possibly cover the early stages of the disease when defects are microscopic. The most adequate resolution of a chondral defect should involve the regeneration into a tissue identical to hyaline cartilage. A simple repair means the filling with a non-identical tissue, which should be able to seal the defective area with good adhesiveness to subchondral bone and total integration within the surrounding cartilage, as well as to resist the mechanic wear and incorporate into the natural turnover of the normal tissue.

All this makes the AC a tissue with almost no healing capacity for intrinsic healing in most *in vivo* situations, although tissue culture demonstrates that *in vitro* conditions may reactivate a significant regenerative potential for juvenile tissue and articular chondrocytes. In partial cartilage defects, the subchondral plate remains intact, without access to the vascular system and they are not usually repaired spontaneously. On the other hand, defects that penetrate the full chondral thickness are generally associated to the violation of the subchondral plate, thus exposing the defect to the vascular system though the marrow area. This leads to the migration of MSCs to the damaged area, where they undergo the chondrogenic differentiation. Nonetheless, in many instances the repair tissue is made up of fibrocartilage that contains mainly collagen type I fibres, and does not meet the criteria precise for a functional tissue. Small defects can be repaired spontaneously with hyaline cartilage production, whereas larger defects will only be able to be repaired with the production of fibrous or fibrocartilaginous tissue, which are biomechanically and biochemically different from the normal hyaline cartilage. As a result, degeneration subsequently appears which can evolve into osteoarthritic change in some cases.

Focal cartilage defects have been detected in up to 63% of patients undergoing arthroscopy of the knee and fortunately most of them remain symptomless for a long time. However, symptomatic lesions can result in significant pain and morbidity and a prospective clinical study has demonstrated that the risk of patients with a cartilage lesion to progress to osteoarthritis is enhanced more than fivefold. Thus, such patients require a treatment filling the gap between palliation and resurfacing via arthroplastia. This need as well as the detailed knowledge we have about the functional elements of cartilage tissue, have turned cartilage repair into a pioneering and very successful area of regenerative medicine.

2.2.1 Therapeutic interventions without active biologics

2.2.1.1 Bone marrow stimulation

First approaches to heal cartilage by *in situ* regeneration date back to 1959. Pridie technique was directed to BM cells recruitment to be used in cartilage defects by drilling small holes into the subchondral BM space underlying the damaged cartilage regions. It was improved later on by reducing the size of the perforations and being then called microfracture technique which is now a frequently performed and well studied procedure (Steadman et al., 1999). This technique is based on the mechanism of mesengenesis or capacity of the non-differentiated mesenchymal cells in choosing a determined phenotype as a response to inducing or GFs. A non-differentiated cell from the BM can be promoted to different cell types such as osteoblasts, with a later maturing to osteocytes, chondroblasts and chondrocytes, but also to endothelial cells, mesothelial cells, fibroblasts or adipocytes. It is a cell signalling process of local cytokines on local cells. In order to achieve all this, the surgical technique is based on drilling the subchondral plate to get bleeding and a superclot that will become a scaffold and supply cells and proteins, starting this way the physiological cascade of the chondrogenic cell differentiation. Other alternative techniques of BM stimulation to regenerate cartilage would be abrasion chondroplasty and in case the articular surface remained untouched, the retrograde stimulation technique. Cartilage defects are repaired only with fibrous tissue or fibrocartilage when using these methods, probably because the number of chondroprogenitors recruited from the BM is too small to promote the hyaline cartilage repair and results are often followed by degeneration of the repair tissue. This was used as an explanation for the observations of other studies that good short term results may be followed by deterioration starting about 18 months after surgery.

Clinical observations and theoretical considerations pointed towards several possible limitations of marrow stimulation techniques. The non-adhesive properties of the cartilage surface and the softness and shrinking of the superclot can lead to only partial defect filling and facilitate an early loss of repair tissue from the cartilage lesion. To avoid this, the treatment has been recently advanced into a matrix-supported technique in which the performed defect was stabilized in an additional way with a biomaterial. The microfractured lesion is covered with a collagen type I/III scaffold and it is called autologous matrix induced chondrogenesis (AMIC) (Kramer et al., 2006; Steinwachs et al., 2008). This technique has been developed to allow the treatment of larger defects by microfracturing and it is used as alternative treatment to autologous chondrocytes transplantation (ACT).

2.2.1.2 Autologous osteochondral transplantation: mosaicplasty

Autologous osteochondral mosaicplasty, sometimes known as osteoarticular transfer system, OATS, is an effective method for the resurfacing of osteochondral defects of the knee. The technique consists in transplantation of many osteocondral autologous plugs obtained from the periphery of the femoral condyle articular surface, which supports less weight and transferring them to create a durable resurfaced area in the defect (Fig. 9). The procedure shows some advantages regarding other repair techniques, such as the viable hyaline cartilage transplantation, a relatively short rehabilitation period and the possibility of carrying out the procedure in one only operation.

However, the OATS limitations are the donor-site morbidity and a limited availability of grafts that can be obtained from the femoropatellar joint or the area adjacent to the intercondylar fossa. Other possible limitations are differences in bearing, thickness and mechanical properties between the donor's and the receiver's cartilages, as well as the graft sinking into the surface due to the support of weight after surgery. Besides, the lack of filling and the possible dead space between cylindrical grafts can limit the repair quality and integrity. Lane et al. transplanted autologous osteochondral grafts into sheep knee joints and reported the lack of integration of the cartilage, which determined the persistence of gaps through the full thickness in all the specimens (Mishima et al., 2008).

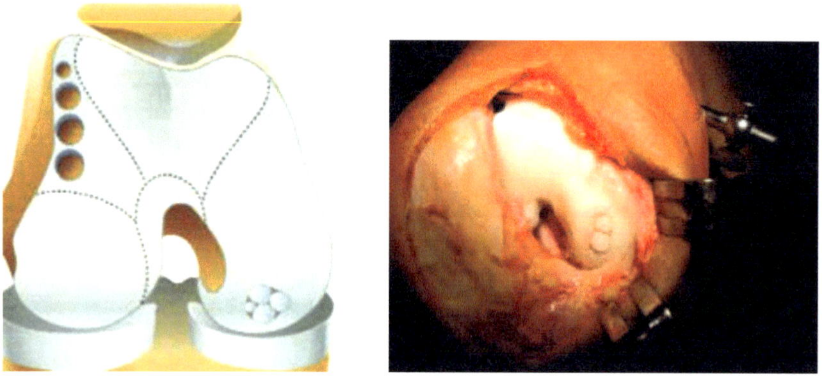

Fig. 9. In mosaicplasty cylindrical osteochondral plugs are harvested from nonload-bearing sites in the affected joint and pressed into place within the osteochondral defect, creating an autograft "mosaic" to fill the lesion

2.2.1.3 Alogenic osteochondral transplantation

Osteochondral allograft transplantation is a procedure for cartilage resurfacing which involves the transplantation into the defect a cadaveric graft composed of viable, intact AC and its underlying subchondral bone. It is a well known resource, especially for tumour surgery. The defect size, its location and its depth are crucial factors for the suitability of the donor graft. Advantages of using osteochondral allografts are the possibility if achieving a precise architecture of the surface, the immediate transplantation of viable hyaline cartilage in a one-time procedure, the possibility to repair large defects, even half-condyles and the donor-site lack of morbidity. Gross et al. have reported results from fresh allografts in 123 patients with good clinical results in 95% of the patients after five years (Gross et al., 2002). There are different possible allografts. Fresh osteochondral allografts are generally used because both freezing and cryopreservation have proved to reduce the chondrocytes viability. Traditionally grafts have been obtained, kept in lactated Ringer's solution at 4 °C and then transplanted in a week. Another alternative for allografts conservation and implantation is cryopreservation, which involves freezing at a controlled speed of specimens within a nutrients rich medium, a cryoprotector agent (glycerol or dymethil sulfoxide), to minimize the cells freezing and keep their viability; finally, there is the possibility of fresh-frozen allografts, with the advantages of lower immunogenic capacity and less transmission of diseases but with lower chondrocyte viability.

2.2.1.4 Soft tissues transplantation

Two main theories support the practicing of covering the cartilage defects with soft tissues, such as perichondrium or periosteum. On one hand, the defect has to be covered mechanically and on the other, we know about the presence of pluripotential stem cells in the perichondrium and the periosteum cambium layer. The different factors able to promote these cells differentiation into active chondrocytes still remain unknown.

2.2.2 Therapeutic interventions with active biologics

2.2.2.1 Autologous chondrocytes implant

The clinical use of the autologous chondrocytes implant (ACI) technique was first reported by Brittberg et al. in 1994, following animal studies which had shown its effectiveness (Grande et al., 1989). In this method, chondrocytes are obtained from a biopsy taken from a non-weight bearing part of the patients cartilage, and are expanded *in vitro*, followed by the injection of a suspension of chondrocytes into cartilage defects, covered with autologous periosteal flap (Fig. 10). This technique premises are based on the capacity of adhesiveness of the cells to certain surfaces, they spread on them and proliferate producing their specific ECM. Although clinical results of the original ACI looked promising (Minas, 2001; Peterson et al., 2000), this procedure has some potential disadvantages, such as leakage of transplanted cells, invasive surgical method, hypertrophy of periosteum (Haddo et al., 2005; Kreuz et al., 2009) and loss of chondrogenic phenotype of expanded chondrocytes in monolayer culture (Benya & Shaffer, 1982). Second generation ACI, named membrane autologous chondrocyte implantation (MACI), has a similar procedure, but a collagen type I and III membrane instead of periosteum. This technique was introduced to improve the ACI problems, and biomaterials such as collagen type I gel (Ochi et al., 2002), hyaluronan-based scaffold (Manfredini et al., 2007) and collagen type I/III membrane (Bartlett et al., 2005) were applied to secure cells in the defect area, to restore chondrogenic phenotype by way of three dimensional cultures (Gigante et al., 2007]) and to replace the periosteum as defect coverage. This is the way MACI technique is created *a posteriori*, by implanting autologous chondrocytes in three dimensional matrices of collagen types I and III, or hyaluronic acid.

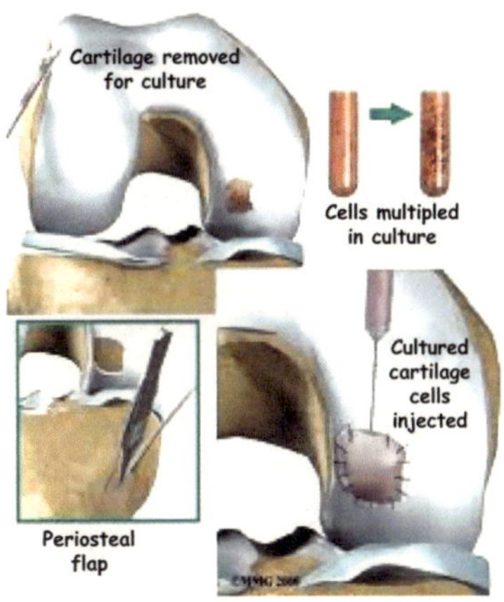

Fig. 10. In autologous chondrocytes implantation (ACI) a chondral biopsy is taken from a donor site at the time of clinical examination to be treated with enzymes in the laboratory to obtain chondrocytes cultures that are re-injected under the periosteal flap

At present, only two prospective studies comparing the original and second generation ACI are available (Bartlett et al., 2005; Manfredini et al., 2007) and both studies show no significant differences in the short term clinical results. As for the first generation ACI, the newly regenerated cartilage often consists of fibrous tissue (Horas et al., 2003; Tins et al., 2005), possibly due to the limited number of chondrocytes and their low proliferation potential. Bone overgrowth that causes thinning of the regenerated cartilage and the violation of the tidemark are also of concern. Moreover, this method still sacrifices healthy cartilage. Thus, these aspects limit ACI in the treatment of large defects and may increase the long-term risk of osteoarthritis development.

2.2.2.2 Mesenchymal stem cells therapy

TE based on cell and genetic therapy offers some of the most promising strategies of tissue repair, including AC repair. It is the science able to create alive tissue to replace, repair or strengthen ill tissue. Thus, the term tissue engineering refers to a wide variety of techniques. Regardless of the used technique, it needs four components (Fig. 2): a growth inducing stimulus (GF), cells that may respond to such influences, a scaffold that might provide a support for the tissue generation, and vascular support (angiogenesis and/or vasculogenesis). This approach involves the use of different cell types acting as chondroprogenitor cells and/or vehicles for supply of genes that synthesize therapeutic proteins. MSCs provide a new potential for cartilage regeneration, as their differentiation to several lineages can be induced, in this case into chondrocytes. Destiny of these cells within tissues is determined by specific cell–to-cell and cell–to-matrix interactions and is controlled by extracellular signalling molecules and their respective receptors and by events that control genetic transcription in a cell-specific way. It has been shown that several differentiation factors are required, such as the BMPs, the FGF and Wnt molecules, although they are not specific for chondrogenesis (Jorgensen et al., 2004). These factors promote the formation of cartilage as well as of bone *in vivo* (Noel et al., 2004), and the precise molecular pathways that rule each specific lineage are being researched.

Beside the characteristics of MSCs exposed before, these cells have self-renewal potential as well as multilineage differentiation potential, including chondrogenesis (Johnstone et al., 1998; Pittenger et al., 1999; Prockop 1997; Sacchetti et al., 2007). MSCs chondrogenesis was first reported by Ashton et al. (1980) and the first ones to describe a defined medium for *in vitro* chondrogenesis of MSCs were Johnstone et al. (1998), who used micromass culture with TGF-β and dexamethasone. Sekiya et al. (2001, 2005) reported that addition to bone BMPs enhanced chondrogenesis under the conditions employed by Johnstone et al. (1998). Nowadays, the micromass culture is widely used to evaluate chondrogenic potential of MSCs *in vitro*. However, this *in vitro* chondrogenesis does not mimic cartilage formation during development. During micromass culture, MSCs increase expressions of both collagen type II (chondrocytes marker) and X (hypertrophic chondrocytes marker) (Barry et al., 2001; Ichinose et al., 2005). Other cytokines such as IGF (Pei et al., 2008) and parathyroid hormone-related peptide (PTHrP) had been tried for better differentiation cocktails, but it is still difficult to obtain *in vitro* MSC-based cartilage formation comparative to native cartilage tissue.

Although BM is considered an acceptable source of MSCs some comparative studies show that MSCs from BM have more chondrogenic potential *in vitro* than those from the adipose tissue (Colter et al. 2001; Huang et al. 2005; Liu et al. 2007; 53, 54). Sakaguchi et al. (47) harvested human BM, synovium, periosteum, muscle and adipose tissue and isolated and expanded MSCs in a similar condition. They demonstrated that MSCs derived from synovium had higher chondrogenic potential than those from other mesenchymal tissues. Yoshimura et al. (71) also demonstrated in a similar way higher chondrogenic differentiation potential of MSCs from synovium in rats. Park et al. (72) showed that MSCs from BM and periosteum are superior to cells isolated from fat with respect to forming hyaline cartilaginous tissue when transplanted into cartilage defects in rats.

To start any regeneration based on MSC activity, first the cells have to be recruited to the damage site. Second step is adhesion to local matrix, followed by activation and extensive proliferation to provide the necessary number of chondroprogenitor cells to build up new tissue. Finally, it is required to switch from expansion to chondrogenic matrix production via chondrogenesis induction, to build up the shock absorbance and gliding characteristics for a proper tissue function. Seamless integration with neighbouring cartilage and bone tissues depends on successful crosstalk between new and old tissues. For durable cartilage repair, the tissue eventually needs to regenerate a tidemark, get adapted to mechanical loading and build up a balanced tissue homeostasis.

Cell migration is a requisite for development from conception to adulthood and plays a major role in regeneration of all tissues. A number of studies demonstrated that chondrocytes migrate under the action of different stimuli, on or within planar and 3D matrices. Attracting factors include BMPs (Frenkel et al., 1996), hepatocyte growth factor (HGF) (Takebayashi et al., 1995), IGF-1 (Chang et al., 2003), TGF-β (Chang et al., 2003), PDGF (Fujita et al., 2004), FGF (Hidaka et al., 2006), fibronectin, fibrin and collagen type I (Maniwa et al., 2001). However, results remain contradictory for some of these factors, as testing of human chondrocytes revealed no effects for BMP-2, BMP-4, BMP-7, IGF and TGF-β in other studies (Mishima &, Lotz, 2008), being IGF-1, PDGF (Fiedler et al., 2004; Ozaki et al., 2007; Ponte et al., 2007) and VEGF the ones that induced higher response (Mishima & Lotz, 2008). Effects of GFs vary *in vitro* and *in vivo*, regarding animal species and specimens' age, among other aspects. But it is generally accepted that GFs stimulate the synthesis of the cartilage ECM components, they inhibit proteases and activate their inhibition systems. Thus, TGF-β illustrates perfectly the complex as well as paradoxical nature of the action of the GFs, which on one hand can increase or decrease the expression and activity of some metalloproteinases (MMP) produced by articular chondrocytes, increasing MMP-9 and reducing MMP-2. Another anabolic factor for chondrocytes is IGF-I, which can keep chondrocytes phenotype *in vitro*, stimulate intensely the synthesis of proteoglycans and type II collagen, as well as block the harmful effects induced by IL1 on the proteoglycans degradation. On the other hand, both IGF-1 and TGF-β stimulate the expression on the cell surface of the subunits of integrins a3- and a5-, as well as adhesion of chondrocytes to fibronectin and collagen type II, being integrins, accepted as the main receptors, the molecules used by the cells (chondrocytes, in cartilaginous tissue case) to adhere to ECM. A new GFs family has been reported, Wnt, with a major role in chondrocyte differentiation. Wnt family members are important regulators of several development processes, including skeletogenesis. After the binding of Wnt to the Frizzled receptors family and the LRP5/6 co-receptors family, the canonical Wnt pathway will stabilize the β-catenin, which translocates to the nucleus and interacts with members of the TCF/LEF (β-catenin-T-cell factor/lymphoid enhancer factor) families to activate target genes. Whereas inactivation of β-catenin causes ectopic formation of chondrocytes at the expense of osteoblasts formation, the canonical Wnt pathway leads to enhanced ossification and suppression of chondrocytes due to the transcriptional downregulation of Sox9 (Boyden et al., 2002; Guo et al., 2004). In fact, Church and colleagues have shown that Wnt4 blocks the chondrogenesis start and accelerates the terminal chondrocyte differentiation *in vitro*, while Wnt5a and Wnt5b promote early chondrogenesis and inhibit terminal differentiation *in vivo* (Church et al., 2002). On the other hand, Wnt7 blocks chondrogenesis (Tufan et al., 2002). These studies illustrate that Wnt/β-catenin signalling play an essential role on MSCs by controlling osteoblastic and chondrocytic differentiations.

Once the process of activation and attraction for the tissue repair has been started, we have just to review MSCs proliferation or expansion. Requisites for human MSCs growth are

different than those for other species and several factors have been identified as potent mitogens, taking PDGF-BB, EGF and TGF-β as the most important ones, leading jointly cell migration and proliferation steps. MSCs have been expanded with foetal bovine serum (FBS), supplemented with GF, for research as well as for clinical use; this supplementation with FBS has several risks such as diseases transmission or immune reactions, which promoted the development of research lines with autologous human serum as substitute for FBS. Human platelet lysate (hPL) has been also proposed as a substitute for culture and expansion of MSCs, and some reports showed that there is a higher proliferation in hPL regarding FBS, keeping their differentiation potential including chondrogenesis.
Finally, as one of the basic columns of the tissue repair we have to highlight the importance of scaffolds, because treatment with MSCs requires cells and scaffold transplantation. Importance of such structures is based on their ability for a fast filling of defects, their raised persistence to carry out repair, they also achieve an uniform distribution of cells in an enhance volume and also because they provide an active environment that may allow the local delivery of local molecules to stimulate repair (Andrades et al., 2010). Many studies have reported that using several scaffolds improve the new tissue quality. Such scaffolds are based mainly on hyaluronic acid, polylactic acid and/or polyglicolic acid, which can help to keep the cells inside the defect and provide a chondroinductive matrix. Natural materials, such as agarose, alginate, gelatine and collagen derivatives are inferior than synthetic and hybrid materials, due to a poor resistance to mechanical stress, so their clinical usefulness is severely limited. We have to take on account that MSCs chondrogenic potential is favoured by hypoxia and that it does not only depend on hydrostatic pressure, but also on cell density within the matrix, as well as on the presence of GFs. It is suggested by the obtained information that scaffolds or matrices allow the cell differentiation and the maintenance of a mature phenotype (Fig. 11); this combined with the use of stem cells, provides us promising perspectives for the regeneration into a functional tissue (Becerra et al., 2010; Reddi et al., 2011).

Fig. 11. Scanning electron microscopy of natural alginate wire mesh containing chondrocytes

From the knowledge of these biological phenomena of in situ repair, adequate conditions for cartilage repair based on MSCs are considered. Although some studies demonstrate that non-differentiated MSCs transplantation in cartilage defects, without scaffolds and not having been chondroinduced provided good, it has been reported that the before mentioned cytokines and particularly the association of TGF-β and BMPs improve the cartilage repair when combined with MSCs. Such stimuli could be obtained by direct administration of recombining GFs in the culture media or via transfer of the respective genes. Thus, the possibility of considering genetic therapy as an applicable measure for the treatment of cartilaginous lesions arises.

2.3 Approaches for tendon regeneration

Tendons make part of a biomechanical chain formed by muscle, bones and tendon itself. Muscle contraction is converted by tendons into bone motion or viceversa; thus, tendon's main function is the transmission of force. Far from being a passive element, thanks to its composition and structure tendon has an adaptive, dynamic behaviour to face special functional requirements.

2.3.1 Tendon composition, structure and mechanical properties

Tendon is a dense connective tissue. Having to withstand basically tensile stresses, tendons are bundles of fibrillar proteins aligned parallelly to the tendon's longitudinal axis, with small amounts of other ECM components, specialized cells (tenocytes, fibroblasts), vessels and neuroreceptors. Between a 75% and a 96% of a tendon's dry weight is constituted by collagen, around a 2% by elastin. These proteins are imbibed in a mucopolysaccharide gel, the ground substance, formed mainly by hyaluronic acid and chondroitin sulfate, proteoglycans (decorin) and glycoproteins (fibronectin), which may account for a 1% of the tendon's dry weight (Crisp, 1972; Hooley, 1977). Collagen comes mainly as type-I collagen, with its characteristic hierarchical organization (Fig. 12) so important to understand the tendon's overall mechanical behaviour (Kastelic et al., 1978; Kastelic & Baer, 1980; Baer et al., 1975; Silver et al., 2003; Wang 2006): tropocollagen molecules assemble into microfibrils, these microfibrils into subfibrils, and several subfibrils give rise to a collagen fibril, with a characteristic 65 nm periodicity visible in scanning electron microscopy. Assemblies of

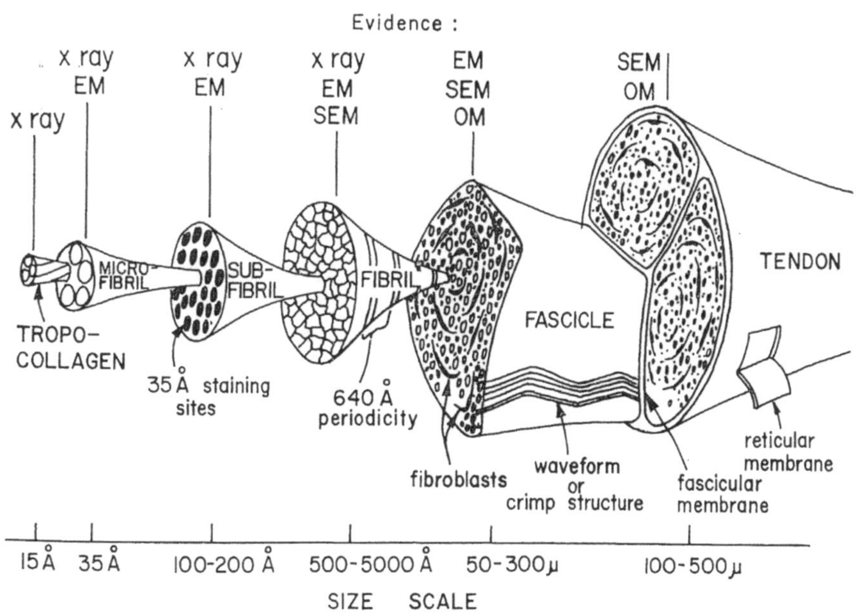

Fig. 12. Type I collagen hierarchical organization in tendon. *Modified from Kastelic J et al. The multicomposite structure of tendon. Connective Tissue Research 1978; 6:11-23*

fibrils bound by a thin layer of connective tissue (the endotenon) form typically crimped planar ribbons, with a periodicity of the order of a hundred microns, visible for the optical microscope. Packages of these ribbons surrounded by the epitenon (a sheath of loose connective tissue) constitute one tendinous bundle or fascicle; several fascicles may constitute larger units, and a tendon may consist of several of these units surrounded, again, by a layer of connective tissue (paratenon). Blood and lymphatic vessels, nerves and cells reside in the endo-, epi- and paratenon. Elastin, as said, is much less abundant in tendon than collagen, and is believed to help recover the original configuration of collagen fibres after cessation of applied stresses thanks to its rubberlike elasticity. The ground substance is a highly hydrated gel that allows for the diffusion of nutrients, signals and metabolytes to and from the cellular component of the tendon, contributing also to its mechanical behaviour as it is sheared during fascicle extension, (Minns et al., 1973; Kastelic & Baer, 1980).

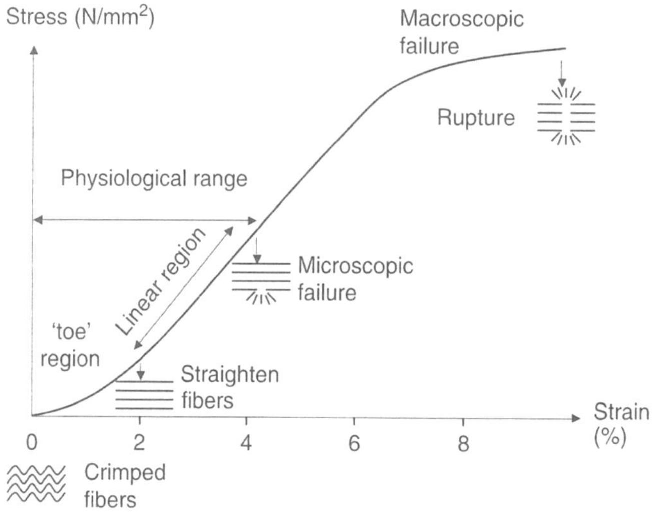

Fig. 13. Mechanism of internal deformation of tendon. *Modified from Wang JH. Mechanobiology of tendon. Journal of Biomechanics 2006; 39:1563-1582*

The crimped geometry of the parallelly aligned collagen-I fibre bundles and the properties of the highly crystalline collagen fibrils are the main responsibles for the tendon's overall mechanical behaviour. The main feature of this behaviour is its nonlinear viscoelasticity, 'nonlinear' meaning that the tendon's stiffness increases as it is progressively stretched, and 'viscoelastic' meaning that the mechanical response of the tissue to an external stimulus does not adjust instantaneously but lags in time, giving rise to the phenomena of stress relaxation and creep (Hooley et al., 1978; Monleón & Díaz, 1990). Thanks to both characteristics tendon is able to damp sudden stresses transmitted to muscles and to achieve a progressive adaptation of the whole biomechanical chain to varying mechanical stimuli. The origin of nonlinearity lies in the mechanism of internal deformation of tendon: the undulation ('crimping') of the collagen ribbons is not uniform, but its angles with the longitudinal axis of the tendon are distributed; thus, as the tendon is progressively deformed a varying number of ribbons first straightens out, and then become stretched (Kastelic & Baer, 1980; Monleón & Díaz, 1990). This gives rise to a typical nonlinear, convex stress-strain relationship, with an initial ('toe') region of low modulus extending up to some

2-3 % of strain, and a subsequent ('linear') high-modulus part, which corresponds to a state in which most of the tendon fascicles have lost their rest wavy crimp and are bearing load (Fig. 13). Physiological regime of tendons lies in the strain interval up to 5-7 %, but this figure may vary considerably depending on age, weight, tendon type and other factors.

2.3.2 Cell response to mechanical stimuli

Even though tendon is a very poorly cellularized tissue, its cells are very important during development and healing, and thus a consideration of their role is critical for cell therapy purposes. Endothelial cells and synovial cells are present in the peritendon, but specialized fibroblasts (tenoblasts and tenocytes) are the dominant cell type. Tendon fibroblasts align in rows between collagen fibre bundles, and are responsible for synthesizing the ECM proteins, organizing them into their characteristic structures, and remodelling the ECM during tendon healing. These processes occur as a consequence of the biochemical and mechanical stimuli transmitted to the cells by their ECM. Tendon cells respond to mechanical forces by altering gene expression, protein synthesis, and cell phenotype. Mechanotransduction (the conversion of mechanical stimuli into biochemical signals leading to cell response) is a not entirely understood complex process which involves the interaction of proteins from the ECM, transmembrane integrins and G proteins, and cytoskeleton proteins such as actin. This interaction triggers signalling pathways that lead to differential gene expression and protein synthesis by the cells (Wang, 2006). Increased ECM protein production occurs as a consequence of GF release stimulated by mechanical loading (Skutek et al., 2001; Kim et al., 2002); matrix remodelling is also influenced by mechanical stimuli, which enhance the secretion of matrix metalloproteinases (Archambault et al., 2002; Tsuzaki et al. 2003). Different mechanical loading conditions lead to varying proportions of ECM proteins produced.

2.3.3 Lesions, standard approaches and challenges posed to cell therapy

Weakly injured tendons tend to heal spontaneously, but the remodelled tissue usually has poorer mechanical properties than the original one. More extensive trauma and degenerative processes, tumors or congenital malformations may require different degrees of surgery, including transplantation. Autologous grafts are in these cases the golden standard, but they may result in donor site morbidity and there may be lack of sufficient supply; artificial prostheses have not until today stood up to the requirements to represent an alternative. Cell therapy and engineered tendon constructs would be of utility in these cases (Thorfinn et al., 2010). The pluripotency discovered in several cell types, embryonic and adult, opens the way to their differentiation into tenocytes and their transplantation into injured or defective sites. A pure cell supply strategy, however, may be appropriate to small defects and injuries, but does not seem to be sufficient in cases of extensive tissue loss or rupture; here cell supply will presumably have to be combined into an engineered tendon construct with biomaterials acting as supporting or guiding structures and maybe GFs enhancing matrix production, remodelling and neovascularization. It is a challenge for this strategy to identify suitable candidates for a scaffold material and architecture, for the cell type to be supplied, for the GFs, their concentration and their release kinetics, and, maybe, for a cell training strategy through convenient *ex vivo* stimuli in bioreactors.

2.3.4 Tissue engineering and cell therapy of tendons

The paradigm of the TE approach is a biodegradable scaffold seeded with cells and bioactive factors, such that the scaffold is progressively replaced by neotissue formed *in situ*, hystologically and functionally normal. The scaffold must initially bear the full mechanical load, but this must soon be gradually transferred to the ingrowing tissue to ensure the mechanical stimuli on the cells necessary for them to differentiate and produce and organize the ECM. A part of this process may take place *ex vivo*, in bioreactors. *In vitro* and *in vivo* animal studies have been undertaken along lines aiming to understand the relative influence and cross-talk of all these factors: *(i)* scaffolds, *(ii)* cell types supplied, *(iii)* biochemical signals, and *(iv)* mechanical stimulation.

i. Three dimensional environments for cells have proved to be necessary to achieve *in vitro* significant tendon-like matrix production (Garvin et al., 2003), and such 3D scaffolds are also likely to make part of any successful *in vivo* strategy. Synthetic materials available for scaffold fabrication belong to the standard families of polyesthers (polylactic and polyglycolic acid and their copolymers, polycaprolactone), and they have been assayed with different morphologies, including knitted and electrospun versions (Ouyang et al., 2003; Guo & Spector, 2006; Thorfinn et al., 2010; Ladd et al., 2011; Jiang et al., 2011). None of these synthetic materials, however, possesses mechanical properties approaching those of natural tendon; the less so when they have a porous or microfibrilar structure. Moreover, the effects of their degradation process have to be taken into account, since they imply local acidification. This is why many researchers show preference for different collagen-based gels, where cells are seeded and reside as in a 3D matrix, and can be tested *in vitro* (Garvin et al., 2003; Kuo & Tuan, 2008); these gels could also serve as vector (transport, localization) for the supply of cells *in vivo*. However, these collagen gels are even less resistant than the synthetic polymers, and thus their utility *in vivo* has to be limited to those cases involving only limited injuries and defects, and where more complex regeneration and extensive reconstruction is not required. Nonetheless, systems based on these gels could be the solution also for these latter cases if they could be engineered *in vitro* to develop a high modulus construct which would be then transplanted, and this is the hope of much current work. All in all, scaffold choice, its composition and structure, remains an open problem for tendon TE.
ii. The cells to be employed in tendon regeneration should be capable of producing and organizing into its functional structure the ECM of tendons; they may be adult or embryonic, differentiated or pluripotent (stem) cells, autologous or not. For obvious reasons, adult autologous cells seem those more probable to enter a clinical solution; and, since BM MSCs and adipose tissue-derived stem cells (ADSC) have been successfully differentiated to functional tenocyte-like cells (Altman et al., 2001; Hoffmann et al., 2006; Juncosa-Melvin et al., 2007; Lee et al., 2007), it is they who are receiving most attention. Bone MSCs, in particular, are cells likely to enter wounded areas, and they have demonstrated a paracrine effect on tendon fibroblasts, secreting factors which enhance cell migration and proliferation (Shimode et al., 2007). Significant results have been obtained with ADSC *in vivo* (de Matos Carvalho et al., 2011). However, the sole transplantation of MSC into tendon defects does not improve significantly the tendon's microstructure, and the cells should be induced into the tenocytic linage by specific molecules (Araque et al., 2011).
iii. Since a natural regeneration accompanies healing processes, an understanding of natural tendon healing and the cascade of signals involved in it is crucial for the design of cell therapy strategies. A number of GFs that play an important role during the inflammatory, repairing and remodelling stages of the healing process has been identified, the most important ones being IGF-1, PDGF, VEGF, bFGF and some of the

TGF-β family (Molloy et al., 2003). Specific molecules which selectively induce the tenocytic differentiation of MSC while supressing their osteogenic and chondrogenic differentiation have been identified (Hoffmann et al., 2006); they may provide a powerful platform for successful tendon regeneration when combined with mechanical stimulation within scaffolds.

iv. Though MSC have been made to express tendon-like phenotype and produce aligned collagen-I, mechanical properties of the neotissue produced could not match those of native tendon unless the cells had been subjected to mechanical stimulation (Juncosa-Melvin et al., 2006; Benhardt & Cosgriff-Hernández, 2009). Several types of bioreactors have been developed which allow to apply load uniaxially, in a static or dynamic (cyclical) mode, to seeded constructs (Guo & Spector, 2006; Garvin et al., 2003; Abousleiman et al., 2009). With their help it has been possible to check hypotheses as to the importance of force magnitude, strain amplitude, time frequency of loading, loading duration, and synergisms with GFs on the proliferation, differentiation and ECM production of cells. An excessive mechanical conditioning has been shown to lead to tendon disorders and matrix degeneration, however; thus, optimal values of those mechanical parameters must be determined (Wang, 2006). Generally, cyclic stress application has resulted in better results than static stresses as regards matrix formation and fibre alignment (Kuo & Tuan, 2008; Benhardt & Cosgriff-Hernández, 2009; Nguyen et al., 2009).

Taking into account those concepts, we are currently investigating the potential of human MSCs to differentiate into functional tenocytes. This cell application is directed to heal tendon and ligament injuries using either biosutures (Fig. 14) or poly-l-lactic acid (PLLA) scaffolds microparticles (Fig. 15) that could represent adequate surfaces for MSCs adhesion, proliferation and differentiation to tenocytes.

3. Conclusion

In the last ten years we have published reports suggesting that MSCs have a promising potential to be utilized for regenerative medicine. For example, our laboratory has accumulated experience working *in vitro* with blood-, fat-, skeletal muscle-, BM-, and now synovial-derived MSCs, performing ectopic implants (in mouse and rat), preclinical trial with small animals (rabbit, using an osteochondral defect model), with big animals (sheep, using a spinal fusion model), and with a pilot clinical trial in a patient affected by osteomyelitis. Cells to be used for an efficient regenerative medicine should be chosen. Undifferentiated versus differentiated or predifferentiated cells will be the choice. A permanent solution will come when the new tissue built in the defect is of the same nature and is perfectly integrated in the whole structure in any pathology and in any age. Only in such as way structure and function will be fully recovered.

When regenerative medicine started several years ago, the main goal was for the implanted cells to directly participate in the reconstruction of the damaged tissue. But now, after the reported paracrine effects in several MSCs therapies we know that MSCs, far from building those tissues, they exert immunomodulatory functions, secreting, in addition, several bioactive molecules that inhibit apoptosis and scarring at sites of injury, and stimulate angiogenesis and mitosis of tissue-specific progenitors (Caplan, 2009; Ito et al., 2010; Park et al., 2010). These actions have been found either when the implanted MSCs coming from BM or adipose derived in adherent cell cultures (García-Olmo et al., 2010) or when the mononuclear fraction of BM was infused. We cannot predict the extent of the paracrine effect, immunomodulatory, or if the effective replenishment of differentiated cells can be assigned, in each case, to the MSCs, or whether these effects have some degree of integration between them. We cannot even know if all those effects can influence the surrounding tissue positively during regeneration, but perhaps negatively towards the pathogenesis of cancer and metastasis (Kuhn & Tuan, 2010). For now, we can only say that in many cases, these actions have a certain synergy to the purpose they claim. In the future, in might be useful to know the responsibility of each action in the regenerative process in order to control it appropriately. All this indicates the necessity to highlight again the importance of a tight control over the stem cell culture method in order to define the cell products for transplantation properly, according to the specific functional outcome sought.

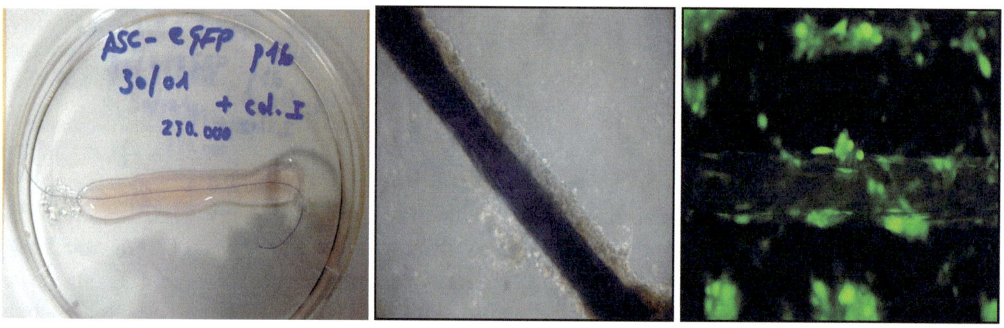

Fig. 14. Fresh collagen type I containing human MSCs embedding a resorbable polyglactic suture in vitro. After 30 min. of incubation at 37 °C, the collagen gelify and cells oriented around and along the suture line, as we can see under a fluorescence microscope (grean: hMSCs transfected with the green fluorescent protein)

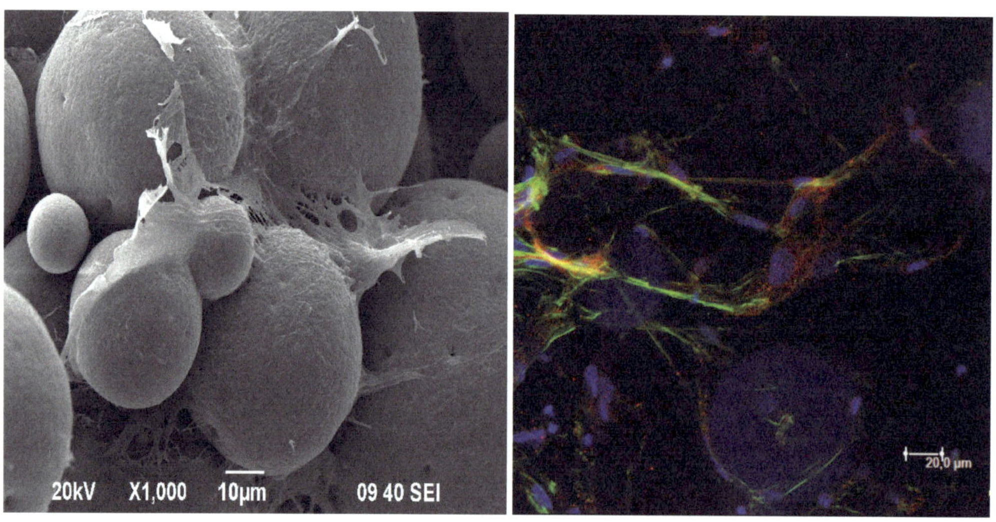

Fig. 15. Scanning electron microscopy of PLLA microparticles containing human MSCs attached to their surfaces after 14 days in culture, and their phenotypic features by confocal microscopy (blue: cell nuclei, grean: actin cytoskeleton, red: collagen type I)

4. Acknowledgments

Laboratory of Bioengineering and Tissue Regeneration-University of Málaga (LABRET-UMA) is supported by grants from the Spanish (BIO2009-13903-C02-01, PLE2009-0163, PI10/02529, and Red de Terapia Celular, RD06/00100014), and the Andalusian Governments (P07-CVI-2781, PI-0729-2010, and PAID BIO217). CIBER-BBN is an initiative funded by the VI National R&D&I Plan 2008-2011, *Iniciativa Ingenio 2010, Consolider Program, CIBER Actions* and financed by the Instituto de Salud Carlos III with assistance from the *European Regional Development Fund.*

5. References

Abousleiman, R.I.; Reyes, Y.; McFetridge, P. & Sikavitsas, V. (2009). Tendon tissue engineering using cell-seeded umbilical veins cultured in a mechanical stimulator. *Tissue Engineering,* 15, 787-795.

Alexanian, A.R.; Kwok, W.M.; Pravdic, D.; Maiman, D.J. & Fehlings, M.G. (2010). Survival of neurally induced mesenchymal cells may determine degree of motor recovery in injured spinal cord rats. Restoration Neurology and Neuroscience, 28(6):761-767.

Alonso, M.; Claros, S.; Becerra, J. & Andrades, J.A. (2008). The effect of type I collagen on osteochondrogenic differentiation in adipose-derived stromal cells in vivo. *Cytotherapy,* 10(6), 597-610.

Altman, G.H.; Horan, R.L.; Martin, I.; Farhadi, J.; Stark, P.R.H.; Volloch, V.; Richmond, J.C.; Vunjak-Novakovic, G. & Kaplan, D.L. (2001). Cell differentiation by mechanical stress. *FASEB J,* 16, 270-272.

Andrades, J.A.; Han, B.; Nimni, M.E.; Ertl, D.C.; Simpkins, R.J.; Arrabal, M.P. & Becerra, J. (2003). A modified rhTGF-beta1 and rhBMP-2 are effective in initiating a chondro-osseous differentiation pathway in bone marrow cells cultured in vitro. *Connective Tissue Research,* 44(3-4), 188-197.

Andrades, J.A.; López-Puertas, J.M.; Cuenca-López, M.D.; Jiménez-Palomo, P. & Becerra, J. (2010). Mesenchymal stem cells and a composed membrane for osteocondral wound treatment. *Patent application* nº 201031016, 29 pages.

Araque, M.C.; Claros, S.; Jiménez-Palomo, P.; Becerra, J.; Monleón, M. & Andrades, J.A. (2011). Differentiation of human mesenchymal stem cells into tenogenic lineage on different designs of microparticles. *Journal of Orthopaedic Research,* in press.

Arrington, E.D.; Smith, W.J.; Chambers, H.G., et al. (1996). Complications of iliac crest bone harvesting. *Clinical Orthopaedic Related Research,* 329, 300-9.

Archambault, J.; Tsuzaki, M.; Herzog, W. & Banes, A.J. (2002). Stretch and interleukin-1beta induce matrix metalloproteinases in rabbit tendon cells in vitro. *Journal of Orthopaedic Research,* 20, 36–39.

Aslan, H.; Zilberman, Y.; Arbeli, V.; Sheyn, D.; Matan, Y.; Liebergall, M. et al. (2006). Nucleofection-based ex vivo nonviral gene delivery to human stem cells as a plataform for tissue regeneration. *Tissue Engineering Part A,* 12(4), 877-889.

Azizi, S.A.; Stokes, D.; Augelli, B.J.; DiGirolamo, C. & Prockop, D.J. (1998). Engraftment and migration of human bone marrow stromal cells implanted in the brains of albino rats-similarities to astrocyte grafts. *Proceedings of National Academic of Sciences USA,* 95(7), 3908-3913.

Baer, E.; Gathercole, L. & Keller, A. (1975). Structure hierarchies in tendon collagen: an interim summary. In: Atkins, E.D.T. & Keller, A. (eds.). Structure of fibrous biopolymers. Butterworth, London.

Baksh, D. & Tuan, R.S. (2007). Canonical and non-canonical Wnts differentially affect the development potential of primary isolate of human bone marrow mesenchymal stem cells. *Journal of Cell Physiology,* 212(3), 817-826.

Banwart, J.C.; Asher, M.A. & Hassanein, R.S. (1995). Iliac crest bone graft harvest donor site morbidity. A statistical evaluation. *Spine,* 20, 1055-60.

Barry, F.; Boynton, R.E.; Liu, B. & Murphy, J.M. (2001). Chondrogenic differentiation of mesenchymal stem cells from bone marrow: differentiation-dependent gene expression of matrix components. *Experimental Cell Research,* 268, 189–200.

Bartlett, W.; Skinner, J.A.; Gooding, C.R.; Carrington, R.W.; Flanagan, A.M.; Briggs, T.W. & Bentley, G. (2005). Autologous chondrocyte implantation versus matrix-induced autologous chondrocyte implantation for osteochondral defects of the knee: a prospective, randomised study. *Journal of Bone and Joint Surgery Brithis,* 87, 640–645.

Becerra J.; Andrades J.A.; Guerado, E.; Zamora-Navas, P.; López-Puertas, J.M. & Reddi, A.H. (2010). Articular cartilage: structure and regeneration. *Tissue Engineering Part B Review,* 16(6), 617-627.

Becerra J.; Santos-Ruiz, L.; Andrades, J.A. & Marí-Beffa, M. (2011). The stem cell niche should be a key issue for cell therapy in regenerative medicine. *Stem Cell Review,* 7, 248-255

Behr, B.; Leucht, P.; Longaker, M.T. & Quarto, N. (2010). Fgf-9 is required for angiogenesis and osteogenesis in long bone repair. *Proceedings of the National Academic of Sciences USA,* 107(26):11853-11858.

Belthur, M.V.; Conway, J.D.; Jindal, G., et al. (2008). Bone graft harvest using a new intramedullary system. *Clinical Orthopaedic and Related Research,* 466, 2973-80.

Bendall, S.C.; Stewart, M.H.; Menendez, P.; George, D.; Vijayaragavan, K.; Werbowetski-Ogilvie, T.; Ramos-Mejia, V.; Rouleau, A.; Yang, J.; Bossé, M.; Lajoie, G. & Bhatia, M. (2007). IGF and FGF cooperatively establish the regulatory stem cell niche of pluripotent human cells in vitro. *Nature,* 448(7157), 1015-1021.

Benhardt, H.A. & Cosgriff-Hernández, E.M. (2009). The role of mechanical loading in ligament tissue engineering. *Tissue Engineering,* 15, 467-475.

Benya, P.D. & Shaffer, J.D. (1982). Dedifferentiated chondrocytes reexpress the differentiated collagen phenotype when cultured in agarose gels. *Cell,* 30, 215–224.

Boatright, K.C. & Boden, S.D. (2002). Biologic enhacement of spinal arthrodesis: past, present, and future. In: Fardon, D.F.; Grafin, S.R.; Abitbol, J.J.; Boden, S.D.; Herkowitz, H.N. & Mayer, T.G. Orthopaedic Knowledge Update, *Spine,* 2, 459-68.

Boyden, L.M.; Mao, J.; Belsky, J. *et al.* (2002). High bone density due to a mutation in LDL-receptor-related protein 5. *New England Journal of Medicine,* 346, 1513–1521.

Boland, G.M.; Perkins, G.; Hall, D.J. & Tuan, R.S. (2004). Wnt 3a promotes proliferation and suppresses osteogenic differentiation of adult human mesenchymal stem cells. *Journal of Cell Biochemistry,* 93(6), 1210-1230.

Boquest, A.C.; Shahdadfar, A.; Fronsdal, K.; Sigurjonsson, O.; Tunheim, S.H.; Collas, P. et al. (2005). Isolation and transcription profiling of purified uncultured human stromal stem cells: alteration of gene expression after in vitro culture. *Molecular Biology and Cell,* 16(3), 1131-1141.

Bostrom, M.P. & Seigerman, D.A. (2005). The clinical use of allografts, demineralized bone matrices, synthetic bone graft substitutes and osteoinductive growth factors: a survey study. *Hospital for Surgery Survery Journal,* 1, 9-18.
Brittberg, M.; Lindahl, A.; Nilsson, A.; Ohlsson, C.; Isaksson, O. & Peterson, L. (1994). Treatment of deep cartilage defects in the knee with autologous chondrocyte transplantation. *New England Journal of Medicine,* 331, 889–895.
Cao, R.; Bråkenhielm, E.; Pawliuk, R.; Wariaro, D.; Post, M.J.; Wahlberg, E.; Leboulch, P. & Cao, Y. (2003). Angiogenic synergism, vascular stability and improvement of hind-limb ischemia by a combination of PDGF-BB and FGF-2. *Nature Medicine,* 9(5), 604-613.
Caplan, A.I. (1991). Mesenchymal stem cells. *Journal Orthopaedic Research,* 9, 641-50.
Caplan, A.I. (2009). Why are MSCs therapeutic? New data: new insight. *Journal of Pathology,* 217, 318-324.
Caplan, A.I. & Dennis, J.E. (2006). Mesenchymal stem cells as trophic mediators. *Journal of Cell Biochemistry,* 98(5), 1076-1084.
Chai, C. & Leong, KW. (2007). Biomaterials approach to expand and direct differentiation of stem cells. *Molecular Therapy,* 15, 467-480.
Chang, C.; Lauffenburger, D.A. & Morales, T.I. (2003). Motile chondrocytes from newborn calf: migration properties and synthesis of collagen II. *Osteoarthritis and Cartilage,* 11, 603–12.
Chiou, M.; Xu, Y. & Longaker, M.T. (2006). Mitogenic and chondrogenic effects of fibroblast growth factor-2 in adipose-derived mesenchymal cells. *Biochemical and Biophysic Research Communication,* 343(2), 644-652.
Church, V.; Nohno, T.; Linker, C. *et al.* (2002). Wnt regulation of chondrocyte differentiation. *Journal of Cell Sciences,* 115, 4809–4818.
Claros, S.; Alonso, M.; Becerra, J. & Andrades, J.A. (2008). Selection and induction of rat skeletal muscle-derived cells to the chondro-osteogenic lineage. *Cell and Molecular Biology,* 54(1), 1-10.
Colter, D.C.; Sekiya, I. & Prockop, D.J. (2001). Identification of a subpopulation of rapidly self-renewing and multypotential adult stem cells in colonies of human marrow stromal cells. *Proceedings of National Academic of Sciences USA,* 98, 7841–7845.
Crigler, L.; Robey, R.C.; Asawachaicharn, A.; Gaupp, D. & Phinney, D.G. (2006). Human mesenchymal stem cell subpopulations express a variety of neuro-regulatory molecules and promote neuronal cell survival and neuritogenesis. *Experimental of Neurology,* 198(1), 54-64.
Crisan, M.; Yap, S.; Casteilla, L., et al. (2008). A perivascular origin for mesenchymal stem cells in multiple human organs. *Cell Stem Cell,* 3, 301-13.
Crisp, J. (1972). Properties of tendon and skin. In: Fung, Y.C.; Perrone, N. & Anliker, M. (eds.). Biomechanics: its foundations and objectives, pp. 141-171. Prentice-Hall, Englewood Cliffs, New Jersey.
Cuenca-López, M.D.; Zamora-Navas, P.; García-Herrera, J.M.; Godino, M.; López-Puertas, J.M.; Guerado, E.; Becerra, J. & Andrades, J.A. (2008). Adult stem cells applied to tissue engineering and regenerative medicine. *Cell and Molecular Biology,* 54(1), 40-51.
Da Silva Meirelles, L.; Caplan, A.I. & Nardi, N.B. (2008). In search of the in vivo identity of mesenchymal stem cells. *Stem Cells,* 26, 2287-99.
Daley, G.Q. & Scadden, D.T. (2008). Prospect for stem cell-based therapy. *Cell,* 132, 544-8.

Day, S.; Ostrum, R.; Clinton, R.; et al. (2000). Bone injury, regeneration, and repair. In: *Biology and Biomechanics of Musculoskeletal System*. Buckwalter, J.; Einhorn, T. & Simon, S.; (Eds.). Chicago, Rosemont IL: American Academy of Orthopaedic Surgeons 388-75.

Deckers, M.M.; Karperien, M.; van der Bent, C.; Yamashita, T.; Papapoulos, S.E. & Löwik, C.W. (2000). Expression of vascular endothelial growth factors and their receptors during osteoblast differentiation. *Endocrinology*, 141(5), 1667-1674.

Delawi, D., Dhert, W.J.; Rillardon, L. et al. (2010). A prospective, randomized, controlled, multicenter study of osteogenic protein-1 in instrumented posterolateral fusions: report on safety and feasibility. *Spine*, 35, 1185-91.

De Mattos Carvalho, A.; Garcia-Alves, A.L.; Galvão Gomes de Oliveira, P.; Cisneros Álvarez, L.E.; Laufer-Amorim, R.; Hussni, C.A. & Deffune, E. (2011). Use of adipose tissue-derived mesenchymal stem cells for experimental tendinitis therapy in equines. *Journal of Equine Veterinary Science*, 31, 26-34.

Dominici, M.; le Blanc, K.; Mueller, I.; Slaper-Cortenbach, I.; Marini, F.; Krause, D. et al. (2006). Minimal criteria for defining multipotent mesenchymal stromal cells. The International Society for Cellular Therapy position statement. *Cytotherapy*, 8(4), 315-317.

Enneking, W.F.; Burchardt, H.; Puhl, J.J., et al. (1975). Physical and biological aspects of repairs in dog cortical bone transplants. *Journal of Bone and Joint Surgery American*, 57, 237-52.

Epstein, N.E. (2008). Efficacy of different bone volume expanders for augmenting lumbar fusions. *Surgical Neurology*, 69, 16-9.

Etheridge, S.L.; Spencer, G.J.; Heath, D.J. & Genever, P.G. (2004). Expression profiling and functional analysis of wnt signaling mechanisms in mesenchymal stem cells. *Stem Cells*, 22(5), 849-860.

Fiedler, J.; Etzel, N. & Brenner, R.E. (2004). To go or not to go: migration of human mesenchymal progenitor cells stimulated by isoforms of PDGF. *Journal of Cell and Biochemistry*, 93, 990–8.

Fiedler, J.; Leucht, F.; Waltenberger, J.; Dehio, C. & Brenner, R.E. (2005). VEGF-A and PIGF-1 stimulate chemotactic migration of human mesenchymal progenitor cells. *Biochemical and Biophysic Research Communication*, 334(2), 561-568.

Frenkel, S.R.; Clancy, R.M.; Ricci, J.L.; Di Cesare, P.E.; Rediske, J.J. & Abramson, S.B. (1996). Effects of nitric oxide on chondrocyte migration, adhesion, and cytoskeletal assembly. *Arthritis and Rheumatism*, 39, 1905–12.

Friedenstein, A.J.; Chailakhyan, R.K. & Gerasimov, U.V. (1987). Bone marrow osteogenic stem cells: in vitro cultivation and transplantation in diffusion chambers. *Cell Tissue Kinetic*, 20(3), 263-272.

Friedenstein, A.J.; Latzinik, N.W.; Grosheva, A.G. & Gorskaya, U.F. (1982). Marrow microenvironment transfer by heterotopic transplantation of freshly isolated and cultured cells in porous sponges. *Experimental Hematology*, 10(2), 217-227.

Fuchs, E.; Tumbar, T. & Guasch, G. (2004). Socializing with the neighbors: stem cell and their niche. *Cell*, 116, 769-778.

Fujita, T.; Azuma, Y.; Fukuyama, R. et al. (2004). Runx2 induces osteoblast and chondrocyte differentiation and enhances their migration by coupling with PI3K-Akt signaling. *Journal of Cell Biology*, 166, 85–95.

García-Olmo, D.; Herreros, D.; De-La-Quintana, P. et al. (2010). Adipose-derived stem cells in Crohn's rectovaginal fistula. *Case Reports in Medicine,* 961758.

Garvin, J.; Qi, J.; Maloney, M. & Banes, A.J. (2003). Novel system for engineering bioartificial tendons and application of mechanical load. *Tissue Engineering,* 9, 967-979.

Gerstenfeld, L.C.; Cullinane, D.M.; Barnes, G.L.; Graves, D.T. & Einhorn, T.A. (2003). Fracture healing as a post-natal developmental process: molecular, spatial, and temporal aspects of its regulation. *Journal of Cell Biochemistry,* 88(5), 873-884.

Giannoudis, P.V.; Einhorn, T.A. & Marsh, D. (2007). Fracture healing: the diamond concept. *Injury,* 38, Suppl 4 S3-6.

Giannoudis, P.V.; Einhorn, T.A.; Schmidmaier, G. & Marsh, D. (2008). The diamond concept: open questions. *Injury,* 39, Suppl 2 S5-8.

Gigante, A.; Bevilacqua, C.; Ricevuto, A.; Mattioli-Belmonte M. & Greco, F. (2007). Membrane-seeded autologous chondrocytes: cell viability and characterization at surgery. *Knee Surgery of Sports and Traumatology Arthroscopyc,* 15, 88–92.

Glassman, S.D.; Howard, J.; Dimar, J. 2nd., et al. (2010). Complications With rhBMP-2 in Posterolateral Spine Fusion: A Consecutive Series of One Thousand Thirty-Seven Cases. *Spine,* 10 doi: 10.1097/BRS.0b013e3181d133d0.

Gnecchi, M.; Zhang, Z.; Ni, A. & Dzau, V.J. (2008). Paracrine mechanisms in adult stem cell signaling and therapy. *Circulation Research,* 103(11), 1204-1219.

Grande, D.A.; Pitman, M.I.; Peterson, L.; Menche, D. & Klein, M. (1989). The repair of experimentally produced defects in rabbit articular cartilage by autologous chondrocyte transplantation. *Journal of Orthopaedic Research,* 7, 208–218.

Greco, S.J. & Rameshwar, P. (2007). Enhancing effect of IL-1alpha on neurogenesis from adult human mesenchymal stem cells: implication for inflammatory mediators in regenerative medicine. *Journal of Immunology,* 179(5), 3342-3350.

Groh, M.E.; Maitra, B.; Szekely, E. & Koç, O.N. (2005). Human mesenchymal stem cells require monocyte-mediated activation to suppress alloreactive T cells. *Experimental of Hematology,* 33(8), 928-934.

Gross, A.E.; Aubin, P.; Cheah, H.K.; Davis, A.M. & Ghazavi, M.T. (2002). A fresh osteochondral allograft alternative. *The Journal of Arthroplasty,* 17, (4 suppl 1).

Guerado, E. (2005). Osteogénesis terapéutica en cirugía del raquis. Bases científicas de la artrodesis vertebral. I: Fundamentos biomecánicos. *Revista Ortopédica de Traumatología,* 49, 29-45.

Guerado, E.; Godino, M.; Andrades, J.A. & Becerra, J.(2005). Osteogénesis terapéutica en cirugía del raquis. Bases científicas de la artrodesis vertebral. II. Fundamentos biológicos. *Revista Ortopédica de Traumatología,* 49, 46-58.

Guo, X.; Day, T.F.; Jiang, X. *et al.* (2004). Wnt/β- catenin signaling is sufficient and necessary for synovial joint formation. *Genes Development,* 18, 2404–2417.

Guo, Ch. & Spector, M. (2006). Tissue engineering of tendons and ligaments. In: Ma, P.X., & Elisseeff, J. (eds.), Scaffolding in tissue engineering. CRC Press, Boca Raton.

Haddo, O.; Mahroof, S.; Higgs, D.; David, L.; Pringle, J.; Bayliss, M.; Cannon, S.R. & Briggs, T.W. (2004). The use of chondroguide membrane in autologous chondrocyte implantation. *Knee,* 11, 51–55.

Hansen-Algenstaedt, N.; Joscheck, C.; Wolfram, L.; Schaefer, C.; Müller, I.; Böttcher, A.; Deuretzbacher, G.; Wiesner, L.; Leunig, M.; Algenstaedt, P. & Rüther W. (2006). Sequential changes in vessel formation and micro-vascular function during bone repair. *Acta Orthopaedic,* 77(3), 429-439.

Hecht, B.P.; Fischgrund, J.S.; Herkowitz, H.N. et al. (1999). The use of recombinant human bone morphogenetic protein 2 (rhBMP-2) to promote spinal fusion in a nonhuman primate anterior interbody fusion model. *Spine,* 24, 629-636.

Heineck, J.; Haupt, C.; Werner, K. et al. (2010). Fracture models in the lumbar sheep spine: a biomechanical investigation. *Journal of Orthopaedic Research,* 28, 773-7.

Helm, G.A. & Gazit, Z. (2005). Future uses of mesenchymal stem cells in spine surgery. *Neurosurgery Focus,* 19(6), E13.

Hidaka, C.; Cheng, C.; Alexandre, D.; Bhargava, M. & Torzilli, P.A. (2006). Maturational differences in superficial and deep zone articular chondrocytes. *Cell and Tissue Research,* 323, 127–35.

Hollinger, J.O.; Hart, C.E.; Hirsch, S.N.; Lynch, S. & Friedlaender GE. Recombinant human platelet-derived growth factor: biology and clinical applications. *Journal of Bone and Joint Surgery American,* 90 Suppl 1, 48-54.

Hoffmann, A.; Pelled, G.; Turgeman, G.; Eberle, P.; Zilberman, Y.; Shinar, H.; Keinan-Adamsky, K.; Winkel, A.; Shahab, S.; Navon, G.; Gross, G. & Gazit, D. (2006). Neotendon formation induced by manipulation of the Smad8 signalling pathway in mesenchymal stem cells. *The Journal of Clinical Investigation,* 116, 940-952.

Hooley, C. (1978). The viscoelastic properties of tendon. PhD Thesis, Oxford University (1977).

Horas, U.; Pelinkovic, D.; Herr, G.; Aigner, T. & Schnettler, R. (2003). Autologous chondrocyte implantation and osteochondral cylinder transplantation in cartilage repair of the knee joint. A prospective, comparative trial. *Journal of Bone and Joint Surgery American,* 85-A, 185–192.

Huang, J.I.; Kazmi, N.; Durbhakula, M.M.; Hering, T.M.; Yoo, J.U. & Johnstone, B. (2005). Chondrogenic potential of progenitor cells derived from human bone marrow and adipose tissue: a patient matched comparison. *Journal of Orthopaedic Research,* 23, 1383–1389.

Huang, P.; Gupta, M.C.; Sarigul-Klijn, N. & Hazelwood, S. (2006). Two in vivo surgical approaches for lumbar corpectomy using allograft and a metallic implant: a controlled clinical and biomechanical study. *Spine Journal,* 6, 648-58.

Ichinose, S.; Yamagata, K.; Sekiya, I.; Muneta, T. & Tagami, M. (2005). Detailed examination of cartilage formation and endochondral ossification using human mesenchymal stem cells. *Clinical Experimental Pharmacology and Physiology,* 32, 561–570.

Ito, T.; Itakura, S.; Todorov, I. et al. (2010). Mesenchymal stem cell and islet co-transplantation promotes graft revascularization and function. *Transplantation,* 89, 1438-1445.

Janssens, K.; ten Dijke, P.; Janssens, S. & Van Hul W. (2005). Transforming growth factor-beta1 to the bone. *Endocrinology Review,* 26(6), 743-774.

Javazon, E.H.; Beggs, K.J. & Flake, A.W. (2004). Mesenchymal stem cells: paradoxes of passaging. *Experimental Hematology,* 32(5), 414-425.

Jiang, Y.; Liu, H.; Li, H.; Wang, F.; Cheng, K.; Zhou, G.; Zhang, W.; Ye, M.; Cao, Y.; Liu, W. & Zou, H.A. (2011). Proteomic analysis of engineered tendon formation under dynamic mechanical loading in vitro. *Biomaterials,* 32, 4085-4095.

Johnstone, B.; Hering, T.M.; Caplan, A.I.; Goldberg, V.M. & Yoo, J.U. (1998). In vitro chondrogenesis of bone marrow-derived mesenchymal progenitor cells. *Experimental Cell Research,* 238, 265–272.

Jones, D.L., & Wagers, A.J. (2008). No place like home: anatomy and function of the stem cell niche. *Nature Review Molecular Cell Biology,* 9, 11-21.

Jorgensen, C.; Gordeladze, J. & Noel, D. (2004). Tissue engineering through autologous mesenchymal stem cells. *Current Opinion in Biotechnology,* 15, 406–410.

Juncosa-Melvin, N.; Shearn, J.T.; Boivin, G.P.; Gooch, C.; Galloway, M.T.; West, J.R.; Nirmalanandhan, V.S.; Bradica, G. & Butler, D.L. (2006). Effects of mechanical stimulation on the biomechanics and histology of stem cell-collagen sponge constructs for rabbit patellar tendon repair. *Tissue Engineering,* 12, 2291-2300.

Juncosa-Melvin, N.; Matlin, K.S.; Holdcraft, R.W.; Nirmalanandhan, V.S. & Butler, D.L. (2007). Mechanical stimulation increases collagen type I and collagen type III gene expression of stem cell-collagen sponge constructs for patellar tendon repair. *Tissue Engineering,* 13, 1219-1226.

Kastelic, J. & Baer, E. (1980). Deformation in tendon collagen. In: Vincent, J. & Currey, J. (eds.), The mechanical properties of biological materials. Cambridge University Press.

Kastelic, J.; Galeski, A. & Baer, E. (1978). The multicomposite structure of tendon. *Connective Tissue Research,* 6, 11-23.

Kawazoe, Y.; Katoh, S.; Onodera, Y.; Kohgo, T.; Shindoh, M. & Shiba T. (2008). Activation of the FGF signaling pathway and subsequent induction of mesenchymal stem cell differentiation by inorganic polyphosphate. *International Journal of Biology Sciences,* 4(1), 37-47.

Kim, S.G.; Akaike, T.; Sasagaw, T.; Atomi, Y. & Kurosawa, H. (2002). Gene expression of type I and type III collagen by mechanical stretch in anterior cruciate ligament cells. *Cell Structure and Function,* 27, 139–144.

Kobbe, P.; Tarkin, I.S.; Frink., et al. (2008). Voluminous bone graft harvesting of the femoral marrow cavity for autologous transplantation. An indication for the "Reamer-Irrigator-Aspirator" (RIA) technique. *Unfallchirurg,* 111, 469-72.

Koob, S.; Torio-Padron, N.; Stark, G.B.; Hannig, C.; Stankovic, Z. & Finkenzeller, G. (2011). Bone formation and neovascularization mediated by mesenchymal stem cells and endothelial cells in critical-sized calvarial defects. *Tissue Engineering Part A,* 17(3-4), 311-321.

Kobbe, P.; Tarkin, I.S. & Pape, H.C. (2008). Use of the "reamer irrigator aspirator" system for non-infected tibial non-union after failed iliac crest grafting. *Injury,* 39, 796-800.

Kreuz, P.C.; Steinwachs, M.; Erggelet, C.; Krause, S.J.; Ossendorf, C.; Maier, D.; Ghanem, N.; Uhl, M. & Haag, M. (2009). Classification of graft hypertrophy after autologous chondrocyte implantation. *Knee Surgery of Sports Traumatology Arthroscopy,* 17, 1289–1297.

Kubo, H.; Shimizu, M.; Taya, Y.; Kawamoto, T.; Michida, M.; Kaneko, E.; et al. (2009). Identification of mesenchymal stem cell (MSC)-transcription factors by microarray and knockdown analyses, and signature molecule-marked MSC in bone marrow by immunohistochemistry. *Genes Cells,* 14(3), 407-424.

Kuhn, N.Z. & Tuan, R.S. (2010). Regulation of stemness and stem cell niche of mesenchymal stem cells: implications in tumorigenesis and metástasis. *Journal of Cellular Physiology,* 222, 268-277.

Kuo, C.K. & Tuan, R.S. (2008). Mechanoactive tenogenic differentiation of human mesenchymal stem cells. *Tissue Engineering Part A,* 14(10), 1615-1627.

Ladd, M.R.; Lee, S.Y.; Stitzel, J.D.; Atala, A. & Yoo, J.J. (2011). Co-electrospun dual scaffolding system with potential for muscle-tendon junction tissue engineering. *Biomaterials*, 32, 1549-1559.

Lee, I.C.; Wang, J.H.; Lee, Y.T. & Young, T.H. (2007). The differentiation of mesenchymal stem cells by mechanical stress or/and co-culture system. *Biochemical and Biophysical Research Communications*, 352, 147-152.

Lee, T. (2010). Host tissue response in stem cell therapy. *World Journal of Stem Cells*, 2(4), 61-66.

Li, H.; Pujic, Z.; Xiao, Y. et al. (2002). Identification of bone morphogenetic proteins 2 and 4 in commercial demineralized freeze-dried allograft preparations: pilot study. *Clinical Implant Dentistry and Related Research*, 2, 110-7.

Li, L.; Zhang, S.; Zhang, Y.; Yu, B.; Xu, Y. & Guan, Z. (2009). Paracrine action mediate the antifibrotic effect of transplanted mesenchymal stem cells in a rat model of global heart failure. *Molecular Biology Reports*, 36(4), 725-731.

Li, Y.; Chen, J.; Chen, X.G.; Wang, L.; Gautam, S.C.; Xu, Y.X.; Katakowski, M.; Zhang, L.J.; Lu, M.; Janakiraman, N. & Chopp, M. (2002). Human marrow stromal cell therapy for stroke in rat: neurotrophins and functional recovery. *Neurology*, 59(4), 514-523.

Ling, L.; Nurcombe, V. & Cool, S.M. (2009). Wnt signaling controls the fate of mesenchymal stem cells. *Gene*, 433(1-2), 1-7.

Linke, A.; Müller, P.; Nurzynska, D.; Casarsa, C.; Torella, D.; Nascimbene, A.; Castaldo, C.; Cascapera, S.; Böhm, M.; Quaini, F.; Urbanek, K.; Leri, A.; Hintze, T.H.; Kajstura, J. & Anversa P. Stem cells in the dog heart are self-renewing, clonogenic, and multipotent and regenerate infarcted myocardium, improving cardiac function. *Proceedings of National Academic of Sciences USA*, 102(25), 8966-8971.

Liu, T.M.; Martina, M.; Hutmacher, D.W.; Hui, J.H.; Lee, E.H. & Lim, B. (2007). Identification of common pathways mediating differentiation of bone marrow- and adipose tissue-derived human mesenchymal stem cells into three mesenchymal lineages. *Stem Cells*, 25, 750–760.

López-Puertas, J.M.; Cuenca-López, M.D.; Jiménez-Palomo, P.; Becerra, J. & Andrades, J.A. (2011). Bone marrow mesenchymal stem cells and a biphasic collagen membrane are effective for articular cartilage regeneration. *Journal of Biomedicine and Biotechnology*, in press.

Ma, T.; Gutnick, J.; Salazar, B.; Larsen, M.D., et al. (2007). Modulation of allograft incorporation by continuous infusion of growth factors over a prolonged duration in vivo. *Bone*, 41, 386-92.

McCulloch, E.A. & Till, J.E. (2005). Perspectives on the properties of stem cells. *Nature Medicine*, 11, 1026-1028.

Mahmood, A.; Lu, D. & Chopp, M. (2004). Marrow stromal cell transplantation after traumatic brain injury promotes cellular proliferation within the brain. *Neurosurgery*, 55(5), 1185-1193.

Makino, T.; Jinnin, M.; Muchemwa, F.C.; Fukushima, S.; Kogushi-Nishi, H.; Moriya, C.; Igata, T.; Fujisawa, A.; Johno, T. & Ihn H. Basic fibroblast growth factor stimulates the proliferation of human dermal fibroblasts via the ERK1/2 and JNK pathways. *Brithish of Journal Dermatology*, 162(4), 717-723.

Manfredini, M.; Zerbinati, F.; Gildone, A. & Faccini, R. (2007). Autologous chondrocyte implantation: a comparison between an open periosteal-covered and an arthroscopic matrix-guided technique. *Acta of Orthopaedic Belgium*, 73, 207–218.

Maniwa, S.; Ochi, M.; Motomura, T.; Nishikori, T.; Chen, J. & Naora, H. (2001). Effects of hyaluronic acid and basic fibroblast growth factor on motility of chondrocytes and synovial cells in culture. *Acta Orthopaedic Scandinavia,* 72, 299–303.

Mayer, H.; Bertram, H.; Lindenmaier, W.; Korff, T.; Weber, H. & Weich, H. (2005). Vascular endothelial growth factor (VEGF-A) expression in human mesenchymal stem cells: autocrine and paracrine role on osteoblastic and endothelial differentiation. *Journal of Cell Biochemistry,* 95(4), 827-839.

Meinel, L.; Zoidis, E.; Zapf, J.; Hassa, P.; Hottiger, M.O.; Auer, J.A.; Schneider, R.; Gander, B.; Luginbuehl, V.; Bettschart-Wolfisberger, R.; Illi, O.E.; Merkle, H.P. & von Rechenberg, B. (2003). Localized insulin-like growth factor I delivery to enhance new bone formation. *Bone;* 33(4), 660-672.

Meirelles, Lda.S.; Fontes, A.M.; Covas, D.T. & Caplan AI. (2009). Mechanisms involved in the therapeutic properties of mesenchymal stem cells. *Cytokine Growth Factor Review,* 20(5-6), 419-427.

Minas, T. (2001). Autologous chondrocyte implantation for focal chondral defects of the knee. *Clinical Orthopaedic and Related Research,* S349–S361.

Minns, R.J.; Soden, P.D. & Jackson, D.S. (1973). The role of the fibrous components and ground substance in the mechanical properties of biological tissues: a preliminary investigation. *Journal of Biomechanics,* 6, 153-165.

Mishima, Y. & Lotz, M. (2008). Chemotaxis of human articular chondrocytes and mesenchymal stem cells. *Journal of Orthopaedic Research,* 26, 1407–12.

Mitsiadis, T.A.; Barrandon, O.; Rochat, A. et al. (2007). Stem cell niches in mammals. *Experimental Cell Research,* 313, 3377-3385.

Molloy, T.; Wang, Y. & Murrell, G. (2003). The roles of growth factors in tendon and ligament healing. *Sports Medicine,* 33, 381–394.

Monleón, M. & Díaz-Calleja, R. (1990). Nonlinear viscoelastic behaviour of the fllexor tendon of the human hand. *Journal of Biomechanics,* 23, 773-781.

Morrison, S.J. & Sprading, A.C. (2008). Stem cells and niches: mechanics that promote stem cell maintenance throughout life. *Cell,* 132, 598-611.

Muschler, G.F. & Midura, R.J. (2002). Connetive tissue progenitors: practical concepts for clinical applications. *Clinical Orthopaedic,* 395, 66-80.

Neufeld, G.; Cohen, T.; Gengrinovitch, S. & Poltorak, Z. (1999). Vascular endothelial growth factor (VEGF) and its receptors. *FASEB Journal,* 13(1), 9-22.

Nimni, M.E. (1997). Polypeptide growth factors: targeted delivery systems. *Biomaterials,* 18(18), 1201-1225.

Nguyen, T.D.; Liang, R.; Woo, S.L.Y.; Burton, S.D.; Wu, Ch.; Almarza, A.; Sacks, M.S. & Abramowitch, S. (2009). Effects of cell seeding and cyclic stretch on the fiber remodeling in an extracellular matrix–derived bioscaffold. *Tissue Engineering Part A,* 15, 957-963.

Niu, T. & Rosen, C.J. (2005). The insulin-like growth factor-I gene and osteoporosis: a critical appraisal. *Gene,* 361, 38-56.

Niu, C.C.; Tsai, T.T.; Fu, T.S.; Lai, P.L.; Chen, L.H. & Chen, W.J. (2009). A comparison of posterolateral lumbar fusion comparing autograft, autogenous laminectomy bone with bone marrow aspirate, and calcium sulphate with bone marrow aspirate: a prospective randomized study. *Spine,* 34, 2715-9.

Noel, D.; Gazit, D. & Bouquet, C. (2004). Short-term BMP-2 expression is sufficient for *in vivo* osteochondral differentiation of mesenchymal stem cells. *Stem Cells,* 22, 74–85.

Ochi, M.; Uchio, Y.; Kawasaki, K.; Wakitani, S. & Iwasa, J. (2002). Transplantation of cartilage-like tissue made by tissue engineering in the treatment of cartilage defects of the knee. *Journal of Bone and Joint Surgery Brithish*, 84, 571–578.

Ouyang, H.W.; Goh, J.C.H.; Thambyah, A.; Teoh, S.H. & Lee, E.H.L. (2003). Knitted poly-lactide-co-glycolide scaffold loaded with bone marrow stromal cells in repair and regeneration of rabbit Achilles tendon. *Tissue Engineering Part A*, 9, 431-439.

Ozaki, Y.; Nishimura, M.; Sekiya, K. et al. (2007). Comprehensive analysis of chemotactic factors for bone marrow mesenchymal stem cells. *Stem Cells Development*, 16, 119–29.

Pape, H.C.; Evans, A. & Kobbe, P. (2010). Autologous Bone Graft: Properties and Techniques. *Journal of Orthophaedic and Trauma*, 24, S36-S40.

Park, K.S.; Kim, Y.S.; Kim, J.H. et al. (2010). Trophic molecules derived from human mesenchymal stem cells enhance survival, function, and angiogenesis of isolated islets after transplantation. *Transplantation*, 89, 509-517.

Pei, M.; He, F. & Vunjak-Novakovic, G. (2008). Synovium-derived stem cell-based chondrogenesis. *Differentiation*, 76(10), 1044-1056.

Peterson, L.; Minas, T.; Brittberg, M.; Nilsson, A.; Sjogren-Jansson, E. & Lindahl, A. (2000). Two- to 9-year outcome after autologous chondrocyte transplantation of the knee. *Clinical Orthopaedic and Related Research*, 212–234.

Pittenger, M.F.; Mackay, A.M.; Beck, S.C.; Jaiswal, R.K.; Douglas, R.; Mosca, J.D.; Morman, M.A.; Ponte, A.L.; Marais, E.; Gallay, N. et al. (2007). The in vitro migration capacity of human bone marrow mesenchymal stem cells: comparison of chemokine and growth factor chemotactic activities. *Stem Cells*, 25, 1737–45.

Porter, R.M.; Liu, F.; Pilapil, C. et al. (2008). Osteogenic potential of reamer irrigator aspirator (RIA) aspirate collected from patients undergoing hip arthroplasty. *Journal of Orthopaedic Research*, 27, 42-9.

Pricola, K.L.; Kuhn, N.Z.; Haleem-Smith, H.; Song, Y. & Tuan, R.S. (2009). Interleukin-6 maintains bone marrow-derived mesenchymal stem cells stemness by an ERK1/2-dependent mechanism. *Journal of Cell Biochemistry*, 108(3), 577-588.

Prockop, D.J. (1997). Marrow stromal cells as stem cells for nonhematopoietic tissues. *Science*, 276, 71–74.

Prockop, D.J. (2007). "Stemness" does not explain the repair of many tissues by mesenchymal stem/multipotent stromal cells (MSCs). *Clinical Pharmacology Therapy*, 82(3), 241-243.

Prockop, D.J. (2009). Repair of tissue by adult stem/progenitor cells (MSCs), controversies, myths, and changing paradigms. *Molecular Therapy*, 17(6), 939-946.

Reddi, A.H.; Becerra, J. & Andrades J.A. (2011). Nanomaterials and hydrogel scaffolds for articular cartilage regeneration. *Tissue Engineering Part B Review*, 20 (pending).

Rodrigues, M.; Griffith, L.G. & Wells A. Growth factor regulation of proliferation and survival of multipotential stromal cells. *Stem Cell Research Therapy*, 1(4), 32.

Sacchetti, B.; Funari, A.; Michienzi, S.; Di Cesare, S.; Piersanti, S.; Saggio, I.; Tagliafico, E.; Ferrari, S.; Robey, P.G.; Riminucci, M. & Bianco, P. (2007). Self-renewing osteoprogenitors in bone marrow sinusoids can organize a hematopoietic microenvironment. *Cell*, 131, 324–336.

Sandhu, H.S.; Grewal, H.S. & Parvataneni, H. (1999). Bone grafting for spinal fusion. *Orthopaedic Clinical North America*, 30, 685-98.

Sato, T.; Vries, R.G.; Snippert, H.J., et al. (2009). Single Lgr5 stem cells build crypt-villus structures in vitro without a mesenchymal Niche. *Nature,* 459, 262-266.

Schmid, G.J., Kobayashi, C.; Sandell, L.J. & Ornitz, D.M. (2009). Fibroblast growth factor expression during skeletal fracture healing in mice. *Developmental Dynamic,* 238(3), 766-774.

Schmidmaier, G.; Hermann, S.; Gree, J. et al. (2006). Quantitative assessment of growth factors in reaming aspirate, iliac crest, and platelet preparation. *Bone,* 39, 1156-63.

Shimode, K.; Iwasaki, N.; Majima, T.; Funakoshi, T.; Sawaguchi, N.; Onodera, T. & Minami, A. (2007). Bone marrow stromal cells act as feeder cells for tendon fibroblasts through soluble factors. *Tissue Engineering Part A,* 13, 333-341.

Shum, L.; Coleman, C.M.; Hatakeyama, Y. & Tuan, R.S. (2003). Morphogenesis and dysmorphogenesis of the appendicular skeleton. *Birth Defects Research C Embryo Today,* 69(2), 102-122.

Silver, F.H.; Freeman, J.W. & Seehra, G.P. (2003). Collagen selfassembly and the development of tendon mechanical properties. *Journal of Biomechanics,* 36, 1529–1553.

Skutek, M.; van Griensven, M.; Zeichen, J.; Brauer, N. & Bosch, U. (2001). Cyclic mechanical stretching modulates secretion pattern of growth factors in human tendon fibroblasts. *European Journal of Applied Physiology,* 86, 48–52.

Slosar, P.J.; Josey, R. & Reynolds, J. (2007). Accelerating lumbar fusions by combining rhBMP-2 with allograft bone: a prospective analysis of interbody fusion rates and clinical outcomes. *Spine Journal,* 7, 301-7.

Simonetti, D.W.; Craig, S. & Marshak, D.R. (1999). Multilineage potential of adult human mesenchymal stem cells. *Science,* 284, 143–147.

Sotiropoulou, P.A.; Perez, S.A.; Gritzapis, A.D.; Baxevanis, C.N. & Papamichail, M. (2006). Interactions between human mesenchymal stem cells and natural killer cells. *Stem Cells,* 24(1), 74-85.

Spees, J.L.; Olson, S.D.; Ylostalo, J.; Lynch, P.J.; Smith, J.; Perry, A. et al. (2003). Differentiation, cell fusion, and nuclear fusion during ex vivo repair of epithelium by human adult stem cells from bone marrow stroma. *Proceedings of National Academic of Sciences USA,* 100(5), 2397-2402.

Steadman, J.R.; Rodkey, W.G.; Briggs, K.K. & Rodrigo JJ (1999). The microfracture technic in the management of complete cartilage defects in the knee joint. *Orthopade,* 28, 26-32.

Takebayashi, T.; Iwato, M.; Jikko, A. et al. (1995). Hepatocyte growth factor / scatter factor BMdulates cell BMtility, proliferation, and proteoglycan synthesis of chondrocytes. *Journal of Cell Biology,* 129, 1411–9.

Thalgott, J.S.; Fogarty, M.E.; Giuffre, J.M.; Christenson, S.D.; Epstein, A.K. & Aprill, C. (2009). A prospective, randomized, blinded, single-site study to evaluate the clinical and radiographic differences between frozen and freeze-dried allograft when used as part of a circumferential anterior lumbar interbody fusion procedure. *Spine,* 34, 1251-6.

Tins, B.J.; McCall, I.W.; Takahashi, T.; Cassar-Pullicino, V.; Roberts, S.; Ashton, B. & Richardson, J. (2005). Autologous chondrocyte implantation in knee joint: MR imaging and histologic features at 1-year follow-up. *Radiology,* 234, 501–508.

Thorfinn, J.; Angelidis, I.; Pridgen, B. & Chang, J. (2010). Tendon engineering. In: Chang, J. & Gupta, G. (eds.). Tissue engineering for the hand. World Scientific Publishers, Singapore.

Tsutsumi, S.; Shimazu, A.; Miyazaki, K.; Pan, H.; Koike, C.; Yoshida, E.; Takagishi, K. & Kato, Y. (2001). Retention of multilineage differentiation potential of mesenchymal cells during proliferation in response to FGF. *Biochemical and Biophysic Research Communication,* 288(2), 413-9.

Tsuzaki, M.; Bynum, D.; Almekinders, L.; Yang, X.; Faber, J. & Banes, A.J. (2003). ATP modulates load-inducible IL-1beta, COX 2, and MMP-3 gene expression in human tendon cells. *Journal of Cellular Biochemistry,* 89, 556–562.

Tufan, A.C.; Daumer, K.M.; DeLise, A.M. *et al.* (2002). AP-1 transcription factor complex is a target of signals from both WnT-7a and N-cadherin-dependent cell-cell adhesión complex during the regulation of limb mesenchymal chondrogenesis. *Experimental Cell Research,* 273, 197–203.

Van der Kraan, P.M.; Blaney-Davidson, E.N.; Blom, A. & van den Berg, W.B. TGF-beta signaling in chondrocyte terminal differentiation and osteoarthritis: modulation and integration of signaling pathways through receptor-Smads. *Osteoarthritis and Cartilage,* 17(12), 1539-1545.

Villars, F.; Bordenave, L.; Bareille, R. & Amédée. J. (2000). Effect of human endothelial cells on human bone marrow stromal cell phenotype: role of VEGF? *Journal of Cell Biochemistry,* 79(4), 672-685.

Wang, J.H.C. (2006). Mechanobiology of tendon. *Journal of Biomechanics,* 39, 1563-1582.

Wang, L.; Fan, H.; Zhang, Z.Y.; Lou, A.J.; Pei, G.X.; Jiang, S.; Mu, T.W.; Qin, J.J.; Chen, S.Y. & Jin. D. (2010). Osteogenesis and angiogenesis of tissue-engineered bone constructed by prevascularized β-tricalcium phosphate scaffold and mesenchymal stem cells. *Biomaterials,* 31(36), 9452-9461.

Wang, Y.; Cheng, Z.; Elalieh, H.Z.; Nakamura, E.; Nguyen, M.T.; Mackem, S.; Clemens, T.L.; Bikle, D.D. & Chang, W. (2011). IGF-1R Signaling in chondrocytes modulates growth plate development by interacting with the PTHrP/Ihh pathway. *Journal of Bone and Mineral Research,* Feb 10.

Weissman, IL. (2000). Translating stem and progenitor cell biology to the clinic: barriers and opportunities. *Science,* 287, 1442-1446.

Yang, Y. (2003). Wnts and wing: Wnt signaling in vertebrate limb development and musculoskeletal morphogenesis. *Birth Defects Research C Embryo Today,* 69(4), 305-317.

Yue, W.M.; Liu, W.; Bi, Y.W.; He, X.P.; Sun W.Y., Pang, X.Y. et al. (2008). Mesenchymal stem cells differentiate into an endothelial phenotype, reduce neointimal formation, and enhance endothelial function in a rat vein grafting model. *Stem Cells Development,* 17(4), 785-793.

Zisa, D.; Shabbir, A.; Suzuki, G., Lee T. Vascular endothelial growth factor (VEGF) as a key therapeutic trophic factor in bone marrow mesenchymal stem cell-mediated cardiac repair. *Biochemical and Biophysical Research Communication,* 390(3), 834-838.

6

Production of Clinical Grade Mesenchymal Stromal Cells

Pytlík[1], Slanař[1], Stehlík[1] and Matějková[2]
[1]First Medical Faculty, Charles University, Prague
[2]National Tissue Center, Brno
Czech Republic

1. Introduction

Mesenchymal stromal cells (MSC, MSCs) are cells firstly described 35 years ago in bone marrow (Friedenstein *et al*, 1976), but present basically in all adult and fetal tissues, where they reside in the vessel wall as part of the population of pericytes (Crisan et al, 2008). These rare cells (10^{-6}-10^{-4} of nucleated cells in various tissues - Werntz, 1996) received special attention of biomedical researchers as they are easy to expand and able to differentiate in various cell and tissue types (Pittenger et al, 1999, Battula et al, 2009, and many others). Later, these cells were found to be little immunogenic and to have immunosuppressive properties, which they exert by action on T cells, B cells, NK cells and dendritic cells (Beyth et al, 2005; Corcione et al, 2006; Spaggiari et al, 2008; Spaggiari et al, 2007). Furthermore, MSCs do not necessarily need to differentiate into tissue of interest, but they can exert their therapeutic effect through secretion of various cytokines (Phinney & Prockop, 2007; Horwitz & Dominici, 2008). Use of these cells therefore appears to be a promising strategy for treatment of various disorders, including orthopedics, heart and vessel, or graft versus host disease (Shenaq et al, 2010; Mathiasen et al, 2009; Le Blanc et al, 2008).

Mesenchymal stromal cells cultivated in vitro are a mixture of cells of various clonogenic and differentiating properties, which are partly dependent on cultivation conditions, partly they are donor specific (Tsai et al, 2011; Friedl et al, 2009). There is no single marker which would distinguish MSCs from other fibroblastoid cells. Therefore, The International Society for Cellular Therapy set in 2006 minimal set of requirements which mesenchymal stromal cells should fulfill. These are: 1. adherence to plastic, 2. expression of CD73, CD90, and CD105 antigens, while being CD14, CD34, CD45, and HLA-DR negative, and 3. their ability to differentiate to osteogenic, chondrogenic and adipogenic lineage (Dominici et al, 2006). Every method used for production of MSCs must be shown to produce cells of above mentioned characteristics. There are several exception from these rules, however - for example MSCs derived from the adipose stromal vascular fraction are CD34 positive (De Ugarte et al, 2003), or can express HLA-DR in certain culture conditions (Tarte et al, 2010).

Though first clinical trial of mesenchymal stromal cells was reported as early as in 1995 (Lazarus et al, 1995), the transfer to clinics has been relatively slow and complicated by several issues, especially in the context of good manufacturing practice preparations of these cells. Among them are issues of cultivation medium, serum and supplementation used, suitable expansion systems and reproducibility of results. Mesenchymal stromal cells can be

freezed and stored (Haack-Sørensen & Kastrup, 2011), but it is not clear if autologous (customer specific) or allogeneic („off the shell or „one size fits all") products are preferable. As a result of these problems, there is still no standard or universally accepted or preferred way how to produce mesenchymal stromal cells for clinical use.
In this review, we will focus on application of good manufacturing practice standards principles to MSC production, with reference to choice of starting material, cultivation media, serum and supplements, cultivation systems, cultivation process, quality control, efficacy and safety concerns. References to universal GMP principles will be made as appropriate and selected choices of clinical grade cultivation components will be provided, as authors knowledge permits. As cellular therapy is a quickly evolving field (both from the practical and regulatory point of view), we have to state that following paragraphs will not be by any means exhaustive, and they may be also subject of changes during the time. On the other hands, some of the outlined principles may apply to other somatic cell products as well.

2. Mesenchymal stromal cells as advanced medicinal products

Mesenchymal stromal cells (with some exceptions mentioned below) belong to advanced therapy medicinal products (ATMPs), according the EC Regulation No 1394/2007, together with other somatic cell therapy medicinal products. According to this regulation, somatic cell therapy medicinal product means a biological medicinal product that has the following characteristics: 1. contains or consists of cells or tissues that have been subject to *substantial manipulation* so that biological characteristics, physiological functions or structural properties relevant for the intended clinical use have been altered, or of cells or tissues *that are not intended to be used for the same essential function(s) in the recipient and the donor* (italics added by authors); 2. is presented as having properties for, or is used in or administered to human beings with a view to treating, preventing or diagnosing a disease through the pharmacological, immunological or metabolic action of its cells or tissues.
There is a significant trend to establish risk-based systems for regulation of ATMPs including mesenchymal stromal cell - based therapy products within the regulatory systems worldwide, although some countries did not follow this route so far (e.g., Australia). Numbers of governments have moved to introduce specific regulations for this sector, while others try to develop the traditional model of regulation for medicines and enable the regulatory authorities to respond to technology changes. The International Conference on Harmonization (ICH) established in 1990 in order to harmonize different regional requirements for registration of pharmaceutical drug products has no specific guidance document for cell products at the moment, although some guidance may be applicable (e.g. ICH S6, ICH Q5A-E). Clearly, there is insufficient worldwide unity of the regulatory approaches to cell based products at the moment.
Recently, the EU, USA, and Canada have implemented new systems for the regulation of ATMPs that are risk based oriented and specific for the geographic areas governed by appropriate local regulation.
The EU system formulates minimum quality and safety standards for harvesting, procurement, testing, processing, preservation, storage and distribution of human cells that need to be documented including donor selection procedures, traceability and adverse event reporting processes, GMP-based quality system, data and confidentiality protection. Within the EU, to assess the quality, safety and efficacy of ATMPs, including mesenchymal

stromal cells, Committee for Advanced Therapies (CAT) was established at the European Medicines Agency in accordance with Regulation (EC) No 1394/2007 on advanced-therapy medicinal products. The main responsibility of the CAT is to prepare a draft opinion on each ATMP application submitted to the European Medicines Agency, before the Committee for Medicinal Products for Human Use (CHMP) adopts a final opinion on the granting, variation, suspension or revocation of a marketing authorization for the medicine concerned. This decision is subsequently formalized by the decision of the European Commission that is binding in all EU member states.

The US FDA review process is conducted by the Office of Cellular, Tissue and Gene Therapies (OCTGT), Center for Biologics Evaluation and Research (CBER), under the Code of Federal Regulations Title 21 Parts 1270 and 1271. Health Canada regulates mesenchymal stromal cell derived products as medicines under Schedule D of the Food and Drugs Act. Health Canada also developed a regulatory framework under the Food and Drugs Act, "The safety of human cells, tissues and organs for transplantation regulations" (CTO Regulations 2007), which specifies requirements for the establishment of licensing and processing quality standards for cells, tissues and organs.

The Australian´s Therapeutic Goods Administration (TGA) plans to introduce new framework for human cell and tissue therapy products with a classification where mesenchymal stromal cells will fit into a Class 3 product - "A cell or tissue processed in a manner that may alter the structure and properties of the cell or tissue but does not purposefully alter the biological activity." This class of products will require: TGA Licensed Manufacturer, Relevant cGMPs, Good Tissue Practice Standards, and TGA pre-market approval.

The regulatory systems distinguish between cell products with substantial manipulation or without. Now, it is clear that in vitro cultivation represents substantial manipulation for cells isolated from human body, making from naturally occurring cells an artifact, changed by the unnatural cell culture environment. Substantial manipulation involves also cell purification or enrichment, for example their selection by monoclonal antibodies against CD34, CD49a, or CD271 antigens, as outlined below. On the other hand, the term „substantial manipulation" does not apply to simple isolation of bone marrow mononuclear cells or adipose tissue stromal vascular fraction, and also can be argued that cells isolated in this way are intended for use for similar purposes as they fulfill in the body (i.e., regeneration of damaged or aging tissues). Therefore, it can be argued that the mentioned EC Regulation does not apply to such products, though this is still under debate. This does not mean, however, that the harvesting, isolation and preparation of these otherwise unmanipulated cellular products for clinical use does not require the adherence to good manufacturing practice (GMP) principles, but regulatory requirements in these cases are similar for blood banking products and therefore these cells may be prepared in suitable transfusion or blood banking facilities. As preparation of crude cell mixtures is less difficult than preparation of better defined cell populations, these were also used in preclinical and clinical trials, sometimes with encouraging results (Hernigou et al, 2005; Chochola et al, 2008; Akita et al, 2010). There is also a frank exception from the EC Regulation No. 1394/2007: custom-made (hospital) ATMPs, which are prepared on a non-routine basis according to specific quality standards, if they are used within the same member state in a hospital under exclusive professional responsibility of a medical practitioner to comply with a medical prescription for a custom-made product for an individual patient (Sensebé, 2010).

The following text focuses mostly on further manipulated MSCs, as production of such cells are much more complex and basic principles of good manufacturing practice as well as national and international regulation requirements have to be followed.

3. Good manufacturing practice principles

It has to be understand that good manufacturing practice guidelines are not instructions on how to manufacture products. Rather, these are series of principles that must be fullfilled during the manufacturing process. Their goal is to obtain a final product from defined materials, by a defined, documented and traceable way, by trained operators. Currently, cell products regulated as medicines must comply with the GMP for Medicinal Products. However, the GMP for Medicinal Products is not yet fully adapted for dealing with the unique circumstances of cell based products. The main principles are as follows:

Materials. Raw materials have to be of a documented quality. They should be certified by their manufacturer and their batches should be registered. This does not necessarily mean that all materials have to be of clinical grade, though this is clearly an advantage. When the clinical grade material is not availlable, sufficient documentation about its production and about composition of individual batches has to be obtained to minimize the risk of its contamination by undesired elements. The documentation about used materials have to be preserved for legally determined time period.

Manufacturing processes. Manufacturing processes have to be clearly defined by a set of instructions known as standard operational procedures. These should be written in clear and unambiguous language and easily availlable for operators.

Documentation. Each part of the manufacturing process have to be documented, beginning from the storage conditions of raw materials (freezer and fridges temperatures) to the final product. These records should demonstrate that the standard operational procedures were in fact followed and that the quality of the product is as expected. Any deviations from standard operational procedures have to be documented.

Validation. National legislations have usually sets of recommended procedures for certain parts of the manufacturing process (e.g., the required tests for bacterial contaminations are described in pharmacopoiea). These procedures are usually designed for conventional drugs and cannot be allways used for somatic cell therapy products (e.g., sterilization, microbial tests of final product). The process called validation means the comparison of alternative procedures to the customary ones and proofs that these deviations from standard procedures bring desired outcomes.

Standardization. For good management of internal quality controls is a must. At present, there are also many programs of external quality controls performed by national authorities or commercial subjects. Manufacturer shall control storage areas to prevent mix-ups, deterioration, contamination, cross-contamination, and improper release or distribution of products. The storage temperature must be validated for each type of product and it is convenient to use devices with appropriate certificates.

Also set of standards have to be adopted for release of the product and these standards have to be followed and release criteria for every batch have to be documented.

Requirements for cellular products are also mentioned in International Standards for Cellular Therapy Product Collection, Processing and Administration (Fourth Edition, Version 4.1, April 2011) made by FACT-JACIE. These Standards are designed to provide minimum guidelines for programs, facilities, and individuals performing cell transplantation and therapy or providing support services for such procedures.

Traceability. Records of manufacture (including distribution) that enable the complete history of a batch to be traced are retained. A system is available for recalling any batch of product from sale or supply. If undesired effect of the product occur, the causes for possible quality defects have to be investigated and appropriate measures have to be taken to exlude the defective batch from further use and to prevent recurrence of possible mistakes. Also, database of undesired drug effect should be established (pharmacovigilance).

Training. Operators have to be fully trained in standard operational procedures and their knowledge should be periodically examined.

GMP requirements are regulated by national and international legislatives and adherence to their principles is controlled by special agencies - in Europe, it is EMA (European Medical Agency), in the United States the FDA (Food and Drug Administration). Other countries, as Australia, Canada, Japan, Singapore or United Kingdom have highly developed GMP requirements. In other countries, especially in the developing world, the World Health Organization (WHO) version of GMP is used by pharmaceutical regulators and the pharmaceutical industry. Control of adherence to the GMP principles is performed by regular inspections by the governmental agencies.

4. GMP facilities

Good manufacturing practice facilities are the basic prerequisites for GMP preparation of medicinal products. They are designed to create the appropriate production environment, to prevent product contamination by raw materials and cross-contamination between batches and to ensure that standard operational procedures may be followed as intended. Again, GMP facilities for somatic cell therapy products may differ from facilities designed for manufacturing of conventional drugs.

Cleanroom desings should in general complie to International Standard ISO 14644 - Cleanrooms and associated controlled environments. ISO 14644 consists of eight parts:

- ISO 14644-1: Classification of air cleanliness
- ISO 14644-2: Specifications for testing and monitoring to prove continued compliance with ISO 14644-1
- ISO 14644-3: Test methods
- ISO 14644-4: Design, construction and start-up
- ISO 14644-5: Operation
- ISO 14644-6: Vocabulary
- ISO 14644-7: Separative devices (clean air hoods, gloveboxes, isolators and mini-environments)
- ISO 14644-8: Classification of airborne molecular contamination

It is above the scope of this chapter to run into details. In following examples, ISO 14644-1 requirements for airborne particulate cleanliness (Table 1), and scheme of contamination control concept following ISO 14644-4 (Figure 1) are shown.

The ascending requirements for cleanliness (from rooms class C to process core class A) is achieved by elaborate air conditioning systems, which ensure the highest pressure in the process core, with pressure gradient descending to peripheral parts of the facility. Air is blowed into the facility by systems of high-effective or ultrahigh-effective particle filters

(HEPA or UEPA). Between spaces of different classes of cleanliness, filters are designed for decontamination of materials as well as operating personell.

As is clear from what was written above, construction and running of GMP facility is very expensive. These expenses are naturally calculated to the final cost of the product. However, for small GMP productions, e.g. university based, or for Phase 1 clinical trials, solutions also exist (Xvivo production systems - www.biospherix.com, and others).

ISO classification number	Maximum concentration limits (particles/m3 of air) for particles equal to or larger than the considered sizes shown below					
	0,1 μm	0,2 μm	0,3 μm	0,5 μm	1 μm	5 μm
ISO Class 1	10	2				
ISO Class 2	100	24	10	4		
ISO Class 3	1 000	237	102	35	8	
ISO Class 4	10 000	2 370	1 020	352	83	
ISO Class 5	100 000	23 700	10 200	3 520	832	29
ISO Class 6	1 000 000	237 000	102 000	35 200	8 320	293
ISO Class 7				352 000	83 200	2 930
ISO Class 8				3 520 000	832 000	29 300
ISO Class 9				35 200 000	8 320 000	293 000

Table 1. Requirements for airborne particulate cleanliness

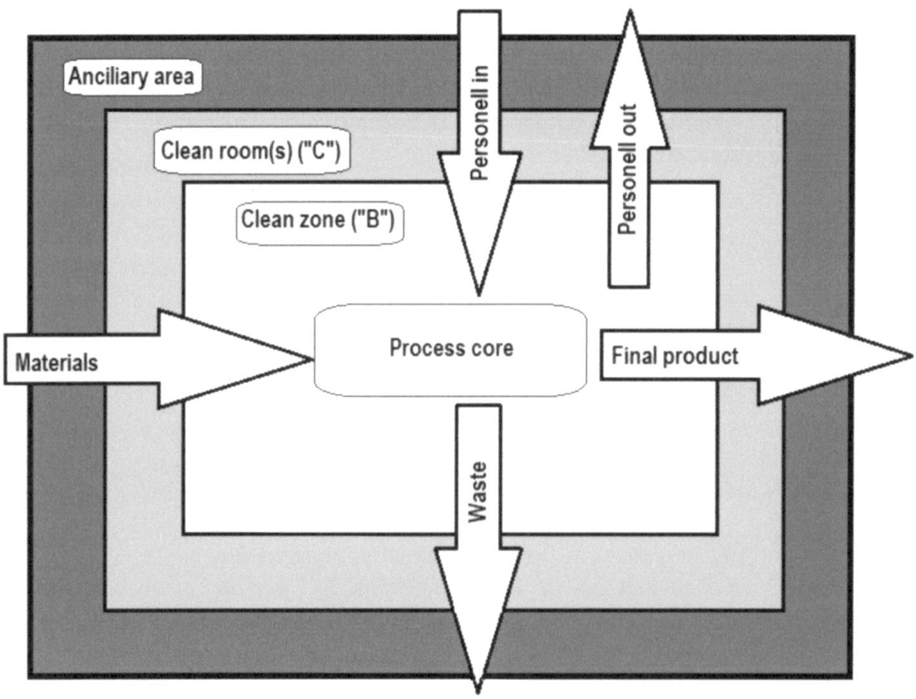

Fig. 1. Scheme of clean facility for GMP. Classification of rooms matches the requirements for airborne particulate cleanliness. For example, room of GMP class C requires cleanliness ISO Class 8 or higher

5. Choice of the starting material

MSCs are most easily obtained from bone marrow or adipose tissue. Other tissues of interest involve cord blood, amniotic fluid, or trophoblast.

5.1 Autologous or allogeneic?

Use of autologous material may be preferable for number of reasons. It eliminates or significantly reduces risks of disease transmission and overcomes the problems with a suitable donor selection. This is true even if we assume that MSCs are non-imunogenic and suitable for use across histocompatibility bariers (Niemeyer et al, 2006, Hare et al, 2009), as cells from different donors do not grow equally well. Also, there are reports that MSCs grow better in autologous plasma than in fetal calf serum (Stute et al, 2004).
On the other hand, the indisputable advantage of allogeneic cells is the possibility of their cryopreservation for later use as "off the shelf" product. This means that for acute or unpredicted indications (trauma, burns) there will be no need to wait several days or weeks untill autologous cells become availlable. Also, the quality tests, including sterility, differentiation or immunosuppressive abilities of cells, and viability of cells after thawing may be performed in full before the use of the product. Use of allogeneic, well characterized and quality-controlled cells also solves the problem of unpredicable growth of MSCs from different donors, as is autologous setting, adequate number of good quality cells may not be allways obtained. Allogeneic MSCs have been used successfully in treatment of graft versus host disease (Le Blanc et al, 2008), or myocardial infarction (Hare et al, 2009). Most donors in Le Blanc´s and all donors in Hare´s work were unrelated and HLA-mismatched individuals.
The choice of autologous versus allogeneic cells will be also guided by the wider settings of their therapeutic use. For example, in academic settings, where small-scale production is sufficient and customarized production of MSCs is feasible, autologous products might be preferrable. Also, it is possible that in certain countries it would be easier to obtain authorisation for autologous rather than for allogeneic products. However, in large-scale production setting, use of larger batches of an allogeneic product is probably inevitable.

5.2 Bone marrow

Bone marrow was the first source of MSCs for experimental and later for clinical use. For experimental use, it is still probably the best and most accessible, as small amounts of bone blood can be easily aspirated in local analgesia. Also, for small-scale experimental MSC cultivation, local hematology department may provide an easy access to bone blood when bone marrow samples from patients with known or suspected hematological disorder are taken for diagnostic purposes. However, one must be careful when MSCs are cultivated from hematological patients, as in certain diseases (acute and chronic leukemias, multiple myeloma, myelodysplastic syndrome) mesenchymal stromal cells may have slightly different properties from MSCs from healthy bone marrow (Garaoya et al, 2009), and may even harbor chromosomal abnormalities (Blau et al, 2007).
For larger scale and therapeutic purposes, where several hundred mililiters of bone marrow blood is necessary, bone marrow must be harvested under general anesthesia. Harvest is usually performed from posterior iliac crests similarly as for bone marrow transplantation

for hematological disorders, i.e., in collection bags of vessels with an anticoagulant, usually heparin. There are several recommendations about the volume that should be taken in one aspiration, but they are somehow contradictory. Fennema recommends to harvest at least 8 ml portions, as in lower volume the yield of nuclear cells and MSCs in unpredictable. Also, second portion of up to 10 ml may be taken from the same site without significant loss of quality due to dilution by peripheral blood (Fennema et al, 2009). On the other hand, in older work Muschler concludes that aspiration of more than 2 ml of bone blood results in smaller number of colony forming units -fibroblast (CFU-F) (Muschler et al, 1997). It may be therefore prudent for each centre to establish its own aspiration protocol, which yields the best results in their settings.

For isolation of mononuclear fraction, several methods are availlable. For smaller volumes, Ficoll centrifugation method is usually used, and GMP-compatible Ficoll is availlable. In larger volumes, bone marrow is usually processed on blood separators, which is similar to harvest of peripheral blood progenitor cells. Sometimes, the older starch or dextran sedimentation methods are still at use. In any case, closed isolation system is preferrable for mononuclear cell isolation. This may be challenging especially with small bone marrow blood volumes. Therefore, the alternative is to seed bone marrow mononuclear cells on plastic without previous enrichment, with or without preceeding red cell lysis (Tarte et al, 2010, online supplement; Horn et al, 2008; Horn et al, 2011).

5.3 Adipose tissue

Multipotent stromal cells in adipose tissue reside in stromal vascular fraction (SVF), which can be easily separated from fat cells after collagenase digestion (Zuk et al, 2002; Gimble & Guilak, 2003). Adipose tissue is richer source of MSCs than bone marrow, as these cells account for almost 2% of cells in SVF (Valle et al, 2009). This means that from 200-500 g of fat, 100-300x10^6 MSCs can be obtained, a number that might be used for treatment even without further expansion (unpublished data). Adipose tissue mesenchymal stromal cells are CD34 positive and they may differ from bone marrow stromal cells a little, but not by their differentiating potential into three main lineages (Kern et al, 2006).

Adipose tissue MSCs may be obtained after lipoexcision or lipoaspiration. If any of these procedures has a clear advantage in number or quality of cells over the other, is still a matter of debate (Torio Padron et al, 2010), but lipoaspiration is far less invasive procedure. Also, lipoaspirates are superior to excisates from the point of view of their stability, as the cell number in aspirates remains stable even after 24 hours, in contrary to rapid decrease of their yield in the excided tissue (Bieback et al, 2010). There is also uncertainity about the part of the body which is the richest in adipose tissue-derived MSCs (Fraser et al, 2007; Jurgens et al, 2008), but this does not seem to be of great clinical importance. On the other hand, it was established that higher aspiration pressures (-350 mm Hg) are preferrable to lower ones (-700 mm Hg), as they lead to higher cell yield (Mojallal et al, 2008).

The greatest disadvantage of adipose tissue in contrast to bone marrow is the necessity to digest the starting material by collagenase. From the technical point of view, besides manual method there are also automated devices which can produce SVF by good-manufacturing practice compatible method (TGI1200, Cytori Celution, Adistem et al). GMP compatible collagenase also exists, but is very expensive (Brooke et al, 2009). However, the advantages of adipose tissue seem to prevail and therefore, it may be expected that it will become more popular for MSC production than bone marrow.

5.4 Trophoblastic tissues

In general, trophoblastic tissues present an excellent source of cells for stem cell therapy, as they are abundant, their use is not connected with any ethical problems, and they contain developmentally young and putatively more plastic stem cells. The use of umbilical cord blood is well established in transplantation medicine (Wagner & Gluckman, 2010). Mesenchymal stromal cells can be retrieved from cord blood, Wharton jelly, and placenta (Wang, 2004; Flynn et al, 2007; Troyer & Weiss, 2008; Brooke et al, 2009). Mesenchymal stromal cells from trophoblastic tissues have been shown to have similar transcriptome, proteome, immunsuppressive and differentiation abilities as MSCs from bone marrow (Jones et al, 2007; Tsai et al, 2007; Barlow et al, 2008). Of these, umbilical cord blood MSCs were first described and probably most extensively characterized. They have higher proliferation capacity than bone marrow mesenchymal cells, but they are quite rare and even in experienced laboratories they can be successfully isolated from only about two thirds of cord blood units (Kern et al, 2006). Isolation of Wharton jelly and placental MSCs is in general similar to isolation of adipose tissue stromal cells in that digestive enzymes have to be used. However, they are more abundant and more easily isolated than MSCs from the cord blood.

One protocol for GMP isolation and expansion of placental MSCs was described recently (Brooke et al, 2009). Briefly, the placenta was aseptically collected and minced in small pieces, which were then digested by collagenase, type 1, and DNA-se I in Dulbecco´s modified Eagle´s medium, low glucose (DMEM, LG). After digestion, centrifugation tubes were pulse spun to remove large particular matter and mononuclear cells from the suspension were retrieved after centrifugation on Ficoll-Paque. Then adherent cells were isolated by cultivation on plastic for three days, in DMEM-LG with 20% FCS and 50 mg/l gentamycin. Cells were propagated for a total of five passages and cells not required for further propagation were cryopreserved after each passage.

After isolation, 74% of cells were found to be CD45+ leukocytes and 0.6% of cells were CD73+CD105+. Initial propagation in eight 175 cm^2 flasks yielded 40-100x10^6 adherent cells, from which still approximately 25% were CD45+ (P0). Percentage of CD45+ cells rose to more than 50% during first passage (P1), but fell quickly under 1% in following passages (P2-P5). In first passage, there was 40% of CD73+CD105+ cells and the percentage was above 90% in successing passages. Cell recoveries after cryopreservation were 96% from P2, 100% from P3, and 60% from P4 and P5. 120x10^6 MSCs from one placental unit was infused to a patient suffering with acute myeloid leukemia, who was co-transplanted with two units of umbilical cord blood (HLA-mismatched with infused MSCs). Though the hematopoietic engraftment could not be fully evaluated because of early death of the recipient, the infusion itself was reported to be uneventful.

This report shows that large numbers of MSCs can be obtained from human placenta. However, at least two passages have to be employed to deplete the final product from hematopoietic cells. The manufacturing procedure was reported to be labour-intensive and time consuming, using the open system of plastic cultivation flasks. However, this report shows that placental tissue can be used as an alternative source to bone marrow or adipose tissue for allogeneic applications, as is engraftment or treatment of graft-versus-host disease in blood progenitor cell transplantation setting.

5.5 Enrichment of starting material for MSCs

Mesenchymal stromal cells do not have any particular markers or antigenes to allow for an easy separation, as in case of CD34+ or CD133+ stem cells, or CD3 lymphocytes. Essentially, two ways of enrichment of mononuclear cells for MSCs in the starting population are depletion of lineage-positive cells, or selection for MSC-containing fraction. RosetteSep© Mesenchymal Enrichment Cocktail (StemCell Technologies) contains monoclonal antibodies against hematopoietic cells (CD3, CD14, CD19, CD38, CD66b and Glycophorin A) and is mixed with bone marrow blood before separation on Ficoll gradient. Lineage positive hematopoietic cells form rosettes with erythrocytes and sediment with them. For selection of MSC-containing fraction, antibodies against CD271 (Jarocha et al, 2008; Poloni et al, 2009), or against CD105 (Jarocha et al, 2008), were used. Besides, while cited studies reported good results, Bierback reported worse results for RosetteSep© and CD271+ separation compared to plastic adherence, at least when MSCs were grown in platelet lysate (Bieback et al, 2009). Furthermore, monoclonal antibodies are expensive and they add another element into altogether complex process of GMP isolation and expansion of MSCs. If no new, revolutionary strategy for MSC enrichment will emerge, it is therefore unlikely that enrichment of starting material will become a standard procedure in GMP mesenchymal stromal cell production in the near future.

6. Cultivation conditions

The choice of cultivation conditions is of uttermost importance, with respect to GMP requirements. All components of the cultivation system should be fully characterized, certified or validated, and their robustness and stability of performance of the whole system have to be assured. However, in practical way, this might be quite challenging. As patented techniques and formulas are used for production of many ingredients, from surface treatment of cultivation vessels to the formulations of medium composition, both the researcher and regulatory agency have sometimes to rely on incomplete information. From this point of view, it is questionable whether the newest solutions just introduced to the market are always preferrable to older and well-tried technologies.

6.1 Cultivation vessels and systems

Traditionally, MSCs were cultivated in open systems. There is a plethora of companies (Corning, Nunc, TPP, to name at least few), that produce plastic flasks suitable for research-grade cultivation of MSCs. However, these do not seem to be optimal for clinical-grade production for several reasons:

1. Though these vessels are manufactured as sterile, tissue culture treated and apyrogenic, they are not certified for GMP-production.
2. They have to be opened before each manipulation. This was overcomed e.g. in RoboFlask™ produced by Corning, which have silicon rubber seal that can be repeatedly penetrated by injection needle without the need to open the vessel.
3. Classical flasks are small and difficult to manipulate. For production of clinically meaningful numbers of MSCs, tens or even more than hundred of these flasks would be needed for every single patient. This may be partially overcomed by use of larger flasks with several cultivation surfaces (e.g., CellSTACK® Culture Chambers, HYPER*Flask*®Culture Vessels, both Corning), on the other hand, the visual assessment of culture grow is very difficult under these conditions. Small bioreactors (CellCube®, Corning) offer as much as 80, 000 cm2 culture surface and accessories, as are setup kits, oxygenators, oxygen probes, etc.

4. There is a need for reseeding of MSCs after they reach critical density. Traditionally, this has been performed by trypsinization, centrifugation and reseeding of cells to new flasks. This is labour-demanding, expensive and increases the risk of microbial contamination. Possible solution of this problem may be the dynamic culture surface expansion, as described by Majd (Majd et al, 2009). The bottom of the cultivation vessel was made from high-extension silicon rubber, which could be mechanically expanded by iris-like device from the initial area of 10 cm^2 to area of 80 cm^2. In this device, cells were grown in constant densities for more than 9 weeks. Quick and hands-free harvest of adherent cells can be performed by several robotic systems, as is Tecan Freedom EVO (Tecan Group LTD).

Solutions mentioned above are mostly still suitable for small-volume production. However, larger robotic systems compatible with good manufacturing practice and good tissue practice principles, are available as well. Tecan Cellerity™ is a fully robotic modular system with several possible configurations, including HEPA filtred clean bench, robotic CO2 incubator, media refrigerator, etc. One possible configuration is shown on Figure 2. As is clear, these are already very complex and expensive solutions for large-scale commercial production. For smaller manufacturers, reasonable compromise between optimal and realistic will have to be achieved.

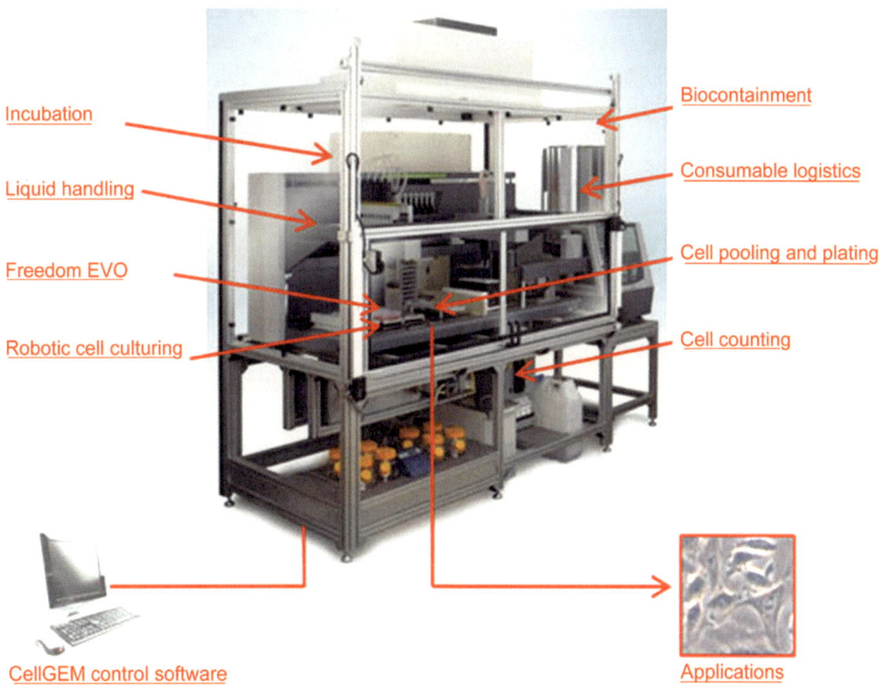

Fig. 2. Tecan Celerity configuration employing laminar box, incubator, liquid handling, cultivation media storage (down, left) and automatic processing of cell cultures

6.2 Choice of cultivation medium

A variety of cultivation media for mesenchymal stromal cells currently exist. Most commonly used are research-grade media DMEM (Dulbecco´s modified Eagle medium)

low-glucose, IMDM (Iscove´s modified Dulbecco´s medium) and alpha-MEM (minimal essential medium). The first of them is most commonly used and is present in EMEA approved GMP-compliant medium (Haack- Sørensen et al, 2008). It was shown that DMEM is preferrable to IMDM with the respect of preservation of MSC "stemness" (Pierri et al, 2011). However, alpha-MEM was found to better preserve osteogenic properties of MSCs than DMEM (Coelho et al, 2000), and in at least one work (Lange et al, 2007) also to lead to higher CFU-F retrieval in primary expansion (P0). Superiority of alpha-MEM over DMEM with respect to MSC expansion was found also in our own unpublished experiments.

There is a number of expansion media claimed to be GMP compliant: the LP02 basic medium (Lange et al, 2007), or the CellGro™ medium for hematopoietic stem cells (Pytlík et al, 2009). However, both of these need to be supplemented with fetal calf serum or some of its human alternatives (see below). Serum-free chemically defined media for mesenchymal stromal cells were also developed. Instead of serum they contain attachment factors for adherent cells and sometimes they have to be supplemented with recombinant cytokines. Of course, these media are all patented, thus the researcher does not know in full what is their exact composition. StemPro® is a serum-free medium pioneered by Invitrogen. While one group (Hartmann et al, 2010) were unable to cultivate MSCs in it without supplementation with 2% human serum (Hartmann et al, 2010), another group was more successfull after pre-coating of cultivation vessels with CELLStart™ xenogeneic free substrate (Invitrogen) or with human fibronectin and adding recombinant PDGF-BB, FGF-2 and TGF-beta (Chase et al, 2010). Another serum-free medium is MesenCult© ACF (StemCell Technologies) (Hartmann et al, 2010), which, however was not found comparable with DMEM and human platelet lysate in our hands (Matějková et al, unpublished data).

6.3 Choice of serum

First published results of clinical trials with MSCs used fetal calf serum (FCS) as a supplement to culture medium (Lazarus et al, 1995; Koç et al, 2000; LeBlanc et al, 2008). EMEA-compliant fetal calf serum does exist (Haack- Sørensen et al, 2008). Such a serum is produced e.g. by PAA Laboratories or Lonza. It origins in bovine spongiform encephalopathy-free countries (Australia, New Zealand) and is treated by irradiation to inactivate possible pathogens. However, fetal calf serum has several disadvantages. The first is great variability among batches with regard to MSC growth support, which necessitates expensive prescreening. The second is possibility of allergic reactions to xenogeneic protein. One group already reported presence of anti-fetal calf serum antibodies in blood of patients treated by MSCs expanded in FCS-containing medium (Sundin et al, 2007), and others reported anaphylatoxic reactions after administration of other cellular products prepared with FCS (Mackensen et al, 2000). Transmission of prion or viral diseases remains a theoretical possibility with fetal calf serum, too, though no zoonosis was reported in several thousands of patients treated with various cellular therapies manufactured with FCS so far.

As mentioned above, experiences with serum-free media are currently limited and reports are controversial. Therefore, before use of such media becomes widespread, human alternatives for FCS should be sought for. Autologous human plasma (AP) was reported to be at least comparable to FCS (Stute et al, 2004). However, given the amount of AP that may be realistically obtained (200-250 ml maximum with one blood donation) and taken into account the amount of medium needed for replacements and serial passagings, AP does not seem to be a realistic option for clinical-scale MSC manufacturing. Experiences with allogeneic human serum (HS) are controversial, too. While several groups reported early senescence of MSCs grown with HS (Stute et al, 2004), others did not observe such a phenomenon (Bierback et al, 2009). In our hands, human serum (unsupplemented) performed worse than FCS and cells grown in human serum frequently underwent early adipogenic differentiation (Pytlík et al, 2009, and unpublished data). Thrombin-activated platelet releasate in plasma (t-PRP) and pooled human platelet lysate (p-HPL) are two other FCS substitutes of human origin. While both of them take advantage of release of platelet-derived cytokines and growth factors in plasma or serum, their manufacturing and subsequently their performance are substantially different. T-PRP is prepared by adding human thrombin to the platelet concentrate, with subsequent centrifugation and filtration through 0.2 μm filter (which also sterilizes the product). It is necessary to freeze t-PRP in small aliquots and thaw it just before preparation of fresh medium. The product may need aditional centrifugation to remove possibly developing clots and heparin have to be added to the complete medium to prevent gel formation.

P-HPL may be produced either from buffy coats or from expired platelet concentrates. These are briefly centrifuged at room temperature and frozen in aliquots in -30 to -80°C. The freeze-thaw cycles may be repeated several times. One team found to be advantageous to adjust the number of residual platelets in platelet-rich plasma (after centrifugation) to $1.5x10^9$/ml, as with these numbers, the performance of p-HPL was found to be optimal (Lange et al, 2007). Before use, the p-HPL should be spun at high speed (4000-8000 g) to remove the cellular debris.

In her seminal work, Bierback et al found that pHPL have better performance than tPRP, but also than HS and FBS. Besides higher yield of MSCs, use of pHPL also led to less contamination with hematopoietic cells (Bierback et al, 2009). P-HPL may be produced in most transfusion departments under GMP conditions, and its sources are not limited as are sources of autologous serum or human plasma. Even if it is not currently known which factors or cytokines in pHPL are responsible for its efficacy, it constitutes a suitable surrogate for fetal calf serum. First clinical experiences with MSCs produced in p-HPL supplemented medium were already reported (von Bonin et al, 2009).

6.4 Supplements

One of the first researchers who studied influence of various growth factors and other supplements on MSCs grown in serum-deprived conditions were Gronthos and Simmons (Gronthos & Simmons, 1995). They studied 25 different growth factors and found that the combination of insulin, platelet-derived growth factor BB (PDGF-BB) and epidermal growth factor (EGF), together with dexamethason and ascorbic acid, led to superior yields of MSCs over other combinations.

PDGFs were first found in platelets and they might be responsible for some of the platelet lysate activity in MSC growth. Some authors described role of PDGF during osteogenic, adipogenic and chondrogenic differentiation, however, the primary effect seems to be mitogenic. PDGF also inhibits differentiation of cells including MSCs. PDGF-BB form can activate all PDGF receptors and therefore is the best choice as a culture supplement. PDGF-BB may be obtained in GMP quality (CellGenix).

EGF has similar mode of action in MSC cultures as PDGF. It acts as a mitogen (Krampera et al, 2005), and in adition it can maintain stem cell properties of hMSCs. Indeed, Kratchmarova found that EGF and PDGF signaling leads to phosphorylation of similar set of proteins (Kratchmarova et al, 2005), with the one exception, the proteins of PI3K pathway, which is phosphorylated by PDGF only. Therefore, synergism of PDGF and EGF in Gronthos & Simmons work seems a little surprising and our experiments have shown that in certain cultivation systems, their action may be redundant (Stehlík, unpublished data).
Subsequently, other growth factors were found to be useful in MSC expansion: Fibroblast growth factor 2 (beta FGF or FGF-2) was found not only to enhance growth of CFU-F colonies, but also to preserve stem cell characteristics of MSC (Tsutsumi et al, 2001; Bianchi et al, 2003). In one work, macrophage colony-stimulating factor (M-CSF) was found also to stimulate expansion of MSCs (Jin-Xiang et al, 2004). FGF-2 factor, clinical grade, is also available from CellGenix. Transforming growth factor beta (TGF-beta) is known to induce so-called epithelial-mesenchymal transition (EMT), i.e., process that enables polarized epithelial cells to acquire a motile fibroblastoid phenotype (Wendt et al, 2009). It also induces chondrogenic differentiation of mesenchymal stromal cells. However, TGF-beta was also found to promote growth of MSCs in serum-free medium, together with PDGF-BB and FGF-2 (Chase et al, 2010).

6.4.1 An example of MSC cultivation with cytokine-suplemented medium

We have developed a rapid cultivation procedure of MSCs grown in CellGro™ for Hematopoietic Stem Cells clinical grade medium supplemented with 10% human serum and five Gronthos & Simmons supplements (insulin, ascorbic acid, dexamethasone, EGF, PDGF-BB), further enhanced with FGF-2 and M-CSF. This medium, though not serum-free, enabled us to expand MSCs significantly in a single step from bone marrow mononuclear cells. Yields of MSCs were consistently above 10^6 MSCs per 10^6 seeded bone marrow mononuclear cells after two weeks of cultivation. Furthermore, medium did not require change and also hematopoietic cells did not require removal. The only manipulation was addition of supplements three times during the two week cultivation period. MSCs cultivated in this medium had phenotype comparable with MSCs cultivated in alpha-MEM + fetal calf serum and were able to differentiate to three mesodermal lineages (Pytlík et al, 2009). During futher development, we successfully transferred this technology to RoboFLASKs™ (Corning) with silicon rubber seal, which was only three times perforated by blunted needle. Cell harvest was also successful without opening the RoboFLASK™.
This cultivation method is very simple, easily transferrable to GMP environment and enables to expand enough MSCs for clinical applications during single two-weeks expansion. After further validation, it may become a solution for MSC production by smaller companies or academic facilities.

6.5 Antibiotics

In preclinical research, cultivation media are often supplemented by antibiotics, usually penicilin-streptomycin combination. However, use of antibiotics, especially beta-lactams, is not advocated for clinical-scale production, as they may mask bacterial contamination. Also, they have allergogenic potential. Aminoglycoside antibiotics may be neutralized by charcoal adsorption (Kielpinski et al, 2005), or on special membrane filters (e.g., TTHVAB210 by Millipore, Steigman et al, 2008). However, the most preferrable option is not to use antibiotics at all and to secure sterility of the product by strict adherence to principles of asepsis, rather than antisepsis.

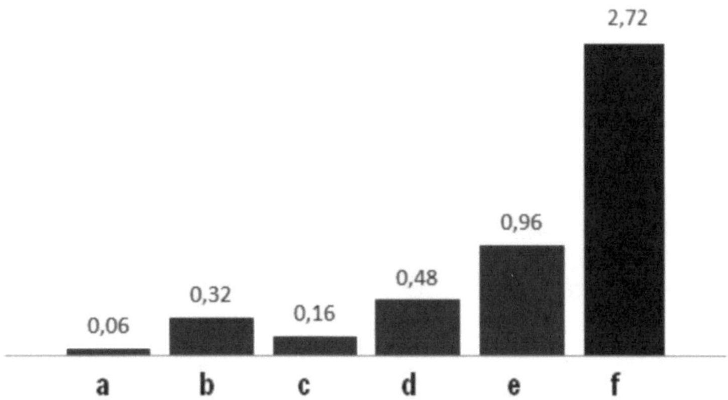

Fig. 3. Comparison of yields of MSCs grown in different media. a – alpha-MEM + FCS, b – alpha-MEM + HS, c – alpha-MEM + HS + 5 Gronthos & Simmons supplements, d – alpha-MEM + HS + 5 Gronthos & Simmons supplements + FGF-2, e – alpha-MEM + HS + 5 Gronthos & Simmons supplements + FGF-2 + M-CSF, f – CellGro™ + HS + 5 Gronthos & Simmons supplements + FGF-2 + M-CSF. Cell yields ($x10^6$) per 10^6 seeded bone marrow mononuclear cells. Adapted from Pytlík et al, 2009

6.6 Harvesting of adherent cells

Harvesting of adherent cells is usually performed by EDTA-trypsin solution. Trypsin is typically porcine, and therefore not optimal for GMP production of MSCs. An alternative, TrypLE™ Select (Invitrogen) is a recombinant bacterial enzyme, produced on dedicated animal origin-free equipment, GMP compliant. It is availlable in two strenghts and the more concentrated (10x) is recommended for MSC harvest. Its performance is at least comparable to classical EDTA-trypsin and it has been already used for clinical-grade preparation of mesenchymal stromal cells (Brooke et al, 2009). Similar performance has also Sigma product from corn, TrypZean (Carvalho et al, 2011).

6.7 Cryopreservation of cellular products

MSCs can be cryopreserved, however, as with their expansion, no method is universally accepted and data differ significantly when recovery rates from different freezing methods and formulas are reported. As most experience in GMP cryopreservation of cellular product have been made with hematopoietic progenitor cells (HPCs), in some centres, protocols derived from HPC-freezing ones are used for MSC cryopreservation as well. Also, bags used for cryopreservation of MSCs are the same as used for HPC freezing and storage. On the other hand, other teams simply extend their experience with research-grade freezing and cryopreservation to clinics.

Most cryopreservation techniques use a mixture of cell culture media, animal sera, and dimethylsulfoxide (DMSO) as a freezing solution. DMSO has been extensively used as a cryoprotectant because of its high membrane permeability. However, despite the protection this cryoprotectant offers, DMSO can be damaging to cells when used in high concentration, especially during the thawing procedure. Also, if not removed, it can cause adverse reactions in patients (nausea, vomiting, tachycardia, bradycardia, hypotension, etc.). Other cryoprotectants, as methylcellulose, sucrose, threhalose, glycerol, hydroxyethylstarch, polyvinylpyrrolidon, or various combinations of these were tested, however, in the end none of them has been found to be superior to DMSO. Therefore, the main issues in development of GMP freezing and cryopreservation protocols is the choice of serum (if any), and the adjustment of the DMSO concentration to lowest possible level.

Haack-Sørensen et al (Haack-Sørensen et al, 2007; Haack-Sørensen & Kastrup, 2011) advocates the use 5% concentrations of DMSO together with 95% fetal calf serum. Control-rate freezing method (freezing at rate of 1°C per minute) is employed, as this was clearly found to be better than uncontrolled freezing (Fuller & Devireddy, 2008). Cell concentration should be between 0.5-1x10^6/ml (Goh et al, 2007). It is essential that all procedures beginning with adding DMSO and ending with thawing, are performed at 4-8°C on ice and after thawing, the cell suspension is quickly diluted to lower the DMSO concentration. While probably not better than other protocols (the cell viability or CFU-F retrieval after thawing is not presented), this method is quite simple and may be quickly adopted for GMP conditions, if use of FCS is plausible. Autologous serum might be the best alternative to FCS, if serum is needed at all, however, its use is limited by the same problems as its use in MSC cultivation (Reuther et al, 2006). Human allogeneic serum is surely an option, too, but it still brings risk of disease transmission.

The question therefore is, if serum is needed at all for MSC cryopreservation. Other researchers found that even 2% DMSO, with culture medium (DMEM) without serum was as good as 10% DMSO with 80% human or fetal calf serum (Thirumala et al, 2010a, 2010b, 2010 c). Defined, serum-free and animal components-free freezing media, as is Cryostor™ CS 10 (StemCell Technologies, Woods et al, 2009), or Plasmalyte-A (Baxter, Steigman et al, 2008) are also availlable.

The results of freezing-thawing procedures with the respect to cell viability are controversial and difficult to compare. For similar protocols, recovery rates are as different as 50% to 90% of viable cells. Comparisons of CFU-F formation from unfrozen and frozen cells from the same passage were not reported, to our knowledge. The greatest problem with viability reports is that most researchers use only the simplest method for its evaluation, which is trypan blue exclusion. Freeze-thawing process may start early apoptosis in cryopreserved cells, while these cells still may appear as viable in the trypan blue exclusion test (Baust et al, 2002). Better tests (e.g., flowcytometric staining by $DiOC_6$ for analysis of mitochondrial transmembrane potential, together with propidiumiodine or 7-AAD to exclude dead cells) therefore should be employed. CFU-F assays, for (at least) post-hoc quality control would be also desirable, as in our hands, only half of colony forming ability could be retreieved after the freezing-thawing procerures (unpublished data). As is evident, cryopreservation of MSCs is still largerly an unresolved issue and further attempts toward its optimalization and standardization have to be performed.

7. Quality issues

7.1 Viability, clonogenicity and senescence

Viability of cells is traditionally performed by trypan blue staining in Burkers chambers. Special counting machines, which evaluate both cell concentration and tolluidin blue permeability (Countess® from Invitrogene, Cellometer® from Nexcelcom Bioscience) are also availlable, but their running is quite expensive. Flow cytometric staining by propidium iodine or 7-AAD may be more advantageous than trypan blue, especially when combined with suitable other fluorochrome to detect still viable, but early apoptotic cells. Clonogenicity of produced cells should be compared with cells produced by standard procedure by CFU-F formation (Colter et al, 2000). This test, which involves seeding cells in densities of 1.5, 3, 5, and 10 cells/cm^2 in a 100 mm Petri dish, is simple, inexpensive and has been shown in our hands to be highly reproducible. Its disadvantage, however, is the lenght of this test, which lasts from 7-14 days. This may be longer than the shelf-live of the final product. However, this testing might be useful as a part of a post-release evaluation of a product quality to ensure that possible failure of the treatment procedure was not due to poor graft quality.

Senescence is an underestimated problem in MSC production. MSCs have only limited number of population doublings, known as Hayflick limit, before senescence growth arrest occur (Hayflick, 1963). In MSC, this is typically from 20 to 50 doublings, depending on cell source and culture conditions (Izadpanah et al, 2006; Suchánek et al, 2007, Cholewa et al, 2011). Senescent cells not only cease to proliferate, but their differentiation properties are also impaired and they can differentiate to osteogenic lineage only. They can display aneuploidy without transformation (Tarte et al, 2010), and exhibit certain mutations, e.g., in p53 gene. P53 mutated MSC can migrate to mammary tissue and form an inductive microenvironment for breast cancer (Houghton et al, 2010). Senescence is easy to evaluate in non-confluent cell cultures by beta-galactosidase staining (Bandyopadhyay et al, 2005), and this test has been recently adapted for flow cytometry as well (Noppe et al, 2009). Senescent MSC have typical secretome and gene expression profiles - they secrete for example interleukin-6, matrix metalloproteinases, hepatocyte growth factor, or FGF2. These molecules can reinforce the senescent arrest or stimulate growth and invasion of established cancer (Coppé et al, 2008). Senescent gene expression profile of MSC - cultivated either with fetal calf serum or human platelet lysate - includes upregulation of hyaluronan and proteoglycan link protein 1 *(HAPLN1)*, keratin 18 *(KRT18)*, brain-derived neurotrophic factor *(BDNF)*, or renal tumor antigen *(RAGE)*, while pleiotropin *(PTN)* is downregulated (Schallmoser et al, 2010). From practical point of view, beta-galactosidase testing should become a routine pre-release quality test of MSC preparations, while rt-PCR testing for selected senescence-associated genes may be a part of post-release quality surveillance (see below).

7.2 Product characterization

The minimal set of requirements for cells to be recognized as MSCs are set in Introduction to this chapter (Dominici et al, 2006). However, testing the full set of these requirements for every batch can be challenging, as e.g. lineage differentiation lasts several weeks, which may be longer than the expiration period of the product. On the other hand, performance of various MSC products, especially in the autologous setting, may be at least partially donor dependent (Friedl et al, 2009). Therefore, a reasonable compromise has to be achieved depending on particular situation.

Full characterization of the product has to be performed in the preclinical phase of its development. These tests should show reproducible profile of surface markers, performed by flow cytometry or immunocytochemistry. Testing for surface markers may also reveal some impurities of the final product, especially admixture of hematopoietic cells. Release standards (percentage of cells positive for given antigen, acceptable amount of undesired cells) should be set at this point.

If the manufacturing procedure differs from accepted standard (e.g., from cultivation of cells in DMEM or alpha-MEM medium with 10% fetal calf serum), full tests for differentiation properties should be performed to prove that the final product complies with the minimal set of requirements for MSCs. This may involve comparison of cells produced by alternative procedure with cells produced by standard way. These comparisons should be not only qualitative (Oil red staining for adipogenic, von Kossa or alizarin red staining for osteogenic, and staining for collagen II for chondrogenic differentiation), but also quantitative (e.g., calcium accumulation or triglyceride synthesis). If desired effect of MSC product is immunosuppression (e.g., for treatment of graft-versus-host disease), allogeneic mixed leukocyte reactions with admixture of various ratios of MSCs should be performed (Le Blanc et al, 2003).

Preclinical characterization of the product may involve also gene expression profiling. In current literature, there is a number of papers describing gene expression profiles of MSC obtained from various tissues or cultivated by different methods (Wagner et al, 2005; Tsai et al, 2007; Secco et al, 2009), and it might not be necessary to repeat these expensive, cumbersome and poorly standardised experiments in full. If it is necessary, Wang et al. provide a detailed protocol how to perform gene expression analysis (Wang et al, 2011). In any way, qualitative or quantitative testing for expression of selected genes of interest (genes related to growth, stemness, differentiation properties or senescence) might be useful. This testing could also apply to final clinical products, either as pre-release or post-release control of quality.

For quality testing of fully developed and approved product intended to clinical use, it must be kept in mind that the donor variability might substantially influence the final quality of the batch. Full testing for clonality, differentiation, immunosuppressive properties or gene expression might not be possible because of the short shelf-life of the product. Even when there is an intention to freeze the product before use, its performance might be different before freezing and after thawing (our unpublished experiments). However, it might be advisable to perform these tests as a part of post-release testing, to ensure the released product was of sufficient quality and to be able to show that possible treatment failure was not due to poor quality of the MSC product.

8. Safety issues

8.1 Donor screening

In general, the same set of examinations as for blood banking purposes should be performed. Testing for hepatitis B, C, HIV and syphilis are mandatory. In certain areas of the world, testing for HTLV-1 and/or Chagas disease may also apply.

8.2 Microbial contamination

Bacterial contamination of classical pharmaceutical products is excluded by standardized tests, as set for example in European Pharmacopoiea (EP, chapter 2.6.1), or US Pharmacopoiea (USP, chapter 71). These growth promotion tests (GPT) involve two different cultivation media – Fluid thyoglicollate medium and soya-bean casein digest medium, and two different temperatures - 22.5°C and 32.5°C - for growth of each tested sample. However, this test takes 14 days to finish and is clearly unsuitable for products with short shelf-live. There are instructions on validation of rapid microbiological tests both in USP (ch. 1223) and EP (section 5.1.6). These require for alternative microbiological testing to ensure following, compared with standard GPT:

Specificity. All microbial strains must be detected and confirmed. Aerobic strains must be detected in the aerobic culture bottles. Anaerobic strains must be detected in anaerobic bottles. It should be confirmed that cell cultures themselves will not generate false positive tests. Microbial strains can be bought e.g. from ATCC (American Type Culture Collection), however, the set should also include isolates from microbiologically positive samples and from environment of the facility.

Limit of detection. Each challenge microorganism must be detected at less than 100 CFU but greater than 0 CFU.

Repeatability. All replicates inoculated with challenge microorganisms are determined to be positive.

Ruggedness. All strains must be detected and confirmed as prepared by different analysts.

Equivalence. Alternative method must detect challenge organisms sooner than the compendium method.

There are several solutions for rapid microbiological testing, but all have their advantages and disadvantages. Best comparable to pharmacopoieal methods are cultivation methods based on CO_2 detection (BACTEC™ - Becton Dickinson, BacT/ALERT® - bioMérieux), and they have already been approved for tissue products (Kielpinski et al, 2005). These are also relatively unexpensive, easy to handle and do not require much space. Results are typically obtained in 48-72 hours. DNA detecting tests (e.g., LightCycler® Septi*Fast* Test - Roche) may be more challenging to be validated, as they may not detect all possible contaminating organisms (especially the environmental isolates) and, on the other hand, they may detect DNA from unviable organisms. Also, the number of gene copies (GC) is not easy to compare directly with the number of colony-forming units (see also 8.3). However, these tests are attractive as they can detect microorganisms in less than 24 hours. Fluorescent cytometry tests (Scan*RDI*® AES Chemunex) provide ultra-rapid detection of microorganisms (90 minutes), but are very expensive and used typically by large pharmacological companies.

It have to be stressed that validation of an alternative microbiological testing method may be very laborious and time consuming and can take several years before successfully completed. This may change, as these tests are getting more widespread. Close cooperation with the regulatory agency from the very time such a method is contemplated, is necessary in any case. For a close introduction to the rapid sterility test implementation, see Gressett (Gressett et al, 2008).

A good question is what to do when final products - especially customized ones - are eventually found to be microbiologically positive. At that time already a lot of work and money have been invested in the product, not to mention a patient who might in the meantime undergo some kind of preparative procedure for cellular treatment. This is similar to situations in hematopoietic progenitor cell transplants, where even microbiologically positive graft cannot be withdrawn and discarded, as this would mean inevitable death of the patient in many cases. Positive grafts are found in wide range of 0-43% of cases (Lowder & Whelton, 2003), but surprisingly they do not appear to present unacceptable risks. In two large studies, (Patah, 2007; Phinney, 2007) the frequency of positive grafts was between 1-2%. While Phinney gave preemptive antibiotic treatment to the graft recipients, Phinney just observed them. The frequency of adverse events was zero in the first study and close to zero in second. Preemptive antibiotic treatment based on tested or presumed microbial sensitivity might be a reasonable strategy for transplantation of microbiologically positive products, under strictly controlled conditions.

8.3 Mycoplasma contamination

Mycoplasmas are microorganisms without cell wall, which may pass through sterilization with 0.2 µm filters. They have quite complex requirements for survival conditions, but cell culture media make good environment for their growth. As such, mycoplasmas present significant thread to cell and tissue cultivation. European, United States or Japanese pharmacopoieas state requirements for mycoplasma testing. Essentially, two types of tests are used: first is inoculation of cell culture samples on a solid agar or in a liquid enrichment medium, from which are mycoplasma cultures after several days transferred on agar. This test is quite sensitive (10 CFU/ml), but takes 28 days to complete. In second method, the indicator cell culture, samples are co-cultured with permissive cell lines (usually Vero cells) and then stained with fluorescent DNA-binding dyes (DAPI or Hoechst). This approach also takes time and is less sensitive than agar cultivation (100 CFU/ml).

Fortunately, there are several tests, based on nucleic acid testing (NAT), which have been already validated, though NAT is not without its problems. First, it does not distinguish dead cells from living ones. Second, the translation of gene copy numbers to colony finding units is problematic. Not only all mycoplasmas detected by NAT are not necessary live ones, but also CFU is not an equivalent to living cell – it is an expression of its ability to form typical colony. Also, cultivation methods work with larger volumes (1 to 10 ml of medium) than NAT tests (tens to hundreds µl). Therefore, enrichment of a starting material (e.g., by high-speed centrifugation) may be necessary. It has to be assured that sequences of all mycoplasmas are covered by single PCR reaction and it also has to be assured that this reaction will not amplify sequences from related microorganisms (Streptococci, Clostridia, Lactobacilli).

MycoTOOL™ (Roche Diagnostics) is a test amplifying a part of the 16S rDNA of Mycoplasmas. It was validated with the European Pharmacopoiea tests (Chapter 2.6.7.) and is able to detect Mycoplasmas with sensitivity of at least 10 CFU/ml (Deutschmann et al, 2010). A quantitative MycoSensor QPCR assay kit was developed by Stratagene and found acceptable in preclinical regulatory validation of amniotic MSC manufacturing protocol (Steigman et al, 2008). For detail description of NAT-based Mycoplasma detection techniques, problems with alternative non-microbial detection and possible other solutions, see Volokhov (Volokhov et al, 2011).

8.4 Endotoxin testing

Endotoxins are lipopolysaccharides from gram-negative bacteria and are the most common cause of toxic reactions resulting from contaminations with pyrogens. Reactions to endotoxin can cause serious health problems, as is diarrhea, septic shock, marrow necrosis and others (Opal & Steven, 2007). Testing for endotoxins is therefore a standard release test for cellular and gene therapy products. The acceptable level of endotoxin in these products is usually 5.0 EU/kg/dose.

Endotoxin is usually tested with the Limulus Amebocyte Lysate (LAL) method. The problem with this test is not the time (results can be usually obtained in 3-4 hours), but its sensitivity to external factors and complexity of its setting. Endosafe® PTSTM is a chromogenic LAL test that provides quantitative results in approximately 15 minutes (Gee et al, 2008). It has been already validated for testing of bone marrow mononuclear cells for cardiac regeneration (Soncin et al, 2009), and is relatively easy to use. For other applications, however, comparison with standard accepted method may still be necessary.

8.5 Tumorigenicity

There were several reports of spontaneous transformation of human MSC in cultures (Rubio et al, 2005; Wang et al, 2005; Rosland et al, 2009). Most, if not all, these results reflect cross-contamination of mesenchymal cultures with exogenous tumor cell lines (Torsvik et al, 2010), which hardly can be a concern in a well-conducted GMP facility. However, transformation of MSC was observed after prolonged cultivation in human telomerase immortalized cells (Serakinci et al, 2004). This should not again cause concern in production of non-manipulated MSC, however, it shows to potential danger in case MSC were genetically manipulated. Furthermore, these immortalized transformed MSCs lost the $p16^{ink4a}$ gene, which was shown to occur occasionally even in non-immortalized MSC cultures (Shibata et al, 2007). In conclusion, risk of spontaneous malignant transformation of human MSC products does not seem to be very high. The question of routine cytogenetic testing of MSC product has to take in account the fact of low sensitivity of classical cytogenetic examination which may easily miss potentially dangerous but still very small clone and certainly will miss most losses of heterozygozity or similar small genetic changes. Also, cytogenetic testing can lead to false-positive results, as it was shown that aneuploidy might in fact be quite common in MSC undergoing senescence, but not transformation (Tarte et al, 2010).

8.6 Clinical safety and surveillance principles

As the experience with somatic cell therapy is still limited, there are no universally applicable principles of clinical safety monitoring. Until sufficient information will be availlable, all recipients of somatic cell therapy, including the treatment with MSC, should be followed indefinitely (for a lifetime), and monitored for possible adverse effects of treatment. Adverse events should be collected in context of the clinical trials in the premarketing phase, and according to general pharmacovigilance principles in the postmarketing phase.

Possible acute complications connected with mesenchymal stromal cell treatment, as perceived from preclinical evaluation, are infusion related complications, immunological reactions (more probable with use of xenogeneic proteins during MSC production and/or after repeated use), and local reactions (with local application). Significant number of MSCs, especially from the Stro-1^{-} fraction, engrafts in lungs (Devine et al, 2003; Bensidhoum et al, 2004), and lungs are the first organ attended by intravenously administered MSCs. Therefore, the possibility of MSC induced lung injury have to be taken seriously. In experimental animals, administration of large numbers of MSCs may cause stroke or even death. Therefore, it is desirable that all systemically treated patients would be closely monitored during the infusion and some time thereafter.

MSCs are little immunogenic and immune reactions caused by their administration therefore should not be problem. However, when cultivated in xenogeneic protein-containing systems, they may internalize and present these proteins to recipient. This problem may become significant especially if repeated administrations of MSCs are planned (e.g., for graft versus host disease treatment), as first exposure of xenogeneic protein may cause priming of the recipient immune system and subsequent administrations may trigger an allergic reaction.

In the long-term follow-up, three issues seem to be particularly important: "maldifferentiation" of MSCs, tumor propagation, and disease transmission.

"Maldifferentiation" of MSCs refer to differentiation in a tissue type not desired in the particular organ. It was shown that MSCs, in contrast to hematopoietic progenitor cells, produce calcifications after local injection to an infarcted heart (Breitbach et al, 2007). In another model of glomerular injury, MSCs prevented progressive renal failure when administered intraarterially to rats, but degraded in kidney to fat cells, surounded by fibrotic tissue (Kunter et al, 2007). To our knowledge, nothing similar was observed after intravenous infusion or in humans, but similar undesired effects of human MSCs cannot be excluded.

For tumor formation, patients should be followed indefinitely. As shown above, the risks of spontaneous transformation of human MSCs are probably very small, however, there are concerns that MSCs may support tumor growth by a variety of mechanisms, involving immunosuppression, transformation of MSCs to CAFs (cancer associated fibroblasts), or tumor vasculature support (Momin et al, 2010; Klopp et al, 2011). Human MSCs have been shown to promote tumor development in several animal models (Zhu et al, 2006; Karnoub et al, 2007). In clinical practice, rather than facilitating growth of previously undiagnosed tumors, MSCs may promote tumor growth when applied to patients with established cancer, for example in hematopoietic cell transplantation setting. There is one report showing that patients who had cotransfused MSCs together with hematopoietic progenitor cells, had less graft versus host disease but more leukemia relapses (Ning et al, 2008). On the other hand, there are also reports that unmanipulated MSCs may also suppress tumor growth (Khakoo et al, 2006; Qiao et al, 2008). Large series of patients, optimally in randomized clinical trials, need to be followed for the frequency of various types of spontaneous tumors before the tumorigenicity of human MSCs may be excluded. The question of tumorigenesis will undoubtebly become even more significant if genetically engineered MSCs will be used for treatment of cancer or metabolic diseases, however, this is beyond the scope of this chapter (reviewed in Aboody et al, 2008 and Momin et al, 2010).

To our knowledge, disease transmission was not reported yet after mesenchymal stem cell therapy. Usual infection surveillance should be sufficient. Infectious origin of any febrile reaction during and after MSC application should be excluded and MSC recipients should be tested for hepatitis or HIV transmission in a fixed time after cellular therapy. If blood-transmitted infection is confirmed in recipient, donor of MSC should be investigated as well, in case of allogeneic therapy. If fetal calf serum is used for MSC expansion, it must be from bovine spongiform encephalopathy-free area, as noted above.

9. Conclusion

Mesenchymal stromal cell therapy offers solutions for a number of currently unmet clinical needs in modern medicine. These solutions might be less than optimal in certain cases, or may not fulfill the expectations at all. Mesenchymal stromal cell therapy may well provide only temporary clinical solutions, before better understanding of underlying principles of diseases and their treatment become availlable and better treatment approaches (e.g., targeted delivery systems for cytokines and gene products, small molecules, etc.) will be developed. At this time, however, it seems that somatic cell therapy is worth exploring, despite the new challenges connected with it.

Because of the complex regulatory requirements, cell therapy will probably be very expensive and during this time, when its safety and efficacy are being tested, it will be difficult to find reimbursement of expenses connected with its development. An inequality in access to new treatments may result on one hand and difficulties with accrual of patients to clinical trials on the other. Therefore, it is crucial that all involved in mesenchymal stromal cell treatment, including funding institutions, regulatory institutions, academic facilities and private subjects, would cooperate closely together, on national or international platforms. On these platforms, fabrication of GMP-compatible facilities and development of GMP prepared products for cellular therapy will undoubtedly prove to be crucial in transferring the experimental knowledge into clinical practice. Falling behind the international level of knowledge and experience may have very undesired effects on health care in underdeveloped countries or regions. The purpose of this chapter was to provide at least partial solutions to challenges in this exciting new area of medicine.

10. Acknowledgment

This work was supported by grant from Czech Ministry of Schools and Education, MSM 0021620808, and TA01010964 from Technology Agency of Czech Republic.

11. References

Aboody, K.S.; Najbauer, J. & Danks, M.K. (2008). Stem and progenitor cell-mediated tumor selective gene therapy. *Gene Therapy*, Vol.15, No.10, pp. 739-752, ISSN 0969-7128

Akita, S.; Akino, K.; Hirano, A.; Ohtsuru, A. & Yamashita, S. (2010). Noncultured autologous adipose-derived stem cells therapy for chronic radiation injury. *Stem Cells International*, 2010 Dec 1;2010:532704, ISSN 1687-9678

Bandyopadhyay, D.; Gatza, C.; Donehower, L.A. & Medrano, E.E. (2005). Analysis of Cellular Senescence in Culture In Vivo: The Senescence-Associated β-Galactosidase Assay. In: *Current Protocols in Cell Biology*, J.S. Bonifacino, (Ed.), 18.9.1–18.9.9., Wiley & Sons, Online ISBN: 9780471143031

Barlow, S.; Brooke, G.; Chatterjee, K.; Price, G.; Pelekanos, R.; Rossetti, T.; Doody, M.; Venter, D.; Pain, S.; Gilshenan, K. & Atkinson, K. (2008). Comparison of human placenta- and bone marrow-derived multipotent mesenchymal stem cells. *Stem Cells and Development*, Vol.17, No.6, pp.1095-1107, ISSN 1547 3287

Battula, V.L.; Treml, S.; Bareiss, P.M.; Gieseke, F.; Roelofs, H.; de Zwart, P.; Müller, I.; Schewe, B.; Skutella, T.; Fibbe, W.E.; Kanz, L. & Bühring HJ. (2009). Isolation of functionally distinct mesenchymal stem cell subsets using antibodies against CD56, CD271, and mesenchymal stem cell antigen-1. *Haematologica*, Vol.94, No.2, pp.173-184, ISSN 0390 6078

Baust, J.M.; Van Buskirk, R. & Baust, J.G. (2002). Gene Activation of the Apoptotic Caspase Cascade Following Cryogenic Storage. *Cell Preservation Technology*, Vol.1, No.1, pp.63-80, ISSN 1538 344X

Bensidhoum, M.; Chapel, A.; Francois, S.; Demarquay, C.; Mazurier, C.; Fouillard, L.; Bouchet, S.; Bertho, J.M.; Gourmelon, P.; Aigueperse, J.; Charbord, P.; Gorin, N.C.; Thierry, D. & Lopez, M. (2004). Homing of in vitro expanded Stro-1- or Stro-1+ human mesenchymal stem cells into the NOD/SCID mouse and their role in supporting human CD34 cell engraftment. *Blood,* Vol.103, No.9, pp. 3313-3319, ISSN 0006 4971

Beyth, S.; Borovsky, Z.; Mevorach, D.; Liebergall, M.; Gazit, Z.; Aslan, H.; Galun, E. & Rachmilewitz, J. (2005). Human mesenchymal stem cells alter antigen-presenting cell maturation and induce T-cell unresponsiveness. *Blood,* Vol.105, No.5, pp. 2214-2219, ISSN 0006 4971

Bianchi, G.; Banfi, A.; Mastrogiacomo, M.; Notaro, R.; Luzzatto, L.; Cancedda, R. & Quarto, R. (2003). Ex vivo enrichment of mesenchymal cell progenitors by fibroblast growth factor 2. Experimental Cell Research, Vol.287, No.1, pp.98-105, ISSN 0014 4827

Bieback, K.; Hecker, A.; Kocaömer, A.; Lannert, H.; Schallmoser, K.; Strunk, D. & Klüter, H. (2009). Human alternatives to fetal bovine serum for the expansion of mesenchymal stromal cells from bone marrow. Stem Cells, Vol. 27, No.9, pp. 2331-2341, ISSN 1066 5099

Blau, O.; Hofmann, W.K.; Baldus, C.D.; Thiel, G.; Serbent, V.; Schümann, E.; Thiel, E. & Blau, I.W. (2007). Chromosomal aberrations in bone marrow mesenchymal stroma cells from patients with myelodysplastic syndrome and acute myeloblastic leukemia. *Experimental Hematology,* Vol.35, No.2, pp. 221-229; ISSN 0301 472X

von Bonin, M.; Stölzel, F.; Goedecke, A.; Richter, K.; Wuschek, N.; Hölig, K.; Platzbecker, U.; Illmer, T.; Schaich, M.; Schetelig, J.; Kiani, A.; Ordemann, R.; Ehninger, G.; Schmitz, M. & Bornhäuser, M. (2009). Treatment of refractory acute GVHD with third-party MSC expanded in platelet lysate-containing medium. *Bone Marrow Transplantation,* Vol.43, No.3, pp.245-251, ISSN 0268 3369

Breitbach, M.; Bostani, T.; Roell, W.; Xia, Y.; Dewald, O.; Nygren, J.M.; Fries, J.W.; Tiemann, K.; Bohlen, H.; Hescheler, J.; Welz, A.; Bloch, W.; Jacobsen, S.E. & Fleischmann, B.K. (2007). Potential risks of bone marrow cell transplantation into infarcted hearts. *Blood,* Vol.110, No.4, pp. 1362-1369, ISSN 0006 4971

Brooke, G.; Rossetti, T.; Pelekanos, R.; Ilic, N.; Murray, P.; Hancock, S.; Antonenas, V.; Huang, G.; Gottlieb, D.; Bradstock, K. & Atkinson, K. (2009). Manufacturing of human placenta-derived mesenchymal stem cells for clinical trials. *British Journal of Haematology,* Vol.144, No.4, pp. 571-579, ISSN 0007 1048

Carvalho, P.P.; Wu, X.; Yu, G.; Dietrich, M.; Dias, I.R.; Gomes, M.E.; Reis, R.L. & Gimble, J.M. (2011). Use of animal protein-free products for passaging adherent human adipose-derived stromal/stem cells. *Cytotherapy,* 2011, Jan 3. [Epub ahead of print]. ISSN 1465 3249

Chochola, M.; Pytlík, R.; Kobylka, P.; Skalická, L.; Kideryová, L.; Beran, S.; Varejka, P.; Jirát, S.; Kriváriek, J.;, Aschermann, M. & Linhart A. (2008). Autologous intra-arterial infusion of bone marrow mononuclear cells in patients with critical leg ischemia. *International Angiology,* Vol.27, No.4, pp. 281-290, ISSN 0392 9590

Cholewa, D.; Stiehl, T.; Schellenberg, A.; Bokermann, G.; Joussen, S.; Koch, C.; Walenda, T.; Pallua, N.; Marciniak-Czochra, A.; Suschek, C.V. & Wagner, W. (2011). Expansion of Adipose Mesenchymal Stromal Cells is affected by Human Platelet Lysate and Plating Density. *Cell Transplantation,* Mar 7 [Epub ahead of print], ISSN 0963 6879

Coelho, M.J.; Cabral, A.T. & Fernande, M.H. (2000). Human bone cell cultures in biocompatibility testing. Part I: osteoblastic differentiation of serially passaged human bone marrow cells cultured in alpha-MEM and in DMEM. *Biomaterials,* Vol.21, No.11, pp.1087-1094, ISSN 0142 9612

Corcione, A.; Benvenuto, F.; Ferretti, E.; Giunti, D.; Cappiello, V.; Cazzanti, F.; Risso, M.; Gualandi, F.; Mancardi, G.L.; Pistoia, V. & Uccelli, A. (2006). Human mesenchymal stem cells modulate B-cell functions. *Blood,* Vol.107, No.1, pp. 367-372, ISSN 0006 4971

Colter, D.C.; Class, R.; DiGirolamo, C.M. & Prockop, D.J. (2000). Rapid expansion of recycling stem cells in cultures of plastic-adherent cells from human bone marrow. *Proceedings of the National Academy of Sciences of the United States of America,* Vol.97, No.7, pp. 3213-3218, ISSN 0027 8424

Committee for Advanced Therapies (CAT); CAT Scientific Secretariat, Schneider, C.K.; Salmikangas, P.; Jilma, B.; Flamion, B.; Todorova, L.R.; Paphitou, A.; Haunerova, I.; Maimets, T.; Trouvin, J.H; Flory, E.; Tsiftsoglou, A.; Sarkadi, B.; Gudmundsson, K.; O'Donovan, M.; Migliaccio, G.; Ancāns, J.; Maciulaitis, R.; Robert, J.L.; Samuel, A.; Ovelgönne, J.H.; Hystad, M.; Fal, A.M.; Lima, B.S.; Moraru, A.S.; Turcáni, P.; Zorec, R.; Ruiz, S.; Akerblom, L.; Narayanan, G.; Kent, A.; Bignami, F.; Dickson, J.G.; Niederwieser, D.; Figuerola-Santos, M.A.; Reischl, I.G.; Beuneu, C.; Georgiev, R.; Vassiliou, M.; Pychova, A.; Clausen, M.; Methuen, T.; Lucas, S.; Schüssler-Lenz, M.; Kokkas, V.; Buzás, Z.; MacAleenan, N.; Galli, M.C.; Linē, A.; Gulbinovic, J.; Berchem, G.; Fraczek, M.; Menezes-Ferreira, M.; Vilceanu, N.; Hrubisko, M.; Marinko, P.; Timón, M.; Cheng, W.; Crosbie, G.A.; Meade, N.; di Paola, M.L.; VandenDriessche, T.; Ljungman, P.; D'Apote, L.; Oliver-Diaz, O.; Büttel, I. & Celis, P. Challenges with advanced therapy medicinal products and how to meet them. (2010). *Nature Reviews Drug Discovery,* Vol.9, No.3, pp. 195-201, ISSN 1474 1776

Coppé, J.P.; Patil, C.K.; Rodier, F.; Sun, Y.; Muñoz, D.P.; Goldstein, J.; Nelson, P.S.; Desprez, P.Y. & Campisi, J. (2008). Senescence-associated secretory phenotypes reveal cell-nonautonomous functions of oncogenic RAS and the p53 tumor suppressor. *PLoS Biology,* Vol.6, No.12, pp. 2853-2868, ISSN 1544 9173

Crisan, M.; Yap, S.; Casteilla, L.; Chen, C.W.; Corselli, M.; Park, T.S.; Andriolo, G.; Sun, B.; Zheng, B.; Zhang, L.; Norotte, C.; Teng, P.N.; Traas, J.; Schugar, R.; Deasy, B.M.; Badylak, S.; Buhring, H.J.H.; Giacobino, J.P.; Lazzari, L.; Huard, J. & Peault, B. (2008). A perivascular origin for mesenchymal stem cells in multiple human organs. *Cell Stem Cell,* Vol.3, No. , pp. 301-313, ISSN 1934 5909

De Ugarte, D.A.; Morizono, K.; Elbarbary, A.; Alfonso, Z.; Zuk, P.A.; Zhu, M.; Dragoo, J.L.; Ashjian, P.; Thomas, B.; Benhaim, P.; Chen, I.; Fraser, J. & Hedrick, M.H. (2003). Comparison of multi-lineage cells from human adipose tissue and bone marrow. *Cells, Tissues, Organs,* Vol.174, No.3, pp. 101-109, ISSN 1422 6405

Deutschmann, S.M.; Kavermann, H. & Knack, Y. (2010). Validation of a NAT-based Mycoplasma assay according European Pharmacopoiea. *Biologicals,* Vol.38, No.2, pp. 238-248, ISSN 1045 1056

Devine, S.M.; Cobbs, C.; Jennings, M.; Bartholomew, A. & Hoffman, R. (2003). Mesenchymal stem cells distribute to a wide range of tissues following systemic infusion into nonhuman primates. *Blood,* Vol.101, No.8, pp. 2999-3001, ISSN 0006 4971

Dominici, M.; Le Blanc, K.; Mueller, I.; Slaper-Cortenbach, I.; Marini, F.; Krause, D.; Deans, R.; Keating, A.; Prockop, D.J. & Horwitz, E. (2006). Minimal criteria for defining multipotent mesenchymal stromal cells. The International Society for Cellular Therapy position statement. *Cytotherapy,* Vol.8, No.4, pp. 315-317, ISSN 1465 3249

Fennema, E.M.; Renard, A.J.; Leusink, A.; van Blitterswijk, C.A. & de Boer, J. (2009). The effect of bone marrow aspiration strategy on the yield and quality of human mesenchymal stem cells. *Acta Orthopaedica,* Vol.80, No.5, pp. 618-621, ISSN 1745 3674

Flynn, A.; Barry, F. & O'Brien, T. (2007). UC blood-derived mesenchymal stromal cells: an overview. *Cytotherapy,* Vol.9, No.8, pp. 717-726, ISSN 1465 3249

Friedenstein, A.J.; Gorskaja, J.F. & Kulagina, N.N. (1976). Fibroblast precursors in normal and irradiated mouse hematopoietic organs. *Experimental Hematology,* Vol.4, No. , pp. 267-274, ISSN 0301 472X

Friedl, G.; Windhager, R.; Schmidt, H. & Aigner, R. (2009). The osteogenic response of undifferentiated human mesenchymal stem cells (hMSCs) to mechanical strain is inversely related to body mass index of the donor. *Acta Orthopaedica,* Vol.80, No.4, pp. 491-498, ISSN 1745 3674

Fraser, J.K.; Wulur, I.; Alfonso, Z.; Zhu, M. & Wheeler, E.S. (2007). Differences in stem and progenitor cell yield in different subcutaneous adipose tissue depots. *Cytotherapy,* Vol.9, No.5, pp. 459-467, ISSN 1465 3249

Fuller, R. & Devireddy, R.V. (2008). The effect of two different freezing methods on the immediate post-thaw membrane integrity of adipose tissue derived stem cells. *International Journal of Heat and Mass Transfer,* Vol.51, No.23-24, pp.5650-5654, ISSN 0017 9310

Garayoa, M.; Garcia, J.L.; Santamaria, C.; Garcia-Gomez, A.; Blanco, J.F.; Pandiella, A.; Hernández, J.M.; Sanchez-Guijo, F.M.; del Cañizo, M.C.; Gutiérrez, N.C. & San Miguel, J.F. (2009). Mesenchymal stem cells from multiple myeloma patients display distinct genomic profile as compared with those from normal donors. *Leukemia,* Vol.23, No.8, pp. 1515-1527, ISSN 0887 6924

Gee, A.P.; Sumstad, D.; Stanson, J.; Watson, P.; Proctor, J.; Kadidlo, D.; Koch, E.; Sprague, J.; Wood, D.; Styers, D.; McKenna, D.; Gallelli, J.; Griffin, D.; Read, E.J.; Parish, B. & Lindblad R. (2008). A multicenter comparison study between the Endosafe PTS rapid-release testing system and traditional methods for detecting endotoxin in cell-therapy products. *Cytotherapy,* Vol.10, No.4, pp.427-435, ISSN 1465 3249

Gimble, J. & Guilak, F. Adipose-derived adult stem cells: isolation, characterization, and differentiation potential. *Cytotherapy,* Vol. 5, No.5, pp. 362-369, ISSN 1465 3249

Goff, L.A.; Boucher, S.; Ricupero, C.L.; Fenstermacher, S.; Swerdel, M.; Chase, L.G.; Adams, C.C.; Chesnut, J.; Lakshmipathy, U. & Hart, R.P. (2008). Differentiating human multipotent mesenchymal stromal cells regulate microRNAs: prediction of microRNA regulation by PDGF during osteogenesis. *Experimental Hematology,* Vol.36, No.10, pp.1354-1369, ISSN 0301 472X

Goh, B.C.; Thirumala, S.; Kilroy, G.; Devireddy, R.V. & Gimble, J.M. (2007). Cryopreservation characteristics of adipose-derived stem cells: maintenance of differentiation potential and viability. *Journal of Tissue Engineering and Regenerative Medicine,* Vol.1, No.4, pp. 322-324, ISSN 1932 6254

Gressett, G.; Vanhaecke, E. & Moldenhauer, J. (2008). Why and how to implement a rapid sterility test. *PDA Journal of Pharmacological Science and Technology,* Vol.62, No.6, pp. 429-444, ISSN 1079 7440

Gronthos, S. & Simmons, P.J. (1995). The growth factor requirements of STRO-1-positive human bone marrow stromal precursors under serum-deprived conditions in vitro. Blood, Vol.85, No.4, pp. 929-940, ISSN 0006 4971

Haack-Sørensen, M.; Bindslev, L.; Mortensen, S.; Friis, T. & Kastrup, J. (2007). The influence of freezing and storage on the characteristics and functions of human mesenchymal stromal cells isolated for clinical use. *Cytotherapy,* Vol.9, No.4, pp. 328-337, ISSN 1465 3249

Haack- Sørensen, M.; Friis, T.; Bindslev, L.; Mortensen, S.; Johnsen, H.E. & Kastrup, J. (2008). Comparison of different culture conditions for human mesenchymal stromal cells for clinical stem cell therapy. *Scandinavian Journal of Clinical and Laboratory Investigation,* Vol.68, No.3, pp. 192-203, ISSN 0036 5513

Haack-Sørensen, M. & Kastrup, J. (2011). Cryopreservation and revival of mesenchymal stromal cells. In: *Mesenchymal Stem Cell Assays and Applications,* Vemuri, M.C.; Chase, LG. & Rao M.S. (eds.), Methods in Molecular Biology, vol. 698, pp. 161-174, Springer Science+Business Media, ISBN 978-1-60761-998-7

Hare, J.M.; Traverse, J.H.; Henry, T.D.; Dib, N.; Strumpf, R.K.; Schulman, S.P.; Gerstenblith, G.; DeMaria, A.N.; Denktas, A.E.; Gammon, R.S.; Hermiller, J.B. Jr; Reisman, M.A.; Schaer, G.L. & Sherman, W. (2009). A randomized, double-blind, placebo-controlled, dose-escalation study of intravenous adult human mesenchymal stem cells (prochymal) after acute myocardial infarction. *Journal of the American College of Cardiology,* Vol.54, No.24, pp. 2277-2286, ISSN 0735 1097

Hayflick, L. (1963). The limited in vitro lifetime of human diploid cell strains. *Experimental Cell Research,* Vol.37, pp. 614-636, ISSN 0014 4827

Hernigou, P.; Poignard, A.; Beaujean, F. & Rouard, H. (2005). Percutaneous autologous bone-marrow grafting for nonunions. Influence of the number and concentration of progenitor cells. *The Journal of Bone and Joint Surgery. American Volume,* Vol.87, No.7, pp. 1430-1437, ISSN 0021 9355

Horn, P.; Bork, S.; Diehlmann, A.; Walenda, T.; Eckstein, V.; Ho, A.D. & Wagner, W. (2008). Isolation of human mesenchymal stromal cells is more efficient by red blood cell lysis. *Cytotherapy,* Vol.10, No.7, pp. 676-685, ISSN 1465 3249

Horn, P.; Bork, S. & Wagner, W. Standardized Isolation of Human Mesenchymal Stromal Cells with Red Blood Cell Lysis. (2011). In: *Mesenchymal Stem Cell Assays and Applications,* Vemuri, M.C.; Chase, LG. & Rao M.S. (eds.), Methods in Molecular Biology, vol. 698, pp. 23-35, Springer Science+Business Media, ISBN 978-1-60761-998-7

Horwitz, E.M. & Dominici M. (2008). How do mesenchymal stromal cells exert their therapeutic benefit? *Cytotherapy,* Vol.10, No.8, pp. 771-774, ISSN 1465 3249

Houghton, J.; Li, H.; Fan, X.; Liu, Y.; Liu, J.H.; Rao, V.P.; Poutahidis, T.; Taylor, C.L.; Jackson, E.A.; Hewes, C.; Lyle, S.; Cerny, A.; Bowen, G.; Cerny, J.; Moore, N.; Kurt-Jones, E.A. & Erdman SE. (2010). Mutations in bone marrow-derived stromal stem cells unmask latent malignancy. *Stem Cells and Development,* Vol.19, No.8, pp. 1153-1166, ISSN 1547 3287

Izadpanah, R.; Trygg, C.; Patel, B.; Kriedt, C.; Dufour, J.; Gimble, J.M. & Bunnell, B.A. (2006). Biologic properties of mesenchymal stem cells derived from bone marrow and adipose tissue. *Journal of Cellular Biochemistry,* Vol.99, No.5, pp. 1285-1297, ISSN 0730 2312

Jarocha, D.; Lukasiewitz, E. & Majka, M. (2008). Adventage of mesenchymal stem cells (MSC) expansion directly from purified bone marrow CD105+ and CD271+ cells. *Folia Histochemica et Cytobiologica,* Vol. 46, No.3, pp. 307-314, ISSN 0239 8508

Jin-Xiang, F.; Xiaofeng, S.; Jun-Chuan, Q.; Yan, G. & Xue-Guang, Z. (2004). Homing efficiency and hematopoietic reconstitution of bone marrow-derived stroma cells expanded by recombinant human macrophage-colony stimulating factor in vitro. *Experimental Hematology,* Vol.32, No.12, pp. 1204-1211, ISSN 0301 472X

Jones, B.J.; Brooke, G.; Atkinson, K. & McTaggart, S.J. (2007). Immunosuppression by placental indoleamine 2,3-dioxygenase: a role for mesenchymal stem cells. *Placenta,* Vol.28, No.11-12, pp.1174-1181, ISSN 0143 4004

Jurgens, W.J.; Oedayrajsingh-Varma, M.J.; Helder, M.N.; Zandiehdoulabi, B.; Schouten, T.E.; Kuik, D.J.; Ritt, M.J. & van Milligen, F.J. (2008). Effect of tissue-harvesting site on yield of stem cells derived from adipose tissue: implications for cell-based therapies. *Cell and Tissue Research,* Vol. 332, No.3, pp. 415-426, ISSN 0302 766X

Karnoub, A.E.; Dash, A.B.; Vo, A.P.; Sullivan, A.; Brooks, M.W.; Bell, G.W.; Richardson, A.L.; Polyak, K.; Tubo, R. & Weinberg RA. (2007). Mesenchymal stem cells within tumour stroma promote breast cancer metastasis. *Nature,* Vol.449, No.7162, pp. 557-563, ISSN 0028 0836

Kern, S.; Eichler, H.; Stoeve, J.; Klüter, H. & Bieback, K. (2006). Comparative analysis of mesenchymal stem cells from bone marrow, umbilical cord blood, or adipose tissue. *Stem Cells,* Vol.24, No.5, pp. 1294-1301, ISSN 1066 5099

Khakoo, A.Y.; Pati, S.; Anderson, S.A.; Reid, W.; Elshal, M.F.; Rovira, I.I.; Nguyen, A.T.; Malide, D.; Combs, C.A.; Hall, G.; Zhang, J.; Raffeld, M.; Rogers, T.B.; Stetler-Stevenson, W.; Frank, J.A.; Reitz, M. & Finkel, T. (2006). Human mesenchymal stem cells exert potent antitumorigenic effects in a model of Kaposi's sarcoma. *Journal of Experimental Medicine,* Vol.203, No.5, pp. 1235-1247, ISSN 0022 1007

Kielpinski, G.; Prinzi, S.; Duguid, J. & du Moulin, G. (2005). Roadmap to approval: use of an automated sterility test method as a lot release test for Carticel, autologous cultured chondrocytes. *Cytotherapy,* Vol.7, No.6, pp. 531-541, ISSN 1465 3249

Klopp, A.H.; Gupta, A.; Spaeth, E.; Andreeff, M. & Marini F 3rd. (2011). Concise review: Dissecting a discrepancy in the literature: do mesenchymal stem cells support or suppress tumor growth? *Stem Cells,* Vol.29, No.1, pp. 11-19, ISSN 1066 5099

Koç, O.N.; Gerson, S.L.; Cooper, B.W.; Dyhouse, S.M.; Haynesworth, S.E.; Caplan, A.I. & Lazarus, H.M. Rapid hematopoietic recovery after coinfusion of autologous-blood stem cells and culture-expanded marrow mesenchymal stem cells in advanced breast cancer patients receiving high-dose chemotherapy. *Journal of Clinical Oncology,* Vol.18, No.2, pp. 307-316, ISSN 0732 183X

Krampera, M.; Pasini, A.; Rigo, A.; Scupoli, M.T.; Tecchio, C.; Malpeli, G.; Scarpa, A.; Dazzi, F.; Pizzolo, G. & Vinante, F. (2005). HB-EGF/HER-1 signaling in bone marrow mesenchymal stem cells: inducing cell expansion and reversibly preventing multilineage differentiation. Blood, Vol.106, No.1, pp.59-66, ISSN 0006 4971

Kratchmarova, I.; Blagoev, B.; Haack-Sorensen, M.; Kassem, M. & Mann M. (2005). Mechanism of divergent growth factor effects in mesenchymal stem cell differentiation. *Science,* Vol.308, No.5727, pp. 1472-1477, ISSN 0036 8075

Kunter, U.; Rong, S.; Boor, P.; Eitner, F.; Müller-Newen, G.; Djuric, Z.; van Roeyen, C.R.; Konieczny, A.; Ostendorf, T.; Villa, L.; Milovanceva-Popovska, M.; Kerjaschki, D. & Floege, J. (2007). Mesenchymal stem cells prevent progressive experimental renal failure but maldifferentiate into glomerular adipocytes. *Journal of American Society of Nephrology,* Vol.18, No.6, pp. 1754-1764, ISSN 1046 6673

Lange, C.; Cakiroglu, F.; Spiess, A.N.; Cappallo-Obermann, H.; Dierlamm, J. & Zander, A.R. (2007). Accelerated and safe expansion of human mesenchymal stromal cells in animal serum-free medium for transplantation and regenerative medicine. *Journal of Cellular Physiology,* Vol.213, No.1, pp.18-26, ISSN 0021 9541

Lazarus, H.M.; Haynesworth, S.E.; Gerson, S.L.; Rosenthal, N.S. & Caplan, A.I. (1995). Ex vivo expansion and subsequent infusion of human bone marrow-derived stromal progenitor cells (mesenchymal progenitor cells): implications for therapeutic use. *Bone Marrow Transplantation,* Vol.16, No.4, pp. 557-564, ISSN 0268 3369

Le Blanc, K.; Tammik, L.; Sundberg, B.; Haynesworth, S.E. & Ringdén, O. (2003) Mesenchymal stem cells inhibit and stimulate mixed lymphocyte cultures and mitogenic responses independently of the major histocompatibility complex. *Scandinavian Journal of Immunology,* Vol.57, No.1, pp. 11-20, ISSN 0300 9475

Le Blanc, K.; Frassoni, F.; Ball, L.; Locatelli, F.; Roelofs, H.; Lewis, I.; Lanino, E.; Sundberg, B.; Bernardo, M.E.; Remberger, M.; Dini, G.; Egeler, R.M.; Bacigalupo, A.; Fibbe, W.; Ringdén, O. & Developmental Committee of the European Group for Blood and Marrow Transplantation. (2008). Mesenchymal stem cells for treatment of steroid-resistant, severe, acute graft-versus-host disease: a phase II study. *Lancet,* Vol.371, No.9624, pp: 1579-1586; ISSN 0140 6736

Lowder, J.N.; Whelton, P. (2003). Microbial contamination of cellular products for hematolymphoid transplantation therapy: assessment of the problem and strategies to minimize the clinical impact. *Cytotherapy,* Vol.5, No.5, pp. 377-390, ISSN 1465 3249

Mackensen, A.; Dräger, R.; Schlesier, M.; Mertelsmann, R. & Lindemann A. (2000). Presence of IgE antibodies to bovine serum albumin in a patient developing anaphylaxis after vaccination with human peptide-pulsed dendritic cells. *Cancer Immunology, Immunotherapy,* Vol.49, No.3, pp.152-156, ISSN 0340 7004

Majd, H.; Wipff, P.J.; Buscemi, L.; Bueno, M.; Vonwil, D.; Quinn, T.M. & Hinz B. (2009). A novel method of dynamic culture surface expansion improves mesenchymal stem cell proliferation and phenotype. *Stem Cells,* Vol.27, No.1, pp. 200-209, ISSN 1066 5099

Mathiasen, A.B.; Haack-Sørensen, M. & Kastrup, J. (2009). Mesenchymal stromal cells for cardiovascular repair: current status and future challenges. *Future Cardiology,* Vol.5, No.6, pp. 605-617, ISSN 1479 6678

Mojallal, A.; Auxenfans, C.; Lequeux, C.; Braye, F. & Damour, O. (2008). Influence of negative pressure when harvesting adipose tissue on cell yield of the stromal-vascular fraction. *Bio-Medical Materials and Engineering,* Vol.18, No.4-5, pp. 193-197, ISSN 0959 2989

Momin, E.N.; Vela, G.; Zaidi, H.A. & Quiñones-Hinojosa, A. (2010). The Oncogenic Potential of Mesenchymal Stem Cells in the Treatment of Cancer: Directions for Future Research. *Current Immunology Reviews,* Vol.6, No.2, pp. 137-148, ISSN 1573 3955

Muschler, G.F.; Boehm, C. & Easley, K. (1997). Aspiration to obtain osteoblast progenitor cells from human bone marrow: the influence of aspiration volume. *Journal of Joint and Bone Surgery. American volume,* Vol.79, No.11, pp. 1699-1709, ISSN 0021 9355

Niemeyer, P.; Krause, U.; Kasten, P.; Kreuz, P.C.; Henle, P.; Südkam, N.P. & Mehlhorn, A. Mesenchymal stem cell-based HLA-independent cell therapy for tissue engineering of bone and cartilage. *Current Stem Cell Research & Therapy,* Vol. 1, No.1, pp. 21-27, ISSN 1574-888X

Ning, H.; Yang, F.; Jiang, M.; Hu, L.; Feng, K.; Zhang, J.; Yu, Z.; Li, B.; Xu, C.; Li, Y.; Wang, J.; Hu, J.; Lou, X. & Chen, H. (2008). The correlation between cotransplantation of mesenchymal stem cells and higher recurrence rate in hematologic malignancy patients: outcome of a pilot clinical study. *Leukemia,* Vol.22, No.3, pp. 593-599, ISSN 0887 6924

Noppe, G.; Dekker, P.; de Koning-Treurniet, C.; Blom, J.; van Heemst, D.; Dirks, R.W.; Tanke, H.J.; Westendorp, R.G. & Maier, A.B. (2009). Rapid flow cytometric method for measuring senescence associated beta-galactosidase activity in human fibroblasts. *Cytometry A,* Vol.75, No.11, pp. 910-916, ISSN 1552 4930

Padley, D.J.; Dietz, A.B. & Gastineau, D.A. (2007). Sterility testing of hematopoietic progenitor cell products: a single-institution series of culture-positive rates and successful infusion of culture-positive products. *Transfusion,* Vol.47, No.4, pp. 636-643, ISSN 0041 1132

Patah, P.A.; Parmar, S.; McMannis, J.; Sadeghi, T.; Karandish, S.; Rondon, G.; Tarrand, J.; Champlin, R.; de Lima, M. & Shpall, E.J. (2007). Microbial contamination of hematopoietic progenitor cell products: clinical outcome. *Bone Marrow Transplantation,* Vol.40, No4, pp. 365-368, ISSN 0268 3369

Phinney, D.G. & Prockop, D.J. (2007). Concise review: mesenchymal stem/multipotent stromal cells: the state of transdifferentiation and modes of tissue repair--current views. *Stem Cells,* Vol.25, No.11, pp. 2896-2902, ISSN 1066 5099

Pieri, L.; Urbani, S.; Mazzanti, B.; Pozzo, S.D.; Santosuosso, M.; Saccardi, R.; Bosi, A.; Faussone-Pellegrini, M.S. & Vannucchi, M.G. Human mesenchymal stromal cells preserve their stem features better when cultured in the Dulbecco's modified Eagle medium. *Cytotherapy,* Jan 3. [Epub ahead of print], ISSN 1465 3249

Pittenger, M.F.; Mackay, A.M.; Beck, S.C.; Jaiswal, R.K.; Douglas, R.; Mosca, J.D.; Moorman, M.A.; Simonetti, D.W.; Craig, S.; & Marshak, D.R. (1999). Multilineage potential of adult human mesenchymal stem cells. *Science,* Vol.284, No.5441, pp. 143-147, ISSN 0036 8075

Poloni, A.; Maurizi, G.; Rosini, V.; Mondini, E.; Mancini, S.; Discepoli, G.; Biasio, S.; Battaglini, G.; Felicetti, S.; Berardinelli, E. Serrani, F. & Leoni, P. (2009) Selection of CD271(+) cells and human AB serum allows a large expansion of mesenchymal stromal cells from human bone marrow. *Cytotherapy,* Vol.11, No.2, pp. 153-162, ISSN 1465 3249

Pytlík, R.; Stehlík, D.; Soukup, T.; Kalbácová, M.; Rypácek, F.; Trc, T.; Mulinková, K.; Michnová, P.; Kideryová, L.; Zivný, J.; Klener, P. Jr.; Veselá, R.; Trnený, M.; & Klener, P. (2009). The cultivation of human multipotent mesenchymal stromal cells in clinical grade medium for bone tissue engineering. *Biomaterials,* Vol.30, No.20, pp. 3415-3427, ISSN 0142 9612

Qiao, L.; Xu, Z.; Zhao, T.; Zhao, Z.; Shi, M.; Zhao, R.C.; Ye, L. & Zhang, X. (2008). Suppression of tumorigenesis by human mesenchymal stem cells in a hepatoma model. *Cell Research,* Vol.18, No.4, pp. 500-507, ISSN 1001 0602

Reuther, T.; Kettmann, C.; Scheer, M.; Kochel, M.; Iida, S. & Kubler, A.C. (2006). Cryopreservation of osteoblast-like cells: viability and differentiation with replacement of fetal bovine serum in vitro. *Cells, Tissues, Organs,* Vol.183, No.1, pp. 32-40, ISSN 1422 6405

Rosland, G.V.; Svendsen, A.; Torsvik, A.; Sobala, E.; McCormack, E.; Immervoll, H.; Mysliwietz, J.; Tonn, J.C; Goldbrunner, R.; Lonning P.E.; Bjerkvig, R.; & Shcichor, C. (2009). Long-term cultures of bone marrow-derived human mesenchymal stem cells frequently undergo spontaneous malignant transformation. *Cancer Research,* Vol.69, No., pp. 5331-5339, ISSN 0008 5472

Rubio, D.; Garcia-Castro, J.; Martin, M.C.; de la Fuente, R.; Cigudosa, J.C.; Lloyd, A.C.; & Bernad, A. (2005). Spontaneous human adult stem cell transformation. *Cancer Research,* Vol. 65, No., pp. 3035-3039, ISSN 0008 5472

Schallmoser, K. & Strunk, D. (2009). Preparation of Pooled Human Platelet Lysate (pHPL) as an Efficient Supplement for Animal Serum-Free Human Stem Cell Cultures. *Journal of Visualised Experiments: JoVE,* http://www.jove.com/details.stp?id=1523 doi: 10.3791/1523, ISSN 1940 087X

Schallmoser, K.; Bartmann, C.; Rohde, E.; Bork, S.; Guelly, C.; Obenauf, A.C.; Reinisch, A.; Horn, P.; Ho, A.D.; Strunk, D. & Wagner, W. (2010). Replicative senescence-associated gene expression changes in mesenchymal stromal cells are similar under different culture conditions. *Haematologica,* Vol.95, No.6, pp. 867-874, ISSN 0390 6078

Secco, M.; Moreira, Y.B.; Zucconi, E.; Vieira, N.M.; Jazedje, T.; Muotri, A.R.; Okamoto, O.K.; Verjovski-Almeida, S. & Zatz, M. (2009). Gene expression profile of mesenchymal stem cells from paired umbilical cord units: cord is different from blood. *Stem Cell Reviews,* Vol.5, No.4, pp. 387-401, ISSN 1550 8943

Sensebé, L.; Krampera, M.; Schrezenmeier, H.; Bourin, P. & Giordano, R. (2010). Mesenchymal stem cells for clinical application. *Vox Sanguinis,* Vol.98, No.2, pp. 93-107, ISSN 0042 9007

Serakinci, N.; Guldberg, P.; Burns, J.S.; Abdallah, B.; Schrodder, H.; Jensen, T.; & Kassem, M. (2004). Adult human mesenchymal stem cell as a target for neoplastic tranformation. *Oncogene,* Vol.23, No., pp. 5095-5098, ISSN 0950 9232

Shenaq, D.S.; Rastegar, F.; Petkovic, D.; Zhang, B.Q.; He, B.C.; Chen, L.; Zuo, G.W.; Luo, Q.; Shi, Q.; Wagner, E.R.; Huang, E.; Gao, Y.; Gao, J.L.; Kim, S.H.; Yang, K.; Bi, Y.; Su, Y.; Zhu, G.; Luo, J.; Luo, X.; Qin, J.; Reid, R.R.; Luu, H.H.; Haydon, R.C. & He, T.C. (2010). Mesenchymal Progenitor Cells and Their Orthopedic Applications: Forging a Path towards Clinical Trials. *Stem Cells International,* Dec 16;2010:519028, ISSN 1687 9678

Soncin, S.; Lo Cicero, V.; Astori, G.; Soldati, G.; Gola, M.; Sürder, D.; Moccetti, T. (2009). A practical approach for the validation of sterility, endotoxin and potency testing of bone marrow mononucleated cells used in cardiac regeneration in compliance with good manufacturing practice. *Journal of Translational Medicine,* Sep 8;7:78, ISSN 1479 5876

Spaggiari, G.M.; Capobianco, A.; Abdelrazik, H.; Becchetti, F.; Mingari, M.C. & Moretta L. (2008). Mesenchymal stem cells inhibit natural killer-cell proliferation, cytotoxicity, and cytokine production: role of indoleamine 2,3-dioxygenase and prostaglandin E2. *Blood,* Vol.111, No.3, pp. 1327-1333, ISSN 0006 4971

Spaggiari, G.M.; Abdelrazik, H.; Becchetti, F. & Moretta L. (2009). MSCs inhibit monocyte-derived DC maturation and function by selectively interfering with the generation of immature DCs: central role of MSC-derived prostaglandin E2. *Blood,* Vol.113, No.26, pp. 6576-6583, ISSN 0006 4971

Steigman, S.A.; Armant, M.; Bayer-Zwirello, L.; Kao, G.S.; Silberstein, L.; Ritz, J. & Fauza, D.O. (2008). Preclinical regulatory validation of a 3-stage amniotic mesenchymal stem cell manufacturing protocol. *Journal of Pediatric Surgery,* Vol.43, No.6, pp. 1164-1169, ISSN 0022 3468

Stute, N.; Holtz, K.; Bubenheim, M.; Lange, C.; Blake, F. & Zander, A.R. (2004). Autologous serum for isolation and expansion of human mesenchymal stem cells for clinical use. *Experimental Hematology,* Vol. 32, No.12, pp. 1212-1225, ISSN 0301 472X

Suchánek, J.; Soukup, T.; Ivancaková, R.; Karbanová, J.; Hubková, V.; Pytlík, R. & Kucerová L. (2007). Human dental pulp stem cells--isolation and long term cultivation. *Acta Medica (Hradec Kralove),* Vol.50, No.3, pp. 195-201, ISSN 1211 4286

Sundin, M.; Ringdén, O.; Sundberg, B.; Nava, S.; Götherström, C. & Le Blanc K. (2007). No alloantibodies against mesenchymal stromal cells, but presence of anti-fetal calf serum antibodies, after transplantation in allogeneic hematopoietic stem cell recipients. *Haematologica,* Vol.92, No.9, pp. 1208-1215, ISSN 0390 6078

Tamama, K.; Kawasaki, H. & Wells, A. (2010). Epidermal growth factor (EGF) treatment on multipotential stromal cells (MSCs). Possible enhancement of therapeutic potential of MSC. *Journal of Biomedicine & Biotechnology,* 2010:795385, Epub 2010 Feb 17, ISSN 1110 7243

Tarte, K.; Gaillard, J.; Lataillade, J.J.; Foulliard, L.; Becker, M.; Mossafa, H.; Tchirkov, A.; Rouard, H.; Henry, C.; Splingard, M.; Dulong, J.; Monnier, D.; Gourmelon, P.; Gorin, N.C. & Sensebé, L.; Société Francaise de Greffe de Moelle et Therapie Cellulaire. (2010). Clinical-grade production of human mesenchymal stromal cells: occurrence of aneuploidy without transformation. *Blood,* Vol.115; No.8, pp. 1549-1553, ISSN 0006 4971

Thirumala, S.; Gimble, J.M. & Devireddy, R.V. (2010a). Cryopreservation of stromal vascular fraction of adipose tissue in a serum-free freezing medium. *Journal of Tissue Engineering and Regenerative Medicine,* Vol.4, No.3, pp. 224-232, ISSN 1932 6254

Thirumala, S.; Wu, X.; Gimble, J.M. & Devireddy, R.V. (2010b). Evaluation of polyvinylpyrrolidone as a cryoprotectant for adipose tissue-derived adult stem cells. *Tissue Engineering. Part C, Methods,* Vol.16, No.4, pp. 783-792, ISSN 1076 3279

Thirumala, S.; Gimble, J.M. & Devireddy, R.V. (2010c). Evaluation of methylcellulose and dimethyl sulfoxide as the cryoprotectants in a serum-free freezing media for cryopreservation of adipose-derived adult stem cells. *Stem Cells and Development,* Vol.19, No.4, pp. 513-522, ISSN 1547 3287

Torsvik, A.; Rosland, G.V.; Swendsen, A.; Molven, A.; Immervoll, H.; McCormack, E.; Lonning, P.E.; Primon, M.; Sobala, E.; Tonn, J.C.; Goldbrunner, R.; Schichor, C.;

Mysliwietz, J.; Lah, T.T. & Motaln, H. (2010). Spontaneous malignant transformation of human mesenchymal stem cells reflect cross-contamination: Putting the research field on track. *Cancer Research,* Vol.70, No.13, pp. 6393-6396, ISSN 0008 5472

Torio-Padron, N.; Huotari, A.M.; Eisenhardt, S.U.; Borges, J. & Stark, G.B. (2010). Comparison of pre-adipocyte yield, growth and differentiation characteristics from excised versus aspirated adipose tissue. *Cells, Tissues, Organs,* Vol. 191, No.5, pp. 365-371, ISSN 1422 6405

Troyer, D.L. & Weiss, M.L. (2008). Wharton's jelly-derived cells are a primitive stromal cell population. *Stem Cells,* Vol.26, No.3, pp. 591-599, ISSN 1066 5099

Tsai, C.C.; Chen, Y.J.; Yew, T.L.; Chen, L.L.; Wang, J.Y.; Chiu, C.H. & Hung, S.C. (2011). Hypoxia inhibits senescence and maintains mesenchymal stem cell properties through down-regulation of E2A-p21 by HIF-TWIST. *Blood,* Vol.117, No.2, pp. 459-469, ISSN 0006 4971

Tsai, M.S.; Hwang, S.M.; Chen, K.D.; Lee, Y.S.; Hsu, L.W.; Chang, Y.J.; Wang, C.N.; Peng, H.H.; Chang, Y.L.; Chao, A.S.; Chang, S.D.; Lee, K.D.; Wang, T.H.; Wang, H.S. & Soong, Y.K. (2007). Functional network analysis of the transcriptomes of mesenchymal stem cells derived from amniotic fluid, amniotic membrane, cord blood, and bone marrow. *Stem Cells,* Vol.25, No.10, pp. 2511-2523, ISSN 1066 5099

Tsutsumi, S.; Shimazu, A.; Miyazaki, K.; Pan, H.; Koike, C.; Yoshida, E.; Takagishi, K. & Kato, Y. (2001). Retention of multilineage differentiation potential of mesenchymal cells during proliferation in response to FGF. *Biochemical and Biophysical Research Communications,* Vol.288, No.2, pp.413-419, ISSN 0006 291X

Volokhov, D.V.; Graham, L.J.; Brorson, K.A. & Chizhikov, V.E. (2011). Mycoplasma testing of cell substrates and biologics: Review of alternative non-microbiological techniques. *Molecular and Cellular Probes,* 2011 Jan 11. [Epub ahead of print]. ISSN 0890 8508

Wagner, J.E. & Gluckman, E. Umbilical cord blood transplantation: the first 20 years. *Seminars in Hematology,* Vol.47, No.1, pp. 3-12, ISSN 0037 1963

Wagner, W.; Wein, F.; Seckinger, A.; Frankhauser, M.; Wirkner, U.; Krause, U.; Blake, J.; Schwager, C.; Eckstein, V.; Ansorge, W. & Ho, A.D. (2005). Comparative characteristics of mesenchymal stem cells from human bone marrow, adipose tissue, and umbilical cord blood. *Experimental Hematology,* Vol.33, No.11, pp. 1402-1416, ISSN 0301 472X

Wang, T.H.; Lee, Y.S. & Hwang, S.M. (2011). Transcriptome Analysis of Common Gene Expression in Human Mesenchymal Stem Cells Derived from Four Different Origins. In: *Mesenchymal Stem Cell Assays and Applications,* Vemuri, M.C.; Chase, LG. & Rao M.S. (eds.), Methods in Molecular Biology, vol. 698, pp. 405-417, Springer Science+Business Media, ISBN 978-1-60761-998-7

Wang, Y.; Huso, D.L.; Harrington, J.; Kellner, J.; Jeong, D.K.; Turney, J. & McNiece, I.K. (2005). Outgrowth of a transformed cell population derived from normal human BM mesenchymal stem cell culture. *Cytotherapy,* Vol.7, No.6, pp. 509-519, ISSN 1465 3249

Wendt, M.K.; Allington, T.M. & Schiemann WP. (2009). Mechanisms of the epithelial-mesenchymal transition by TGF-beta. Future Oncology, Vol.5, No.8, pp. 1145-1168, ISSN 1479 6694

Werntz, J.R.; Lane, J.M.; Burstein, A.H.; Justin, R.; Klein, R. & Tomin, E. (1996). Qualitative and quantitative analysis of orthotopic bone regeneration by marrow. *Journal of Orthopaedics Research*, Vol.14, No.1, pp. 85-93, ISSN 0736 0266

Woods, E.J.; Perry, B.C.; Hockema, J.J.; Larson, L.; Zhou, D. & Goebel, W.S. Optimized cryopreservation method for human dental pulp-derived stem cells and their tissues of origin for banking and clinical use. *Cryobiology*, Vol.59, No.2, pp. 150-157, ISSN 0011 2240

Zhu, W.; Xu, W.; Jiang, R.; Qian, H.; Chen, M.; Hu, J.; Cao, W.; Han, C. & Chen Y. (2006). Mesenchymal stem cells derived from bone marrow favor tumor cell growth in vivo. *Experimental and Molecular Pathology*, Vol.80, No.3, pp. 267-274, ISSN 0014 4800

Zuk, P.A.; Zhu, M.; Ashjian, P.; De Ugarte, D.A.; Huang, J.I.; Mizuno, H.; Alfonso, Z.C.; Fraser, J.K.; Benhaim, P. & Hedrick, M.H. (2002). Human adipose tissue is a source of multipotent stem cells. *Molecular Biology of the Cell*, Vol.13, No.12, pp. 4279-4295, ISSN 1059 1524

7

Adipose-Derived Stem Cells (ASCs) for Tissue Engineering

Mathias Tremp[1], Souzan Salemi[1], Rita Gobet[2],
Tullio Sulser[1] and Daniel Eberli[1]
[1]*University Hospital Zürich, Laboratory for Tissue Engineering and Stem Cell Therapy, Department of Urology, Zürich,*
[2]*Division of Pediatric Urology, Department of Pediatric Surgery, University Children's Hospital, Zürich*
Switzerland

1. Introduction

A current challenge in the field of tissue engineering is soft tissue replacement. Techniques for soft tissue reconstruction include use of autologous fat transplantation, alloplastic implants and autologous tissue flaps. However, these approaches have disadvantages, including donor-site morbidity, implant migration, foreign body reaction and immune system rejection. The use of autologous stem cells expanded *in vitro* and combined with novel biomaterials for organ reconstruction offers a potential solution for replacement of tissue or whole organs. Stem cells, first identified in embryonic tissue and later in numerous adult tissues, possess the unique capacity to differentiate into wide range of tissue types. However, although embryonic cells are the most flexible of stem cell lines, they raise the problem of ethical issues. For tissue engineering, candidates of stem cells include embryonic stem cells (Ahn, et al.), induced pluripotent stem cells (iPS) (Crisan, et al.;2008, Parker, et al.;2007) and adult stem cells. The ability of adult stem cells to divide or self-renewal make them attractive source of stem cells for use in tissue engineering. A significant amount of current interest has focused on the possibility that adult human stem cells are the therapeutic alternative to embryonic stem cells because of their plasticity (Aoki, et al.;2010).
The presence of self-renewing cells within the bone marrow of mice was reported in 1963 which was later known as hematopoietic stem cells (HSCs) (Becker, et al.;1963, Zhang, et al.;1999). Several years later, HSCs were identified in umbilical cord blood by other investigators (Aust, et al.; 2004, Dellavalle, et al.;2007). Furthermore, several other adult stem cell types such as neural stem cells (Dellavalle, et al.; 2007, Guilak, et al.;2010), were isolated and identified. Moreover, a population of plastic adherent cells were isolated from collagenase digests of adipose tissue.
Adipose tissue derived stem cells were termed as: adipose derived stem/stromal cells (ASCs), adipose-derived adult stem (ADAS) cells, adipose-derived adult stromal cells, adipose-derived stromal cells (ADSCs), adipose stromal cells (ASCs), adipose mesenchymal stem cells (AdMSCs), lipoblast, pericyte, preadipocyte, and processed lipoaspirate (PLA) cells. To prevent the confusion in the literature, the International Fat Applied Technology

Society reached a consensus to adopt the term "adipose-derived stem cells" (ASCs) to identify the isolated, plastic-adherent, multipotent cell population (Gimble, et al.;2007). Adipose tissue derives from the mesodermal layer of the embryo and develops both pre- and postnatally (Gonda, et al.;2008, Kakudo, et al.;2008). Researchers studying the development of adipose tissue have long worked with a fibroblastic cell line, known as preadipocytes, isolated from adipose tissue (Cowherd, et al.;1999, Gregoire, et al.;1998, Kirkland and Hollenberg;1998, Nam and Lobie;2000, Sorisky;1999).
The reason why adipose tissue would contain a stem cell population is not still clear. There is some discussion whether the cells are subpopulation of fibroblasts reside within the fat tissue or are perhaps mesenchymal or peripheral blood stem cells passing through the fat tissue (Crisan, et al.; 2008, Dellavalle, et al.; 2007).
The ASCs represent a readily available source for isolation of potentially useful stem cells (Sterodimas, et al.; 2010). In culture, they have shown to have an impressive developmental plasticity, including the ability to undergo multilineage differentiation and self-renewal (Liu, et al.; 2009). When ASCs are compared with BM-MSCs, further similarities have been demonstrated in regards to their growth kinetics, cell senescence, gene transduction efficiency (De Ugarte, et al.; 2003), as well as CD surface marker expression (Gronthos, et al.; 2001, Katz, et al.; 2005, Zuk, et al.; 2002) and gene transcription (Katz, et al.; 2005). Compared to bone marrow MSCs, ASCs have potential advantages for tissue engineering application, because of the tissue accessibility, multipotency and ease of isolation without painful procedures or donor site injury.
In this chapter we will discuss the potential use of adipose-derived stem cells in the field of tissue engineering.

2. Isolation and expansion of ASCs

The initial methods to isolate cells from adipose tissue were pioneered by Rodbell and colleagues in the 1960s (Rodbell; 1966, Rodbell; 1966, Rodbell and Jones; 1966) using rat fat tissue. These methods were further adapted for human tissues by several other groups (Deslex, et al.; 1987, Engfeldt, et al.; 1980, Ho, et al.; 2010). The current methods for isolating ASCs rely on a collagenase digestion followed by centrifugal separation to isolate the stromal/vascular cells from primary adipocytes. The pellet is resuspended with a basal medium containing 10% foetal bovine serum (Estes, et al.; 2008). The cell suspension is filtered through 100 µm cell strainer and the cells are plated and incubated at 37°C in the presence of 5% CO_2. The medium is changed every second day until the cells reach 80-90% confluence. A large number of ASCs can be harvested in this manner, with yields of approximately 250,000 cells per gram of tissue (Aust, et al.; 2004, Guilak, et al.; 2010). In order to remove the use of animal products in human ASC cultures, a very low human serum expansion medium and a completely serum-free medium have been recently reported (Parker, et al.; 2007). Furthermore it was reported that use of platelet- rich plasma can enhance the proliferation of human ASCs. These results can support the clinical application of platelet-rich plasma for cell based, soft-tissue engineering and wound healing (Kakudo, et al.; 2008).
ASCs should be harvested at 80% confluence for freezing. Cryopreservation medium consists of 80% fetal bovine serum, 10% dimethylsulfoxide (DMSO) and 10% DMEM/Ham's F-12. The cells should be stored in a final concentration of 1–2 million viable cells per milliliter of cryopreservation medium. Aliqouted vials are first frozen in an alcohol freezing

container and store and are stored at -80°C overnight. On the next day, the frozen vials can be transferred to a liquid nitrogen container for long-term storage. Successful storage of ASCs more than 6 months has been shown. This ensures the availability of autologous banked ASCs for clinical applications in the future (De Rosa, et al.;2009, Gonda, et al.;2008).

2.1 Characterization of ASCs

In order to characterize the undifferentiated animal or human ASC cells cultured in vitro flow cytometric and immunohistochemical methods are widely used (Gronthos, et al.; 2001, Zuk, et al.; 2001). The cell surface phenotype of ASC is quite similar to MSCs (mesenchymal stem cells). Both ASC and MSC cells express CD29, CD44, CD71, CD90, and CD105 (Zuk, et al.; 2002). In contrast, no expression of the hematopoietic lineage markers CD31, CD34 and CD45 was observed in either of the cultures. In addition the ASC cells express the neutral endopeptidase (CD10 or common acute lymphocytic leukemia antigen CALLA), aminopeptidase (CD13), and ecto nucleotidase (CD73). Furthermore, ASC cells produce Type I and Type III collagens, osteopontin, ostenectin, Thy-1 (CD90), and MUC-18 (CD146) (Gimble and Guilak; 2003).

Different investigator have reported different pattern of expression. For example, while Gronthos et al. (Gronthos, et al.;2001) detected CD34 and VCAM (CD106) on ASC cells, Zuk et al. (Zuk, et al.;2002) did not. Likewise, while Zuk et al. (Zuk, et al.;2002) detected Stro-1, Gronthos et al. did not. These discrepancies could be due to the differences in cell isolation methods, how long the cells were cultured prior to analysis and sensitivity differences between immunohistochemical and flow cytometric detection methods (Zuk, et al.;2002, Zuk, et al.;2001).

3. ASCs applications

Adipose tissue has proven to serve as an abundant source of adult stem cells with multipotent properties suitable for tissue engineering and regenerative medical applications. ASCs can be differentiated into variety of cell types. Differentiation is commonly induced by insulin, dexamethasone, cyclic AMP agonist, β-glycerophosphate, heparin, ascorbate and different cytokines depending on the lineage type.

ASCs like BM-MSCs, differentiate *in vitro* towards soft tissue such adipocytes, smooth muscle and cardiac myocytes when treated with established lineage-specific factors. In addition they can differentiate toward musculoskeletal tissues such as osteocytes, myocytes and chondrocytes. Furthermore neurogenic differentiation of these cells is reported by several investigators.

3.1 ASCs differentiation

3.1.1 Adipogenesis

ASCs in response to inductive compounds including glucocorticoid receptor ligands (dexamethasone), insulin, cyclic AMP agonist (forskolin) and peroxisome proliferator-activated receptor gamma (PPARγ) undergo adipogenic differentiation (Farmer; 2006, Hauner, et al.; 1989, Lazar; 2005, Zuk, et al.; 2001). During the differentiation process ASCs reduce their proliferation rate and undergo morphological changes. ASCs are induced in the adipocyte differentiation medium containing biotin, d-pantothenate, dexamethasone, methylisobutylxanthine, insulin and equivalent PPARγ agonist. After induction for 2 weeks in adipogenic medium the human ASC contain vacuoles filled with neutral lipid cells which

can be further stained for intracellular lipid droplets accumulation using an Oil Red O stain (Preece;1972).
In addition, these cells secrete increased amounts of the adipocyte protein leptin, and transcribe adipogenic mRNAs such as the fatty acid binding protein, aP2 and lipoprotein lipase (Halvorsen, et al.; 2001, Hauner, et al.; 1989, Sen, et al.; 2001). Some of these parameters such as leptin, aP2 mRNA levels were quantified and found to be increased by several hundred-fold during the differentiation process (Halvorsen, et al.; 2001, Sen, et al.; 2001). It is reported that ASCs harvested from female mice differentiate more efficiently into adipocytes than those from male mice (Ogawa, et al.; 2004).
One of the most important uses of ASCs is for the replacement of adipose tissue itself. Large soft tissue defects are a common problem following trauma, burns and oncological resections. Several studies demonstrated the *in vitro* differentiation of ASCs along adipogenic lineages, including the accumulation of intracellular lipid droplets, and the expression of characteristic proteins and enzymes (Ogawa, et al.; 2004, Tchkonia, et al.; 2002, Zuk, et al.; 2002). ASCs were used to seed artificial scaffolds and were further implanted subcutaneously in mice and rats (von Heimburg, et al.; 2001, von Heimburg, et al.; 2001). The cell-seeded grafts showed significant neovascularisation of the implant, as well as penetration of the preadipocytes or ASCs into the scaffolding, and their differentiation into mature lipid-laden adipocytes.

3.1.2 Smooth muscle

ASCs can be differentiated to smooth muscle cells (SMCs) and might offer a cell source for hollow organ engineering. For myogenic differentiation ASCs at passages 3 through 5 are cultured in smooth muscle inductive medium consisting of MCDB131 supplemented with 1% FBS and 100u/ml of heparin for up to 6 weeks at 37°C with 5% CO_2. The media is changed every 3 days and cell splitting is not required (Jack, et al.; 2009).
The cellular changes after differentiation can be investigated by real- time PCR at mRNA level. As reported the expression of muscle actin (SMA), calponin and myosin heavy chain showed an increase after growth in differentiation medium (Jack, et al.; 2009). The same was observed at protein levels, induction media induced differentiation of the ASCs into a smooth muscle phenotype in which the expression of smooth muscle specific proteins SMA, caldesmon, and myosin heavy chain (MHC) was increased (Jack, et al.; 2009).
Differentiation is a complex process and has a dramatical effect on cell size, shape, membrane potential, metabolic activity and responsiveness to external signals. One of the main characteristic of SMCs is their contractility which plays important roles in angiogenesis, blood vessel maintenance, and mechanical regulation of hollow organs such as bladder.
Differentiated SMCs can show two specific phenotypes, which vary from synthetic and proliferative to contractile and quiescent (Beamish, et al.; 2010). SMCs exhibit a contractile phenotype characterized by high expression of specific contractile proteins including SMA, calponin, SM22, smoothelin, *h*-caldesmon and smooth muscle myosin heavy chain (SM-MHC) (Owens, et al.; 2004, Shanahan, et al.; 1993).
Since differentiated human ASCs express smooth muscle specific proteins they may prove to be of value in the repair of smooth muscle defects in the gastrointestinal and urinary tracts. Juan et al. reported that the ASCs from different sites show different myogenic differentiation abilities *in vitro*. ASCs from the adipose tissues of the nape of the neck and

vicinity of epididymis can be used as ideal seed cells for tissue engineering of lower urinary tract (Yuan, et al.; 2010). Similar study was performed by other group using human subcutaneous and omental adipose tissues. They could show that subcutaneous adipose tissue has higher differentiation capacity than omental adipose tissue which can be a suitable cell source for use in regenerative medicine (Toyoda, et al.; 2009).

3.1.3 Osteogenesis. Bone defect repair

In the past decade, several groups isolated cells from the adipose tissue of humans and other species capable of differentiating into osteoblasts *in vitro* (Dragoo, et al.; 2003, Mizuno, et al.; 2002, Zuk, et al.; 2001). ASC cells differentiate into osteoblast-like cells in the presence of ascorbate, b-glycerophosphate, dexamethasone and vitamin D3.

For osteogenic differentiation confluent ASCs cells are incubated for 3 weeks in DMEM containing 10% FBS, 100 nM dexamethasone, 10 mM β-glycerophosphate and 50 μM L-ascorbic acid-2-phosphate. After fixation cells are incubated at 37°C for 1 hour with 0.16% naphthol AS-TR phosphate and 0.8% Fast Blue BB dissolved in 0.1 M tris buffer (pH 9.0). For osteogenic differentiation cells were also incubated in 1% alizarin red S for 3 minutes to detect calcium deposition (Sakuma, et al.; 2009).

Over a 2- 4 week period *in vitro* , both human and rat ASC cells deposit calcium phosphate mineral within their extracellular matrix, and express osteogenic genes. Under osteogenic conditions ASCs are observed to express genes and proteins associated with osteoblasts phenotypes such as osteopontin, osteonectin, osteocalcin collagen type I, BMP-2 and BMP-4 (Halvorsen, et al.; 2001, Zuk, et al.; 2001). In addition ASCs are able to form mineralized matrix *in vitro* in both long term 2-D or 3-D osteogenic cultures.

In vivo, ASC cells embedded in porous cubes of hydroxyapatite/tricalcium phosphate form bone were used as implants in immunodeficient mice (Hicok, et al.; 2004). New osteoid, derived from the human ASC cells, is present within a 6-week incubation period (Hicok, et al.; 2004). This finding indicates that ASCs cells will have therapeutic applications in bone repair. The first case of autologous ASC use for osseous repair has been reported in the treatment of a calvarial defect in a7-year-old girl (Lendeckel, et al.; 2004). Using different type of scaffolds, human ASC can form bone in immunodefficient mice (Hicok, et al.; 2004, Lee, et al.; 2003).

3.1.4 Myogenesis: skeletal muscle repair

There is several line of evidence that ASC cells can differentiate along each of the myocyte lineage pathways when cultured in myogenic induction medium containing 0.1 *m*M dexamethasone, 50 *m*M hydrocortisone, 10% FBS and 5% horse serum. ASCs express MyoD and myogenin, transcription factors regulating skeletal muscle differentiation (Pittenger, et al.; 1999, Zuk, et al.; 2001).

Skeletal myogenesis is characterized by a period of myoblast proliferation, followed by the expression of muscle-specific proteins and fusion to form multinucleated myotubules. Early myogenic differentiation is characterized by the expression of several myogenic regulatory factors including myogenic determination factor MyoD1 (Weintraub, et al.; 1991). Terminally differentiated myoblasts can be characterized by the expression of myosin and the presence of multiple nuclei (Silberstein, et al.; 1986).

In the first *in vivo* report, F Bacou et al injected ASCs into the anterior tibialis muscle of rabbits following cardiotoxin-induced injury. Consistent with prior work using satellite cells treated muscles were found to be heavier, have an increased fibre area cross-section and exert greater maximal force (Boubaker el Andalousi, et al.; 2002).

3.1.5 Chondrogenesis

ASCs display chondrogenic characteristics following induction with ascorbate, dexamethasone and transforming growth factor-β (Awad, et al.; 2003, Huang, et al.; 2004, Zuk, et al.; 2001). Under inductive conditions ASCs express aggrecan, chondroitin sulphate, collagen type II and IV and proteoglycans associated with chondrogenic phenotype.(Awad, et al.; 2004, Erickson, et al.; 2002, Wickham, et al.; 2003, Zuk, et al.; 2001)
For chondrogenic differentiation, ASCs cells are grown to confluency in 30-mm dishes and incubated for 3 weeks in DMEM containing 1% FBS, 50 mM L-ascorbic acid-2-phosphate, 40 mg/ml proline, 100 mg/ml pyruvate, 10 ng/ml transforming growth factor (Harriman, et al.)-b3, and 1x ITS. Induction medium is replaced every 3 days. At the indicated time points, differentiated cells are fixed for 1h with 4% paraformaldehyde and rinsed with PBS. Accumulation of chondrocyte matrix is detected with alcian blue staining (pH 2.5, Wako) (Matsumoto, et al.; 2008).

3.1.6 Neuronal differentiation

There is preliminary evidence suggesting that human ASCs can display neuronal and/or oligodendrocytic markers. ADSC at passages 2–5 are seeded in six-well plates at 40%–60% confluence. After three washes with PBS, the cells are induced with NIM (DMEM supplemented with 500 mM IBMX, 200 mM INDO, and 5 mg/ml insulin) for 1 hr. The cells are then examined for the expression of neuronal markers S100, NF70, and nestin followed by hematoxylin-eosin (HE) staining (Ning, et al.; 2006).
The *in vivo* test for the therapeutic potential of ASCs looked at their effects when injected intraventricularly in rats. ASCs survived with increased engraftment at the site of injury compared with controls. Neural lineage markers microtubule-associated protein-2 and glial fibrillary acidic protein were expressed in some engrafted cells. Behavioural tests of the motor and sensory systems showed clear improvements in those treated with ASCs after infarct (Kang, et al.; 2003). It is not clear whether transplanted cells replaced the lost neurons or provided a support role for existing stem cells and injured neurons. Furthermore, in a co-culture model, Kang *et al.* studied the interactions between neural stem cells (NSCs) and ASCs. In comparison to laminin-coated dishes, ASC feeder layers showed ability to support the differentiation and survival of NSCs over 14 days in culture.

4. Biomaterials in tissue engineering with ASCs

Currently, autologous and allogenic adipose tissues represent a ubiquitous source of material for fat reconstructive therapies. However, these approaches are limited, and often accompanied by a 40–60% reduction in graft volume following transplantation. A number of factors including a stable scaffold support structure and vascularisation is necessary to support *de novo* adipogenesis and long-term maintenance of adipose tissue formation within adipose tissue engineered constructs (Patrick, et al.; 2002). Recently, cell-based approaches utilizing adipogenic progenitor cells in combination with biomaterial carriers for fat tissue engineering have been developed and were reported to promote both short-term *in vivo* adipogenesis and to repair defect sites (Borges, et al.; 2003, Patrick, et al.; 1999). To date, however the efficacy of exogenously delivered stem cell populations to support the generation of long-term volume stable adipose tissue *in vivo* is restricted by suboptimal properties of their biomaterial carriers including insufficient biocompatibility and rapid scaffold degradation rates (Patrick, et al.; 2002).

For functional tissue replacement such as bone, Silk-based biomaterials have previously been demonstrated to offer exceptional benefits over conventional synthetic (e.g. poly-glycolic and lactic acid copolymers) and natural (e.g. collagen type I) biomaterials (Meinel, et al.; 2004). The slow degradation and mechanical integrity of silk scaffolds in comparison with other conventional biomaterials such as collagen and PLA, above all for long-term *in vivo* studies, suggest that silk fibroin-based scaffolds would be an optimal biomaterial for long-term adipose tissue growth and function (Mauney, et al.; 2007). Also, previous studies have demonstrated the ability of adipocytes to secrete various paracrine factors which can positively influence both the migration and differentiation of preadipocytes (Shillabeer, et al.; 1989). Mauney et al. studied biomaterials derived from silk fibroin prepared by aqueous (AB) and organic (HFIP) solvent-based processes, along with collagen (COL) and poly-lactic acid (PLA)-based scaffolds *in vitro* and *in vivo* for their utility in adipose tissue engineering strategies (Mauney, et al.; 2007).

For *in vitro* studies, they used adipose-derived mesenchymal stem cells (hASCs) and seeded them on the various biomaterials and cultured them for 21 days in the presence of adipogenic stimulants (AD). In their study, hASCs (and hMSCs) cultured on all biomaterials in the presence of AD showed significant upregulation of adipogenic mRNA transcript levels (e.g. GLUT4) to similar extents when compared to noninduced controls (Mauney, et al.; 2007). Also, oil-red O analysis of hASC-seeded scaffold displayed substantial amounts of lipid accumulating adipocytes following cultivation with AD. Following a 4-week implantation period in a rat muscle pouch defect model, both AB and HFIP scaffolds supported *in vivo* adipogenesis either alone or seeded with hASCs (Mauney, et al.; 2007). On the other hand, COL and PLA scaffolds underwent rapid scaffold degradation and were irretrievable following the implantation period. The authors concluded that macroporous 3D AB and HFIP silk fibroin scaffolds offer an important platform for cell-based adipose tissue engineering applications, and in particular, provide longer-term structural integrity to promote the maintenance of soft tissue *in vivo* (Mauney, et al.; 2007).

Tissue-specific scaffolds and signalling systems are essential to differentiate stem cells into the required cells and use them effectively to construct three-dimensional (3D) tissues (Sterodimas, et al.; 2010). It has also been proved that adipose tissues engineered with ASCs and type I collagen scaffolds can serve *in vivo* for the replacement of damaged tissue (Lu, et al.; 2006). This has been confirmed by Zhang et al., where collagen I scaffold exhibited excellent cellular compatibility and can be used as a vehicle for adipose tissue engineering (Zhang, et al.; 2007).

Porous collagenous microbeads can be useful as injectable cell delivery vehicles for adipose-derived stem cells, allowing *ex vivo* proliferation and differentiation on particles that are small enough to be injected into a defect and molded into the desired shape without migration of the cells. The cell-seeded microbeads can be injected through a needle into the wound site, and agglomeration of the microbeads can retain the cells and microbeads in the site (Rubin, et al.; 2007). Furthermore, the use of natural material hold promises in tissues engineering.

Placental decellular matrix (PDM) holds potential as a scaffold for adipose tissue engineering applications. The placenta is a rich source of human extracellular matrix (ECM) components that can be harvested without harm to the donor. Constructs derived from the ECM may mimic the native environment of the body, promoting normal cellular organization and behavior. Natural materials also have advantages in terms of ease of processing, biodegradability and biocompatibility (Schmidt and Baier; 2000). Cell-adhesive placental decellular matrix scaffolds facilitate proliferation and viability, while differentiation is augmented when the cells are encapsulated in non-adhesive cross-linked hyaluronan (XLHA) scaffolds (Flynn, et al.; 2008). Incorporation of XLHA into the PDM scaffolds may improve the construct bulking properties and may influence cellular infiltration, differentiation and wound healing (Shu XZ; 2004).

Other candidate for suitable scaffolds is non-woven polyglycolic acid (PGA) and hyaluronic acid gel. In a recent study it has been shown that more adipose-tissue-like construct is regenerated when using type I collagen sponge than when the non-woven polyglycolic acid or hyaluronic acid gel are used (Itoi, et al.; 2010). In addition, significant evidence has been shown that ASCs and PLGA spheres can be used in a clinical setting to generate adipose tissue as a noninvasive soft tissue filler (Choi, et al.; 2006).

Altman et al. could show that human adipose-derived stem cells seeded on a silk fibroin-chitosan scaffold enhance wound healing and show differentiation into fibrovascular, endothelial, and epithelial components of restored tissue (Altman, et al.; 2009). In addition, it has been shown that transfection of human ASCs with liposome- enveloped xenogenic protein from a neonatal rat tissue preparation can induce differentiation of stem cells along the directed lineage (Gaustad, et al.; 2004). These observations support the hypothesis that the inductive biochemical and structural cues of the microenvironment are conserved across species and that a silk fibroin-chitosan delivery vehicle can provide a beneficial niche in supporting migration, proliferation, and differentiation of the applied cells (Altman, et al.; 2009).

Another complex biomolecule which has sparked great interest for tissue engineering is Hyaluronic acid which has been stated to support the growth and development of progenitor cells (Brun, et al.; 1999, Solchaga, et al.; 1999). The material has a progressive rate of biodegradation, lacks cytotoxicity and does not induce a systemic immune response or chronic inflammation in a human *in vivo* model (Stillaert, et al.; 2008). Hyaluronic acid-based (HA) scaffolds were demonstrated to be suitable materials for soft-tissue regeneration; they maintain volume when seeded with preadipocytes. Hemmrich et al. evaluated, *in vitro* and *in vivo,* human preadipocytes seeded onto plain hyaluronan benzyl ester (HYAFF®11) or HYAFF®11 coated with the extracellular matrix glycosaminoglycan hyaluronic acid and they found extensive formation of new vessels throughout the construct but with only minor adipose tissue (Hemmrich, et al.; 2005).

Long-standing, 3D predefined-shape adipose tissue from hAD-MSCs of human adipose tissue remains a challenge. Lin et al. cultured scaffolds (Gelatin sponges, monofilament polypropylene and polyglycolic acid meshes) with hAD-MSCs in adipogenic medium for 2 weeks before implantation, and implanted scaffolds were harvested after 2, 4, and 6 months *in vivo.* All of the successfully harvested scaffolds were filled with newly formed adipose tissue and had retained their predefined shape and dimensions (Lin, et al.; 2008). It has been shown that hAD-MSCs are not successful soft tissue filler if used alone (Lin, et al.; 2008, Moseley, et al.; 2006).

There are numerous adipose tissue engineering culture strategies in which the core tissue engineering principles comprising appropriate cells, scaffold, and microenvironment are optimized (Patrick; 2001). Specifically, there are static versus dynamic culture, co-cultivation, and addition of growth factors, vascularization, and long-term sustainability of engineered constructs. The advantage of dynamic culture includes increased nutrient and oxygen delivery to cells within a 3D construct (Frye and Patrick; 2006). Improved bioreactor designs to address direct perfusion conditions are necessary to advance dynamic culture techniques (Choi, et al.; 2010).

It has been shown that hASCs express pericyte lineage markers *in vivo* and *in vitro*, exhibit increased migration in response to PDGF-BB *in vitro*, exhibit perivascular morphology when injected *in vivo*, and contribute to increases in microvascular density during angiogenesis by migrating toward vessels (Amos, et al.; 2008). In cell-assisted lipotransfer (CAL), autologous ASCs are used in combination with lipoinjection (e.g. Parry-Romberg syndrome). A stromal vascular fraction (SVF) containing ASCs is freshly isolated from half of the aspirated fat and recombined with the other half. The preliminary results suggest that CAL is effective and safe for soft-tissue augmentation and superior to conventional lipoinjection and microvasculature can be detected more prominently in CAL fat (Yoshimura, et al.; 2008). Recently, adipose stem cells have proved to selectively induce neovascularisation and increase the viability of random-pattern skin flaps. This mechanism might be both due to the direct differentiation of ASCs into endothelial cells and the indirect effect of angiogenic growh factors released from ASCs (Lu, et al.; 2008).

5. Conclusion

ASCs provide an abundant and readily accessible source of multipotent stem cells. The use of autologous stem cells expanded *in vitro* and combined with novel selected biomaterials for organ reconstruction offers a potential solution for replacement of tissue or whole organs. ASC does have one important advantage over the other sources of stem cells namely easy availability. There is no human tissue as expendable as adipose tissue, making it relatively easy to isolate adequate numbers of ASCs for possible human therapies. Human ASCs can be ideal cell source for tissue engineering. They are available in large quantities of cells per individual, multipotent, are transplantable in an autologous setting.

However, further studies are needed before ASCs can be used clinically. In particular, investigators need to demonstrate the safety and efficacy of ASCs cells in animal models, either alone or in combination with novel biomaterial scaffolds.

6. Acknowledgment

Both the authors Dr. Mathias Tremp and Dr. Souzan Salemi contributed equivalently in writing this chapter.

7. References

Ahn, H. H., Lee, J. H., Kim, K. S., Lee, J. Y., Kim, M. S., Khang, G., Lee, I. W., and Lee, H. B. (2008). Polyethyleneimine-mediated gene delivery into human adipose derived stem cells. *Biomaterials* Vol.29, No.15, pp. 2415-22, ISSN 0142-9612

Altman, A. M., Yan, Y., Matthias, N., Bai, X., Rios, C., Mathur, A. B., Song, Y. H., and Alt, E. U. (2009). IFATS collection: Human adipose-derived stem cells seeded on a silk fibroin-chitosan scaffold enhance wound repair in a murine soft tissue injury model. *Stem Cells* Vol.27, No.1, pp. 250-8, ISSN 1549-4918

Amos, P. J., Shang, H., Bailey, A. M., Taylor, A., Katz, A. J., and Peirce, S. M. (2008). IFATS collection: The role of human adipose-derived stromal cells in inflammatory microvascular remodeling and evidence of a perivascular phenotype. *Stem Cells* Vol.26, No.10, pp. 2682-90, ISSN 1549-4918

Aoki, T., Ohnishi, H., Oda, Y., Tadokoro, M., Sasao, M., Kato, H., Hattori, K., and Ohgushi, H. (2010). Generation of induced pluripotent stem cells from human adipose-derived stem cells without c-MYC. *Tissue Eng Part A* Vol.16, No.7, pp. 2197-206, ISSN 1937-335X

Aust, L., Devlin, B., Foster, S. J., Halvorsen, Y. D., Hicok, K., du Laney, T., Sen, A., Willingmyre, G. D., and Gimble, J. M. (2004). Yield of human adipose-derived adult stem cells from liposuction aspirates. *Cytotherapy* Vol.6, No.1, pp. 7-14, ISSN 1465-3249

Awad, H. A., Halvorsen, Y. D., Gimble, J. M., and Guilak, F. (2003). Effects of transforming growth factor beta1 and dexamethasone on the growth and chondrogenic differentiation of adipose-derived stromal cells. *Tissue Eng* Vol.9, No.6, pp. 1301-12, ISSN 1076-3279

Awad, H. A., Wickham, M. Q., Leddy, H. A., Gimble, J. M., and Guilak, F. (2004). Chondrogenic differentiation of adipose-derived adult stem cells in agarose, alginate, and gelatin scaffolds. *Biomaterials* Vol.25, No.16, pp. 3211-22, ISSN 0142-9612

Beamish, J. A., He, P., Kottke-Marchant, K., and Marchant, R. E. (2010). Molecular regulation of contractile smooth muscle cell phenotype: implications for vascular tissue engineering. *Tissue Eng* Part B Rev Vol.16, No.5, pp. 467-91, ISSN 1937-3376

Becker, A. J., Mc, C. E., and Till, J. E. (1963). Cytological demonstration of the clonal nature of spleen colonies derived from transplanted mouse marrow cells. *Nature* Vol.197, pp. 452-4, ISSN 0028-0836

Borges, J., Mueller, M. C., Padron, N. T., Tegtmeier, F., Lang, E. M., and Stark, G. B. (2003). Engineered adipose tissue supplied by functional microvessels. *Tissue Eng* Vol.9, No.6, pp. 1263-70, ISSN 1076-3279

Boubaker el Andalousi, R., Daussin, P. A., Micallef, J. P., Roux, C., Nougues, J., Chammas, M., Reyne, Y., and Bacou, F. (2002). Changes in mass and performance in rabbit muscles after muscle damage with or without transplantation of primary satellite cells. *Cell Transplant* Vol.11, No.2, pp. 169-80, ISSN 0963-6897

Brun, P., Cortivo, R., Zavan, B., Vecchiato, N., and Abatangelo, G. (1999). In vitro reconstructed tissues on hyaluronan-based temporary scaffolding. *J Mater Sci Mater Med* Vol.10, No.10/11, pp. 683-8, ISSN 0957-4530

Choi, J. H., Gimble, J. M., Lee, K., Marra, K. G., Rubin, J. P., Yoo, J. J., Vunjak-Novakovic, G., and Kaplan, D. L. (2010). Adipose tissue engineering for soft tissue regeneration. *Tissue Eng* Part B Rev Vol.16, No.4, pp. 413-26, ISSN 1937-3376

Choi, Y. S., Cha, S. M., Lee, Y. Y., Kwon, S. W., Park, C. J., and Kim, M. (2006). Adipogenic differentiation of adipose tissue derived adult stem cells in nude mouse. *Biochem Biophys Res Commun* Vol.345, No.2, pp. 631-7, ISSN 0006-291X

Cowherd, R. M., Lyle, R. E., and McGehee, R. E., Jr. (1999). Molecular regulation of adipocyte differentiation. *Semin Cell Dev Biol* Vol.10, No.1, pp. 3-10, ISSN 1084-9521

Crisan, M., Yap, S., Casteilla, L., Chen, C. W., Corselli, M., Park, T. S., Andriolo, G., Sun, B., Zheng, B., Zhang, L., Norotte, C., Teng, P. N., Traas, J., Schugar, R., Deasy, B. M., Badylak, S., Buhring, H. J., Giacobino, J. P., Lazzari, L., Huard, J., and Peault, B.

(2008). A perivascular origin for mesenchymal stem cells in multiple human organs. *Cell Stem Cell* Vol.3, No.3, pp. 301-13, ISSN 1875-9777

De Rosa, A., De Francesco, F., Tirino, V., Ferraro, G. A., Desiderio, V., Paino, F., Pirozzi, G., D'Andrea, F., and Papaccio, G. (2009). A new method for cryopreserving adipose-derived stem cells: an attractive and suitable large-scale and long-term cell banking technology. *Tissue Eng Part C Methods* Vol.15, No.4, pp. 659-67, ISSN 1937-3392

De Ugarte, D. A., Morizono, K., Elbarbary, A., Alfonso, Z., Zuk, P. A., Zhu, M., Dragoo, J. L., Ashjian, P., Thomas, B., Benhaim, P., Chen, I., Fraser, J., and Hedrick, M. H. (2003). Comparison of multi-lineage cells from human adipose tissue and bone marrow. *Cells Tissues Organs* Vol.174, No.3, pp. 101-9, ISSN 1422-6405

Dellavalle, A., Sampaolesi, M., Tonlorenzi, R., Tagliafico, E., Sacchetti, B., Perani, L., Innocenzi, A., Galvez, B. G., Messina, G., Morosetti, R., Li, S., Belicchi, M., Peretti, G., Chamberlain, J. S., Wright, W. E., Torrente, Y., Ferrari, S., Bianco, P., and Cossu, G. (2007). Pericytes of human skeletal muscle are myogenic precursors distinct from satellite cells. *Nat Cell Biol* Vol.9, No.3, pp. 255-67, ISSN 1465-7392

Deslex, S., Negrel, R., Vannier, C., Etienne, J., and Ailhaud, G. (1987). Differentiation of human adipocyte precursors in a chemically defined serum-free medium. *Int J Obes* Vol.11, No.1, pp. 19-27, ISSN 0307-0565

Dragoo, J. L., Samimi, B., Zhu, M., Hame, S. L., Thomas, B. J., Lieberman, J. R., Hedrick, M. H., and Benhaim, P. (2003). Tissue-engineered cartilage and bone using stem cells from human infrapatellar fat pads. *J Bone Joint Surg* Br Vol.85, No.5, pp. 740-7, ISSN 0301-620X

Engfeldt, P., Arner, P., and Ostman, J. (1980). Influence of adipocyte isolation by collagenase on phosphodiesterase activity and lipolysis in man. *J Lipid Res* Vol.21, No.4, pp. 443-8, ISSN 0022-2275

Erickson, G. R., Gimble, J. M., Franklin, D. M., Rice, H. E., Awad, H., and Guilak, F. (2002). Chondrogenic potential of adipose tissue-derived stromal cells in vitro and in vivo. *Biochem Biophys Res Commun* Vol.290, No.2, pp. 763-9, ISSN 0006-291X

Estes, B. T., Diekman, B. O., and Guilak, F. (2008). Monolayer cell expansion conditions affect the chondrogenic potential of adipose-derived stem cells. *Biotechnol Bioeng* Vol.99, No.4, pp. 986-95, ISSN 1097-0290

Farmer, S. R. (2006). Transcriptional control of adipocyte formation. *Cell Metab* Vol.4, No.4, pp. 263-73, ISSN 1550-4131

Flynn, L. E., Prestwich, G. D., Semple, J. L., and Woodhouse, K. A. (2008). Proliferation and differentiation of adipose-derived stem cells on naturally derived scaffolds. *Biomaterials* Vol.29, No.12, pp. 1862-71, ISSN 0142-9612

Frye, C. A., and Patrick, C. W. (2006). Three-dimensional adipose tissue model using low shear bioreactors. *In Vitro Cell Dev Biol Anim* Vol.42, No.5-6, pp. 109-14, ISSN 1071-2690

Gaustad, K. G., Boquest, A. C., Anderson, B. E., Gerdes, A. M., and Collas, P. (2004). Differentiation of human adipose tissue stem cells using extracts of rat cardiomyocytes. *Biochem Biophys Res Commun* Vol.314, No.2, pp. 420-7, ISSN 0006-291X

Gimble, J., and Guilak, F. (2003). Adipose-derived adult stem cells: isolation, characterization, and differentiation potential. *Cytotherapy* Vol.5, No.5, pp. 362-9, ISSN 1465-3249

Gimble, J. M., Katz, A. J., and Bunnell, B. A. (2007). Adipose-derived stem cells for regenerative medicine. *Circ Res* Vol.100, No.9, pp. 1249-60, ISSN 1524-4571

Gonda, K., Shigeura, T., Sato, T., Matsumoto, D., Suga, H., Inoue, K., Aoi, N., Kato, H., Sato, K., Murase, S., Koshima, I., and Yoshimura, K. (2008). Preserved proliferative capacity and multipotency of human adipose-derived stem cells after long-term cryopreservation. *Plast Reconstr Surg* Vol.121, No.2, pp. 401-10, ISSN 1529-4242

Gregoire, F. M., Smas, C. M., and Sul, H. S. (1998). Understanding adipocyte differentiation. *Physiol Rev* Vol.78, No.3, pp. 783-809, ISSN 0031-9333

Gronthos, S., Franklin, D. M., Leddy, H. A., Robey, P. G., Storms, R. W., and Gimble, J. M. (2001). Surface protein characterization of human adipose tissue-derived stromal cells. *J Cell Physiol* Vol.189, No.1, pp. 54-63, ISSN 0021-9541

Guilak, F., Estes, B. T., Diekman, B. O., Moutos, F. T., and Gimble, J. M. (2010). 2010 Nicolas Andry Award: Multipotent adult stem cells from adipose tissue for musculoskeletal tissue engineering. *Clin Orthop Relat* Res Vol.468, No.9, pp. 2530-40, ISSN 1528-1132

Halvorsen, Y. D., Bond, A., Sen, A., Franklin, D. M., Lea-Currie, Y. R., Sujkowski, D., Ellis, P. N., Wilkison, W. O., and Gimble, J. M. (2001). Thiazolidinediones and glucocorticoids synergistically induce differentiation of human adipose tissue stromal cells: biochemical, cellular, and molecular analysis. *Metabolism* Vol.50, No.4, pp. 407-13, ISSN 0026-0495

Halvorsen, Y. D., Franklin, D., Bond, A. L., Hitt, D. C., Auchter, C., Boskey, A. L., Paschalis, E. P., Wilkison, W. O., and Gimble, J. M. (2001). Extracellular matrix mineralization and osteoblast gene expression by human adipose tissue-derived stromal cells. *Tissue Eng* Vol.7, No.6, pp. 729-41, ISSN 1076-3279

Harriman, G. R., Bogue, M., Rogers, P., Finegold, M., Pacheco, S., Bradley, A., Zhang, Y., and Mbawuike, I. N. (1999). Targeted deletion of the IgA constant region in mice leads to IgA deficiency with alterations in expression of other Ig isotypes. *J Immunol* Vol.162, No.5, pp. 2521-9, ISSN 0022-1767

Hauner, H., Entenmann, G., Wabitsch, M., Gaillard, D., Ailhaud, G., Negrel, R., and Pfeiffer, E. F. (1989). Promoting effect of glucocorticoids on the differentiation of human adipocyte precursor cells cultured in a chemically defined medium. *J Clin Invest* Vol.84, No.5, pp. 1663-70, ISSN 0021-9738

Hemmrich, K., von Heimburg, D., Rendchen, R., Di Bartolo, C., Milella, E., and Pallua, N. (2005). Implantation of preadipocyte-loaded hyaluronic acid-based scaffolds into nude mice to evaluate potential for soft tissue engineering. *Biomaterials* Vol.26, No.34, pp. 7025-37, ISSN 0142-9612

Hicok, K. C., Du Laney, T. V., Zhou, Y. S., Halvorsen, Y. D., Hitt, D. C., Cooper, L. F., and Gimble, J. M. (2004). Human adipose-derived adult stem cells produce osteoid in vivo. *Tissue Eng* Vol.10, No.3-4, pp. 371-80, ISSN 1076-3279

Ho, J. H., Ma, W. H., Tseng, T. C., Chen, Y. F., Chen, M. H., and Lee, O. K. (2010). Isolation and Characterization of Multi-Potent Stem Cells from Human Orbital Fat Tissues. *Tissue Eng* Part A, pp., ISSN 1937-335X

Huang, J. I., Zuk, P. A., Jones, N. F., Zhu, M., Lorenz, H. P., Hedrick, M. H., and Benhaim, P. (2004). Chondrogenic potential of multipotential cells from human adipose tissue. *Plast Reconstr Surg* Vol.113, No.2, pp. 585-94, ISSN 0032-1052

Itoi, Y., Takatori, M., Hyakusoku, H., and Mizuno, H. (2010). Comparison of readily available scaffolds for adipose tissue engineering using adipose-derived stem cells. *J Plast Reconstr Aesthet Surg* Vol.63, No.5, pp. 858-64, ISSN 1878-0539

Jack, G. S., Zhang, R., Lee, M., Xu, Y., Wu, B. M., and Rodriguez, L. V. (2009). Urinary bladder smooth muscle engineered from adipose stem cells and a three dimensional synthetic composite. *Biomaterials* Vol.30, No.19, pp. 3259-70, ISSN 1878-5905

Kakudo, N., Minakata, T., Mitsui, T., Kushida, S., Notodihardjo, F. Z., and Kusumoto, K. (2008). Proliferation-promoting effect of platelet-rich plasma on human adipose-derived stem cells and human dermal fibroblasts. *Plast Reconstr Surg* Vol.122, No.5, pp. 1352-60, ISSN 1529-4242

Kang, S. K., Lee, D. H., Bae, Y. C., Kim, H. K., Baik, S. Y., and Jung, J. S. (2003). Improvement of neurological deficits by intracerebral transplantation of human adipose tissue-derived stromal cells after cerebral ischemia in rats. *Exp Neurol* Vol.183, No.2, pp. 355-66, ISSN 0014-4886

Katz, A. J., Tholpady, A., Tholpady, S. S., Shang, H., and Ogle, R. C. (2005). Cell surface and transcriptional characterization of human adipose-derived adherent stromal (hADAS) cells. *Stem Cells* Vol.23, No.3, pp. 412-23, ISSN 1066-5099

Kirkland, J. L., and Hollenberg, C. H. (1998). Inhibitors of preadipocyte replication: opportunities for the treatment of obesity. *Prog Mol Subcell Biol* Vol.20, pp. 177-95, ISSN 0079-6484

Lazar, M. A. (2005). PPAR gamma, 10 years later. *Biochimie* Vol.87, No.1, pp. 9-13, ISSN 0300-9084

Lee, J. A., Parrett, B. M., Conejero, J. A., Laser, J., Chen, J., Kogon, A. J., Nanda, D., Grant, R. T., and Breitbart, A. S. (2003). Biological alchemy: engineering bone and fat from fat-derived stem cells. *Ann Plast Surg* Vol.50, No.6, pp. 610-7, ISSN 0148-7043

Lendeckel, S., Jodicke, A., Christophis, P., Heidinger, K., Wolff, J., Fraser, J. K., Hedrick, M. H., Berthold, L., and Howaldt, H. P. (2004). Autologous stem cells (adipose) and fibrin glue used to treat widespread traumatic calvarial defects: case report. *J Craniomaxillofac Surg* Vol.32, No.6, pp. 370-3, ISSN 1010-5182

Lin, S. D., Wang, K. H., and Kao, A. P. (2008). Engineered adipose tissue of predefined shape and dimensions from human adipose-derived mesenchymal stem cells. *Tissue Eng* Part A Vol.14, No.5, pp. 571-81, ISSN 1937-3341

Liu, Z. J., Zhuge, Y., and Velazquez, O. C. (2009). Trafficking and differentiation of mesenchymal stem cells. *J Cell Biochem* Vol.106, No.6, pp. 984-91, Issn 1097-4644

Lu, F., Gao, J. H., Ogawa, R., Mizuro, H., and Hykusoku, H. (2006). Adipose tissues differentiated by adipose-derived stem cells harvested from transgenic mice. *Chin J Traumatol* Vol.9, No.6, pp. 359-64, ISSN 1008-1275

Lu, F., Mizuno, H., Uysal, C. A., Cai, X., Ogawa, R., and Hyakusoku, H. (2008). Improved viability of random pattern skin flaps through the use of adipose-derived stem cells. *Plast Reconstr Surg* Vol.121, No.1, pp. 50-8, ISSN 1529-4242

Matsumoto, T., Kano, K., Kondo, D., Fukuda, N., Iribe, Y., Tanaka, N., Matsubara, Y., Sakuma, T., Satomi, A., Otaki, M., Ryu, J., and Mugishima, H. (2008). Mature adipocyte-derived dedifferentiated fat cells exhibit multilineage potential. *J Cell Physiol* Vol.215, No.1, pp. 210-22, ISSN 1097-4652

Mauney, J. R., Nguyen, T., Gillen, K., Kirker-Head, C., Gimble, J. M., and Kaplan, D. L. (2007). Engineering adipose-like tissue in vitro and in vivo utilizing human bone marrow and adipose-derived mesenchymal stem cells with silk fibroin 3D scaffolds. *Biomaterials* Vol.28, No.35, pp. 5280-90, ISSN 0142-9612

Meinel, L., Karageorgiou, V., Hofmann, S., Fajardo, R., Snyder, B., Li, C., Zichner, L., Langer, R., Vunjak-Novakovic, G., and Kaplan, D. L. (2004). Engineering bone-like tissue in vitro using human bone marrow stem cells and silk scaffolds. *J Biomed Mater Res* A Vol.71, No.1, pp. 25-34, ISSN 1549-3296

Mizuno, H., Zuk, P. A., Zhu, M., Lorenz, H. P., Benhaim, P., and Hedrick, M. H. (2002). Myogenic differentiation by human processed lipoaspirate cells. *Plast Reconstr Surg* Vol.109, No.1, pp. 199-209; discussion 210-1, ISSN 0032-1052

Moseley, T. A., Zhu, M., and Hedrick, M. H. (2006). Adipose-derived stem and progenitor cells as fillers in plastic and reconstructive surgery. *Plast Reconstr Surg* Vol.118, No.3 Suppl, pp. 121S-128S, ISSN 1529-4242

Nam, S. Y., and Lobie, P. E. (2000). The mechanism of effect of growth hormone on preadipocyte and adipocyte function. *Obes Rev* Vol.1, No.2, pp. 73-86, ISSN 1467-7881

Ning, H., Lin, G., Lue, T. F., and Lin, C. S. (2006). Neuron-like differentiation of adipose tissue-derived stromal cells and vascular smooth muscle cells. *Differentiation* Vol.74, No.9-10, pp. 510-8, ISSN 0301-4681

Ogawa, R., Mizuno, H., Watanabe, A., Migita, M., Hyakusoku, H., and Shimada, T. (2004). Adipogenic differentiation by adipose-derived stem cells harvested from GFP transgenic mice-including relationship of sex differences. *Biochem Biophys Res Commun* Vol.319, No.2, pp. 511-7, ISSN 0006-291X

Owens, G. K., Kumar, M. S., and Wamhoff, B. R. (2004). Molecular regulation of vascular smooth muscle cell differentiation in development and disease. *Physiol Rev* Vol.84, No.3, pp. 767-801, ISSN 0031-9333

Parker, A. M., Shang, H., Khurgel, M., and Katz, A. J. (2007). Low serum and serum-free culture of multipotential human adipose stem cells. *Cytotherapy* Vol.9, No.7, pp. 637-46, ISSN 1465-3249

Patrick, C. W., Jr. (2001). Tissue engineering strategies for adipose tissue repair. *Anat Rec* Vol.263, No.4, pp. 361-6, ISSN 0003-276X

Patrick, C. W., Jr., Chauvin, P. B., Hobley, J., and Reece, G. P. (1999). Preadipocyte seeded PLGA scaffolds for adipose tissue engineering. *Tissue Eng* Vol.5, No.2, pp. 139-51, ISSN 1076-3279

Patrick, C. W., Jr., Zheng, B., Johnston, C., and Reece, G. P. (2002). Long-term implantation of preadipocyte-seeded PLGA scaffolds. *Tissue Eng* Vol.8, No.2, pp. 283-93, ISSN 1076-3279

Pittenger, M. F., Mackay, A. M., Beck, S. C., Jaiswal, R. K., Douglas, R., Mosca, J. D., Moorman, M. A., Simonetti, D. W., Craig, S., and Marshak, D. R. (1999). Multilineage potential of adult human mesenchymal stem cells. *Science* Vol.284, No.5411, pp. 143-7, ISSN 0036-8075

Preece, A. (1972). A Manual for Histologic Technicians. Boston, MA: Little, Brown and Co.

Rodbell, M. (1966). Metabolism of isolated fat cells. II. The similar effects of phospholipase C (Clostridium perfringens alpha toxin) and of insulin on glucose and amino acid metabolism. *J Biol Chem* Vol.241, No.1, pp. 130-9, ISSN 0021-9258

Rodbell, M. (1966). The metabolism of isolated fat cells. IV. Regulation of release of protein by lipolytic hormones and insulin. *J Biol Chem* Vol.241, No.17, pp. 3909-17, ISSN 0021-9258

Rodbell, M., and Jones, A. B. (1966). Metabolism of isolated fat cells. 3. The similar inhibitory action of phospholipase C (Clostridium perfringens alpha toxin) and of insulin on lipolysis stimulated by lipolytic hormones and theophylline. *J Biol Chem* Vol.241, No.1, pp. 140-2, ISSN 0021-9258

Rubin, J. P., Bennett, J. M., Doctor, J. S., Tebbets, B. M., and Marra, K. G. (2007). Collagenous microbeads as a scaffold for tissue engineering with adipose-derived stem cells. *Plast Reconstr Surg* Vol.120, No.2, pp. 414-24, ISSN 1529-4242

Sakuma, T., Matsumoto, T., Kano, K., Fukuda, N., Obinata, D., Yamaguchi, K., Yoshida, T., Takahashi, S., and Mugishima, H. (2009). Mature, adipocyte derived, dedifferentiated fat cells can differentiate into smooth muscle-like cells and contribute to bladder tissue regeneration. *J Urol* Vol.182, No.1, pp. 355-65, ISSN 1527-3792

Schmidt, C. E., and Baier, J. M. (2000). Acellular vascular tissues: natural biomaterials for tissue repair and tissue engineering. *Biomaterials* Vol.21, No.22, pp. 2215-31, ISSN 0142-9612

Sen, A., Lea-Currie, Y. R., Sujkowska, D., Franklin, D. M., Wilkison, W. O., Halvorsen, Y. D., and Gimble, J. M. (2001). Adipogenic potential of human adipose derived stromal cells from multiple donors is heterogeneous. *J Cell Biochem* Vol.81, No.2, pp. 312-9, ISSN 0730-2312

Shanahan, C. M., Weissberg, P. L., and Metcalfe, J. C. (1993). Isolation of gene markers of differentiated and proliferating vascular smooth muscle cells. *Circ Res* Vol.73, No.1, pp. 193-204, ISSN 0009-7330

Shillabeer, G., Forden, J. M., and Lau, D. C. (1989). Induction of preadipocyte differentiation by mature fat cells in the rat. *J Clin Invest* Vol.84, No.2, pp. 381-7, ISSN 0021-9738

Shu XZ, P. G. (2004). Therapeutic biomaterials from chemically modified hyaluronan. In: Garg HG, Hales CA, editors *Chemistry and biology of hyaluronan*. Amsterdam: Elsevier Press, pp. 475-504

Silberstein, L., Webster, S. G., Travis, M., and Blau, H. M. (1986). Developmental progression of myosin gene expression in cultured muscle cells. *Cell* Vol.46, No.7, pp. 1075-81, ISSN 0092-8674

Solchaga, L. A., Dennis, J. E., Goldberg, V. M., and Caplan, A. I. (1999). Hyaluronic acid-based polymers as cell carriers for tissue-engineered repair of bone and cartilage. *J Orthop Res* Vol.17, No.2, pp. 205-13, ISSN 0736-0266

Sorisky, A. (1999). From preadipocyte to adipocyte: differentiation-directed signals of insulin from the cell surface to the nucleus. *Crit Rev Clin Lab Sci* Vol.36, No.1, pp. 1-34, ISSN 1040-8363

Sterodimas, A., de Faria, J., Nicaretta, B., and Pitanguy, I. (2010). Tissue engineering with adipose-derived stem cells (ADSCs): current and future applications. *J Plast Reconstr Aesthet Surg* Vol.63, No.11, pp. 1886-92, ISSN 1878-0539

Stillaert, F. B., Di Bartolo, C., Hunt, J. A., Rhodes, N. P., Tognana, E., Monstrey, S., and Blondeel, P. N. (2008). Human clinical experience with adipose precursor cells seeded on hyaluronic acid-based spongy scaffolds. *Biomaterials* Vol.29, No.29, pp. 3953-9, ISSN 0142-9612

Tchkonia, T., Giorgadze, N., Pirtskhalava, T., Tchoukalova, Y., Karagiannides, I., Forse, R. A., DePonte, M., Stevenson, M., Guo, W., Han, J., Waloga, G., Lash, T. L., Jensen, M. D., and Kirkland, J. L. (2002). Fat depot origin affects adipogenesis in primary cultured and cloned human preadipocytes. *Am J Physiol Regul Integr Comp Physiol* Vol.282, No.5, pp. R1286-96, ISSN 0363-6119

Toyoda, M., Matsubara, Y., Lin, K., Sugimachi, K., and Furue, M. (2009). Characterization and comparison of adipose tissue-derived cells from human subcutaneous and omental adipose tissues. *Cell Biochem Funct* Vol.27, No.7, pp. 440-7, ISSN 1099-0844

von Heimburg, D., Zachariah, S., Heschel, I., Kuhling, H., Schoof, H., Hafemann, B., and Pallua, N. (2001). Human preadipocytes seeded on freeze-dried collagen scaffolds investigated in vitro and in vivo. *Biomaterials* Vol.22, No.5, pp. 429-38, ISSN 0142-9612

von Heimburg, D., Zachariah, S., Low, A., and Pallua, N. (2001). Influence of different biodegradable carriers on the in vivo behavior of human adipose precursor cells. *Plast Reconstr Surg* Vol.108, No.2, pp. 411-20; discussion 421-2, ISSN 0032-1052

Weintraub, H., Davis, R., Tapscott, S., Thayer, M., Krause, M., Benezra, R., Blackwell, T. K., Turner, D., Rupp, R., Hollenberg, S., and et al. (1991). The myoD gene family: nodal point during specification of the muscle cell lineage. *Science* Vol.251, No.4995, pp. 761-6, ISSN 0036-8075

Wickham, M. Q., Erickson, G. R., Gimble, J. M., Vail, T. P., and Guilak, F. (2003). Multipotent stromal cells derived from the infrapatellar fat pad of the knee. *Clin Orthop Relat Res*, No.412, pp. 196-212, ISSN 0009-921X

Yoshimura, K., Sato, K., Aoi, N., Kurita, M., Hirohi, T., and Harii, K. (2008). Cell-assisted lipotransfer for cosmetic breast augmentation: supportive use of adipose-derived stem/stromal cells. *Aesthetic Plast Surg* Vol.32, No.1, pp. 48-55; discussion 56-7, ISSN 0364-216X

Yuan, Q., Zeng, X., Chen, L., Peng, E., and Ye, Z. (2010). [Comparison of myogenic differentiation ability of adipose-derived stem cells from different sites in rabbit]. Zhongguo Xiu Fu Chong Jian Wai Ke Za Zhi Vol.24, No.10, pp. 1228-32, ISSN 1002-1892

Zhang, J., Shehabeldin, A., da Cruz, L. A., Butler, J., Somani, A. K., McGavin, M., Kozieradzki, I., dos Santos, A. O., Nagy, A., Grinstein, S., Penninger, J. M., and Siminovitch, K. A. (1999). Antigen receptor-induced activation and cytoskeletal rearrangement are impaired in Wiskott-Aldrich syndrome protein-deficient lymphocytes. *J Exp Med* Vol.190, No.9, pp. 1329-42, ISSN 0022-1007

Zhang, Y. S., Gao, J. H., Lu, F., Zhu, M., and Liao, Y. J. (2007). [Cellular compatibility of type collagen I scaffold and human adipose-derived stem cells]. *Nan Fang Yi Ke Da Xue Xue Bao* Vol.27, No.2, pp. 223-5, ISSN 1673-4254

Zuk, P. A., Zhu, M., Ashjian, P., De Ugarte, D. A., Huang, J. I., Mizuno, H., Alfonso, Z. C., Fraser, J. K., Benhaim, P., and Hedrick, M. H. (2002). Human adipose tissue is a source of multipotent stem cells. *Mol Biol Cell* Vol.13, No.12, pp. 4279-95, ISSN 1059-1524

Zuk, P. A., Zhu, M., Mizuno, H., Huang, J., Futrell, J. W., Katz, A. J., Benhaim, P., Lorenz, H. P., and Hedrick, M. H. (2001). Multilineage cells from human adipose tissue: implications for cell-based therapies. *Tissue Eng* Vol.7, No.2, pp. 211-28, ISSN 1076-3279

8

In Vitro Culture Methods of Skin Cells for Optimal Skin Reconstruction by Tissue Engineering

Moulin VJ, Mayrand D, Laforce-Lavoie A, Larochelle S and Genest H
Centre LOEX de l'Université Laval, Génie tissulaire et régénération: LOEX du Centre de recherche FRSQ du Centre hospitalier affilié universitaire de Québec, Département de Chirurgie, Faculté de Médecine, Université Laval, Québec, QC Canada

1. Introduction

Replacement of a wounded or destroyed tissue is now technically possible using an in vitro method of tissue engineering. This method involves isolating and culturing human cells (autologous or not) in optimal conditions to form a reconstructed tissue with similar characteristics to its in vivo counterpart. Burn victims were the first patients to benefit from this method in 1981, wherein reconstituted human epidermal tissues were used to treat the burns (O'Connor et al., 1981). The pioneered technique comprises epidermal cell isolation, culturing and grafting as a unique layer of cells organized in a sheet. The differentiation of the cells into a pluristratified epidermis is performed in vivo after grafting.

Numerous improvements, most notably the addition of a dermal part in skin tissue reconstruction have been made since the technique's inception. However, the dermis is more difficult to reconstitute because of the complexity of its organization. Briefly, the dermis is a mix of matrix and cells, which are primarily fibroblasts. Moreover, the dermis is also composed of several structures such as the epidermal annexes (e.g., hair and sebaceous and sweat glands) and a network of capillaries. Several others cells like lymphocytes, or neurons are also more difficult to add on dermis despite their crucial roles in acquiring all the functions of the skin. Furthermore, the third part of the skin, the hypodermis is vital in vivo and addition of this tissue will increase the function of the grafted skin.

In classical cutaneous grafting with split- or full-thickness skin, the amount of dermis that is grafted to a wound bed inversely correlates with the degree of scarring and wound contracture which impacts the functional and cosmetic outcome (Bombaro et al., 2003). Similarly, dermal cells have been fundamental in the efficacy and quality of cultured keratinocyte grafts (Moulin et al., 2000) (El Ghalbzouri et al., 2002) (Kirfel and Herzog, 2004; Gallant-Behm et al., 2011) (Robert et al., 1997).

Currently clinics widely use dermis that is constituted of matrix and fibroblasts. The methods used to obtain cultured living dermis, which can be grouped into three different categories (Table 1) represent the three main methods of tissue engineering. These categories are based on the materials used to originate the matrix, such as biomaterials, biological materials and dermal fibroblasts. Biomaterials are materials that are not present in skin but that are compatible with cells. Biological materials such as collagens are primarily derived from non-human sources. Lastly, the matrix may be created by human dermal fibroblasts themselves during culture. We refer to this last method as the "self-assembly method" in our lab. The cell source is primarily allogeneic, which allows for a high quantity production of tissues with a quick delay before grafting. Using this method, the cells are available to stimulate healing and are replaced by the patient's cells with time. In a few cases (Hyalograft 3D™ or "LOEX" skin), cells are isolated directly from the patient's skin. Using the approach, reconstituted tissues comprise a true graft and do not disappear with time. However, this method is time consuming and costly.

Matrix protein origin	Description
Biomaterial	
Biodegradable polyglactin mesh scaffold	Dermagraft™: Cryopreserved allogeneic fibroblasts are cultured on the mesh
Biological	
Esterified hyaluronic acid matrix	Hyalograft 3D™: Autologous fibroblasts are cultured in the scaffold
Bovine collagen sponge	Orcel™: Allogenic keratinocytes are seeded over a dermal scaffold containing allogeneic fibroblasts.
Bovine collagen sponge	Apligraft™: Allogenic keratinocytes are seeded over a dermal scaffold containing allogeneic fibroblasts.
Bovine collagen matrix	Permaderm™: Autologous keratinocytes are seeded on collagen gels made with autologous fibroblasts
Cellular	
Fibrin matrix at the beginning replaced by fibroblast-secreted matrix	ICX-SKN™: Matrix is freeze dried prior to repopulation with allogeneic fibroblasts
Fibroblast-secreted matrix	VCT01™: Allogenic keratinocytes are seeded on top of a dermis made of allogenic fibroblasts secreting their own ECM
Fibroblast-secreted matrix	LOEX: Autologous keratinocytes are seeded on top of a dermis made of autologous fibroblasts secreting their own ECM

Table 1. Matrix types in commercial and pre-commercial tissue-engineered dermis or skins

2. Fibroblast functionality

While simplified dermis comprised of matrix and fibroblasts are presently used to treat patients, several improvements in this tissue reconstruction method are necessary to obtain a skin with better functional acceptability.

The capacity of the skin fibroblasts to grow, remodel the matrix, secrete and degrade proteins, and produce numerous growth factors or cytokines is crucial for the regulation of the tissue structure and its cellular microenvironment. During reconstitution of dermal substitutes, the cells must retain their capacity to regulate their environment. Several parameters must be monitored to ensure proper fibroblast functions in growth capacity, matrix deposition and remodeling or differentiation state maintenance, all of which are inherent to these cells in normal human skin. There is a paucity of published studies dedicated to the analysis of the cell phenotype changes in vitro versus in vivo. This phenotype can change depending on several parameters, such as the cell duration of the cell culture (Endt et al., 2011), the location of the skin biopsies (Chang et al., 2002) and the age of the donor (Dumas et al., 1994). However, these results are highly controversial due to a lack of reproductibility by other groups (Ng et al., 2009) (Falanga et al., 1991).

2.1 Cell passage number

The replicative potential of a human cell population must be considered when those cells are used to reconstruct an engineered tissue. Indeed, because the replicative potential of most cells is limited, tissues reconstructed from cells at lower passage yield tissues of better quality. However, using cells promptly after their isolation and limited expansion is not always possible due to the requirement for several passages to obtain the quantity of cells needed. Therefore, knowing the cell passage range that yields the optimal results when using fibroblasts in tissue reconstruction is important.

We studied three fibroblast populations isolated from human skin biopsies at different times of culture, which were assigned by passage number. Fibroblasts were isolated using a collagenase solution (Rompré et al., 1990) and cultured with DME+10% fetal calf serum. When cells reached confluency, they were trypzinized and seeded until they reached the twelfth passage. Population doubling times, which were indicative of the cell growth capacity, the diameter of the cells, and the contractile capacity were evaluated from the first passage until the 12th passage.

Doubling times were calculated using the following formula: $DT = \ln 2 * t / (\ln C1 - \ln C0)$, where t is the culture duration, $C1$ is the number of cells at the end of the culture and $C0$ is the number of seeded cells. Doubling times of three cell populations isolated from human skin biopsies were studied, and no statistical difference was detected from Passage 1 to 12 (Figure 1A). This is in agreement with results reported by Endt et al. (2011), wherein analysis of only one fibroblast population did not reveal any doubling time differences before the 15th passage. The cell diameter variation has also been investigated (Figure 1B) as a marker of a cell phenotype modification. This parameter remained constant with time (17 $\mu m \pm 0.83$) and lacked any apparent modifications.

Cell senescence has been shown to increase with time and has been detected in fibroblasts from the 30th passage (Endt et al., 2011). However, a higher sensitivity to apoptosis is also indicative of cellular aging. We monitored the apoptotic rate of cells using propidium iodide incorporation (Moulin et al., 2004) in the presence of 10% FBS, an additive that induces cell growth. A very low rate of apoptotic cells can be detected (<4%) at all passages and cell populations (Figure 2A). When FBS was removed from the culture medium, the absence of a growth factor induced a slight increase in apoptosis (<4%) during the initial passages. However, passages 10 and 12 exhibited a marked increase in the apoptotic rate, which is indicative of a change in the cell's response to apoptotic stimuli (Figure 2B).

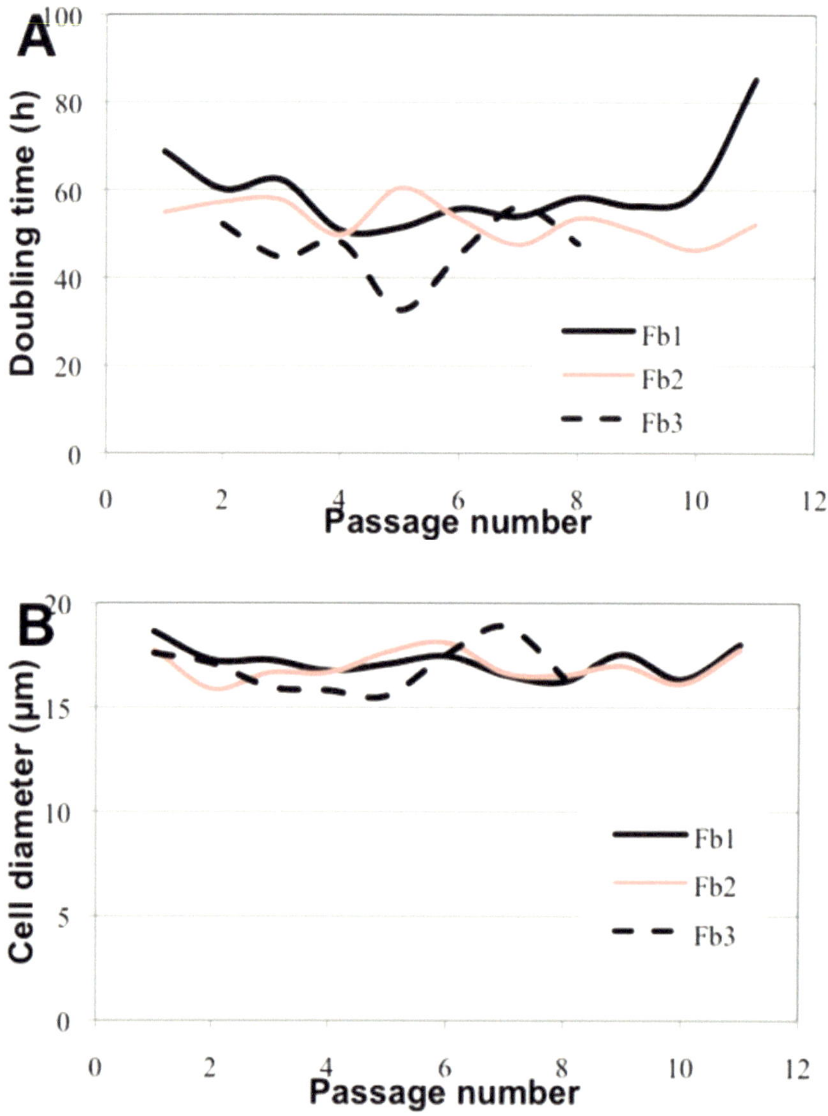

Fig. 1. The doubling times and diameters of fibroblasts cultured for different durations

The contractile capacity of cells has been studied to evaluate one of their function during healing (Finesmith et al., 1990; Delvoye et al., 1991). Cells were seeded into a collagen gel according to Moulin et al. (1998), and the gel diameter was evaluated daily (Figure 3). Differences in the gel contraction speed, which reflects the contractile capacity of the cells, were not observed when cells were cultured from passages 3 to 12.

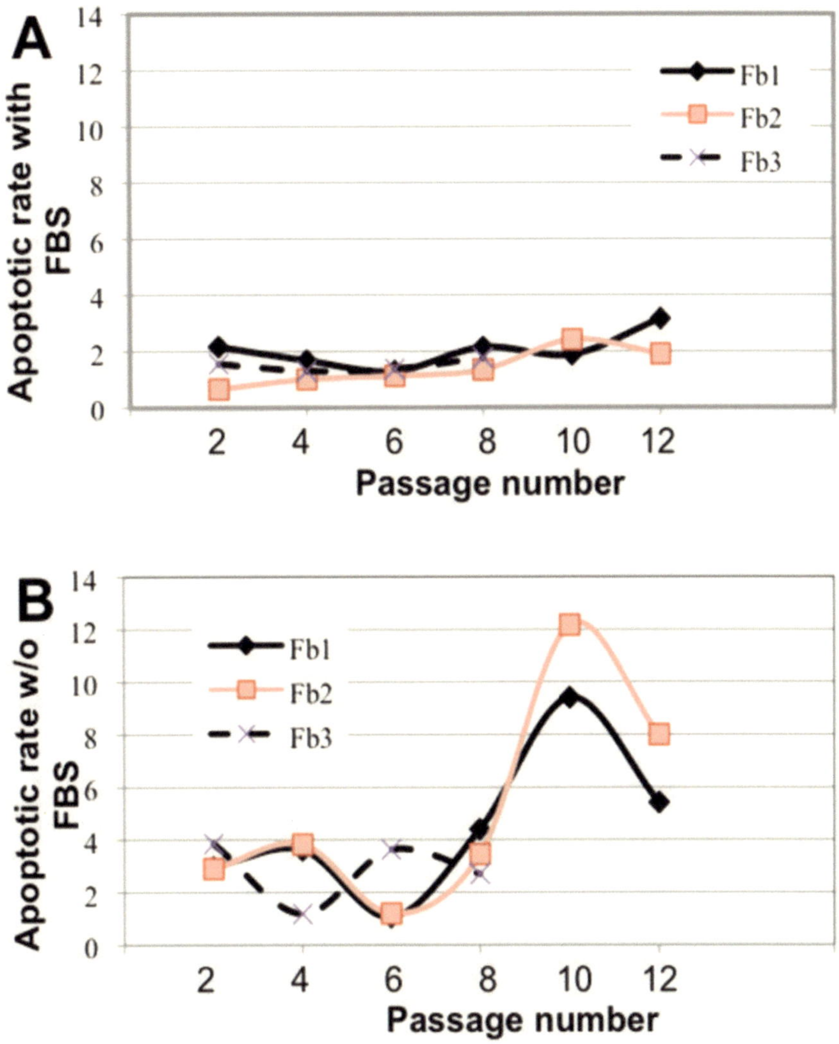

Fig. 2. The apoptotic rate of fibroblasts (A) in presence of 10% FBS or (B) without FBS. The rate of apoptosis was estimated from particle numbers containing less than 2n of DNA after propidium iodide incorporation

In summary, fibroblasts isolated from the dermis can be cultured without any major changes for at least 8 passages. After this time, the occurrence of several minor stresses, such as a transient lack of nutrients, can interfere with the phenotype of the cells and, thus, their response to stress in vivo. After grafting, cells placed in an engineered, reconstructed tissue must respond to several stresses, such as an absence of nutrients before revascularization, trauma, or infection, for the entirety of the patient's life. Thus, cells must be as similar to the original cells as possible. For a skin graft, sub-culturing of fibroblasts for 8 passages, corresponding to at least 25 doubling populations where one isolated cell will generate 8.4 million cells, permits the accumulation of enough cells to reconstruct a large surface area of the dermis.

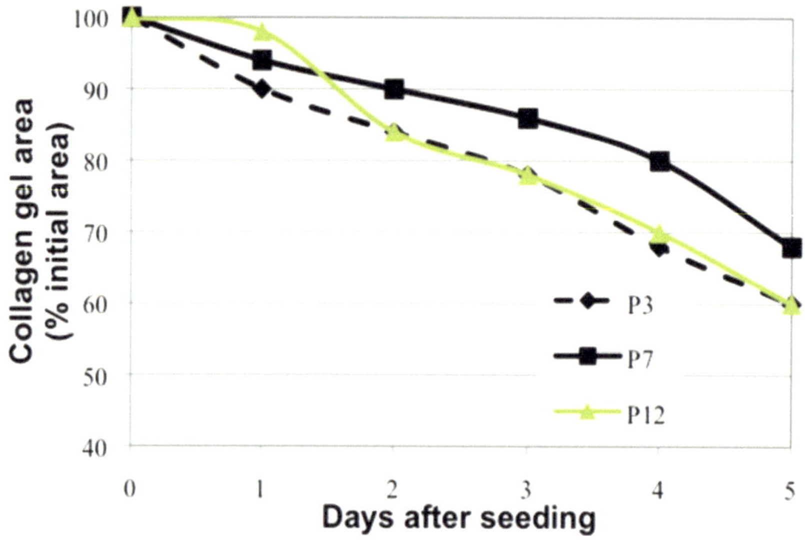

Fig. 3. The contractile capacity of fibroblasts (Fb8) at different passages. Representative results obtained with 3 different populations are presented

2.2 Age and sex of the donor

For autologous grafting, the age of the donor should not change. However, many tissue-engineered derma are produced using allogeneic cells. Therefore, the age of the donors' cells is a valuable parameter to estimate. We have calculated the doubling times for 17 different cell populations isolated from the human dermis at passage 5. Biopsies were taken from 20- to 64-year-old donors. Variations in the doubling time were not detected with donor age. In our experimental conditions, the doubling time mean of the fibroblasts was 47.2 ± 7.5 hours (Figure 4).

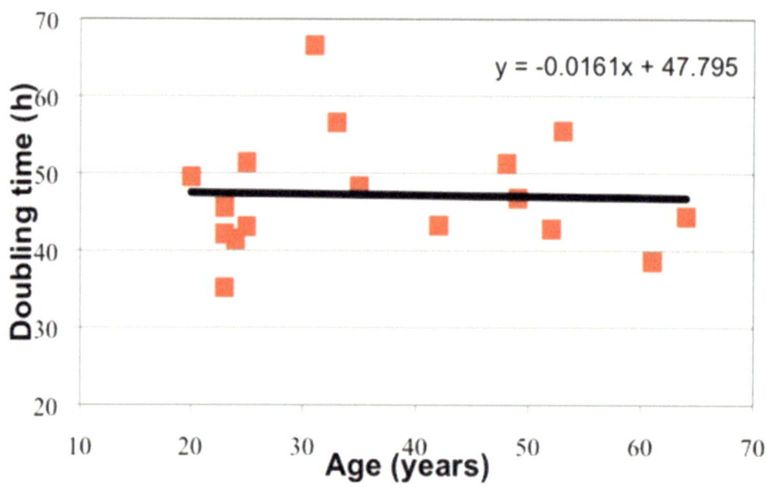

Fig. 4. The doubling time of fibroblasts from skin donors of different ages. The formula of the line was calculated using linear regression analysis

A comparison of cell populations from female and male donors of similar age and biopsy location did not reveal a statistically different doubling time (Student's t-test, p=0.63, N=3 for each category) (data not shown).

2.3 Location of the biopsy

The importance of the location of the biopsy at the origin of the fibroblasts in modulating in vitro cellular phenotypes has been addressed with mixed results (Chang et al., 2002; Chipev and Simon, 2002; Falanga et al., 1991). We did not find a statistical difference in the doubling time of fibroblasts from different biopsy locations, including the forearm, scalp, abdomen, and breast (Figure 5).

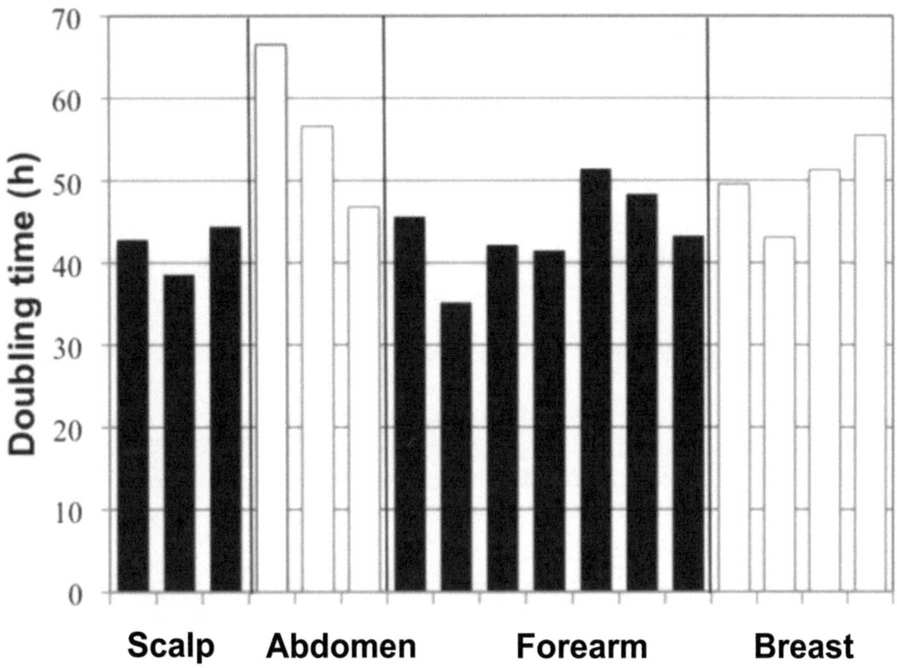

Fig. 5. The doubling time of the fibroblasts from different biopsy locations. Student-Newman-Keul multiple range test: *, p<0.05

One of the most important properties of fibroblasts is to secrete extracellular matrix to create the dermis. To evaluate this parameter, we used the self-assembly approach developed at the LOEX (Michel et al., 1999). Fibroblasts that are cultured for 4 weeks in the presence of ascorbate secrete and remodel extracellular proteins, a measure of the matrix remodeling capacity of the cells (Bellemare et al., 2005). A statistical difference in the thickness of the reconstructed dermis was observed for the various biopsy locations tested. When cells were isolated from the scalp or abdomen, the dermis was thicker than when fibroblasts were isolated from the breast or forearm (Figure 6). However, the thickness of the dermis was not dependent on the age of the patient (data not shown). Because the biopsy site may dictate the quality of the reconstructed dermis and, thus, the viability of its graft, the need for accurately choosing the biopsy site from which to collect cells to reconstruct a tissue is crucial.

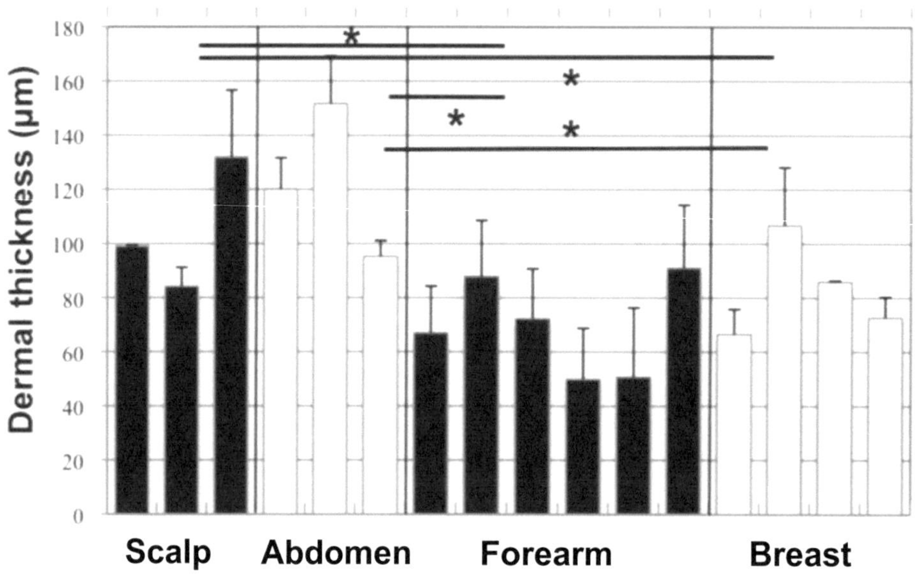

Fig. 6. The thickness of the dermis reconstructed using the self-assembly approach with fibroblasts isolated from different location of the body. Student-Newman-Keul multiple range test: *: $p<0.05$

3. Capillary network

The graft of thick tissues is always a challenge due to the absence of a capillary network, which impedes nourishment of the tissue during the first days after grafting. Malnourishment is often responsible for the increase in the percentage of rejected grafts observed with these types of tissues. Different approaches are currently used to improve skin substitute vascularization before or after transplantation. The addition of growth factors, including VEGF and FGF2, via absorption or incorporation into scaffolds has been shown to trigger a localized and sustained delivery that promotes vascularization (Bouhadir and Mooney, 2001). Transplanted cells can also be genetically modified to produce angiogenic proteins that promote better and faster vascularization (Supp et al., 2000). Forming a capillary network into the tissues before grafting has been reported as an effective method to bypass this problem. Of several described techniques, seeding endothelial cells in biomaterial or on a dermal sheet that is then superimposed has been used most often (Black et al., 1998).

3.1 Isolation of human capillary endothelial cells

The isolation of the needed capillary cells can be performed from the same small tissue biopsy (less than 1cm^2) used to isolate the keratinocytes and the fibroblasts, that allows for the reconstitution of an autologous skin with a differentiated epithelium and a vascularized dermis.

The skin biopsy was cut into 0.5 cm^2 pieces and transferred into a thermolysin solution (500µg/ml) in HEPES buffer (Germain et al., 1993) then incubated at 4°C overnight. The epidermis was gently peeled from the dermis with forceps and constantly agitated in a trypsin/EDTA (Valeant Canada Limited) solution for 15 min. The epidermal cells were centrifuged, and plated in the presence of irradiated 3T3 feeder layer cells as previously described (Moulin et al., 2000).

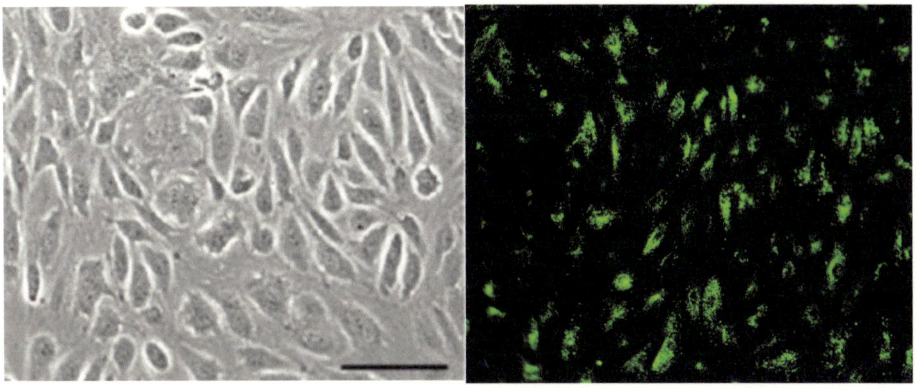

Fig. 7. Microvascular endothelial cells were isolated from human dermis; (left) phase contrast and (right) von Willebrand factor immunostaining (bar: 200μm)

The capillaries were extruded by pressing the dermal layer and were plated on gelatin-coated tissue culture flasks with EGM-2 medium (Cambrex Bio Science Baltimore, Inc., Baltimore, MD). After 24h, the medium was replaced to eliminate non-adherent cells and the cells were cultured for one week. The cells were removed from the culture flask by trypsin/EDTA and processed with CD31 antibody-coupled magnetic bead (Dynabeads, Invitrogen) for 30 min, allowing for further purification of the cells (Richard et al., 1998). Immunostaining was performed to validate the presence of the von Willebrand factor, a specific marker of endothelial cells, in the whole population (Figure 7). After the extrusion of the capillaries, the dermis was incubated in a collagenase H (0.125U/ml) solution for 4-5 hours at 37°C then plated and cultured as in (Moulin et al., 2001).

3.2 Reconstruction of a tissue-engineered endothelialized skin

The self-assembly approach is a tissue engineering method based on the in vitro production of mesenchymal sheets devoid of biomaterial or exogenous matrix proteins that were previously described for the production of human skin and blood vessels (L'Heureux et al., 1998; Michel et al., 1999). Dermal fibroblasts are cultured on petri dishes with the fibroblast medium supplemented with ascorbic acid (50μg/ml). Cells secrete and remodel matrix that form a manipulable sheet after 4 weeks. Endothelial cells are then seeded on the sheet and cultured for an additional week in EGM-2 medium with ascorbic acid. Two of these sheets are then superimposed to form an endothelialized tissue-engineered dermal substitute. Subsequently, keratinocytes can then be added onto the dermis to form an epidermis after differentiation at the air-liquid interface (Rochon et al., 2010) (Figure 8).

We have evaluated the formation of the capillary structure with time after endothelial cell seeding on dermal sheets and we observed an increase of capillary like structure over time. The formation was evaluated using CD31, a protein that localizes to junctions between endothelial cells. The formation of capillary-like structures reached a maximum at 21 days and remained stable until at least day 28 (Figures 9 and 10).

This endothelialized tissue-engineered skin has been grafted onto mice. Complete re-vascularization of non-vascularized skin substitutes usually takes 14 days. Authors have demonstrate that endothelial network inosculates with the host's own vascular system within 4 days allowing to a quick revascularization of the tissue (Gibot et al., 2010). Furthermore, an active invasion of the dermis by the sprouting of host capillaries from the wound bed has been detected. The authors conclude that the microvascular network constructed in vitro is an interesting method to quickly vascularize a thick tissue. This finding will facilitate the graft take of thick engineered-tissues that, currently, die due to a lack of oxygen and nutriments.

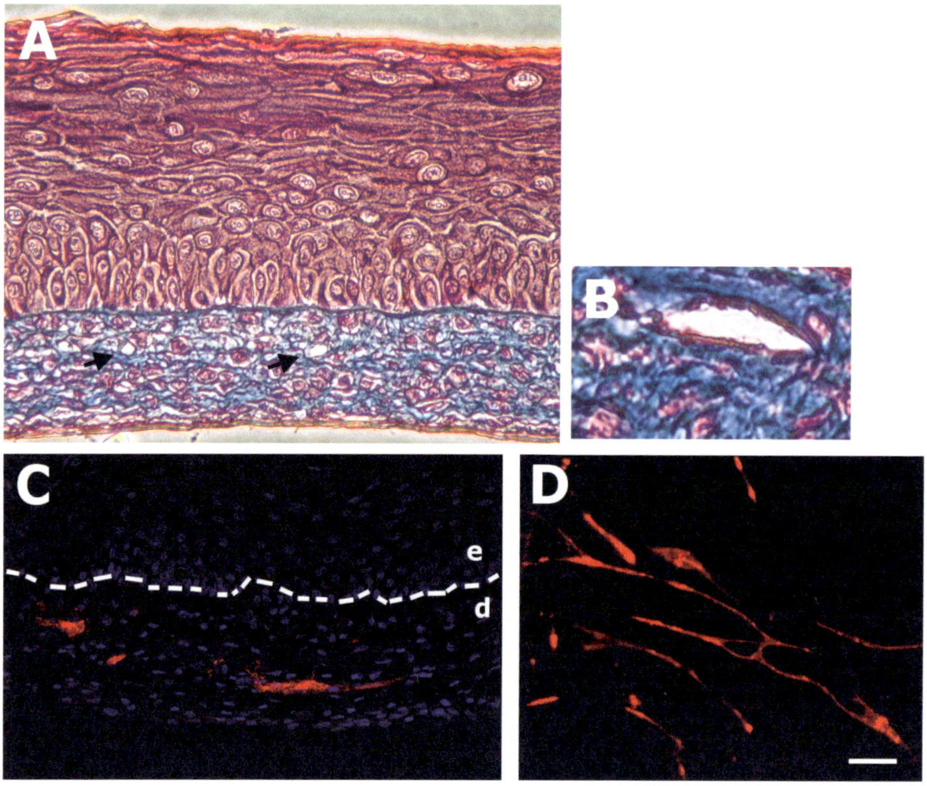

Fig. 8. Endothelialized tissue-engineered skin performed with three cell populations (i.e., keratinocytes, fibroblasts and endothelial cells) isolated from the same skin biopsy (A) Masson's Trichrome staining. Arrows indicate capillary-like structures. (B) Higher magnification of a capillary-like structure. (C) Transversal section of CD31-immunolabeled endothelialized tissue-engineered skin (dotted line: basal membrane between epidermis (e) and dermis (d)). (D) Bird's-eye view of CD-31-labeled endothelialized tissue-engineered dermis. Bar: 50µm for A, C and D and 7µm for B

4. Other possible improvements

The skin is a complex organ that cannot be pared down to a bilayer tissue comprised simply of an epidermis and a dermis. The third part of the skin, the hypodermis, is often ignored; however, it is crucial in controlling the temperature and nutriment storage, as well as secreting important hormones such as leptin that are important during healing (Frank et al., 2000). The addition of this skin moiety to a tissue-engineered reconstructed skin should increase the functions of the grafted skin. Researchers can now reconstruct an autologous hypodermis using human adipose-derived stem/stromal cells and incorporate it to a reconstructed skin (Trottier et al., 2008).

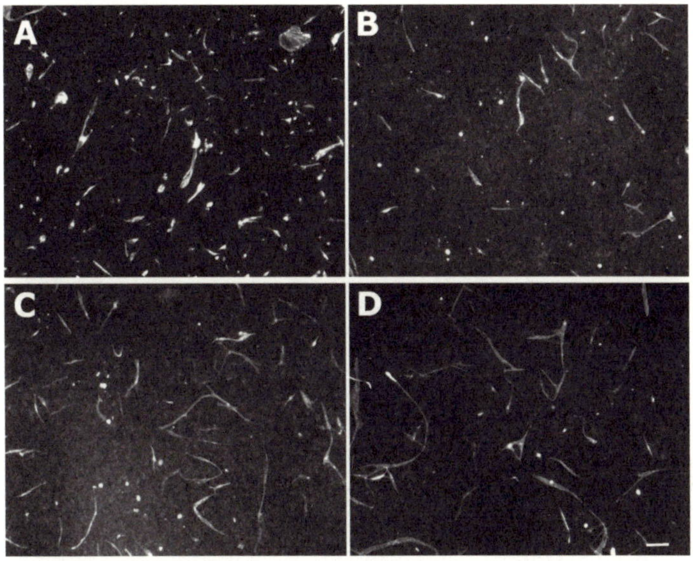

Fig. 9. Immunohistochemistry with CD31 antibodies on endothelialized-tissues after 7 (A), 14 (B), 21 (C) and 28 (D) days of culture. Bar: 200μm

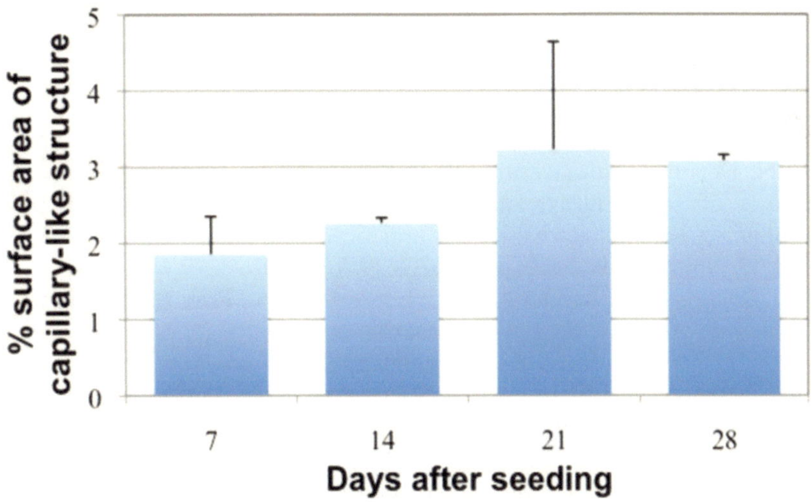

Fig. 10. The percent surface area of capillary-like structure evaluated from pictures of immunohistostaining with CD31-antibodies as in Figure 9

The absence of sebaceous and sweat glands in the reconstructed tissue is a recurrent problem for patients with large burn areas that have been grafted with tissue-engineered skin. This absence induces important thermoregulation problems and causes a dry skin that needs to be constantly moisturized. The addition of these glands to a tissue-engineered skin is not trivial but recent results offer promise (Huang et al., 2010).

Hair follicle regeneration was thought to be impossible during adult life. Wound stimulus and wnt pathway activation have been recently shown to trigger de novo hair follicle formation from epidermal stem cells (Ito et al., 2007). This observation suggests that tissue-engineered skin with normal hair follicles is plausible.

If added, these complex skin structures can improve the functionality of the tissue. Furthermore, several other cell populations have also been added such as melanocytes, improving UV protection of the skin (Scuderi et al., 2008); Langerhans and dendritic cells, immunological cells present in the skin (Bechetoille et al., 2007) or nerves to improve skin sensation (Blais et al., 2009). These additions, if possible in laboratories for experiments, have now to be added in tissue-engineered skin for routine use as human skin replacement on patient.

5. Conclusion

The possibilities of the tissue engineering method are wide and an increase in the functionality of the grafted tissues will enhance their use. The skin has been the first tissue to be reconstructed and used in clinics to cure patients including large burn victims and for ulcer therapy. Techniques that have been developed for the skin reconstitution are currently used to reconstitute other tissues with more complex structures. However, numerous improvements are needed before obtaining an identical tissue.

6. Acknowledgments

The authors are supported by the Canadian Institutes of Health Research, Fonds de la Recherche en Santé du Québec (FRSQ), Fondation du CHA-Hôpital Enfant-Jesus/Saint-Sacrement and Réseau ThéCell, FRSQ.

7. References

Bechetoille, N., Dezutter-Dambuyant, C., Damour, O., Andre, V., Orly, I. & Perrier, E. (2007) Effects of solar ultraviolet radiation on engineered human skin equivalent containing both Langerhans cells and dermal dendritic cells. *Tissue Engineering*, Vol. 13, No. 11, (Nov 2007), pp.:2667-2679, ISSN 1076-3279

Bellemare, J., Roberge, C.J., Bergeron, D., Lopez-Valle, C.A., Roy, M. & Moulin, V.J. (2005) Epidermis promotes dermal fibrosis: role in the pathogenesis of hypertrophic scar. *Journal of Pathology*, Vol. 206 (may 2005), pp.:1-8

Black, A.F., Berthod, F., L'Heureux, N., Germain, L. & Auger, F.A. (1998) In vitro reconstruction of a human capillary-like network in a tissue-engineered skin equivalent. *FASEB Journal*, Vol. 12, No. 13, (Oct 1998), pp.:1331-1340, ISSN 0892-6638

Blais, M., Grenier, M. & Berthod, F. (2009) Improvement of nerve regeneration in tissue-engineered skin enriched with schwann cells. *Journal of Investigative Dermatology*, Vol. 129, No. 12, (Dec 2009), pp.:2895-2900, ISSN 1523-1747

Bombaro, K.M., Engrav, L.H., Carrougher, G.J., Wiechman, S.A., Faucher, L., Costa, B.A., Heimbach, D.M., Rivara, F.P. & Honari, S. (2003) What is the prevalence of hypertrophic scarring following burns? *Burns*, Vol. 29, No. 4, (Jun 2003), pp.:299-302

Bouhadir, K.H. & Mooney, D.J. (2001) Promoting angiogenesis in engineered tissues. *Journal of Drug Targeting*, Vol. 9, No. 6, pp.:397-406, ISSN 1061-186X

Chang, H., Chi, J.-T., Dudoit, S., Bondre, C., van de Rijn, M., Botstein, D. & Brown, P. (2002) Diversity, topographic differentiation and positional memory in human fibroblasts. *Proceeding of National Academy of Science of U S A*, Vol. 99, No. 20, pp.:12877-12882,

Chipev, C. & Simon, M. (2002) Phenotypic differences between dermal fibroblasts from different body sites determine their responses to tension and TGFß1. *BMC Dermatology*, http://www.biomedcentral.com/1471-5945/2/13

Delvoye, P., Wiliquet, P., Leveque, J.-L., Nusgens, B.V. & Lapière, C.M. (1991) Measurement of mechanical forces generated by skin fibroblasts embedded in a three-dimensional collagen gel. *Journal of Investigative Dermatology*, Vol. 97, No. pp.:898-902

Dumas, M., Chaudagne, C., Bonte, F. & Meybeck, A. (1994) In vitro biosynthesis of type I and III collagens by human dermal fibroblasts from donors of increasing age. *Mechanisms of Ageing & Development*, Vol. 73, No. 3, (Mar 1994), pp.:179-187

El Ghalbzouri, A., Gibbs, S., Lamme, E., Van Blitterswijk, C.A. & Ponec, M. (2002) Effect of fibroblasts on epidermal regeneration. *British Journal of Dermatology*, Vol. 147, No. 2, (Aug 2002), pp.:230-243

Endt, H., Sprung, C.N., Keller, U., Gaipl, U., Fietkau, R. & Distel, L.V. (2011) Detailed analysis of DNA repair and senescence marker kinetics over the life span of a human fibroblast cell line. *Journals of Gerontology. Series A, Biological Sciences & Medical Sciences*, Vol. 66, No. 4, (Apr 2011), pp.:367-375, ISSN 1758-535X

Falanga, V., Katz, M.H. & Goodless, D.R. (1991) Proliferation and collagen synthesis by human dermal fibroblasts cultured from different body sites. *Journal of Dermatological Science*, Vol. 2, No. 3, (May 1991), pp.:166-170, ISSN 0923-1811

Finesmith, T.H., Broadley, K.N. & Davidson, J.M. (1990) Fibroblasts from wounds of different stages of repair vary in their ability to contract a collagen gel in response to growth factors. *Journal of Cellular Physiology*, Vol. 144, No. pp.:99-107

Frank, S., Stallmeyer, B., Kampfer, H., Kolb, N. & Pfeilschifter, J. (2000) Leptin enhances wound re-epithelialization and constitutes a direct function of leptin in skin repair. *Journal of Clinical Investigation*, Vol. 106, No. 4, (Aug 2000), pp.:501-509, ISSN 0021-9738

Gallant-Behm, C.L., Du, P., Lin, S.M., Marucha, P.T., Dipietro, L.A. & Mustoe, T.A. (2011) Epithelial Regulation of Mesenchymal Tissue Behavior. *Journal of Investigative Dermatology*, Vol. 131, (Jan 2011), pp: 892-899, ISSN 1523-1747

Germain, L., Rouabhia, M., Guignard, R., Carrier, L., Bouvard, V. & Auger, F.A. (1993) Improvement of human keratinocyte isolation and culture using thermolysin. *Burns*, Vol. 19, No.2 pp.:99-104

Gibot, L., Galbraith, T., Huot, J. & Auger, F.A. (2010) A preexisting microvascular network benefits in vivo revascularization of a microvascularized tissue-engineered skin substitute. *Tissue Engineering Part A*, Vol. 16, No. 10, (Oct 2010), pp.:3199-3206, ISSN 1937-335X

Huang, S., Xu, Y., Wu, C., Sha, D. & Fu, X. (2010) In vitro constitution and in vivo implantation of engineered skin constructs with sweat glands. *Biomaterials*, Vol. 31, No. 21, (Jul 2010), pp.:5520-5525, ISSN 1878-5905

Ito, M., Yang, Z., Andl, T., Cui, C., Kim, N., Millar, S.E. & Cotsarelis, G. (2007) Wnt-dependent de novo hair follicle regeneration in adult mouse skin after wounding. *Nature*, Vol. 447, No. 7142, (May 2007), pp.:316-320, ISSN 1476-4687

Kirfel, G. & Herzog, V. (2004) Migration of epidermal keratinocytes: mechanisms, regulation, and biological significance. *Protoplasma*, Vol. 223, No. 2-4, (Jun 2004), pp.:67-78

L'Heureux, N., Paquet, S., Labbé, R., Germain, L. & Auger, F.A. (1998) A completely biological tissue-engineered human blood vessel. *FASEB Journal*, Vol. 12, pp.:47-56

Michel, M., L'Heureux, N., Pouliot, R., Xu, W., Auger, F.A. & Germain, L. (1999) Characterization of a new tissue-engineered human skin equivalent with hair. *In Vitro Cellular & Developmental Biology. Animal*, Vol. 35, pp.:318-326

Moulin, V., Auger, F., Garrel, D.&Germain, L. (2000) Role of wound healing myofibroblasts on reepithelialization of skin. *Burns*, Vol. 26, No. 1, pp.:3-12

Moulin, V., Castilloux, G., Auger, F.A., Garrel, D., O'Connor-McCourt, M. & Germain, L. (1998) Modulated response to cytokines of human wound healing myofibroblasts compared to dermal fibroblasts. *Experimental Cell Research*, Vol. 238, pp.:283-293

Moulin, V., Larochelle, S., Langlois, C., Thibault, I., Lopez-Valle, C.A. & Roy, M. (2004) Normal skin wound and hypertrophic scar myofibroblasts have differential responses to apoptotic inductors. *Journal of Cellular Physiology*, Vol. 198, No. 3, (Mar 2004), pp.:350-358

Moulin, V., Tam, B.Y.Y., Castilloux, G., Auger, F.A., O'Connor-McCourt, M.D., Philip, A. & Germain, L. (2001) Fetal and adult human skin fibroblasts display intrinsic differences in contractile capacity. *Journal of Cellular Physiology*, Vol. 188, pp.:211-222

Ng, M.H., Aminuddin, B.S., Hamizah, S., Lynette, C., Mazlyzam, A.L. & Ruszymah, B.H. (2009) Correlation of donor age and telomerase activity with in vitro cell growth and replicative potential for dermal fibroblasts and keratinocytes. *Journal of Tissue Viability*, Vol. 18, No. 4, (Nov 2009), pp.:109-116, ISSN 0965-206X

O'Connor, N., Mulliken, J., Banks-Schlegel, S., Kehinde, O. & Green, H. (1981) Grafting of burns with cultured epithelium prepared from autologous epidermal cells. *Lancet*, Vol. 1, No. 8211, (Jan 1981), pp.:75-78, ISSN 0140-6736

Richard, L., Velasco, P. & Detmar, M. (1998) A simple immunomagnetic protocol for the selective isolation and long-term culture of human dermal microvascular endothelial cells. *Experimental Cell Research*, Vol. 240, No. 1, (Apr 1998), pp.:1-6, ISSN 0014-4827

Robert, M., Dusser, I., Muriel, M.-P., Noël-Hudson, M.-S., Aubery, M. & Wepierre, J. (1997) Barrier function of reconstructed epidermis at the air-liquid interface: influence of dermal cells and extracellular components. *Skin Pharmacology*, Vol. 10, pp: 247-260

Rochon, M.H., Fradette, J., Fortin, V., Tomasetig, F., Roberge, C.J., Baker, K., Berthod, F., Auger, F.A.&Germain, L. (2010) Normal human epithelial cells regulate the size and morphology of tissue-engineered capillaries. *Tissue Engineering Part A*, Vol. 16, No. 5, (May 2010), pp.:1457-1468, ISSN 1937-335X

Rompré, P., Auger, R.A., Germain, L., Bouvard, V., Lopez Vallé, C.A., Thibault, J. & Le Duy, A. (1990) Influence of initial collagen and cellular concentrations on the final surface area of dermal and skin equivalents: a Box-Behnken analysis. *In Vitro Cellular & Developmental Biology. Animal*, Vol. 26, pp.:983-990

Scuderi, N., Onesti, M.G., Bistoni, G., Ceccarelli, S., Rotolo, S., Angeloni, A. & Marchese, C. (2008) The clinical application of autologous bioengineered skin based on a hyaluronic acid scaffold. *Biomaterials*, Vol. 29, No. 11, (Apr 2008), pp.:1620-1629, ISSN 0142-9612

Supp, D.M., Supp, A.P., Bell, S.M. & Boyce, S.T. (2000) Enhanced vascularization of cultured skin substitutes genetically modified to overexpress vascular endothelial growth factor. *Journal of Investigative Dermatology*, Vol. 114, No. 1, (Jan 2000), pp.:5-13, ISSN 0022-202X

Trottier, V., Marceau-Fortier, G., Germain, L., Vincent, C. & Fradette, J. (2008) IFATS collection: using human adipose-derived stem/stromal cells for the production of new skin substitutes. *Stem Cells*, Vol. 26, No. 10, (Oct 2008), pp.:2713-2723, ISSN 1549-4918

9

Mesenchymal Stem Cells for Cell Therapy and Tissue Regeneration in Urology

Yingai Shi[1,2] and Yuanyuan Zhang[2]
[1]*Key Laboratory of Pathobiology, Ministry of Education, Jilin University, Changchun*
[2]*Wake Forest University, Institute for Regenerative Medicine (WFIRM), Winston-Salem, North Carolina*
[1]*China*
[2]*U.S.A.*

1. Introduction

Adult stem cells are defined as clonogenic, self-renewing progenitor cells that reside in adult tissues and can generate one or more specialized types of cells required for the function of that tissue. The majority of adult tissues contain their own stem/ progenitor cells that are capable of maintaining, generating, and replacing terminally differentiated cells within the tissue in response to physiologic cell turnover or tissue damage resulting from injury. Stem cell populations in the bone marrow were the first adult stem cell populations to be described, but in recent years adult stem populations have been demonstrated in the brain,[1, 2] skin, [3] and muscle [4], hair follicle and the gastrointestinal tract. Adult stem cells, especially hematopoietic stem cells, are the best understood cell type in stem cell biology [5], yet they remain an area of intense study, as their potential for therapy may be applicable to a myriad of degenerative disorders. These cells are a particularly attractive option for cell therapy and tissue engineering applications because they can be used in autologous therapies, thus avoiding any complications associated with immune rejection.

Research into adult stem cells has progressed slowly in the past, mainly because true stem cells are present in extremely low numbers in adult tissue[6-8], and because adult non-mesenchymal stem cells have been challenging to isolate, expand and maintain in culture. Some cells, such as those of the liver, pancreas and nerve, have very low proliferative capacity *in vitro*, and the functionality of some cell types is reduced after the cells are cultivated. These issues have limited the use of adult stem cells in tissue engineering and cell therapy research. However, the discovery of native targeted progenitor cells has allowed some of these limitations to be overcome. Native targeted progenitor cells are tissue specific unipotent cells derived from most organs. These cells are already programmed to become a specific cell type, and as with adult stem cells, native progenitor cells can be obtained from the specific organ to be regenerated, expanded, and used in the same patient without rejection, in an autologous manner [9-26]. By studying the niche in which the progenitor cells reside, as well as by exploring conditions that promote the differentiation of these cells, it has been possible to overcome some of the problems facing cell expansion *in vitro*. Major advances in cell culture techniques have been made within the past decade, and these techniques make the use of autologous cells possible for clinical application. In this chapter, we will focus on the use of autologous cells for regeneration and/or repair of the urinary tract.

2. Stem / progenitor cells derived from the genitourinary tract

Bladder

Adult human stem/progenitor cells from urinary tract system have been recently described and characterized [6,27]. It has been shown experimentally that the bladder neck and trigone area have a higher density of urothelial progenitor cells [28], and these cells are localized in the basal region [29]. In the past, it was possible to grow urothelial cells in the laboratory setting, but only with limited success. However, several protocols have been developed over the last two decades that have improved urothelial growth and expansion by enhancing culture conditions to support proliferation and differentiation of urothelial progenitor cells [17, 30-32]. It is now possible to expand a urothelial strain from a single surgical specimen that initially covers a surface area of 1 cm^2 to one covering a surface area of 4202 m^2 (the equivalent area of one football field) within 8 weeks[17]. Now, normal human bladder epithelial and muscle cells can be efficiently harvested from surgical material, extensively expanded in culture, and their differentiation characteristics, growth requirements, and other biologic properties can be studied [17, 19, 20, 31-40]. In addition, human urothelial and muscle cells can attach and form sheets of cells when seeded onto polymer scaffolds. The cell-polymer scaffold can then be implanted for repairing urological tissue defects. Histological analysis indicates that, within the cell-polymer construct, viable cells are able to self assemble back into their respective tissue types, and they retain their native phenotype [11].

To determine whether these engineered tissues could be implanted in continuity with the urinary tract, large animal models of bladder augmentation were used [22]. Partial cystectomies were performed in dogs and the animals were divided into 2 experimental groups. In one group, the bladder was augmented with a non-seeded bladder-derived collagen matrix, and in the second group, the bladder was augmented with a cell-seeded construct. The bladders augmented with seeded matrices demonstrated a 100% increase in capacity compared with bladders augmented with cell-free matrices, which only generated a 30% increase in capacity.

It has been known for a number of years that the bladder is able to regenerate generously over cell-free scaffolds, because urothelium has a high reparative capacity [41]. However, bladder muscle tissue is less likely to regenerate in the same fashion as urothelium, which leads to contracture or resorption of the graft. In addition, the inflammatory response toward the materials used to form the graft matrix may contribute to the resorption of cell-free grafts as well. The dog study demonstrated a major difference between matrices used with autologous cells (tissue-engineered matrices) and those used without cells [22]. The matrices seeded with cells prior to use in bladder augmentation retained most of their implanted diameter, as opposed to the matrices implanted without cells, in which graft contraction and shrinkage occurred. As in previous studies, the histomorphology demonstrated a lack of muscle cells and a more aggressive inflammatory reaction in the unseeded matrices.

The results of initial studies showed that the creation of tissue engineered bladders using autologous urothelial and smooth muscle cells could be achieved; however, it could not be determined whether the improvement in functional parameters noted was due to the implanted segment or to the remaining native bladder tissue. To better address the functional parameters of tissue-engineered bladders, additional animal model of subtotal cystectomy with subsequent replacement with a tissue-engineered organ was created[25]. In this model cystectomy-only and non-seeded controls maintained average capacities of 22% and 46% of preoperative values, respectively. An average bladder capacity of 95% of the original precystectomy volume was achieved with the cell-seeded tissue engineered bladder replacements. These findings were confirmed radiographically. The subtotal cystectomy reservoirs that were not reconstructed and the reservoirs reconstructed with unseeded grafts showed a marked decrease in bladder compliance (10% and 42% total compliance, respectively). In contrast, the compliance of the tissue-engineered bladders showed almost no difference from preoperative values (106%). Histologically, the non-seeded scaffolds resulted in a structure composed of normal urothelial cells with a thickened fibrotic submucosa and a thin layer of muscle fibers. The retrieved tissue-engineered bladders showed a normal cellular organization, consisting of a trilayer of urothelium, submucosa, and muscle. Immunocytochemical analyses confirmed the muscle and urothelial phenotypes, and indicated the presence of neural structures [25]. These studies, performed with polyglocolic acid based-scaffolds, have been repeated by other investigators, showing similar results in large numbers of animals long-term [42, 43]. Subsequent studies indicated that biodegradable scaffolds seeded with cells can be used without concerns for local or systemic toxicity [44].

Bladder tissue engineered using autologous cells has been used clinically. A small clinical study was conducted starting in 1998. Seven patients were treated using a collagen scaffold seeded with cells taken from biopsies of their own bladders, either with or without omental coverage, or a combined PGA-collagen scaffold seeded with cells and omental coverage. The patients reconstructed with the engineered bladder tissue created with the PGA-collagen cell-seeded scaffolds with omental coverage showed increased compliance, decreased end-filling pressures, increased capacities and longer dry periods over time [45]. It is clear from this small study that the engineered bladders continued their improvement with time, suggesting continued development in vivo. Although the experience is promising, it is just a start and the technology is not yet ready for wide dissemination, as further experimental and clinical studies are required.

Kidney

Kidney has long been considered an organ that is incapable of true regeneration. Furthermore, the question of whether or not the kidney contains adult stem cells remains controversial. However, increasing evidence of a regenerative response in the kidney has been observed following the injury resulting from both toxic and ischemic insults. These observations include evidence of renal progenitors of specific cell types involved in the formation of new renal tubular cells and the recovery of renal function recovery after ischemic injury. The presence of injured or dead cells following ischemia causes denudation of the tubular basement membrane, and sloughed cells and cellular debris fills tubular lumens. The kidney responds to the ischemic injury with a prompt regenerative response, resulting in regenerating tubules and improving kidney function. Although they remain elusive, the cells participating in renal regeneration are likely from pools of both exogenous and endogenous stem cells. The exogenous stem cells are probably largely derived from bone marrow, and may be both hematopoietic and mesenchymal stem cells. In some studies, these cells appear to home to damaged sites in the injured kidney and form tubular epithelial cells following acute renal injury [46-49]. These MSC might also produce growth factors such as IGF-1 to promote renal repair [50]. The endogenous stem cells are resident kidney stem cells found in the renal tubules and the papilla[51, 52]. They are inactivated under physiological conditions. These stem cells posses the capacity to give rise renal tubule cells following injury repair [53, 54]. A recent study showed that repopulation of damaged renal tubules occurs primarily from proliferation of tubular epithelial cells and resident renal-specific stem cells, with some contribution of paracrine factors from bone marrow-derived mesenchymal stem cells [55-56].

Testis

Primordial germ cells (PGC) are the embryonic progenitor cells of the gametes (spermatogonial stem cells and ova). *In vivo*, PGC colonize the gonadal ridge during early embryonic development and are then restricted to producing the gametes. However, if PGC are cultured *in vitro* in the presence of specific growth factors, they are able to form pluripotent embryonic germ cells (EGC) through a process that is relatively similar to that of nuclear reprogramming and generation of induced pluripotent (iPS) cells in the laboratory (discussed later in this review) [57]. These cells can contribute to all cellular lineages in chimeric embryos, including the germline. They also form teratomas when injected into immunocompromised animals.

For regenerative medicine purposes, PGC are not ideal, because they are derived from embryos and there are a number of controversial ethical issues surrounding the manipulation of human embryos. However, there has been much interest in isolating and describing spermatogonial stem cells (SSC) in recent years. The presence of SSC, which are derived from PGC in the testis, was originally inferred from the presence of ongoing spermatogenesis in the adult male. Recently, the cells believed to be the actual SSC were isolated from the adult testis of both mice and humans through selection of the markers STRA-8, GPR125, CD49fm, CD133 and others[58, 59 60]. Interestingly, when SSC from both mice and humans are cultured in specified media containing growth factors known to be required for maintenance of pluripotency of other types of stem cells, such as leukemia inhibitory factor (LIF) and glial cell line-derived neurotrophic factor (GDNF), they appear to convert to an embryonic stem (ES) cell like state.[60] These converted cells have been termed adult germline stem cells (aGSC) and they can differentiate into a number somatic cell types encompassing all three embryonic germ layers when they are exposed to the same conditions used to differentiate ES cells. They also form teratomas when implanted *in vivo*. These results suggest that SSC, which can be obtained through a small testicular biopsy, may be useful for the development of cell-based, autologous organ regeneration strategies. However, more research is required to overcome additional hurdles before this technology can be used clinically. In addition, since autologous regeneration strategies based on SSC would only benefit males, researchers are working to identify and describe similar pluripotent cells that may reside in an ovarian niche for use in females.

Urine

We recently demonstrated that it is possible to isolate and expand stem/progenitor cells from human urine[6]. Approximately, 0.2% of cells collected from urine express markers characteristic of mesenchymal stem cells (MSC), can expand extensively in culture, and can differentiate towards multiple bladder cell lineages as identified by the expression of urothelial, smooth muscle, endothelial and interstitial cell markers. We initially referred to these cells as urine progenitor cells. However, our more recent experiments indicated that urine-derived cells can give rise to additional specialized types, including osteocytes, chondrocytes, and adipocytes. Furthermore, these cells have self-renewal capability consistent with stem cells. There is now sufficient evidence to provisionally designate them as urine-derived stem cells (USC).

Three types of cells exist in urine: differentiated, differentiating and progenitor cells. Most cells in urine are fully differentiated. They do not attach to tissue culture plates. About 0.1% of cells in urine are differentiating cells, which do attach to plates and display the morphology and protein markers of various bladder cell lineages. However, these cells do not expand further after subculture. About 0.2% of the cells in urine have a phenotype consistent with multipotent stem cells. USC are easily cultured, appear genetically stable after a number of passages, and maintain the ability to give rise to more differentiated progeny.

USC comprise an average of about 7 cells/100ml urine (from 5 to 10 cells/100 ml urine). We have shown that, a few days after being placed in a tissue culture well, a single cell forms a cluster of cells which appeared small, compact and uniform. A consistently high yield of cells was achieved from each of these clonal lines. The cells reached confluence in about two weeks when placed in a 3-cm diameter well at passage one. At passage 2, cells were plated in 10 cm culture dishes and a cell number of approximately one million was reached in 3.5 weeks. Finally, in six to seven weeks, the cultures expanded to approximately 100 million cells at passage four. These cells displayed normal exponential cell growth patterns, with a steady increase in cell numbers during a 10-day culture period. The average population doubling time was 31.3 hours in mixed media. These urine-derived cells also showed the ability to differentiate into various cell lineages as described below, and were capable of growing for at least 14 passages *in vitro*.

Cells from human urine specimens could be consistently cultured long-term using a medium that we originally developed for culture of rat urothelium[61]. However, the phenotype of the cultured human urine-derived cells was not that of primary urothelial cells. The primary cultures from urine did not show expression of the cytokeratins (CK7, CK13, and CK19/20), which are characteristic of epithelial cells, nor did the cells express the urothelial-specific protein uroplakin. After growth in medium containing higher levels of epidermal growth factor (EGF), the cells were induced efficiently to express the cytokeratin proteins and uroplakin[6]. However, after growth in myogenic medium, the cultured cells expressed markers consistent with smooth muscle, including alpha-smooth muscle actin (α-SM actin), desmin, calponin, and myosin[6]. This led us to conclude that the urine-derived cells were progenitors (initially designated UPC) capable of giving rise to both urothelium and bladder smooth muscle. Furthermore, we found that the cells displayed a surface marker phenotype consistent with mesenchymal stem cells (MSC). Specifically, they expressed CD 44, CD73, CD90, CD105, and CD 146, and they were negative for both hematopoietic markers and endothelial markers including CD45, CD34, and CD31. We concluded that the urine-derived progenitors were at least bipotential for the major bladder cell types. This result was surprising, because it was generally believed that muscle and epithelial cells in bladder represent separate cell lineages derived from mesoderm and endoderm, respectively. We have recently observed that the urine-derived cells express markers typical of mesenchymal stem cells (MSC) and pericytes, and that they can differentiate to yield the characteristic cell lineages obtained from MSC, namely, osteocytes, adipocytes and chondrocytes [62].

In our recent study, USC clones could be obtained from 85% of urine sample tested. Fresh urine gave the highest rate of colony formation (67%) and urine stored at 4° C the lowest

(30%). Urine from volunteers aged 13-40 gave the highest rate of clone recovery. Catheterization significantly enhanced the number of USC in urine compared to spontaneously voided urine, possibly because catheterization resulted in cells being scraped off the inner bladder wall. Collecting triple urine samples increased the rate of clone formation.

There are many potential advantages to using USC as a cell source for urological tissue engineering. First, cells can be easily harvested and grown in culture. USC do not require enzyme digestion or culture on a layer of feeder cells to support cell growth. Second, since invasive surgical biopsy procedures are not necessary to harvest cells from urine, patient morbidity and potential complications, such as urethral or bladder trauma and urinary tract infections, are avoided. As USC are autologous somatic cells, no ethical issues are involved in their use for tissue reconstruction, and no immune reaction to engineered implants should occur.

The quality of cells obtained from urine is similar to that of the biopsy-derived cells described above. When differentiated, USC express all proteins characteristic of the various bladder cell lineages. Karyotype analysis has demonstrated that these cells are genetically stable. Importantly, there is a major cost advantage to using USC – it costs about $50 to obtain cells from urine, versus about $5,000 to isolate cells from a biopsy procedure[6]. About 1.4 $\times 10^9$ urothelial and smooth muscle cells (SMC) are required for bladder tissue regeneration[63]. We estimate that 3-4 urine samples (about 40-45 USC/600 ml urine) expanded for 4-5 weeks would yield a sufficient quantity of low passage, healthy cells for clinical tissue engineering applications. This time frame is comparable to that required for expansion from a tissue biopsy (7-8 weeks)[64]. USC and the cells obtained through urological tissue biopsies come from the same urinary tract systems and have similar biological features. Therefore, collecting cells from urine could be an attractive alternative to the standard urological tissue biopsies currently used in cell therapy and tissue engineering.

3. Stem/progenitor cells derived from non-urological tissues

Despite the convenience of using differentiated cells in tissue engineering applications, these cells have several shortcomings. These cells have a limited ability to grow in culture and they tend to dedifferentiate *in vitro*, which may lead to insufficient numbers of cells. In addition, autologous bladder cells cannot be taken from patients with urinary tract malignancies. One solution to these problems is to prepare engineered tissues using stem cells from various sources. These types of stem/progenitor cells from non-urological tissue have been studied as cell sources for bladder regeneration and cell therapy for stress urinary incontinence.

Mesenchymal stem cells

Mesenchymal stem cells (MSC), isolated from bone marrow, skeletal muscle, and adipose tissue, possess the capacity to differentiate into cells of connective tissue lineages, including muscle. Isolation and characterization of MSC, and control of their myogenic differentiation derived from both pre-clinical and clinical studies have attracted attention to their potential use in urological regenerative medicine and tissue engineering.

Currently, the most effectively characterized types of multipotent stem cells are from bone marrow. Bone marrow stem cells (BMSC) have been shown to differentiate into specialized cells, including hepatocytes[65-67], neural cells[68-71] and mainly mesodermal derivatives such as bone, cartilage, cardiac muscles, skeletal muscle, and fat. If BMSC are placed on a proper bio-degradable scaffold and implanted, they can act as anti-fibrotic, angiogenic, anti-apoptotic, and mitotic agents. Recently, BMSC were evaluated as an alternative cell type for use in replacement of bladder SMC when native bladder muscle tissue is unavailable. The potential of BMSC to differentiate into cells with bladder SMC characteristics was assessed *in vitro*[72] and in different animal models [73-78]. Kanematsu et al[75] showed that *in vitro,* both supernatants from cultured rat bladder cells (conditioned media) or media containing TGF-β and VEGF induced bone marrow cells to adopt a SMC phenotype. Recently, we have investigated the impacts of soluble growth factors, bladder extracellular matrix (ECM), and 3D dynamic culture on cell proliferation and differentiation of human BMSC into smooth muscle cells (SMC)[72]. Myogenic growth factors (PDGF-BB and TGF-β1) alone, or combined either with bladder ECM or dynamic cultures, induced BMSC to express smooth muscle specific genes and proteins. Either ECM or the dynamic culture alone promoted cell proliferation but did not induce myogenic differentiation of BMSC. A highly porous nanofibrous poly-L-lactic acid (PLLA) scaffold provided a 3D structure for maximizing the cell-matrix penetration, maintained myogenic differentiation of the induced BMSC, and promoted tissue remolding with rich capillary formation *in vivo*. This study demonstrates that myogenic-differentiated BMSC seeded on a nanofibrous PLLA scaffold can be used for cell-based tissue engineering for bladder cancer patients requiring cystoplasty.

In order to test this *in vivo,* bone marrow cells expressing green fluorescent protein were transplanted into lethally irradiated rats. Eight weeks following transplantation, bladder domes were replaced with acellular matrix grafts. Two weeks after the graft procedure, GFP expression in the matrices indicated that the transplanted marrow cells had repopulated the graft. By 12 weeks, these cells reconstituted the smooth muscle layer, with native SMC also infiltrating the graft. In another rat study[74], rapid regeneration of bladder SMC and urothelium occurred on BMSC seeded collagen matrices, whereas fibrotic changes were observed in the non-seeded matrix group 3 months after bladder augmentation. In a large animal study[73], BMSC proliferated at the same rate as primary cultured bladder SMC *in vitro,* and they had a similar histological appearance and contractile phenotype as primary cultured bladder SMC. BMSC had a significant contractile response to calcium-ionophore *in vitro,* and this response was similar to that seen in bladder SMC but markedly different from fibroblasts. Immunohistochemical staining and Western blotting indicated that BMSC expressed α -smooth muscle actin, but did not express desmin or myosin. *In vivo,* small intestinal submucosa (SIS) grafts seeded with BMSC developed solid smooth-muscle bundle formations throughout the grafts, as did bladder cell-seeded SIS grafts. However, bladder tissue regeneration did not occur in animals that received cell-free scaffolding. These results indicate that BMSC may provide an alternative cell source for bladder tissue engineering. This is relevant for patients with bladder malignancies who require bladder augmentation or replacement but do not have enough normal, non-malignant bladder cells to use in tissue engineering applications.

Other MSC such as skeletal muscle-derived progenitor cells [79-88] [89-92] and adipose stem cells[80, 90, 91, 93-98] have been investigated as potential candidates for cell-based tissue engineering and injection therapy stress urinary incontinence (SUI), and these studies are further described in Section 5.

Induced pluripotent stem cells

iPS cells are a type of pluripotent stem cell that is artificially derived from a patient's own somatic cells (a non-pluripotent cell) by inducing a "forced" expression of certain genes. iPS cells were first produced in 2006 from mouse cells and then in 2007 from human cells. iPS cells are typically derived by transfecting stem cell-associated genes into non-pluripotent cells, such as adult fibroblasts. Transfection is typically achieved through viral vectors, such as retroviruses. Transfected genes include the master transcriptional regulators Oct-3/4 (Pouf51) and Sox2, although it is suggested that other genes may enhance the efficiency of induction. After 3–4 weeks, small numbers of transfected cells begin to become morphologically and biochemically similar to pluripotent stem cells, and these cells are typically isolated through morphological selection or through a reporter gene and/or antibiotic selection. This has been cited as an important advancement in stem cell research, as it may allow researchers to obtain pluripotent stem cells, which are important in research and potentially have therapeutic uses in urology, without the controversial use of embryos.

iPS cells are believed to be similar to natural pluripotent stem cells, such as embryonic stem (ES) cells in many respects, including expression of certain stem cell genes and proteins, chromatin methylation patterns, doubling time, embryoid body formation, teratoma formation, viable chimera formation, potency and differentiability, but the full extent of their relation to natural pluripotent stem cells is still being assessed.

However, depending on the methods used, reprogramming of adult cells to obtain iPS cells may pose significant risks that currently limit the use of this technique in human therapy. For example, if viruses are used to genetically alter the cells, expression of oncogenes may potentially be triggered. In February 2008, a report published in the journal *Cell* announced the discovery of a technique that removed the need for oncogenes such as c-myc in induction of pluripotency, thereby increasing the potential use of iPS cells in human diseases. Even more recently, in April 2009, Sheng Ding in La Jolla, California, showed that the generation of iPS cells was possible without any genetic alteration of the adult cell [99]. Repeated treatment of the cells with certain proteins channeled into the cells via poly-arginine anchors was sufficient to induce pluripotency. The cells generated by this process are known as protein-induced pluripotent stem cells (piPS cells).

Human Amniotic Fluid Stem Cells

Human amniotic fluid cells are commonly used clinically as a diagnostic tool for the prenatal diagnosis of fetal genetic anomalies. Recently, increasing evidence demonstrated that fetal-derived stem cells can be isolated from amniotic fluid. These cells represent a novel class of pluripotent stem cells with intermediate characteristics between embryonic and adult stem cells, as they are capable of giving rise to lineages representative of all three germ layers but do not form teratomas when implanted *in vivo*[100]. These features, in addition to the absence of ethical concerns about their use, indicate that amniotic fluid stem (AFS) cells might be a promising cell source for tissue engineering and stem cell therapy. Perin et al [101-103] have recently reported that AFS cells may be useful for kidney regeneration. In a series of studies, this group demonstrated that these pluripotent cells are able to differentiate into *de novo* kidney structures during organogenesis *in vitro*. Human male amniotic fluid cells were isolated between 12 and 18 weeks of gestation. AFS cells were isolated from these cultures and labeled with green fluorescent protein and Lac-Z protein. Labeled human AFS cells were then microinjected into murine embryonic kidneys (12.5–18 days gestation) and these were maintained in a co-culture system for 10 days. Histological analysis revealed that human AFS cells were able to contribute to the development of elemental kidney structures including renal vesicles, and C- and S-shaped bodies.

Expression of the early kidney markers zona occludens-1, glial-derived neurotrophic factor and claudin were confirmed by RT-PCR. Therefore, it is possible that amniotic fluid stem cells represent a potential cell source for future renal cell therapies.

4. Approaches for inducing myogenic differentiation of stem cells

When stem cells are used as a cell source for urological tissue engineering and regeneration, they can be used three ways: 1) stem cells can be induced to differentiate into the target cells/tissue–like cells *in vitro* before cell implantation; 2) stem cells are implanted directly into the tissues where repair is needed, and the surrounding cell- and tissue-based signals induce the stem cells to differentiate into the specific cells required for regeneration, and 3) a cell-free scaffold is implanted and recruits the host's own stem/progenitor cells, which then differentiate into the proper cell type required for tissue repair.

Current research tends to focus largely on the first strategy described above to allow control over the signals that the cells receive so that differentiation into the tissue type required progresses without problems. For this strategy, it is necessary to mimic the physiological conditions that guide stem cells to differentiate into the desired target cells before implantation. For example, several factors have been shown to enhance autologous adult stem cell differentiation into functional SMC, including:

i. Growth factors, such as vessel endothelial growth factor (VEGF), platelet-derived growth factor (PDGF-BB), transforming growth factor-β (TGF-β) and insulin-like growth factor (IGF)[104],
ii. Components of the extracellular matrix (ECM). Cellular interactions with the ECM play an important role in cell adhesion, growth, migration, apoptosis and differentiation [105]. It consists of compounds such as collagen, laminin [106], and fibronectin. Collagen IV can promote embryonic stem cells to differentiate into stem cell antigen-1-positive ($Sca\text{-}1^+$) progenitor cells and SMC[107]. However, ECM can promote adult cell proliferation or improve the yield of SMC derived from adult MSC, but ECM alone cannot induce adult MSC to differentiate into SMC[108, 109]. Additionally, culture on a three-dimensional ECM scaffold in a dynamic culture system can improve cell proliferation, maintain cell phenotypes and lead to a more homogenous distribution of cells on the scaffold [110, 111] when compared to 2-D static culture.
iii. *In vitro* co-culture of fully-differentiated SMC and stem cells appears to improve stem cell differentiation into muscle cells, most likely because the SMC secrete specific factors into the culture medium. Baskin et al demonstrated that mature urothelium can induce urological embryonic tissue or stem cells to differentiate into smooth muscle cells *in vivo* through epithelial-stromal cell interaction or cell-cell interaction. In contrast, the embryonic tissue failed to differentiate into SMC when urothelium was not present[112]. Because of this, conditioned medium is commonly used for stem cell differentiation. Conditioned medium is essentially culture medium that is partially used by cells, and it is enriched with cell-derived material including small amounts of growth factors.
iv. The application of cyclic mechanical strain to cell cultures has been demonstrated to increase the expression of smooth muscle cell markers in stem cells[113]. Periodic stretching occurs *in vivo* as a part of the natural function of hollow organs; for example, as the bladder fills and empties. Differentiated SMC easily lose their contractile function in static culture once the cells leave the body, but the use of mechanical strain in culture

can prevent this spontaneous loss of phenotype *in vitro* and maintain SMC functional characteristics[114].

5. Autologous stem cells for endoscopic therapies

Another exciting area of clinical urologic investigation is the use of various autologous cells to treat vesicoureteral reflux (VUR) and SUI. An increasing number of clinical trials using tissue engineering approaches have been reported (Table 1). All of these clinical applications of urological tissue engineering are based on a series of successful animal experiments[115, 116, 117].

Cell types	Require Tissue biopsy	Highly expandable	Potential applications	Clinical trail
Skeleton muscle derived progenitor cells	Yes		Cell therapy for VRU and urinary incontinent, Bladder regeneration	Yes
Bone marrow stem cells	Yes	Depend on age	Tissue engineered bladder; Cell therapy for VRU; urinary incontinent	Yes
Adipose stem cells	Yes		Tissue engineered bladder;	Yes
Urine progenitor cells	No	no depend on age	Tissue engineered urological organs	No

Table 1. Potential use of autologous stem/progenitor cells for urological tissue engineering and cell therapy

Endoscopic therapy offers a simple method for definitive treatment in SUI and VUR. Two types of injectable substances have been investigated. First, natural and synthetic biomaterials that serve as bulking agents, such as silicone, fibrin, bioglass, polyvinyl alcohol foam, alginate gel, a small-intestinal submucosal suspension and Defulx have been used[118]. Currently, injectable therapy based on bulking agents is used for only about two-thirds of patients with SUI, with even lower cure rates. Potential problems of these injectable substances include decreasing volume of the injectable substance over time *in vivo*, the need for multiple injections to obtain and maintain optimal efficacy, potential antigenicity of the injectable and related allergic reactions, migration of the injected material, and urethral pain both at the time of injection and afterward. The ideal bulking agent should remain efficacious over time and have few side effects, but so far, none of the substances in use have met these criteria for success and the search for a superior injectable therapy for SUI continues.

Cell-based therapy is a promising alternative in urological procedures for VUR and USI. Autologous cells that can be used for this purpose include chondrocytes[115, 119], adipose-derived stem cells (ASC) [120-122], BMSC [121, 123], and skeletal muscle derived progenitor cells [85, 124-127]. In 1994, Atala et al proposed the use of injectable autologous chondrocytes to correct VUR via endoscopy [12, 128]. Using a minipig model, they noted that autologous chondrocytes injected around the ureter to stop reflux did not migrate and the cartilage bead produced by this technique maintained its volume with time. Reflux was corrected in all animals treated endoscopically with autologous chondrocytes. After these successful animal experiments, several clinical trials have begun[115, 119]. One study [129] was conducted in a total of 29 children (46 ureters) with grades II to IV reflux. Chondrocytes were harvested from a biopsy of each patient's ear cartilage and were grown in culture for 6 weeks. Patients then returned to the clinic for transurethral injection of chondrocytes into the bladder trigone to correct reflux. Ultrasound was performed 1 month after this procedure and radionuclide cystography was done 3 months postoperatively to confirm reflux resolution. When reflux persisted, repeat treatment with stored chondrocytes was offered. In this study, a single chondrocyte injection corrected reflux in 26 of the 46 ureters (57%), while secondary injection was successful in 12 of 19 (63%). Overall, reflux was corrected in 38 of the 46 ureters (83%) and in 24 of the 29 patients (83%). There were no significant complications, and transurethral injection of autologous chondrocytes to correct VUR in children appears to be an effective and safe technique. The only limitation of this therapy is the high cost.

The use of myocyte- and stem cell-based injection therapy has also been tested in VUR cases. Primary VUR is a congenital anomaly of the ureter-vesical junction that creates a deficiency of the longitudinal muscle of the intravesical ureter. This leads to an inadequate valvular mechanism and allows urine to flow backward from the bladder to the kidney. Thus, myocyte-based therapies are attractive options for the recovery of this muscle defect at the ureteral orifice.

Autologous progenitor cell-based therapy has also made significant progress in treatment of SUI. This cell therapy could soon become a standard procedure. The objective of this therapy is to improve or cure the sphincter dysfunction via periurethral endoscopic injection. Currently, myoblasts obtained from skeletal muscle biopsies and adipose-derived cells are the most commonly used cells for therapy for SUI. Recently, autologous myoblasts and fibroblasts have been evaluated as a potential injectable therapy for SUI. One group[130-133] has studied a combination therapy consisting of autologous myoblasts injected into the rhabdosphincter and fibroblasts injected into the urethral submucosa. A year follow-up study of 123 women was performed from 2004 to 2005. A cure rate of about 79% with improvements in quality-of-life scores, rhabdosphincter contractility, and urethral closure pressures has been achieved. All patients were continent 1 year after receiving this therapy and maintained their good outcome at further follow-up visits. Ultrasound images before treatment clearly revealed poor periurethral integrity of the sphincteric mechanism; postinjection images revealed a completely normal-appearing urethra.

Rodriguez *et al* recently reported that adipose-derived stem cells (ASC) have the potential to differentiate into functional SMC[89] . ASC expressed a series of contractile proteins, including α-SM actin, desmin, myosin heavy chain, calponin, caldesomon, smoothlin, and SM22 following aspiration from fat tissue and culture in SM differentiation medium. One important advantage of using ASC is that adipose tissue can be harvested in large quantities with minimal morbidity. Autologous fat tissues were used for cell injection therapy for vesicoureteral reflux in a clinical study [134]. Two out of 11 patients had a reduction in grade of reflux, including one ureter that ceased refluxing altogether[135]. One recent study[136] showed ASC could correct neurogenic erectile dysfunction in rats as effectively as bone marrow stem cells did. More research is underway to determine whether ASC can differentiate into Leydig, Sertoli and male germ cells. The eventual goal of the research is to use ASC to treat male infertility and testosterone deficiency.

6. Conclusions

Current advances in urological tissue engineering and stem cell-based therapy demonstrate that bladder and urethral tissues can be regenerated using autologous cells seeded onto biodegradable scaffolds. VUR and SUI can be corrected with injections of autologous stem cells contained in a hydrogel. However, many issues must be elucidated before these techniques can become widely used in the clinic. For example, the role of donor cells in tissue regeneration remains unclear, and it is not known whether the seeded stem cells proliferate and populate scaffold materials themselves, or if they stimulate to the activation, migration, proliferation, and differentiation of the local progenitor cells to complete the tissue regeneration. Additionally, an approach to promote angiogenesis and to facilitate innervation with a functional network of regenerated nerves will greatly improve tissue regeneration strategies to create a *de novo* urological organ. (6,015 words)

7. References

[1] Jiao, J., Chen, D. F.: Induction of neurogenesis in nonconventional neurogenic regions of the adult central nervous system by niche astrocyte-produced signals. *Stem Cells*, 26: 1221, 2008

[2] Taupin, P.: Therapeutic potential of adult neural stem cells. Recent Patents CNS Drug Discov, 1: 299, 2006

[3] Jensen, U. B., Yan, X., Triel, C. et al.: A distinct population of clonogenic and multipotent murine follicular keratinocytes residing in the upper isthmus. *Journal of Cell Science*, 121: 609, 2008

[4] Crisan, M., Casteilla, L., Lehr, L. et al.: A reservoir of brown adipocyte progenitors in human skeletal muscle. *Stem Cells*, 26: 2425, 2008

[5] Ballas, C. B., Zielske, S. P., Gerson, S. L.: Adult bone marrow stem cells for cell and gene therapies: implications for greater use. *Journal of Cellular Biochemistry - Supplement*, 38: 20, 2002

[6] Zhang, Y., McNeill, E., Tian, H. et al.: Urine derived cells are a potential source for urological tissue reconstruction. *J Urol*, 180: 2226, 2008

[7] Mimeault, M., Batra, S. K.: Recent progress on tissue-resident adult stem cell biology and their therapeutic implications. *Stem Cell Reviews*, 4: 27, 2008

[8] Hristov, M., Zernecke, A., Schober, A. et al.: Adult progenitor cells in vascular remodeling during atherosclerosis. *Biological Chemistry*, 389: 837, 2008

[9] Atala, A., Vacanti, J. P., Peters, C. A. et al.: Formation of urothelial structures in vivo from dissociated cells attached to biodegradable polymer scaffolds in vitro. *Journal of Urology*, 148: 658, 1992

[10] Atala, A., Cima, L. G., Kim, W. et al.: Injectable alginate seeded with chondrocytes as a potential treatment for vesicoureteral reflux. *Journal of Urology*, 150: 745, 1993

[11] Atala, A., Freeman, M. R., Vacanti, J. P. et al.: Implantation in vivo and retrieval of artificial structures consisting of rabbit and human urothelium and human bladder muscle. *Journal of Urology*, 150: 608, 1993

[12] Atala, A., Kim, W., Paige, K. T. et al.: Endoscopic treatment of vesicoureteral reflux with a chondrocyte-alginate suspension. *Journal of Urology*, 152: 641, 1994

[13] Atala, A., Schlussel, R. N., Retik, A. B: Renal cell growth in vivo after attachment to biodegradable polymer scaffolds. *J Urol*, 153, 1995

[14] Atala, A., Guzman, L., Retik, A. B.: A novel inert collagen matrix for hypospadias repair. *Journal of Urology*, 162: 1148, 1999

[15] Atala, A.: Tissue engineering in the genitourinary system. In: *Tissue Engineering*. Edited by A. Atala and D. J. Mooney. Boston, MA: Birkhauser Press, p. 149, 1997

[16] Atala, A.: Autologous cell transplantation for urologic reconstruction. Journal of Urology, 159: 2, 1998

[17] Cilento, B. G., Freeman, M. R., Schneck, F. X. et al.: Phenotypic and cytogenetic characterization of human bladder urothelia expanded in vitro. *Journal of Urology*, 152: 665, 1994

[18] Yoo, J. J., Atala, A.: A novel gene delivery system using urothelial tissue engineered neo-organs. *Journal of Urology*, 158: 1066, 1997

[19] Fauza, D. O., Fishman, S. J., Mehegan, K. et al.: Videofetoscopically assisted fetal tissue engineering: skin replacement. *Journal of Pediatric Surgery*, 33: 357, 1998

[20] Fauza, D. O., Fishman, S. J., Mehegan, K. et al.: Videofetoscopically assisted fetal tissue engineering: bladder augmentation. *Journal of Pediatric Surgery*, 33: 7, 1998

[21] Machluf, M., Atala, A.: Emerging concepts for tissue and organ transplantation. *Graft*, 1: 31, 1998

[22] Yoo, J. J., Meng, J., Oberpenning, F. et al.: Bladder augmentation using allogenic bladder submucosa seeded with cells. *Urology*, 51: 221, 1998

[23] Amiel, G. E., Atala, A.: Current and future modalities for functional renal replacement. Urologic Clinics of North America, 26: 235, 1999

[24] Kershen, R. T., Atala, A.: New advances in injectable therapies for the treatment of incontinence and vesicoureteral reflux. *Urologic Clinics of North America*, 26: 81, 1999

[25] Oberpenning, F., Meng, J., Yoo, J. J. et al.: De novo reconstitution of a functional mammalian urinary bladder by tissue engineering.[see comment]. *Nature Biotechnology*, 17: 149, 1999

[26] Park, H. J., Yoo, J. J., Kershen, R. T. et al.: Reconstitution of human corporal smooth muscle and endothelial cells in vivo. *Journal of Urology*, 162: 1106, 1999

[27] Oottamasathien, S., Williams, K., Franco, O. E. et al.: Urothelial inhibition of transforming growth factor-beta in a bladder tissue recombination model. *Journal of Urology*, 178: 1643, 2007

[28] Nguyen, M. M., Lieu, D. K., deGraffenried, L. A. et al.: Urothelial progenitor cells: regional differences in the rat bladder. *Cell Proliferation*, 40: 157, 2007

[29] Kurzrock, E. A., Lieu, D. K., Degraffenried, L. A. et al.: Label-retaining cells of the bladder: candidate urothelial stem cells. *American Journal of Physiology - Renal Physiology*, 294: F1415, 2008

[30] Scriven, S. D., Booth, C., Thomas, D. F. et al.: Reconstitution of human urothelium from monolayer cultures. *Journal of Urology*, 158: 1147, 1997

[31] Liebert, M., Hubbel, A., Chung, M. et al.: Expression of mal is associated with urothelial differentiation in vitro: identification by differential display reverse-transcriptase polymerase chain reaction. *Differentiation*, 61: 177, 1997

[32] Puthenveettil, J. A., Burger, M. S., Reznikoff, C. A.: Replicative senescence in human uroepithelial cells. *Advances in Experimental Medicine & Biology*, 462: 83, 1999

[33] Liebert, M., Wedemeyer, G., Abruzzo, L. V. et al.: Stimulated urothelial cells produce cytokines and express an activated cell surface antigenic phenotype. *Seminars in Urology*, 9: 124, 1991

[34] Tobin, M. S., Freeman, M. R., Atala, A.: Maturational response of normal human urothelial cells in culture is dependent on extracellular matrix and serum additives. *Surg Forum*, 45: 786, 1994

[35] Harriss, D. R.: Smooth muscle cell culture: a new approach to the study of human detrusor physiology and pathophysiology. *British Journal of Urology*, 75 Suppl 1: 18, 1995

[36] Freeman, M. R., Yoo, J. J., Raab, G. et al.: Heparin-binding EGF-like growth factor is an autocrine growth factor for human urothelial cells and is synthesized by epithelial and smooth muscle cells in the human bladder. *Journal of Clinical Investigation*, 99: 1028, 1997

[37] Solomon, L. Z., Jennings, A. M., Sharpe, P. et al.: Effects of short-chain fatty acids on primary urothelial cells in culture: implications for intravesical use in enterocystoplasties.[see comment]. *Journal of Laboratory & Clinical Medicine*, 132: 279, 1998

[38] Lobban, E. D., Smith, B. A., Hall, G. D. et al.: Uroplakin gene expression by normal and neoplastic human urothelium. *American Journal of Pathology*, 153: 1957, 1998

[39] Nguyen, H. T., Park, J. M., Peters, C. A. et al.: Cell-specific activation of the HB-EGF and ErbB1 genes by stretch in primary human bladder cells. *In Vitro Cellular & Developmental Biology. Animal*, 35: 371, 1999

[40] Rackley, R. R., Bandyopadhyay, S. K., Fazeli-Matin, S. et al.: Immunoregulatory potential of urothelium: characterization of NF-kappaB signal transduction. *Journal of Urology*, 162: 1812, 1999

[41] de Boer, W. I., Schuller, A. G., Vermey, M. et al.: Expression of growth factors and receptors during specific phases in regenerating urothelium after acute injury in vivo. *American Journal of Pathology*, 145: 1199, 1994

[42] Jayo, M. J., Jain, D., Ludlow, J. W. et al.: Long-term durability, tissue regeneration and neo-organ growth during skeletal maturation with a neo-bladder augmentation construct. *Regenerative Medicine*, 3: 671, 2008

[43] Jayo, M. J., Jain, D., Wagner, B. J. et al.: Early cellular and stromal responses in regeneration versus repair of a mammalian bladder using autologous cell and biodegradable scaffold technologies.[see comment]. *Journal of Urology*, 180: 392, 2008

[44] Kwon, T. G., Yoo, J. J., Atala, A.: Local and systemic effects of a tissue engineered neobladder in a canine cystoplasty model. *Journal of Urology*, 179: 2035, 2008

[45] Atala, A., Bauer, S. B., Soker, S. et al.: Tissue-engineered autologous bladders for patients needing cystoplasty.[see comment]. *Lancet*, 367: 1241, 2006

[46] Lin, F.: Renal repair: role of bone marrow stem cells. *Pediatr Nephrol*, 23: 851, 2008

[47] Lin, F.: Stem cells in kidney regeneration following acute renal injury. *Pediatr Res*, 59: 74R, 2006

[48] Krause, D., Cantley, L. G.: Bone marrow plasticity revisited: protection or differentiation in the kidney tubule? *J Clin Invest*, 115: 1705, 2005

[49] Kale, S., Karihaloo, A., Clark, P. R. et al.: Bone marrow stem cells contribute to repair of the ischemically injured renal tubule. *J Clin Invest*, 112: 42, 2003

[50] Imberti, B., Morigi, M., Tomasoni, S. et al.: Insulin-like growth factor-1 sustains stem cell mediated renal repair. *J Am Soc Nephrol*, 18: 2921, 2007

[51] Oliver, J. A., Maarouf, O., Cheema, F. H. et al.: The renal papilla is a niche for adult kidney stem cells. *J Clin Invest*, 114: 795, 2004

[52] Maeshima, A., Sakurai, H., Nigam, S. K.: Adult kidney tubular cell population showing phenotypic plasticity, tubulogenic capacity, and integration capability into developing kidney. *J Am Soc Nephrol*, 17: 188, 2006

[53] Duffield, J. S., Park, K. M., Hsiao, L. L. et al.: Restoration of tubular epithelial cells during repair of the postischemic kidney occurs independently of bone marrow-derived stem cells. *J Clin Invest*, 115: 1743, 2005

[54] Duffield, J. S., Bonventre, J. V.: Kidney tubular epithelium is restored without replacement with bone marrow-derived cells during repair after ischemic injury. *Kidney Int*, 68: 1956, 2005

[55] Liu, K. D., Brakeman, P. R.: Renal repair and recovery. *Crit Care Med*, 36: S187, 2008

[56] Lin, F., Moran, A., Igarashi, P.: Intrarenal cells, not bone marrow-derived cells, are the major source for regeneration in postischemic kidney. *J Clin Invest*, 115: 1756, 2005

[57] Nayernia, K.: Stem cells derived from testis show promise for treating a wide variety of medical conditions. *Cell Res*, 17: 895, 2007

[58] Kanatsu-Shinohara, M., Inoue, K., Lee, J. et al.: Generation of pluripotent stem cells from neonatal mouse testis. *Cell*, 119: 1001, 2004

[59] Seandel, M., James, D., Shmelkov, S. V. et al.: Generation of functional multipotent adult stem cells from GPR125+ germline progenitors. *Nature*, 449: 346, 2007

[60] Conrad, S., Renninger, M., Hennenlotter, J. et al.: Generation of pluripotent stem cells from adult human testis. *Nature*, 456: 344, 2008

[61] Zhang, Y. Y., Ludwikowski, B., Hurst, R. et al.: Expansion and long-term culture of differentiated normal rat urothelial cells in vitro. *In Vitro Cell Dev Biol Anim*, 37: 419, 2001

[62] Bharadwaj B, W. S., Rohozinski J, Furth M, Atala A, Zhang Y: Multipotential Differentiation of Human Urine-Derived Stem Cells. *Tissue Engineering and Regenerative Medicine 2nd World Congress*: S293, 2009

[63] Atala, A., Bauer, S. B., Soker, S. et al.: Tissue-engineered autologous bladders for patients needing cystoplasty. *Lancet*, 367: 1241, 2006

[64] Oberpenning, F., Meng, J., Yoo, J. J. et al.: De novo reconstitution of a functional mammalian urinary bladder by tissue engineering. *Nat Biotechnol*, 17: 149, 1999

[65] Yang, Y., Qu, B., Huo, J. H. et al.: Serum from radiofrequency-injured livers induces differentiation of bone marrow stem cells into hepatocyte-like cells. *J Surg Res*, 155: 18, 2009

[66] Saulnier, N., Lattanzi, W., Puglisi, M. A. et al.: Mesenchymal stromal cells multipotency and plasticity: induction toward the hepatic lineage. *Eur Rev Med Pharmacol Sci*, 13 Suppl 1: 71, 2009

[67] Mizuguchi, T., Hui, T., Palm, K. et al.: Enhanced proliferation and differentiation of rat hepatocytes cultured with bone marrow stromal cells. *J Cell Physiol*, 189: 106, 2001
[68] Yaghoobi, M. M., Mowla, S. J.: Differential gene expression pattern of neurotrophins and their receptors during neuronal differentiation of rat bone marrow stromal cells. *Neurosci Lett*, 397: 149, 2006
[69] Xin, H., Li, Y., Chen, X. et al.: Bone marrow stromal cells induce BMP2/4 production in oxygen-glucose-deprived astrocytes, which promotes an astrocytic phenotype in adult subventricular progenitor cells. *J Neurosci Res*, 83: 1485, 2006
[70] Shichinohe, H., Kuroda, S., Yano, S. et al.: Improved expression of gamma-aminobutyric acid receptor in mice with cerebral infarct and transplanted bone marrow stromal cells: an autoradiographic and histologic analysis. *J Nucl Med*, 47: 486, 2006
[71] Scintu, F., Reali, C., Pillai, R. et al.: Differentiation of human bone marrow stem cells into cells with a neural phenotype: diverse effects of two specific treatments. *BMC Neurosci*, 7: 14, 2006
[72] Tian, H., Bharadwaj, S., Liu, Y. et al.: Myogenic differentiation of human bone marrow mesenchymal stem cells on a 3D nano fibrous scaffold for bladder tissue engineering. *Biomaterials*, 31: 870
[73] Zhang, Y., Lin, H. K., Frimberger, D. et al.: Growth of bone marrow stromal cells on small intestinal submucosa: an alternative cell source for tissue engineered bladder. *BJU Int*, 96: 1120, 2005
[74] Chung, S. Y., Krivorov, N. P., Rausei, V. et al.: Bladder reconstitution with bone marrow derived stem cells seeded on small intestinal submucosa improves morphological and molecular composition. *J Urol*, 174: 353, 2005
[75] Kanematsu, A., Yamamoto, S., Iwai-Kanai, E. et al.: Induction of smooth muscle cell-like phenotype in marrow-derived cells among regenerating urinary bladder smooth muscle cells. *Am J Pathol*, 166: 565, 2005
[76] Sharma, A. K., Hota, P. V., Matoka, D. J. et al.: Urinary bladder smooth muscle regeneration utilizing bone marrow derived mesenchymal stem cell seeded elastomeric poly(1,8-octanediol-co-citrate) based thin films. *Biomaterials*, 31: 6207, 2010
[77] Sharma, A. K., Fuller, N. J., Sullivan, R. R. et al.: Defined populations of bone marrow derived mesenchymal stem and endothelial progenitor cells for bladder regeneration. *J Urol*, 182: 1898, 2009
[78] Sharma, A. K., Bury, M. I., Marks, A. J. et al.: A Non-Human Primate Model for Urinary Bladder Regeneration Utilizing Autologous Sources of Bone Marrow Derived Mesenchymal Stem Cells. *Stem Cells*, 2010
[79] Strasser, H., Marksteiner, R., Margreiter, E. et al.: Transurethral ultrasonography-guided injection of adult autologous stem cells versus transurethral endoscopic injection of collagen in treatment of urinary incontinence. *World J Urol*, 25: 385, 2007
[80] Smaldone, M. C., Chen, M. L., Chancellor, M. B.: Stem cell therapy for urethral sphincter regeneration. *Minerva Urol Nefrol*, 61: 27, 2009
[81] Peyromaure, M., Sebe, P., Praud, C. et al.: Fate of implanted syngenic muscle precursor cells in striated urethral sphincter of female rats: perspectives for treatment of urinary incontinence. *Urology*, 64: 1037, 2004

[82] Nolazco, G., Kovanecz, I., Vernet, D. et al.: Effect of muscle-derived stem cells on the restoration of corpora cavernosa smooth muscle and erectile function in the aged rat. *BJU Int*, 101: 1156, 2008

[83] Lu, S. H., Yang, A. H., Wei, C. F. et al.: Multi-potent differentiation of human purified muscle-derived cells: potential for tissue regeneration. *BJU Int*, 2009

[84] Lu, S. H., Yang, A. H., Chen, K. K. et al.: Purification of human muscle-derived cells using an immunoselective method for potential use in urological regeneration. *BJU Int*, 2009

[85] Lu, S. H., Wei, C. F., Yang, A. H. et al.: Isolation and characterization of human muscle-derived cells. *Urology*, 74: 440, 2009

[86] Lu, S. H., Cannon, T. W., Chermanski, C. et al.: Muscle-derived stem cells seeded into acellular scaffolds develop calcium-dependent contractile activity that is modulated by nicotinic receptors. *Urology*, 61: 1285, 2003

[87] Hoshi, A., Tamaki, T., Tono, K. et al.: Reconstruction of radical prostatectomy-induced urethral damage using skeletal muscle-derived multipotent stem cells. *Transplantation*, 85: 1617, 2008

[88] Cannon, T. W., Lee, J. Y., Somogyi, G. et al.: Improved sphincter contractility after allogenic muscle-derived progenitor cell injection into the denervated rat urethra. *Urology*, 62: 958, 2003

[89] Rodriguez, L. V., Alfonso, Z., Zhang, R. et al.: Clonogenic multipotent stem cells in human adipose tissue differentiate into functional smooth muscle cells. *Proc Natl Acad Sci U S A*, 103: 12167, 2006

[90] Montzka, K., Heidenreich, A.: Application of mesenchymal stromal cells in urological diseases. *BJU Int*, 2009

[91] Jack, G. S., Zhang, R., Lee, M. et al.: Urinary bladder smooth muscle engineered from adipose stem cells and a three dimensional synthetic composite. *Biomaterials*, 30: 3259, 2009

[92] Jack, G. S., Almeida, F. G., Zhang, R. et al.: Processed lipoaspirate cells for tissue engineering of the lower urinary tract: implications for the treatment of stress urinary incontinence and bladder reconstruction. *J Urol*, 174: 2041, 2005

[93] Sakuma, T., Matsumoto, T., Kano, K. et al.: Mature, adipocyte derived, dedifferentiated fat cells can differentiate into smooth muscle-like cells and contribute to bladder tissue regeneration. *J Urol*, 182: 355, 2009

[94] Smaldone, M. C., Chen, M. L., Chancellor, M. B.: Stem cell therapy for urethral sphincter regeneration. *Minerva Urol Nefrol*, 2008

[95] Liu, J., Huang, J., Lin, T. et al.: Cell-to-cell contact induces human adipose tissue-derived stromal cells to differentiate into urothelium-like cells in vitro. *Biochem Biophys Res Commun*, 2009

[96] Lin, G., Wang, G., Banie, L. et al.: Treatment of stress urinary incontinence with adipose tissue-derived stem cells. *Cytotherapy*, 2009

[97] Lin, G., Banie, L., Ning, H. et al.: Potential of adipose-derived stem cells for treatment of erectile dysfunction. *J Sex Med*, 6 Suppl 3: 320, 2009

[98] Cavarretta, I. T., Altanerova, V., Matuskova, M. et al.: Adipose Tissue-derived Mesenchymal Stem Cells Expressing Prodrug-converting Enzyme Inhibit Human Prostate Tumor Growth. *Mol Ther*, 2009

[99] Lin, T., Ambasudhan, R., Yuan, X. et al.: A chemical platform for improved induction of human iPSCs. *Nat Methods*, 6: 805, 2009

[100] De Coppi, P., Bartsch, G., Jr., Siddiqui, M. M. et al.: Isolation of amniotic stem cell lines with potential for therapy. *Nat Biotechnol*, 25: 100, 2007

[101] Perin, L., Sedrakyan, S., Da Sacco, S. et al.: Characterization of human amniotic fluid stem cells and their pluripotential capability. *Methods Cell Biol*, 86: 85, 2008

[102] Perin, L., Giuliani, S., Sedrakyan, S. et al.: Stem cell and regenerative science applications in the development of bioengineering of renal tissue. *Pediatr Res*, 63: 467, 2008

[103] Perin, L., Giuliani, S., Sedrakyan, S. et al.: Stem Cell and Regenerative Science Applications in the Development of Bioengineering of Renal Tissue. *Pediatr Res*, 2008

[104] Hayashi, K., Takahashi, M., Kimura, K. et al.: Changes in the balance of phosphoinositide 3-kinase/protein kinase B (Akt) and the mitogen-activated protein kinases (ERK/p38MAPK) determine a phenotype of visceral and vascular smooth muscle cells. *J Cell Biol*, 145: 727, 1999

[105] Hirst, S. J., Twort, C. H., Lee, T. H.: Differential effects of extracellular matrix proteins on human airway smooth muscle cell proliferation and phenotype. *Am J Respir Cell Mol Biol*, 23: 335, 2000

[106] Hubschmid, U., Leong-Morgenthaler, P. M., Basset-Dardare, A. et al.: In vitro growth of human urinary tract smooth muscle cells on laminin and collagen type I-coated membranes under static and dynamic conditions. *Tissue Eng*, 11: 161, 2005

[107] Chen, X. D., Dusevich, V., Feng, J. Q. et al.: Extracellular matrix made by bone marrow cells facilitates expansion of marrow-derived mesenchymal progenitor cells and prevents their differentiation into osteoblasts. *J Bone Miner Res*, 22: 1943, 2007

[108] Tian, H., Bharadwaj, S., Liu, Y. et al.: Differentiation of human bone marrow mesenchymal stem cells into bladder cells: potential for urological tissue engineering. *Tissue Eng Part A*, 16: 1769, 2010

[109] Tian, H., Bharadwaj, S., Liu, Y. et al.: Myogenic differentiation of human bone marrow mesenchymal stem cells on a 3D nano fibrous scaffold for bladder tissue engineering. Biomaterials, 31: 870, 2010

[110] Arrigoni, C., Chitto, A., Mantero, S. et al.: Rotating versus perfusion bioreactor for the culture of engineered vascular constructs based on hyaluronic acid. B*iotechnol Bioeng*, 100: 988, 2008

[111] Gomes, M. E., Sikavitsas, V. I., Behravesh, E. et al.: Effect of flow perfusion on the osteogenic differentiation of bone marrow stromal cells cultured on starch-based three-dimensional scaffolds. *J Biomed Mater Res A*, 67: 87, 2003

[112] DiSandro, M. J., Li, Y., Baskin, L. S. et al.: Mesenchymal-epithelial interactions in bladder smooth muscle development: epithelial specificity. *J Urol*, 160: 1040, 1998

[113] Kobayashi, N., Yasu, T., Ueba, H. et al.: Mechanical stress promotes the expression of smooth muscle-like properties in marrow stromal cells. *Exp Hematol*, 32: 1238, 2004

[114] Akhyari, P., Fedak, P. W., Weisel, R. D. et al.: Mechanical stretch regimen enhances the formation of bioengineered autologous cardiac muscle grafts. *Circulation*, 106: I137, 2002

[115] Paltiel, H. J., Diamond, D. A., Zurakowski, D. et al.: Endoscopic treatment of vesicoureteral reflux with autologous chondrocytes: postoperative sonographic features. *Radiology*, 232: 390, 2004

[116] Carr, L. K., Steele, D., Steele, S. et al.: 1-year follow-up of autologous muscle-derived stem cell injection pilot study to treat stress urinary incontinence. *International Urogynecology Journal*, 19: 881, 2008

[117] Capozza, N., Patricolo, M., Lais, A. et al.: Endoscopic treatment of vesico-ureteral reflux: twelve years' experience. *Urol Int*, 67: 228, 2001

[118] Puri, P., Chertin, B., Velayudham, M. et al.: Treatment of vesicoureteral reflux by endoscopic injection of dextranomer/hyaluronic Acid copolymer: preliminary results. *J Urol*, 170: 1541, 2003

[119] Diamond, D. A., Caldamone, A. A.: Endoscopic correction of vesicoureteral reflux in children using autologous chondrocytes: preliminary results. *Journal of Urology*, 162: 1185, 1999

[120] Zhao, W., Zhang, C., Jin, C. et al.: Periurethral Injection of Autologous Adipose-Derived Stem Cells with Controlled-Release Nerve Growth Factor for the Treatment of Stress Urinary Incontinence in a Rat Model. *Eur Urol*, 2010

[121] Furuta, A., Jankowski, R. J., Honda, M. et al.: State of the art of where we are at using stem cells for stress urinary incontinence. *Neurourol Urodyn*, 26: 966, 2007

[122] De Girolamo, L., Sartori, M. F., Arrigoni, E. et al.: Human adipose-derived stem cells as future tools in tissue regeneration: osteogenic differentiation and cell-scaffold interaction. *Int J Artif Organs*, 31: 467, 2008

[123] Furuta, A., Jankowski, R. J., Pruchnic, R. et al.: The promise of stem cell therapy to restore urethral sphincter function. *Curr Urol Rep*, 8: 373, 2007

[124] Stangel-Wojcikiewicz, K., Majka, M., Basta, A. et al.: Adult stem cells therapy for urine incontinence in women. *Ginekol Pol*, 81: 378, 2010

[125] Nikolavsky, D., Chancellor, M. B.: Stem cell therapy for stress urinary incontinence. Neurourol Urodyn, 29 Suppl 1: S36, 2010

[126] Tedesco, F. S., Dellavalle, A., Diaz-Manera, J. et al.: Repairing skeletal muscle: regenerative potential of skeletal muscle stem cells. *J Clin Invest*, 120: 11, 2010

[127] Kim, D. K., Jankowski, R. J., Pruchnic, R. et al.: In vitro and in vivo effect of lidocaine on rat muscle-derived cells for treatment of stress urinary incontinence. *Urology*, 73: 437, 2009

[128] Atala, A., Kavoussi, L. R., Goldstein, D. S. et al.: Laparoscopic correction of vesicoureteral reflux. *Journal of Urology*, 150: 748, 1993

[129] Diamond, D. A., Zurakowski, D., Atala, A. et al.: Is adolescent varicocele a progressive disease process? *Journal of Urology*, 172: 1746, 2004

[130] Mitterberger, M., Pinggera, G. M., Marksteiner, R. et al.: Functional and histological changes after myoblast injections in the porcine rhabdosphincter.[see comment][erratum appears in Eur Urol. 2008 Nov;54(5):1208 Note: Bartsch, Georg [removed]]. *European Urology*, 52: 1736, 2007

[131] Mitterberger, M., Pinggera, G. M., Marksteiner, R. et al.: Adult stem cell therapy of female stress urinary incontinence. *Eur Urol*, 53: 169, 2008

[132] Mitterberger, M., Marksteiner, R., Margreiter, E. et al.: Myoblast and fibroblast therapy for post-prostatectomy urinary incontinence: 1-year followup of 63 patients. *J Urol*, 179: 226, 2008

[133] Mitterberger, M., Marksteiner, R., Margreiter, E. et al.: Autologous myoblasts and fibroblasts for female stress incontinence: a 1-year follow-up in 123 patients. *BJU International*, 100: 1081, 2007

[134] Chancellor, M. B., Rivas, D. A., Liberman, S. N. et al.: Cystoscopic autogenous fat injection treatment of vesicoureteral reflux in spinal cord injury. *J Am Paraplegia Soc*, 17: 50, 1994

[135] Arena, S., Fazzari, C., Implatini, A. et al.: Dextranomer/hyaluronic Acid copolymer implant for vesicoureteral reflux: role of myofibroblast differentiation. *J Urol*, 181: 2695, 2009

[136] Lin, C. S., Xin, Z. C., Deng, C. H. et al.: Recent advances in andrology-related stem cell research. *Asian J Androl*, 10: 171, 2008

10

Glandular Stem Cells A New Source for Myocardial Repair?

Norbert W. Guldner[1], Charli Kruse[2] and Hans - H. Sievers[1]
[1]*Clinic of Cardiac Surgery, University of Lübeck,*
[2]*Fraunhofer EBM, Cellular Biotechnology, Lübeck*
Germany

1. Introduction

For more than 95% of the patients with end-stage heart failure there is no definitive treatment option up to now (AHA, 2009). This fact is caused by a severe shortage in donor hearts on the one hand and on the other hand by technical and economic limitations of cardiac assist devices and artificial hearts. Thus there is a big need for alternative treatement options. Basically heart failure is due to the inability of adult cardiomyocytes to divide and repair damaged heart muscle. For myocardial regenerative medicine stem cells which show the ability to differentiate into functional cardiomyocytes might be a promising source. Although human embryonic stem cells can differentiate into beating cardiomyocytes (Xu et al, 2002), however the therapeutic use of these cells is not without legal and ethical problems.

Previous studies in animal models have demonstrated the therapeutical potential of intra-myocardial injection of adult murine stem cells. In a cell-cell contact dependent manner in mammals, mesenchymal stem cells aquired a cardiomyocyte phenotype (Orlic&Kocher 2001, Kawamoto&Yeh 2003, Wang et. al., 2006). Furthermore spontaneously beating cardiomyocytes were derived from isolated cardiomyogenic cell lines of murine bone marrow stromal cells (Makino et al., 1999) adipose tissue stroma cells (Planat-Bernard et. al., 2004) and from spermatogonial stem cells from the adult mouse testis (Guan et. al., 2006). In human myocardium however, adult stem cells from different origins applied intramyocardially did not show a differentiation into cardiomyocytes (Yoon&Wollert 2005). Thus, the search for an appropriate source of cells being able to differentiate into functional cardiomyocytes in human or to contribute to a contractile myocardial patch is still continuing.

From adult rat pancreatic tissue multipotential stem cells were isolated and differentiated into the endodermal pancreatic and hepatic cells (Zulewski, 2001). Human pancreatic stem cells showed also a differentiation into mesodermal structures, like adipocytes, chondrocytes and osteocytes (Seeberger et. al., 2006). Furthermore rat and human pancreatic stem cells gave rise to cellular aggregates containing cell types of all three germ layers (Kruse et.al, 2004) including ectodermal lineages also shown in further recent publications (Seaberg et. al., 2004; Choi et. al., 2004). But up to now neither in animals nor in humans cardiomyocytes were generated from pancreatic stem cells. Due to the clinical need of cardiomyocytes, we investigated the differentiation of human pancreatic stem cell cultures

into cardiomyocytes, a process potentially promoted by co-culture with human myocardial biopsies.
Myocardial regeneration with artificially applied cardiomyocytes is emerging to a promising issue of significant scientific and clinical impact. Nevertheless the source of cells for human cardiomyocyte differentiation especially from adult tissue is still unclear. We hypothesized that human pancreatic stem cells may differentiate into cardiomyocyte-like cells enhanced when co- cultured with myocardial tissue.

2. Harvesting, isolating, culturing

From four patients undergoing surgery after an abdominal injury, pancreatic tissue was obtained with informed consent (ethical accreditation of the Ethics Committees, University Hospital of Lübeck, AZ: 03-065). Stem cells were selected, cultured within fetal calf serum and passaged more than 21 times as described elsewhere (Kruse C et al., 2004). Briefly, the pancreatic tissue was treated with digestion medium containing HEPES-Eagle-medium (pH 7.4), 0.1mM HEPES-buffer (pH 7.6), 70% (v/v) modified Eagle-medium, 0.5% (v/v) Trasylol (Bayer AG, Leverkusen, Germany), 1% (w/v) bovine serum albumin, 2.4 mM $CaCl_2$ and collagenase (0.63 PZ/mg, Serva, Heidelberg, Germany). After digestion the acini were dissociated, centrifuged and further purified by washing in Dulbecco's modified Eagle's medium (DMEM, Gibco, Germany) supplemented with 20 % fetal calve serum (FCS). The washing procedure was repeated 5 times. The acini were resuspended in DMEM and cultured at 37°C in a 5% CO_2 humidified atmosphere. After 1-2 days of culture spindle-shaped stem cells were observed on the bottom of the cell culture flask. After reaching confluence pancreatic stem cells were subcultured by trypsinization, counted and reseeded with a density of 2-4 x 10^5 cells/cm^2. This procedure was repeated until sufficient cells were available. For differentiation we used cells from passages 4 and 14.
Pancreatic stem cells as a source of cardio-myocytes can be obtained from patients using minimally invasive procedure during laparoscopy. Due to their retroperitoneal topography this cell harvesting might not be as easy as obtaining cells from blood, bone marrow or fatty tissue. Nevertheless current investigations show that stem cells from salivary glands easier accessible than pancreatic tissue might have a comparable potential generating cardiomyocytes. On the other hand, if these pancreatic stem cells could react as a general donor on the basis of a recently postulated immunological compatibility of adult stem cells (Chiu RCJ, 2005; Puissant et. al, 2005) sophisticated harvesting might not be necessary.
When cells became confluent in culture dishes netlike clusters could be detected. The cell layer was washed with the less nutritive phosphate buffered saline (PBS) and mechanically lifted partially from the culture bottom with a scraper.
Irrespective of whether adult human pancreatic stem cells were cryo-conserved or freshly isolated, their phenotype built netlike cell clusters in different passages. After development of these netlike cell clusters, changing culture conditions and partial mobilization from the bottom, very few cell clusters showed distinct cellular autonomous contractions with about 20 beats per minute. Contracting regions were then video documented in real time and reproduced in five settings.

3. Co-culture with myocardium

To promote self-differentiation into cardiomyocytes, human pancreatic stem cells were co-cultured with biopsies of human myocardium. This myocardial stimulation practically was achieved by co- culture of the primary cells with each 5 pieces of myocardium (4x4x4mm) for 2 days. The tissue (left ventricular wall, mitral papillary muscle) was received during heart surgery (ethical accreditation of the Ethics Committees, University hospital of Lübeck, AZ 05-206). Heart muscle pieces were adhered to the bottom of the culture dishes for 3 hours until primary pancreatic stem cells (1x 10^6) were applied. After 48h heart muscle pieces were removed and the stem cells were further cultured as described above. Cells were passaged every time after reaching confluence. Immunocytochemical analyses were performed directly 48 hours after treatment. To investigate the long term effects of differentiation, cells were collected 17 days after treatment for PCR- analyses.

When cells became confluent in culture dishes netlike clusters could be detected. The cell layer was washed with the less nutritive phosphate buffered saline (PBS) and mechanically lifted partially from the culture bottom with a scraper. Contracting regions were then video documented in real time and reproduced in five settings.

To test whether cardio-myocytes grow out of biopsies from heart tissues, we cultured them as described above but without pancreatic stem cells. After two days no outgrowing cells could be detected, thus cell fusion of cardio-myocytes from biopsies with pancreatic stem cells seems very unlikely.

After having been in contact with human myocardium for 48 hours and growing for further 14-40 days in co-culture, cells were partially mobilized from the bottom of the culture flask and treated with a less nutritive culture medium. Many contracting areas (4-6 fold) were found in comparison to the unstimulated cells. However, the structure of a contracting area itself was comparable with that already observed before having been developed spontaneously without contact to myocardium. These autonomous contractions with about 20 beats per minute as well, were found in all five co-cultures investigated.

The number of contracting areas was enhanced by co-cultured human myocardial biopsies.

The influence from cardio-myocytes on glandular stem cells could become already demonstrated by an immunocytochemical visualization of sarcomeres (red) like shown in figure1 seeing a clearly detectable gradient from M, where the myocardium was located, to the periphery. This cellular stimulation could also be documented as described as follows.

These findings of the stem cell transformation in contact with myocardium in-vitro may become repeated in-vivo after an intra-myocardial injection,

4. Analysis of co-cultured glandular stem cells

After co-culturing and breeding, the influence of the added myocardium to the glandular stem cells was examined by immunocytochemistry of sarcomere-related myosin, immuno-cytochemistry of cardiac specific troponin, at the electrone-microscopic level and phenotyped with respect to RNA, protein and cardiomyocyte specificity.

4.1 Immunocytochemistry of sarcomere-related myosin

Both, the stimulated as well as the non stimulated stem cells were seeded on chamber slides and cultured for at least two days before they were fixed with methanol:acetone (7:3) containing 1 g/ml DAPI (Roche, Switzerland) and washed 3 times in PBS. After incubation in 10% normal goat serum at room temperature for 15 min the specimens were incubated with the primary antibody overnight at 4°C in a humid chamber. Primary monoclonal antibody was directed against sarcomere myosin MF 20 (DSHB, USA). After rinsing 3 times with PBS, slides were incubated for 45 minutes at 37°C with Cy3-labelled anti-mouse IgG, diluted 1:200. Slides were washed 3 times in PBS, covered in Vectashield mounting medium (Vector, USA) and analyzed with a fluorescence microscope (Axioskop Zeiss, Germany). Excluding these detected sarcomeres were released from the biopsy and attached to the stem cells, controls with fibroblasts and endothelial cells were cocultured with myocardium. Within these controls, tested cells were negative for sarcomeres by immunocytochemistry.

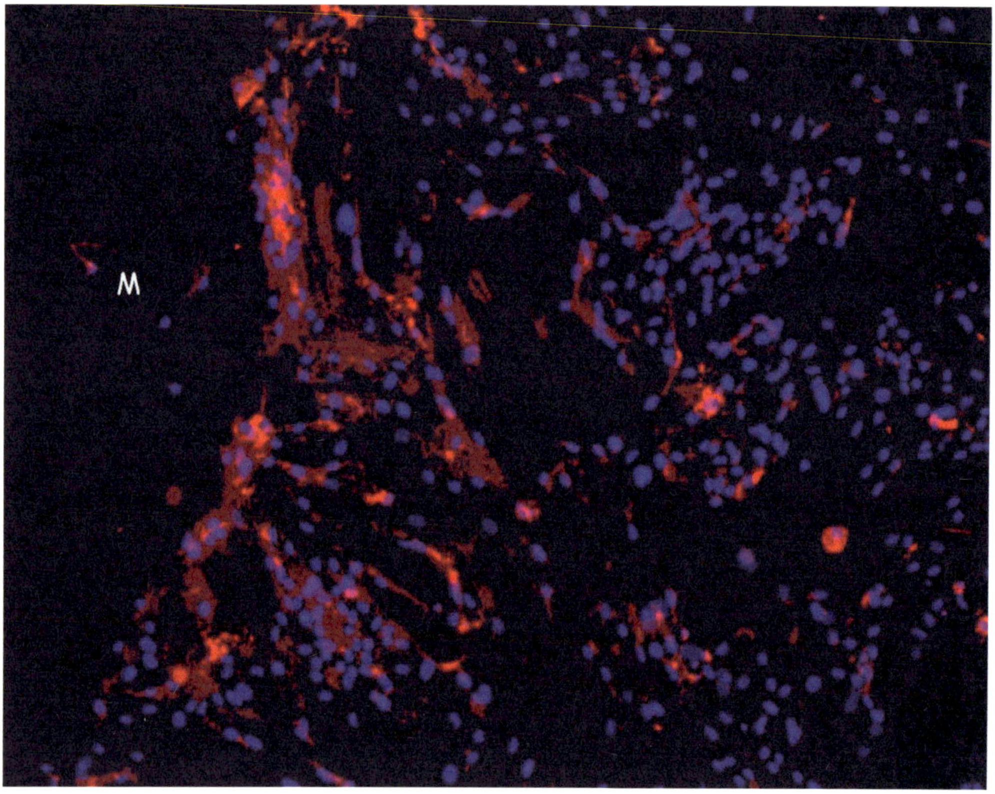

Fig. 1. Immunocytochemical visualization of sarcomeres (red) in transformed adult pancreatic stem cells (blue nuclei) which had been co-cultured with human myocardium (M) for two days . They were clearly positive for the tested antibody compared to the untreated pancreatic stem cells . A decreasing gradient of myosin containing cells from M to the periphery is to be seen

4.2 Immunocytochemistry of cardiac specific troponin I

Stem cells were co-cultured with myocardial biopsies for 48 hours and breeded for further 2 to 4 days after removing the myocardium. Thereafter probes were rinsed 2 times with PBS and dried for 24 hours on air by RT and then fixated by pure acetone for 10 minutes and -20C°, rinsed again for 2X5 minutes with tris-buffered-saline (TBS) and preincubated with RPMI 1640 (Sigma, USA) with 10% AB serum (Biochrom AG, Germany). Monoclonal anti-troponin I antibody (Clone 2d5, Biozal 1:25) was used as the primary antibody for 60 minutes. Secondary antibody administration (rabbit anti-mouse; DAKO, Denmark; 1:25 for 30 minutes) followed by alkaline phosphatase anti-alkaline phosphatase complex incubation (DAKO, Denmark; 1:50, 30 minutes) was repeated for a total of times each. Finally, substrate incubation (naphthole/neofuchsine) and counterstaining with hemalaun were performed before microscopic evaluation. Additionally an isotype-control was carried out with mouse IgG 1 (DAKO, Denmark) and for further negative control skeletal muscle was stained. Myocardium was used for a positive control. An isotype-control with mouse IgG 1 (DAKO, Denmark) was negative as well. Additional controls applying on skeletal muscles were negative too. As expected, a control on human myocardium showed positive results (data not shown).

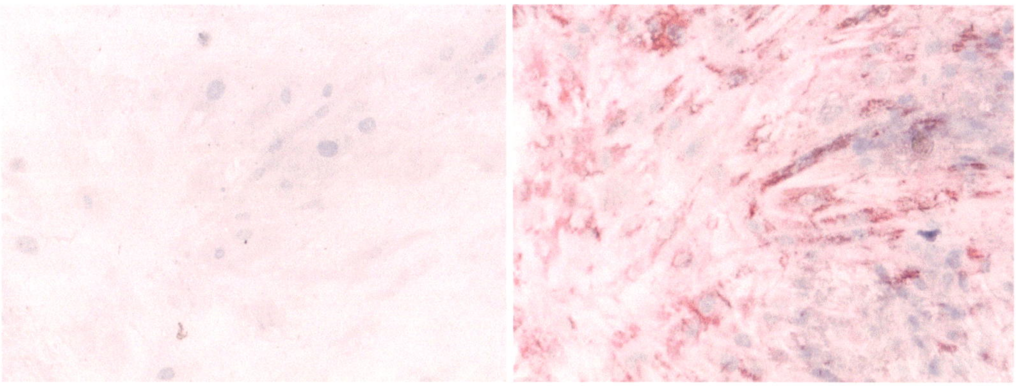

Fig. 2. Human pancreatic adult stem cells with immunocytochemical staining of cardiospecific troponin I (left) and after a two day co-culture with human myocardium (right). The existance of cardiospecific troponin I in transformed cells is clearly demonstrated

4.3 Electron microscopic evaluation

Cells grown on coverslips were fixed with 2.5% glutaraldehyde in 0.1 M cacodylate buffer for 1 h. Postfixation was performed with 1% OsO_4 in 0.1 M cacodylate buffer for 2 h; samples were dehydrated with ethanol and embedded in araldite (Fluka, Switzerland). Ultrathin sections were stained with uranyl acetate and lead citrate (Ultrostainer Carlsberg System, LKB, Sweden) and were examined with a Philips electron microscope EM 400 at 60 kV (Philips, The Netherlands).

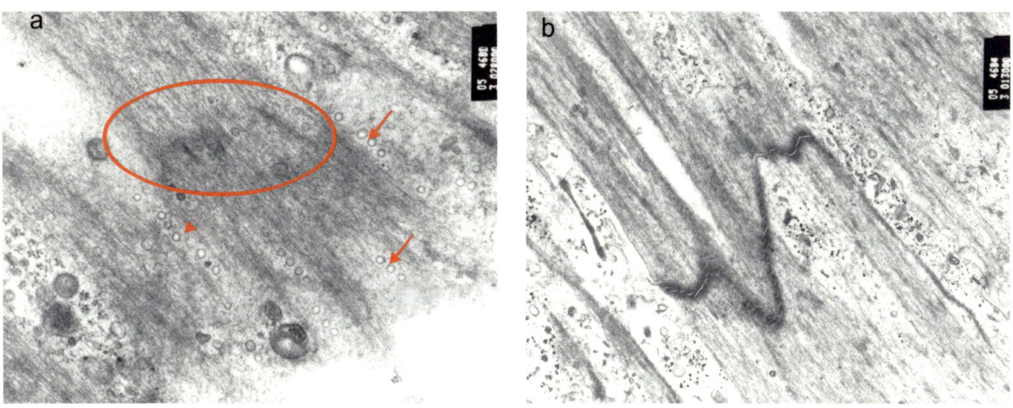

Fig. 3. Electron micrographs (magnification 13 000X) four days after a 48-hour contact with biopsies of human myocardium. Myofilaments and structures of less (left) and complete developed intercalated disc (right) are demonstrated

Electron microscopy (Fig. 3 left) showed differentiated cardiomyocytes containing a number of contractile fibrils. Furthermore, different developmental stages of intercalated discs were observed. While intercalated discs are slightly but clearly recognizable (encircled, Fig. 3 left) in some cells, they are as well differentiated as in mature tissue in others (Fig. 3 right).

4.4 Semiquantative RT-PCR analysis

Total cellular RNA was isolated using NucleoSpin®RNA II-Kit (Macherey-Nagel, Germany). 0.5 µg of total RNA was reverse transcribed into cDNA using Superscript II Reverse Transcriptase RNase H- (RT, Invitrogen, The Netherlands) and oligo dT-Primers (Invitrogen, The Netherlands) according to the manufacturer's instructions. The PCR's were performed in 50 µl reaction volume using Taq DNA Polymerase (MBI Fermentas GmbH, Germany). The reactions were carried out for 38 cycles. Control run of RNA without reverse transcriptions were performed to avoid contamination with genomic DNA and produced no bands. To normalize cDNA concentration in different RT-probes, we measured relative expression of GAPDH as representative for an internal housekeeping gene control.

The expected fragment sizes and optimal PCR annealing temperatures used were as follows: GAPDH , 5':gagtcaacggatttggtcgt, 3':ggaagatggtgatgggattt (213bp, 58,8°C), troponin T2, 5':gattctggctgagaggagga, 3':tggagactttctggttatcgttg (197bp, 62.6°C), alpha-actin, 5':gtgtgacgacgaggagacca, 3':cttctgacccatacccacca (154bp, 62,6°C), desmin, 5':ctgtccctcccacctctgt, 3':agcccctgctttctaagtcc (250bp, 62,6°C). A purified human heart RNA (Ambion, USA) and a carcinogenic cell line (HEp-2) served as functional controls for PCR-primer. PCR-analyses were carried out in a total of four settings.

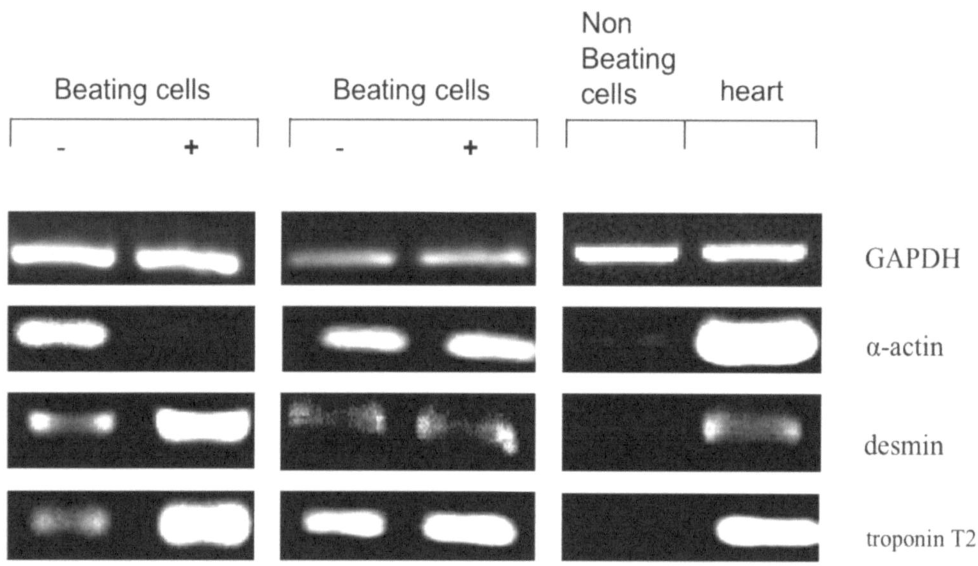

Fig. 4. Gene expression analysis with the specific PCR primers for the target genes α-actin, desmin and troponin T2 isoform1 demonstrated a greater increase of muscle cell-specific molecules in co-cultured beating cells (+) than in untreated beating cells (-). Non beating carcinogenic cells did not show any muscle specific marker. As a positive control served human heart cDNA (heart)

To test the long-term effect of differentiation PCR-analyses were performed two weeks after co-culture. The target genes for α-actin, desmin and troponin T were detected to a somewhat greater extent in pancreatic stem cells cocultured with myocardium as compared to untreated cells (Fig. 4). A-actin could partially not been amplified, whereas desmin and troponin T showed greater amounts in most of the experiments. It remains unclear why α-actin is suppressed in some cultures. One explanation could be that the differentiation progression of single cell lines follows different roads. We could find a myocardial differentiation within 14 days, but also a shutdown of genes, e.g. α-actin, in between. Fig.4 demonstrates typical results of these two possibilities.
However, as the differences between co-culture and untreated cells were slightly, the differentiation seemed to abate during time. Probably a permanent stimulus might be necessary to keep the cells in differentiation processes.

5. Homing

Applying stem cell therapy in a failing myocardium, the dimension of an intra-myocardial cell homing is significant. Thus a comparison of the homing potential between glandular (GSCs) and mesenchymal stem cells (MSCs) was performed within the myocardium of a big animal model.
In 6 African Bore Goats the intra- myocardial homing of glandular stem cells and MSCs ($CD133^+$) was evaluated. Glandular stem cells were characterized by red PKH26 respectively green PKH67 (MSCs) makers. Myocardial samples were taken after one resp. three hours (n=3), others were harvested 6 weeks after injection in additional three goats. Frozen tissue slices were generated and examined for the marked cells.

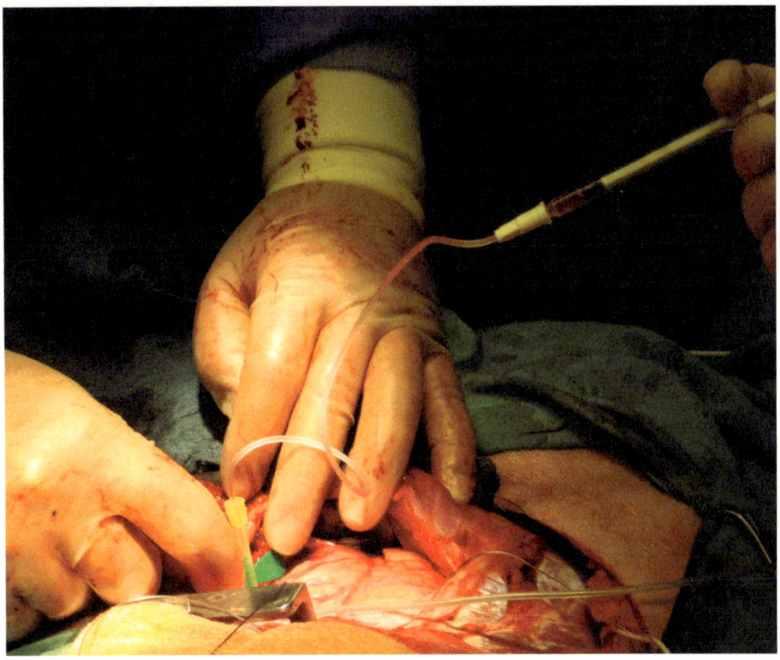

Fig. 5. Through a left lateral thoracotomy and exposure of the left heart a mix of one million of each cell type was injected into three locations of the goat's myocardium of the left ventricle

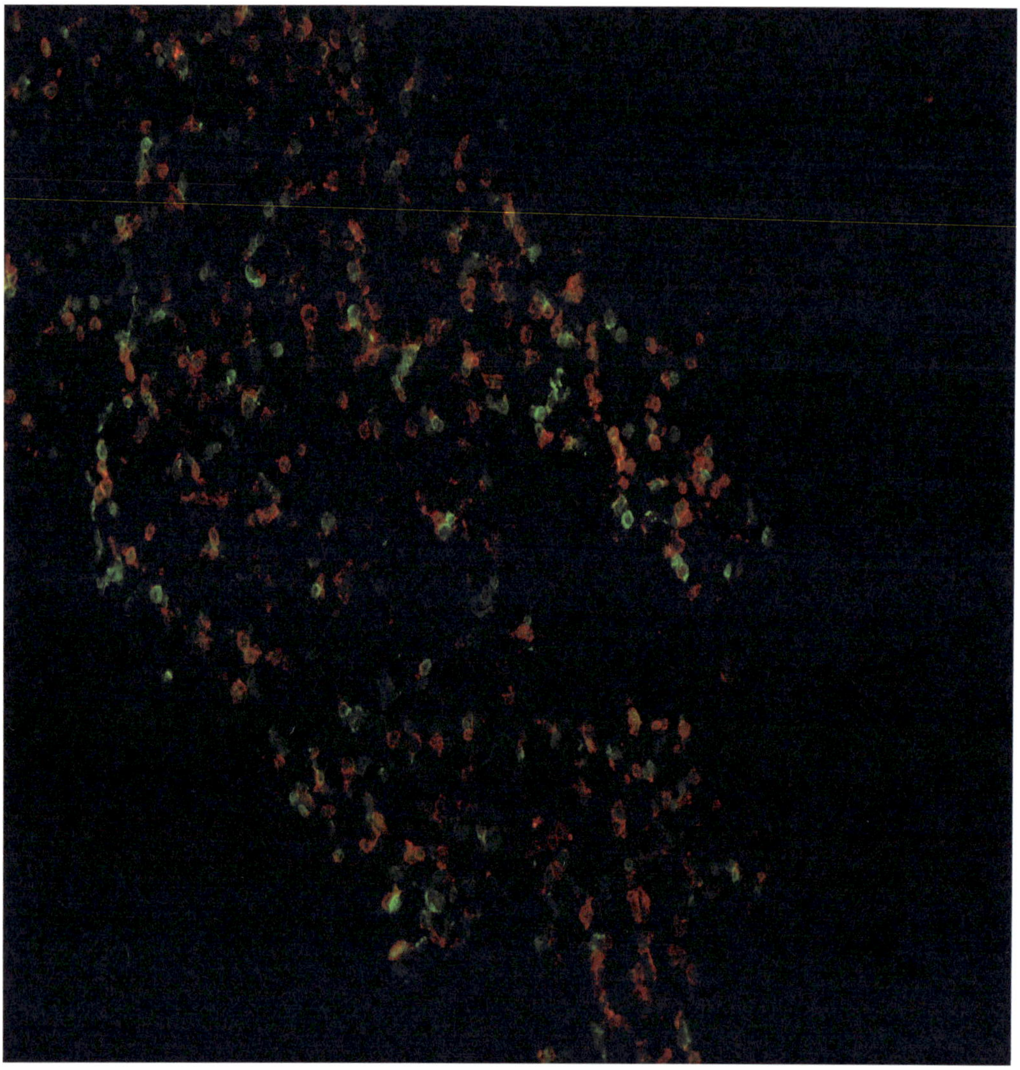

Fig. 6. a A mix of GSCs and MSCs ($CD133^+$) are shown before intra- myocardial injection . GSCs are characterized by red PKH26 respectively MSCs by green PKH67 (MSCs) makers. They stay within the cell membrane for about three months. Cell counts were performed after one and three hours and 6 weeks after the intra-myocardial injection

Red PKH26 markers for glandular stem cells (GSCs) respectively green PKH67 markers for mesenchymal stem cells (MSCs) will enable to detect the intramyocardial injected stem cell within a time frame of six months. Within this time frame it should be possible to visualize connexin anti-bodies for the detection of gap junctions.

Using a mix of an intra- myocardial injection of GSCs and MSCs, solely in MSCs (green) a significant cell migration into the surrounding myocardium (n=3) was observed. After 6 weeks nearly all GSCs remained within the myocardium while the MSCs disappeared almost completely. Within the frozen myocardial slices >90% of the marked stem cells were identified as GSCs (red) but <10% as green MSCs.

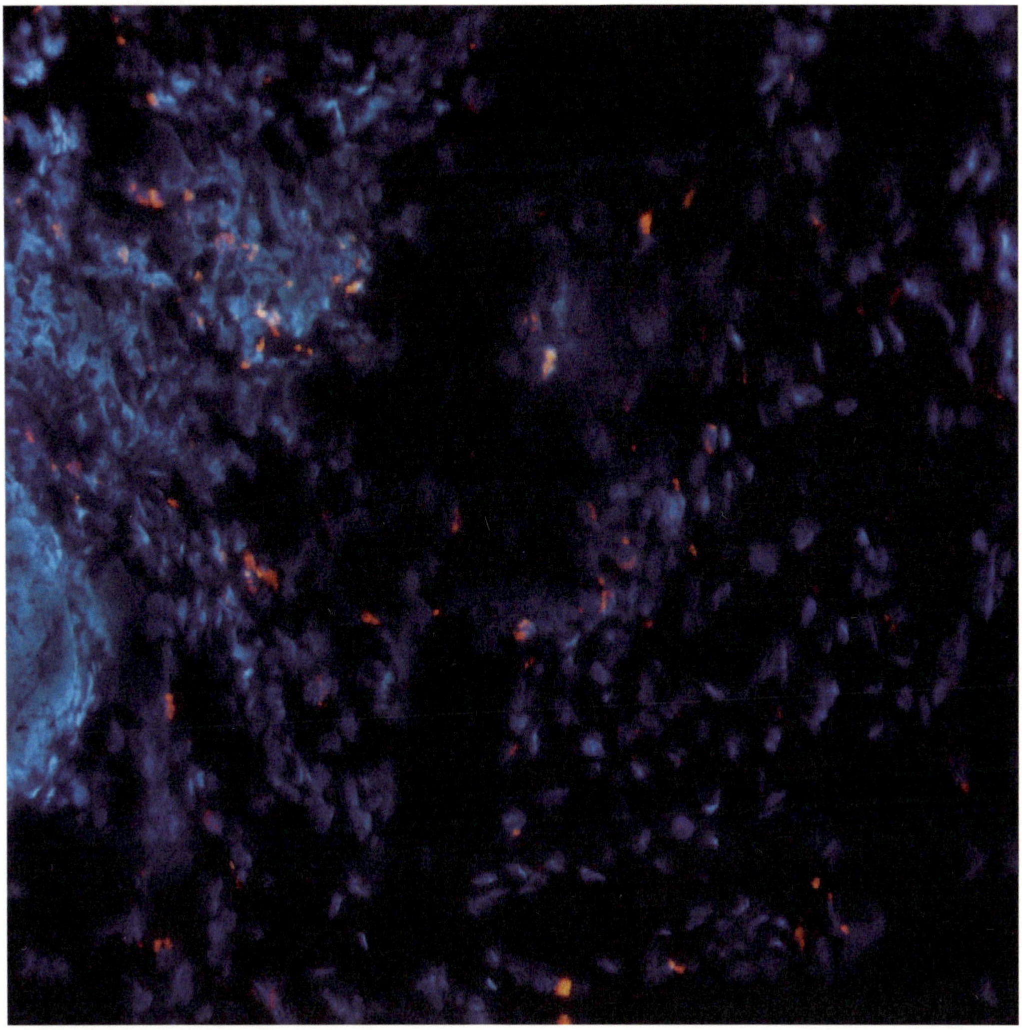

Fig. 6. b Mainly glandular stem cells (GSCs) stained by red PKH26 were found six weeks after intramyocardial injection in the myocardium (blue) of the Bore Goats (n=3). Due to a >90% homing of GSCs combined with the ability developing cardio- myocyte like cells, glandular stem cells might become a very promising treatment option in the therapy of a failing myocardium

6. Clinical application

Adult bone marrow derived adult stem cells i. e. mesenchymal stem cells (MSCs) were successfully applied clinically to restore the myocardium solely and in combination with trans-myocardial laser (channels) for myocardial revascularization because MSCs mainly transform into capillaries) but not significantly into cardio-myocytes. (Stamm C et al.,2003 and Steinhoff G et al., 2006).

Adult glandular stem cells (pancreatic, parotic and submandibular) however are able to form tissue with the distinct ability to generate cell types of all three germ layers (Kruse C et al., 2004). We showed that mesenchymal cells from this tissue differentiate into cardiomyocytes promoted in co-cultures with human myocardial biopsies (Guldner et al., 2006).

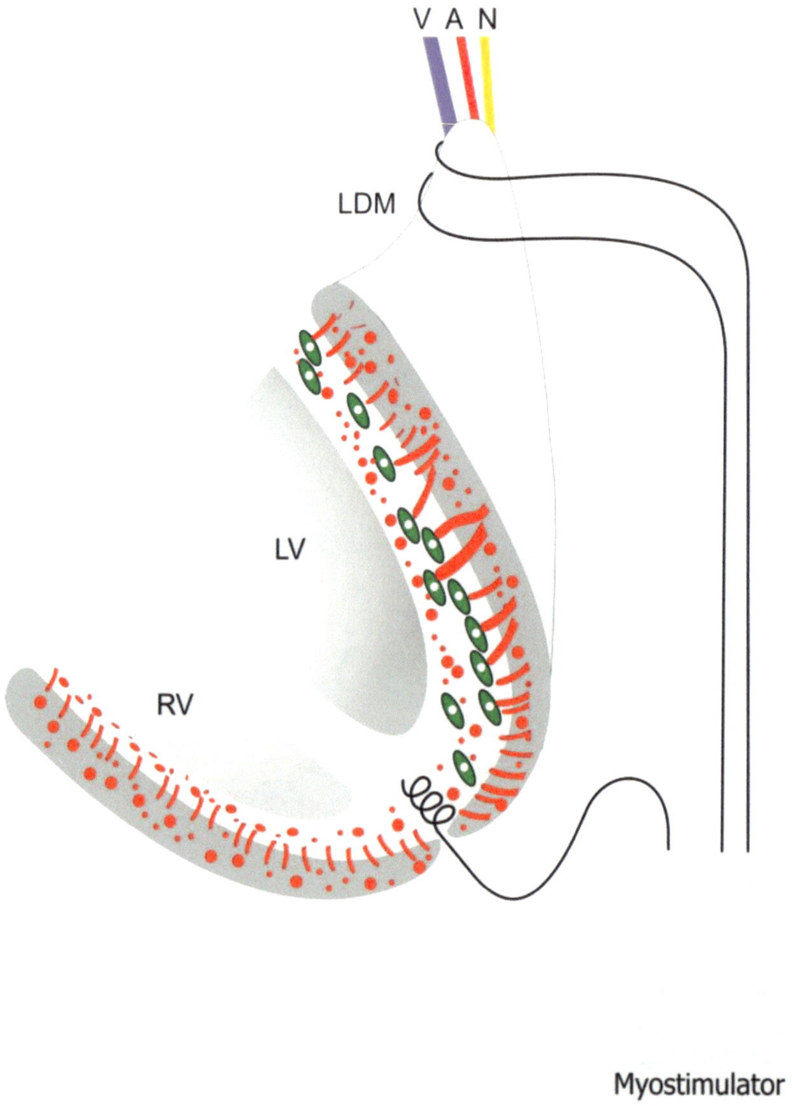

Fig. 7. The concept of a glandular stem cell cardiomyoplasty is designed as a treatment option for end-stage heart failure. It combines an expected regenerative potential of transformed adult glandular stem cells into cardiomyocytes within the myocardium or onto the myocardium and the potential of a hypercapillarized latissimus dorsi muscle (LDM) wrapped around the heart for stem cell nutrition and girdling. Muscle transformation from glandular stem cells into cardiomyocytes is documented. A hypercapillarization (capillary to fiber ratio) of 36% within the LDM was found after an intermittent electrical stimulation over 14 days by an implanted myostimulator delivering a controlled stimulation pattern by a new designed devise (Microstim myostimulator, Germany)

Harvesting pancreatic stem cells may cause severe complications due to pancreatic fistulas accompanied with a peritonitis. Biopsies of the glandula parotis may injure the facialis nerve with facialis paresis. The authors assume minor complications are to expect using the submandibular gland by harvesting cells. An amount of more than 100 mio. stem cells are expected from each submandibular gland, which can become harvested in the same operation injecting them into the myocardium.

Glandular stem cells have the potential transforming into cardio myocytes (Guldner et al., 2006). Furthermore GSCs showed a more than 90% homing in comparison with MSCs after 6 weeks injected in a goat's myocardium (n=3). Combined with their ability developing cardio-myocyte like cells, glandular stem cells are expected to be superior to MSCs for myocardial repair.

In goats and humans electrical continuous and intermittent stimulation of the latissimus dorsi muscle (LDM) has been shown to enhance capillary density of skeletal muscle tissue likewise in small animals as shown by Hudlicka O et al., 1984; Dawson JM et al., 1989; Mathieu-Costello et al., 1996; and Skorjanc D et al., 1998. Additionally known are data of capillary density in intermittent stimulated human sized animals, correlated with functional data like blood flow at rest and under exercise which were evaluated elsewhere (Guldner et al., 2008).The recent investigations in goats demonstrated a 36% higher capillary to fiber ratio in comparison with the non stimulated control after 14 days of electrical pacing of the in-situ LDM (Guldner et al., 2008). An electrically stimulated and therefore hyper-capillarized latissimus dorsi muscle (LDM) and wrapped around the heart could serve for GSC's nutrition (Mannion et al., 1992, 1993, 1996 ;Salmons et al., 1998; Guldner et.al.2008).

A stem cell cardio-myoplasty combines a cellular cardio-myoplasty (Stamm et. al, 2003; Steinhoff&Stamm, 2006) with a muscular dynamic cardio-myoplasty (Carpentier& Chachques 1985). The additional elastic girdling from LDM around the heart (Figure 7) reduces left ventricular's wall tension and therefore reduces the oxygen consumption of the myocardium (Hagège A A et al., 1995).

7. Conclusion

These are first investigations demonstrating the feasibility of generating autonomously contracting cardio-myocyte-like cells from adult human glandular stem cells and their transformation in contact with myocardium. Injecting them into the myocardium is an in vivo co-culturing. We expect in opposite to the injection of bone marrow derived stem cells, resulting in capillaries, a substantial increase of cardiomyocytes. This mechanism might be helpful because of the inability of adult cardiomyocytes to divide and repair damaged heart muscle.

Harvesting glandular stem cells intra-operatively might become feasible in the same operative procedure as the intramyocardial implantation likewise it has been practiced already with bone-marrow-derived stem cells (Steinhoff&Stamm, 2006). The estimated amount of harvested glandular stem cells from one glandular submandibularis might be many-fold higher than the amount of MSCs applied nowadays clinically.

Regarding a superior intra-myocardial homing of GSCs in comparison to MSCs stem cell therapy with GSCs might become an advanced cellular treatment option in the regenerative medicine for end-stage heart failure.

Glandular stem cells might enable intramyocardial injections solely or in combination with a pre-stimulated cardiomyoplasty, a muscle powered cardiac assist procedure with an increased capillary to fiber ratio, which is to perform by adequate electrical stimulation patterns. This muscular indirect myocard revascularisation may not only increase myocardial`'s blood supply supporting the implanted GSCs but will additionally decrease the oxygen consumption by a girdling effect. This is due to a deceases wall tension of the myocardium. Thus glandular stem cells should enable a glandular stem cell cardiomyoplasty which combines an expected regenerative potential of transformed adult glandular stem cells into cardiomyocytes within the myocardium with the potential of a hypercapillarized latissimus dorsi muscle (LDM) wrapped around the heart for stem cell nutrition and girdling.

8. Acknowledgment

The authors thank T. Hardel, P.M. Rumpf, B. Keding, J. KajanE. Klink and E. Theißing for generously culturing cells and performing immunocytochemistry and M Klinger performing the electron microscopy. This study was supported by a grant from the European Union (CellPROM). We thank the Microstim GmbH, Germany for the technical supply and the provision of myostimulators for the hypercapillarization studies.

9. References

American Heart Association, Heart Disease and Stroke Statistics- (2009) Updates, Dallas, Texas

Beyer M, (1997). Cardiomyopexy-current status of an indirect revascularization method. *Z Kardiol*; 86:Suppl 1:125-132

Beyer M, Beyer U, Mierdl S, Sirch J, von Behren H, Hannekum A, (1994). Indirect myocardial revascularization--an experimental study in the dog. *Eur J Cardiothorac Surg*; 8:557-562

Carpentier A, Chachques JC,(1985). Myocardial substitution with stimulated skeletal muscle:first successful clinical case. *Lancet*;1:1267.

Chiu, R.C.J. (2005) "Stealth Immune Tolerance" in stem cell Transplantation: Potential for "universa Donors" in myocardial regenerative therapy. *J of Heart and Lung Transplant*; 24: 511-516.

Choi Y, Ta M, Atouf F, Lumelsky N. (2004) Adult pancreas generates multipotent stem cells and pancreatic and nonpancreatic progeny. *Stem Cells*; 22: 1070-1084.

Dawson JM, Hudlicka O. (1989). The effect of long-term activity on the microvasculature of rat glycolytic muscle. *Int J Microcirc*; 8 53-69.

Guan, K. et al. (2006) Pluripotency of spermatogonial stem cells from adult mouse testis. *Nature* [online March]1-5

Guldner NW, , Klapproth P, Schwarz PO et al. (2009)Bio-technologies for glandular stem cell cardiomyopexy. *Ann.Anatomy*;191:45-50.

Guldner NW, Eichstaedt HC, Klapproth P et al., (1994). Dynamic training of skeletal muscle ventricles. A method to increase muscular power for cardiac assistance. *Circulation*; 89:1032-1040

Guldner NW, Kajahn, Klinger M, Sievers HH, Kruse C, (2006). Autonomously contracting human cardiomyocytes generated from adult pancreatic stem cells and enhanced in co-cultures with myocardial biopsies. *Int J Artif Organs*; 29:1158-66.

Guldner NW, Klapproth P, Grossherr M et al., (2001). Biomechanical hearts: muscular blood pumps, performed in a 1-step operation, and trained under support of clenbuterol. *Circulation*;104:717-722

Guldner NW, Klapproth P, Grossherr M et al.,(2000). Clenbuterol-supported dynamic training of skeletal muscle ventricles against systemic load: a key for powerful circulatory assist? *Circulation*;101:2213-2219.

Hagège A.A., Desnos M., Fernandez F., Besse B., Mirochnik N., Castaldo M., Chachques J.C., Carpentier A., Guérot C, (1995). Clinical study of the effects of latissimus dorsi muscle flap stimulation after cardiomyoplasty. *Circulation*,; 92(9 Suppl): II210-215.

Hescheler J, Fleischmann BK, Lentini S, Maltsev VA, Rohwedel J, Wobus AM, et.al. (1997) Embryonic stem cells: a model to study structural and functional properties in cardiomyogenesis. *Cardiovasc Res.*; 36:149-162.

Hudlicka O, Tyler KR,(1984). The effect of long-term high frequency stimulation on capillary density and fibre types in rabbit fast muscles. *J Physiol*; 353:435-445.

Kawamoto A, Tkebichava T, Yamaguchi J(2003). Intramyocardial transplantation of autologous endothelial progenitor cells for therapeutic neo-vascularization of myocardial ischemia. *Circulation*; 107: 461-468.

Kocher AA, Schuster MD, Szaboles MJ, Takuma S, Gurkhoff D, Wang J, et al. (2001) Neovascularization of ischemic myocardium by human bone-marrow-derived angioblasts prevents cardiomyocyte apoptosis, reduces remodelling and improves cardiac function. *Nat Med*,; 7: 430-436.

Kruse C, Birth M, Rohwedel I, Assmuth K, Goepel A, Wedel T(2004). Pluripotency of adult stem cells derived from human and rat pancreas. *Appl Physics* A; 1-15.

Kruse C, Birth M, Rohwedel I, Assmuth K, Goepel A, Wedel T.(2004) Pluripotency of adult stem cells derived from human and rat pancreas. *Appl Physics* A; 1-15.

Makino S, Fukuda K, Miyoshi S, Komada FK, Pan J, Sano M, Takahashi T, Hori S, Abe H, Hata J, Umezawa A, Ogawa S (1999). Cardiomyocytes can be generated from marrow stromal cells in vitro *J Clin Invest*; 103: 697-705.

Mannion JD, Blood V, Bailey W et al., (1996). The effect of basic fibroblast growth factor on the blood flow and morphologic features of a latissimus dorsi cardiomyoplasty. *J Thorac Cardiovasc Surg*; 111:19-28.

Mannion JD, Buckman PD, Magno MG, DiMeo F, (1992). Collateral blood flow from skeletal muscle to normal myocardium. *J Surg Res*; 53:578-587.

Mannion JD, Magno M, Buckman P et al., (1993) Techniques to enhance extramyocardial collateral blood flow after a cardiomyoplasty. *Ann Surg*; 218:544-553.

Mannion JD, Magno MG, Buckman PD et al., (1993). Acute electrical stimulation increases extramyocardial collateral blood flow after a cardiomyoplasty. *Ann Thorac Surg*; 56: 1351-8.

Mathieu-Costello et al., (1996). Capillary-to-fibre surface ratio in rat fast-twitch hindlimb muscles after chronic electrical stimulation. *J Appl Physiol*; 80(3):904-909.

Orlic D, Kajsturaj J, Chimenti S, Anderson SM, Li B, Pickel J, McKay R, et: al. (2001). Bone marrow cells regenerate infarcted myocardium. *Nature*.;410:701-705.

Planat-Benard V, Menard C, Andre M, Puceat M,Perez A, Garcia- Verdugo J, Penicaud, Casteilla L(2004). Spontaneous cardiomyocyte differentiation from adipose tissue stroma cells. *Circ Res*; 94: 223-229.

Puissant B, Barreau C., Bourin Ph, Clavel C., Bousquet Ch, Taureau Ch, et.al. (2005)Immunmodulary effect of human adipose tissue-derived adult stem cells: comparison with bone marrow mesenchymal stem cells. *Br J Haematol,*); 129: 118-129.

Salmons S, Tang AT, Jarvis JC, Degens H, Hastings M, Hooper TL, (1998). Morphological and functional evidence, and clinical importance, of vascular anastomoses in the latissimus dorsi muscle of the sheep. *J Anat*; 193(1):93-104.

Seaberg M; (2004). Clonal identification of multipotent precursors from adult mouse pancreas that generate neural and pancreatic lineages. *Nat biotech* [online March] 1-10

Seeberger KL, Dufour JM, Shapiro AM, Lakey JR, Rajotte RV, Korbutt GS (2006). Expansion of mesenchymal stem cells from human pancreatic ductal epithelium. *Lab Invest*; 86:141-53.

Skorjanc D, Jaschinski F, Heine G, Pette D,(1998). Sequential increases in capillarization and mitochondrial enzymes in low-frequency-stimulated rabbit muscle. *Am J Physiol*; 274:810-818

Stamm C, Westphal B, Kleine HD et al. (2003).Autologous bone marrow stem-cell transplantation for myocardial regeneration. *Lancet*; 361:45–46.

Steinhoff G, Choi Y-H, Stamm C. (2006). Intramyocardial bone marrow stem cell treatment for myocardial regeneration. *Eur Heart J*; Suppl:32-39.

Wang T, Xu Z, Jiang W, Ma A(2005). Cell-to-cell contact induces mesenchymal stem cell to differenciate into cardiomyocyte and smooth muscle cell. *Int J Cardiol*. Aug 22; [Epup ahead of print].

Wollert KC and Drexler H(2005). Clinical applications of stem cells for the heart, *Circ. Res*; 96: 151-163.

Xu C, Police S, Rao N, Carpenter MK(2002). Characterization and enrichment of cardiomyocytes derived from human embryonic stem cells. *Circ Res*; 91:501-8.

Yeh ETH, Zhang S, Wu HD, Körbling M, Willerson JT, Estroy Z(2003). Transdifferentiation of human peripheral blood CD34-enriched cell population into cardiomyocytes, endothelial cells, and smooth muscle cells in vivo. *Circulation*; 108: 2070-75.

Yoon Y., Wecker A., Heyed L., Park JS, Lorsodo DW(2005). Clonally expanded novel multipotent stem cells from human bone marrow regenerate myocardium after myocardial infarction. *J Clin Invest* :115: 326-338.

Zulewski, H(2001). Multipotential nestin-positive stem cells isolated from adult pancreatic islets differentiate ex vivo into pancreatic endocrine, exocrine, and hepatic phenotypes. *Diabetes* 50, 521-533.

Part 3

Biomaterials for Regenerative Medicine

11

Preparation and Characterization Urea-Solubilized Sol-Gel Type I Collagen and Its Possible Use in Applications

Xin Xiong[1,2], Herwig Brunner[1,2] and Robin Ghosh[3]
[1]*Institute of Interfacial Engineering, University of Stuttgart, Stuttgart,*
[2]*Fraunhofer Institute of Interfacial Engineering and Biotechnology, Stuttgart,*
[3]*Department of Bioenergetics, Institute of Biology, University of Stuttgart, Stuttgart*
Germany

1. Introduction

The extracellular matrix (ECM) is composed of a mixture of glue-like macromolecules in the extracellular spaces which give a tissue its characteristic form and maintains its integrity [Exposito, 2002; Gelse, 2003]. The ECM also regulates a variety of cellular behaviors such as proliferation, differentiation, motion and communication [Kadler, et al., 2007; Tsuchiya, et al., 2007]. Collagens are the most abundant proteins in the animal kingdom and have been highly conserved in evolution [Exposito, 2002; Kadler, et al., 2007]. These proteins constitute one third of the human proteome and contribute three-quarters of dry weight of human skin [Kotch & Raines, 2006]. Increasingly, this family of proteins is gaining attention from scientists in different fields, ranging from biochemistry to tissue engineering.

In the last century, collagen has been studied extensively by researchers from different disciplines including chemistry, physics, medicine and biology. Despite of this intensity, the understanding of many collagen properties is still limited and the number of collagen types or collagen-like proteins (CLP) discovered are increasing continuously. Type I collagen is the most abundant protein in the animal body and located in all ECM of connective tissues like bone, skin, tendon and blood vessels. The primary function of Type I collagen is to give mechanical support for the structural organization of the ECM of connective tissues, therefore giving the tissues their characteristic form and maintaining the cellular integrity in the ECM. Possibly there are still many unknown functions to be found. Type I collagen represents the characteristic structure very well and has the characteristic large domains comprised of peptides with varied length of [Gly-X-Y], where X and Y are generally proline (Pro), repeats with a length about 300 nm and folded into unique triple helix structure as its native form [Prockop, & Kivirikko, 1995]. Every chain from the three polypeptide-chains $(\alpha_1)_2(\alpha_2)$ form a left-handed helix then fold together to form a right-handed triple helix.

1.1 Self-assembly of collagen Type I and role of proline hydroxylation in collagen aggregation

After the biosynthesis of collagen, it is post-translationally modified by different enzymes, including some modifications (particularly hydroxylation) which are unique to this class of proteins. Enzyme-mediated hydroxylation introduces a hydroxyl (–OH) group into prolyl or lysyl residues, thereby oxidizing them to 4-hydroxyproline (4-Hyp), 3-hydroxyproline (3-Hyp) and hydroxylysine (Hyl). The function of 4-Hyp is generally accepted to stabilize the collagen triple helix structure, while the function of 3-Hyp is still not clearly known [Jenkins, et al., 2003; Berg & Prockop, 1973; Wick, et al., 1978]. Hydroxyproline can be determined by chemical methods, as described by [Jamall et al., 1981]. Whereas many studies have suggested that the 4-Hyps enhance the stability of collagen strongly, in contrast, 3-Hyp has been suggested to exhibit a *destabilizing* function [Jenkins, et al., 2003].

Since collagen contains very few aliphatic or aromatic residues, it is unlikely that the assembly of the collagen chains is entropy-mediated as a consequence of the hydrophobic association of amino acids. In fact, extensive calorimetric studies by Privalov in the early 1980s (see [Privalov, 1982]) showed that the thermodynamic behavior of Type I collagen melting was exceptional. In particular, the melting transition was exceptionally sharp compared with those observed for other proteins, and the magnitude of the specific heat change for the melting transition was also significantly larger than for most proteins. Privalov [Privalov, 1982] also concluded, from a limited data set, that 4–Hyp *at the third position* of the characteristic amino acid triplet made the largest contribution to Hyp-mediated collagen stabilization. Finally, Privalov concluded that the every large enthalpy change upon collagen denaturation was due to extensive hydrogen bonding, and that the evidence strongly favored one of the original Ramachandran models [Ramachandran et al., 1973], where an additional water bridge can be formed between Hyp at the third position of the triplet and a strongly-bound water molecule. This unusual bonding leads to an extensive hydrogen-bonding network, which strongly stabilizes the structure. From the thermodynamic data, and also the exceptionally high value (determined experimentally) for the strongly bound water, about 7-8 water molecules must be tightly bound per collagen triplet. Brodsky and Persikov [Brodsky, & Persikov, 2005] have also noted that 4-Hyp on the Y position can accept a hydrogen bond from the neighboring Gly, therefore contributing to the conformational stability of the triple helix. Reports also indicate that the Hyp is rarely found to be at the X position thus preventing steric hindrance, which would lead to structural destabilization.

Collagen fibril formation is a self-assembly process that has been investigated *in vitro* for over four decades. Although it is now clear that the major driving force of the fibril building is from the amino acids residing in the sequence, many studies showed other factors could affect or participate in this process [Kadler, et al., 1986; Ioozo, 1998]. Not only the kinetics of fibril formation but also the diameter of the fibril are controlled by these factors. Studies have shown that proteoglycan is involved in control of the shape and size of Type I collagen in echinoderm. The presence of the protein decorin has also been suggested to influence the fibril fusion [Kadler, et al., 1996]. Although different assembly forms were observed, the aggregation of collagen molecules into unbranched fibrils has been conserved in evolution. The fibril formation *in vivo* in vertebrates is either bipolar or unipolar, thus the spatial arrangement of the fibrils determine the mechanical and physical properties of the tissues.

Fibril-forming collagens including Type I-III, V, XI, XXIV, and XXVII are all similar in size and share the same basic sequence in that they contain a large uninterrupted triple-helical (collagenous) region. The collagenous region is flanked by N- and C-terminal non-collagenous domains and both propeptides [Khoshnoodi, et al., 2006]. The propeptides are synthesized first in the cytoplasm as the collagen precursor, procollagen, and then processed to collagen in the extracellular space. At this step, the N- and C-terminal propeptides are cleaved off to yield mature collagen. However, the assembly to triple helical structure begins already in the ER, with collagen single chains forming homo- or heterotrimers, even though the homotrimer of $(\alpha_1)_3$ is the thermodynamically preferred form of Type I collagen assembly. *In vitro* studies suggest that collagen trimerization is controlled by the non-collagenous (NC) regions [Kuznetsova, et al., 2003; Leikina, et al., 2002]. It has been shown that the homotrimer of α_1 is the default form which assembles in absence of the α_2 chain, whereas the homotrimer of α_2 is only formed if the C-terminal non-collagenous domain is replaced with synthetic peptides. Thus the C-NC-domain of α_2 is recognized as a key domain for the heterotrimer formation. Early studies have also shown that the disulfide bridges in the C-NC domain of the α_2 are crucial for heterotrimer assembly and folding [Koivu, 1987]. Energy minimization and molecular dynamic simulations have indicated that the trimerization of both α_1 C-NC domains with the α_2 C-NC domains is a key step in folding, docking and assembly of the three heterotrimer chains [Malone, et al., 2005]. Thus, folding of collagen is initiated from the C-termini by association of α_2 and $(\alpha_1)_2$ and then proceeds sequentially toward the N-termini. Trimerization begins after the interchain disulfide bonds have been formed between α_2 and $(\alpha_1)_2$ C-NC domains.

At the same time, new recognition sites, encoded by the sequence information, allow formation of oligomers and fibril according to the tissue specific requirements [Khoshnoodi, et al., 2006]. These recognition sites determine the enzymatic cleavage sites, connections of collagen molecules, lateral association and interaction with other macromolecules from cells such as integrin and proteoglycans in the ECM. Although the understanding of the mechanisms for the trimerization and self-assembly has been increased significantly in last decades, several features are still open for future studies, e.g. the recognition motif for trimerization, differential regulation of different C-NC domains. These could be very useful for engineering proteins with a superstructure which is more suitable for application as a biomaterial. Furthermore, the understanding of N-NC domain is still limited [Bornstein, 2002]. It is known that the N-NC peptides in procollagen α_1 are involved in numerous processes, including the maturation of collagen and function of Type I collagen. Strangely, however, the deletion of these peptides appears to have no significant phenotype in the mouse. Interestingly, the N-NC domains are cysteine-rich peptides (10 Cys in N-NC of α_1) and are very well-conserved in almost all known procollagen types and all animals. No other amino acids in the N-NC domains are seen to be conserved so strongly as cysteine. However, the function of the disulfide bonds in this region is still an open question [Bornstein, 2002].

Whether purified collagen can refold to triple-helical structure after the denaturation or unfolding is still unclear. Classical results in the last 40 years have indicated that unfolded collagen cannot assemble to form native triple helix. In these studies, although UV-CD-spectroscopy showed a "significant" positive peak around 225 nm, which is characteristic for triple helix, the electron microscopic images revealed that only misaligned helices were present. Some studies have also shown so-called segment-long-spacing aggregations. This conclusion is supported by the above theory that the C-NC domain is a key regulator for triple helix folding. Surprisingly, all studies so far have indicated that collagen at body temperature (37 °C) is only marginally stable [Privalov, 1982]. For instance, the most

modern studies show RTT Type I collagen to be only about 30% in the folded (fibrillar) state at 37 °C [Privalov, 1982]. Possibly, collagens are a subgroup of the class of "unfolded proteins" which maintain a high degree of conformational flexibility at physiological temperatures. It is also reasonable to suppose that the presence of proteoglycans, glycosylation, heat shock proteins (HSP) and Hyps serve to stabilize the collagen structure.
Recent studies have introduced some new results to understand the folding of the collagen which are based upon model peptides. Interestingly, these model peptides are from the collagenous region consisting only of [GPP] repeats. These researchers observed that the Hyp is not essential in stabilization of the collagen model *in vitro* [Mohs, et al., 2007], which is contradictory to the other studies. The question how collagen molecules are stabilized *in vivo* is still unresolved.

1.2 Experimental studies using synthetic peptides

There are two major methods to study collagen: using isolated natural collagen [Kadler, et al., 1988] and synthetic model peptide/protein [Kotch & Raines, 2006; Malone, et al., 2005]. Due to its high complexity and high molecular weight, most structural studies have relied on model peptides. Unfortunately, most model peptides studies have focused on the trimer repeats of [Gly-X-Y] and have been mostly limited to $[\text{Gly-Pro-Hyp}]_{10}$ [Mohs, et al., 2002], which is only a short peptide much shorter than natural collagen (≈300 nm). Although these model peptides yield much information about the dynamic process of the assembly and triple helix building, the information is just part of that necessary for the understanding of real collagen. Studies have shown that polymerization of these model peptides yield the triple helix form but with a high polydispersity [Kotch & Raines, 2006; Paramonov, et al., 2005]. Furthermore, it must be taken into account, the α_1 and α_2 chains, containing about 1000 amino acids each, show only 43 and 28 [Gly-Pro-Hyp] (rat Type I collagen, SP:02454 and 02466) repeats, respectively. For this reason, an understanding of the structure of the whole collagen protein is still required [Orgel, et al., 2001]. However, this task is always hampered by the low quality of the collagen derived from conventional procedures, including neutral salt and acetic acid-extraction [Becker, & Timpl, 1972; Miller, et al., 1982; Chandrakasan, et al., 1976; Rajan, et al., 2007] (A comparison of the three procedures is shown in Fig. 1).

1.3 Laboratory scale preparation of highly purified natural Type I collagen, using urea-extraction, and its characterization

In our previous study [Xiong, et al., 2009], we reported a new procedure for obtaining highly purified native like Type I collagen from rat tail tendon with a high yield. This urea-extracted collagen (UC) exhibits a very high "solubility" in water, and thus, using current physicochemical terminology, can be considered to be a sol-gel preparation. The collagen sol-gel can be directly employed as matrix for cell culture without any pre-treatment such as neutralization and gelling. Another new finding we reported concerned the differential hydroxylation pattern of the proline residues at the X positions, and the existence of the differential hydroxylation among the population of collagen α chains. Studies of the population structure of differential hydroxylations, may contribute to an understanding of the tissue-specific differences of Type I collagen as well as the different states of the aging process.
Finally, we further characterized the UC. In particular, we improved the ESI-MS/MS technique by means of combinational digestion, resulting in strongly increased sequence coverage from an initial α_1-chain (95.8%)/α_2-chain (88.1%) coverage to 99% coverage for both

α-chains. Moreover, we developed a new method based on UV-spectroscopy for fast determination of the urea collagen concentration facilitating the application of the UC. A special focus was also directed to understand the reversible self-assembly of the UC, which differed strongly from other collagen isolates. We hypothesized, and later proved experimentally, that the specific carbamylation of the lysine residues in the C-terminus of the α_1 chains might play a decisive role.

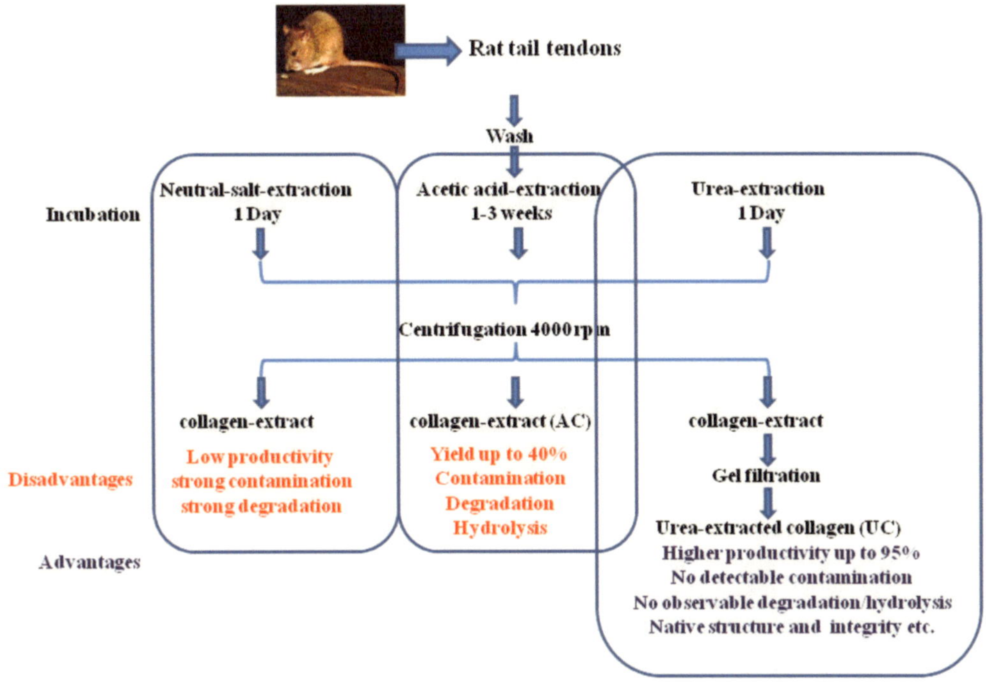

Fig. 1. Brief chart of major differences between the conventional method and newly established urea-extraction

Summary and comparison of the conventional procedures for collagen purification to urea extraction. In all three procedures, the tail tendons were solubilized into respective solutions within the indicated time frames. After centrifugation, the crude extracts are either stored as final products (neutral/acidic extracted collagen) or further purified (urea-extracted collagen). As the neutral salt extract has some known drawbacks, such as low yield and high contamination with other proteins, as well as often showing significant degradation, in our studies, only AC and UC have been directly compared using either commercially available AC or AC produced in our lab. Yield of the UC is calculated from independent experiments from dry weight of the tendons and isolated purified urea extracted collagen. The yield of the acetic acid procedure has been calculated using the wet weight of the tendons and the dried weight of the acetic collagen [Kadler, et al., 1988; Paramonov, et al., 2005; Becker, & Timpl, 1972; Orgel, et al., 2001; Miller & Gay, 1987]

In 2010, Eyre and coworkers published two articles describing a hypothesis based on the MS/MS studies of different types of collagens from a variety of tissues including rat tail tendon, human skin and bone. Thereby, they identified unusual 3-hydroxylations on the first Pro-residues in the C-terminal [GPP] repeats while the second Pro-residues are conventionally 4-hydroxylated [Eyre, et al., 2011; Weis, et al., 2010]. Further, they speculated that the 3-Hyp is a tissue-specific feature and can be very important for defining the structure and function as well as assembly status of the Type I collagen in different tissues. This hypothesis is consistent with observations of our previous studies, where we demonstrated an unusual differential hydroxylation pattern in the RTT-Type I collagen. This aspect has not been studied previously in great detail, largely due to the severe limitations of the classical procedures for protein sequence analysis. The new findings are based on more modern and highly discerning mass spectrometric methods, in combination with our newly established UC preparation method. We believe that the combination of these two factors opens the door for a new understanding of this class of protein.

1.4 Goal of this chapter

In the following part, we will show some major results from the urea extraction procedure and the characterization which involved the AFM, SEM, UV-CD-spectroscopy and MS/MS as well as cell culture techniques. We also compare our data to those published recently by other groups.

2. Methods and results

2.1 Comparison of the isolation methods and the yield

Generally, the neutral salt and acetic methods have been established since 1960s. Later many groups modified the methods to improve the quality of the isolated collagens and to reduce the working time [Miller, & Rhodes, 1982; Miller, & Gay, 1987; Bailey, et al., 2011]. Briefly, different tissues such as rat tail tendons, skin from human, porcine and bovine sources are defatted and disinfected to germ free (e.g. with an ethanol wash) before collagen isolation. To increase the solubility of the collagen and avoid potential immunogenicity, the tissue is often treated with pepsin so that the telopeptides can be removed from the collagen molecules. The pre-treated tissues are then incubated in either neutral salt or acetic acid solution for several days to weeks, until a gel-like solution is formed. Finally, the raw extract is clarified by centrifugation to remove the non-soluble tissue residue.

In our newly-established urea-extraction method [Xiong, et al., 2009], we demonstrated that soluble sol-gel collagen can be obtained using a simple two step purification procedure: (1) PBS-washed rat tail tendons are solubilized in 10 M urea solution for up to 16 h at 25 °C (room temperature), and (2) followed by Sepharose 12 gel filtration. The collagen-containing fractions are then dialyzed against water, and the highly purified Type I collagen can be stored in water or concentrated by lyophilization. The most interesting aspect with this method is that, Type I collagen in different formats (solution, sol-gel-like preparation and powder) as well as different concentrations is now readily available. Up to 100 mg/ml UC can be redissolved in water or culture medium as required.

2.2 Alternative digestion of UC for mass spectrometric characterization

In order to increase the sequence coverage of the Type I collagen using the MS/MS technique, we applied an alternative digestion approach whereby the samples were digested separately with CNBr, trypsin and GluC. We also compared the results from the MS/MS analysis with those obtained by amino-acid composition analysis.

Lyophilized UC from gel filtration was dissolved in formic acid containing 30% ethanol (v/v) at a collagen concentration of 3 mM, as determined by the ΔA220 value. For the

cleavage of methionine, cyanogen bromide (CNBr) in a 75-fold molar excess of the His amount (which can be estimated from the ΔA220 (unpublished data)) was added to the solution in a Reacti-Vial (Pierce) and the reaction was performed for 6 h in the dark at room temperature under stirring. The reaction mix was centrifuged at 16,000 xg for 10 min and then the supernatant was transferred to an Eppendorf tube for drying in a Speed-Vac (Univapo-100H, Uniequip GmbH, Germany) for 2 h.
About 0.5 mg/ml of protein was dissolved in a solution containing 8 M urea and 50 mM NH_4HCO_3, then reduced and alkylated using standard protocols. The sample was diluted and tryptic digestion was carried out at 37 °C overnight. After digestion the solution was adjusted to a pH of 2–3 with concentrated formic acid.

2.3 UV-spectroscopy and UV-CD-spectroscopy

UV-spectra were obtained using a JASCO V-560 spectrometer with a scanning speed of 200 nm/min and slit width of 2 nm. Generally, 2 mm path-length quartz cuvettes were used for the measurements. The whole measurement was performed at 25 °C using a thermostatted cuvette holder.
Further, for determination of the UC concentration, we developed a rapid UV-spectroscopic measurement. As mature Type I collagen lacks the aromatic residues Trp, Phe and Tyr, but contains large amounts of His, we employed this amino acid as a spectroscopic indicator. In control experiments, 1 mM His (Serva) was dissolved in 10 mM sodium phosphate buffer adjusted to either pH 5.5 (protonated, charged form of His) or pH 7.5, and measured in both the presence and absence of 1 M urea. The same measurement was performed using mature Type I collagen dissolved in the same buffers. A comparison of the molar concentration of UC collagen obtained after a dry weight determination following lyophilization with that determined from the His extinction coefficient obtained above, showed the two measurements to be in close agreement.
To study the refolding of the UC after dialysis, UV-CD spectra were acquired with a JASCO J-715 CD spectrometer (scanning speed of 50 nm/min, 2 mm cuvette, slit width 2 nm, averaging over 9 scans). Spectra were obtained in the range of 320 nm to 190 nm. For the spectroscopic measurements, all buffers were degassed under vacuum prior to use.

2.4 Atomic force microscopy (AFM) of the UC-matrix and comparison to native rat tail tendon

The dialyzed urea extract was characterized by AFM. Images were captured using a CP microscope (Park Scientific Instrument Auto probe microscope CP/Veeco) in the tapping mode by using silicon-etched RTESPA cantilevers (Veeco Nanoprobes), which have a nominal tip radius smaller than 10 nm and a spring constant of 20-80 N/m. The drive frequency was set at 200-400 kHz with integral and proportional gains of 0.5-1.5. The scan rate was 1.0 Hz and the scan size was 2-5 and 20-50 µm for collection of heights. All images were flattened and zoomed off-line with ProScan software.

2.5 Cell culture assays for applications as a biomatrix for tissue engineering

The goal here was to characterize the UC collagen and compare its properties with the more conventional AC. The fibroblast 3T3 cell line was grown in Dulbecco's modified Eagle medium (DMEM) (GIBCO) in tissue culture flask (Greiner) at 37 °C, 5% CO2, until reaching confluence. This DMEM contains 10% fetal calf serum (FCS) (Clonetics), 2 mM L-glutamine,

0.1% gentamicin, 0.45% glucose and 0.1% penicillin-streptomycin. The same amount of UC and AC were dissolved in DMEM (1 mg/ml) and dropped into a 24 well-plate with inserts for expression analysis and into a 6-well plate with cover slips for phenotypical studies by SEM. The collagen solutions were incubated under a sterile bench for 2 h to allow gelling. Then, 1 x 10^4 and 2.5 x 10^4 cells in 100 µl DMEM were dropped onto the collagen surface in the inserts. For the collagen-coated cover slips only 5000 cells were employed. For both studies, the cells were cultivated for 14 days with changes of medium every 2 days. Cell vitality was judged by the live/dead cytotoxicity test. The samples were washed twice in phosphate-buffered saline (PBS) and stained for 60 min with 200 µl calcein/ethidium homodimer-1 solution. Subsequently, the samples were washed in PBS and fixed in 4% glutaraldehyde. To study the difference in cell behavior on different collagens, specific genes were chosen to be compared at a transcriptional level using real-time polymerase chain reactions (RT-PCR). Three culture conditions were compared cells growing on a plastic surface used as reference culture. Cells grew on UC and AC were compared to the reference culture.

3. Results and discussion

3.1 Comparison of the preparation methods for Type I collagen

As mentioned above and shown in Fig. 1, in the conventional isolation procedures, there are several drawbacks that cannot be ignored: (1) the contamination from other tissue proteins such as albumin and other type of collagens; (2) uncontrolled degradation of the collagens during pepsin treatment and the extraction, as the tissue own enzyme cannot be totally removed by washing the tissue prior to the extraction; (3) the acid treatment also causes some partial hydrolysis during the extraction, which is largely uncontrollable (Fig. 2); (4) the longer the tissue is incubated in the acid solution and the more collagen is solubilized, the higher the viscosity of the solution, which reduces the further solubilization of the remaining tissue collagen and complicates further purification steps e.g. chromatography. The last point creates severe limitations for up-scaling of the procedure for industrial high-throughput production; (5) the isolated collagen is normally kept in acidic solution which limits the shelf life and handicaps the delivery and application in the laboratory or hospital. To obviate these difficulties, some sources provide freeze-dried collagen products. However, as mentioned above, the collagen powder cannot be redissolved rapidly, and in the normal case, only part of the dried powder is soluble. Several other problems with the AC, such as the pH adjustment prior to usage in cell culture with medium and buffers, which is also batch-dependent, makes the application of such collagen very time- and labor-consuming. most importantly, this reduced the reproducibility of the collagen based experiments as indeed, the collagens produced from the neutral salt or acidic procedure are mixture from differentially hydrolyzed/degraded collagen peptides and intact collagen [Miller, et al., 1982; Chandrakasan, et al., 1976; Bailey, et al., 2011; Bann, et al., 2003; Mizuno, et al., 2004; Rajan, et al., 2006]. In Fig. 2, the isolated UC was analyzed by a 4% SDS-PAGE gel and the characteristic bands for α_1 and α_2 polypeptides are indicated with arrows. During the incubation in 9 M urea several samples were taken after 1 h, 3 h, 5 h, and 24 h respectively to test for degradation. As shown in Fig. 2A, the longer the incubation the more collagen could be dissolved in the 9 M urea. The amounts of the high molecular weight fractions were increased during the incubation. A smear across by the high molecular weight fractions was observed in all samples (Fig. 2A). Even though tendons were incubated for 2 weeks in 9 M urea solution, no degradation was observed. There was no difference in the SDS-PAGE if the extracted collagen was treated with DTT prior to loading, indicating that the occurrence high molecular weight fractions is not due to the formation of disulfide bridges (Fig. 2B). Furthermore, the UC was treated in 0.5 M acetic acid for 7 days. The collagen was degraded continuously, indicating the obvious advantage of UC-process if compared to the acidic procedures with the regard to the integrity of the isolated collagen.

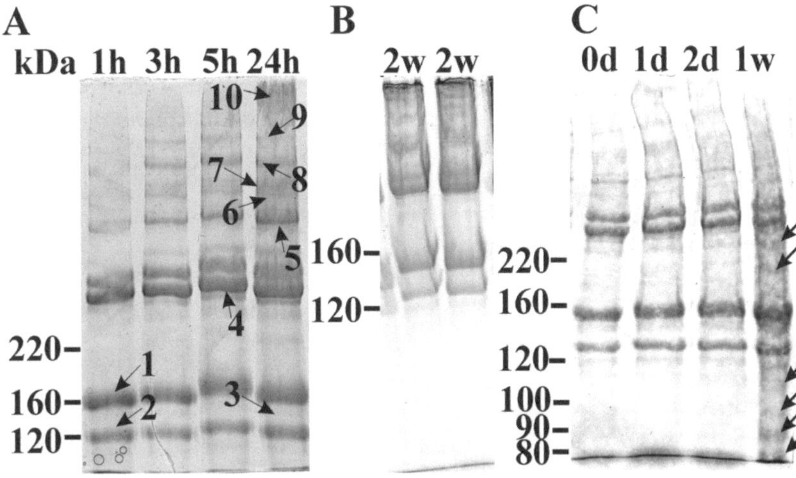

Fig. 2. Urea-extraction and acetic acid partial degradation of UC.
(A) Extraction of collagen from rat tail tendons. Lanes 1 to 4 were the samples taken at 1h, 3h, 5h and 24h during the incubation with 9 M urea. Arrows numbered from 1 to 10 correspond to bands excised for ESI-MS/MS protein identification. The bands 1 and 2 in lane 1 correspond to the α_1 and α_2 chains of the Type I collagen, respectively. All the protein bands indicated by arrows were identified using ESI-MS/MS and these bands contain only Type I collagen. (B) Tendons were incubated in 9 M urea for 2 weeks. Lane 1 shows the untreated sample (no DTT or heating prior to loading). Lane 2 shows the DTT-reduced and heated (at 100 °C for 5 min) sample. (C) The lyophilized extract from urea extraction was incubated in 0.5 M acetic acid. Lane 1 shows the input sample at the beginning, Lane 2 and 3 are the sample taken after one and two days respectively; Lane 4 is sample after 1 week incubation. The arrows indicate the bands which appeared during the incubation

In our lab, we have demonstrated that the yield of UC from the starting tissue is about 93%, significantly higher than the value we obtain for AC (compared to a value of about 60%, see [Xiong, et al., 2009] for a more detailed discussion)

3.2 UV-determination of the UC concentration

Surprisingly, a continuing problem is the need for a fast and accurate method for the determination of collagen protein concentration. Due to the low amount of aromatic amino acid residues present in Type I collagen, widely-used protein assays such as the Lowry-Ciocalteau reaction[Lowry, et al., 1951; Peterson, et al., 1977], or UV-Vis determinations at 280 nm, and the Bradford-type determinations [Bradford, 1976], which rely the binding of a chromophore to hydrophobic binding sites (almost not present in collagen), are largely useless in this context. In addition, for these latter methods, the collagen should be highly purified and strongly diluted to prevent precipitation. This is often achieved by purification using acid extraction, and storing in dilute acid. The acetic acid storage method often limits the accuracy of these protein assays. The most accurate method is dry weight determination, whereby the collagen in solution is directly dried in oven or lyophilized to constant water content then weighed. The disadvantage here is that lyophilization requires long drying times (typically longer than 1 h), which is very inconvenient for routine measurements. Other groups use chemical or immunological reactions, which are based either upon a chemical reaction followed with photometric detection [Shormanov, & Bulatnikov, 2006] or anti-collagen antibodies. In principle, the dry weight methods are the most accurate but time-consuming whereas photometric methods are complicated due to the chemical reaction, which need an optimum condition for detection, and antibody reactions are only accurate when purified monoclonal antibodies are employed. Unfortunately, most commercial antibodies against Type I collagen show high cross-reactivity (up to 10%) to other fibril collagens such as Type III, so that their use in quantitative immunoblotting is often not accurate and also time- consuming.

Histidine (His) is present in a high quantity in collagen molecules. In the course of recent work, we examined the possibility of using UV-VIS spectrum of His to determine collagen protein concentration. His absorption spectroscopy has rarely, if ever, been used to determine protein concentration, as its extinction is much weaker than those of the aromatic amino acids. However, Type I collagen contains large amounts of His and very low amounts of aromatic amino acids. From a total of 30 His residues/mol (pro-α_1)$_2$(pro-α_2) procollagen (SP: P02454 and P02466, respectively), 13 are located in the "core" collagen sequences (2 in α_1 at position 256 and 1099; 9 His in α_2 at positions 152, 188, 518, 950, 972, 981, 1031, 1044 and 1076). In contrast to core region, 7 His are found in the pro-α_1 propeptides and only 3 His in pro-α_2 propeptides region. Of particular interest was whether the His spectrum could be observed in the presence of 1-4 M urea, which is often employed as a solubilizing agent, and whether the ionization state (His has pK_a 6.0) affects the measurement.

To test our procedure, we employed Type I UC. The final preparation is homogenous and corresponds to the "core" sequences described above, as indicated by mass spectrometry (data not shown). This UC can be dissolved completely in water following urea removal (by dialysis) and lyophilization, and provides an ideal test object for testing the His determination method here. We showed that 1 mM His dissolved in 4 M urea yielded a differential absorbance value at 220 nm (ΔA_{220}) of 0.64. 1.3 mg dried collagen was dissolved in 1 ml buffer containing 4 M urea and from this sample 200 µl was added to 2 ml for spectroscopic determination.

We determined the amount of His using UV-spectroscopy. In the α_1 chain contains 2 His residues, whereas the α_2 chain, contains 9 His residues. Assuming an $(\alpha_1)_2\,\alpha_2$ stoichiometry, these values imply that a single native mature collagen molecule contains 13 His residues. Using the extinction coefficient at 220 nm, we calculated the predicted concentration of weighed 1.3 mg UC sample (4.6 nmol collagen) to be 1.6 mg (5.6 nmol collagen). Although the 20% error margin is still significant, the method is very accurate compared to those mentioned above. This measurement is also in good agreement with the data from amino acid composition analysis and the ESI-MS/MS data, which identified all the His present in UC. This measurement is highly reproducible and can be used even in presence of different salts and urea. Although amounts of phenylalanine are also very high in Type I collagen (26 in α_1, 22 in α_2), these were not visible in a UV-absorption spectrum, indicating that these phenylalanine residues are most likely modified.

We found out that the His absorption at 220 nm is very useful and reliable for this purpose even in presence of urea at high concentrations (up to 8 M). A full description of the method will be reported elsewhere.

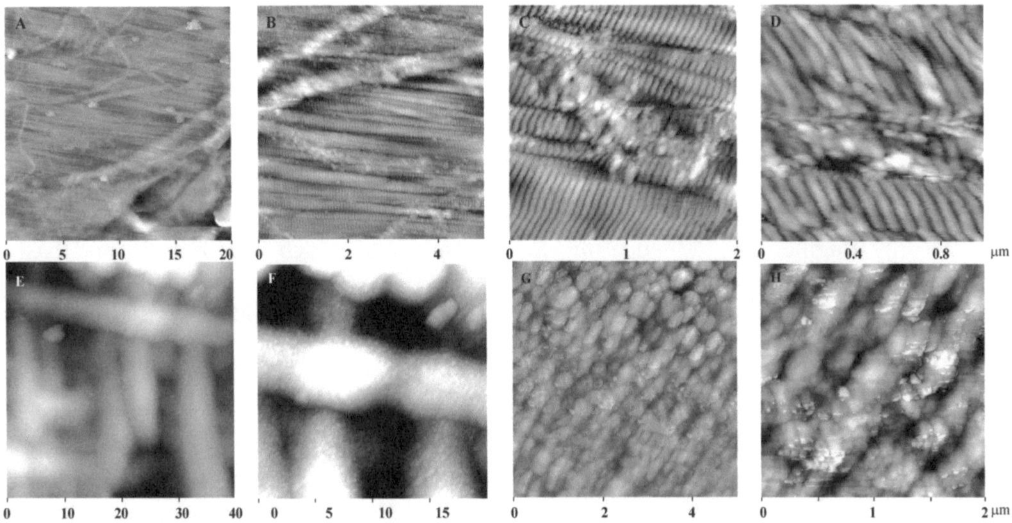

Fig. 3. AFM image comparison of RTT and the UC samples
(A-D) are the images taken directly from the lyophilized RTTs without any further purification steps to show the native fibril structure of the RTT in different scales from 20 to 0.4 μm. (E-H) are the correspondent scaled images from the UC samples after dialysis against water without lyophilization. The sol-gel-like UC, showed significant random fibril formation at this stage, which could be seen to be highly similar to native RTT fibrils shown in (E). However, the fibril has another dimension (large diameter) if compared to native RTT. Additionally, the sol-gel UC has a high water content and no fine D-periods structure

3.3 Discovery of differential hydroxylations of proline residues and its application as identifying feature for collagen

The differential hydroxylation of collagen is a topical subject, which until only recently, has been very difficult to address experimentally. An illuminating example of the use of this phenomenon is illustrated by recent work by Mary Schweitzer and colleagues, reported recently in two seminal Science papers, who have demonstrated that soft tissue, purported to be collagen, can be isolated from fossils which are between 68-80 million years old [Schweitzer, et al., 2007; Schweitzer, et al., 2009]. In both works, Schweitzer and co-workers were able to extract and sequence several small polypeptides from the fossil material, some of which was assigned to arise from collagen. If correct, the Schweitzer work would represent a huge breakthrough in molecular paleontology. However, the first paper, where only 4 small collagen fragments could be isolated from a 68 million year-old *Tyranosaurus rex* fossil, was heavily criticized. In particular, the mass spectrometric sequence determination was only slightly above the noise level, and at least one of the sequences may have arisen from bacterial contamination (see [Schweitzer, et al., 2009] for a detailed discussion). In the latest paper by Schweitzer and collaborators, however, 8 collagen fragments (six arising from the Type I collagen α_1 chain, and two arising from the Type I collagen α_2 chain), extracted from soft tissue arising from an 80 million year-old hadrosaur (*Brachylophosaurus canadiensis*) fossil, could be obtained and sequenced with a high level of precision. In this work, significant effort was taken to minimize contamination problems, redressing many of the criticisms of the *T. rex* work. Furthermore, in an impressive technical *tour de force*, Schweitzer and colleagues were able to determine the hydroxylation profile of the isolated polypeptides, confirming the assignments by comparison to synthetic hydroxylated polypeptides.

However, an unsatisfying aspect of the *B. canadiensis* work is the exclusive reliance upon synthetic polypeptides. The confirmation of hydroxylated amino acids at the same homologous positions of a modern-day native collagen would support the *B.canadiesis* assignments significantly. So far, this has been difficult to perform, since the databases contain only fragmentary sequence information, much of which is rather old and based upon the contiguous assembly of CNBr fragments sequenced by Edman analysis. Usually, Edman analysis does not provide easy and unambiguous access to peptide hydroxylation data.

Recently, we have published the first (to our knowledge) almost complete tandem mass spectrometric (MS/MS) analysis of the hydroxylation profile of a native (rat) collagen I molecule [Xiong, et al., 2009]. Our present (improved since [Xiong, et al., 2009]) coverage for the α_1 and α_2 chains is 99.2% and 89.8%, respectively (Fig. 1a, for methods see [Xiong, et al., 2009]). Given the high precision of our data, we thought it might be interesting to compare it to the hydroxylation assignments of the *B. canadiensis* Type I collagen polypeptides. A BLASTP analysis of the *B. canadiensis* collagen Type I-derived polypeptides against those of the full-length rat α_1 and α_2 polypeptides yielded a single unambiguous homology in all cases (Fig. 4b). For five of the six *B. canadiensis* α_1-derived polypeptides the hydroxylation assignment is essentially identical to that found for rat tail-derived α_1 polypeptide, as determined by MS/MS (Fig. 4b). For the sixth polypeptide the hydroxylated proline residue in the rat sequence has been replaced by an isoleucine in the *B. canadiensis* sequence (Fig. 4b). Interestingly, and in stark contrast to the α_1 polypeptides, one of the two α_2-derived polypeptides from *B. canadiensis* does not show the same hydroxylation pattern as in the rat α_2 sequence, since the hydroxylable proline in the rat sequence has been replaced by alanine in the *B. canadiensis* sequence (Fig. 4b, lower). Nevertheless, for this *B. canadiensis* α_2 sequence, the distal hydroxylation sites neighbour those found for the rat α_2 sequence. For the second α_2 sequence, our partial sequence shows a hydroxylation difference at the third position. Although the rat distal sequences were not detectable, the hydroxylable proline in *B. canadiensis* is replaced by isoleucine in rat α_2. In summary, the hydroxylation assignments of the *B. canadiensis* are consistent with their experimental observation in a modern-day species (rat), but also show significant differences consistent with a long evolutionary divergence.

Mass spectrometry is now becoming a standard method in many biological and medical institutions. The results presented here, showed that the MS/MS based characterization of collagen, including mass determination and identification, can be easily applied as a very accurate procedure for quality control in the production and application of collagens. The method is of high accuracy and yields results in a very short time frame if compared to conventional methods. Moreover, for study of the collagens, in particular, for the same type of collagen in different tissues and different aging status, the MS/MS technique shows a unique advantage in being able to distinguish the above mentioned PTMs unambiguously.

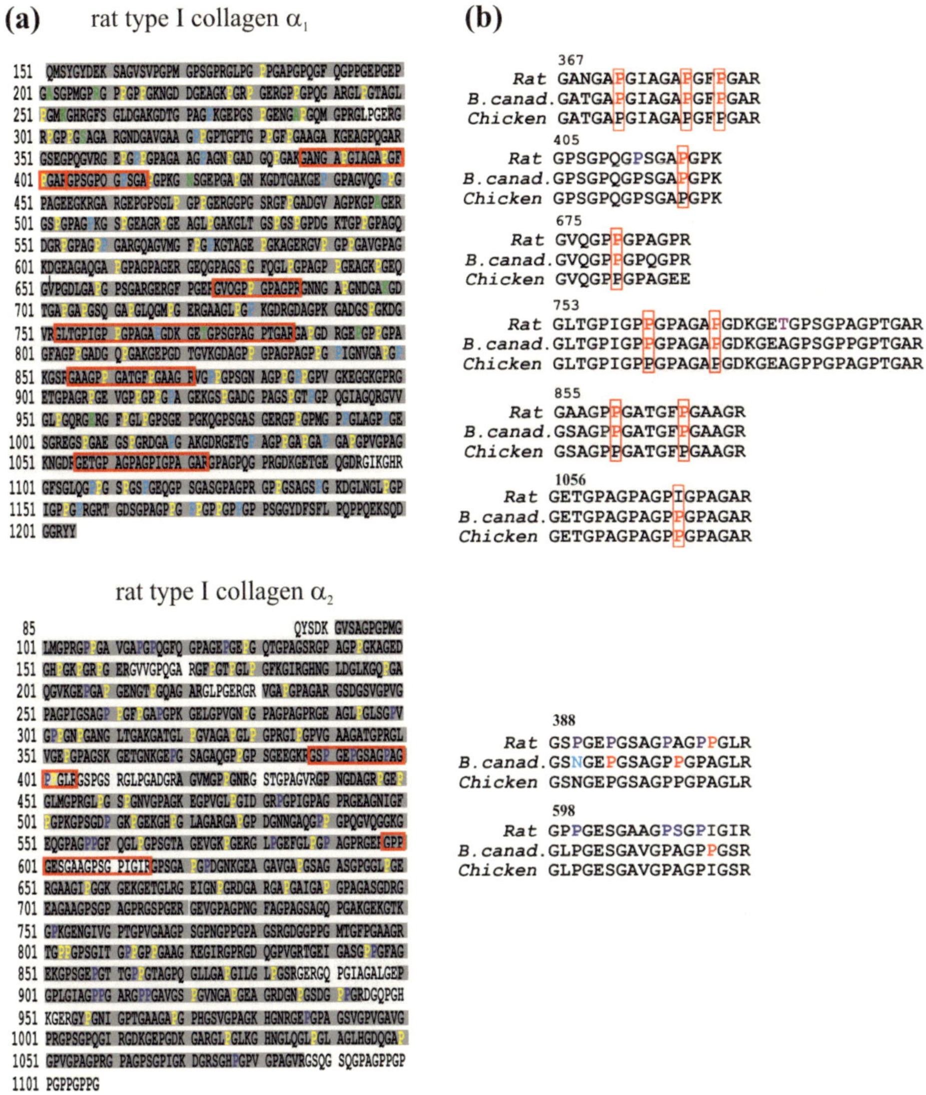

Fig. 4. MS/MS Sequence coverage with emphasis on the positional hydroxylation of proline as identifying feature for mammalian Type I collagen, as compared with published data from [Schweitzer, M.H., et al., 2007 & 2009]

(a) The amino acid sequences of the full-length mature Type I collagen α_1 (top) and α_2 (bottom) polypeptides (SP: P02454 (α_1) and P02466 (α_2), respectively. The gray shading indicates the sequences derived by MS/MS (see (3) for methods), and the positions of the homologous polypeptides from *B. canadiensis* are highlighted as red boxes. Persistently hydroxylated and partially modified proline residues shown by MS/MS are highlighted in yellow and blue, respectively. Residues differing from the SwissProt sequences were highlighted in green. (b) Detailed comparison of the homologous sequences from rat, *B. canadiensis* and chicken α_1 (top) and α_2 (bottom) polypeptides. The chicken sequence was chosen as this was considered extensively in (2) and because birds are thought to be evolutionarily closer related to dinosaurs than mammals. The red boxes indicate persistently hydroxylated proline residues

3.4 Applications of UC as biomatrix in tissue engineering

Type I collagen is one of the most widely-used biomaterials as a cell culture matrix and as a biomaterials in the clinic. However, as mentioned above, these applications have been hampered by many drawbacks, most of which are due to the production procedure. We demonstrated in our previous work that the UC-cultured 3T3 fibroblasts showed 3D-like cell morphology and all cell assays of gene regulation employed showed that the UC shows a significant benefit to the cultured cells if compared to plastic surface and acetic acid-extracted collagen. Moreover, we introduced the sol-gel concept to define our UC as a sol-gel like materials in high concentration in aqueous solution at neutral pH (water or culture media with pH 7.4). This product makes the long-term storage of collagen as a powder format possible and simplifies the delivery and application, as the powder can be easily redissolved on demand, even at high concentrations, without any detectable precipitation and or other change of properties. A very important property of the sol-gel UC is the avoidance of contraction of matrix during the cell culture time frame, thereby eliminating an important negative effect of the conventional collagen matrix. In the latter case, the collagen must be used in relative low concentrations and pre-gelled to give a solid gel-like structure. (for detailed results see [Xiong, et al., 2009])

In our study we also used the SEM technique to compare possible fibril formation of UC to that of the native structure of RTT. The SEM images showed random formation of large fibrils during/soon after the dialysis in water with a length more than 10 μm and an average diameter of approximately 2.4 μm. These dimensions are much larger than those of natural collagen (<100 nm). However, the arrangement of the large fibrils, examined using AFM, consists of many small fibrils and showed a visual similarity to those of native tissue (Fig. 3). The difference in fibril diameter is possibly attributable to the lack of additional components (bound carbohydrate, glycoproteins or lipid) present in the starting material and subsequently removed by gel filtration [Brodsky, & Eikenberry, 1982; Brodsky, & Persikov, 2005].

4. Outlook

Although collagen has been studied extensively during last decades, the understanding of its structure and biological functions are still limited, in particular due to the low quality of available collagen for highly accurate analysis. Based upon our newly-established urea-extraction procedure, we have not only demonstrated the high potential of UC as a biomatrix for tissue engineering but also showed indicated some new insights into understanding the structure, assembly, and function as well. We first showed a reversible assembly of UC from urea-solubilized α_1 and α_2 polypeptides in urea and water [Xiong, et al., 2009]. We then turned to the interesting question of whether this preparation can be used to design a new stable collagen-like/collagen containing biomatrix for tissue engineering. Throughout this review, we have tried to present some new ideas for the characterization of collagen, using modern highly accurate methods such as mass spectrometry and variety of newly established UV-spectroscopic methods for collagen quality control. We believe that there are still a great deal of information to be obtained concerning the post-translational modification of collagen, and that this is relevant for many commercial applications.

5. Acknowledgment

This work was supported by a fellowship of the Peter und Traudl Engelhorn Foundation to X.X.

6. References

Bailey, J.L., et al., Collagen oligomers modulate physical and biological properties of three-dimensional self-assembled matrices. *Biopolymers*, 2011. 95: 77-93.

Bann, J.G., Bachinger, H.P. and Peyton, D.H., Role of carbohydrate in stabilizing the triple-helix in a model for a deep-sea hydrothermal vent worm collagen. *Biochemistry*, 2003. 42: 4042-4048.

Becker, U. and Timpl, R., Cyanogen bromide peptides of the rabbit collagen a1-chain. *FEBS Lett*, 1972. 27: 85-88.

Berg, R.A. and Prockop, D.J., The thermal transition of a non-hydroxylated form of collagen. Evidence for a role for hydroxyproline in stabilizing the triple-helix of collagen. *Biochem Biophys Res Commun*, 1973. 52: 115-120.

Bornstein, P., The NH(2)-terminal propeptides of fibrillar collagens: highly conserved domains with poorly understood functions. *Matrix Biol*, 2002. 21: 217-226.

Bradford, M.M., A rapid and sensitive method for the quantitation of microgram quantities of protein utilizing the principle of protein-dye binding. *Anal Biochem*, 1976. 72: 248-254.

Brodsky, B. and Eikenberry, E.F., Characterization of fibrous forms of collagen. In *Meth Enzymol*, Academic Press, Paris 1982. 82: 127-174.

Brodsky, B. and Persikov, A.V., Molecular structure of the collagen triple helix. *Adv Protein Chem*, 2005. 70: 301-339.

Chandrakasan, G., Torchia, D.A. and Piez, K.A., Preparation of intact monomeric collagen from rat tail tendon and skin and the structure of the nonhelical ends in solution. *J Biol Chem*, 1976. 251: 6062-6067.

Exposito, J.Y., et al., Evolution of collagens. *Anat Rec*, 2002. 268: 302-316.

Eyre, D.R., et al., A novel 3-hydroxyproline (3HYP)-rich motif marks the triple-helical C-terminus of tendon Type I collagen. *J Biol Chem.*, 2011. 286: 7732-7736.

Gelse, K., E. Poschl, and T. Aigner, Collagens--structure, function, and biosynthesis. *Adv Drug Deliv Rev*, 2003. 55: 1531-1546.

Iozzo, R.V., Matrix proteoglycans: from molecular design to cellular function. *Ann Rev Biochem*, 1998. 67: 605-652.

Jamall, I.S., Finelli, V.N. and Que Hee, S.S., A simple method to determine nanogram levels of 4-hydroxyproline in biological tissues. *Anal Biochem*, 1981. 112: 70-75.

Jenkins, C.L., et al., Effect of 3-hydroxyproline residues on collagen stability. *J Am Chem Soc*, 2003. 125: 6422-6427.

Kadler, K.E., Hojima, Y. and Prockop, D.J., Assembly of Type I collagen fibrils de novo. Between 37 and 41 degrees C the process is limited by micro-unfolding of monomers. *J Biol Chem*, 1988. 263: 10517-10523.

Kadler, K.E., et al., Collagen fibril formation. *Biochem J*, 1996. 316: 1-11.

Kadler, K.E., et al., Collagens at a glance. *J Cell Sci*, 2007. 120: 1955-1958.

Khoshnoodi, J., et al., Molecular recognition in the assembly of collagens: Terminal noncollagenous domains are key recognition modules in the formation of triple helical protomers. *J Biol Chem*, 2006. 281: 38117-38121.

Koivu, J., Identification of disulfide bonds in carboxy-terminal propeptides of human type I procollagen. *FEBS Lett*, 1987. 212: 229-232.

Kotch, F.W. and Raines, R.T., Self-assembly of synthetic collagen triple helices. *Proc Natl Acad Sci USA*, 2006. 103: 3028-3033.

Kuznetsova, N.V., McBride, D.J. and Leikin, S., Changes in thermal stability and microunfolding pattern of collagen helix resulting from the loss of alpha2(I) chain in osteogenesis imperfecta murine. *J Mol Biol*, 2003. 331: 191-200.

Leikina, E., et al., Type I collagen is thermally unstable at body temperature. *Proc Natl Acad Sci USA*, 2002. 99: 1314-1318.

Lowry, O.H., et al., Protein measurement with the folin phenol reagent. *J Biol Chem*, 1951. 193: 265-275.

Malone, J.P., Alvares, K. and Veis, A., Structure and assembly of the heterotrimeric and homotrimeric C-propeptides of Type I collagen: Significance of the 2(I) chain. *Biochemistry*, 2005. 44: 15269-15279.

Miller EJ, Gay, S & Leon, WC, The collagens: An overview and update. In *Meth Enzymol*, Academic Press, Paris 1982. 82: 3-32.

Miller, E.J. and Rhodes, R.K., Preparation and characterization of the different types of collagen. In *Meth Enzymol*, Academic Press, Paris 1982. 82: 33-64.

Miller, E.J. and Gay, S., The collagens: an overview and update. In *Meth Enzymol*, Academic Press, Paris 1987. 144: 3-41.

Mizuno, K., et al., Hydroxylation-induced stabilization of the collagen triple helix. Acetyl-(glycyl-4(R)-hydroxyprolyl-4(R)-hydroxyprolyl)(10)-NH(2) forms a highly stable triple helix. *J Biol Chem*, 2004. 279: 38072-38078.

Mohs, A., et al., Mechanism of stabilization of a bacterial collagen triple helix in the absence of hydroxyproline. *J Biol Chem*, 2007. 282: 29757-29765.

Orgel, J.P., et al., The in situ supermolecular structure of Type I collagen. *Structure* (Camb), 2001. 9: 1061-1069.

Paramonov, S.E., Gauba, V. and Hartgerink, J.D., Synthesis of collagen-like peptide polymers by native chemical ligation. *Macromolecules*, 2005. 38: 7555-7561.

Peterson, G.L., A simplification of the protein assay method of Lowry et al. which is more generally applicable. *Anal Biochem*, 1977. 83: 346-356.

Privalov, P. L., Stability of proteins. *Adv Prot Chem*, 1982. 35, 1-104.

Prockop, D.J. and Kivirikko, K.I., Collagens: molecular biology, diseases, and potentials for therapy. *Ann Rev Biochem*, 1995. 64: 403-434.

Rajan, N., et al., Preparation of ready-to-use, storable and reconstituted Type I collagen from rat tail tendon for tissue engineering applications. *Nat Protocols*, 2007. 1: 2753-2758.

Ramachandran et al., A hypothesis on the role of hydroxyproline in stabilizing collagen structure. *Biochim Biophys Acta*, 1973 322: 166-171.

Schweitzer, M.H., et al., Analyses of soft tissue from Tyrannosaurus rex suggest the presence of protein. *Science*, 2007. 316: 277-280.

Schweitzer, M.H., et al., Biomolecular characterization and protein sequences of the Campanian hadrosaur *B. canadensis*. *Science*, 2009. 324: 626-631.

Shormanov, V.K. and Bulatnikov, G.G., Photometric determination of collagen. *J Anal Chem*, 2006. 61: 320-324.

Tsuchiya, S., et al., Collagen type I matrix affects molecular and cellular behavior of purified porcine dental follicle cells. *Cell Tissue Res*, 2008. 331: 447-459.

Weis, M.A., et al., Location of 3-hydroxyproline residues in collagen types I, II, III, and V/XI implies a role in fibril supramolecular assembly. *J Biol Chem*, 2010. 285: 2580-2590.

Wick, G., et al., The diagnostic application of specific antiprocollagen sera. II. Analysis of liver biopsies. *Int Arch Allergy Appl Immunol*, 1978. 56: 316-324.

Xiong, X., et al., A new procedure for rapid, high yield purification of Type I collagen for tissue engineering. *Process Biochem*, 2009. 44: 1200-1212.

12

Direct Use of Resorbable Collagen-Based Beads for Cell Delivery in Tissue Engineering and Cell Therapy Applications

Veronica Glattauer[1], Wei-Bor Tsai[2], Jacinta F. White[1], Julie Nigro[1], Tracy A. Tebb[1], Jerome A. Werkmeister[1] and John A.M. Ramshaw[1]
[1]*CSIRO Materials Science and Engineering, Bayview Avenue, Clayton,*
[2]*Department of Chemical Engineering, National Taiwan University, Taipei*
[1]*Australia*
[2]*Taiwan*

1. Introduction

Tissue engineering and cell therapy represent significant emerging technologies in medicine. Tissue engineering is, "the persuasion of the body to heal itself, through the delivery to the appropriate sites of molecular signals, cells and supporting structures" (Williams, 1999). Various approaches have been described which are labelled tissue engineering, which form a continuum from complex regeneration of scaffold-based tissues to more simple cell-based therapies, where these supporting structures are usually absent or are present in the form of gels simply to aid cell delivery or encapsulation. In tissue engineering the implant to be delivered may frequently be a near mature functional product or a cell-seeded tissue scaffold, both derived in a bioreactor (Haycock, 2011). For cell-based therapies, a bioreactor is not necessary but may be used to amplify the cell numbers for the therapy. Indeed, a key element for both approaches is that they normally require significant expansion of the replacement cells, which have typically been derived in low numbers from small biopsy samples (Brittberg et al., 1994; Chiang et al., 2005; Thissen et al., 2005). A consequence of this expansion is that there can be a significant loss in the quality of cells available for the procedure.

In many cases, monolayer cell culture has been the method of choice for cell expansion (Brittberg, 1999; Brittberg et al., 1994; Henderson et al., 2007). However, while this method is well established and simple it has problems. One of these is that certain cells may de-differentiate and lose their phenotype and biochemical characteristics (Arterburn et al., 1995; Mallein-Gerin et al., 1991). For example, in extended culture, chondrocytes lose their rounded morphology (Fig. 1A) and develop an extended, fibroblast like appearance (Fig. 1B). Also, accompanying this morphological change, matrix biosynthesis switches from type II collagen to type I collagen (Aulthouse et al., 1989; Lefebvre et al., 1990; von der Mark et al., 1977). It is clear that the collagen produced changes from type II collagen (Fig. 2A), which is the characteristic collagen of articular, hyaline cartilage (Mayne, 1989; Miller & Matukas, 1969) to type I collagen (Fig. 2B), which is a mechanically-inferior fibrocartilage (Chiang et al., 2005; Wegener et al., 2009). Type I collagen is the principal collagen of many tissues, including skin, bone, ligament and tendons. It is also present in poorly functioning fibrocartilage which can be observed in some arthritic conditions (Aulthouse et al., 1989; Lefebvre et al., 1990; von der Mark et al., 1977). Depending on the cell type, variations in media or other conditions can potentially be used to extend the control on phenotype and biochemical characteristics. For example, lowering of the amount of oxygen has been suggested for control of chondrocyte phenotype in monolayer cultures (Malda et al., 2003), but the positive effect may be limited, possibly to around 10 days (Malda et al., 2003, Oesser & Seifert, 2003), which is less time than would typically be needed for cell expansion for cell therapy applications (Brittberg et al., 1994). Stem cells, especially embryonic stem cells have particular difficulties in maintaining stemness in culture and current trends are aimed at developing specialised media or scaffold-based bioreactors to expand these types of cells.

Chondrocytes have been a cell type of particular interest as they are the cell used in a major cell therapy application, autologous chondrocyte implantation (ACI). The development of ACI has proved to be a particularly useful innovation for certain types of cartilage defects (Brittberg et al., 1994; Brittberg, 1999), and is now an established cell-based therapy for regeneration of certain types of articular cartilage damage. Defects of articular cartilage in the knee do not have the capacity to self repair (Hippocrates, cited by Hunter, 1742-1743). A range of surgical techniques have been proposed for repair of damage and pain relief. These include debridement, abrasion, mosaicplasty and microfracture (Bedi et al., 2010). A range of grafting options have also recently been developed, including osteochondral autografts and allografts and periosteal grafting (Hunziker, 2001; Shah et al., 2007). In ACI a cartilage biopsy sample is taken from the patient that yields low numbers of chondrocytes, which then must be expanded several fold to give sufficient cells for the regeneration procedure. In ACI, the cell expansion takes place in monolayer culture and may take 21 days or longer (Brittberg et al., 1994; Chiang et al., 2005; Henderson et al., 2005). As a consequence, the majority, if not all, of the implanted cells are likely to have lost their chondrocyte characteristics. This compromises the regeneration of a true hyaline cartilage after implant (Brittberg, 1999; Hunziker, 2001; Temenoff & Mikos, 2000). Several other variants on this chondrocyte-based techniques have subsequently been proposed to give tissue regeneration constructs that are placed and secured in the defect (Bartlett et al., 2005; Marlovits et al., 2005). These methods, including ACI, provide methods for pain relief and extend the time period before a total knee reconstruction is required.

There are clear deficiences in using monolayer culture for cell expansion. An alternative culture system that has been developed is based on spinner culture for cells in suspension (McLimans et al., 1957), which was subsequently adapted to use beads as cell microcarriers (van Wezel, 1967). Microcarriers have have been shown to provide clear advantages for cell expansion, as the rate of cell expansion can be considerably greater, and for chondrocytes, the cell phenotype, including the biochemical characteristics, is substantially, if not fully retained (Frondosa et al., 1996; Malda et al., 2003, 2006). Compared to monolayer culture, the mechanical effects that are present in spinner culture may lead to positive effects on certain cell types (Freeman et al., 1994). A wide range of microcarriers have been developed, predominantly based on synthetic materials, which provide a wide range of surface chemistries, such as negatively and positively charged and capable of further modification to introduce biological functions (Jacobson & Ryan, 1982).

The benefits of using spinner culture for chondrocyte expansion for potential application to ACI were initially demonstrated using non-resorbable beads (Frondoza et al., 1996; Lahiji et al., 2000), and spinner culture was shown to be significantly superior to monolayer culture, with excellent proliferation rates achieved while the cells retained their phenotype (Frondosa et al., 1996). In addition, chondrocytes cultured on a variety of bead types in spinner culture systems have a greater ability to re-differentiate to the differentiated phenotype (Lee et al., 2011; Malda et al., 2003).

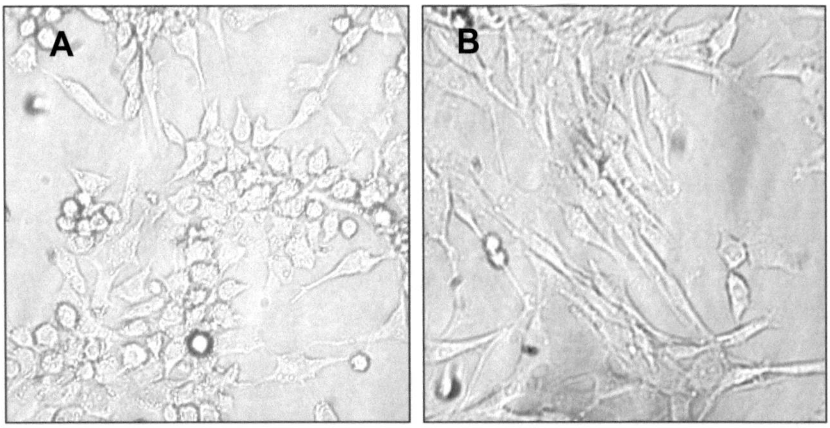

Fig. 1. De-differentiation of chondrocytes in monolayer culture to fibroblast-like phenotype. (A) Cells after 3 days in culture, starting to show some elongated characteristics, (B) Cells after 6 days in culture, showing a predominately elongated, fibroblast-like shape

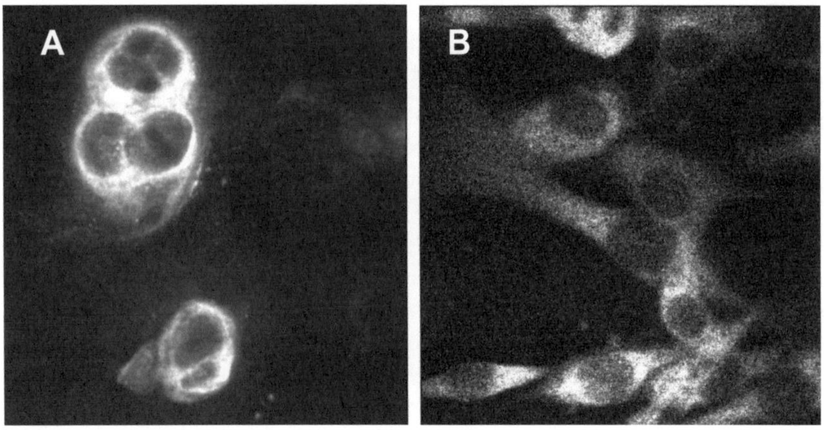

Fig. 2. De-differentiation of chondrocytes in monolayer culture to fibroblast-like phenotype, with collagen production identified by immune-histological staining. (A) Immunostaining for type II collagen production by cells after 3 days in culture. (B) Immunostaining for type I collagen production by cells after 6 days in culture

Various approaches are available for making beads that would potentially be suitable for cell therapy, and several types, for example Cytodex™ beads (Fig. 3A), are commercially available for cell culture. In some cases the beads are biodegradable, for example, resorbable synthetic polymeric (PLGA) beads (Fig. 3B) have been described, including those with surface modifications to enhance cell attachment (Chen et al., 2006; Hong et al., 2008; Thissen et al., 2005).

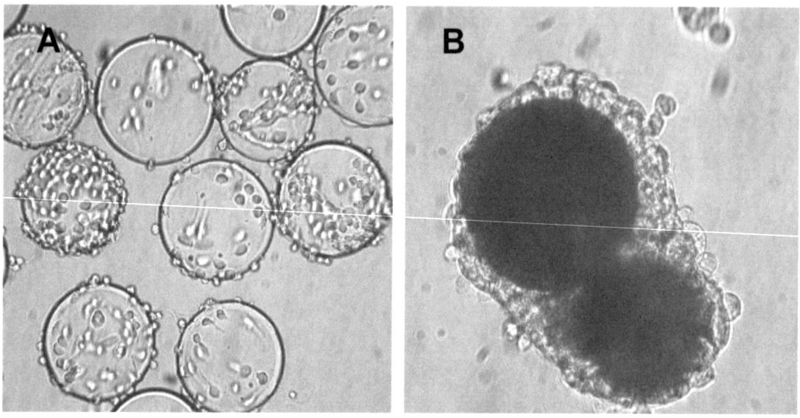

Fig. 3. Chondrocytes in culture on synthetic beads. (A) Culture for 6 days on Cytodex™ beads coated by absorption with type II collagen, showing retention of a round cell morphology. (B) Culture for 12 days on PLGA beads (Thissen et al., 2005) coated by absorption with type II collagen, showing proliferation of cells with rounded morphology

Natural, extracellular matrix tissues can also be used to make beads for spinner culture, and some, for example highly crosslinked gelatin, Cultispher™, are commercially available. Natural tissue-based beads have a range of advantages compared with synthetic beads, including the presence of natural cell binding and growth factor binding sites, a three dimensional environment including variation in surface topography, in vivo resorption using normal biochemical pathways, and when crosslinking is introduced, control of the resorption rate can be achieved (Glattauer et al., 2010). These beads can often be readily processed for cell isolation; for example, mildly crosslinked gelatin beads can be readily dispersed using trypsin or collagenase allowing easy separation of cells by filtration. A wide range of possible beads has been described, along with various approaches for manufacture. Beads can be made of natural tissue components (Glattauer et al., 2010), including those based on powdered acellular dermis (Cymetra™) (Maloney et al., 2004), porous collagen beads that are formed after removal of an alginate carrier (Tebb et al., 2007), or mildly, <0.5% (Glattauer et al., 2010) or heavily, >>1%, cross-linked gelatin (Tao et al., 2003).

If cells are expanded for cell therapy using an in vivo biodegradable, resorbable bead, then there is no need to isolate the cells from the beads prior to administration (Glattauer et al., 2010; Werkmeister et al., 2006). Inclusion of the resorbable beads as an integral part of the therapy procedure provides the advantage of minimizing the extent of cell handling and eliminating a final trypsin treatment to detach cells from the beads. Delivery of a cell/bead construct is also preferably performed using a gel system so that an appropriate distribution of the material is obtained and the cell/bead construct can be readily administered. A gel delivery system could be a synthetic polymer, preferably one that could be cured in situ to prevent migration after delivery of the cell/bead materials, while maintaining sufficient porosity to allow cell respiration and growth (Adhikari et al., 2010; Gunatillake et al., 2006; Werkmeister et al., 2010). Another good material for a delivery system is collagen. When cold solutions of soluble collagen are warmed to body temperature, 37 °C, the soluble collagen forms into fibrils and a gel then forms (Gross and Kirk, 1958). This gel is biocompatible, and has been the basis for soft tissue augmentation technology (Knapp et al., 1978; Kaplan et al., 1983). Hence, with use of cell/bead constructs, where the cells do not require final removal after the expansion phase and can be implanted directly as a cell/bead/gel delivery system, the "curing" of the collagen delivery gel will hold the cell/bead constructs in place. In addition, the potential presence of some extracellular matrix (ECM) that had accumulated during culture (Frondosa et al., 1996; Tebb et al., 2007; Thissen et al., 2005) could assist in retaining the cells at the desired implant location. When resorbable beads are used for cell expansion and then direct delivery of cells, it is possible to design suitable carrier beads with a range of stabilities that matches the implant requirements.

2. Materials and methods

2.1 Preparation of beads

2.1.1 Gelatin beads

Solid gelatin-based beads with cross-linking throughout the bead were produced as previously described (Glattauer et al., 2010). Briefly, 20% w/v porcine A-type gelatin, 270-300 g Bloom (Sigma, St Louis, MO) in 50 mM acetic acid was heated to 50 °C and dispersed at 10% v/v in olive oil at 37 °C by rapid stirring. After 90 min, the emulsion was transferred to 4 °C and stirred for 30 min. Beads were separated from oil by 3 extractions with 0.2% Triton X-100 in phosphate buffered saline, and then cross-linked by addition of selected concentrations of glutaraldehyde (GA) ranging from 0.005% to 0.1%. Cross-linked beads were further extracted with 0.2% Triton X-100 in phosphate buffered saline, water and finally EtOH. Beads were collected by filtration and freeze dried.

2.1.2 Gelatin/collagen beads

For gelatin/collagen beads containing type I or II collagen, collagen was added to cooled, 37 °C, gelatin to give a 10% w/w collagen to gelatin mixture immediately prior to emulsion formation, as for gelatin only beads. Soluble bovine hide type I collagen and soluble bovine articular cartilage type II collagens were prepared using pepsin digestion (see below).

2.1.3 Collagen beads

Collagen beads, based on co-formation with alginate followed by alginate removal were prepared as previously described (Tebb et al., 2007).
Collagen beads can also be formed by an emulsion method, similar to that used to prepare gelatin beads (see above). Type I collagen in phosphate buffered saline, pH 7.3, at 8 mg/ml and 4 °C was dispersed at 10% v/v in olive oil at 12 °C by rapid stirring. After 15 min, the emulsion was heated quickly to 37 °C and stirred very gently. The collagen was allowed to gel for 3 h at 37 °C. The collagen gel beads were then stabilised by addition of 0.1% w/v glutaraldehyde and were crosslinked overnight. Beads were washed, as above for the gelatin beads, collected and freeze dried, followed by ethylene oxide sterilisation prior to use. The typical size range prior to fractionation, eg: by sieving, was 100-200 μm.

2.1.4 Bone based beads

Powdered bovine bone particles were obtained from Waitaki Biosciences (Christchurch, NZ) and were sieved to give a 70 to 150 μm fraction. For production of demineralised bone (DMB) particles, the bovine bone particles were suspended in either 0.6 M HCl, or 0.5 M EDTA at pH 7.4, and stirred for 16 h at 4°C. After settling, solutions were removed by decanting. Particles were then further extracted each day for 6 days using the same conditions, before being washed and freeze dried (Glattauer et al., 2010). A 70 to 150 μm fraction was collected by sieving. The loss of Ca was confirmed as previously described (Glattauer et al., 2010).

2.1.5 Endosteal particles

Endosteal particles from bovine bone marrow were prepared from fresh sternum from young calves, ~4 weeks old (Nigro et al., 2010). Briefly, dissected pieces of sternum were crushed in liquid nitrogen and the frozen bone marrow powder was freeze-dried. To decellularise the dried particles, they were suspended in water and mixed vigorously for 30 min, and then allowed to settle. The water was replaced with 4% ethanol/0.1% peracetic acid and mixed vigorously for 2 h, and then allowed to settle. The solution was then replaced and the particles mixed with phosphate buffered saline for 15 min, and then allowed to settle. This wash was repeated and then the settled particles were treated with a cocktail of DNase I (50 U/ml) and RNase A (1 U/ml) in 10 mM Tris-HCl/2.5 mM $MgCl_2$/0.5 mM $CaCl_2$, pH 7.6 for 24 h at 37 °C, with gentle agitation. The particles were washed with phosphate buffered saline and then water, and then exchanged into 80% v/v ethanol and sieved to provide a fraction between 40 - 230 μm. To remove any residual calcification, particles were treated with 0.6 N HCl for 16 h at 4 °C, washed three times in sterile water and then stored in 80% (v/v) ethanol.

2.1.6 Basement membrane particles

Fresh bovine testis was obtained from an abattoir. After removal of the external membranes, the tissue comprising the seminiferous tubules was sliced into approximately 5 x 5 x 5 mm pieces and was macerated on a stainless steel wire mesh to break the tissue and release cells with continual irrigation by phosphate buffered saline, including a protease inhibitor cocktail. The tubules were retained on the mesh. Samples were suspended in excess phosphate buffered saline containing protease inhibitor cocktail, and washed and collected by settling 3 times. Washed samples were examined by immunohistology to examine the presence of basement membrane components using antibodies to laminin (Sigma) and collagen type IV (Biodesign, ME) following the methods previously described (Glattauer et al., 2007). Basement membrane samples were fragmented further using Ultra Turrax blender (IKA Werke, Germany). The resulting suspension was examined by microscopy to determine the particle size that was present. The particles were observed to have a maximal dimension of 500 μm. The basement membrane particles can be fractionated if required to give particles of the desired size by use of mesh sieves of appropriate size.
All collagen-based particle types were sieved to collect 70 to 150 μm fractions and were sterilised by ethylene oxide, except the seminiferous tubule-derived particles which were sanitised by soaking in 80% EtOH. Bead size distributions were verified by microscopy and image analysis. Unmodified Cytodex™ beads (Amersham Biosciences, Sweden) were used as a synthetic control bead and were sterilised prior to use by autoclaving.

2.2 Culture of chondrocytes

2.2.1 Isolation of chondrocytes

Ovine chondrocytes were isolated from fresh articular cartilage by digestion of 1 mm^3 pieces with 10 % trypsin (2 ml/g) (MP Biomedicals), for 1 h at 37 °C, followed by 300 units/ml bacterial collagenase (354 U/ml *Clostridium histolyticum*, type IAS, Sigma, St Louis, MO) and

760 units/ml hyaluronidase (2 ml/ g) (760 U/ml bovine testicular type IV-S, Sigma, St Louis, MO), for 16 h at 37 °C. Cells were collected through a 70 µm filter, and washed and counted prior to seeding. Human articular chondrocytes were obtained from Edward Keller (Cambrex BioScience, Mount Waverley, Australia), (EK 23-7-02, KN8823) and maintained in complete chondrocyte medium (F12:DMEM, 1:1, containing 10% FCS, plus 1 mM glutamine and 10 mM 2-mercaptoethanol); comparable results were obtained with both the ovine and human cells.

2.2.2 Chondrocyte seeding and culture on beads

Cells were seeded onto the various microcarrier beads and cultured in spinner bottles, with an initial density of 5 x 10^5 cells/150 mg particles in 50ml DMEM/F12/10% FBS containing 100 µg/ml penicillin and streptomycin, at 37 °C in 5% CO_2. Cell seeding was performed with initial 25 rpm intermittent stirring for 2 min every 30 min. After 3 h, the stirring speed was increased to 45 rpm, intermittently as above, for 1.5 h, followed by continuous stirring at 15 min intervals at 45, 50, 55 and finally 60 rpm. After 24 h in culture, the efficiency of cell attachment onto the beads was monitored.

2.2.3 Mesenchymal stem cell seeding and culture on particles

Mesenchymal stem cells were obtained and seeded as previously described (Nigro et al., 2010). A settled volume (0.45 ml) of particles was exchanged into 50 ml MEM/20% FCS/5 ng/ml FGF-2 before adding 2.5 x 10^5 mesenchymal stem cells in a spinner flask. The flask was stirred at 25 rpm for 2 min every 30 min at 37 °C with 5%CO_2. After 24 h, the spinner flask was set to stir at 25 rpm continuously. After 7 days, a sample of the mesenchymal stem cells/particle construct was taken and cell viability was assessed using Calcein-AM. Images were captured with a fluorescence microscope (Olympus).

2.3 Collagen materials

2.3.1 Isolation and preparation of collagen

Type I bovine collagen was isolated and purified from bovine dermis by pepsin digestion (1 mg/ml) in 50 mM acetic acid at 4 °C as previously described (Ramshaw, 1986). Type I collagen was collected by precipitation by 0.7 M NaCl, and was the further purified by differential salt precipitation at neutral pH (Trelstad et al., 1976). Type II collagen from bovine articular cartilage was isolated in a similar manner. The purified bovine collagens were dialysed exhaustively against 20 mM acetic acid and then freeze dried and stored at -20 °C.

2.3.2 Preparation of collagen gels

To prepare collagen gels, purified collagen was dissolved at 3 mg/ml in 50 mM acetic acid. Once dissolved, the solution was taken to 1 M acetic acid and sealed in a dialysis bag and held against 1 M acetic acid for 24 h. It was then dialysed exhaustively against water followed by 20 mM acetic acid and then taken to about 25 mg/ml by air evaporation at room temperature. Finally, while still maintaining the seal of the dialysis tube, it was taken to 4 °C and dialysed against water and then phosphate buffered saline, prior to removal from the dialysis bag under sterile conditions in a biohazard cabinet.

2.3.3 Preparation of collagen gel and cell-bead composite plugs

Cells were taken after culture on beads in a spinner culture system. Typically, 14 x 10^6 cells on 0.5 ml beads were used. This was to mimic the concentration of cells in natural hyaline cartilage. The beads/cells were incorporated by mixing with 0.4 ml of 20 mg/ml type I collagen gel in phosphate buffered saline containing 0.1 ml of 10x concentration cell culture medium. The mixture was then taken into a syringe and centrifuged briefly at low speed to remove any air bubbles and 200 μl aliquots dispensed into 96 well tissue culture plate, which was then incubated at 37 °C for 1 h to allow collagen fibril formation and collagen gelation to occur (Gross and Kirk, 1958). The cell-bead-collagen plugs were then transferred into 24 well culture plates, covered with cell culture medium, and kept in culture for up to 21 days. Ascorbate was added daily to a final concentration 50 μg/ml.

2.4 Evaluation of tissue samples

2.4.1 Histological evaluation

Samples were processed for conventional histology to examine the extent of bead and particle degradation as well as the extent of cellular infiltration into the collagen/bead (particle) plugs. Stains used were standard Haematoxylin and Eosin and Alcian Blue at pH 2.5 and pH 5.8.

2.4.2 Immunohistological evaluation

Immunohistology staining used specific antibodies against collagen types I, II, VI and other extracellular matrix components. Prior to staining with antibodies, cultured cells on beads were washed in warm phosphate buffered saline, and then pre-fixed in ice cold methanol. A range of antibodies was used to assess the type of matrix: goat anti-collagen type II (Southern Biotechnology, Birmingham, AL), goat anti-collagen type I (Southern Biotechnology, Birmingham, AL), mouse monoclonal anti-collagen type VI (1D8-F8/Col6) (Werkmeister et al., 1993) and mouse monoclonal anti-keratan sulphate (5D4, Seikagaku Corp., Tokyo, Japan). Beads were incubated with antibodies for 1 h, washed and then reacted with corresponding FITC-conjugated secondary antibodies, rabbit anti-goat IgG (Southern Biotechnology, Birmingham, AL) or sheep anti-mouse Ig (Silenus, Melbourne, Australia). As controls in all studies, primary antibodies were omitted and replicate slides were stained with only secondary FITC-labeled antibodies.

3. Results and discussion

3.1 Production of collagen-based beads

We have successfully prepared a wide range of biological, collagen-based beads from a variety of starting tissues. These include beads based solely on gelatin, the denatured form of collagen, and on gelatin mixed with non-denatured collagen (Fig. 4) which are readily formed by an emulsion method. Further stabilisation by crosslinking is then used, as otherwise the gelatin beads would not be stable at the 37 °C required for the subsequent cell culture. In the present case, glutaraldehyde has been used as the crosslinking agent. Variation in the extent of the glutaraldehyde crosslinking allows control of the rate of turnover of the beads (Glattauer et al., 2010). A wide range of other crosslinking agents are also possible (Ramshaw et al., 2000) which have also been used successfully in biomedical materials for clinical applications. The gelatin beads formed regular spheres with a reasonably smooth surface (Fig. 4A), although at high power, using SEM, some texture is present (Glattauer et al., 2000) The gelatin beads containing either 10% w/w collagen type I or type II were again uniform spheres, but in this case both showed some surface texture

(Fig. 4C, Fig. 4E). When the beads were dehydrated in EtOH, the gelatin beads retained their spherical shape (Fig. 4B), while the collagen containing beads became irregular in shape (Fig. 4D, Fig. 4F). However, when these beads were re-hydrated for cell culture they generally returned to their previous spherical shape.

Beads can also be made from collagen alone, without the gelatin component, by various approaches. For example, the emulsion approach used for gelatin (Fig. 4A) can be adapted for use with collagen. In this case the collagen solution at neutral pH must be kept cold during the emulsification stage, so as to prevent fibril formation. Once an emulsion has been formed, it is then warmed to 37 °C which causes the collagen to form fibrils (Gross & Kirk, 1958) and then stable gel particles. Like the gelatin beads, these can then be further stabilised by crosslinking using glutaraldehyde or other reagents (Fig. 5).

Alternatively, collagen beads can be formed by the use of a carrier during the bead forming stage, which is subsequently removed. For example (Fig. 6) alginate has been used as an easy and effective carrier system. In this case a mixture of alginate with collagen that contains a high proportion, ~30-40%, of collagen can be formed, and beads are produced by dropwise (aerosol) addition to 1.5% $CaCl_2$ (Tebb et al., 2007). The $CaCl_2$ leads to the precipitation of the alginate to form stable beads, and the collagen can be stabilised by fibril formation and crosslinking as required. The alginate phase can be removed by treatment with Na citrate, giving a collagen only bead (Tebb et al., 2007). An advantage of this method is that the collagen beads are quite porous, and cell infiltration readily occurs (Tebb et al., 2007). Other materials could be used as carriers, including gelatin and other proteins. The collagen triple-helix structure is generally highly stable to proteolytic action, and pepsin treatment is often used during its preparation (Miller & Rhodes, 1982). So a protein carrier, such as gelatin, can be removed by proteolysis, leaving porous collagen beads.

Collagen beads can also be formed directly from tissue. For example, Cymetra™ is prepared from acellular human dermis by fragmentation (Maloney et al., 2004) and is used clinically for tissue augmentation and equivalent procedures. The particles can also be used readily as a cell growth substrate.

An alternate source of collagen beads or particles is from natural collagen containing tissues. The principle protein component of bone is type I collagen. Hence bone particles themselves may present surface collagen to cells to enhance binding (Fig. 7A). Bone particles can be refined using demineralisation, for example with EDTA or HCl, to produce beads enriched of collagen (Glattauer et al., 2010) (Fig. 6B). There are also low levels of growth factors and other biological macromolecules that can add to the effectiveness of the collagen particles in promoting cell adhesion and proliferation. The type and quantity of these molecules may vary depending on the method used to demineralise the tissue and may vary between batches (Wildermann et al., 2007). However, when chondrocytes were grown on particles prepared by both these methods no significant difference was seen in the proliferation rate for the cells (Glattauer et al., 2010). The bone and demineralised bone particles differ significantly from the previous gelatin and collagen beads in that they are irregular in shape, have an irregular surface topography, and do not need additional crosslinking as they are already fully stabilised by the naturally occuring collagen crosslinks. Any soluble collagen would be removed during the demineralisation process. The demineralised bone particles are preferred for spinner culture as they are more bouyant and hence easier to use. Whereas the area of smooth spherical beads in an aliquot, such as synthetic beads or gelatin beads, can be readily calculated from the size distribution of the beads, this is more difficult for irregular shaped beads, and beads with variations in surface topology.

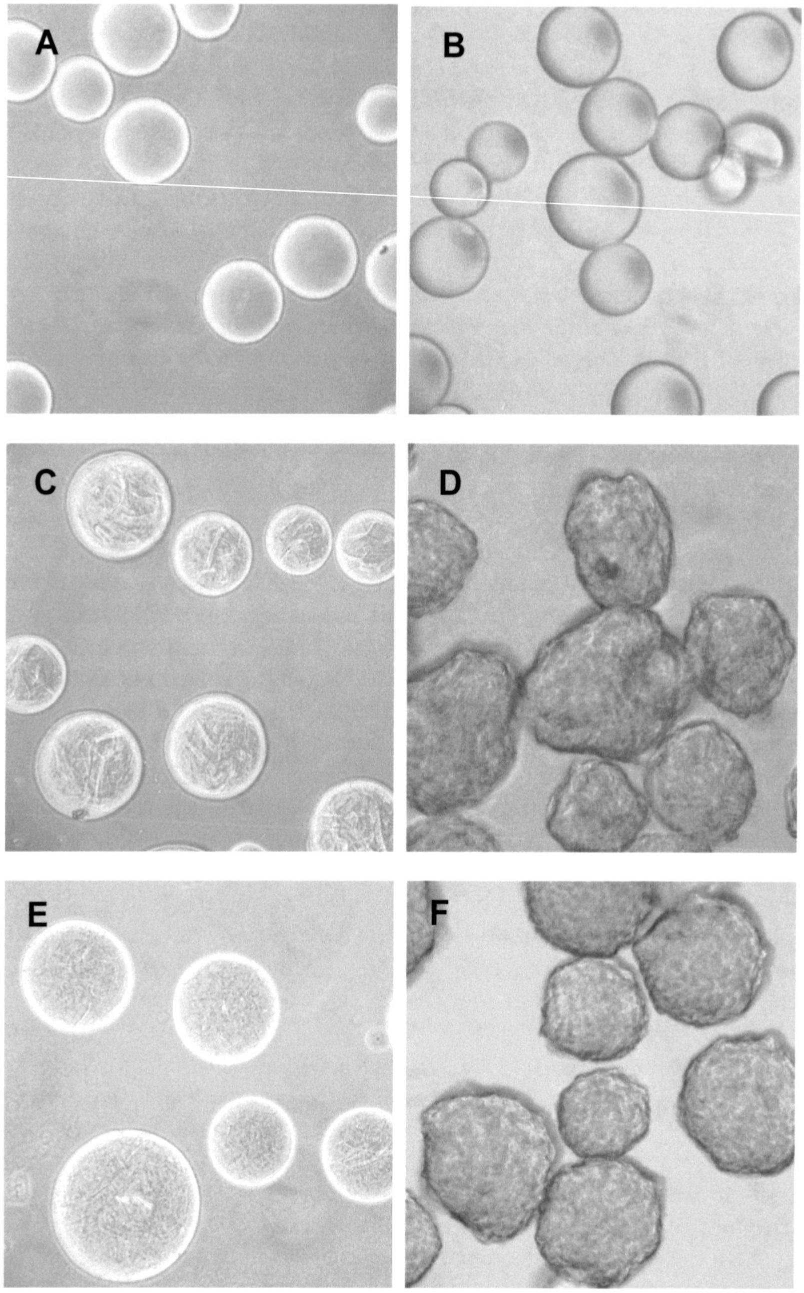

Fig. 4. Production of gelatin and gelatin/collagen beads. Beads in phosphate buffered saline (A, C, E) and after dehydration in EtOH (B, D, F). After rehydration, these beads returned to their original size and shape. (A, B) Gelatin beads, average hydrated bead size prior to fractionation, 106 ± 20μm. (C, D) Gelatin/10% w/w type I collagen beads, average hydrated bead size prior to fractionation, 169 ± 53μm. (E, F) Gelatin/10% w/w type II collagen beads, average hydrated bead size prior to fractionation, 159 ± 53μm

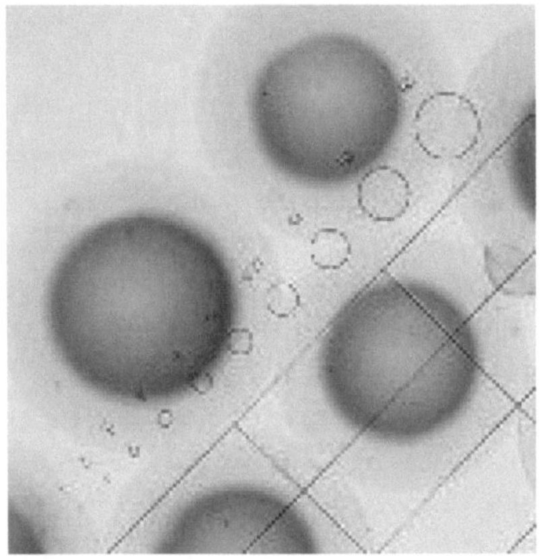

Fig. 5. Collagen beads prepared by an emulsion method

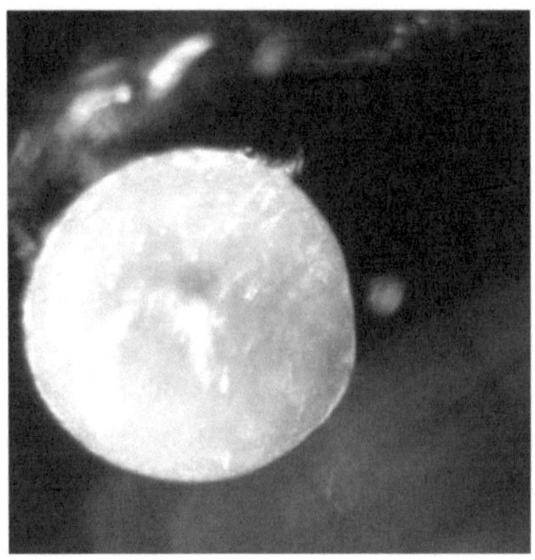

Fig. 6. Production of type II collagen beads using an alginate carrier (Tebb et al., 2007), with type II collagen at 6.6 mg/ ml and alginate at 0.3% during initial bead formation. Bead size ~ 0.7 mm

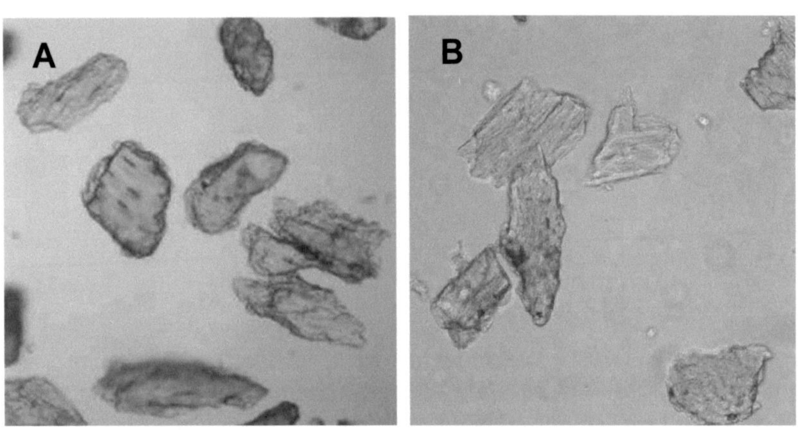

Fig. 7. Examples of (A) bone particles, and (B) demineralised bone particles

The gelatin, collagen and bone particles are all based principally on type I collagen. Other collagens are abundant in different tissues. For example, type II collagen is the collagen found in normal hyaline cartilage (Mayne, 1989; Miller and Rhodes, 1982) and which could be used to make type II collagen particles by tissue fragmentation and further processing, if required, to remove other components such as proteoglycan. Similarly, type III collagen is a major component of blood vessels, along with type I collagen and elastin (Hanson & Bentley, 1983; McCloskey and Cleary, 1974). Hence, particles rich in type III collagen could be prepared by processing arterial tissue. Type IV collagen is the major collagen component of basement membranes. Basement membranes are unique, highly organised supportive sheet-like structures in the extracellular matrix that are formed at the interface between parenchymal cells and their surrounding tissues (Yurchenco et al., 2004). There are different types of basement membranes in the body. Some act as a tissue boundary where certain cells can attach; some can act as filters with selective permeability, and some support very selective cellular differentiation of a number of different cell types, including stem cells. The basement membrane comprises a range of specific macromolecular components. As well as type IV collagen, which may be important for mechanical properties (Yurchenco et al., 2004), the other components are laminin, nidogen and heparan sulphate proteoglycans, such as perlecan, which are believed to interact in a highly ordered manner with type IV collagen (Yurchenco et al., 2004).

Thus, we have shown that by using natural tissues from different origins, it is also possible to make other tissue derived particles. By having different protein and other macromolecules present, these have the potential to provide different interactions with various cell types. An example is the production of particles from the endosteum of bone (Fig. 8) (Nigro et al., 2010). This tissue is proximal to the location of haematopoetic stem cells (Haylock et al., 2007) and so may be useful in the proliferation of this cell type. The particle composition is distinct; examination of the composition using immunohistology showed that they contain perlecan and some laminin, suggesting similarity to a basement membrane, but collagen type IV was not detected (Nigro et al., 2010).

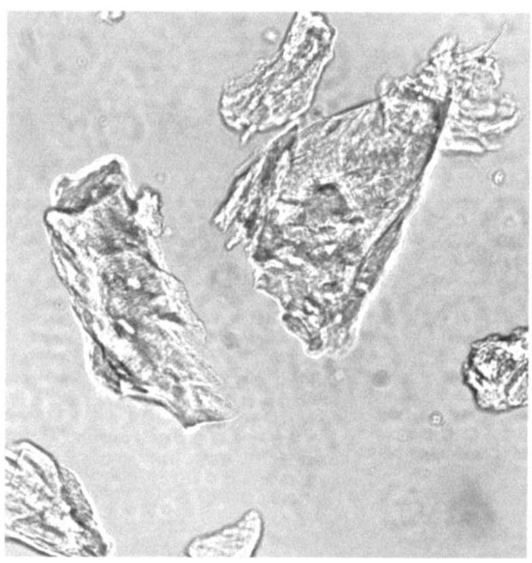

Fig. 8. Decalcified endosteal particles

Basement membrane particles can also be formed from a range of other tissues, including placenta, muscle and kidney (Werkmeister et al., 2006). In the present example, particles have been prepared from seminiferous tubules. The tubules are readily isolated from testis tissue (Fig. 9A), and immunohistological analysis has shown that the major components of collagen IV (data not shown) and laminin (Fig. 9B) have been retained during the processing. Further fragmentation can be used to prepare particles (Fig. 9C). The composition of this basement membrane is distinct (Glattauer et al., 2007) showing the presence of various laminin types based on the positive staining for the α3, α4, α5 and α6 chain of collagen type IV and the positive staining of various laminin chains, including the α2 and β2 chains. This composition is distinct from that of the commercial basement membrane material, Matrigel™, which is based on $(\alpha 1)_2\alpha 2$ type IV collagen and α1β1γ1 laminin (laminin 111)(Kleinman & Martin, 2005). Matrigel™ has been used extensively for addressing cell attachment so a different composition basement membrane may provide a useful alternative option for cell culture.

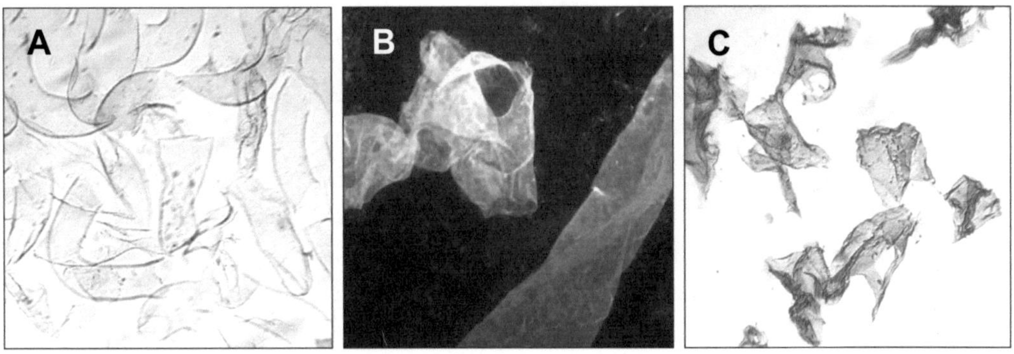

Fig. 9. Seminiferous tubule-based particles. (A) Fragments of isolated seminiferous tubules. (B) Immunostaining of isolated seminiferous tubules, showing the retention of laminin during the processing. (C) Seminiferous tubules particles after further fragmentation of isolated tubules

3.2 Cell compatibility of beads

To be useful, any bead must clearly not be cytotoxic and must allow ready attachment and growth of cells. All the beads and particles described above (Figs. 3-9) meet these criteria. There was no evidence of any cytotoxicity due to residual glutaraldehyde that had been used to crosslink some of the beads.

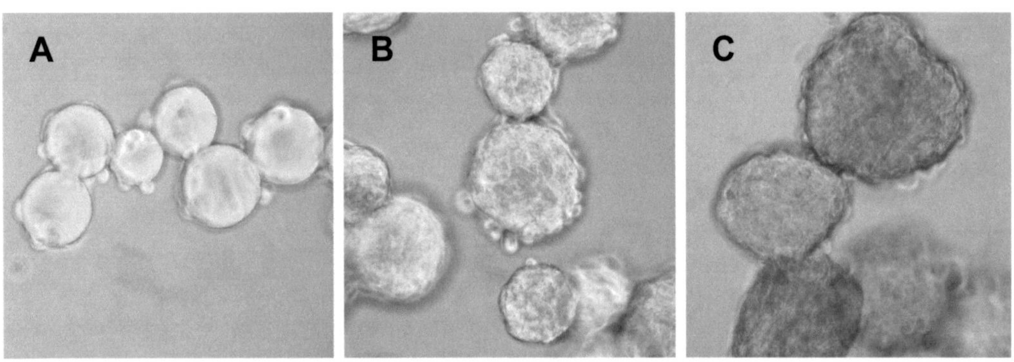

Fig. 10. Gelatin-based beads with chondrocytes after 3 days in spinner culture, showing good attachment and retention of the round chondrogenic phenotype. (A) Gelatin beads, (B) Gelatin/Type I collagen beads, (C) Gelatin/Type II collagen beads

As an example, chondrocytes readily attach to gelatin and gelatin/collagen beads and are viable and starting to proliferate after 3 days in culture (Fig. 10). Similarly, cells readily attach and start to proliferate on bone (Fig. 11A) and demineralised bone particles (Fig. 11B). Attachment and growth of chondrocytes was good on the seminiferous tubule particles, even after only 2 days in culture (Fig. 12).

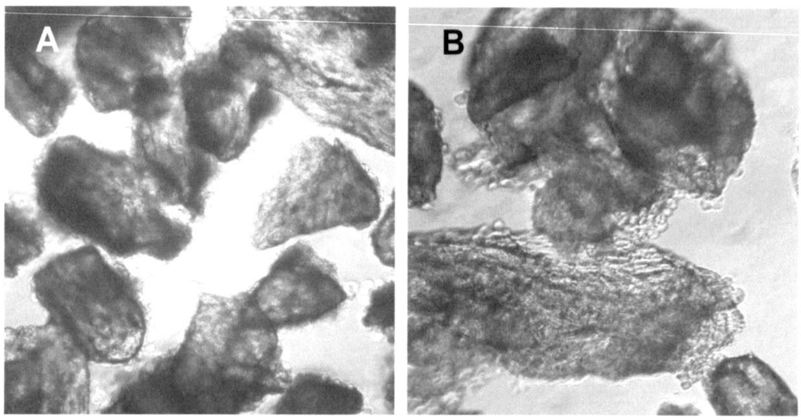

Fig. 11. Chondrocytes growing on bone particles. (A) Sheep chondrocytes on bone particles after 3 days in spinner culture, (B) Human chondrocytes on demineralised bone particles after 4 days in spinner culture

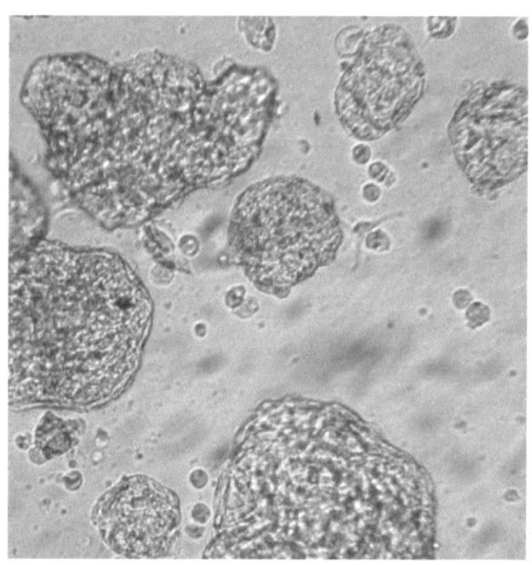

Fig. 12. Human chondrocytes growing on bovine seminiferous tubule particles for 2 days

3.3 Cell growth rate on beads

Previous studies (Frondoza et al., 1996) had suggested that cell proliferation of chondrocytes on synthetic microcarrier beads was superior to that found in monolayer culture. The present work using biologically based beads supports this finding (Fig. 13). Thus, over a 21 day period, with cell collection and re-seeding after 7 and 14 days, the spinner culture using both gelatin and demineralised bone particles was significantly better than that found using monolayer cultures. In all cases, the bead experiments were established so that the surface area of the carriers was equivalent to that used in the monolayer cultures. The cell yield for the gelatin beads was almost one order of magnitude greater than with monolayer cell culture (Fig. 13). The performance of demineralised bone particles was even better, being better than 2 orders of magnitude greater than for monolayer culture. As similar culture conditions were used for both the gelatin and demineralised bone particles, this reflects the differences in surface compositions and the surface topography of the two bead types as well as the potential for the demineralised bone particles to contain some active growth factors that may boost cell proliferation (Wildemann et al., 2007). Comparisons in an additional experiment including other bead systems (Table 1) showed that proliferation rates were consistently better for biological, collagen-based beads when compared to monolayer culture and a synthetic bead.

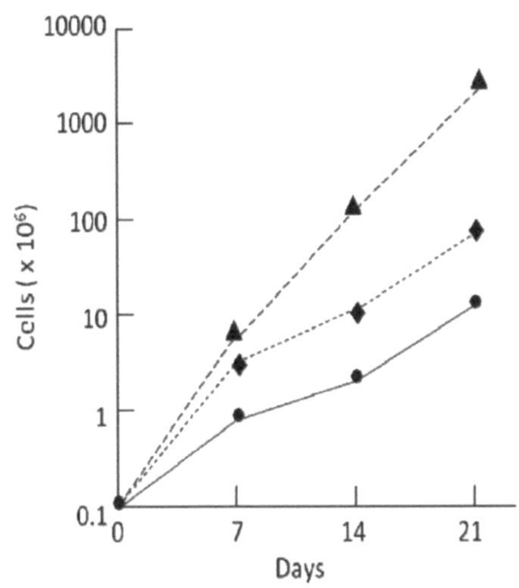

Fig. 13. Proliferation of human chondrocytes in monolayer culture and in spinner culture using different microcarrier systems. The initial loading was 0.25 x 10^6 cells on beads that gave equivalent surface areas of ~125 cm^2 (equivalent to 5 T25 monolayer flasks. • monolayer culture, ♦ spinner culture using gelatin beads, ▲ spinner culture using demineralised bone particles

Culture system	Bead/particle type	Proliferation rate after 21 days
Monolayer	n/a	5.5
Spinner Culture	Cytodex	6.5
Spinner Culture	Gelatin	38.4
Spinner Culture	Bone	37.2
Spinner Culture	Demineralised bone	47.2

Table 1. Comparison of the proliferation rate for human chondrocytes obtained using different culture systems. The proliferation rate was determined relative to the initial number of cells that were seeded

3.4 Extracellular matrix production in culture

In the present study we have focussed principally on chondrocytes. In these cases, with a variety of different beads in spinner culture, newly formed extracellular matrix was found to accumulate around the beads and cells, causing clumps of tissue-like material to form in the cultures. This was clearly evident by 7 days, for example, with gelatin and gelatin/collagen beads (Fig. 14), where a matrix of cells and collagen can be seen holding most beads in clumps.

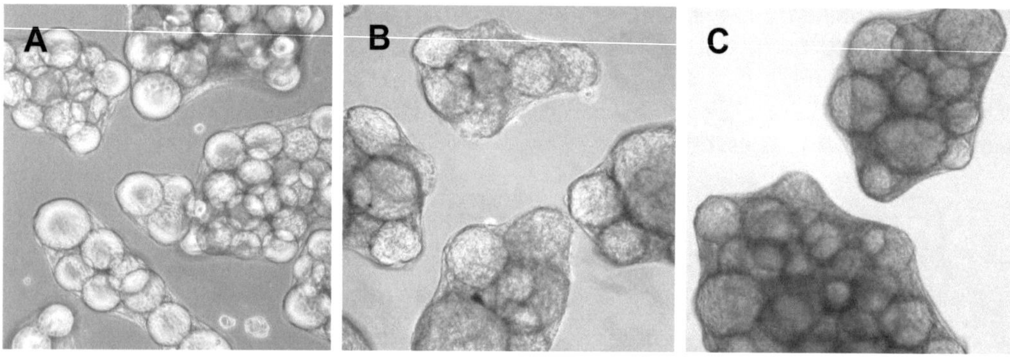

Fig. 14. Growth of chondrocytes on various beads after 7 days in spinner culture, showing the clumping of beads due to the formation of new extracellular matrix. (A) Gelatin beads, (B) Gelatin/Type I collagen beads, (C) Gelatin/Type II collagen beads

Histology, using Haematoxylin and Eosin staining (Fig. 15) shows the presence of this new connective tissue with cells throughout. Not all cells have migrated from the beads into this new matrix and some are still stained on the bead surfaces. A similar result is seen for chondrocytes grown in spinner culture on bone particles (Fig. 15C).

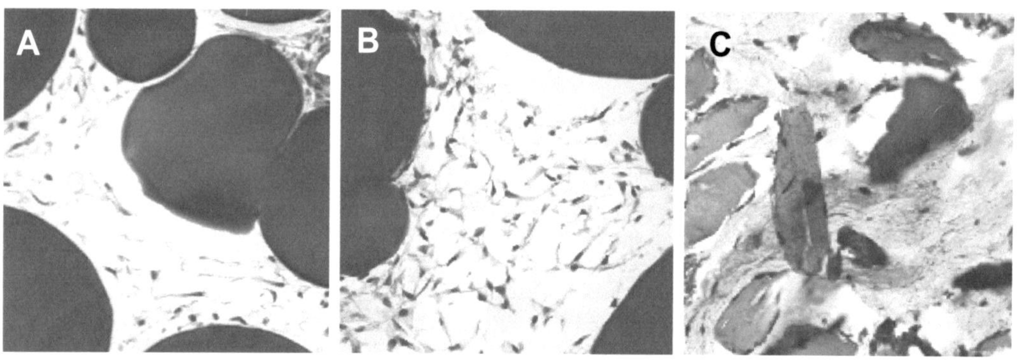

Fig. 15. Haematoxylin and Eosin staining of the growth of chondrocytes on various beads after 7 days in spinner culture, showing the new extracellular matrix that is formed (A) Gelatin beads, (B) Gelatin/Type I collagen beads, (C) Bone particles

With the porous collagen beads, the initial phase of cell proliferation was mainly infiltration of the collagen bead and less accumulation of cells and extracellular matrix external to the particles and hence less clumping was observed than with the gelatin beads (Fig. 16). On the other hand, a small amount of clumping of particles was also observed when human mesenchymal stem cells were grown on decalcified endosteal particles for 7 days. Cell attachment was good and proliferation had commenced, but was not as advanced as seen for chondrocytes on other bead systems (Fig. 17).

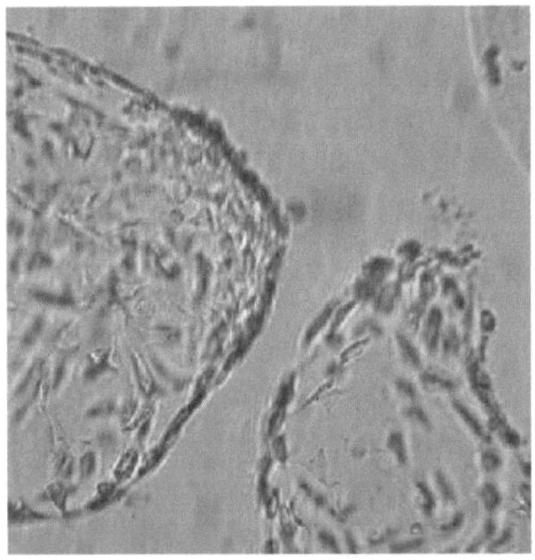

Fig. 16. Growth of chondrocytes for 7 days on porous collagen beads prepared through alginate removal (Tebb et al., 2007), showing infiltration of the beads by cells

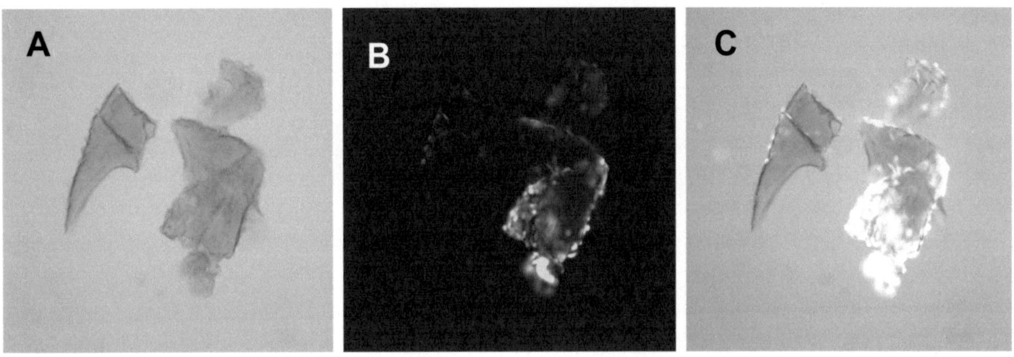

Fig. 17. Growth of human MSCs for 7 days on decalcified endosteal particles. (A) Cells attached to particles, (B) staining with Calcein-AM, which stains cell cytoplasm, and (C) Merging of the previous 2 images

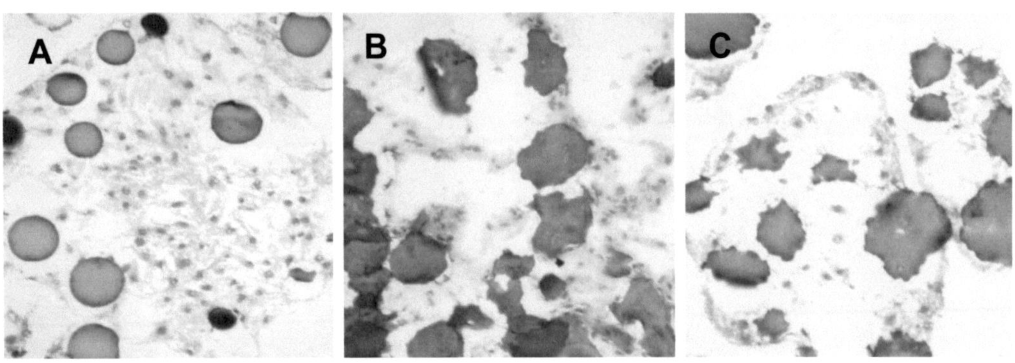

Fig. 18. Haematoxylin and Eosin staining of the growth of chondrocytes on various beads after 21 days in spinner culture, showing the formation of new extracellular matrix. (A) Gelatin beads, (B) Gelatin/Type I collagen beads, (C) Gelatin/Type II collagen beads

After longer periods of culture, 14-21 days, continued extracellular matrix production was seen when ascorbate was included, and cell proliferation also occured. After 21 days,

Haematoxylin and Eosin staining of materials containing gelatin and gelatin/collagen beads showed extensive matrix between particles (Fig. 18), which were now more separated than in the 7 day clusters (Figs. 14 & 15). In addition, the particles containing collagens were now showing erosion and degradation (Fig. 18B, Fig. 18C), while the gelatin alone particles were still substantially intact (Fig. 18A). Alcian Blue staining of these samples confirmed that the gelatin alone beads were more stable than those containing collagen (Fig. 19) and showed that the extracellular matrix contained proteoglycans/glycosaminoglycans which are key products of chondrocytes and are indicators of proper cell function and cartilage formation. The staining was strongest around the periphery of the particles where cells were still attached. The chondrocyte cell phenotype was also examined by immunohistology (Fig. 20). This showed that after 14 days in spinner culture the chondrocytes were still producing type II collagen, the correct collagen for articular (hyaline) cartilage formation, and the amount of type I collagen being produced was very low (Fig. 20).

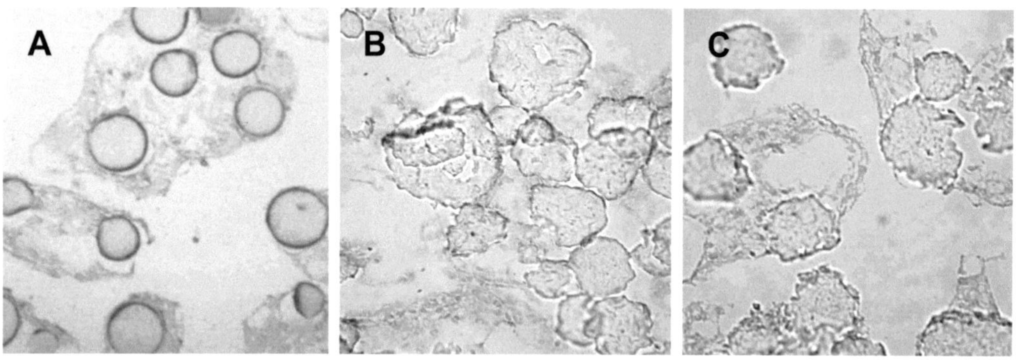

Fig. 19. Alcian Blue staining of the growth of chondrocytes on various beads after 21 days in spinner culture, showing the formation of proteoglycans within the new extracellular matrix. (A) Gelatin beads, (B) Gelatin/Type I collagen beads, (C) Gelatin/Type II collagen beads

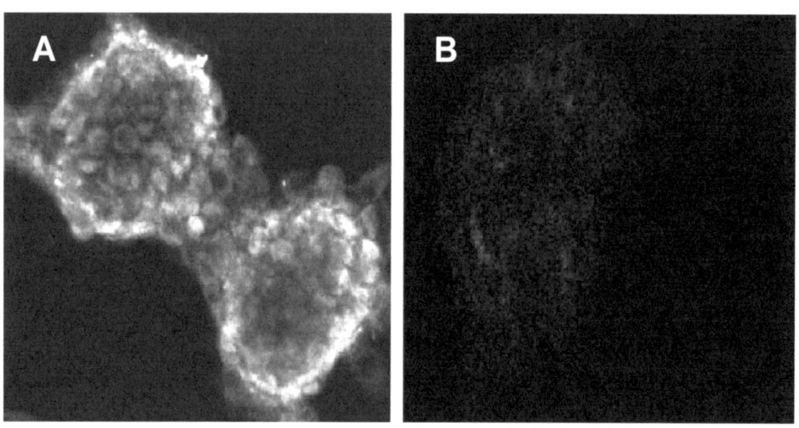

Fig. 20. Immunohistological staining of chondrocytes grown on gelatin beads in spinner culture for 14 days, showing continued production of type II collagen and little, if any type I collagen production. (A) Immunostaining for type II collagen, (B) Immunostaining for type I collagen

3.5 Bead culture in a collagen carrier gel

It was noted earlier, in the Introduction, that if cells are expanded for cell therapy on biodegradable, resorbable beads, then there is no need to isolate the cells from the beads prior to administration (Glattauer et al., 2010; Werkmeister et al., 2006), and that this has the advantage of minimizing the extent of cell handling that is required. In such a system, delivery of the cells/bead in a gel or gel forming material has further advantages. The presence of the gel stabilises the cell implant, and aids in the delivery of the cell/bead constructs.

Collagen type I gel is an ideal FDA approved protein that can be used as the delivery vehicle. When mixed with the cell/bead constructs the mixture can still be readily delivered through a 22-gauge needle, while the particles with bound cells remain uniformly dispersed and do not settle out rapidly (see Fig. 21A, where the particles have been stained to highlight the distribution in the gel). On warming to 37 °C for 1 h, the collagen formed a fibrous, solid gel that was able to be picked up, if required (Fig. 21B). This would not be necessary for in situ application of the mixture by injection into tissue. Indeed, we have injected a solution of cell/bead constructs in collagen gels and found that the mixture gelled in situ and gel contraction was markedly reduced by the presence of the beads (data not shown).

The cell/bead constructs embedded in a collagen gel can be maintained in culture for greater than 3 weeks. During the initial stages of culture, within 3 days, if synthetic, unmodified Cytodex™ beads are used, the cells show a greater affinity for the collagen of the gel than the bead and migrate from the beads into the surrounding gel. Cell migration is much slower from gelatin and other biologically-seeded beads where there seems a possible better affinity for the bead.

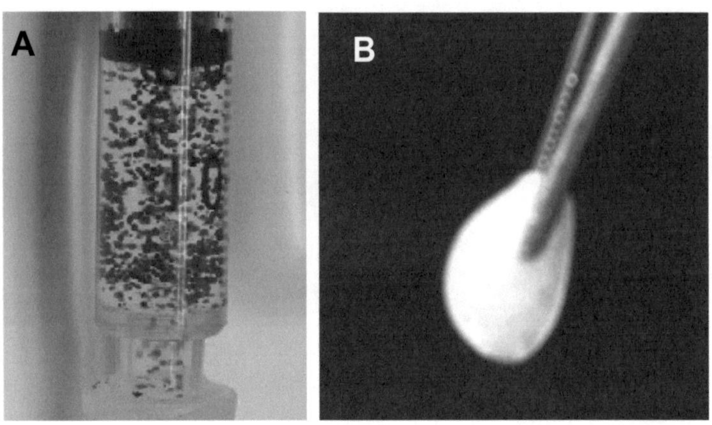

Fig. 21. Delivery of cell/bead constructs using a collagen gel. (A) Stable dispersion of gelatin beads, after staining, in a 1% collagen type I gel. (B) Collagen gel with beads, after heating to 37 °C forms a mechanically stable gel

After 3 weeks in static culture, the cell/gelatin beads/collagen gel construct had a tissue-like appearance (Fig. 22), and the loose structure of the original gel construct had formed a firmer, more compact structure, which showed very minimal contraction. When cells are seeded into collagen gels alone, the gel contracts significantly, by ~50% or more, within the first 7 days of culture (Ramshaw et al., 1991) and again forms a more compact tissue-like format.

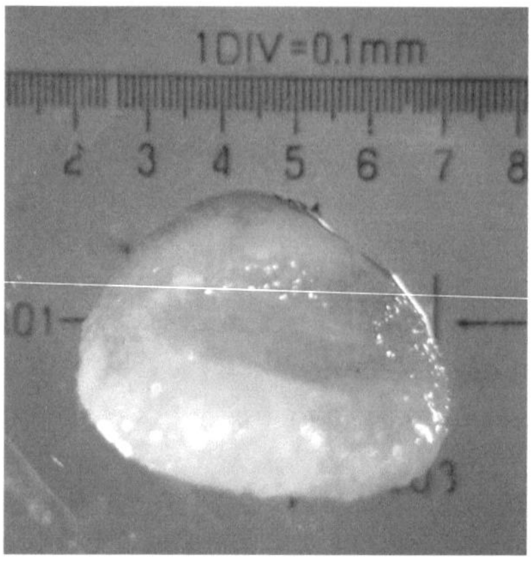

Fig. 22. A collagen type I gel with cells/gelatin beads after 3 weeks in static culture, showing the tissue-like material that is formed

The presence of the beads appears to retard any contraction of the collagen gel, while at the same time the synthesis of extracellular matrix augments the structure. Chondrocytes in culture can produce matrix metalloproteinases (Lefebvre et al., 1990) which could degrade the construct. The presence of the beads, where the cells have formed an established matrix prior to incorporation in the collagen gel may minimise the production of these proteinases. However, after 2 to 3 weeks of culture some erosion of gelatin beads was seen (Fig. 23A, Fig. 23B). The extent of this erosion was greater than when the collagen delivery gel was absent (Fig. 18A).

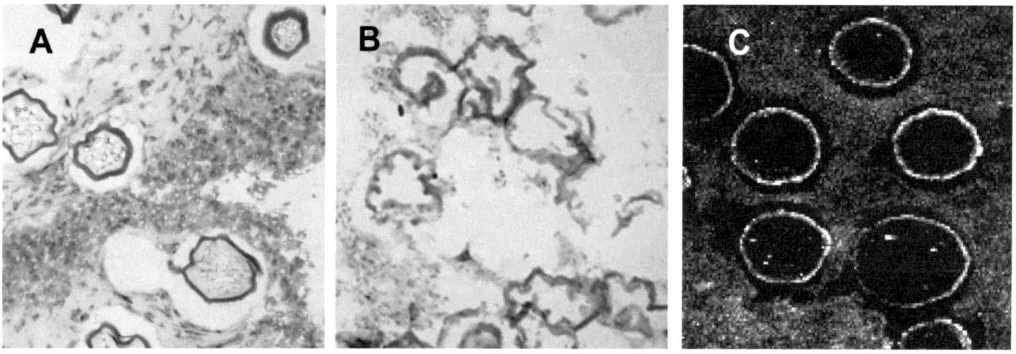

Fig. 23. Histological examination of cell-bead constructs in a collagen gel after 2-3 weeks in culture. (A) Haematoxylin and eosin staining after 2 weeks of culture, (B) Alcian Blue staining after 3 weeks in culture, and (C) Immunostaining of Type II collagen after 2 weeks in culture

The presence of new extracelluar matrix can be seen by histochemical and specific immunohistological staining. Heamatoxylin and Eosin staining after 2 weeks of culture (Fig. 23A) shows extensive cell infiltration of the collagen gel, and new collagen synthesis around the beads and within the gel. Alcian Blue staining after 3 weeks of culture (Fig. 23B) shows significant deposition of proteoglycan/glycosaminoglycan within the gel surrounding the beads, as well as very strong staining around the beads themselves. This staining was not present in the original collagen gel and is therefore indicative of new matrix production.

This is also the case for type II collagen production. Immunohistology after 2 weeks of culture (Fig. 23C) shows that there is type II collagen accumulation, possibly at low levels, throughout the matrix, while there is strong staining around the beads.

4. Conclusions

The data presented in this Chapter show that a wide variety of collagen-based beads and particles can be made which are useful for spinner culture. The beads can be made with a range of tissue-like conformations, ranging from denatured collagen (gelatin) through to beads that are based around basement membrane and collagen type IV structures. The beads, including those that have been further crosslinked using glutaraldehyde, were not cytotoxic and readily supported cell growth. The cell growth in spinner culture was excellent and within 7 days there was normally accumulation of new extracellular matrix. For cell therapy applications, the proliferation rates of chondrocytes were excellent, several fold better than monolayer cultures and synthetic beads. While most biological tissue based particles or biologically fabricated beads were excellent cell carriers, demineralised bone particles and gelatin beads were particularly good, the former possibly augmented by residual growth factors. In a therapeutic application the demineralised bone-based beads could have a fairly long resorption rate. If a shorter resorption rate was preferred, then gelatin beads are ideal, and the rate of resorption can be controlled by varying the extent of glutaraldehyde stabilisation (Glattauer et al., 2010). After longer times in culture, exceeding 21 days, the chondrocytes maintained their phenotype and were still producing macromolecules characteristic of native articular cartilage, such as proteoglycans and type II collagen. For therapeutic applications a superior approach is to deliver the cell/bead constructs in a delivery gel; in the present study a gel of type I collagen has been used. In this case the choice and design of the beads can be selected to match the requirements with respect to cell type and bead resorption rate. These gel constructs were shown to be stable, retaining their size and becoming more robust with new matrix accumulation. Again the chondrocytes maintained their phenotype over 21 days, as shown by proteoglycan and collagen type II production.

Together, these data show the utility of expanding cells on biological, collagen-based beads in spinner culture for tissue engineering and cell therapy applications. Further, these data also show that an ideal approach is to use resorbable beads for cell expansion and then direct delivery of cells still attached to the beads, where it is possible to design carrier beads with a range of stabilities that matches the implant requirements. An ideal way of achieving this implant is to combine the cells/beads together with a delivery gel. This proposed approach has been used in an animal trial for articular cartliage repair, where it was compared to a standard ACI approach (Chiang et al., 2005). The results strongly confirmed the validity of the proposed method, with an independent multifactor scoring system showing that the cell/bead/gel approach was significanly better than a standard ACI procedure and a non-cell control in a mini-pig model (Chiang et al., 2005). The approach can be further improved by examining delivery systems that allow the delivery and stabilisation of the cell/bead construct by arthroscopy, removing the need for open knee surgery. Equally, although the present techniques use cartilage biopsies obtained from non-load bearing cartilage, recent developments in cell biology, for example the ready acquisition of mesenchymal stem cells which can be differentiated into chondrocytes, opens up further opportunities that may minimise invasive surgery for the patient.

5. References

Adhikari, R.; Danon, S.J.; Bean, P.; Le, T.; Gunatillake, P.A.; Ramshaw, J.A.M. & Werkmeister, J.A. (2010). Evaluation of in situ curable biodegradable polyurethanes containing zwitterion components. *Journal of Materials Science: Materials in Medicine*, Vol.21, No.4, (April), pp. 1081–1089, ISSN 0957-4530

Arterburn, L.M.; Zurlo, J.; Yager, J.D.; Overton, R.M. & Heifetz A.H. (1995). A morphological study of differentiated hepatocytes in vitro. *Hepatology*, Vol.22, No.1, (July), pp. 175-187, ISSN 0270-9139

Aulthouse, A.L.; Beck, M.; Griffey, E.; Sanford, J.; Arden, K.; Machado, M.A. & Horton, W.A. (1989). Expression of the human chondrocyte phenotype in vitro. *In Vitro Cellular and Developmental Biology*, Vol.25, No.7, (July), pp. 659–668, ISSN 0883-8364

Bartlett, W.; Skinner, J.A.; Gooding, C.R.; Carrington, R.W.; Flanagan, A.M.; Briggs, T.W. & Bentley G. (2005). Autologous chondrocyte implantation versus matrix-induced autologous chondrocyte implantation for osteochondral defects of the knee: A prospective, randomised study. *The Journal of Bone and Joint Surgery. British Volume*, Vol.87, No.5, (May), pp. 640-645, ISSN 0301-620X

Bedi, A.; Feeley, B.T. & Williams, R.J. (2010). Management of articular cartilage defects of the knee. Journal of Bone and Joint Surgery, American Volume, Vol.92, No.4, (April), pp. 994-1009, ISSN 0021-9355

Brittberg, M. (1999). Autologous chondrocyte transplantation. *Clinical Orthopaedics and Related Research*, Vol.367S, pp. S147-S155, ISSN 0009-921X

Brittberg, M.; Lindahl, A.; Nilsson, A.; Ohlsson, C.; Isaksson, O. & Peterson, L. (1994). Treatment of deep cartilage defects in the knee with autologous chondrocyte transplantation. *The New England Journal of Medicine*, Vol.331, No.14, (October), pp. 889-895, ISSN 0028-4793

Chen, R.; Curran, S.J.; Curran, J.M. & Hunt, J.A. (2006). The use of poly(L-lactide) and RGD modified microspheres as cell carriers in a flow intermittency bioreactor for tissue engineering cartilage. *Biomaterials*, Vol.27, No.25, (September), pp. 4453-4460, ISSN 0142-9612

Chiang, H.; Kuo, T-F.; Tsai, C-C.; She, B-R.; Huang, Y-Y.; Lee, H-S.; Shieh, C-S.; Chen, M-H.; Ramshaw, J.A.; Werkmeister, J.A.; Tuan, R.S. & Jiang, C-C. (2005). Repair of porcine articualr cartilage defect with autologous chondrocyte transplantation. *Journal of Orthopaedic. Research*, Vol.23, No.3, (May), pp. 584-593, ISSN 0736-0266

Freeman, P.M.; Natarajan, R.N.; Kimura, J.H. & Andriacchi, T.P. (1994). Chondrocyte cells respond mechanically to compressive loads. *Journal of Orthopaedic Research*, Vol.12, No.3, (May), pp. 311–320, ISSN 0736-0266

Frondoza, C.G.; Sohrabi, A. & Hungerford, D. (1996). Human chondrocytes proliferate and produce matrix components in microcarrier suspension culture. *Biomaterials*, Vol.17, No.9, (May), pp. 879–888, ISSN 0142-9612

Glattauer, V.; Irving-Rodgers, H.F.; Rodgers, R.J.; Stockwell, S.; Brownlee, A.; Werkmeister, J.A. & Ramshaw, J.A.M. (2007). Examination of the basement membrane components associated with the bovine spermatogonial tubule basal lamina. *Reproduction, Fertility & Development*, Vol.19, No.3, pp. 473-481, ISSN 1031-3613

Glattauer, V.; White, J.F.; Tsai, W.-B.; Tsai, C.-C.; Tebb, T.A.; Werkmeister, J.A. & Ramshaw, J.A.M. (2010). Preparation and properties of extra-cellular matrix-based beads for spinner culture. *Journal of Biomedical Materials Research A,* Vol.92A, No.4, (March), pp. 1301-1309, ISSN 1549-3296

Gross, J. & Kirk, D. (1958). The heat precipitation of collagen from neutral salt solutions: some rate-regulating factors. *The Journal of Biological Chemistry,* Vol.233, No.2, (August), pp. 355-360, ISSN 0021-9258

Gunatillake, P.; Mayadunne, R. & Adhikari, R. (2006). Recent developments in biodegradable synthetic polymers. *Biotechnology Annual Review,* Vol.12, pp. 301-347, ISSN 1387-2656

Hanson, A.N. & Bentley, J.P. (1983). Quantitation of type I to type III collagen ratios in small samples of human tendon, blood vessels, and atherosclerotic plaque. *Analytical Biochemistry,* Vol.130, No.1, (April), pp. 32-40, ISSN 0003-2697

Haycock, J.W. (2011). 3D cell culture: a review of current approaches and techniques. *Methods in Molecular Biology,* Vol.695, pp. 1-15, ISSN 1064-3745

Haylock, D.N.; Williams, B.; Johnston, H.M.; Liu, M.C.; Rutherford, K.E.; Whitty, G.A.; Simmons, P.J.; Bertoncello, I. & Nilsson, S.K. (2007). Hemopoietic stem cells with higher hemopoietic potential reside at the bone marrow endosteum. *Stem Cells,* Vol.25, No.4, (April), pp. 1062-1069, ISSN 1536-2302

Henderson, I.; Lavigne, P.; Valenzuela, H. & Oakes, B. (2007). Autologous chondrocyte implantation: superior biologic properties of hyaline cartilage repairs. *Clinical Orthopaedics and Related Research,* Vol.455, pp. 253-261 ISSN 0009 921X

Hong, Y.; Gong, Y.; Gao, C. & Shen, J. (2008). Collagen-coated polylactide microcarriers/chitosan hydrogel composite: Injectable scaffold for cartilage regeneration. *Journal of Biomedical Materials Research A,* Vol.85A, No.3, (June), pp. 628-637, ISSN 1549-3296

Hunter, W. (1742-1743). Of the structure and diseases of articulating cartilages. *Philosophical Transactions of The Royal Society B,* Vol.42, pp. 514-521, ISSN 0962-8436

Hunziker, E.B. (2001). Articular cartilage repair: Basic science and clinical progress. A review of the current status and prospects. *Osteoarthritis and Cartilage,* Vol.9, No.1, (January), pp. 22-32, ISSN 1063-4584

Jacobson, B.S. & Ryan, U.S. (1982). Growth of endothelial and HeLa cells on a new multipurpose microcarrier that is positive, negative or collagen coated. *Tissue & Cell,* Vol.14, No.1, pp. 69-83, ISSN 040-8166

Kaplan, E.N.; Falces, E. & Tolleth, H. (1983). Clinical utilization of injectable collagen. *Annals of Plastic Surgery,* Vol.10, No.6, (June), pp. 437-451, ISSN 0148-7043

Kleinman, H.K. & Martin, G.R. (2005). Matrigel: basement membrane matrix with biological activity. *Seminars in Cancer Biology,* Vol.15, No.5, (October), pp. 378-386, ISSN 1044-579X

Knapp, T.R.; Kaplan, E.N. & Daniels, J.R. (1977). Injectable collagen for soft tissue augmentation. *Plastic and Reconstructive Surgery,* Vol.60, No.3, (September), pp. 398-405, ISSN 0032-1052

Lahiji, A.; Sohrabi, A.; Hungerford, D.S. & Frondoza, C.G. (2000). Chitosan supports the expression of extracellular matrix proteins in human osteoblasts and chondrocytes. *Journal of Biomedical Materials Research,* Vol.51, No.4, (September), pp. 586-595, ISSN 1549-3296

Lee, T.J.; Bhang, S.H.; La, W.G.; Yang, H.S.; Seong, J.Y.; Lee, H.; Im, G.I.; Lee, S.H. & Kim, B.S. (2011). Spinner-flask culture induces redifferentiation of de-differentiated chondrocytes. *Biotechnology Letters,* Vol.33, No.4, (April), pp. 829-836, ISSN 0141-5492

Lefebvre, V.; Peeters-Joris, C. & Vaes, G. (1990). Production of collagens, collagenase and collagenase inhibitor during the dedifferentiation of articular chondrocytes by serial subcultures. *Biochimica et Biophysica Acta,* Vol.1051, No.3, (March), pp. 266-275, ISSN 0006-3002

Malda, J. & Frondoza, C.G. (2006). Microcarriers in the engineering of cartilage and bone. *Trends in Biotechnology,* Vol.24, No.7, (July), pp. 299-304, ISSN 0167-7799

Malda, J.; Martens, D.E.; Tramper, J.; van Blitterswijk, C.A. & Riesle, J. (2003). Cartilage tissue engineering: Controversy in the effect of oxygen. *Critical Reviews in Biotechnology,* Vol.23, No.3, pp. 175-194, ISSN 0738-8551

Malda, J.; van Blitterswijk, C.A.; Grojec, M.; Martens, D.E.; Tramper, J. & Riesle, J. (2003). Expansion of bovine chondrocytes on microcarriers enhances redifferentiation. *Tissue Engineering,* Vol.9, No.5, (October), pp. 939-948, ISSN 1076-3279

Malemud, C.J.; Norby, D.P.; Sapolsky, A.I.; Matsuta, K.; Howell, D.S. & Moskowitz, R.W. (1981). Neutral proteinases from articular chondrocytes in culture. I. A latent collagenase that degrades human cartilage type II collagen. *Biochimica et Biophysica Acta, Vol.657, No.2, (February), pp.* 517-529, ISSN 0006-3002

Mallein-Gerin, F.; Garrone, R. & van der Rest, M. (1991). Proteoglycan and collagen synthesis are correlated with actin organization in dedifferentiating chondrocytes. *European Journal of Cell Biology,* Vol.56, No.2, (December), pp. 364-373, ISSN 0171-9335

Maloney, B.P.; Murphy, B.A. & Cole, H.P. (2004). Cymetra. *Facial Plastic Surgery,* Vol.20, No.2, (May), pp. 129-134, ISSN 0736-6825

Marlovits, S.; Striessnig, G.; Kutscha-Lissberg, F.; Resinger, C.; Aldrian, S.M.; Vécsei, V. & Trattnig, S. (2005). Early postoperative adherence of matrix-induced autologous chondrocyte implantation for the treatment of full-thickness cartilage defects of the femoral condyle. *Knee Surgery, Sports Traumatology, Arthroscopy,* Vol.13, No.6, (September), pp. 451-457, ISSN 0942-2056

Mayne, R. (1989). Collagens: what is their function, are they involved in articular disease? *Arthritis and Rheumatism,* Vol.32, No.3, (March), pp. 241-246, ISSN 0004-3591

McCloskey, D.I. & Cleary, E.G. (1974). Chemical composition of the rabbit aorta during development. *Circulation Research,* Vol.36, No.6, (June), pp. 828-835, ISSN 0009-7330

McLimans, W.F.; Davis, E.V.; Glover, F.L. & Rake G.W. (1957). The submerged culture of mammalian cells; the spinner culture. *Journal of Immunology,* Vol.79, No.5, (November), pp. 428-433, ISSN 0022-1767

Miller, E.J. & Matukas, V.J. (1969). Chick cartilage collagen: a new type of alpha 1 chain not present in bone or skin of the species. *Proceedings of the National Academy of Sciences of the United States of America,* Vol.64, No.4, (December), pp. 1264-1268, ISSN 0027-8424

Miller, E.J. & Rhodes, R.K. (1982). Preparation and characteriztion of the different types of collagen. *Methods in Enzymology,* Vol.82, Pt.A, pp. 33-64, ISSN 0076-6879

Nigro, J.; White, J.F.; Ramshaw, J.A.M.; Haylock, D.N.; Nilsson, S.K. & Werkmeister, J.A. (2010). The effect of bovine endosteum-derived particles on the proliferation of human mesenchymal stem cells. *Biomaterials,* Vol.31, No.21, (July), pp. 5689-5699, ISSN 1042-9612

Oesser, S. & Seifert, J. (2003). Stimulation of type II collagen biosynthesis and secretion in bovine chondrocytes cultured with degraded collagen. *Cell and Tissue Research*, Vol.311, No.3, (March), pp. 393-399, ISSN 0302-766X

Ramshaw, J.A.M. (1986). Distribution of type III collagen in bovine skin of various ages. *Connective Tissue Research*, Vol.14, No.4, pp. 307-314, ISSN 0300-8207

Ramshaw, J.A.M.; Glattauer, V. & Werkmeister, J.A. (2000). Stabilisation of collagen in clinical applications. In: *Biomaterials and Bioengineering Handbook*, D.L. Wise, (Ed.), 717-738, Marcel Dekker Inc., ISBN 0-8247-0318-9, New York, NY, USA

Ramshaw, J.A.M.; Mitrangas, K. & Bateman, J.F. (1991). Heterogeneity in dermatosparaxis is shown by contraction of collagen gels. *Connective Tissue Research*, Vol.25, No.3-4, pp. 295-300, ISSN 0300-8207

Shah, M.R.; Kaplan, K.M.; Meislin, R.J. & Bosco, J.A. (2007). Articular cartilage restoration of the knee. *Bulletin of the NYU Hospital for Joint Diseases*, Vol.65, No.1, pp. 51-60, ISSN 1936-9719

Tao, X.; Shaolin, L. & Yaoting, Y. (2003). Preparation and culture of hepatocyte on gelatin microcarriers. *Journal of Biomedical Materials Research A*, Vol.65A, No.2, (May), pp. 306-310, ISSN 1549-3296

Tebb, T.A.; Tsai, S.-W.; Glattauer, V.; White, J.F.; Ramshaw, J.A.M. & Werkmeister, J.A. (2006). Development of porous collagen beads for chondrocyte culture. *Cytotechnology*, Vol.52, No.2, (October), pp. 99-106, ISSN 0920-9069

Thissen, H.; Chang, K.-Y.; Tebb, T.A.; Tsai, W.B.; Glattauer, V.; Ramshaw, J.A.M. & Werkmeister, J.A. (2005). Synthetic biodegradable microparticles for articular cartilage tissue engineering. *Journal of Biomedical Materials Research A*, Vol.77A, No.3, (June), pp. 590-598, ISSN 1549-3296

Trelstad, R.L.; Catanese, V.M. & Rubin, D.F. (1976). Collagen fractionation: separation of native types I, II and III by differential precipitation. *Analytical Biochemistry*, Vol.71, No.1, (March), pp. 114-118, ISSN 0003-2697

van Wezel, A. L. (1967). Growth of cell-strains and primary cells on micro-carriers in homogeneous culture. *Nature (London)*, Vol.216, No.5110, (October), pp. 64–65, ISSN 0028-0836

von der Mark, K.; Gauss, V.; von der Mark, H. & Muller, P. (1977). Relationship between cell shape and type of collagen synthesised as chondrocytes lose their cartilage phenotype in culture. *Nature (London)*, Vol.267, No.5611, (June), pp. 531-532, ISSN 0028-0836

Wegener, B.; Schrimpf, F.M.; Pietschmann, M.F.; Milz, S.; Berger-Lohr, M.; Bergschmidt, P.; Jansson, V. & Müller, P.E. (2009). Matrix-guided cartilage regeneration in chondral defects. *Biotechnology and Applied Biochemistry*, Vol.53, Pt.1, pp. 63-70, ISSN 0885-4513

Werkmeister, J.; Glattauer, V. & Ramshaw J.A.M. (2006). Preparation and use of basement membrane particles. *Patent Application*, WO2006128216-A1

Werkmeister, J.A.; Adhikari, R.; White, J.F.; Tebb, T.A.; Le, T.P.T.; Taing, H.C.; Mayadunne, R.; Gunatillake, P.A.; Danon, S.J. & Ramshaw, J.A.M. (2010). Biodegradable and injectable cure-on-demand polyurethane scaffolds for regeneration of articular cartilage. *Acta Biomaterialia*, Vol.6, No.9, (September), pp. 3471-3481, ISSN 1742-7061

Wildemann, B.; Kadow-Romacker, A.; Haas, N.P. & Schmidmaier, G. (2007). Quantification of various growth factors in different demineralized bone matrix preparations.

Journal of Biomedical Materials Research A, Vol.81, No.2, (May, pp. 437-442, ISSN 1549-3296

Williams, D.F. (1999). *The Williams Dictionary of Biomaterials,* Liverpool University Press, ISBN 0-85323-734-4, Liverpool, UK

Yurchenco, P.D.; Amenta, P.S. & Patton, B.L. (2004). Basement membrane assembly, stability and activities observed through a developmental lens. *Matrix Biology,* Vol.22, No.7, (January), pp. 521-538, ISSN 0945-053X

13

Collagen: Applications of a Natural Polymer in Regenerative Medicine

Umber Cheema, Michael Ananta and Vivek Mudera
UCL Division of Surgery & Interventional Science,
Institute of Orthopaedics & Musculoskeletal Science,
Stanmore Campus, Brockley Hill,
UK

1. Introduction

Collagen is a naturally occurring matrix polymer which is highly conserved across species. It is the predominant extra-cellular matrix component of most connective tissues within the mammalian body, comprising one third of all protein found within tissues. Collagens are extracellular and have a mainly structural role. Critical parameters including density, packing and orientation (or direction) results in distinctively varying mechanical properties in tissues such as bone, skin, tendon and cartilage.

In connective tissue lost to trauma or disease, replacement tissue strategies, have to consider mechanical implications. Synthetic polymers can be designed to have the mechanical integrity of the native structure to be replaced but eventually this will be degraded and replaced by the host. The major target protein that will be replaced in connective tissues is collagen. The other alternative is to start with collagen as a natural polymer substrate and tailor its mechanical properties *in vitro*. Given the critical role this protein plays in tissue structure there have been and continue to be efforts into extracting this protein, reforming 3D scaffolds for tissue engineering as well as controlling density and direction parameters to form tissues in vitro. The main difference in the building of bulk tissues is the cell-rich or matrix-rich nature of the tissue being engineered. Where the matrix dominates a tissue, the mechanical properties of the matrix are critical, i.e in connective tissue. In tendon for example, the alignment of collagen fibrils along the principle axis of strain application, and the nature of a 'dual' size of fibril diameters, provides incredible strength to this tissue. Compared directly to a tissue like dermis, where collagen is interlocked in a basket-weave formation with elastin to provide tensile strength in multiple axis to ensure stretching of this tissue does not compromise its integrity. Thus, when engineering tissues in vitro, technologies and processing to control parameters of collagen architecture have been developed to mimic those found in tissues in situ. This field of controlled processing is growing, as the sophistication of methods employed to create biomimetic scaffolds advances.

Type I collagen scaffolds are widely used in clinical practice and the collagen for these materials are generally obtained either from cultured cells or extracted from native tissues. Extraction encompasses the entire range from decellularisation of collageneous tissues preserving the native architecture to the complete break down into collagen molecules which can later be reconstituted into their native fibrillar structure. One of the most common collagen scaffolds in clinical use is de-cellularised dermal tissue. By decellularising an entire piece of connective tissue, the immunogenicity of the tissue is removed whilst retaining the collagen architecture and other matrix components in the native form. These scaffolds induce and guide tissue repair when implanted in a full-thickness skin defect as a template for dermal regeneration. Methods to enhance the in vivo persistence of such decellurarised scaffolds, involve dispersion of the collagen matrix. The dispersion of collagen can result in the degradation of the collagen into polymers, oligomers or monomers dependent upon the treatment methods applied. This is followed by co-precipition with a glycosaminoglycan. In these cases the inherent architecture of the native tissue is lost and reformed *in vitro*.

The formation of living collagenous tissue equivalents rely upon the cell seeding of decellularised native tissues or cell-seeding into prefabricated, porous collagen scaffolds. The next generation of skin equivalents are bio-engineered cell based technologies using cell produced collagen. These include the next generation bilayered skin equivalents produced by Organogenesis. The use of solubilised collagen as dermal and lip fillers for the correction of contour deformaties, is a growing cosmetic procedure. This relies upon collagen dispersion from either human cadaver or bovine tissue, which can then be stably re-injected to 'plump' to fill minor defects. The question is whether the fibrosis response by the injection of these products causes the therapeutic effect or the products themselves.

The main focus of this chapter is to address the importance of collagen protein structure and its relation to normal mechanical function in matrix-rich tissues. This chapter will overview the established, clinically used and new novel processing technologies being researched to improve and control bulk collagen processing for applications in regenerative medicine, and new directions needed to control collagen architecture.

2. Collagen- the dominant extra-cellular protein

Collagen is a naturally occurring matrix polymer which is well conserved across species. It is the predominant extra-cellular matrix component of most connective tissues within the mammalian body, comprising one third of all protein found within tissues, particularly musculo-skeletal tissues. There are 27 known types of Collagen which are extracellular and have a mainly structural role. The configuration of this protein greatly affects its role in tissue architecture. Parameters including density, packing, degree of cross-linking and orientation (direction) result in distinctively varying mechanical properties in tissues such as bone, skin, tendon and cartilage.

An example is the bi-modal distribution of fibril diameters found in tendons, which are aligned parallel to the direction in which strain is applied (Morgan et al. 2006). This confers strength in one direction, which is precisely the axis of strain generation required for a tendon.In comparison to this, the collagen architecture of skin varies considerably. Collagen fibrils in this tissue are weaved in to a more 'random' structure, along with components like elastin, to confer a 'stretchy' property on this tissue, which is constantly being deformed due to the structure and function of skin. In fact, parallel alignment of collagen fibrils is more apparent in scarred skin tissue compared to normal (Verhaegen et al. 2009). The architecture of this protein is, therefore, tissue-specific, and mimicking this architecture will be important for scaffolds design and tissue engineering.

Although Collagen is the major extra-cellular component of most connective tissues in the body, additional ECM components contribute to the mechanical properties, cell-attachment properties and regulation of architecture. An example of this is the presence of collagen II in conjunction with proteoglycans in cartilage, which provide compressive load strength to this tissue. Although mainly collagen type I, Hydroxyapatite is the major mineral component of bone. There is currently active research in making composite matrices containing both of these components, to create biomimetic scaffolds for bone engineering. For the tissue engineering of vascular grafts, the elasticity of vascular structures (such as the pulmonary aorta) is critical to mimic *in vitro*. This mechanical feature is dependent upon the very high presence of elastin within these structures, and for vascular graft engineering the incorporation of elastin protein, along with the other component proteins is necessary. Vessels contain multiple cell types and specific basement membrane proteins are necessary for the correct phenotype of these cells, particularly endothelial cells, which line the lumen. One of the most successful approaches for tissue engineering remains de-cellularisaiton of xenograft tissue, followed by cell-seeding using bioreactor culture.

3. Matrix-rich tissues

The matrix component of different tissue types is dependent upon whether these tissues are cell-rich (central nervous system, skeletal muscle, organs like heart, liver, kidneys), versus matrix-rich (tendon, ligament, cartilage and bone). Collagen type I plays an important role in most matrix rich tissues, but there are numerous other protein components critical to the unique architecture of tissues.

Matrix-rich tissues are often very hierarchical in structure, because the physical nature of the tissue is mainly as a support system for the skeleton. Both tendon and skeletal muscle have distinct 'bundle within a bundle' structure, where mainly collagen (in the case of tendon) and myofibrils (in the case of skeletal muscle) are bundled in graduating structures, along the principal axis of strain the tissue will have to bear. The packing of protein components thus becomes critical to how load is distributed within a tissue, and ultimately results in whether a tissues is a successful component for locomotion.

The architecture of collagen type I is particularly crucial to matrix-rich tissues, as the strength required of these tissues can only really be achieved by recapitulating these precise features, including orientation of fibrils (direction), diameter and density. If we take bone as an example, this tissue is intrinsically linked to the skeletal muscle system, and itself through connections with tendons and ligaments. It is mainly composed of collagen type I and a mineral component. The conformation of collagen varies dependent on which bone we study, and where we study it. The most common 'surface' feature of bone is composed of alternate lamellae of longitudinal and transverse bundles of collagen fibres, running parallel to the surface of the bone (Smith 1960). The mineral component within bone is interspersed within these lamellae structures. The intense load endured by bone tissue is therefore appropriated by the particular structure of the extra-cellular components. As bone is subject to load in varying axis of strain, a simple parallel bundled array of fires (as seen in tendon) is not observed.

Whilst engineering specific tissues, mimicking the correct matrix components as well as the specific architecture will result in increased success in modeling of a tissue. Functionality of any engineered tissue is likely to be critically dependent on the biomimicry, particularly of the structure of the protein component. For replacement of matrix-rich tissues, the mechanical functionality of any tissue engineered scaffold will be optimal if the architecture and packing of the matrix components closely mimics that found *in vivo* naturally. Collagen to collagen attachment is primarily through increased cross-linking, and cell-collagen interaction is primarily through the integrins α1β1 and α2β1. The majority of replacement tissues engineered implants and constructs currently either rely upon synthetic grafts, which provide the adequate mechanical load-bearing properties, or on compositions of native proteins which do not necessarily provide the appropriate mechanical properties.

4. Collagen used *in vivo*

The majority of collagen scaffolds used for clinical purposes are made from extracting collagen from tissues. In the process of extraction very often collagen and other proteins are highly denatured. This is a result of the highly effective methods of cross-linking *in vivo*. The limited understanding of the protein Elastin, for example, is mainly due to our inability to extract it in its native form from tissues. The efficient mechanism by which proteins are built into tissue architectures by the body are in part hindering our complete understanding of those very mechanisms and proteins.

The main use of collagen for clinical application are as replacement scaffolds used as tissue fillers, and as support matrices for matrix rich tissues. These collagen scaffolds are now widely used in clinical practice, primarily as skin substitutes and dermal fillers. Because of the naturally high composition of collagen in matrix rich tissues, collagen scaffolds are the natural protein to use as a scaffold, however in the process of engineering or assembling such scaffolds, much of the native protein structure, and thus inevitably function, is lost.

The extraction method varies for each scaffold type and resulting scaffolds properties therefore vary. The majority of these scaffolds are composite materials, often with synthetic components, primarily to provide strength to the scaffold, necessary to hold sutures and withstand the mechanical loads at the implant site and/or tissue. Here, we overview the products available as collagen type I based tissue equivalents and substitutes.

4.1 Collagen dermal tissue equivalents

The most widely used tissue equivalent scaffolds clinically, is for replacement dermal equivalents. The majority of products available rely upon a fabricated collagen mesh, mainly without a cellular component. Generally collagen is isolated from tissue samples, and in the process the collagen is greatly disaggregated and denatured, often resulting in complete loss of the original architecture and packing, and comprising mainly of fragments of the original protein. The majority of these scaffold equivalents are composed solely of collagen, without the addition of other extracellular matrix components.

Integra (Yannas et al. 1981)is the first FDA-approved skin substitute consisting of a suturable, semi-permeable silicone elastomer (polysiloxane, Dow Corning Liquid Silastic Medical Adhesive Type A) cured to an underlying dermal component made of a degradable crosslinked (vacuum dehydration and gluteraldehyde) coprecipitation of bovine collagen and (8%) chondroitin 6-sulfate (a shark cartilage derived glycosaminoglycan). Integra is used to reconstruct the skin in a two stage procedure in surgically excised burn injuries (Burke et al. 1981;Heimbach et al. 1988) or in excised benign or malignant lesions (Prystowsky et al. 2001). After it has integrated, generally 2-3 weeks after implantation the silicone membrane is removed and the neodermis is grafted with a split thickness skin graft. Histologically the dermal matrix disappears after one month (Stern et al. 1990) and only elicits a small transient immune response (Michaeli and McPherson 1990). Graft take is similar to allografts, but not as good as autografts. One of the major benefits being is that it is cosmetically superior over meshed autografts (Heimbach, Luterman, Burke, Cram,

Herndon, Hunt, Jordan, McManus, Solem, Warden, & . 1988) and the reconstructed skin has elastic properties matching that of normal skin (Nguyen et al. 2010).
Alloderm (LifeCell Corp, Branchburg, NJ) is decellularised dermal matrix derived from human cadaver skin by the tonic removal of epidermis with NaCl and decellularisation with sodium dodecyl sulphate (a detergent) of the dermal component. For storage, the resultant acellular dermal matrix was freeze-dried in a cryoprotectant solution (dextran, sucrose, raffinose) followed by a two-step drying procedure. For use the dermal matrix is rehydrated in two changes of sterile normal saline. Clinical studies have shown it to be useful as a dermal substitute in full thickness burns without signs of rejection, showing revascularisation, cellular repopulation, incorporation into the wound and cosmetic results similar to intermediate thickness skin grafts (Wainwright et al. 1996).
Dermagraft (Cooper et al. 1991) is fabricated by seeding and maintaining fibroblasts on a Vicryl mesh for a period of 2 to 3 weeks to achieve a sufficient amount of tissue formation for a therapeutic effect. The product is currently FDA approved for the treatment of venous (Omar et al. 2004), neuropathic and diabetic ulcers (Marston et al. 2003). Dermagraft is more effective than conventional treatments of chronic (persisting longer than 6 months) venous (Omar, Mavor, Jones, & Homer-Vanniasinkam 2004), neuropathic and diabetic ulcers (Marston, Hanft, Norwood, & Pollak 2003) and the product is FDA-approved for these applications. The material 'takes' in (immune-compromised) animal wounds, but integration in human (immuno-competent) subjects has not been reported and Dermagraft requires multiple applications to achieve a clinical effect.
Dermagraft is not to be confused with Dermagraft-TC (Dermagraft Transient Cover, currently marketed as TransCyte) which is fabricated by culturing allogeneic human neonatal fibroblasts 17 days on Biobrane (Bertek Pharmaceuticals Inc., WV, USA) a semi-permeable polymer (silicone) membrane with partially embedded woven monofilament nylon mesh onto which porcine collagen peptides are chemically bound to form a hydrophilic surface (Hansbrough et al. 1994). The product is devitalised by freezing and the cell produced ECM contains fibronectin, type I collagen, proteoglycans and growth factors. The product is a temporary cover for partial thickness burns and is spontaneously ejected by healing wounds. It has shown to, decrease hospital stay compared to conventional treatment (silvazine) (Amani et al. 2006), prevent frequent and painful dressing changes, autografting, and decreases the time to wound closure compared to alternative treatments (Biobrane alone, silvazine) (Kumar et al. 2004).
Apligraf (Organogenesis Inc., MA and Novartis Pharmaceutical Corp., NJ, US) (Wilkins et al. 1994) is a commercially available skin equivalent consisting of a bovine collagen (type I) hydrogel seeded with allogeneic human neonatal foreskin fibroblasts and keratinocytes (Bell et al. 1981a;Bell et al. 1981b). The mechanical properties of the dermal component are improved by allowing collagen to contract for 6 days after which keratinocytes are seeded and cultured for 4 days under submerged conditions and terminally differentiated into a stratified epidermis for an additional 7 days at the air/liquid interface. Apligraf is currently FDA approved for the treatment of venous leg (Falanga 2005) and diabetic foot ulcers (Veves et al. 2001). Apligraf is more effective than conventional treatment (i.e. compression therapy and saline-moistened gauze, respectively) in achieving wound closure in chronic (lasting longer than 6 months) non-healing venous leg (Falanga 2005) and diabetic foot ulcers (Veves, Falanga, Armstrong, & Sabolinski 2001) and is FDA-approved for the treatment of these conditions. As previously mentioned, Apligraf does not persist in human acute full thickness skin defects (cells remain in the wound up to 4 weeks) (Griffiths et al.

2004) and requires multiple applications for it to be effective. It is noteworthy that Organogenesis is currently in late stage development of a next generation allogeneic skin equivalent, VCTO1, which similar to Apligraf with the only difference being that the dermal matrix will be human fibroblast derived making its components all human.
OrCel (Forticell Bioscience, Inc., formerly Ortec International, Inc., NY, US) (Eisenberg and Llewelyn 1998) is currently under investigation for the treatment of chronic wounds. The product is currently FDA-approved for the treatment of split-thickness skin graft donor sites (Still et al. 2003) and for use in the surgical release of hand syndactyly secondary to epidermolysis bullosa. This commercially available skin equivalent is fabricated by seeding allogeneic fibroblasts into a preformed lyophilised bovine collagen (type I) sponge which is cultured for 2 days, inverted and seeded with keratinocytes on its non-porous side and cultured for an additional 7-14 days (Bell, Ehrlich, Sher, Merrill, Sarber, Hull, Nakatsuji, Church, & Buttle 1981b)to form a confluent monolayer epidermis.

4.2 Dermal fillers

Bovine collagen

Zyderm I, introduced in 1977 (Knapp et al. 1977) and FDA approved in 1981, is an injectable soluble, bovine collagen implant for the treatment of fine, superficial lines. The acid-soluble collagen is extracted from bovine dermal tissue with acetic-acid and is then pepsin-treated to yield telopeptide-poor collagen with low antigenicity. The collagen is 96% type I and 4% type III collagen and it is purified to over 99% and dispersed in phosphate buffered saline (PBS) containing 0.3% lidocaine hydrochloride to a concentration of 35 mg/ml. The monomeric collagen solution spontaneously polymerises at 37 °C (Knapp, Luck, & Daniels 1977). Zyderm II, FDA approved in 1983, is similar to Zyderm I but contains 65 mg/ml of collagen and is used to treat mild to moderate lines. Both Zyderm I and II only provide a short clinical effect from two to three months (Kligman and Armstrong 1986). Zyplast, FDA approved in 1985, is similar to Zyderm I collagen, but is chemically cross-linked with 0.0075% glutaraldehyde, making it less susceptible to degradation by collagenases (McPherson et al. 1986) and extending *in vivo* persistence from three to six months (Kligman & Armstrong 1986). Multiple applications, however, are still required for an extended effect.
Resoplast (Rofil Medical International, Breda, The Netherlands) is another bovine collagen suspended in a solution of phosphate buffer, sodium chloride, and lidocaine hydrochloride at a concentration of 35 and 65 mg/ml. This dermal filler persists *in vivo* for a period of three to six months (Naoum and Dasiou-Plakida 2001).
Arteplast, introduced in 1991, is an injectable material composed of microspheres 20 - 40 μm of polymethylmethacrylate (PMMA) suspended in a solution of denatured collagen (Lemperle et al. 1991). Due to issues with foreign body granulomas in reaction to the microspheres, in 1994, Arteplast was replaced by Artecoll (Lemperle et al. 1995) which features larger PMMA microspheres (30 - 50 μm) suspended in Resoplast (35 mg/ml). The product was subsequently FDA approved in 2006 and marketed as Artefill (Matarasso 2006). Artefill is a permanent filler for the correction of nasolabial folds and the effect of the collagen carrier is sustained by the microspheres through local induced fibrosis which replaces the carrier with host tissue (Lemperle et al. 2010).

Porcine collagen

Small intestinal submucosa (SIS) is derived from the small intestine of pigs. Following de-cellularisation the remaining matrix, which is mainly composed of collagen as well as a complex mixture of functional and structural molecules, is sterilised and the intact piece of tissue is ready to be used as a replacement scaffold. The benefit of decellularising an intact piece of tissue, whilst retaining the unique 3D ultrastructure, is the low immunogenicity issues which occur when implanting such a scaffold, and the ability of the body to repopulate such a scaffold with its own cells (Badylak et al. 2010). The precise architecture can never be replicated 100%, as even the actions of decellularisation and sterilization affect protein structure, however for matrix-rich tissue replacement this approach is of particular importance, as some elements of the mechanical features of load-bearing tissues can be replicated. There are continuing efforts to use this model to de-cellularise entire organs, and use these 'templet' scaffold architectures to re-seed cells into a more biomimetic environment (Badylak, Taylor, & Uygun 2010)

Evolence (or Dermicol-P35, ColBar LifeScience, Herzliya, Israel) consists of collagen suspended in a phosphate-buffered saline at a concentration of 35 mg/ml (Narins et al. 2007). The collagen is derived from pepsin-treated porcine tendons. The purified, telopeptide-poor collagen monomers are polymerised and then cross-linked with D-ribose. The product was FDA approved in 2008 for the treatment of mid to deep facial wrinkles and the cosmetic effect persists up to one year after administration (Narins et al. 2010). The risk of a hypersensitivity response to the product is less than 1% and the product does not require skin testing prior to use.

Human collagen

Cosmoderm (Inamed Corporation, Santa Barbara, CA), FDA approved in 2003, consists of collagen extracted and purified from tissue cultures of human foreskin fibroblast (Bauman 2004). Because the product is human derived there is virtually no risk of an immune response and patients therefore do not require pre-procedural skin testing. The collagen is suspended in PBS containing 0.3% lidocaine hydrochloride and the product is marketed in concentrations of 35 and 65 mg/ml (Cosmoderm I and II, respectively). Also a longer lasting variant is available in the form of Cosmoplast is cross-linked with glutaraldehyde for greater stability.

Cymetra (Life Cell Corp., Branchburg, NJ) is an intradermal injection product that is composed of cryofractured AlloDerm (Sclafani et al. 2002a). Cymetra is provided as a freeze-dried powder with an average particle size of 123 µm and is reconstituted through the rehydration of 330 mg in 1 ml of 1% lidocaine hydrochloride. Its effect last longer than Zyplast but shorter than Alloderm, possibly due to a larger surface area.

Autologen (Collagenesis, Inc., Beverly, Mass.) is a dermal matrix dispersion obtained from autologous skin obtained from the patient during previous elective surgery (Fagien and Elson 2001). Dermalogen, from the same company, is a similar product except that the skin is obtained from tissue banks. The dermis is mechanically pulverised and solubilised with chemical modifiers to form a dispersion of predominantly native, type I collagen, traces of types III and VI collagen, elastin, fibronectin, chondroitin sulfate, and other proteoglycans. Autologen and Dermalogen persistence is comparable to Zyplast (Sclafani et al. 2000;Sclafani et al. 2002b). Due to the human origin both products do not cause an immune response and do not require pre-procedural skin testing. The major drawbacks, however, are the requirement of a previous surgical procedure to acquire skin and a processing time of 3 to 4 weeks (Autologen) and the theoretical risk of disease transmission (Dermalogen). Both products are currently not available.

Isologen (Isologen Technologies, Metuchen, NJ) is not a collagen filler but uses autologous fibroblast transplantation to provide collagen at the site of interest. Fibroblasts are obtained from a 3 mm punch biopsy and grown for a period of 2 months to obtain a sufficient number of cells for a clinical effect (Watson et al. 1999).

5. Methods to control collagen architecture *in vivo and in vitro*: Engineering collagen conformation: density, direction, diameter

There are a wide variety of collagen substitutes, as have been documented, however to engineer a functional tissue with collagen, controlling the behaviour and architectural features of collagen *in vitro* will be necessary. Some of these mechanisms have been inspired by mechanisms *in vivo*, but the majority have derived from engineering principles derived from our understanding of the protein in question . Here we overview some novel approaches to controlling collagen architecture.

Collagen type I is used for 3D culture of cells, providing a biomimetic environment in which to study cell behaviour (Grinnell and Petroll 2010). Typically collagen scaffolds comprise of collagen hydrogels, which as the name suggests are mainly water. This generally means that the density of such scaffolds is inappropriate for modeling tissue matrix densities. These hydrogels are useful, however, in studying the interactions of cell-matrix, as cells are able to remodel the matrix into which they have been seeded, to orientate collagen fibrils, to control alignment of cells where strain is applied along an axis, and encourage specific behaviours of cells in response to alignment e.g. fusion of single myoblasts to form multi-nucleated fibres (Cheema et al. 2003). During this process the matrix is remodeled, and this tends to result in loss of water from the hydrogel, thus increasing the density of collagen.

Advances have been made to controllably increase the density of collagen scaffolds, with a recent application of controlled load to plastically compress (PC) standard collagen hydrogels to expel excess fluid and increase the collagen density (Brown et al. 2005). Application of this PC technology is very useful for *in vitro* tissue modeling as cell viability is retained in cells embedded within scaffolds undergoing PC, and the density increase (both for matrix and cells) is controllable and results in densities more biomimetic. At 11% collagen density, standard PC techniques bring scaffolds to *in vivo* levels of matrix density, however the mechanical properties of such scaffolds still fall far from those found in tissues. Further methods utilising PC technology have pushed these densities up to 30% and even higher (Abou-Neel et al. 2006). But what it lacks is the specific architecture for which a multi-disciplined approach to mimicking collagen architecture is required.

Despite the importance of collagen fibril diameter to the material properties of tissues, our basic understanding of its control is poor. Control of fibril diameter is distinct from fibrillogenesis, which is the emergence of the tertiary collagen protein structure.

Fibril modifying molecules, such as collagen types V and IX and proteoglycans such as decorin, are the main suggested mechanisms by which collagen fibril diameter is controlled in vivo, and these elements limit how large fibrils can grow in tissues such as cornea or tendon (Ameye and Young 2002;Ezura et al. 2000;Scott 1984). There is also an emerging understanding of how fibrils diameters can be increased, by using mechanical forces to apply cyclical loading of collagen containing gels, encouraging lateral fusion of fibrils, which is plausible when the quarter stagger patterns of the fibrils of collagen are in perfect register (Figure 1) (Cheema et al. 2007). This 'register' is most commonly identified as the banding pattern seen in transmission electron micrographs of native collagen fibrils. The need for banding pattern registration again lies in the short-range, non-covalent bonds presented between adjacent molecules, which drive fibril polymer formation. The bonding involved during this proposed fusion of fibrils is likely to be identical to the ionic and hydrogen bonding thought to stabilise the quarter stagger molecular packing in the original fibril.

Although this appears to be an engineering trick *in vitro*, it is highly likely that such mechanical forces occur in any new tissue *in vivo* under load. And importantly, the ability to control collagen fibril diameter without cells shows for the first time that mechanical forces in vivo may help determine fibril diameter and that cell-free engineering of native *collagen materials* is possible. Using technologies and strategies to manipulate and control fibril diameter will be critical to engineering collagen proteins for suitable use as a scaffold.

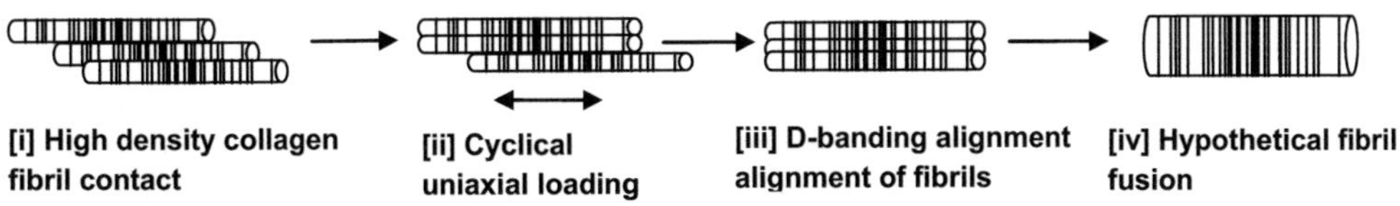

Figure 1a

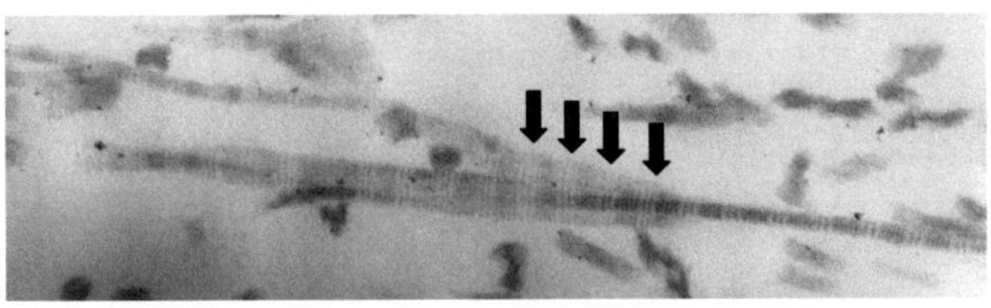

Figure 1b

Fig. 1. a) Hypothesis for the mechanically mediated mechanism for collagen fibril anastomosis. [i] Collagen fibrils in contact. At points of fibril lateral contact there will be very few where adjacent fibrils in 'surface charge' register. [ii] Application of cyclical uniaxial tensile strain inevitably brings a proportion of fibrils into register. [iv] Fibrils in contact and in register will form stable anastomoses, resulting in thicker fibril populations. [v] Bonding for this would comprise non-covalent, ionic, hydrogen and hydrophobic linkages which also drive fibrillogenesis. Figure 1b. Collagen fibrils in Longitudinal section, showing banding pattern register in the parallel elements in a multiple fibril cluster, as the fibril elements appear to anastomose. Arrows indicate alignment of banding pattern between different fibrils, over this length there are three separate fibrils in register, flanked by only one fibril (right) and three fibrils out of register (left) (magnification bar = 700 nm). Figure adapted from Cheema *et al.* 2007

The alignment of collagen fibrils in scaffolds is a critical parameter for control of architectural features. Without mimicking this alignment found in tissues, it is not possible to build a biomimetic tissue. Methods used *in vitro* to control alignment of collagen type I include magnetic alignment, interstitial directional fluid flow to control alignment and flow of collagen solution through microfluidic chambers during gelation (Elsdale and Bard 1972;Girton et al. 1999;Guo and Kaufman 2007;Lee et al. 2006;Ng and Swartz 2003;Ng and Swartz 2006). Elsdale and Bard were amongst the first groups able to align collagen. By simply setting a gel in a slanted chamber, and allowing interstitial fluid to flow downwards, collagen fibrils were observed to align along this fluid flow (Elsdale & Bard 1972). Tranquillo and colleagues have applied magnetic field to type I collagen scaffold, during gelation, and found that they collagen fibrils aligned along the plane in which the magnet was aligned (Girton, Dubey, & Tranquillo 1999).The majority of these alignment strategies have been applied to collagen hydrogels, however it is possible to align collagen fibrils in dense scaffolds using a similar fluid flow mechanism (Kureshi et al. 2010). Recently both the

magnetic alignment and fluid-flow alignment methods have been applied together to controllable align collagen fibrils (Guo & Kaufman 2007).
Electro-spinning, which is a method used to spin nano-and micro- diameter fibres into 3D meshes, is cuurently employed as a mechanism to compose 3D scaffolds using collagen and other natural proteins (Matthews et al. 2002). Collagen specifically retains its composition of polypeptide chains and even exhibits superfifical D-Banding of fibrils following the electro-spinning process (Jhu et al. 2011). Such electro-spun scaffolds have been employed as tissue engineered substitutes for dermal tissue and muscle reconstruction (Jhu et al. 2011). The electro-spun material is still deficient in some of its mechanical properties, including tensile strength, and the cell-interaction with such materials still requires further investigation.
The methods described within this section will need to be critically used to generate the meso and micro-scale architecture required to mimic tissues. There are limitations to each of these methods and further research into how to finely control collagen protein architecture is required. An example is how to control the bimodal distribution of fibril diameters within native tissues. Currently strategies rely mainly on increasing fibril diameter, for example by application of cyclical load, we are as yet unable to restrict fibril diameter in 3D collagen gels. Even with cyclical loading, we are only able to apply a broad loading regime to all the fibrils, and not select a population to increase, to allow for the typical bi-modal distribution. Further complications arise as cells often need to be embedded within 3D scaffolds as they are being cast, this means that making such engineering parameters cell-friendly will be critical to the ultimate success of an engineered tissue.

6. Problems with matrix-cell-interaction

One of the major obstacles in using scaffold materials into which cells are seeded, is the loss of cell directed control of matrix architecture, specifically protein architecture. Culturing cells in 3D collagen hydro-gels allows for cells to interact and remodel the matrix, but there are limitations to how accurately cell-matrix interaction *in vivo* can be mimicked (Grinnell & Petroll, 2010). It is difficult to simulate the force applied *in vivo* as a tissue develops or repairs, often multiple cell types and multiple matrix proteins are involved in the process, and this is time consuming to recapitulate *in vitro*. By engineering architecture is a speedy manner, we remove the cell-interactive component, and we may also be eliminating some of the cues necessary for cells to also mature and/or differentiate along with the protein architecture. As our models become progressively more tissue like, this interaction will need to be addressed.
Cell loading itself has an affect on cell-matrix interaction and matrix remodelling (Cheema et al. 2003). Compliant 3D collagen hydro-gels translate force to cells embedded within the scaffold very well, however as the collagen matrix stiffens and remodels, less force is translated to the embedded cells, primarily due to stress-shielding by the stiffer matrix. Further studies are required to dissect out this iterative process of matrix re-modelling with cell response, and while possible to study in simple single cell type and single matrix cultures *in vitro*, is much more difficult to decipher in a tissue.
Overarching all of this is the adaptive response of tissues, which renders the tissue able to adapt to changing mechanical requirements, environment, repair etc. In an animal, growth and repair of tissues requires a constant degree of adaptation, which is mainly orchestrated by the resident cells. These cells control protein deposition, and the architecture is then directed by the imposition of mechanical loading by the tissue and resident cells. This iterative process is difficult to define within the tissue, due to the presence of multiple cell

types, multiple proteins, interaction with surrounding tissues, complex mechanical loading and stress, and variance with age and degree of re-modelling.
Engineering functional collagen-based tissues, whilst trying to emulate all of these processes, and allowing for how they may further re-model such a scaffold is a challenge. Further research into the biological mechanisms controlling tissue architecture will help us develop more successful 3D implants.

7. Conclusions

Being the dominant protein of the majority of tissues of the body, the architecture and packing of collagen needs to be controlled is we are to successfully engineer tissues *in vitro*. There are many currently available collagen scaffold materials, for use *in vivo*, as replacement skin, tissue fillers and artificial vascular structures, but the best of these have relied upon retaining the original architecture of the collagen from tissues. By understanding how this protein is modified and packed *in vivo*, by cells, mechanical load and other matrix proteins, we can apply these procedures to predictably control collagen architecture. Without this level of controlled engineering of collagen protein for scaffold manufacture, it is unlikely we will successfully engineer scaffolds with the appropriate mechanical properties, to mimic native tissue mechanics.
There has been a significant drive to engineer the more dominant protein components found within matrix-rich tissues, but controlling the architectural parameters of more minor protein components is also critical, and getting these processes to work in synchrony remains a challenge. This control over tissue architecture needs to be done without compromising on cell viability, cell action and appropriate cell differentiation.
Using native proteins has both benefits, due to the biomimicry of protein motifs and structural elements, and pitfalls compared to manufacturing synthetic polymers as scaffold materials. Synthetic polymers are man-made and engineered to precisely fulfil the required mechanical properties of any implant, but they lack the native cell-attachment motifs. Controlling the architecture, *in vitro*, of native proteins, including predominantly collagen, holds the key to the successful engineering of biomimetic, native protein based implants and constructs.

8. References

Abou-Neel, E.A., Cheema, U., Knowles, J.C., Brown, R.A., & Nazhat, S.N. 2006. Use of multiple unconfined compression for fine control of collagen gel scaffold density and mechanical properties. *Soft Matter*, 2, 986-992

Amani, H., Dougherty, W.R., & Blome-Eberwein, S. 2006. Use of Transcyte and dermabrasion to treat burns reduces length of stay in burns of all size and etiology. *Burns*, 32, (7) 828-832 available from: PM:16997480

Ameye, L. & Young, M.F. 2002. Mice deficient in small leucine-rich proteoglycans: novel *in vivo* models for osteoporosis, osteoarthritis, Ehlers-Danlos syndrome, muscular dystrophy, and corneal diseases. *Glycobiology*, 12, (9) 107R-16R

Badylak, S.F., Taylor, D., & Uygun, K. 2010. Whole-Organ Tissue Engineering: Decellularization and Recellularization of Three-Dimensional Matrix Scaffolds. *Annu.Rev.Biomed Eng* available from: PM:21417722

Bauman, L. 2004. CosmoDerm/CosmoPlast (human bioengineered collagen) for the aging face. *Facial.Plast.Surg.*, 20, (2) 125-128 available from: PM:15643578

Bell, E., Ehrlich, H.P., Buttle, D.J., & Nakatsuji, T. 1981a. Living tissue formed in vitro and accepted as skin-equivalent tissue of full thickness. *Science*, 211, (4486) 1052-1054 available from: PM:7008197

Bell, E., Ehrlich, H.P., Sher, S., Merrill, C., Sarber, R., Hull, B., Nakatsuji, T., Church, D., & Buttle, D.J. 1981b. Development and use of a living skin equivalent. *Plast.Reconstr.Surg.*, 67, (3) 386-392 available from: PM:7232576

Brown, R.A., Wiseman, M., Chuo, C.B., Cheema, U., & Nazhat, S.N. 2005. Ultrarapid Engineering of Biomimetic Materials and Tissues: Fabrication of Nano- and Microstructures by Plastic Compression. *Adv Funct Mat*, 15, (11) 1762-70

Burke, J.F., Yannas, I.V., Quinby, W.C., Jr., Bondoc, C.C., & Jung, W.K. 1981. Successful use of a physiologically acceptable artificial skin in the treatment of extensive burn injury. *Ann Surg.*, 194, (4) 413-428 available from: PM:6792993

Cheema, U., Chuo, C.B., Sarathchandra, P., Nazhat, S.N., & Brown, R.A. 2007. Engineering Functional Collagen Scaffolds: Cyclical loading increases material strength and fibril aggregation'. *Adv Funct Mat*, 17, 2426-31

Cheema, U., Yang, S.Y., Mudera, V., Goldspink, G., & Brown, R.A. 2003. 3-D *In Vitro* Model of Early Skeletal Muscle Development. *Cell Motil Cytoskeleton*, 54, 226-236

Cooper, M.L., Hansbrough, J.F., Spielvogel, R.L., Cohen, R., Bartel, R.L., & Naughton, G. 1991. In vivo optimization of a living dermal substitute employing cultured human fibroblasts on a biodegradable polyglycolic acid or polyglactin mesh. *Biomaterials*, 12, (2) 243-248 available from: PM:1652296

Eisenberg, M. & Llewelyn, D. 1998. Surgical management of hands in children with recessive dystrophic epidermolysis bullosa: use of allogeneic composite cultured skin grafts. *Br.J Plast.Surg.*, 51, (8) 608-613 available from: PM:10209464

Elsdale, T. & Bard, J. 1972. Collagen substrata for studies on cell behavior. *J Cell Biol*, 54, (3) 626-37

Ezura, Y., Chakravarti, S., Oldberg, A., Chervoneva, I., & Birk, D.E. 2000. Differential expression of lumican and fibromodulin regulate collagen fibrillogenesis in developing mouse tendons. *J Cell Biol*, 151, (4) 779-88

Fagien, S. & Elson, M.L. 2001. Facial soft-tissue augmentation with allogeneic human tissue collagen matrix (Dermalogen and Dermaplant). *Clin.Plast.Surg.*, 28, (1) 63-81 available from: PM:11248870

Falanga, V. 2005. Wound healing and its impairment in the diabetic foot. *Lancet*, 366, (9498) 1736-1743 available from: PM:16291068

Girton, T.S., Dubey, N., & Tranquillo, R.T. 1999. Magnetic-induced alignment of collagen fibrils in tissue equivalents. *Methods Mol Med*, 18, 67-73

Griffiths, M., Ojeh, N., Livingstone, R., Price, R., & Navsaria, H. 2004. Survival of Apligraf in acute human wounds. *Tissue Eng*, 10, (7-8) 1180-1195 available from: PM:15363174

Grinnell, F. & Petroll, W.M. 2010. Cell motility and mechanics in three-dimensional collagen matrices. *Annu.Rev.Cell Dev.Biol*, 26, 335-361 available from: PM:19575667

Guo, C. & Kaufman, L.J. 2007. Flow and magnetic field induced collagen alignment. *Biomaterials*, 28, (6) 1105-1114 available from: PM:17112582

Hansbrough, J.F., Morgan, J., Greenleaf, G., & Underwood, J. 1994. Development of a temporary living skin replacement composed of human neonatal fibroblasts cultured in Biobrane, a synthetic dressing material. *Surgery*, 115, (5) 633-644 available from: PM:8178264

Heimbach, D., Luterman, A., Burke, J., Cram, A., Herndon, D., Hunt, J., Jordan, M., McManus, W., Solem, L., Warden, G., & . 1988. Artificial dermis for major burns. A multi-center randomized clinical trial. *Ann Surg.*, 208, (3) 313-320 available from: PM:3048216

Jha, B.J. Ayres, C.E. Bowman, J.R. Telemeco, T.A. Sell, S.A. Bowlin, G.L. and SimpsoN, D.E. 2011. Electrospun Collagen: A Tissue Engineering Scaffold with Unique Functional Properties in a Wide Variety of Applications. *Journal of Nanomaterials*, vol. 2011, Article ID 348268. doi:10.1155/2011/348268

Kligman, A.M. & Armstrong, R.C. 1986. Histologic response to intradermal Zyderm and Zyplast (glutaraldehyde cross-linked) collagen in humans. *J Dermatol.Surg.Oncol.*, 12, (4) 351-357 available from: PM:3082955

Knapp, T.R., Luck, E., & Daniels, J.R. 1977. Behavior of solubilized collagen as a bioimplant. *J Surg.Res.*, 23, (2) 96-105 available from: PM:886847

Kumar, R.J., Kimble, R.M., Boots, R., & Pegg, S.P. 2004. Treatment of partial-thickness burns: a prospective, randomized trial using Transcyte. *ANZ.J Surg.*, 74, (8) 622-626 available from: PM:15315558

Kureshi, A., Cheema, U., Alekseeva, T.A., Cambrey, A., & Brown, R.A. 2010. Alignment hierarchies: engineering architecture from the nanometre to the micrometre scale. *J R Soc Interface*, 6, S107-16

Lee, P., Lin, R., Moon, J., & Lee, L.P. 2006. Microfluidic alignment of collagen fibers for in vitro cell culture. *Biomed Microdevices*, 8, 35-41

Lemperle, G., Hazan-Gauthier, N., & Lemperle, M. 1995. PMMA microspheres (Artecoll) for skin and soft-tissue augmentation. Part II: Clinical investigations. *Plast.Reconstr.Surg.*, 96, (3) 627-634 available from: PM:7638287

Lemperle, G., Knapp, T.R., Sadick, N.S., & Lemperle, S.M. 2010. ArteFill permanent injectable for soft tissue augmentation: I. Mechanism of action and injection techniques. *Aesthetic Plast.Surg.*, 34, (3) 264-272

Lemperle, G., Ott, H., Charrier, U., Hecker, J., & Lemperle, M. 1991. PMMA microspheres for intradermal implantation: Part I. Animal research. *Ann Plast.Surg.*, 26, (1) 57-63

Marston, W.A., Hanft, J., Norwood, P., & Pollak, R. 2003. The efficacy and safety of Dermagraft in improving the healing of chronic diabetic foot ulcers: results of a prospective randomized trial. *Diabetes Care*, 26, (6) 1701-1705

Matarasso, S.L. 2006. The use of injectable collagens for aesthetic rejuvenation. *Semin.Cutan.Med Surg.*, 25, (3) 151-157

Matthews,J.A. WJ. G.E. Matthews, A. Wnek, G.E. Simpson, D.G. and G. L. Bowlin. 2002 Electrospinning of collagen nanofibers. *Biomacromolecules*, vol. 3, no. 2, pp. 232–238

McPherson, J.M., Sawamura, S., & Armstrong, R. 1986. An examination of the biologic response to injectable, glutaraldehyde cross-linked collagen implants. *J Biomed Mater.Res.*, 20, (1) 93-107 available from: PM:3949825

Michaeli, D. & McPherson, M. 1990. Immunologic study of artificial skin used in the treatment of thermal injuries. *J Burn Care Rehabil.*, 11, (1) 21-26 available from: PM:2107179

Naoum, C. & Dasiou-Plakida, D. 2001. Dermal filler materials and botulin toxin. *Int.J Dermatol.*, 40, (10) 609-621 available from: PM:11737419

Narins, R.S., Brandt, F.S., Lorenc, Z.P., Maas, C.S., Monheit, G.D., Smith, S.R., & McIntyre, S. 2007. A randomized, multicenter study of the safety and efficacy of Dermicol-P35 and non-animal-stabilized hyaluronic acid gel for the correction of nasolabial folds. *Dermatol.Surg.*, 33 Suppl 2, S213-S221 available from: PM:18086061

Narins, R.S., Coleman, W.P., III, Rohrich, R., Monheit, G., Glogau, R., Brandt, F., Bruce, S., Colen, L., Dayan, S., Jackson, I., Maas, C., Rivkin, A., Sclafani, A., & Spivak, J.C. 2010. 12-Month controlled study in the United States of the safety and efficacy of a permanent 2.5% polyacrylamide hydrogel soft-tissue filler. *Dermatol.Surg.*, 36 Suppl 3, 1819-1829 available from: PM:20969659

Ng, C.P. & Swartz, M.A. 2003. Fibroblast alignment under interstitial fluid flow using a novel 3-D tissue culture model. *Am J Physiol-Heart C*, 284, H1771-H1777

Ng, C.P. & Swartz, M.A. 2006. Mechanisms of interstitial flow-induced remodeling of fibroblast–collagen cultures. *Ann Biomed Eng*, 34, 446-454

Nguyen, D.Q., Potokar, T.S., & Price, P. 2010. An objective long-term evaluation of Integra (a dermal skin substitute) and split thickness skin grafts, in acute burns and reconstructive surgery. *Burns*, 36, (1) 23-28 available from: PM:19864070

Omar, A.A., Mavor, A.I., Jones, A.M., & Homer-Vanniasinkam, S. 2004. Treatment of venous leg ulcers with Dermagraft. *Eur.J Vasc.Endovasc.Surg.*, 27, (6) 666-672 available from: PM:15121121

Prystowsky, J.H., Siegel, D.M., & Ascherman, J.A. 2001. Artificial skin for closure and healing of wounds created by skin cancer excisions. *Dermatol.Surg.*, 27, (7) 648-653 available from: PM:11442617

Sclafani, A.P., Romo, T., III, & Jacono, A.A. 2002a. Rejuvenation of the aging lip with an injectable acellular dermal graft (Cymetra). *Arch.Facial.Plast.Surg.*, 4, (4) 252-257 available from: PM:12437432

Sclafani, A.P., Romo, T., III, Parker, A., McCormick, S.A., Cocker, R., & Jacono, A. 2000. Autologous collagen dispersion (Autologen) as a dermal filler: clinical observations and histologic findings. *Arch.Facial.Plast.Surg.*, 2, (1) 48-52 available from: PM:10925424

Sclafani, A.P., Romo, T., III, Parker, A., McCormick, S.A., Cocker, R., & Jacono, A. 2002b. Homologous collagen dispersion (dermalogen) as a dermal filler: persistence and histology compared with bovine collagen. *Ann Plast.Surg.*, 49, (2) 181-188 available from: PM:12187346

Scott, J.E. 1984. The periphery of the developing collagen fibril. *Biochem.J.*, 195, 229-33

Smith, J.W. 1960. Collagen fibre patterns in mammalian bone. *J Anat*, 94, (3) 329-44

Stern, R., McPherson, M., & Longaker, M.T. 1990. Histologic study of artificial skin used in the treatment of full-thickness thermal injury. *J Burn Care Rehabil.*, 11, (1) 7-13 available from: PM:2179224

Still, J., Glat, P., Silverstein, P., Griswold, J., & Mozingo, D. 2003. The use of a collagen sponge/living cell composite material to treat donor sites in burn patients. *Burns*, 29, (8) 837-841 available from: PM:14636761

Veves, A., Falanga, V., Armstrong, D.G., & Sabolinski, M.L. 2001. Graftskin, a human skin equivalent, is effective in the management of noninfected neuropathic diabetic foot ulcers: a prospective randomized multicenter clinical trial. *Diabetes Care*, 24, (2) 290-295 available from: PM:11213881

Wainwright, D., Madden, M., Luterman, A., Hunt, J., Monafo, W., Heimbach, D., Kagan, R., Sittig, K., Dimick, A., & Herndon, D. 1996. Clinical evaluation of an acellular allograft dermal matrix in full-thickness burns. *J Burn Care Rehabil.*, 17, (2) 124-136 available from: PM:8675502

Watson, D., Keller, G.S., Lacombe, V., Fodor, P.B., Rawnsley, J., & Lask, G.P. 1999. Autologous fibroblasts for treatment of facial rhytids and dermal depressions. A pilot study. *Arch.Facial.Plast.Surg.*, 1, (3) 165-170 available from: PM:10937098

Wilkins, L.M., Watson, S.R., Prosky, S.J., Meunier, S.F., & Parenteau, N.L. 1994. Development of a bilayered living skin construct for clinical applications. *Biotechnol.Bioeng.*, 43, (8) 747-756 available from: PM:18615798

Yannas, I.V., Burke, J.F., Warpehoski, M., Stasikelis, P., Skrabut, E.M., Orgill, D., & Giard, D.J. 1981. Prompt, long-term functional replacement of skin. *Trans.Am Soc Artif.Intern.Organs*, 27, 19-23 available from: PM:7036496

14

Tissue-Engineered Extracellular Matrices (ECMs) as Adjuvant Scaffolds for Endovascular Aneurysmal Repair (EVAR)

Anthony Callanan, Niall F. Davis,
Michael T. Walsh and Tim M. McGloughlin
*Centre for Applied Biomedical Engineering Research (CABER) and MSSI,
University of Limerick, Castletroy, Co. Limerick,
Ireland*

1. Introduction

Abdominal aortic aneurysms (AAA) are permanent, irreversible, localised dilatations of they aorta. Usually, they develop as a result of a progressive localised weakness within the vessel wall. They typically occur below the level of the renal arteries and have a high propensity for rupture. In fact, ruptured AAAs account for approximately 8,000 and 15,000 deaths in the United Kingdom (UK) and United States of America (USA) respectively on an annual basis (Sakalihasan et al. 2005, Thompson 2003, Vorp and Vande Geest 2005). Risk factors for their development include male gender, age >65 and a history of smoking. Other associated risk factors are connective tissue disorders that typically have a genetic predisposition, syphilitic infections and cystic medial necrosis.

Currently, there are two surgical treatments for AAA; the traditional open repair and a minimally invasive procedure known as endovascular aneurysm repair (EVAR) (Kamineni and Heuser 2004, Parodi et al. 1991, Sakalihasan et al. 2005). The endovascular technique has been widely applied in clinical practice, however important limitations persist (Egelhoff et al. 1999, Kamineni and Heuser 2004, Parodi et al. 1991). Among these limitations are device migration, endoleaks, and thrombotic occlusion. It has been suggested that tissue-engineered xenografts may play a role for preventing these complications in EVAR. In the present chapter we discuss limitations of stent-grafts deployed in the EVAR procedure. We place particular emphasis on tissue-engineered extracellular matrices (ECMs) as adjuvant scaffolds for optimisation of the EVAR procedure.

2. Aetiology of aneurysms

The arterial wall is primarily composed of 3 layers (or tunicae) that surround the luminal cavity as illustrated in Fig. 1. The inner layer (or tunica intima) consists of a monolayer of endothelial cells. A thin membrane known as the elastica interna separates the tunica intima from the tunica media and the tunica media itself consists of concentric layers of smooth muscle cells interwoven between networks of connective tissue. The tunica media is separated from the outer tunica adventitia by the elastica externa and adventitial constituents include collagen and interspersed fibroblasts. Elastin is the predominant tissue within the aorta and it functions as the principal load bearing element of the aortic wall. During aneurysm formation degradation of elastin occurs along the walls of the aorta (Raghavan et al. 2005). It is widely believed that degradation of elastin may promote an inflammatory response within the wall leading to weakened tissue, abnormal remodelling responses and subsequent AAA development.

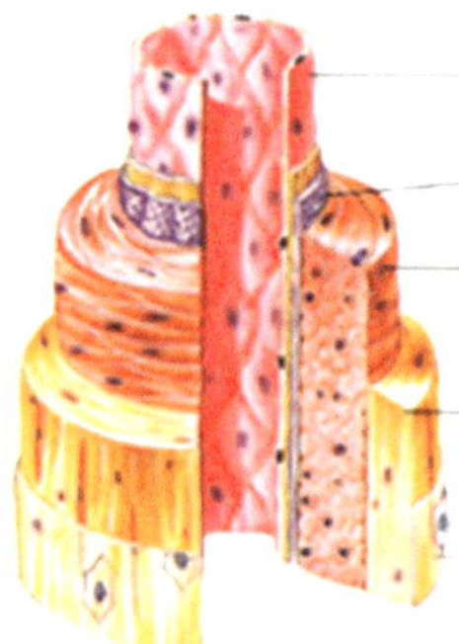

Fig. 1. Histological structure of arterial wall

Currently, aneurysms are classified relative to their shape and location. Fusiform aneurysms are the most common and they typically occur due to a circumferential weakness along an extended portion of the aorta with the weakened portion appearing as a symmetrical bulge. In contrast saccular aneurysms frequently form on one side of the aorta and are asymmetrical in their nature. Finally, pseudoaneurysms usually occur as a result of trauma to the aortic wall that causes all 3 layers of the vessel to separate.

3. Surgical treatment options for AAA

3.1 Open repair

Currently there are two vascular procedures available for the treatment of AAA; an open procedure and minimally invasive surgery. Open surgery involves a midline incision to gain access to the aneurysmal site. Intraoperatively, the aorta and iliac arteries are exposed and

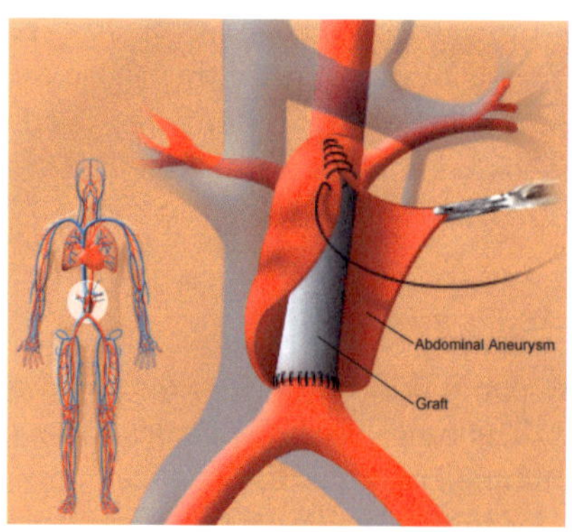

Fig. 2. Open repair of abdominal aortic aneurysm (AAA)
(http://www.musc.edu/radiology/interventional/index.htm)

cross-clamped prior to incising the wall of the AAA. Intraluminal thrombi are removed and a synthetic graft is sutured *in situ* before the aneurysm is closed over the graft as illustrated in Fig. 2. Complications associated with the open method include infection, increased inpatient stay and a predisposition to acute renal failure.

3.2 Minimally invasive approaches

The minimally invasive approach involves the use of a stent graft device as illustrated in Fig 3 and this method is referred to as 'EndoVascular Aneurysm Repair (EVAR)'. Importantly, EVAR is associated with a significant decrease in mortality and reduced duration of inpatient stay when compared to open repair; however EVAR has additional procedural risks.

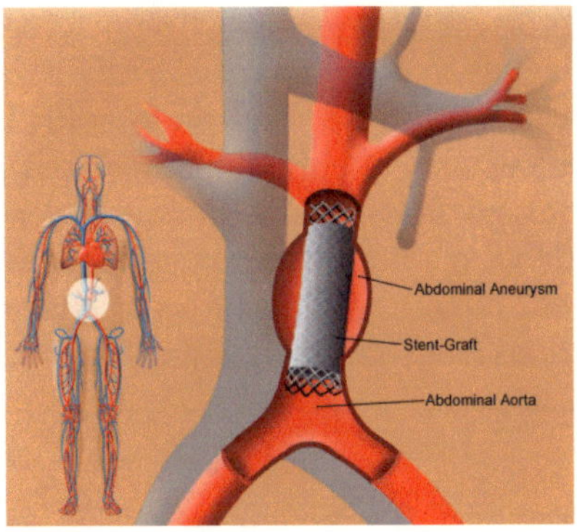

Fig. 3. Endovascular aneurysmal (EVAR) repair of AAA
(http://www.musc.edu/radiology/interventional/index.htm)

3.2.1 Complications associated with EVAR

Intraoperative complicatons include vascular injury during initial catheterisation and stent deployment and postoperative complications associated with EVAR include persistent blood flow outside the graft that can result in increased pressure on the aneurysmal sac and a subsequent endoleak. To date, endoleaks have been classified into 5 different subtypes and types 1-5 are illustrated in table. 1

Endoleak Type	Characteristics
1	Occurs at proximal or distal end of the stent-graft where it attaches to vessel wall
2	Precipitated by collateral flow from mesenteric or lumbar arteries
3	Tear in graft fabric and blood leaks between modular components of stent graft
4	Leak occurs through pores within the graft's fabric
5	Occurs as a result of 'intra-sac' pressurisation and is commonly referred to as endotension

Table 1. Endoleaks type 1-5 and their associated clinical features (Greenhalgh and Powell 2008)

Typically, endoleaks that are classified as type 3 or type 4 resolve spontaneously; however endoleaks classified as type 1 or type 2 often require surgical intervention (Greenhalgh and Powell 2008). Treatment options for type 5 remain controversial as surgeons remain divided on advocating immediate surgical repair versus a more conservative surveillance approach (Mennander et al. 2005, Veith et al. 2002).

Migration of the stent-graft following the EVAR procedure is another complication associated with considerable postoperative morbidity. Clinically, migration can be defined as ≥5mm of distal movement of the stent-graft from its attachment site. Usually migration can be caused by inadequate attachment of the graft to the proximal neck of the aneurysm or by morphological changes within the neck of the vessel. Postoperative complications include a widening of the aneurysm that may result in decreased radial force exerted by the proximal portion of the stent *in vivo*. Radial force is an important fixation method in stent-grafts without hooks or barbs and widening of the neck inevitably predisposes these devices to migration. Less frequent complications include stenosis or occlusion of the graft or distal vessels.

On account of these complications only 5 stent-grafts with FDA approval are currently available. These are the AneuRx® and Talent® from Medtronic, the Zenith® from Cook, the Gore Excluder® and the Endologix Powerlink® (Endovascular Today, 2009). Their success is also limited by a high incidence of endoleaks and stent migration as illustrated in Table 2. Typically, the majority of stent migration failures occur after one year. A reliable method that prevents these complications from occurring is an attractive option.

Device	Type	Stent Material	Graft Fabric	Stent type	Hooks or Barbs	Documented Endoleak	Documented Migration	Reference
AneuRx	Modular external	Nitinol external	Dacron	Nitinol skeleton	No	Yes	Yes	(Zarins et al. 2001)
Talent	Modular self-expanding internal	Nitinol	Dacron	Multi nitinol stents (Bare)	No	Yes	Yes	(Criado et al. 2003)
Zenith	Modular	Stainless steel	Dacron	Z-stents	Yes	Yes	Yes	(Greenberg 2003)
Gore Excluder	Unitary external tube type and internal	Nitinol	Teflon ePTFE	Spiral shape	No	Yes	Yes	(Bush et al. 2001)
Endologix PowerLink	Unitary Body internal stent type	Cobalt-Chromium Alloy	ePTFE	Single wire Z-shaped	No	Yes	Yes	(Wang et al. 2008)

Table 2. Stent graft devices for EVAR: Their structural properties and associated complications

3.3 Possible method for improvement of EVAR

Previously, investigations for improving the stent-graft design have predominantly focussed on the mechanical aspects of the stent. In general, most of these investigations have failed and a reliable solution remains elusive. Recent investigations (Brown et al. 2006, Schoder et al., 2004, Niyyati et al. 2005, Yavuz et al. 2006) suggest that tissue-engineered extracellular matrix (ECM) scaffolds derived from xenogenic sources may have the potential to overcome limitations that are associated with traditional mechanical solutions.

4. Tissue-engineered Extracellular matrix (ECM) scaffolds

Extracellular matrices (ECMs) are biological scaffolds usually derived from xenogenic sources. They are acellular in nature and induce a host derived tissue-remodelling response after implantation while undergoing simultaneous degradation processes (Davis et al., 2010). Therefore, they may provide an attractive alternative as suitable biomaterials for improving EVAR treatment of AAA. Urinary bladder matrix (UBM) and small intestine submucosa (SIS) are two common ECM scaffolds of porcine origin that have had good clinical outcomes after surgical implantation across numerous subspecialties. These biomaterials are prepared via numerous physical, chemical and enzymatic processes.

Author (Year)	Application	Animal Model	Result
S.F. Badylak (1989)	Large diameter vascular graft	Canine	No thrombus formation and no intimal hyperplasia after 44 weeks. No endothelialisation of the scaffold and greater stiffness within the scaffold compared to artery
D.J. Schultz (2002)	Reparation of enterocutaneous fistula	Human	No data available
S.G. de la Fuente (2002)	Gastric reparation	Rat	Regeneration of gastric mucosa after 3 weeks
M. Chen (2001)	Small intestine	Canine	Tubular failure secondary to obstruction and leakage
M.A. Cobb (1999)	Dura mater substitute	Canine	Complete resorption after 60 days
M. Rosen (2002)	Regeneration of biliary system	Canine	Infiltration with fibroblasts after 2 weeks. Biliary epithelium replaced with native collagen after 3 months.
S.F. Badylak (2001)	Body wall repair	Canine	Preoperative tensile strength achieved after 24 months
T.G. Smith (2002)	Reparation of ureteral defect	Porcine	SIS graft replaced with urothelium and smooth muscle after 9 weeks
B.P. Kropp (1995)	Urinary bladder regeneration	Rat	Urothelium, lamina propria and smooth muscle replaced within 3 weeks

Table 3. Multisystemic applications for SIS and associated remodelling duration after *in vivo* implantation

Extensive experimental evaluation of SIS and UBM materials has been undertaken; examining sterilisation effects, cell interactions, cell growth effects, gene expression, mechanical properties, processing effects, suture retention effects and repeatability issues associated with their clinical applications (Ahn et al. 2007, Cimini et al. 2005, Freytes et al. 2005, Freytes et al. 2008, Gilbert et al. 2006, Hodde et al. 2002, Roeder et al. 1999, Sellaro et al. 2007, Teebken et al. 2000). These characterisation methods have led to a greater understanding of an ECM's biological, structural and mechanical properties. Clinical applications of ECM scaffolds are described in Table 3. Influences such as stent interaction, compliance differences, aortic endothelial cell interactions and flow effects, have not been fully characterised on ECMs using *in vitro* or *in vivo* experimental approaches.

Although UBM has been utilised to a lesser extent compared to SIS, it has been successfully applied for the treatment of dysplastic oesophygeal tissue with excellent patency rates during the follow up period (Badylak et al., 2005). In addition, UBM has been applied for effectively treating strictures of the trachea with no evidence of stenosis or tracheomalacia during the follow up period (Gilbert et al. 2008). Finally, UBM has also been effectively applied for reparation of the thoracic wall in a canine model (Gilbert et al. 2007).

4.1 Constituents and preparation of ECMs

The major constituents found within mammalian ECMs are collagen, glycoproteins, glycosaminoglycans (GAGs) and growth factors as illustrated in Table 4. These constituents provide structural, functional, adhesive and stimulatory functions to their surrounding cells enabling them to survive and proliferate (Badylak et al. 2009, Baldwin 1996, Laurie et al. 1989). Naturally, the transition phase of an ECM scaffold from intact mammalian tissue to viable donor xenograft material requires several processing steps. Initially, the native tissue is manually separated from unwanted tissue structures. Tissue decellularisation is achieved through a combination of sonication, agitation, freezing and thawing processes (Badylak et al. 2009). These treatments disrupt the cell membrane and facilitate the removal of intracellular remnants. During the decellularisation process it is of paramount importance to preserve as many mechanical and biological properties of the donor ECM as possible. Disruption of collagen architecture can decrease the mechanical strength of the scaffold, removal of GAGs adversely affect its viscoelastic behaviour and the absence of growth factors will decrease the scaffold's bioinductive properties (Lovekamp et al. 2006). After the xenograft is decellularised it is then sterilised by exposure to irradiation or ethylene oxide (Rosario et al. 2008).

To date, porcine SIS and porcine UBM have been strongly favoured as potential donor scaffolds for many different surgical subspecialties. Their harvesting sites differ; however both have had a considerable degree of success when applied clinically. SIS is harvested from the small intestine and UBM originates from the urinary bladder. Their collagen components also differ to a small extent after decellularisation and sterilisation. The decellularised SIS scaffold is primarily composed of collagen type 1 (Badylak et al. 2009) with smaller amounts of collagen types 3, 4, 5, 6 also present (Badylak SF 1995). In contrast, UBM is rich in collagen types 3, 4 and 7. UBM possess an intact basement membrane which has many characteristics that favour its application to the vasculature. Importantly, a basement membrane may support the growth and differentiation of a confluent endothelial cell layer on the luminal surface of the scaffold. Different characteristics of SIS and UBM are compared in Table 5.

Constituents	Features
Collagen	• Most abundant protein within ECMs • More than 20 different types have been identified • Provides distinct mechanical and physical properties to the ECM
Laminin	• Large adhesion glycoprotein • Involved in cell and tissue differentiation • Promotes tissue development and angiogenesis
Fibronectin	• Extracellular glycoprotein • Promotes host biocompatibility • Induces cell adhesion by binding to membrane-spanning receptor proteins known as integrins
Glycosaminoglycans	• Mucopolysaccharides • Bind covalently to a protein core to form a proteoglycan molecule • Act as a reservoir when cells stop growth factor production • Enables ECMs to store growth factors that may be used during tissue regeneration
Growth Factors	• Present in small quantities within ECMs • Naturally occurring substances capable of stimulating cellular growth, proliferation and differentiation

Table 4. Constituents of ECM scaffolds and their associated biological features

Extracellular Matrix	Characteristics
Small Intestinal Submucosa (SIS)	• Derived from porcine jejunum • Primarily composed of the submucosal layer and the stratum compactum of the tunica mucosa • Mainly consists of type 1 collagen
Urinary Bladder Matrix (UBM)	• Derived from the porcine urinary bladder • Possesses an intact basement membrane consisting of collagen types 4 and 7

Table 5. Differentiating the characteristics of SIS from UBM

4.2 Immunogenic response after implantation

Theoretically, an implanted ECM should not elicit an immediate or delayed immune response due to its acellular and avascular nature (Allman et al. 2001, Ho et al. 2004, Sandusky et al. 1992). However, we know that elimination of all nuclear materials and cell membrane products is almost impossible despite extensive measures taken during the decellularisation process. Therefore, it is expected that the recipient should mount an immune response against the graft's cell remnants and arguably, against the intact xenogenic proteins. This hypothesis has been extensively studied by assessing the host's cell-mediated T-helper 1 (rejection) and T-helper 2 (accommodation) immune responses to implanted xenografts (Strom et al. 1996, Zhai et al. 1999) .

Results from preliminary studies on mice are favourable, as the implanted SIS scaffold elicits an immune lymphocytic response that is predominately Th2-like (Allman et al. 2001). The Th-2 pathway stimulates the production of interleukins IL-4, IL-5, IL-6 and IL- 10. These interleukins promote graft acceptance and prevent the activation of neighbouring inflammatory macrophages (Bach et al. 1997, Chen and Field 1995) . Activation of the Th2 pathway also promotes effective tissue remodelling, structural repair and functional recovery of the injured tissue after graft acceptance (Piterina AV 2009). Undoubtedly, activation of this humoral response is encouraging as activation of the alternate lymphocytic pathway (i.e. Th1) produces an acute inflammatory reaction. Cytokines such as IL-2, interferon (IFN) gamma and tumour necrosis factor (TNF) beta activate neighbouring macrophages and stimulate the differentiation of CD 8+ cells to a cytotoxic phenotype. This host derived inflammatory response ultimately leads to xenogenic graft rejection (Abbas et al. 1996, Matsumiya et al. 1994).

The terminal alpha 1,3 galactose epitope (Gal-epitope) is present in cell membranes of all mammals except humans and concerns over the epitope's inflammatory potential exist (Galili 1993, Koren et al. 1994). In humans this epitope (i.e. antigenic determinant) is recognised by IgM, IgG and IgA antibodies that mediate hyperacute or delayed graft rejection through complement fixation and antibody dependent cell mediated cytotoxicity (Good et al. 1992, Koren et al. 1994, Schussler et al. 2001). The potential for complement activation has been investigated with results suggesting that it does not occur when the graft is implanted (McPherson et al. 2000). Researchers have attributed the absence of host immune responses to the distribution of the epitope within the xenograft and to the minute quantities that are present (McPherson et al. 2000). In whole organ transplantation levels of the Gal-epitope are expectantly higher and these high levels have been linked to chronic graft rejection (Schussler et al. 2001). Currently, methods of eliminating the epitope prior to scaffold implantation are under investigation and it has been suggested that treatment of the xenogenic scaffold with alpha galactosidase during the decellularisation process is a potential solution. Should clinical concerns persist it might also be possible to harvest the graft material from transgenic Gal-knockout pigs that are bred specifically for tissue engineering purposes.

The graft's response to potential host derived pathogenic micro-organisms has also raised concerns among vascular surgeons as graft infection is associated with considerably morbidity. The xenograft's response to Gram-positive and Gram-negative bacteria has been evaluated and compared to conventional synthetic graft materials (i.e. polytetrafluoroethylene) in animal studies (Badylak et al. 2003). Interestingly, xenogenic ECM materials were resistant to persistant bacterial infection after deliberate contamination at the graft implantation site. This has been attributed to the presence of multiple low-molecular weight peptides that survive the decellularisation and sterilisation processes (Brennan et al. 2006, Sarikaya et al. 2002). These peptides demonstrate bacteriostatic activity against micro-organisms and inhibit bacterial proliferation for up to 12 hours after initial exposure. Their antimicrobial activity protects the remodelling site from circulating pathogens (Brennan et al. 2006). However, their origin and structural homology to natural antimicrobial peptides (AMP) and defensins are important aspects that have not been clarified to date. Their spectrum of activity and pathways of incorporation are also poorly understood. These factors need to be thoroughly investigated so the extent of their antibacterial role can be clearly established.

4.3 Remodelling and degradation

Biological growth factors found within SIS and UBM are key contributors to cell growth and tissue regeneration (Babensee et al. 2000, Tabata 2004, Tabata 2005). Proteoglycans facilitate their survival during matrix decellularisation and sterilisation by functioning as storage vessels (Hodde et al. 2005). As the matrix is implanted growth factors are released stimulating angiogenesis, host cell infiltration and mitogenesis (Table 6). Matrix degradation coincides with this and the degradation process is influenced by host derived enzymatic and cellular processes (Badylak 2007). During the degradation process growth factors dissociate from their binding proteins and are activated to promote tissue neovascularisation. Matrix degradation and growth factor activation continues until the ECM scaffold is completely replaced by host cells (Clyne and Edelman 2009).

Growth Factor	Function
Vascular Endothelial Growth Factor (VEGF)	Regulates angiogenesis by controlling blood vessel formation and growth
Platelet Derived Growth Factor (PDGF)	Deposition of granulation tissue and stimulation of angiogenesis
Bone Morphogenetic Protein (BMP)	Stimulates formation of bone and cartilage
Keratinocyte Growth Factor (KGF)	Epithelialisation of wounds during healing
Fibroblast Growth Factor (FGF)	Induces growth of fibroblasts and endothelial cells during wound healing
Transforming Growth Factor (TGF-beta)	Reorganization of matrix molecules to improve dermal architecture and reduce scarring

Table 6. Bioinductive growth factors found within the ECMs and their functions

Common growth factors that influence tissue remodelling responses include 'Vascular Endothelial Cell Growth Factor' (VEGF) and 'basic Fibroblast Growth Factor (bFGF)' as illustrated in Table 6. VEGF has been shown to stimulate angiogenesis, vascular permeability and endothelial cell proliferation and migration while bFGF encourages wound healing (Ferrara et al. 1992). Other retained growth factors within the ECM include keratinocyte growth factor (KGF) which mediates epithelial cell proliferation and differentiation (Alpdogan et al. 2006) and platelet-derived growth factor-beta-polypeptide (PDGF-BB) which promotes chemotaxis, proliferation, angiogenesis and tissue remodelling. It appears that the strong remodelling effect exerted by biological growth factors is accentuated by cryptic peptides that are also released from the implanted scaffold during the degradation process. These peptides are involved in recruiting circulating bone-marrow derived cells that can partake in long-term tissue remodelling processes (S. F. Badylak et al. 2001, Zantop et al. 2006).

An ability to be completely degraded while stimulating a native remodelling response over a relatively short period of time is perhaps ECM's most attractive feature (Badylak et al.

2000, Davis et al. 2011, Gilbert et al. 2007). These impermanent properties were investigated by determining the rates of *in vivo* graft degradation and excretion in canine models during the nineties (Badylak et al. 1998, Kropp et al. 1995, Kropp et al. 1996a, Kropp et al. 1996b, Vaught et al. 1996). Studies show that xenogenic ECMs are rapidly degraded and absorbed when implanted in the genitourinary tract with up to 90% of the scaffold being replaced by host tissue within 28 days (Badylak et al. 1998, Record et al. 2001). Generally, excretion rates of all ECMs vary between 28 and 90 days depending on the type of tissue that is being remodelled (Allman et al. 2001, Badylak et al. 1998, Record et al. 2001). Naturally, the process may be prolonged when multiple layers of xenogenic scaffold are implanted (e.g. 90-120 days). Shortly after the degradation process the ECM briefly enters the blood stream and is excreted via the kidneys through glomerular filtration. This has been shown by measuring quantitative studies of ^{14}C- labelled SIS after augmentation cystoplasty in canine models (Record et al. 2001). More than 50% of the scaffold was removed from the implantation site at 28 days and almost 100% of the scaffold was replaced by 100 days. During the follow up period 95% of degradation products were found in the host's urine (Badylak et al. 2000, Record et al. 2001).

4.4 Mechanical properties

ECMs' remodelling capacity is dependent on the preservation of bioinductive growth factors during the sterilisation process. Similarly, its mechanical effectiveness is largely dependent on preserving intact collagenous arrangements and adhesive glycoproteins during this process. This is highlighted by the scaffold's mechanical response to different sterilisation techniques. Studies have shown that the graft's uniaxial and biaxial mechanical properties are significantly reduced after exposure to gamma irradiation, electron beam irradiation and ethylene oxide (Freytes et al. 2008) . The reduction in mechanical strength is dose-dependent and this emphasises the preparation difficulties encountered between graft sterilisation and constituent preservation techniques (Gouk et al. 2008).

Short-term mechanical limitations are also present during the initial remodelling response (Davis et al. 2011). Typically, both SIS and UBM show a decrease in mechanical strength after implantation that is caused by a temporal imbalance between the rate of scaffold degradation and the rate of infiltrating host cell deposition (Gilbert et al. 2007). While the rapid degradation rate of implanted genitourinary ECMs is often lauded, one must consider the temporal mismatch that occurs between xenograft degradation and host-derived matrix deposition. One study demonstrated a 30-fold decrease in bladder compliance (in comparison to the pre-operative status) after canine urinary bladder was replaced with SIS (Kropp et al. 1996b).

Short-term strength limitations have been addressed by increasing the number of layers within the implanted scaffold as a single layer of implanted SIS has proved insufficient for most load bearing organs. The graft's mechanical strength increases by 150% simply by increasing the number of layers of SIS layers from 2 to 4 (Freytes et al. 2004). Importantly, the imbalance between matrix degradation and deposition is temporary in nature and is only relevant until the host's remodelling capability equates-to or surpasses the ECM's degradation rate. A rapid remodelling response can occur once infiltrating host cells self-organise and begin producing their own ECM. This results in a time dependent return to expected mechanical strength and site-appropriate mechanical behaviour after xenogenic implantation (S. Badylak et al. 2001, Badylak et al. 2005, Liang et al. 2006).

5. ECMs as potential vascular grafts

A number of studies have investigated SIS's potential as a tissue-engineered vascular substitute. Initially, Badylak *et al.* replaced a segment of canine aorta in 1989 with a tubularised SIS scaffold (Badylak et al. 1989). Results from this preliminary study demonstrated patency of the aorta during thc follow up period. Importantly, adverse effects such as infection, thrombosis, intimal hyperplasia and hypertension were avoided. Histological assessment after follow up revealed organised, dense non-thombogenic collagenous connective tissue. However, there was no evidence of endothelial cell growth on the luminal surface of the SIS graft after 44 weeks. The authors concluded by suggesting that SIS merits further investigation as a large diameter graft for aortic replacement purposes. Consequentially, SIS was subsequently investigated as a potential small diameter graft in a canine model (Lantz et al., 1990) where the biological scaffold replaced the carotid and femoral artery (Lantz et al. 1990). Like Badylak *et al.*, results from this study indicated no evidence of endothelial growth on the scaffold's luminal surface. Luminal and abluminal surfaces of the scaffold were comprised of dense organised collagenous tissue, with no evidence of infection, propagating thrombus or intimal hyperplasia.

A more comprehensive study by Sandusky *et al.* also evaluated SIS as a small calibre vascular graft for carotid arteries in canine models in 1992 (Sandusky et al. 1992). A sample size of 24 canine models was included in this study over a period of 180 days where gold standard saphenous vein grafts were directly compared with SIS scaffolds. Results from this study revealed endothelialisation of the SIS scaffold with transmural growth of capillaries and infiltrating smooth muscle cells from the host. In addition, no significant differences were noted between saphenous vein grafts and SIS scaffolds when intimal thickening was compared in this study. In 1995 Hiles *et al.* assessed the mechanical properties of SIS as a potential aortic graft in a canine model (Hiles et al. 1995). Their study demonstrated that host tissue completely replaced the SIS scaffold and that the remodelled scaffold had appropriate physical and mechanical properties to adequately function in a vascular system. However, compliance values from the remodelled SIS construct were three-fold lower than compliance values of a normal thoracic aorta (Hiles et al. 1995).

Another study by Roeder *et al.* also investigated the compliance, burst pressure and remodelling effects of SIS as a small diameter vascular graft (Roeder et al. 2001). Findings from their study demonstrated a degree of tissue remodelling with improved compliance at the site of the implanted SIS scaffold. The authors concluded by suggesting that mechanical properties of the remodelled SIS scaffolds were similar to the vessel of the animal model that was replaced. Ovine models have also been utilised to assess SIS's potential as a small diameter vascular graft (Pavcnik et al. 2009). One study, evaluated the implanted SIS scaffolds with angiography during their follow up period. Angiographic assessment of implanted SIS scaffold revealed a multitude of complications that included stenosis of the anastamotic site, aortic dissections, recurrent aneurysmal formation and diffuse dilatations of the implanted scaffold.

Although SIS has been extensively investigated as a potential vascular replacement scaffold it is interesting to note that UBM has never been previously investigated for this purpose. Encouragingly, more recent studies have suggested that UBM merits further investigation as a potential vascular substitute (Badylak 2005, Brown et al. 2006). After the preparation process UBM can be manipulated into many different physiological configurations. Its malleable nature in conjunction with its biocompatibility may provide researchers with an alternative ECM scaffold for vascular replacement purposes.

6. Justification of ECMs in EVAR

In EVAR failure of the implanted stent most frequently occurs one year after surgical implantation as discussed in section 3.2.1. However, ECMs take approximately 3 months to induce a constructive tissue remodelling effect (Badylak 2005, Gilbert et al. 2008). Therefore, a tissue-engineered ECM is likely to reabsorb within this 12 month timeframe and provide a secure seal that could potentially prevent the complication of stent migration.

6.1 Disadvantages of biodegradable polymers

The potential for biodegradable polymers as potential vascular replacements in EVAR stent-grafts has previously been investigated. Poly D, L-lactic–glycolic acid co-polymer (PLGA) and Poly ε-caprolactone (PCL) are 2 polymers that have been assessed *in vivo* and *in vitro* as potential vascular substitutes. In one study the degradation rates of both polymers were assessed and compared *in vivo* and *in vitro*. Results demonstrated that degradation rates of both scaffolds occurred at a more rapid rate *in vivo* compared to *in vitro*. Hydrolysis of PLGA and PCL polymers were influenced by the concentration of carboxylic acid end-groups. Therefore, it appears that degradation products of both polymers may serve as catalysts for reactions in static conditions, which are likely to accelerate degradation. This study suggests that by-products of the initial degradation influence the effect of cell infiltration (Sung et al. 2005). Therefore, these initial studies imply that biodegradable polymers are unsuitable in the setting of EVAR as they are associated with undesirable by-products that adversely affect their degradation rate (Sung et al. 2005).

6.2 Configurations of ECM stent grafts for EVAR

SIS has been investigated in a stented environment on the abdominal aorta of ovine models where remodelling of the SIS scaffold onto the aortic wall was assessed by Yamada *et al.* (Yamada et al. 2001). In this study the SIS stent graft was manufactured by sandwiching the stent between two sheets of SIS as illustrated in Fig. 4. Results demonstrated no evidence of stenosis and no evidence of endoleak formation around the implanted stent grafts. Histological assessment showed incorporation of the graft into the wall of the aorta with a dense neo-intima replacing the SIS scaffold. Endothelialisation occurred in areas where the graft was in direct contact with the aortic wall and central portions of the graft were partially endothelialised after the 12 week follow up period. A similar study by Noishiki et al., 2001 reported comparable results to Yamada *et al.* with a partial endothelium forming on the implanted SIS hybrid stent graft (Noishiki et al. 2001).

After these promising results the performance of SIS covered endografts (stent devices) implanted into ovine femoral arteries was investigated by Nakata *et al.* in 2003. The study compared the performance of the SIS covered endografts to non-covered nitinol stents and PTFE covered endografts. In their conclusion the authors suggest that SIS and bare metal nitinol stents display similar attachment features to the aortic wall and performed superior to a poly-tetra-fluoro-ethylene (PTFE) covered stent also included in the study (Nakata et al. 2003). It should also be noted that both SIS endografts and bare nitinol stent exhibited eccentric intimal hyperplasia with eventual occlusion of the stented vessel during the follow up period. A study by Schoder *et al.* also investigated SIS in the setting of EVAR repair (Schoder et al. 2004). In this study the SIS scaffold was suspended against the wall of the aorta as illustrated in Fig. 5. Results from this study suggest that this deployment method is promising for the prevention of type 2 endoleaks. In addition, results also demonstrated evidence of a host derived tissue remodelling response during the follow up period. A detailed analysis of the remodelling response revealed an established endothelium along the distal and proximal regions of the scaffold with poor endothelialisation of its central portion.

Fig. 4. Sandwiched stent composed of SIS (Yamada et al., 2001)

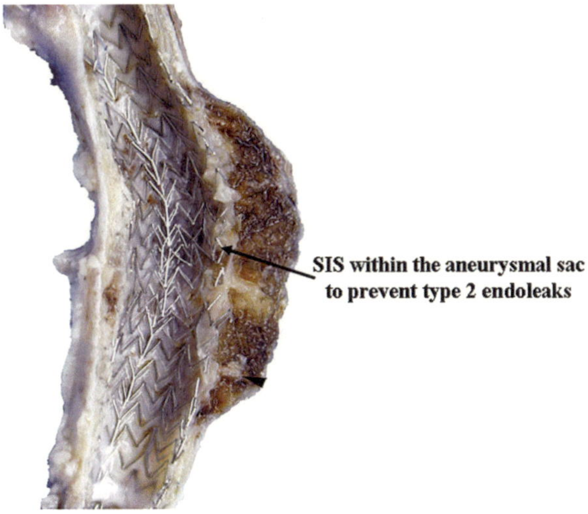

Fig. 5. SIS stent graft deployed inside an AAA (Schoder et al., 2004)

In 2005 Niyyati *et al.* assessed the potential for SIS as an intrahepatic protocaval shunt. Only one device remained functional in 6 animals after a 14 day experimental time period (Niyyati et al. 2005). Intuitively, the authors discouraged SIS in this setting. The configured SIS stent device and the implanted SIS stent are illustrated in Fig. 6 A and B. Histological assessment of the luminal surface after follow up demonstrated a smooth neointima on the surface of the functional stent.

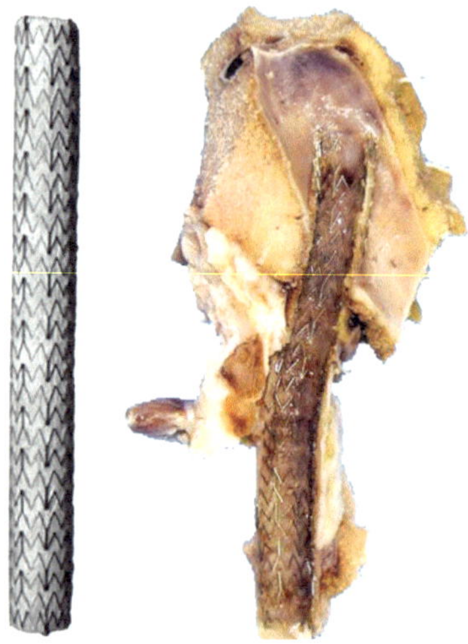

Fig. 6. (A) Stent SIS configuration, (B) Gross section through excised graft implantation (Niyyati et al. 2005)

A more recent study investigated the effects of different stent grafts on the portal vein of canine models (Ishii et al., 2005). In this study, four different stents were assessed, bare metal stent, PTFE covered stent, Dacron covered stent and an SIS covered stent. The study concluded that SIS covered stents confer no advantages in comparison to other conventional stent grafts. In fact PTFE consistently outperformed the other 3 stents and was recommended as the most suitable stent for implantation into the portal vein. In 2006 the endothelialisation of an implanted SIS stent graft was compared with Dacron and PTFE in an ovine model. In this study the stent grafts were inserted into the thoracoabdominal aorta and endothelialisation of the stent graft was assessed during the follow up period (Yavuz et al. 2006). Results showed that Dacron exhibited the greatest and most progressive amount of endothelialisation. In comparison, SIS demonstrated progressive tissue remodelling and a moderate amount of neointimal formation.

7. Concept solution

A potential mechanism for improving the performance of the EVAR stent-graft is the insertion of a tissue-engineered stabilisation collar at the proximal and distal ends of the device as illustrated in Fig. 7.

A tissue-engineered 'stent-collar' may prevent common complications such as endoleaks and graft migration. Intuitively, a number of important critical issues need to be addressed prior to implementation of this possible solution. Compliance of the scaffold in tubular structures and the reduction in the scaffold's strength caused by interactions with the 'stent-graft' should be investigated. Compliance issues may arise due to fluid flow and elastic characteristics of the arterial wall exerted on the tissue-engineered scaffold. Furthermore, radial forces exerted by the stent-graft induce stress loadings on the surface of the tissue-engineered scaffold. Structural properties of the scaffold also need to be accurately characterised. The scaffold's potential to induce cellular attachment and host derived tissue-remodelling responses need to be explored. Contact between the tissue-engineered material and arterial wall may result in cell infiltration from the host's endothelium (Fig. 8). To date the majority of these questions remain unanswered and require further research to adequately develop ECM scaffolds into AAA endovascular treatment.

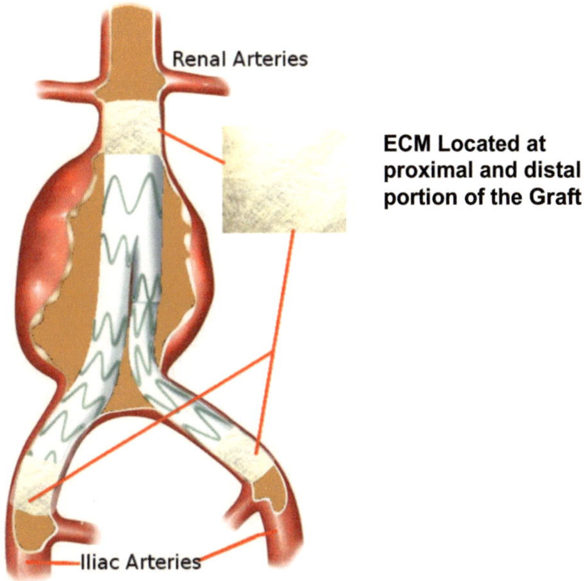

Fig. 7. Possible solution for optimisation and stabilisation of the stent graft during EVAR. (Adapted from http://www.medtronic.com/your-health/abdominal-aortic-aneurysm/getting-a-device/surgery)

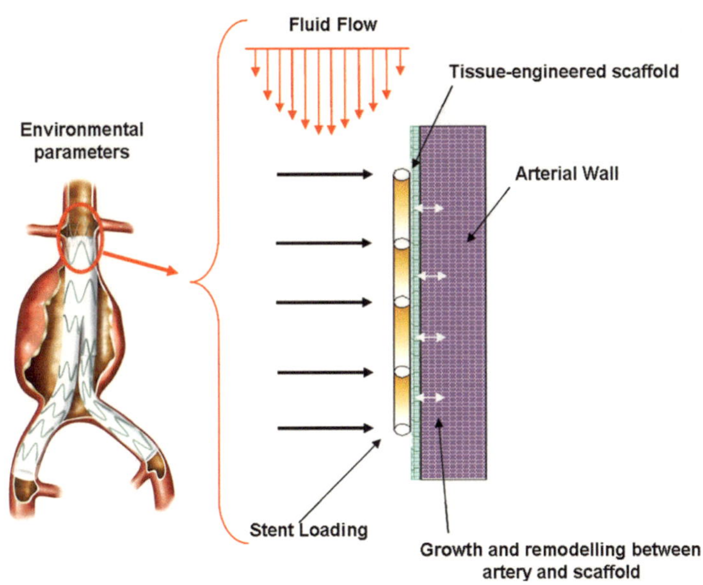

Fig. 8. ECM collar interacting with aortic wall to anchor the stent-graft and prevent migration of the device

8. Conclusions

In this chapter we have examined key issues associated with medium-term failure of endovascular stents used in the EVAR procedure. Common complications associated with EVAR include endoloeaks and migration of the deployed stent. Although tissue-engineered xenografts offer an attractive alternative for improving the EVAR procedure, it is notable that implantation of ECM scaffolds into stented environments have shown conflicting results to date. Encouragingly, the advent of alternative types of biological ECMs, such as UBM, has opened up new avenues for researchers with an interest in optimizing the EVAR procedure. Development of a tissue-engineered scaffold that optimizes the performance of the stent-graft remains a valuable possibility and exciting option for the future.

9. Acknowledgements

This was supported by Enterprise Ireland Commercialisation Funds with a Proof of Concept Phase and Technology Development Phase grants, and also by the Irish Research Council for Science, Engineering & Technology (IRCSET) - Marie Curie International Mobility Fellowship.

10. References

Abbas, A. K., Murphy, K. M. and Sher, A. (1996) 'Functional diversity of helper T lymphocytes', *Nature*, 383(6603), 787-93.

Ahn, H. H., Kim, K. S., Lee, J. H., Lee, M. S., Song, I. B., Cho, M. H., Shin, Y. N., Kim, M. S., Khang, G. and Lee, H. B. (2007) 'Porcine small intestinal submucosa sheets as a scaffold for human bone marrow stem cells', *Int J Biol Macromol*, 41(5), 590-6.

Allman, A. J., McPherson, T. B., Badylak, S. F., Merrill, L. C., Kallakury, B., Sheehan, C., Raeder, R. H. and Metzger, D. W. (2001) 'Xenogeneic extracellular matrix grafts elicit a TH2-restricted immune response', *Transplantation*, 71(11), 1631-40.

Alpdogan, O., Hubbard, V. M., Smith, O. M., Patel, N., Lu, S., Goldberg, G. L., Gray, D. H., Feinman, J., Kochman, A. A., Eng, J. M., Suh, D., Muriglan, S. J., Boyd, R. L. and van den Brink, M. R. (2006) 'Keratinocyte growth factor (KGF) is required for postnatal thymic regeneration', *Blood*, 107(6), 2453-60.

Babensee, J. E., McIntire, L. V. and Mikos, A. G. (2000) 'Growth factor delivery for tissue engineering', *Pharm Res*, 17(5), 497-504.

Bach, F. H., Ferran, C., Hechenleitner, P., Mark, W., Koyamada, N., Miyatake, T., Winkler, H., Badrichani, A., Candinas, D. and Hancock, W. W. (1997) 'Accommodation of vascularized xenografts: expression of "protective genes" by donor endothelial cells in a host Th2 cytokine environment', *Nat Med*, 3(2), 196-204.

Badylak, S., Kokini, K., Tullius, B. and Whitson, B. (2001) 'Strength over time of a resorbable bioscaffold for body wall repair in a dog model', *J Surg Res*, 99(2), 282-7.

Badylak, S., Meurling, S., Chen, M., Spievack, A. and Simmons-Byrd, A. (2000) 'Resorbable bioscaffold for esophageal repair in a dog model', *J Pediatr Surg*, 35(7), 1097-103.

Badylak, S. F. (2005) 'Regenerative medicine and developmental biology: the role of the extracellular matrix', *Anat Rec B New Anat*, 287(1), 36-41.

Badylak, S. F. (2007) 'The extracellular matrix as a biologic scaffold material', *Biomaterials*, 28(25), 3587-93.

Badylak, S. F., Freytes, D. O. and Gilbert, T. W. (2009) 'Extracellular matrix as a biological scaffold material: Structure and function', *Acta Biomater,* 5(1), 1-13.

Badylak, S. F., Kropp, B., McPherson, T., Liang, H. and Snyder, P. W. (1998) 'Small intestional submucosa: a rapidly resorbed bioscaffold for augmentation cystoplasty in a dog model', *Tissue Eng,* 4(4), 379-87.

Badylak, S. F., Lantz, G. C., Coffey, A. and Geddes, L. A. (1989) 'Small intestinal submucosa as a large diameter vascular graft in the dog', *J Surg Res,* 47(1), 74-80.

Badylak, S. F., Park, K., Peppas, N., McCabe, G. and Yoder, M. (2001) 'Marrow-derived cells populate scaffolds composed of xenogeneic extracellular matrix', *Exp Hematol,* 29(11), 1310-8.

Badylak, S. F., Vorp, D. A., Spievack, A. R., Simmons-Byrd, A., Hanke, J., Freytes, D. O., Thapa, A., Gilbert, T. W. and Nieponice, A. (2005) 'Esophageal reconstruction with ECM and muscle tissue in a dog model', *J Surg Res,* 128(1), 87-97.

Badylak SF, V. S., Kokini K, et al. (1995) *'The use of xengenic small intestinal submucoas as a biomaterial for Achilles tendon repair in a dog model ',* *J Biomed Mater Res,,* 29, 977-85.

Badylak, S. F., Wu, C. C., Bible, M. and McPherson, E. (2003) 'Host protection against deliberate bacterial contamination of an extracellular matrix bioscaffold versus Dacron mesh in a dog model of orthopedic soft tissue repair', *J Biomed Mater Res B Appl Biomater,* 67(1), 648-54.

Baldwin, H. S. (1996) 'Early embryonic vascular development', *Cardiovasc Res,* 31 Spec No, E34-45.

Brennan, E. P., Reing, J., Chew, D., Myers-Irvin, J. M., Young, E. J. and Badylak, S. F. (2006) 'Antibacterial activity within degradation products of biological scaffolds composed of extracellular matrix', *Tissue Eng,* 12(10), 2949-55.

Brown, B., Lindberg, K., Reing, J., Stolz, D. B. and Badylak, S. F. (2006) 'The basement membrane component of biologic scaffolds derived from extracellular matrix', *Tissue Eng,* 12(3), 519-26.

Bush, R. L., Najibi, S., Lin, P. H., Weiss, V. J., MacDonald, M. J., Redd, D. C., Martin, L. G., Chaikof, E. L. and Lumsden, A. B. (2001) 'Early experience with the bifurcated Excluder endoprosthesis for treatment of the abdominal aortic aneurysm', *J Vasc Surg,* 34(3), 497-502.

Chen, N. and Field, E. H. (1995) 'Enhanced type 2 and diminished type 1 cytokines in neonatal tolerance', *Transplantation,* 59(7), 933-41.

Cimini, M., Boughner, D. R., Ronald, J. A., Johnston, D. E. and Rogers, K. A. (2005) 'Dermal fibroblasts cultured on small intestinal submucosa: Conditions for the formation of a neotissue', *J Biomed Mater Res A,* 75(4), 895-906.

Clyne, A. M. and Edelman, E. R. (2009) 'Vascular growth factor binding kinetics to the endothelial cell basement membrane, with a kinetics-based correction for substrate binding', *Cytotechnology.*

Criado, F. J., Fairman, R. M. and Becker, G. J. (2003) 'Talent LPS AAA stent graft: results of a pivotal clinical trial', *J Vasc Surg,* 37(4), 709-15.

Davis, N. F., Callanan, A., McGuire, B. B., Flood, H. D. and McGloughlin, T. M. (2011) 'Evaluation of Viability and Proliferative Activity of Human Urothelial Cells Cultured Onto Xenogenic Tissue-Engineered Extracellular Matrices', *Urology.*

Davis, N. F., Callanan, A., McGuire, B. B., Mooney, R., Flood, H. D. and McGloughlin, T. M. (2011) 'Porcine extracellular matrix scaffolds in reconstructive urology: An ex vivo comparative study of their biomechanical properties', *J Mech Behav Biomed Mater,* 4(3), 375-82.

Davis, N. F., McGuire, B. B., Callanan, A., Flood, H. D. and McGloughlin, T. M. (2010) 'Xenogenic extracellular matrices as potential biomaterials for interposition grafting in urological surgery', *J Urol,* 184(6), 2246-53.

Egelhoff, C. J., Budwig, R. S., Elger, D. F., Khraishi, T. A. and Johansen, K. H. (1999) 'Model studies of the flow in abdominal aortic aneurysms during resting and exercise conditions', *J Biomech,* 32(12), 1319-29.

Ferrara, N., Houck, K., Jakeman, L. and Leung, D. W. (1992) 'Molecular and biological properties of the vascular endothelial growth factor family of proteins', *Endocr Rev,* 13(1), 18-32.

Freytes, D. O., Badylak, S. F., Webster, T. J., Geddes, L. A. and Rundell, A. E. (2004) 'Biaxial strength of multilaminated extracellular matrix scaffolds', *Biomaterials,* 25(12), 2353-61.

Freytes, D. O., Rundell, A. E., Vande Geest, J., Vorp, D. A., Webster, T. J. and Badylak, S. F. (2005) 'Analytically derived material properties of multilaminated extracellular matrix devices using the ball-burst test', *Biomaterials,* 26(27), 5518-31.

Freytes, D. O., Stoner, R. M. and Badylak, S. F. (2008) 'Uniaxial and biaxial properties of terminally sterilized porcine urinary bladder matrix scaffolds', *J Biomed Mater Res B Appl Biomater,* 84(2), 408-14.

Galili, U. (1993) 'Interaction of the natural anti-Gal antibody with alpha-galactosyl epitopes: a major obstacle for xenotransplantation in humans', *Immunol Today,* 14(10), 480-2.

Gilbert, T. W., Gilbert, S., Madden, M., Reynolds, S. D. and Badylak, S. F. (2008) 'Morphologic assessment of extracellular matrix scaffolds for patch tracheoplasty in a canine model', *Ann Thorac Surg,* 86(3), 967-74; discussion 967-74.

Gilbert, T. W., Sacks, M. S., Grashow, J. S., Woo, S. L., Badylak, S. F. and Chancellor, M. B. (2006) 'Fiber kinematics of small intestinal submucosa under biaxial and uniaxial stretch', *J Biomech Eng,* 128(6), 890-8.

Gilbert, T. W., Stewart-Akers, A. M., Simmons-Byrd, A. and Badylak, S. F. (2007) 'Degradation and remodeling of small intestinal submucosa in canine Achilles tendon repair', *J Bone Joint Surg Am,* 89(3), 621-30.

Good, A. H., Cooper, D. K., Malcolm, A. J., Ippolito, R. M., Koren, E., Neethling, F. A., Ye, Y., Zuhdi, N. and Lamontagne, L. R. (1992) 'Identification of carbohydrate structures that bind human antiporcine antibodies: implications for discordant xenografting in humans', *Transplant Proc,* 24(2), 559-62.

Gouk, S. S., Lim, T. M., Teoh, S. H. and Sun, W. Q. (2008) 'Alterations of human acellular tissue matrix by gamma irradiation: histology, biomechanical property, stability, in vitro cell repopulation, and remodeling', *J Biomed Mater Res B Appl Biomater,* 84(1), 205-17.

Greenberg, R. (2003) 'The Zenith AAA endovascular graft for abdominal aortic aneurysms: clinical update', *Semin Vasc Surg,* 16(2), 151-7.

Greenhalgh, R. M. and Powell, J. T. (2008) 'Registries and RCTs for new interventional procedures', *Lancet,* 372(9642), 891-2.

Hiles, M. C., Badylak, S. F., Lantz, G. C., Kokini, K., Geddes, L. A. and Morff, R. J. (1995) 'Mechanical properties of xenogeneic small-intestinal submucosa when used as an aortic graft in the dog', *J Biomed Mater Res,* 29(7), 883-91.

Ho, K. L., Witte, M. N. and Bird, E. T. (2004) '8-ply small intestinal submucosa tension-free sling: spectrum of postoperative inflammation', *J Urol*, 171(1), 268-71.

Hodde, J., Record, R., Tullius, R. and Badylak, S. (2002) 'Fibronectin peptides mediate HMEC adhesion to porcine-derived extracellular matrix', *Biomaterials*, 23(8), 1841-8.

Hodde, J. P., Ernst, D. M. and Hiles, M. C. (2005) 'An investigation of the long-term bioactivity of endogenous growth factor in OASIS Wound Matrix', *J Wound Care*, 14(1), 23-5.

Kamineni, R. and Heuser, R. R. (2004) 'Abdominal aortic aneurysm: a review of endoluminal treatment', *J Interv Cardiol*, 17(6), 437-45.

Koren, E., Kujundzic, M., Koscec, M., Neethling, F. A., Richards, S. V., Ye, Y., Zuhdi, N. and Cooper, D. K. (1994) 'Cytotoxic effects of human preformed anti-Gal IgG and complement on cultured pig cells', *Transplant Proc*, 26(3), 1336-9.

Kropp, B. P., Eppley, B. L., Prevel, C. D., Rippy, M. K., Harruff, R. C., Badylak, S. F., Adams, M. C., Rink, R. C. and Keating, M. A. (1995) 'Experimental assessment of small intestinal submucosa as a bladder wall substitute', *Urology*, 46(3), 396-400.

Kropp, B. P., Rippy, M. K., Badylak, S. F., Adams, M. C., Keating, M. A., Rink, R. C. and Thor, K. B. (1996a) 'Regenerative urinary bladder augmentation using small intestinal submucosa: urodynamic and histopathologic assessment in long-term canine bladder augmentations', *J Urol*, 155(6), 2098-104.

Kropp, B. P., Sawyer, B. D., Shannon, H. E., Rippy, M. K., Badylak, S. F., Adams, M. C., Keating, M. A., Rink, R. C. and Thor, K. B. (1996b) 'Characterization of small intestinal submucosa regenerated canine detrusor: assessment of reinnervation, in vitro compliance and contractility', *J Urol*, 156(2 Pt 2), 599-607.

Lantz, G. C., Badylak, S. F., Coffey, A. C., Geddes, L. A. and Blevins, W. E. (1990) 'Small intestinal submucosa as a small-diameter arterial graft in the dog', *J Invest Surg*, 3(3), 217-27.

Laurie, G. W., Horikoshi, S., Killen, P. D., Segui-Real, B. and Yamada, Y. (1989) 'In situ hybridization reveals temporal and spatial changes in cellular expression of mRNA for a laminin receptor, laminin, and basement membrane (type IV) collagen in the developing kidney', *J Cell Biol*, 109(3), 1351-62.

Liang, R., Woo, S. L., Takakura, Y., Moon, D. K., Jia, F. and Abramowitch, S. D. (2006) 'Long-term effects of porcine small intestine submucosa on the healing of medial collateral ligament: a functional tissue engineering study', *J Orthop Res*, 24(4), 811-9.

Lovekamp, J. J., Simionescu, D. T., Mercuri, J. J., Zubiate, B., Sacks, M. S. and Vyavahare, N. R. (2006) 'Stability and function of glycosaminoglycans in porcine bioprosthetic heart valves', *Biomaterials*, 27(8), 1507-18.

Matsumiya, G., Shirakura, R., Miyagawa, S., Izutani, H., Nakata, S. and Matsuda, H. (1994) 'Assessment of T-cell subsets involved in antibody production and cell-mediated cytotoxicity in rat-to-mouse cardiac xenotransplantation', *Transplant Proc*, 26(3), 1214-6.

McPherson, T. B., Liang, H., Record, R. D. and Badylak, S. F. (2000) 'Galalpha(1,3)Gal epitope in porcine small intestinal submucosa', *Tissue Eng*, 6(3), 233-9.

Mennander, A., Pimenoff, G., Heikkinen, M., Partio, T., Zeitlin, R. and Salenius, J. P. (2005) 'Nonoperative approach to endotension', *J Vasc Surg*, 42(2), 194-9.

Nakata, M., Pavcnik, D., Uchida, B. T., VanAlstine, W., Timmermans, H. A., Toyota, N., Terada, M., Brountzos, E., Kaufman, J. A., Keller, F. S. and Rosch, J. (2003)

'Comparison of small intestinal submucosa-covered and noncovered nitinol stents with PTFE endografts in injured ovine femoral arteries: a pilot study', *Cardiovasc Intervent Radiol*, 26(5), 459-67.

Niyyati, M., Petersen, B. D., Pavcnik, D., Uchida, B. T., Timmermans, H. A., Hiraki, T., Wu, R. H., Brountzos, E., Keller, F. S. and Rosch, J. (2005) 'A flexible stent with small intestinal submucosa covering for direct intrahepatic portocaval shunt: experimental pilot study in swine', *Cardiovasc Intervent Radiol*, 28(2), 215-20.

Noishiki, Y., Ichikawa, Y., Kosuge, T., Yamazaki, I., Yamamoto, K., Manabe, T., Lu, Y. and Yamane, Y. (2001) 'Introduction of tissue engineering concepts into the field of endovascular grafts: an attempt to solve endoleakage problems of endovascular grafts implanted in aortic aneurysms', *Artif Organs*, 25(3), 228-35.

Parodi, J. C., Palmaz, J. C. and Barone, H. D. (1991) 'Transfemoral intraluminal graft implantation for abdominal aortic aneurysms', *Ann Vasc Surg*, 5(6), 491-9.

Pavcnik, D., Obermiller, J., Uchida, B. T., Van Alstine, W., Edwards, J. M., Landry, G. J., Kaufman, J. A., Keller, F. S. and Rosch, J. (2009) 'Angiographic evaluation of carotid artery grafting with prefabricated small-diameter, small-intestinal submucosa grafts in sheep', *Cardiovasc Intervent Radiol*, 32(1), 106-13.

Piterina AV, C. A., Meaney CL, Davis LM, Callanan A, Walsh MT, McGloughlin TM (2009) *Int J. Mol. Sci 2009* Submitted.

Raghavan, M. L., Kratzberg, J. A. and Golzarian, J. (2005) 'Introduction to biomechanics related to endovascular repair of abdominal aortic aneurysm', *Tech Vasc Interv Radiol*, 8(1), 50-5.

Record, R. D., Hillegonds, D., Simmons, C., Tullius, R., Rickey, F. A., Elmore, D. and Badylak, S. F. (2001) 'In vivo degradation of 14C-labeled small intestinal submucosa (SIS) when used for urinary bladder repair', *Biomaterials*, 22(19), 2653-9.

Roeder, R., Wolfe, J., Lianakis, N., Hinson, T., Geddes, L. A. and Obermiller, J. (1999) 'Compliance, elastic modulus, and burst pressure of small-intestine submucosa (SIS), small-diameter vascular grafts', *J Biomed Mater Res*, 47(1), 65-70.

Roeder, R. A., Lantz, G. C. and Geddes, L. A. (2001) 'Mechanical remodeling of small-intestine submucosa small-diameter vascular grafts--a preliminary report', *Biomed Instrum Technol*, 35(2), 110-20.

Rosario, D. J., Reilly, G. C., Ali Salah, E., Glover, M., Bullock, A. J. and Macneil, S. (2008) 'Decellularization and sterilization of porcine urinary bladder matrix for tissue engineering in the lower urinary tract', *Regen Med*, 3(2), 145-56.

Sakalihasan, N., Limet, R. and Defawe, O. D. (2005) 'Abdominal aortic aneurysm', *Lancet*, 365(9470), 1577-89.

Sandusky, G. E., Jr., Badylak, S. F., Morff, R. J., Johnson, W. D. and Lantz, G. (1992) 'Histologic findings after in vivo placement of small intestine submucosal vascular grafts and saphenous vein grafts in the carotid artery in dogs', *Am J Pathol*, 140(2), 317-24.

Sarikaya, A., Record, R., Wu, C. C., Tullius, B., Badylak, S. and Ladisch, M. (2002) 'Antimicrobial activity associated with extracellular matrices', *Tissue Eng*, 8(1), 63-71.

Schoder, M., Pavcnik, D., Uchida, B. T., Corless, C., Timmermans, H. A., Yin, Q., Brountzos, E., Nakata, M., Hiraki, T., Niyyati, M., Kaufman, J. A., Keller, F. S. and Rosch, J. (2004) 'Small intestinal submucosa aneurysm sac embolization for endoleak prevention after abdominal aortic aneurysm endografting: a pilot study in sheep', *J Vasc Interv Radiol,* 15(1 Pt 1), 69-83.

Schussler, O., Shen, M., Shen, L., Carpentier, S. M., Kaveri, S. and Carpentier, A. (2001) 'Effect of human immunoglobulins on the immunogenicity of porcine bioprostheses', *Ann Thorac Surg,* 71(5 Suppl), S396-400.

Sellaro, T. L., Ravindra, A. K., Stolz, D. B. and Badylak, S. F. (2007) 'Maintenance of hepatic sinusoidal endothelial cell phenotype in vitro using organ-specific extracellular matrix scaffolds', *Tissue Eng,* 13(9), 2301-10.

Strom, T. B., Roy-Chaudhury, P., Manfro, R., Zheng, X. X., Nickerson, P. W., Wood, K. and Bushell, A. (1996) 'The Th1/Th2 paradigm and the allograft response', *Curr Opin Immunol,* 8(5), 688-93.

Sung, H. J., Su, J., Berglund, J. D., Russ, B. V., Meredith, J. C. and Galis, Z. S. (2005) 'The use of temperature-composition combinatorial libraries to study the effects of biodegradable polymer blend surfaces on vascular cells', *Biomaterials,* 26(22), 4557-67.

Tabata, Y. (2004) 'Tissue regeneration based on tissue engineering technology', *Congenit Anom (Kyoto),* 44(3), 111-24.

Tabata, Y. (2005) 'Significance of release technology in tissue engineering', *Drug Discov Today,* 10(23-24), 1639-46.

Teebken, O. E., Bader, A., Steinhoff, G. and Haverich, A. (2000) 'Tissue engineering of vascular grafts: human cell seeding of decellularised porcine matrix', *Eur J Vasc Endovasc Surg,* 19(4), 381-6.

Thompson, M. M. (2003) 'Controlling the expansion of abdominal aortic aneurysms', *Br J Surg,* 90(8), 897-8.

Vaught, J. D., Kropp, B. P., Sawyer, B. D., Rippy, M. K., Badylak, S. F., Shannon, H. E. and Thor, K. B. (1996) 'Detrusor regeneration in the rat using porcine small intestinal submucosal grafts: functional innervation and receptor expression', *J Urol,* 155(1), 374-8.

Veith, F. J., Baum, R. A., Ohki, T., Amor, M., Adiseshiah, M., Blankensteijn, J. D., Buth, J., Chuter, T. A., Fairman, R. M., Gilling-Smith, G., Harris, P. L., Hodgson, K. J., Hopkinson, B. R., Ivancev, K., Katzen, B. T., Lawrence-Brown, M., Meier, G. H., Malina, M., Makaroun, M. S., Parodi, J. C., Richter, G. M., Rubin, G. D., Stelter, W. J., White, G. H., White, R. A., Wisselink, W. and Zarins, C. K. (2002) 'Nature and significance of endoleaks and endotension: summary of opinions expressed at an international conference', *J Vasc Surg,* 35(5), 1029-35.

Vorp, D. A. and Vande Geest, J. P. (2005) 'Biomechanical determinants of abdominal aortic aneurysm rupture', *Arterioscler Thromb Vasc Biol,* 25(8), 1558-66.

Yamada, K., Pavcnik, D., Uchida, B. T., Timmermans, H. A., Corless, C. L., Yin, Q., Yamakado, K., Park, J. W., Rosch, J., Keller, F. S., Sato, M. and Yamada, R. (2001) 'Endoluminal treatment of ruptured abdominal aortic aneurysm with small intestinal submucosa sandwich endografts: a pilot study in sheep', *Cardiovasc Intervent Radiol,* 24(2), 99-105.

Yavuz, K., Geyik, S., Pavcnik, D., Uchida, B. T., Corless, C. L., Hartley, D. E., Goktay, A., Correa, L. O., Timmermans, H., Hodde, J. P., Kaufman, J. A., Keller, F. S. and Rosch, J. (2006) 'Comparison of the endothelialization of small intestinal submucosa, dacron, and expanded polytetrafluoroethylene suspended in the thoracoabdominal aorta in sheep', *J Vasc Interv Radiol,* 17(5), 873-82.

Zantop, T., Gilbert, T. W., Yoder, M. C. and Badylak, S. F. (2006) 'Extracellular matrix scaffolds are repopulated by bone marrow-derived cells in a mouse model of achilles tendon reconstruction', *J Orthop Res,* 24(6), 1299-309.

Zarins, C. K., Xu, C. and Glagov, S. (2001) 'Atherosclerotic enlargement of the human abdominal aorta', *Atherosclerosis,* 155(1), 157-64.

Zhai, Y., Ghobrial, R. M., Busuttil, R. W. and Kupiec-Weglinski, J. W. (1999) 'Th1 and Th2 cytokines in organ transplantation: paradigm lost?', *Crit Rev Immunol,* 19(2), 155-72.

15

Elastin Based Constructs

Lisa Nivison-Smith and Anthony Weiss
*School of Molecular Bioscience, University of Sydney
Australia*

1. Introduction

Elastin is a key structural protein found in the extracellular matrix (ECM) of all mammals. As the dominant part of the elastic fiber, elastin confers the mechanical properties of resilience and elasticity essential to the function of elastic tissues. Elastin interacts with cells through specific biochemical mechanisms. This chapter considers the (1) mechanical and biochemical roles of elastin in elastic tissues and the subsequent disease phenotypes that result from the degradation and loss of elastin, (2) development and success of current elastin based biomaterials including sources of elastin for tissue engineering and their application, and (3) vascular constructs that our laboratory has developed from recombinant human tropoelastin. These constructs mimic the physical and biochemical properties of native elastin.

2. Elastin formation *in vivo*

Elastin is formed in the process of elastogenesis through the assembly and cross-linking of the protein tropoelastin (Figure 1). The tropoelastin monomer is produced from expression of the elastin gene during perinatal development by elastogenic cells such as smooth muscle cells (SMCs), endothelial cells, fibroblasts and chondroblasts (Uitto, Christiano et al. 1991). The tropoelastin transcript undergoes extensive alternative splicing leading to the removal of entire domains from the protein. In humans, this splicing results in several tropoelastin isoforms, the most common of which lacks exon 26A (Indik, Yeh et al. 1987). Mature, intracellular tropoelastin associates with the elastin binding protein (EBP) and this complex is secreted to the cell surface (Hinek 1995). Competition from galactosides results in the dissociation of EBP from tropoelastin and the return of EBP to the cell (Mecham 1991). Released tropoelastin on the cell surface subsequently aggregates by coacervation. During this process, the hydrophobic domains of tropoelastin associate and tropoelastin molecules become concentrated and increasingly aligned allowing for subsequent formation of cross-links (Vrhovski, Jensen et al. 1997).

Coacervated tropoelastin is deposited onto microfibrils which probably serve as a scaffold to direct tropoelastin cross-linking and consequential elastic fiber formation. Cross-linking is facilitated by the enzyme lysyl oxidase, which deaminates lysine side chains in tropoelastin to form allysine sidechains that can subsequently react with adjacent allysine or lysine side chains to form cross-links (Kagan and Sullivan 1982). These cross-links can then further react to form desmosine and isodesmosine cross-links between tropoelastin molecules

(Umeda, Nakamura et al. 2001). Multiple cross-links result in the mature insoluble elastic fiber.

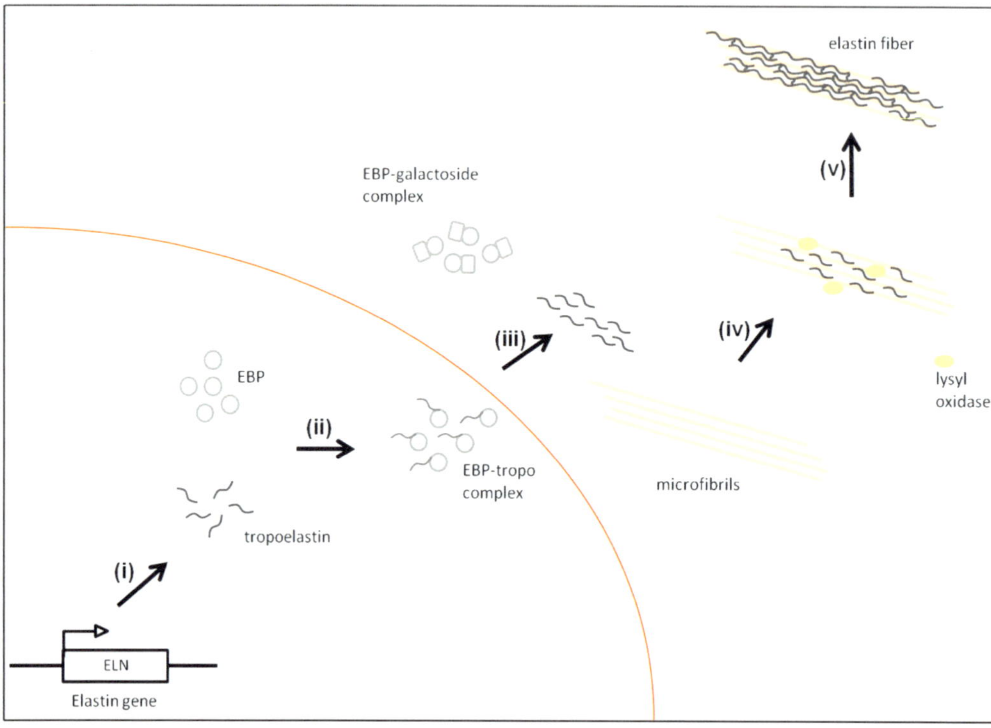

Fig. 1. Schematic of the stages of elastogenesis.
(i) Tropoelastin is transcribed and translated from the elastin (ELN) gene and (ii) transported to the plasma membrane in association with EBP. (iii) Tropoelastin is released and aggregates on the cell surface, while EBP disassociates to form a complex with available galactosides. (iv) Tropoelastin aggregates are oxidized by lysyl oxidase leading to cross-linked elastin that accumulates on microfibrils which help to direct elastin deposition. (v) The process of deposition and cross-linking continues to give rise to mature elastic fibers

3. The role of elastin *in vivo*

Elastin plays a key structural role in elastic tissues including arteries, skin, ligament, cartilage and tendons (Sandberg, Soskel et al. 1981). As the dominant part of the elastic fiber, elastin confers resilience and elasticity essential to the function of these tissues. The arrangement of elastin in the ECM varies between different tissues to yield a wide range of structures with tailored elastic properties. For example, elastin in the form of thin lamina in the arterial wall is mostly responsible for the strength and elasticity necessary for vessel expansion and regulation of blood flow (Glagov, Vito et al. 1992). In the lung, elastin is arranged as a latticework that helps to support the opening and closing of the alveoli (Starcher 2001). In skin, elastin fibers are enriched in the dermis where they impart skin flexibility and extensibility (Roten, Bhat et al. 1996; Pasquali-Ronchetti and Baccarani-Contri 1997).

3.1 Mechanical properties of elastin

Elastin is extremely durable protein with a mean residence time of 74 years (Shapiro, Endicott et al. 1991). It comprises almost 90% of the elastic fiber where it dominates its elastic, mechanical properties. The Young's modulus for elastic fibers typically ranges from 300 - 600 kPa although it can measure as low as 100 kPa for arterial elastin, highlighting the versatile nature of these structures within the ECM (Mithieux and Weiss 2005; Zou and Zhang 2009). Although the mechanism for elasticity has not been fully elucidated, elastic recoil likely to be entropically driven whereby extension of the protein results in a more ordered structure and thus recoil occurs so the protein can return to a disorder state (reviewed by (Rosenbloom, Abrams et al. 1993; Vrhovski and Weiss 1998). This elasticity is due to the inherent elastic properties of the monomer (Holst, Watson et al. 2010; Baldock, Oberhauser et al. 2011).

3.2 Biological properties of elastin

Elastin plays key biological roles in the regulation of cells native to elastic tissues. Studies of elastin knockout mice reveal a crucial role for elastin in arterial morphogenesis through regulation of SMC proliferation and phenotype (Li, Brooke et al. 1998). This model is supported by *in vitro* studies showing that elastin can inhibit SMC proliferation in a dose dependent manner (Ito, Ishimaru et al. 1998). Elastin can also mediate the attachment and proliferation of endothelial cells from several vascular origins (Ito, Ishimaru et al. 1998; Williamson, Shuttleworth et al. 2007; Wilson, Gibson et al. 2010). Similar effects have been observed for dermal fibroblasts (Bax, Rodgers et al. 2009; Rnjak, Wise et al. 2011). Additionally elastin is a chemoattractant for SMCs, endothelial cells and monocytes (Senior, Griffin et al. 1980; Wilson, Gibson et al. 2010).

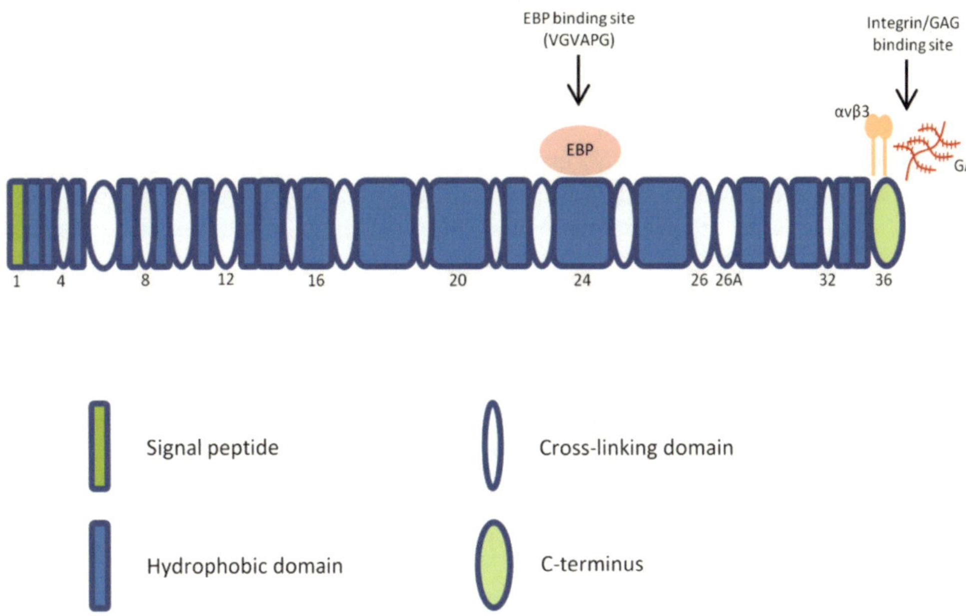

Fig. 2. Schematic of human tropoelastin primary organization and binding partners. All domains are shown. Exons 13, 22, 23, 26A and 32 are subject to alternate splicing

Several cell receptors have been identified for elastin (Figure 2). The most well documented of these receptors is EBP, which binds to multiple sites including the VGVAPG sequence on exon 24 of tropoelastin (Rodgers and Weiss 2005). Upon binding elastin, this receptor activates intracellular signaling pathways involved in cell proliferation, chemotaxis, migration and cell morphology for a range of cell types including SMCs, endothelial cells, fibroblasts, monocytes, leukocytes and mesenchymal cells (Senior, Griffin et al. 1980; Indik, Abrams et al. 1990; Faury, Ristori et al. 1994; Faury, Ristori et al. 1995; Kamisato, Uemura et al. 1997; Jung, Rutka et al. 1998; Mochizuki, Brassart et al. 2002). Other cell receptors, including a less documented glycoprotein termed elastonectin and G protein-coupled receptor can bind elastin through the VGVAPG sequence (Hornebeck, Tixier et al. 1986). Interactions of vascular cells with elastin via these receptors have been shown to dictate focal adhesion formation, cell proliferation and migration (Hornebeck, Tixier et al. 1986; Karnik, Brooke et al. 2003; Karnik, Wythe et al. 2003). Glycosaminoglycans on the SMC and chondrocyte cell surface dominate binding to the C-terminus of bovine tropoelastin (Broekelmann, Kozel et al. 2005; Akhtar, Broekelmann et al. 2011). Cell interactions with human tropoelastin C-terminus specifically occur through the integrin $\alpha_v\beta_3$ (Rodgers and Weiss 2004; Bax, Rodgers et al. 2009). Elastin binding for some cell types is likely to occur through multiple receptors (Bax, Rodgers et al. 2009; Wilson, Gibson et al. 2010; Akhtar, Broekelmann et al. 2011).

4. Elastin and disease

Disease phenotypes manifest due to the degradation and loss of elastin through injury, genetic mutation or age. For example, autosomal dominant and recessive forms of cutis laxa mutations can arise from genetic modifications to the elastin gene and impaired vesicular trafficking, and have been reviewed elsewhere (e.g. (Hucthagowder, Morava et al. 2009; Callewaert, Renard et al. 2011). In skin, the loss of elastin in the dermal layers in severe burns leads significant physical injuries including scarring, wound contraction and loss of skin extensibility (Rnjak, Wise et al. 2011). In the vasculature, genetic mutations in the elastin gene or genes associated with elastic fiber formation result in severe, debilitating diseases (reviewed by (Kielty 2006)). Supravalvular aortic stenosis can arise from point mutations, deletions or translocations within the elastin gene that typically lead to haploinsufficiency and an altered organization of elastic lamellae in the artery, SMC hyperproliferation, increased media thickness and obstruction of the aorta (Urban, Zhang et al. 2001). Elastin is also associated with several vascular pathologies. Damage and fragmentation of elastin in the artery have been linked with deregulation of SMC phenotype, SMC hyperproliferation and invasion which cause vessel occlusion and cardiovascular complications (Brooke, Bayes-Genis et al. 2003). The failure of inelastic materials as arterial replacements further indicate the essential need for intact elastin in functional arteries (Abbott, Megerman et al. 1987).

5. Elastin biomaterials

Common to all elastin diseases is the catalogued *in vivo* inability to adequately regenerate and repair dysfunctional elastic fibers leading to subsequent failure of tissue function. This deficiency is mostly attributed to exclusive expression of elastin during early development

(Mecham 1991), which results in poor renewal of elastin in adult tissues. Materials that can serve as elastin replacements in adult tissues are in demand. This demand is most apparent in vascular tissue engineering as cardiovascular disease is the major contributor to adult mortality worldwide (Lloyd-Jones, Adams et al. 2010). Current synthetic vascular biomaterials, particularly expanded polyterafluoroethylene (ePTFE) and polyethylene terephthalate (Dacron) are poorly mismatched to native arteries in terms of mechanical properties, endothelial cell and SMC interactions and thrombogenecity which lead to a high failure rate in patients (Chlupac, Filova et al. 2009). Elastin can restore properties that are deficient in current grafts, including compliance and strength to match native vessels and regulation of endothelial and smooth muscle cells.

6. Decellularized tissues as elastin biomaterials

Decellularized tissues, generated by the removal of the cellular components of tissue explant are useful as biomaterials as they *a priori* possess much of the complex architecture of the native ECM. Elastic tissues are particular amenable to this method as the stability and insolubility of the elastin protein means it is resistant to many treatments used during decellularization processes.

Decellularizing elastin-rich tissues have been proposed as a path towards the potential replacement of artery, heart valves, bladder skin and lung (Daamen, Veerkamp et al. 2007; Petersen, Calle et al. 2010; Price, England et al. 2010). Enriched elastin vascular grafts generated by decellularization and removal of collagen with proteases from porcine carotid arteries can support fibroblasts *in vitro* (Chuang, Stabler et al. 2009). Cell infiltration has also been observed for other decellularized vascular constructs *in vitro* and *in vivo* (Schmidt and Baier 2000; Conklin, Richter et al. 2002; Dahl, Koh et al. 2003; Uchimura, Sawa et al. 2003). Skin replacements formed from decellularized porcine dermis containing 30% elastin show vascularization and support of cultured keratinocytes when examined in a rat excision model. Degradation of the collagen component of the material also occurs (Hafemann, Ensslen et al. 1999). Transplant of a repopulated decellularized human trachea demonstrates the feasibility of acquiring functionality and improved mechanical capabilities in a patient after 4 months (Macchiarini, Jungebluth et al. 2008).

Despite these advantages, decellularized tissue sources are generally animal derived and are therefore restricted in shape, size and supply. Additionally, decellularization methods involve chemical, physical or enzymatic treatments that can individually or collectively compromise mechanical and biological properties (Gilbert, Sellaro et al. 2006). The common use of detergents can limit the degree of cell repopulation. Decellularization methods are highly specific to a particular tissue thus their broader application to different tissues yields viable results in terms of remaining ECM structure and degree of decellularization (Gilbert, Sellaro et al. 2006). Lack of uniformity and versatility can limit the use of decellularized materials as commercial tissue replacements.

7. Tissue derived elastin constructs

7.1 Insoluble elastin materials

Elastin used for *in vitro* work is generally obtained by purifying the protein directly from elastin-rich tissues. Tissues are treated with chemicals such as NaOH or guanidine-HCl and/or high heat to remove other proteins and cellular material and leave insoluble elastic fibers. However extensive cross-linking and the consequential insolubility of elastin makes it difficult to manipulate *in vitro* (Daamen, Veerkamp et al. 2007).

Freeze-dried scaffolds of insoluble elastin fibers and purified collagen fibers present mechanical properties consistent with those of elastic tissues (Buttafoco, Engbers-Buijtenhuijs et al. 2006). Furthermore, these scaffolds appear to be compatible with SMCs (Buijtenhuijs, Buttafoco et al. 2004; Engbers-Buijtenhuijs, Buttafoco et al. 2005; Buttafoco, Engbers-Buijtenhuijs et al. 2006), endothelial cells (Wissink, van Luyn et al. 2000) and platelets (Koens, Faraj et al. 2010) pointing to potential vascular applications. Also, insoluble elastin/collagen scaffolds have been explored as possible dermal replacements as these materials can support fibroblasts (Daamen, van Moerkerk et al. 2003) and keratinocytes (Lammers, Tjabringa et al. 2009). Other insoluble elastin composites such as elastin/fibrin biomaterials have been generated but characterization of these materials is limited to mechanical capacity (Barbie, Angibaud et al. 1989).

7.2 Hydrolyzed elastin materials

The solubility of tissue-derived elastin can be improved by partial hydrolysis. A fragmented elastin preparation termed α-elastin is obtained by hydrolysis with oxalic acid and is often used in *in vitro* studies of elastin (Partridge, Davis et al. 1955). Hydrolysis can be performed with potassium hydroxide to yield κ-elastin or through mild digestion with proteinases (Partridge, Davis et al. 1955; Jacob and Hornebeck 1985). Hydrolyzed preparations of elastin display various properties that are similar to the native protein including temperature-induced aggregation (coacervation) and regulation of SMC and fibroblast phenotype (De Vries, Zeegelaar et al. 1995; Ito, Ishimaru et al. 1998). Fragmentation of elastin is associated with reduced protein structural integrity and altered cellular signaling properties (Daamen, Veerkamp et al. 2007; Bax, Rodgers et al. 2009).

Multiple vascular materials have been synthesized from hydrolyzed elastin preparations (Table 1). Hydrogels, cross-linked films and electrospun fibers containing hydrolyzed α-elastin all show preferable vascular material properties including regulation of SMC phenotype and increased mechanical elasticity. Electrospun materials are of particular interest as architecturally, these materials closely mimic the dimensions of elastic fibers *in vivo* (Li and Xia 2004).

Hydrolyzed elastin materials have also been proposed for use in the repair of elastic cartilage. In porous PCL scaffolds, infusion of α-elastin demonstrates enhanced scaffold elasticity and attachment and proliferation of articular cartilage chondrocytes *in vitro* (Annabi, Fathi et al. 2011). Replication of auricular-like cartilage has also been explored using alginate, collagen type I and κ-elastin containing hydrogels with auricular cartilage chondrocytes (de Chalain, Phillips et al. 1999). When these materials were implanted in mice and harvested after 12 weeks, matrix components including collagen and elastic fibers were present.

Dermal replacements containing hydrolyzed elastin demonstrate improved properties over elastin-free materials in regards to wound contraction and tissue regeneration (Rnjak, Wise et al. 2011). For example, MatriDerm, a collagen based scaffold with α-elastin shows improved skin elasticity (Ryssel, Gazyakan et al. 2008). Hydrogels formed exclusively from α-elastin (Figure 3) favorably support attachment and proliferation of dermal fibroblasts *in vitro* (Annabi, Mithieux et al. 2009; Annabi, Mithieux et al. 2009).

Scaffold	Advantages	Limitations	Reference
α-elastin film	-Low elastic modulus -attachment & proliferation of SMCs	-reduced SMC proliferation compared to TCPS	(Leach, Wolinsky et al. 2005)
elastin/gelatin gel	-Young's modulus matched to native artery -proliferation & infiltration of SMCs	-reduced SMC growth compared to TCPS	(Lamprou, Zhdan et al. 2010)
Collagen type I gels containing α-elastin	-SMC proliferation inhibited	- EC proliferation inhibited at high α-elastin concentrations	(Ito, Ishimaru et al. 1997)
α-elastin & collagen electrospun blended conduit	-attachment & proliferation of SMCs	-no mechanical testing	(Buttafoco, Kolkman et al. 2006)
α-elastin electrospun sheet	-SMC proliferation inhibited -α-SMA expression observed	-no mechanical testing	(Miyamoto, Atarashi et al. 2009)
α-elastin electrospun fibers	-attachment & proliferation of embryonic mesenchymal cells	-complete 3D constructs not created	(Li, Mondrinos et al. 2005)
α-elastin, PLGA & gelatin electrospun sheet	-mechanical properties tuned to artery through polymer content -proliferation of ECs on scaffold surface & infiltration of SMCs. -expression of functional EC molecules	-mechanical properties tested on electrospun sheets, not tubes	(Han, Lazarovici et al. 2011)
α-elastin, collagen type I & PLGA electrospun conduit	-matched compliance to bovine iliac artery -proliferation of ECs on inner & SMCs on outer surface of conduit -no immune reaction when implanted in mice	-scaffold contraction *in vitro*	(Stitzel, Liu et al. 2006; Lee, Yoo et al. 2007)
α-elastin, collagen type I & PLLA, PCL or PLCL blended electrospun conduit	-growth of bovine ECs -infiltration and α-SMA expression of SMCs	-scaffold contraction of PLCL blends *in vitro*	(Lee, Yoo et al. 2007)

Scaffold	Advantages	Limitations	Reference
α-elastin & PDO blended electrospun conduit	-mechanical properties matched to femoral artery with increased elastin content -increased cell infiltration with increased elastin content -increased graft burst pressure with suture reinforcement	-suture reinforcement lowers compliance	(Sell, McClure et al. 2006; Smith, McClure et al. 2008)
Elastin, collagen type I & collagen type III tri-layered electrospun conduit	-growth of EC, SMC and Fb in separate layers	-delamination of layers -no mechanical testing	(Boland, Matthews et al. 2004)
α-elastin, gelatin & PDS blended electrospun conduit	-matched tensile properties & elastic modulus to femoral artery	-loss of tensile properties due to *in vitro* degradation -no cell studies performed	(Thomas, Zhang et al. 2009)
α-elastin, gelatin & Maxon multi-layered electrospun conduit	-comparable mechanical properties to femoral artery	-no cell studies performed	(Thomas, Zhang et al. 2007)
bovine elastin, PGC, PCL & gelatin bi-layered electrospun conduit	-tensile strength matched to native artery -attachment & proliferation of EC & EPCs	-no SMC characterization	(Zhang, Thomas et al. 2010; Zhang, Thomas et al. 2010; Zhang, Xu et al. 2011)
α-elastin, collagen, PCL tri-layered electrospun conduit	-mechanical properties matched to native artery by modulation of elastin & PCL content	-no cell studies performed	(McClure, Sell et al. 2010)

Table 1. Scaffolds for vascular constructs synthesized using hydrolyzed elastin. Abbreviations: SMC: smooth muscle cell, EC: endothelial cell, EPC: endothelial progenitor cells, Fb: fibroblast, TCPS: tissue culture polystyrene, PLGA: poly(D,L-lactide-*co*-glycolide), PLLA: poly(L-lactide), PCL: polycaprolactone, PLCL: poly(L-lactide-*co*-ε-caprolactone), PDO: polydioxanone, PDS: polydioxanone, PGC: poliglecaprone

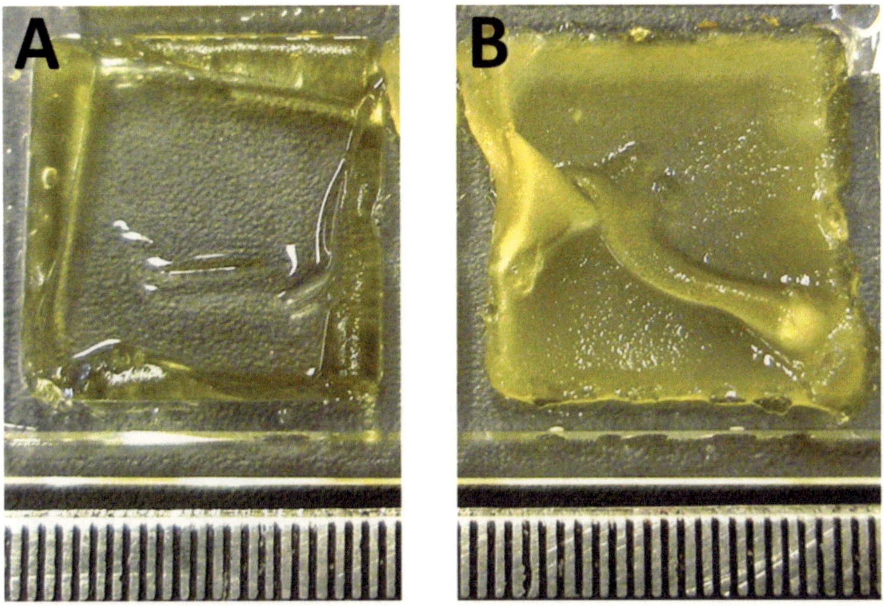

Fig. 3. Examples of α-elastin hydrogels that were synthesized by cross-linking with (A) 0.05% and (B) 0.1% (w/v) glutaraldehyde

8. Elastin-sequence based materials

8.1 Synthetic elastin-based peptides

Synthetic peptides based on key elastin sequences present elastin-like properties including self-assembly, cross-linking and cell interactions (Long, King et al. 1989; Faury, Garnier et al. 1998; Bellingham, Woodhouse et al. 2001; Karnik, Brooke et al. 2003; Karnik, Wythe et al. 2003). Coating of materials with elastin peptides can improve biocompatibility by providing protein sequences required for cell binding (reviewed by (Almine, Bax et al. 2010)). Some three dimensional materials formed from elastin-based peptides demonstrate elastin-like properties, including hydrogels that support cell growth and possess high degrees of elasticity (Keeley, Bellingham et al. 2002; Trabbic-Carlson, Setton et al. 2003). However as with hydrolyzed elastin preparations, synthetic peptides can lack the full repertoire of properties of the fully intact protein and are associated with inflammation (Faury, Ristori et al. 1995).

8.2 Recombinant human tropoelastin

Recombinant human tropoelastin (rhTE) is expressed and purified can be made as a recombinant protein in *Escherichia coli.* rhTE exhibits many properties of native tropoelastin including the ability to coacervate under physiological conditions and be cross-linked *in vitro* to form insoluble elastin fibers (Vrhovski, Jensen et al. 1997; Muiznieks, Jensen et al. 2003). rhTE promotes endothelial cell and fibroblast attachment, spreading and proliferation when used as a surface coating (Bax, Rodgers et al. 2009; Rnjak, Li et al. 2009; Wise, Byrom et al. 2011) and improves the biocompatibility of implanted devices (Yin, Wise et al. 2009; Wilson, Gibson et al. 2010).

Three dimensional biomaterials are produced by cross-linking rhTE to form synthetic human elastin. Synthetic elastin has advantages over decellularized tissue and hydrolyzed elastin preparations as it utilizes human protein avoiding potential problems arising from species differences while benefiting from homogeneity to improve reproducibility and uniformity.

These types of synthetic elastin hydrogels can be made by chemical cross-linking (Mithieux, Rasko et al. 2004), enzyme treatment (Mithieux, Wise et al. 2005) or raising the pH (Mithieux, Tu et al. 2009) of rhTE solutions (Figure 4). The hydrogels demonstrate mechanical properties that are consistent with native elastin including low elastic moduli, support of attachment and proliferation of dermal fibroblasts (Mithieux, Rasko et al. 2004; Rnjak, Li et al. 2009; Annabi, Mithieux et al. 2010). Increases in hydrogel porosity using high pressure CO_2 or the incorporation of glycosaminoglycans improve cell infiltration into hydrogels (Annabi, Mithieux et al. 2010; Tu, Mithieux et al. 2010) where the maintenance of fibroblasts within these scaffolds present them as candidate dermal substitutes.

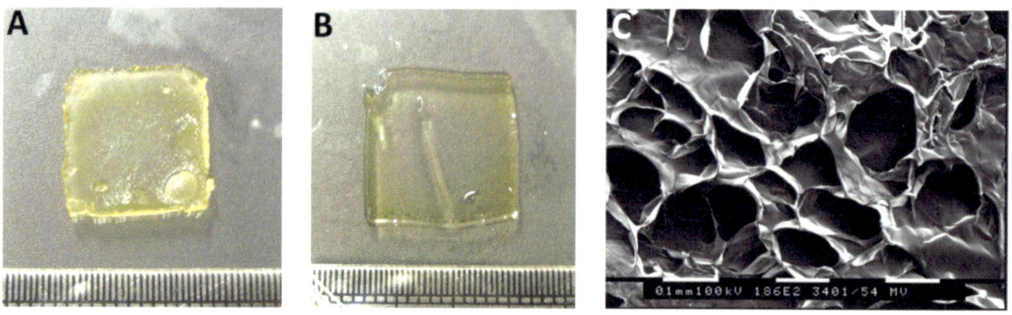

Fig. 4. Synthetic human elastin hydrogels (A) formed from the cross-linking of rhTE and (B) after hydration in phosphate buffered saline. (C) Hydrogel surface porosity shown by scanning electron microscopy

Electrospun synthetic elastin allows for the formation of highly organized biomaterials with tunable mechanical biological properties. Electrospun synthetic elastin is formed by the electrospinning and chemical cross-linking of rhTE to yield ribbon-like microfibers (Figure 5) whose dimensions match those of native elastin fibers (Nivison-Smith, Rnjak et al. 2010). Highly porous electrospun synthetic elastin scaffolds, generated by using high flow rates facilitate the infiltration of dermal fibroblasts *in vitro* and present an alternative to synthetic elastin hydrogels as a dermal replacement (Rnjak, Li et al. 2009).

As a potential vascular material, electrospun synthetic elastin shows attractive characteristics including internal mammary artery-matched elastic mechanical properties, low platelet adhesion (Wise, Byrom et al. 2011) and support of growing human vascular cells including SMCs, endothelial cells (Figure 6) and embryonic palatal mesenchymal stem cells (Li, Mondrinos et al. 2005; Nivison-Smith, Rnjak et al. 2010). Synthetic human elastin fibers can also direct cell spreading to resemble cell organization *in vivo*. For example, the radial alignment of SMC in the arterial media is mimicked by culture of these cells on parallel synthetic elastin fibers (Nivison-Smith & Weiss 2011, *submitted*). Blended conduits of synthetic elastin and silk or polycaprolactone display elasticity and cell adhesion properties courtesy of the rhTE component while the composite component confers additional mechanical strength (Hu, Wang et al. 2010; Wise, Byrom et al. 2011).

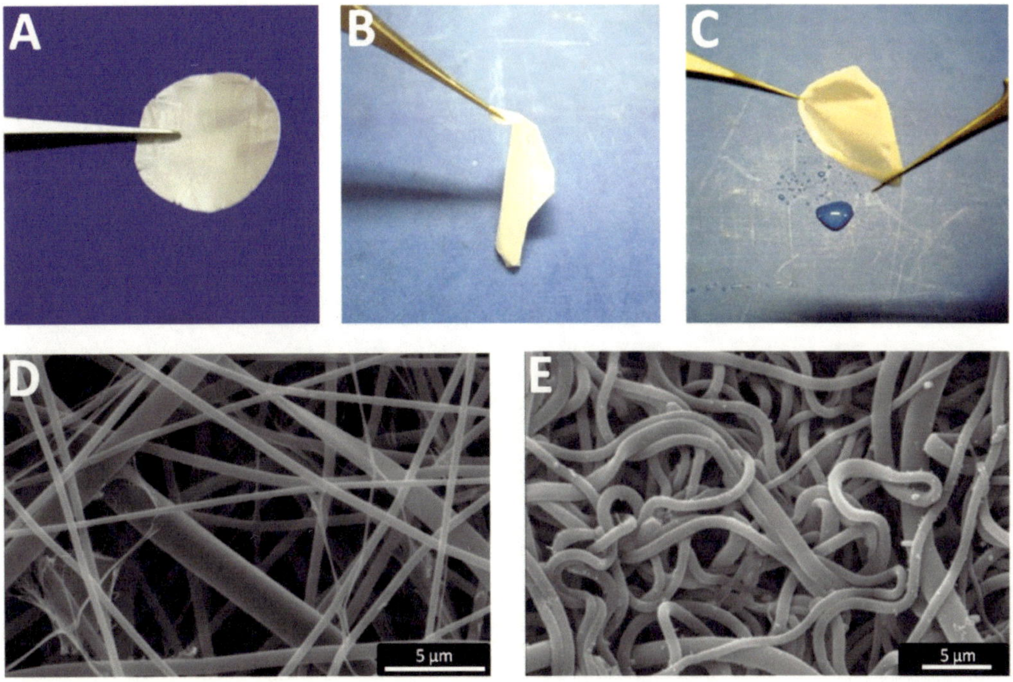

Fig. 5. Synthetic elastin electrospun materials.
Representative photographs of a synthetic human elastin scaffold (A) before cross-linking and (B-C) after cross-linking with hexamethylene diisocyanate and wetting with phosphate buffered saline. Uncross-linked scaffolds are stiff and inflexible while cross-linked scaffolds are highly flexible and collapse when not supported. Scanning electron micrographs of (D) uncross-linked electrospun fibers and (E) cross-linked electrospun fibers reveal the ribbon-like morphology of fibers

9. Conclusion

Elastin is an essential matrix protein, so it is logical that biomaterials designed for elastic tissues should incorporated elastin. Difficulties in sourcing pure intact elastin preparations, particularly those that reflect human sequences, has limited the generation of these materials. Synthetic human elastin that is made from rhTE presents a versatile and stable component of vascular and dermal materials. Elastin-based constructs demonstrate mechanical and biological properties consistent with native elastin and have potential for a wider range of applications.

10. Acknowledgements

We acknowledge grant support from the Australian Research Council and the National Health and Medical Research Council. We thank Dr. Anna Waterhouse for the images related to elastin hydrogels. We acknowledge the contributions of the staff and the facilities made available at the AMMRF (Australian Microscopy & Microanalysis Research Facility) at the Australian Center for Microscopy and Microanalysis, The University of Sydney.

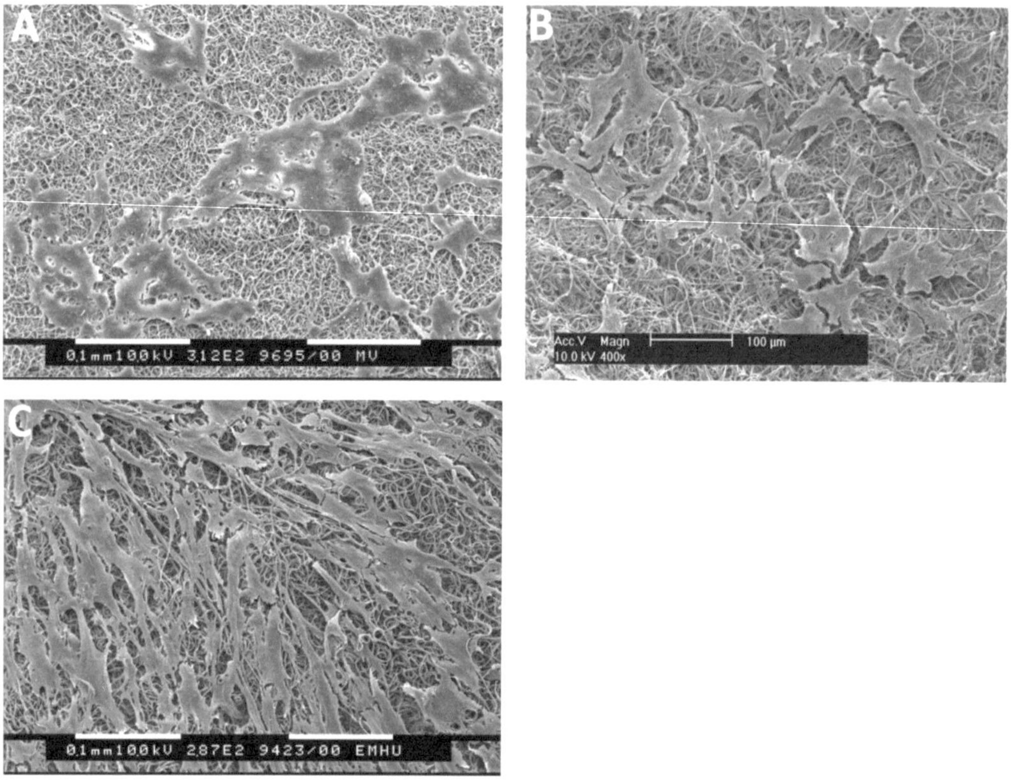

Fig. 6. Examples of synthetic elastin electrospun materials.
Scanning electron micrographs of (A) human umbilical vein endothelial cells, (B) SMCs and (C) dermal fibroblasts cultured on synthetic elastin fibers

11. References

Abbott, W. M., J. Megerman, et al. (1987). Effect of compliance mismatch on vascular graft patency. *J Vasc Surg* 5(2): 376-382.

Akhtar, K., T. J. Broekelmann, et al. (2011). Oxidative modifications of the C-terminal domain of tropoelastin prevent cell binding. *J Biol Chem*.

Almine, J. F., D. V. Bax, et al. (2010). Elastin-based materials. *Chem Soc Rev* 39(9): 3371-3379.

Annabi, N., A. Fathi, et al. (2011). The effect of elastin on chondrocyte adhesion and proliferation on poly (varepsilon-caprolactone)/elastin composites. *Biomaterials* 32(6): 1517-1525.

Annabi, N., S. M. Mithieux, et al. (2009). Synthesis of highly porous crosslinked elastin hydrogels and their interaction with fibroblasts in vitro. *Biomaterials* 30(27): 4550-4557.

Annabi, N., S. M. Mithieux, et al. (2009). The fabrication of elastin-based hydrogels using high pressure CO(2). *Biomaterials* 30(1): 1-7.

Annabi, N., S. M. Mithieux, et al. (2010). Cross-linked open-pore elastic hydrogels based on tropoelastin, elastin and high pressure CO2. *Biomaterials* 31(7): 1655-1665.

Baldock, C., A. F. Oberhauser, et al. (2011). Shape of tropoelastin, the highly extensible protein that controls human tissue elasticity. *Proc Natl Acad Sci U S A* 108(11): 4322-4327.

Barbie, C., C. Angibaud, et al. (1989). Some factors affecting properties of elastin-fibrin biomaterial. *Biomaterials* 10(7): 445-448.

Bax, D. V., U. R. Rodgers, et al. (2009). Cell adhesion to tropoelastin is mediated via the C-terminal GRKRK motif and integrin alphaVbeta3. *J Biol Chem* 284(42): 28616-28623.

Bellingham, C. M., K. A. Woodhouse, et al. (2001). Self-aggregation characteristics of recombinantly expressed human elastin polypeptides. *Biochim Biophys Acta* 1550(1): 6-19.

Boland, E. D., J. A. Matthews, et al. (2004). Electrospinning collagen and elastin: preliminary vascular tissue engineering. *Front Biosci* 9: 1422-1432.

Broekelmann, T. J., B. A. Kozel, et al. (2005). Tropoelastin interacts with cell-surface glycosaminoglycans via its COOH-terminal domain. *J Biol Chem* 280(49): 40939-40947.

Brooke, B. S., A. Bayes-Genis, et al. (2003). New insights into elastin and vascular disease. *Trends Cardiovasc Med* 13(5): 176-181.

Buijtenhuijs, P., L. Buttafoco, et al. (2004). Tissue engineering of blood vessels: characterization of smooth-muscle cells for culturing on collagen-and-elastin-based scaffolds. *Biotechnol Appl Biochem* 39(Pt 2): 141-149.

Buttafoco, L., P. Engbers-Buijtenhuijs, et al. (2006). First steps towards tissue engineering of small-diameter blood vessels: preparation of flat scaffolds of collagen and elastin by means of freeze drying. *J Biomed Mater Res B Appl Biomater* 77(2): 357-368.

Buttafoco, L., N. G. Kolkman, et al. (2006). Electrospinning of collagen and elastin for tissue engineering applications. *Biomaterials* 27(5): 724-734.

Callewaert, B., M. Renard, et al. (2011). New insights into the pathogenesis of autosomal-dominant cutis laxa with report of five ELN mutations. *Hum Mutat* 32(4): 445-455.

Chlupac, J., E. Filova, et al. (2009). Blood vessel replacement: 50 years of development and tissue engineering paradigms in vascular surgery. *Physiol Res* 58 Suppl 2: S119-139.

Chuang, T. H., C. Stabler, et al. (2009). Polyphenol-stabilized tubular elastin scaffolds for tissue engineered vascular grafts. *Tissue Eng Part A* 15(10): 2837-2851.

Conklin, B. S., E. R. Richter, et al. (2002). Development and evaluation of a novel decellularized vascular xenograft. *Med Eng Phys* 24(3): 173-183.

Daamen, W. F., H. T. van Moerkerk, et al. (2003). Preparation and evaluation of molecularly-defined collagen-elastin-glycosaminoglycan scaffolds for tissue engineering. *Biomaterials* 24(22): 4001-4009.

Daamen, W. F., J. H. Veerkamp, et al. (2007). Elastin as a biomaterial for tissue engineering. *Biomaterials* 28(30): 4378-4398.

Dahl, S. L., J. Koh, et al. (2003). Decellularized native and engineered arterial scaffolds for transplantation. *Cell Transplant* 12(6): 659-666.

de Chalain, T., J. H. Phillips, et al. (1999). Bioengineering of elastic cartilage with aggregated porcine and human auricular chondrocytes and hydrogels containing alginate, collagen, and kappa-elastin. *J Biomed Mater Res* 44(3): 280-288.

De Vries, H. J., J. E. Zeegelaar, et al. (1995). Reduced wound contraction and scar formation in punch biopsy wounds. Native collagen dermal substitutes. A clinical study. *Br J Dermatol* 132(5): 690-697.

Engbers-Buijtenhuijs, P., L. Buttafoco, et al. (2005). Analysis of the balance between proliferation and apoptosis of cultured vascular smooth muscle cells for tissue-engineering applications. *Tissue Eng* 11(11-12): 1631-1639.

Faury, G., S. Garnier, et al. (1998). Action of tropoelastin and synthetic elastin sequences on vascular tone and on free Ca2+ level in human vascular endothelial cells. *Circ Res* 82(3): 328-336.

Faury, G., M. T. Ristori, et al. (1994). Role of the elastin-laminin receptor in the vasoregulation. *C R Acad Sci III* 317(9): 807-811.

Faury, G., M. T. Ristori, et al. (1995). Effect of elastin peptides on vascular tone. *J Vasc Res* 32(2): 112-119.

Gilbert, T. W., T. L. Sellaro, et al. (2006). Decellularization of tissues and organs. *Biomaterials* 27(19): 3675-3683.

Glagov, S., R. Vito, et al. (1992). Micro-architecture and composition of artery walls: relationship to location, diameter and the distribution of mechanical stress. *J Hypertens Suppl* 10(6): S101-104.

Hafemann, B., S. Ensslen, et al. (1999). Use of a collagen/elastin-membrane for the tissue engineering of dermis. *Burns* 25(5): 373-384.

Han, J., P. Lazarovici, et al. (2011). Co-electrospun blends of PLGA, gelatin, and elastin as potential nonthrombogenic scaffolds for vascular tissue engineering. *Biomacromolecules* 12(2): 399-408.

Hinek, A. (1995). The 67 kDa spliced variant of beta-galactosidase serves as a reusable protective chaperone for tropoelastin. *Ciba Found Symp* 192: 185-191; discussion 191-186.

Holst, J., S. Watson, et al. (2010). Substrate elasticity provides mechanical signals for the expansion of hemopoietic stem and progenitor cells. *Nat Biotechnol* 28(10): 1123-1128.

Hornebeck, W., J. M. Tixier, et al. (1986). Inducible adhesion of mesenchymal cells to elastic fibers: elastonectin. *Proc Natl Acad Sci U S A* 83(15): 5517-5520.

Hu, X., X. Wang, et al. (2010). Biomaterials derived from silk-tropoelastin protein systems. *Biomaterials* 31(32): 8121-8131.

Hucthagowder, V., E. Morava, et al. (2009). Loss-of-function mutations in ATP6V0A2 impair vesicular trafficking, tropoelastin secretion and cell survival. *Hum Mol Genet* 18(12): 2149-2165.

Indik, Z., W. R. Abrams, et al. (1990). Production of recombinant human tropoelastin: characterization and demonstration of immunologic and chemotactic activity. Arch *Biochem Biophys* 280(1): 80-86.

Indik, Z., H. Yeh, et al. (1987). Alternative splicing of human elastin mRNA indicated by sequence analysis of cloned genomic and complementary DNA. *Proc Natl Acad Sci U S A* 84(16): 5680-5684.

Ito, S., S. Ishimaru, et al. (1997). Inhibitory effect of type 1 collagen gel containing alpha-elastin on proliferation and migration of vascular smooth muscle and endothelial cells. *Cardiovasc Surg* 5(2): 176-183.

Ito, S., S. Ishimaru, et al. (1998). Effect of coacervated alpha-elastin on proliferation of vascular smooth muscle and endothelial cells. *Angiology* 49(4): 289-297.

Jacob, M. P. and W. Hornebeck (1985). Isolation and characterisation of insoluble and kappa-elastins. *Methods of connective tissue research* L. Robert, M. Moczar and E. Moczar. Basel, Karger. 4: 92-129.

Jung, S., J. T. Rutka, et al. (1998). Tropoelastin and elastin degradation products promote proliferation of human astrocytoma cell lines. *J Neuropathol Exp Neurol* 57(5): 439-448.

Kagan, H. M. and K. A. Sullivan (1982). Lysyl oxidase: preparation and role in elastin biosynthesis. *Methods Enzymol* 82 Pt A: 637-650.

Kamisato, S., Y. Uemura, et al. (1997). Involvement of intracellular cyclic GMP and cyclic GMP-dependent protein kinase in alpha-elastin-induced macrophage chemotaxis. *J Biochem* 121(5): 862-867.

Karnik, S. K., B. S. Brooke, et al. (2003). A critical role for elastin signaling in vascular morphogenesis and disease. *Development* 130(2): 411-423.

Karnik, S. K., J. D. Wythe, et al. (2003). Elastin induces myofibrillogenesis via a specific domain, VGVAPG. *Matrix Biol* 22(5): 409-425.

Keeley, F. W., C. M. Bellingham, et al. (2002). Elastin as a self-organizing biomaterial: use of recombinantly expressed human elastin polypeptides as a model for investigations of structure and self-assembly of elastin. *Philos Trans R Soc Lond B Biol Sci* 357(1418): 185-189.

Kielty, C. M. (2006). Elastic fibres in health and disease. *Expert Rev Mol Med* 8(19): 1-23.

Koens, M. J., K. A. Faraj, et al. (2010). Controlled fabrication of triple layered and molecularly defined collagen/elastin vascular grafts resembling the native blood vessel. *Acta Biomater* 6(12): 4666-4674.

Lammers, G., G. S. Tjabringa, et al. (2009). A molecularly defined array based on native fibrillar collagen for the assessment of skin tissue engineering biomaterials. *Biomaterials* 30(31): 6213-6220.

Lamprou, D., P. Zhdan, et al. (2010). Gelatine and Gelatine/Elastin Nanocomposites for Vascular Grafts: Processing and Characterization. *J Biomater Appl.*

Leach, J. B., J. B. Wolinsky, et al. (2005). Crosslinked alpha-elastin biomaterials: towards a processable elastin mimetic scaffold. *Acta Biomater* 1(2): 155-164.

Lee, S. J., J. J. Yoo, et al. (2007). In vitro evaluation of electrospun nanofiber scaffolds for vascular graft application. *J Biomed Mater Res A* 83(4): 999-1008.

Li, D. and Y. N. Xia (2004). Electrospinning of nanofibers: Reinventing the wheel? *Advanced Materials* 16(14): 1151-1170.

Li, D. Y., B. Brooke, et al. (1998). Elastin is an essential determinant of arterial morphogenesis. *Nature* 393(6682): 276-280.

Li, M., M. J. Mondrinos, et al. (2005). Electrospun protein fibers as matrices for tissue engineering. *Biomaterials* 26(30): 5999-6008.

Lloyd-Jones, D., R. J. Adams, et al. (2010). Heart disease and stroke statistics--2010 update: a report from the American Heart Association. *Circulation* 121(7): e46-e215.

Long, M. M., V. J. King, et al. (1989). Elastin repeat peptides as chemoattractants for bovine aortic endothelial cells. *J Cell Physiol* 140(3): 512-518.

Macchiarini, P., P. Jungebluth, et al. (2008). Clinical transplantation of a tissue-engineered airway. *Lancet* 372(9655): 2023-2030.

McClure, M. J., S. A. Sell, et al. (2010). A three-layered electrospun matrix to mimic native arterial architecture using polycaprolactone, elastin, and collagen: a preliminary study. *Acta Biomater* 6(7): 2422-2433.

Mecham, R. P. (1991). Elastin synthesis and fiber assembly. *Ann N Y Acad Sci* 624: 137-146.

Mithieux, S. M., J. E. Rasko, et al. (2004). Synthetic elastin hydrogels derived from massive elastic assemblies of self-organized human protein monomers. *Biomaterials* 25(20): 4921-4927.

Mithieux, S. M., Y. Tu, et al. (2009). In situ polymerization of tropoelastin in the absence of chemical cross-linking. *Biomaterials* 30(4): 431-435.

Mithieux, S. M. and A. S. Weiss (2005). Elastin. *Adv Protein Chem* 70: 437-461.

Mithieux, S. M., S. G. Wise, et al. (2005). A model two-component system for studying the architecture of elastin assembly in vitro. *J Struct Biol* 149(3): 282-289.

Miyamoto, K., M. Atarashi, et al. (2009). Creation of cross-linked electrospun isotypic-elastin fibers controlled cell-differentiation with new cross-linker. *Int J Biol Macromol* 45(1): 33-41.

Mochizuki, S., B. Brassart, et al. (2002). Signaling pathways transduced through the elastin receptor facilitate proliferation of arterial smooth muscle cells. *J Biol Chem* 277(47): 44854-44863.

Muiznieks, L. D., S. A. Jensen, et al. (2003). Structural changes and facilitated association of tropoelastin. *Arch Biochem Biophys* 410(2): 317-323.

Nivison-Smith, L., J. Rnjak, et al. (2010). Synthetic human elastin microfibers: stable cross-linked tropoelastin and cell interactive constructs for tissue engineering applications. *Acta Biomater* 6(2): 354-359.

Partridge, S. M., H. F. Davis, et al. (1955). The chemistry of connective tissues. 2. Soluble proteins derived from partial hydrolysis of elastin. *Biochem J* 61(1): 11-21.

Pasquali-Ronchetti, I. and M. Baccarani-Contri (1997). Elastic fiber during development and aging. *Microsc Res Tech* 38(4): 428-435.

Petersen, T. H., E. A. Calle, et al. (2010). Tissue-engineered lungs for in vivo implantation. *Science* 329(5991): 538-541.

Price, A. P., K. A. England, et al. (2010). Development of a decellularized lung bioreactor system for bioengineering the lung: the matrix reloaded. *Tissue Eng Part A* 16(8): 2581-2591.

Rnjak, J., Z. Li, et al. (2009). Primary human dermal fibroblast interactions with open weave three-dimensional scaffolds prepared from synthetic human elastin. *Biomaterials* 30(32): 6469-6477.

Rnjak, J., S. G. Wise, et al. (2011). Severe Burn Injuries and the Role of Elastin in the Design of Dermal Substitutes. *Tissue Eng Part B Rev.*

Rodgers, U. R. and A. S. Weiss (2004). Integrin alpha v beta 3 binds a unique non-RGD site near the C-terminus of human tropoelastin. *Biochimie* 86(3): 173-178.

Rodgers, U. R. and A. S. Weiss (2005). Cellular interactions with elastin. *Pathol Biol (Paris)* 53(7): 390-398.

Rosenbloom, J., W. R. Abrams, et al. (1993). Extracellular matrix 4: the elastic fiber. *FASEB J* 7(13): 1208-1218.

Roten, S. V., S. Bhat, et al. (1996). Elastic fibers in scar tissue. *J Cutan Pathol* 23(1): 37-42.

Ryssel, H., E. Gazyakan, et al. (2008). The use of MatriDerm in early excision and simultaneous autologous skin grafting in burns--a pilot study. *Burns* 34(1): 93-97.

Sandberg, L. B., N. T. Soskel, et al. (1981). Elastin structure, biosynthesis, and relation to disease states. *N Engl J Med* 304(10): 566-579.
Schmidt, C. E. and J. M. Baier (2000). Acellular vascular tissues: natural biomaterials for tissue repair and tissue engineering. *Biomaterials* 21(22): 2215-2231.
Sell, S. A., M. J. McClure, et al. (2006). Electrospun polydioxanone-elastin blends: potential for bioresorbable vascular grafts. *Biomed Mater* 1(2): 72-80.
Senior, R. M., G. L. Griffin, et al. (1980). Chemotactic activity of elastin-derived peptides. *J Clin Invest* 66(4): 859-862.
Shapiro, S. D., S. K. Endicott, et al. (1991). Marked longevity of human lung parenchymal elastic fibers deduced from prevalence of D-aspartate and nuclear weapons-related radiocarbon. *J Clin Invest* 87(5): 1828-1834.
Smith, M. J., M. J. McClure, et al. (2008). Suture-reinforced electrospun polydioxanone-elastin small-diameter tubes for use in vascular tissue engineering: a feasibility study. *Acta Biomater* 4(1): 58-66.
Starcher, B. (2001). A ninhydrin-based assay to quantitate the total protein content of tissue samples. *Anal Biochem* 292(1): 125-129.
Stitzel, J., J. Liu, et al. (2006). Controlled fabrication of a biological vascular substitute. *Biomaterials* 27(7): 1088-1094.
Thomas, V., X. Zhang, et al. (2007). Functionally graded electrospun scaffolds with tunable mechanical properties for vascular tissue regeneration. *Biomed Mater* 2(4): 224-232.
Thomas, V., X. Zhang, et al. (2009). A biomimetic tubular scaffold with spatially designed nanofibers of protein/PDS bio-blends. *Biotechnol Bioeng* 104(5): 1025-1033.
Trabbic-Carlson, K., L. A. Setton, et al. (2003). Swelling and mechanical behaviors of chemically cross-linked hydrogels of elastin-like polypeptides. *Biomacromolecules* 4(3): 572-580.
Tu, Y., S. M. Mithieux, et al. (2010). Synthetic elastin hydrogels that are coblended with heparin display substantial swelling, increased porosity, and improved cell penetration. *J Biomed Mater Res A* 95(4): 1215-1222.
Uchimura, E., Y. Sawa, et al. (2003). Novel method of preparing acellular cardiovascular grafts by decellularization with poly(ethylene glycol). *J Biomed Mater Res A* 67(3): 834-837.
Uitto, J., A.M. Christiano, V.M. Kahari, M.M. Bashir, and J. Rosenbloom, *Molecular biology and pathology of human elastin.* Biochem Soc Trans, 1991. 19(4): p. 824-9
Umeda, H., F. Nakamura, et al. (2001). Oxodesmosine and isooxodesmosine, candidates of oxidative metabolic intermediates of pyridinium cross-links in elastin. *Arch Biochem Biophys* 385(1): 209-219.
Urban, Z., J. Zhang, et al. (2001). Supravalvular aortic stenosis: genetic and molecular dissection of a complex mutation in the elastin gene. *Hum Genet* 109(5): 512-520.
Vrhovski, B., S. Jensen, et al. (1997). Coacervation characteristics of recombinant human tropoelastin. *Eur J Biochem* 250(1): 92-98.
Vrhovski, B. and A. S. Weiss (1998). Biochemistry of tropoelastin. *Eur J Biochem* 258(1): 1-18.
Williamson, M. R., A. Shuttleworth, et al. (2007). The role of endothelial cell attachment to elastic fibre molecules in the enhancement of monolayer formation and retention, and the inhibition of smooth muscle cell recruitment. *Biomaterials* 28(35): 5307-5318.

Wilson, B. D., C. C. Gibson, et al. (2010). Novel Approach for Endothelializing Vascular Devices: Understanding and Exploiting Elastin-Endothelial Interactions. *Ann Biomed Eng*.

Wise, S. G., M. J. Byrom, et al. (2011). A multilayered synthetic human elastin/polycaprolactone hybrid vascular graft with tailored mechanical properties. *Acta Biomater* 7(1): 295-303.

Wissink, M. J., M. J. van Luyn, et al. (2000). Endothelial cell seeding on crosslinked collagen: effects of crosslinking on endothelial cell proliferation and functional parameters. *Thromb Haemost* 84(2): 325-331.

Yin, Y., S. G. Wise, et al. (2009). Covalent immobilisation of tropoelastin on a plasma deposited interface for enhancement of endothelialisation on metal surfaces. *Biomaterials* 30(9): 1675-1681.

Zhang, X., V. Thomas, et al. (2010). Two ply tubular scaffolds comprised of proteins/poliglecaprone/polycaprolactone fibers. *J Mater Sci Mater Med* 21(2): 541-549.

Zhang, X., V. Thomas, et al. (2010). An in vitro regenerated functional human endothelium on a nanofibrous electrospun scaffold. *Biomaterials* 31(15): 4376-4381.

Zhang, X., Y. Xu, et al. (2011). Engineering an antiplatelet adhesion layer on an electrospun scaffold using porcine endothelial progenitor cells. *J Biomed Mater Res A*.

Zou, Y. and Y. Zhang (2009). An experimental and theoretical study on the anisotropy of elastin network. *Ann Biomed Eng* 37(8): 1572-1583.

16

The Use of a Hydrogel Matrix as a Cellular Delivery Vehicle in Future Cell-Based Therapies: Biological and Non-Biological Considerations

Thomas I. Zarembinski, William P. Tew and Sarah K. Atzet
Glycosan BioSystems, Inc, A Division of Orthocyte Corp.
USA

1. Introduction

Cell therapy is defined as the minimally-invasive delivery of therapeutic cells into a human host to repair damaged or diseased tissue(s). Its long-term goal is to reduce the current expense of 1% of US GDP on organ replacement therapies (Lysaght, et al., 2008). Hematopoietic stem cell transplants are the most popular cell therapy and have been crucial in treating a variety of hematological diseases (Cutler and Antin, 2001; Horowitz, 2004). For novel cellular therapies (i.e. using non-hematopoietic cells directed against non-hematological diseases), the numbers fall off precipitously. Of these, only cell-based cartilage repair is the only novel cell therapy so far used with significant frequency (Martin, et al., 2010). Since this therapy additionally requires the expansion of and the use of autologous chondrocytes, this data suggests a crucial bottleneck for a wide variety of novel cellular therapies will be both in producing enough rare cells for therapeutic effect as well as for obviating immunologic concerns.

The emergence of stem cell biology has reshaped the cell therapy and tissue engineering landscape because the quantity and breadth of therapeutic cell has been dramatically expanded. Stem cells are typically derived from embryos or adult tissues and have two abilities: to self-renew (i.e. one stem cell can divide and make more stem cells) and to differentiate into specialized cells. These characteristics allow stem cells to be used to produce large quantities of a wide variety of cell types, including rare or difficult to harvest cell types (Takahashi, et al., 2007; Thomson, et al., 1998; Yu, et al., 2007). The therapeutic value of these cells can captured by either using them to directly produce new tissue (Keirstead, et al., 1999) or as a source of bioactive agents such as cytokines and growth factors inducing host cells and tissues to regenerate themselves (Caplan and Dennis, 2006). In addition, a wide variety of stem cells can be harvested or derived from adult tissues such as bone marrow, adipose tissue, or cord blood (Ingram, et al., 2004; Lennon and Caplan, 2006; Zuk, et al., 2001). Adult stems can potentially lower the hurdle of immunologic incompatibility due to either their autologous nature or immunomodulatory effects (Caplan and Dennis, 2006).

While stem cells obviate the shortcomings of using a patient's differentiated cells, the rate-limiting step in successful cell therapy is not only the number of transplanted cells but their survival rate post-transplantation. In short, the transplanted stem cells need help to stay alive long enough for their therapeutic effect to be seen. The majority of stem and progenitor cells in the transplanted bolus die shortly after transplantation (Bliss, et al., 2007; Snyder, et al., 2010; Terrovitis, et al., 2008; Zhong, et al., 2010). In some cases, more than 95% of transplanted stem cells die within two weeks of transplantation (Snyder, et al., 2010; Terrovitis, et al., 2008; Zhong, et al., 2010). Since tissues contain cells encapsulated in a carbohydrate and protein-rich extracellular matrix (ECM), one approach to significantly improve stem cell survival is to include a biomaterial carrier that acts as an ECM mimic upon *in vivo* delivery. These carriers have been prepared from synthetic or naturally sourced polymers and provide an adhesion surface which not only localizes cells but also provides a template for new tissue formation (Mooney and Vandenburgh, 2008).

While there are excellent review articles describing the design criteria for clinically useful matrices (Lutolf and Hubbell, 2005; Prestwich, 2008), this chapter focuses on their translation to the clinic—that is, the development of these matrices as FDA-approved injectable cell carriers, or cellular delivery vehicles, and how both biological as well as non-biological considerations (i.e. physician requirements, intellectual property, regulatory, manufacturing) must be satisfied before the biomaterial can reach the medical marketplace. In the first part, we present an overview of some of the basic biomaterials which have the potential to be developed as FDA-approved cellular delivery vehicles. Next, a more in-depth discussion of these biological and non-biological considerations follows. Finally, we describe one matrix, HyStem®, and its potential use in three areas: stroke, cartilage repair, and gene therapy.

2. Injectable matrices for stem cell therapies

2.1 Lessons from Hematopoietic Stem Cell Transplantation—the first stem cell therapy

The first stem cell therapy developed was hematopoietic stem cell transplantation (HSCT) reported in 1968 (Bach, et al., 1968; Gatti, et al., 1968) and is used to treat a variety of hematological malignancies (Cutler and Antin, 2001). It is also by far the most popular: there are in excess of 45,000 HSCTs performed annually worldwide (Horowitz, 2004). While there is still much research to do to understand the entire mechanism of how HSCT works, a review of HSCT is informative since many lessons can be learned and applied to novel non-hematological stem cell therapies.

HSCT can be summarized as follows: HSCT typically begins with the harvest of either bone marrow or the leukocyte fraction of peripheral blood either autologously or allogeneically from an immunologically (or human leukocyte antigen (HLA)) matched donor. The therapeutic fraction of these cells are the CD34+ (early hematopoietic cell marker) cells and contains both multipotent hematopoietic stem cells (CD34+/CD38-) (Cutler and Antin, 2001) and more mature hematopoietic precursor cells (Duran-Struuck and Dysko, 2009). A minimum of 2×10^6 $CD34^+$ cells are required per kg recipient body weight, or 300 million cells per 150 kg adult (Cutler and Antin, 2001; Mavroudis, et al., 1996). Importantly, the mixture of the young pluripotent and more mature hematopoietic stem cells is crucial since the older cells act as escorts to provide temporary immunological restoration and host survival while the younger cells engraft and generate mature cells to replace the former (Duran-Struuck and Dysko, 2009).

After harvest from the donor, these cells are then transplanted intravenously into a recipient who may require prior myeloablation of the host bone marrow. Myeloablative treatment

destroys the host hematopoietic cellular population while leaving many aspects of the hematopoietic niche intact (Dominici, et al., 2009; Duran-Struuck and Dysko, 2009; Slayton, et al., 2007). The niche includes cells such as osteoblasts and mesenchymal stromal cells to which the HSCs attach to as well as communicate with via soluble and insoluble signals (Scadden, 2006; Taichman, 2005). While HSCs find the bone marrow randomly (Cui, et al., 1999), a significant number of cells travel to the marrow where engraftment occurs 2-4 weeks post-transplantation (Cutler and Antin, 2001).
From this example, it can be inferred that the key steps in successful stem cell engraftment include 1) cellular and sample integrity, 2) cellular travel to and anchoring within the niche (homing and lodgment) (Cui, et al., 1999; Lam and Adams, 2010), 3) niche remodeling (Sands and Mooney, 2007), 4) cellular proliferation and differentiation (engraftment) (Cutler and Antin, 2001), and 5) an appropriate space and nourishment to do so (Scadden, 2006). HSCT only requires a 2-4 week post-transplantation time period for engraftment since the transplanted cells and the host cell niche are ready to be merged. The transplanted cells are a mixture of gently-processed cells and they only require less than two days for bone marrow localization after IV administration (Cui, et al., 1999). Once in the bone marrow, the cells should lodge efficiently since the irradiated bone marrow niche mostly retains its native matrix structure and supporting cells (Dominici, et al., 2009; Slayton, et al., 2007) with little remodeling required. In HSCT, both space and access to blood supply is present for the cells since bone has significant open regions for the cells in its interior and is highly vascularized (Gentry-Steele and Bramblett, 1988).

2.2 Challenges for novel stem cell-based therapies

Unlike HSCT, novel stem cell therapies face a more difficult road primarily because neither the stem cell nor its host microenvironment has been prepared in advance of their arrival in their new home. Since stem cells have little function outside of their niche (Scadden, 2006), they must construct their own among their other duties under transplantation duress. Initially, the stem/progenitor cells are extracted from its self-synthesized ECM either from a solid tissue organ or from the surface of a tissue culture plastic plate. Cellular integrity is the first obstacle these cells face. Since harsh enzymatic methods such as trypsinization are typically used, integrins required for cellular attachment are cleaved and need to be resynthesized by the cell (Harrison and Rae, 1997; Wu, et al., 2005). The next major obstacles are lodgment and niche remodeling. While the cells are then injected adjacent to the host target tissue of interest, they have few surface receptors to attach to the surrounding ECM. If the target tissue is diseased or damaged, its niche may also need significant remodeling. For example, the tissue ECM may have significant scar tissue with dramatically different mechanical properties and function from those of the native target tissue (Laflamme, et al., 2007; Martens, et al., 2009; Reilly and Engler, 2009; Scadden, 2006). Altered microenvironment compliance could provide more resistance to remodeling by affecting the desired stem cell differentiation path (Engler, et al., 2006). The final obstacle is the inadequate conditions for engraftment since there is no pre-established vasculature present to nourish the transplanted cells (Martens, et al., 2009). In some cases, there will be little or no space for the cells to divide into since the target tissue has no cavity (Darabi, et al., 2009; Laflamme, et al., 2007; Terrovitis, et al., 2008). It is no surprise that most transplanted cells die within 24 hours of transplantation (Bliss, et al., 2007; Guerette, et al., 1997; Snyder, et al., 2010; Suzuki, et al., 2004; Tate, et al., 2009; Terrovitis, et al., 2008; Zhong, et al., 2010).

2.3 The hydrogel cell delivery matrix

A solution is to borrow principles from the field of Tissue Engineering by providing a temporary, more native niche for the cells until they can synthesize their own. Tissue engineering is the *ex vivo* cultivation of cells on polymeric scaffolds in order to generate tissues or organs for transplantation (Kaihara and Vacanti, 1999). The use of degradable polymer scaffolds is crucial since it provides an initial remodelable substrate for the cells, provides space for the cells to reorganize into more complicated tissues, and potentially can be designed to provide an initial template to guide subsequent structure formation (Kaihara and Vacanti, 1999). Hydrogels made from naturally occurring biopolymers are one kind of degradable scaffold that has four key characteristics: first, their high water content simulates the microenvironment for soft tissues and allows for transfer of gases, small molecules, and proteins (Tibbitt and Anseth, 2009). Second, the starting biopolymers frequently have chemically-available side groups which can functionalized for altering its mechanical properties, degradation time, and cellular adhesion surfaces (Serban and Prestwich, 2008; Tibbitt and Anseth, 2009). Third, they can be constructed to encourage neovascularization for the cells (Cai, et al., 2005; Liu, et al., 2007; Phelps, et al., 2010). Finally, they are injectable when in liquid form, fulfilling the minimal invasive surgery criterion favored by physicians (Miles, et al., 2004).

2.3.1 Types of hydrogels currently available

While there are currently no hydrogel matrices FDA-approved for use specifically for stem cell therapies, there are a number of biomaterials to choose from with varying degrees of regulatory and manufacturing hurdles (Table 1). These include those that 1) have been recently developed in academic laboratories, 2) are commercially available for research use only and whose regulatory status can be extended for human clinical use (e.g. HyStem-C, MT-3D Q-gel, RGD-alginate, etc), or 3) are in fact FDA-approved for acellular indications and would need to be extended for stem cell delivery indications (e.g. Fibrin glue), (Figure 1, Table 1).

All of these matrices roughly fall into one of three classifications: natural, synthetic, or semi-synthetic (Fig. 1). While natural matrices are typically well-tolerated by the host and cells due to their mimicking the natural ECM in terms of backbone and microstructure, they generally suffer from lot-to-lot variability, high degradation rates, and poor tunability (Tibbitt and Anseth, 2009). In addition, their complexity and poor definition can portend difficult manufacture pursuant to current Good Manufacturing Practices (cGMPs) required for clinical application. For example, Matrigel is arguably the most popular hydrogel matrix used in preclinical studies since it is in fact a native ECM with a variety of matricellular proteins and growth factors. Despite these benefits, Matrigel will likely never be placed in humans due to regulatory and manufacturing concerns. In particular, its mouse tumor origin and ill-defined, variable composition are problematic with nearly two-thousand unique proteins present (Hughes, et al., 2010; Nagaoka, et al., 2010).

Synthetic matrices are the opposite: while very reproducible, tunable, and amenable to more facile regulatory and manufacturing protocols, they generally require engineering to provide both cellular attachment sites and degradation rates comparable to those for native ECMs (Tibbitt and Anseth, 2009). Semi-synthetic matrices share characteristics of both classes and can be constructed either by modifying purified natural biopolymers such as HA or Alginate (Alsberg, et al., 2001; Darr and Calabro, 2009; Zheng Shu, et al., 2004) or by engineering synthetic polymers with integrin and/or growth factor binding sites and degradation signals to more closely mimic the natural ECM (Raeber, et al., 2005; Tibbitt and Anseth, 2009).

Name	Base Biopolymer	Crosslinker	Cell Attachment sites	References
Natural				
Matrigel™ (BD Biosciences)	Laminin and Collagen IV	None	Yes	(Kleinman and Martin, 2005)
Purecol ® (Advanced BioMatrix)	Collagen I	None	Yes	(Beckman, et al., 2008)
Tisseel ® (Baxter)	Fibrin	Glutamyl/lysine covalent bonds	FN	(Ahmed, et al., 2008)
Alginate (FMC Biopolymer)	Alginic acid	Calcium	none	(Alsberg, et al., 2001)
Semi-Synthetic				
HyStem-C (Glycosan BioSystems)	Thiolated HA	PEGDA	Thiolated gelatin	(Zheng Shu, et al., 2004)
RGD-alginate (FMC Biopolymer)	Alginic acid	Calcium	RGD	(Alsberg, et al., 2001)
Corgel ™ (Lifecore)	Tyraminated HA	Dityramine covalent bond	none	(Darr and Calabro, 2009)
Synthetic				
Puramatrix ™ (BD Biosciences)	$(RADA)_{16}$	None	RAD	(Semino, 2008)
PEGDA (Glycosan BioSystems)	PEGDA	Covalent bond	none	(Moon, et al., 2009)
MT-3D (Qgel)	PEGDA	Di-cysteine containing peptide	RGD	(Raeber, et al., 2005)

Table 1. Commonly-used commercially available injectable hydrogel matrices. Abbreviations: PEGDA, polyethylene glycol diacrylate; HA, hyaluronic acid; FN, fibronectin; RADA, arginine-alanine-aspartate-alanine tetrapeptide; RGD, arginine-glycine-aspartate tripeptide. Suppliers are shown in parentheses below product name

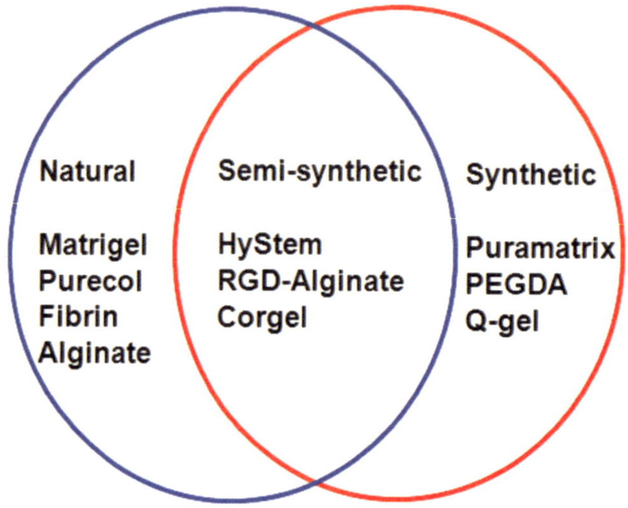

Fig. 1. Different types of hydrogel matrices

A common misconception is that a variety of FDA-approved hydrogel matrices are simple to transfer from its approved indication to one involving stem cell delivery. One case in point is natural biopolymer-derived dermal fillers which are currently approved for use as acellular space fillers in cosmetic indications. Examples include collagen-based Zyplast® and HA-based Juvederm® and Restylane® (Newman, 2009). Many of these biomaterials, however, are pre-crosslinked within prefilled syringes and preclude the addition and homogenous distribution of cells prior to injection. In addition, these HA-based biomaterials have no cellular attachment sites, causing cellular apoptosis or anoikis.

2.4 Matrix choice

Even though there a wide variety of hydrogel matrices to choose from, there is still doubt as to whether a delivery vehicle is even needed and if so, how one might choose the best matrix for an application. For many researchers and physicians, the widespread death of stem cells after transplantation is thought to be unavoidable and provides the rationale for implanting a massive number of cells. The hope is that enough cells survive post-transplantation for there to be a therapeutic effect. For those who see value in including a matrix carrier, its selection is usually an afterthought. For some, the choice is made out of convenience due to collaboration with a bioengineer in the same institution. For others, it is made from a scan of the literature or from collaborator recommendations. Without considering their therapeutic stem cell microenvironment after transplantation, the range of options, and requirements for translation, all matrices appear comparable. However, there are at least three distinct lenses through which a researcher or physician must peer to decide on the best matrix for the therapeutic stem cell and indication. These viewpoints provide the basis for three sets of criteria that must a matrix must fulfill.

2.4.1 Biological criteria for the researcher

From a researcher's perspective, the matrix choice will be driven by maintaining consistency with a) the therapeutic stem cell mode of action b) the properties of the ECM from both the transplanted cell and the target tissue. There are two primary modes of action for a therapeutic stem cell after transplantation: *direct*, or transplanted cell engraftment into the host tissue and *indirect*, or secretion of trophic bioactive factors which induce host tissue repair (Caplan and Dennis, 2006). The former requires the matrix to be degradable in concert with the remodeling and cell proliferation of the transplanted cells (Mooney and Vandenburgh, 2008). Alternatively, the latter requires a release of soluble factors as well as a longer-term protective environment from the host immune system (Luca, et al., 2007; Penolazzi, et al., 2010). Since alginate and PEGDA degrade very slowly (Liao, et al., 2008; Shikanov, et al., 2009), they are well suited for the latter indication. Indeed, both have already been used for non-stem cell based therapeutics; for example encapsulating pancreatic beta cells or islets in alginate or PEGDA prior to transplantation promotes cellular immunoprotection and survival for these cells and extends the window of insulin production (Lin and Anseth, 2009; Wilson and Chaikof, 2008).

Matrix choice is also dependent on the properties of the therapeutic cell ECM since it can provide cues for proper cellular function. In particular, the matrix biopolymer backbone is an essential property to consider since it can contain a great deal of biological information crucial to the cell type. For example, HA has a dual role in cellular biology: it plays a key structural role in the ECM through its interaction with members of the lectican family of proteoglycans while affecting cellular signaling important for development (Camenisch, et al., 2000; Ruoslahti, 1996). In cartilage, HA interacts with the lectican aggrecan which stabilizes the cartilage ECM (Aszodi, et al., 2006; Fraser, et al., 1997; Knudson and Knudson, 1993; Laurent and Fraser, 1992). In addition, HA also interacts with the chondrocyte CD44 receptor to induce genes involved in matrix degradation (Schmitz I et al., 2010). Hence, the use of an HA-based scaffold for implanting chondrogenic cells to heal osteochondral cartilage defects is logical and has in fact been shown to be effective for *in vitro* chondrocyte culture and for animal models of cartilage injury (Chung and Burdick, 2009; Liu, et al., 2006; Toh, et al., 2010). In cardiac development, HA interacts with another lectican, versican, in the cardiac ECM while playing a crucial signaling role in endothelial cell migration and transformation (Camenisch, et al., 2000). Successful use of HA-based hydrogels in the study of cardiac development logically follows (Young and Engler, 2010). Importantly, since all lecticans including neurocan and brevican are also expressed in the nervous system (Yamaguchi, 2000), HA plays a crucial role in the brain extracellular matrix and can provide an excellent substrate for neural tissue engineering (Zhong, et al., 2010).

Matrix choice also depends on the target cell ECM since its matrix compliance and cellular attachment sites present provide differentiation cues for the cells (Engler, et al., 2006; Flaim, et al., 2005; Soen, et al., 2006). With respect to ECM compliance, stem cells are like chameleons—they assess their current substrate compliance and figure out which cells they need to differentiate into so that the compliance of their new ECM matches. For example, a matrix stiffness of less than 1 kPa induces mesenchymal stem cells to become neural-like ((Engler, et al., 2006); Fig. 2), suggesting softer hydrogels are more appropriate for CNS applications (Zhong, et al., 2010). For this reason, it is important to use a matrix whose compliance matches that of the target tissue/cell ECM. A comparison of different matrix compliances is shown in Figure 2. Like matrix stiffness, the appropriate mixture of polypeptide sequences representing cellular attachment sites will likely need to be matched to the cell type and application. While there are hydrogels engineered to contain the RGD peptide for cellular attachment (Alsberg et al., 2001; Raeber et al., 2005), this peptide will likely not be sufficient for each application to maximize differentiation down an specific path (Flaim et al., 2005; Soen et al.,2006).

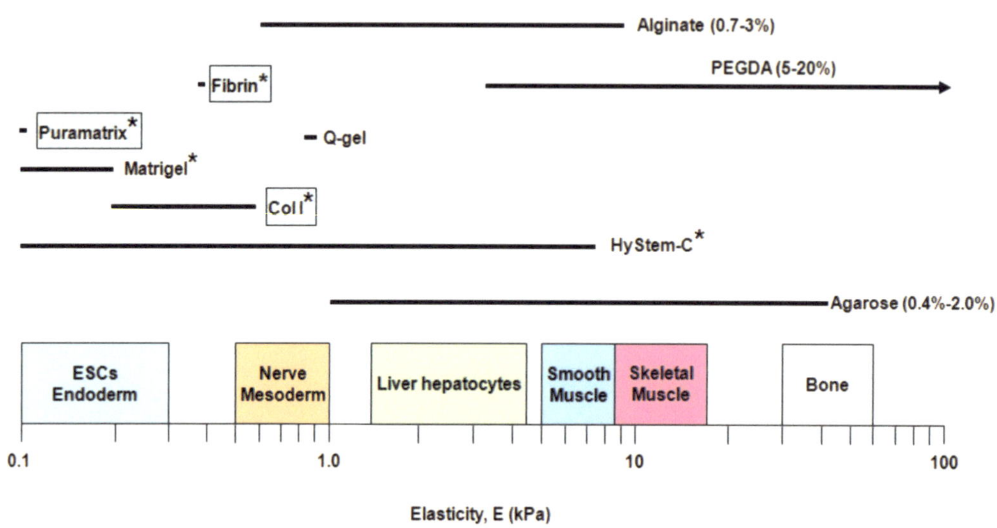

Fig. 2. Elasticities of tissues and of commercially available hydrogel matrices. The Young's modulus of various vertebrate tissues (colored boxes (Reilly and Engler, 2009); ESCs, embryonic stem cells) compared to the ranges attainable with various hydrogel matrices (solid lines; alginate (West, et al., 2007), PEGDA (Patel, et al., 2005), Fibrin (Weisel, 2004), Q-gel (Raeber, et al., 2005), all remaining hydrogels (Doty, 2011)). Hatched boxes indicate which matrices are fibrillar. Asterisk (*) indicates which matrices have basic cellular attachment sites

While not a property of the cell or target tissue, a final criterion from a practical standpoint to consider is the use of potentially noxious crosslinking mechanisms since both the reagents used and/or the conditions to catalyze gelation can cause aberrant cell behavior or cell death. For example, UV light is used to photoinitiate radical chain polymerization of macromolecules functionalized with vinyl groups (e.g. PEGDA, HA methacrylate). Cytotoxicity must be evaluated in the presence of specific photoinitiator concentrations and UV light intensities (Bryant and Anseth, 2006). An additional example is the use of calcium as a crosslinking agent for alginate-based hydrogels. While alginate has been successfully used for culturing a variety of cell types (Alsberg, et al., 2001; Xu, et al., 2006), calcium has also been shown to reduce cellular viability in Sertoli cells with extended exposure during encapsulation (Luca, et al., 2007).

2.4.2 Physician criteria

For a doctor to adopt a new technology, it must satisfy the needs of all stakeholders involved in a patient's care. These include the surgeon transplanting the cells, the assisting nurse who prepares the hydrogel and cell mixture, the patient, and the payer. While all stakeholders have a vested interest in the hydrogel matrix improving cellular survival, arguably the most important are the physician and assisting nurse who find convenience of use crucial (Reiner, 2009). Critical variables affecting convenience include speed of hydrogel dissolution in clinically relevant buffers (e.g. lactated ringer's solution), number of vials required to reconstitute the hydrogel, and speed of gelation of the hydrogel/cell mixture prior to injection. Of these, speed of gelation may be the most important since too fast of gelation precludes facile injection of the hydrogel/cell mixture. Too slow of gelation may not properly localize the incorporated cells since they may either settle to the bottom of the injected bolus, the bolus may begin to disperse, or the very motion of an organ may cause the bolus to ooze out (Martens, et al., 2009). A comparison of the different matrices' gelation times are shown in Table 2

Fast (< 5 min)	Medium (10-20 min)	Slow (60+ min)
Corgel Fibrin PEGDA Alginate RGD-alginate	Matrigel HyStem-C Q-gel	Purecol Puramatrix

Table 2. Gelation times of hydrogel matrices

2.4.3 Non-biological criteria

2.4.3.1 Introduction

While the first two sets of criteria have focused on the science behind cell delivery vehicles, this third set of criteria is a dramatic departure representing key concerns surrounding the development of any human therapeutic. That is, the question must be answered: can it be made profitably and in compliance with regulatory standards? With the increasing levels of research activity in stem cell therapy and tissue engineering, it is expected that many novel stem cell delivery vehicles in addition to different injectable hydrogels will emerge. Indeed, there are already numerous reports in the medical literature of the use of a wide variety of matrices and scaffolds with stem cells and range from synthetic polymers (James, et al., 2011), cell extracts (Rajasingh, et al., 2008), to de-cellularized tissues such as small intestinal mucosa, urinary bladder basement membrane and mucosa, skin, heart, and fat (Badylak and Gilbert, 2008; Flynn, 2010). Nevertheless, all of them must overcome the last and perhaps the most difficult hurdle: commercialization and regulatory approval for human use.

Commercialization is the conversion of a technology into a profitable product and this is usually done by a for-profit institution. Its financial support and guidance are required to fund and navigate an efficient path to the clinic and then to the marketplace. The first step in this process is typically the licensing of the technology's rights from the academic medical institution (Poltorak and Lemer, 2004). However, the majority of these matrices and scaffolds will never be licensed since they cannot be patented or be made profitably. From an intellectual property standpoint, the lack of commercial viability for these materials typically arises from the absence new or novel intellectual property that can be protected by patent. Without patent protection, the ability to maintain a competitive advantage in the marketplace is absent and hence the potential to make a highly profitable product is compromised. If the technology survives IP landscape scrutiny, the cost in manufacture must be significantly less than the price acceptable to the marketplace. In general, there is a high manufacturing cost for biomaterials derived from animal or human sources (including costs for both raw materials and for cGMP manufacture especially in the beginning when low volumes of materials are prepared (Thompson, 2006)). Finally, the path to regulatory approval can be expensive and treacherous.

Successful translation of basic research in tissue engineering and regenerative medicine to clinical applications demands a thorough understanding of a number of regulatory issues for the therapeutic cell as well as the delivery matrix or scaffold. The remainder of this section will focus on describing the regulatory approval path for combination stem cell/matrix products as well as general considerations surrounding cGMP manufacturing.

2.4.3.2 Regulatory

Since a matrix will be combined with the stem cell, two distinct centers within the FDA share the regulatory responsibilities. Cell-based products are regulated by the Center for Biologics Evaluation and Research (CBER) (FDA, 1997). Therapeutic products consisting of stem or progenitor cells delivered to the body in a matrix or implanted in a scaffold however are considered combination products where the matrix or scaffold is a medical device and the cells the new therapeutic agent. CBER will request consultation from the Center for Devices and Regulatory Health (CDRH) for guidance in device approval issues. CBER's final approval will be for a specific delivery matrix and cell type for a clearly defined indication. As a result there is no pathway, at this time, for approval of a matrix or scaffold as a medical device for general cell delivery for stem cell therapy applications.

From the cellular standpoint, the regulatory path for somatic cells and tissues and for stem cells is similar since the former serves as the basis for the latter. Since 1993, the FDA has been developing guidelines and regulatory pathways to regulate the development, manufacture and distribution of somatic cell therapy products. Over the last decade these guidelines have been broadened to include stem and progenitor cell therapeutic products. These guidelines are designed to ensure that such products meet defined safety requirements and have the identity, strength, quality and purity characteristics as those represented to the FDA. In addition, the FDA has recently mandated that any procedure in which human cells manipulated for clinical use are subject to federal manufacturing standards and oversight (18 21 C.F.R. § 1271.3(d) (2009); see 42 U.S.C. § 264). Specifically, Part 1271 of Chapter 21 of the Code of Federal Register unifies the registration and listing system for establishments that manufacture human cells, tissues, and cellular and tissue-based products (HCT/Ps) and establishes current good tissue practices, and other relevant procedures. These regulations, known as the Good Tissue Practice (GTP) requirements, encompass the minimal manipulation of cells for clinical use; i.e. for processes that do not alter the biologic characteristics of the cells (21 CFR Parts 16, 1270, 1271). For procedures in which the biologic nature of the cells is altered to affect a clinical outcome, termed "more than minimal manipulation," Part 211 pharmaceutical cGMP will apply, as well as relevant aspects of Parts 210, 600 and 1271. In addition to these base requirements, any stem-cell–based product that contains cells or tissues that "are highly processed, are used for other than their normal function, are combined with non-tissue components, or are used for metabolic purposes would also be subject to the Public Health Safety Act, Section 351, which regulates the licensing of biologic products and requires the submission of an investigational new drug application (IND) to the FDA before studies involving humans can begin. (21 C.F.R. Part 312 (2009); (Carpenter, et al., 2009))

From the matrix standpoint, its regulatory path depends on which medical device classification it is assigned within the FDA. If the device has not been previously approved or cleared for its intended use, detailed device information must be included in the investigational new drug (IND) submission for a Center for Devices and Radiological Health (CDRH) consult review. It is also recommended that a Device Master File (DMF) be submitted to CDRH (a DMF provides confidential information surrounding manufacture and can be referenced by a sponsor in support of an IND application (FDA, 2011; Read and Khuu, 2009)). In applications where the matrix and cells are used topically, such as wound healing, the delivery matrix is usually classified as a class II medical device. However, when the cells are injected or *implanted* within the body in a matrix to promote attachment and proliferation of the therapeutic cells, these delivery vehicles are classified as class III medical devices (21 CFR, 860.93; 21 U.S.C., 360c (c)(2)(C)). The scope of the information required in a DMF varies only slightly between class II and class III medical devices with the class II devices requiring less biocompatibility testing. For class III, permanently implanted or resorbable, medical devices, biocompatibility of the device alone is assessed through the prescribed *in vitro* and *in vivo* animal testing asset set forth in the ISO-10993 guides. These tests encompass cytotoxicity, sensitization, irritation, acute and chronic systemic toxicity, genotoxicity, and long term implant with histopathology.

In addition to the demonstration of biocompatibility for the matrix alone as described above, the interaction of the therapeutic cells and their delivery matrix or scaffold must also be characterized for safety and toxicology. Such studies should demonstrate that the matrix or scaffold does not alter the function of the cells in such as way as to raise safety and toxicity concerns. In addition to a thorough characterization of the delivery matrix and its interaction with the therapeutic cells, CBER will request complete documentation on the source and manufacture of the cells, dosing studies, and clear evidence of efficacy. While cell specific issues are beyond the scope of this discussion, they are a major part of any IND submission.

2.4.3.3 cGMP manufacturing

Of equal importance to biocompatibility, safety, and toxicity, is the requirement for manufacture and testing of these cell delivery devices with appropriate quality assurance and quality control to meet FDA standards. To this end, manufacturing and testing in compliance with current Good Manufacturing Practices (cGMP) regulations following validated production and analytical testing protocols is required. What is cGMP? While this is the subject of a book chapter onto itself, at its heart it is a quality system which pervades every step of the product and process development. Validation is a critical part of cGMP manufacture and provides the basis for a program which provides documents assuring both proper systems functioning and final product which meets required specifications. It requires analytical methods development and validation, as well as production of engineering batches, process validation batches, and clinical trial material (Beckloff, 2008). Proper validation specifically for manufacturing procedures and analytical testing protocols for the Chemistry, Manufacturing, and Controls section of a DMF is arguably the major commitment of time and money in cGMP manufacture and should be addressed early the device development strategy.

As a final note, a question that frequently arises is: when a supplier claims their products are "cGMP quality", what does this mean? Suppliers of matrices or scaffolds for basic research often represent their products as being made in a cGMP compliant facility. Does this mean that their products are "cGMP quality"? Such statements should be viewed with caution.

As with other industries where quality is crucial, a rigorous audit from an independent regulatory body is required to assure an institution is in substantial compliance with the quality system that the institution follows. Hence, until such time as the FDA has inspected a manufacturing facility and issued a Form 483 listing inspectional observations in response to a specific IND application, there can be no assurance that the manufacture of the device is indeed in substantial compliance with cGMP requirements.

3. HyStem applications in stem cell therapy

While a number of hydrogel matrices will substantially fulfill the criteria described, those matrices which are customizable have a distinct advantage since all stem cells are not created equal. To maximize its utility across a wide range of stem cells and indications, a hydrogel matrix whose composition can be easily modified by the end user is an important feature. The HyStem hydrogel platform is well suited for cell delivery of numerous stem cells since it not only mimics a variety of microenvironments with its basic HA formulation (*vida supra*), but it can be easily adapted to add functionality. The HyStem family of HA-based hydrogels are based on crosslinking HA, gelatin, and heparin with PEGDA using Michael addition chemistry (Zheng Shu, et al., 2004). These three components are modified with thiol groups and used as modules to make three different HA-based hydrogels: HyStem® (thiolated HA), HyStem-C (same as HyStem plus thiolated gelatin for cellular attachment), HyStem-HP (same as HyStem-C plus thiolated heparin for slow growth factor release) (Serban and Prestwich, 2008). Each component is thiolated using EDC/NHS chemistry followed by crosslinking at physiological pH and temperatures via Michael addition using the acrylate groups in PEGDA (Zheng Shu, et al., 2004). Importantly, HyStem's formulation can be further customized by covalent introduction of a variety of molecules compatible with HyStem's thiol-based chemistry. For example, molecules such as cellular attachment peptides with an acrylate or a free cysteine thiol group can be covalently crosslinked into the matrix (Zheng Shu, et al., 2004). In addition, matrix compliance can be modulated by the concentration PEGDA used (Hanjaya-Putra, et al., 2010; Vanderhooft, et al., 2009). Below we highlight three of the newest stem cell applications using HyStem technology.

3.1 Stroke

Stroke is highly prevalent with 550,000 hospitalizations and 150,000 deaths annually in the US alone (Taylor, et al., 1996). While current treatments do little to recover lost function due to cerebral damage, direct implantation of neural stem or progenitors into the infarct cavity may be effective in repair and eventual recovery (Zhong, et al., 2010). The challenge in this approach is that stem and progenitor cells die *en masse* shortly after transplantation (Bliss, et al., 2007). A solution is to deliver the therapeutic cells in an HA-based cellular delivery vehicle. HA provides a biomimic of the brain microenvironment since HA is abundant in the brain ECM (Fraser, et al., 1997; Ruoslahti, 1996), shares brain mechanical properties (Hou et al.,2005) and is conducive to neural growth (Wei, et al., 2007).

HyStem-HP hydrogel was recently used to deliver neural progenitors into an brain infarct cavity (Zhong, et al., 2010). One week after infarct transplantation in HyStem-HP hydrogel, cellular survival of mouse neural progenitor cells doubled, cell distribution was highly localized, and activated microglia/macrophages were excluded from access to the cells

(Zhong, et al., 2010). Importantly, HyStem-HP's use is congruent with current stroke therapy procedures since it can be administered via a catheter or cannula through a burr hole using computed tomographic (CT) guidance (Montes, et al., 2000).

3.2 Cartilage repair

Degenerative joint diseases affect approximately 20% of the adult population and cost nearly US $30 Billion dollars annually to treat (CDC, 2007; Grayson, et al., 2008). Two current treatments involving transplantation of autologous cartilage from other parts of the joint (mosaicplasty) or transplantation of expanded chondrocytes (Carticel) are non-ideal solutions since the transplanted tissue poorly integrates while causing damage to the donor area of joint (Grayson, et al., 2008). One solution is the transplantation of hESC-derived chondrocyte progenitor cells since the large quantities of the cells can be produced. In addition, they have the ability to fully differentiate into chondrocytes which induces cartilage ECM synthesis and hence better integration (Toh, et al., 2010). The delivery of these cells benefit from a hyaluronic acid based matrix since HA plays a key structural and biological role in cartilage ECM (*vida supra*).

Recently two successful approaches using HyStem-C have been reported: the first uses autologous bone-marrow derived mesenchymal stem cells which are injected and localized with HyStem-C in an full-thickness osteochondral defect in rabbits (Liu, et al., 2006); the second, while not an injectable therapy, also uses HyStem-C and serves to show its utility (hESC-derived chondrogenic cells are precultured in HyStem-C before implantation in rats (Toh, et al., 2010)). In both cases, healing was rapid and significant: In 4-6 weeks, a new smooth regenerated cartilage surface begins to form; by 8-12 weeks, smooth, hyaline-like cartilage has completely filled the defect with excellent integration (Liu, et al., 2006; Toh, et al., 2010). Importantly, the regenerated tissue abundantly expresses Collagen type II and proteoglycans indicative of hyaline cartilage (Toh, et al., 2010).

3.3 Stem cells as therapeutic carriers

Neural stem cells (NSCs), mesenchymal stem cells (MSCs), and endothelial progenitor cells can be used to selectively target and eradicate invasive and inoperable cancers because they migrate towards cancerous cells and can be engineered to overexpress and secrete therapeutic payloads. These therapeutics include apoptosis-inducing proteins (Sasportas, et al., 2009), suicide proteins such as cytosine deaminase in the presence of 5-fluorocytosine (Aboody, et al., 2006), and specific monoclonal antibodies (Dudek, 2010; Frank, et al., 2010). In addition to minimizing side effects due to systemic chemotherapy treatments and their side effects, this approach obviates the need to contend with crossing the blood-brain barrier for CNS applications and repeated treatments of potentially unstable proteins (Frank, et al., 2010; Roth, et al., 2008). Much like stem cells used for tissue regeneration, these engineered stem cells benefit greatly from extended survival time with encapsulation in a matrix since the temporal window of therapeutic treatment is extended.

In a recent study, 10^6 MSCs were genetically modified to overexpress a recombinant bispecific antibody (MSC^{dAb}) and injected subcutaneously within Extracel-X® in mice (Extracel-X is based on HyStem technology and specifically designed for mouse tumor xenografts). MSC^{dAb} implanted with Extracel-X survive at least 12 days longer, secrete up to 145 ng antibody/ml plasma over 42 days, and halts HCT-116 tumor xenograft growth (Compte, et al., 2009). In theory, this approach can be extended to improving tissue repair and regeneration since stem cells can be engineered to express a variety of growth factors and/or signaling molecules to encourage orthopedic tissue repair for example (Hoffmann, et al., 2006; Sheyn, et al., 2008).

4. The future of cell delivery vehicles

While the discussion so far has focused on hydrogels as injectable vehicles congruent with minimally invasive surgery, it would be arrogant and limiting to assert that hydrogel matrices are the only type and form of matrix that can be used. In fact, a number of biomaterials can be incorporated into these procedures and offer significant advantages over traditional scaffolds and *in vitro* established structures (Burg and Boland, 2003; Ochi, et al., 2004; Thornton, et al., 2004). One area where additional biomaterial options will play a significant role is in minimizing stem cell manipulation prior to implantation. Cellular manipulation is proportional to the extent to which cells have been changed *ex vivo*. In addition, cell expansion and encapsulation constitute more than minimal manipulation (Hellman and Smith, 2006). The level of cellular manipulation is becoming more heavily scrutinized and can influence the practicability of implementing a proposed cell-based therapy. Currently, the FDA guidelines regulate "highly processed" more stringently than "minimally manipulated bone marrow", so it is advantageous to reduce processing of cellular components (Burger, 2003; Halme and Kessler, 2006). Cell delivery vehicles that can address both requirements for minimally invasive surgical techniques and reduction of cellular manipulation will clearly play a significant role in future of cell based therapies.

Researchers have investigated both synthetic and natural biomaterials that can be adapted to relevant surgical techniques. Few, however, minimize cellular manipulation. Thornton et al, reported a macroporous alginate hydrogel that can be temporarily deformed for delivery through a small catheter and then expanded in situ to their original physical dimensions, these types of materials are also known as shape-memorizing or defining scaffolds (Thornton, et al., 2004). While these can be utilized in minimally invasive techniques they are unable to fill irregularly shaped defects and do not address the biological requirements of numerous stem cells. Elisseeff et al, have developed photopolymerizable synthetic polymers that can be injected as a monomer solution and then cured using transdermal ultra-violet light (Elisseeff, et al., 1999). These concepts allow for catheter deliver and irregular defects but again do not address the biological needs of cells nor the issue of reducing cell manipulation. Finally, numerous types of microcarriers have been proposed as potential cell delivery vehicles. These small spheres on the order of 10-500 mm are often used to encapsulate cells as with alginate, or are highly porous structures that provide high surface area to attach cells, as with PLGA microcarriers. Burg and Boland have reported a composite system that combines an injectable gel delivery vehicle with polymeric microspheres for additional support and cell attachment sites (Burg and Boland, 2003). While novel, this idea misses a crucial benefit of microcarriers role in cell therapy - the ability to reduce cellular processing by expansion on the microcarriers themselves followed by minimally invasive delivery to targeted sites.

A microcarrier culture system that doubles as an expansion substrate and delivery vehicle for human MSCs (hMSC) would be ideal. This concept permits seamless expansion and transplantation of cells for therapy. Besides substantially decreasing the need for tissue culture space (compared with traditional 2D tissue culture plastic expansion) and the possibility of contamination risks (no need for enzymatic passaging), a biologically relevant microcarrier would provide the requisite scaffold to improve a cell's survivability and localization post-transplantation.

There are a wide variety of microcarriers commercially available such as polystyrene-based MicroHex (Nunc) and Plastic Plus (SoloHill) which is compatible with hMSC, however few are biodegradable and even fewer can provide the biological cell requirements. Commercially available biodegradable microcarriers are suboptimal as delivery vehicles for hMSC as their compositions are not physiologically relevant for this stem cell type. Poly(d,l-lactic- co-glycolic acid) (PLGA) and poly (l)-lactic acid (PLLA) microcarriers do not provide a supportive microenvironment after implantation and they degrade to the acidic by-products, lactic acid and glycolic acid. Cultispher microcarriers (Percell Biolytica AB) - crosslinked, porous porcine gelatin microcarriers - have limited utility for regenerative medicine applications using hMSC. Neither collagen nor gelatin bind to the abundant hMSC CD44 (hyaluronic acid, HA) receptors required for function and the introduction of animal-derived components can lead to additional FDA regulations.

A solution is to prepare microcarriers from a physiologically relevant biomaterial for stem cells which can serve as both as an expansion substrate and delivery vehicle. This dual functionality is advantageous as it can reduce the cellular manipulation by eliminating enzymatic treatments to recover cells and can then be delivered either alone or within additional gel via minimally invasive catheter techniques. HyStem-based microcarriers have been recently produced using reverse emulsion techniques, i.e. water in oil polymerization methods (Figure 3). Unlike Cultispher microcarriers, HyStem offers users control over composition and matrix elasticity for ease in customizing future microcarriers. A variety of growth factors, cellular attachment domains, or peptides may be added to customize its formulation (Serban and Prestwich, 2008; Ventura, et al., 2004). In addition, matrix elasticity is easily modulated (Vanderhooft, et al., 2009) and is likely to be of paramount importance for applications involving hMSC transplantation (Engler, et al., 2006).

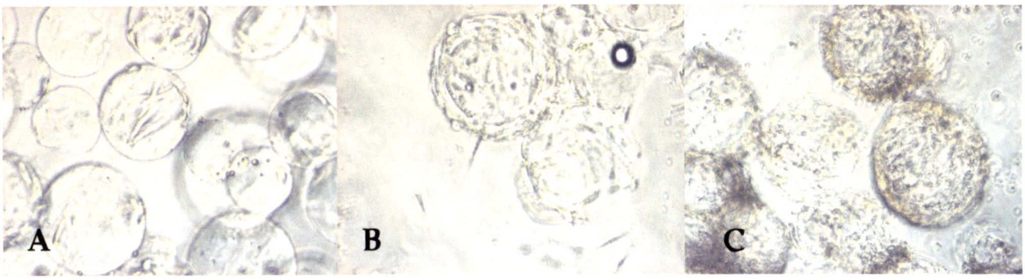

Fig. 3. Mouse fibroblast cells (NIH 3T3 line) seeded on HyStem microcarriers at day 1 (A), 4 (B), and 6(C)

5. Conclusion

In summary, a cellular delivery matrix in regenerative medicine extends the survival of the transplanted cells so that their chances of remodeling their microenvironment or delivering a meaningful therapeutic payload are significantly enhanced. In the context of tissue repair, the use of degradable polymer scaffolds is crucial since it provides a tractable substrate for cells to remodel while providing space and a temporary home. For cell delivery matrices to be useful, however, they must also satisfy non-biological related concerns: From the physician's point of view, injectability and ease of use are of paramount importance. From the commercial point of view, it must be patentable and its surrounding patent landscape must also be favorable (i.e. the intellectual property (IP) has to be novel, non-obvious, and useful as well as unencumbered by potential blocking IP). The market must also be substantial, and the product has to have a significant profit margin in part due to economical manufacture (i.e. raw materials and cGMP costs are low) and the marketplace's acceptance of a fair price.

As a final thought, while there is a great deal of academic excitement and publication activity surrounding the development of ever-increasingly complex cellular delivery matrices that can replicate a new aspect of a stem cell niche (Engler, et al., 2006; Phillips, et al., 2008; Reilly and Engler, 2009; Stern, et al., 2009; Wosnick and Shoichet, 2008), the question is: after a point, when is more complexity in a cellular delivery vehicle too much? The resulting products may become highly unique and patentable from an IP standpoint but from a manufacturing point of view, as the number of components or complexity in manufacturing process increases, manufacturing cost may increase dramatically and block its commercialization. More importantly, from the developmental point of view, do we really need to recreate what nature does by adding every last detail of intricacy into an ECM for a stem cell whose residence time in a developmental intermediate state may be fleeting and only require the intricacy for a moment in time? (Sands and Mooney, 2007).

One strategy for designing an optimal stem cell delivery vehicle is to mimic what nature does during early embryonic development---simply provide a simple, malleable substrate that the embryonic cells can tailor according to their needs. An embryonic cell is in a constant tug-o-war with its surroundings where the cell is exerting a specific force on the ECM or neighboring cell via its acto-myosin cytoskeleton (Ingber, 2006). However, the embryonic organs in which these cells reside likely decide what that specific tension will be based on its physical demands at each of its developmental stage (Mammoto and Ingber, 2010). It is these developmental cues which likely drive the trajectory of the ECM remodeling and cellular differentiation. As an example, let's examine the embryo heart development: initially the embryo is highly compliant (< 10 Pa) into which mesoderm cells penetrate (Reilly and Engler, 2009). There is an eight-fold increase in the developing heart tissue stiffness (Young and Engler, 2010), suggesting that its ECM is constantly being remodeled. In the end, the fully differentiated heart has a stiffness of 10 kPa which provides enough stiffness and stability to allow myocardial cells to generate enough traction to contract against for normal heart beating. In essence, the cells and the ECM are co-developing. Since a variety of stem cells are derived from embryos at various stages, the possibility exists that a simple hydrogel which the stem cells like and can remodel is all that is needed, leaving the building of complexity to nature.

6. Acknowledgements

We thank Dr. Mariluz Henshaw for critical reading of the manuscript.

7. References

Aboody, K.S., Najbauer, J., Schmidt, N.O., Yang, W., Wu, J.K., Zhuge, Y., Przylecki, W., Carroll, R., Black, P.M. and Perides, G., (2006). Targeting of melanoma brain metastases using engineered neural stem/progenitor cells. *Neuro Oncol*, 8 (2):119-126.

Ahmed, T.A., Dare, E.V. and Hincke, M., (2008). Fibrin: A Versatile Scaffold for Tissue Engineering Applications. *Tissue Eng Part B Rev.*

Alsberg, E., Anderson, K.W., Albeiruti, A., Franceschi, R.T. and Mooney, D.J., (2001). Cell-interactive alginate hydrogels for bone tissue engineering. *J Dent Res*, 80 (11):2025-2029.

Aszodi, A., Legate, K.R., Nakchbandi, I. and Fassler, R., (2006). What mouse mutants teach us about extracellular matrix function. *Annu Rev Cell Dev Biol*, 22:591-621.

Bach, F.H., Albertini, R.J., Joo, P., Anderson, J.L. and Bortin, M.M., (1968). Bone-marrow transplantation in a patient with the Wiskott-Aldrich syndrome. *Lancet*, 2 (7583):1364-1366.

Badylak, S.F. and Gilbert, T.W., (2008). Immune response to biologic scaffold materials. *Semin Immunol*, 20 (2):109-116.

Beckloff, M. (ed), (2008). *Validation and Contract Manufacturing*. New York, NY: Informa Healthcare USA.

Beckman, M., Shields, K. and Diegelmann, R. (eds), (2008). *Collagen*: Informa Healthcare USA.

Bliss, T., Guzman, R., Daadi, M. and Steinberg, G.K., (2007). Cell transplantation therapy for stroke. *Stroke*, 38 (2 Suppl):817-826.

Bryant, S. and Anseth, K.S. (eds), (2006). *Photopolymerization of Hydrogel Scaffolds*. Boca Raton, FL: CRC Press.

Burg, K.J. and Boland, T., (2003). Minimally invasive tissue engineering composites and cell printing. *IEEE Eng Med Biol Mag*, 22 (5):84-91.

Burger, S.R., (2003). Current regulatory issues in cell and tissue therapy. *Cytotherapy*, 5 (4):289-298.

Cai, S., Liu, Y., Zheng Shu, X. and Prestwich, G.D., (2005). Injectable glycosaminoglycan hydrogels for controlled release of human basic fibroblast growth factor. *Biomaterials*, 26 (30):6054-6067.

Camenisch, T.D., Spicer, A.P., Brehm-Gibson, T., Biesterfeldt, J., Augustine, M.L., Calabro, A., Jr., Kubalak, S., Klewer, S.E. and McDonald, J.A., (2000). Disruption of hyaluronan synthase-2 abrogates normal cardiac morphogenesis and hyaluronan-mediated transformation of epithelium to mesenchyme. *J Clin Invest*, 106 (3):349-360.

Caplan, A.I. and Dennis, J.E., (2006). Mesenchymal stem cells as trophic mediators. *J Cell Biochem*, 98 (5):1076-1084.

Carpenter, M.K., Frey-Vasconcells, J. and Rao, M.S., (2009). Developing safe therapies from human pluripotent stem cells. *Nat Biotechnol*, 27 (7):606-613.

CDC, (2007). Targeting Arthritis: Reducing Disability for Nearly 19 Million Americans. In: *Prevention, C.f.D.C.a.* (ed): US Department of Health and Human Services.

Chung, C. and Burdick, J.A., (2009). Influence of three-dimensional hyaluronic acid microenvironments on mesenchymal stem cell chondrogenesis. *Tissue Eng Part A*, 15 (2):243-254.

Compte, M., Cuesta, A.M., Sanchez-Martin, D., Alonso-Camino, V., Vicario, J.L., Sanz, L. and Alvarez-Vallina, L., (2009). Tumor immunotherapy using gene-modified human mesenchymal stem cells loaded into synthetic extracellular matrix scaffolds. *Stem Cells*, 27 (3):753-760.

Cui, J., Wahl, R.L., Shen, T., Fisher, S.J., Recker, E., Ginsburg, D. and Long, M.W., (1999). Bone marrow cell trafficking following intravenous administration. *Br J Haematol*, 107 (4):895-902.

Cutler, C. and Antin, J.H., (2001). Peripheral blood stem cells for allogeneic transplantation: a review. *Stem Cells*, 19 (2):108-117.

Darabi, R., Baik, J., Clee, M., Kyba, M., Tupler, R. and Perlingeiro, R.C., (2009). Engraftment of embryonic stem cell-derived myogenic progenitors in a dominant model of muscular dystrophy. *Exp Neurol*, 220 (1):212-216.

Darr, A. and Calabro, A., (2009). Synthesis and characterization of tyramine-based hyaluronan hydrogels. *J Mater Sci Mater Med*, 20 (1):33-44.

Dominici, M., Rasini, V., Bussolari, R., Chen, X., Hofmann, T.J., Spano, C., Bernabei, D., Veronesi, E., Bertoni, F., Paolucci, P., Conte, P. and Horwitz, E.M., (2009). Restoration and reversible expansion of the osteoblastic hematopoietic stem cell niche after marrow radioablation. *Blood*, 114 (11):2333-2343.

Doty, N., (2011). Rheological Properties of Extracellular Matrices. Glycosan BioSystems.

Dudek, A.Z., (2010). Endothelial lineage cell as a vehicle for systemic delivery of cancer gene therapy. *Transl Res*, 156 (3):136-146.

Duran-Struuck, R. and Dysko, R.C., (2009). Principles of bone marrow transplantation (BMT): providing optimal veterinary and husbandry care to irradiated mice in BMT studies. *J Am Assoc Lab Anim Sci*, 48 (1):11-22.

Elisseeff, J., Anseth, K., Sims, D., McIntosh, W., Randolph, M. and Langer, R., (1999). Transdermal photopolymerization for minimally invasive implantation. *Proc Natl Acad Sci U S A*, 96 (6):3104-3107.

Engler, A.J., Sen, S., Sweeney, H.L. and Discher, D.E., (2006). Matrix elasticity directs stem cell lineage specification. *Cell*, 126 (4):677-689.

FDA, (1997).'Proposed Approach to Regulation of Cellular and Tissue-Based Products. Food and Drug Administration.

FDA, (2011). Guideance Compliance Regulatory Information. Food and Drug Administration.

Flaim, C.J., Chien, S. and Bhatia, S.N., (2005).'An extracellular matrix microarray for probing cellular differentiation. *Nat Methods*, 2 (2):119-125.

Flynn, L.E., (2010). The use of decellularized adipose tissue to provide an inductive microenvironment for the adipogenic differentiation of human adipose-derived stem cells. *Biomaterials*, 31 (17):4715-4724.

Frank, R.T., Najbauer, J. and Aboody, K.S., (2010). Concise review: stem cells as an emerging platform for antibody therapy of cancer. *Stem Cells*, 28 (11):2084-2087.

Fraser, J.R., Laurent, T.C. and Laurent, U.B., (1997). Hyaluronan: its nature, distribution, functions and turnover. *J Intern Med*, 242 (1):27-33.

Gatti, R.A., Meuwissen, H.J., Allen, H.D., Hong, R. and Good, R.A., (1968). Immunological reconstitution of sex-linked lymphopenic immunological deficiency. *Lancet*, 2 (7583):1366-1369.

Gentry-Steele, D. and Bramblett, C., (1988). *The Anatomy and Biology of the Human Skeleton*. College Station, TX: Texas A&M Univ Press.

Grayson, W.L., Chao, P.H., Marolt, D., Kaplan, D.L. and Vunjak-Novakovic, G., (2008). Engineering custom-designed osteochondral tissue grafts. *Trends Biotechnol*, 26 (4):181-189.

Guerette, B., Skuk, D., Celestin, F., Huard, C., Tardif, F., Asselin, I., Roy, B., Goulet, M., Roy, R., Entman, M. and Tremblay, J.P., (1997). Prevention by anti-LFA-1 of acute myoblast death following transplantation. *J Immunol*, 159 (5):2522-2531.

Halme, D.G. and Kessler, D.A., (2006). FDA regulation of stem-cell-based therapies. *N Engl J Med*, 355 (16):1730-1735.

Hanjaya-Putra, D., Yee, J., Ceci, D., Truitt, R., Yee, D. and Gerecht, S., (2010). Vascular endothelial growth factor and substrate mechanics regulate in vitro tubulogenesis of endothelial progenitor cells. *J Cell Mol Med*, 14 (10):2436-2447.

Harrison, M. and Rae, I., (1997). *General Techniques of Cell Culture*. New York, NY: Cambridge University Press.

Hellman, K. and Smith, D. (eds), (2006). *The Regulation of Engineered Tissues: Emerging Approaches*. Boca Raton, FL: CRC Press.

Hoffmann, A., Pelled, G., Turgeman, G., Eberle, P., Zilberman, Y., Shinar, H., Keinan-Adamsky, K., Winkel, A., Shahab, S., Navon, G., Gross, G. and Gazit, D., (2006). Neotendon formation induced by manipulation of the Smad8 signalling pathway in mesenchymal stem cells. *J Clin Invest*, 116 (4):940-952.

Horowitz, M. (ed), (2004). *Thomas's Hematopoietic cell transplantation*. Malden, MA: Blackwell Publishing Inc.

Hughes, C.S., Postovit, L.M. and Lajoie, G.A., (2010). Matrigel: a complex protein mixture required for optimal growth of cell culture. *Proteomics*, 10 (9):1886-1890.

Ingber, D.E., (2006). Mechanical control of tissue morphogenesis during embryological development. *Int J Dev Biol*, 50 (2-3):255-266.

Ingram, D.A., Mead, L.E., Tanaka, H., Meade, V., Fenoglio, A., Mortell, K., Pollok, K., Ferkowicz, M.J., Gilley, D. and Yoder, M.C., (2004). Identification of a novel hierarchy of endothelial progenitor cells using human peripheral and umbilical cord blood. *Blood*, 104 (9):2752-2760.

James, R., Kumbar, S.G., Laurencin, C.T., Balian, G. and Chhabra, A.B., (2011). Tendon tissue engineering: adipose-derived stem cell and GDF-5 mediated regeneration using electrospun matrix systems. *Biomed Mater*, 6 (2):025011.

Kaihara, S. and Vacanti, J.P., (1999). Tissue engineering: toward new solutions for transplantation and reconstructive surgery. *Arch Surg*, 134 (11):1184-1188.

Keirstead, H.S., Ben-Hur, T., Rogister, B., O'Leary, M.T., Dubois-Dalcq, M. and Blakemore, W.F., (1999).'Polysialylated neural cell adhesion molecule-positive CNS precursors generate both oligodendrocytes and Schwann cells to remyelinate the CNS after transplantation. *J Neurosci*, 19 (17):7529-7536.

Kleinman, H.K. and Martin, G.R., (2005). Matrigel: basement membrane matrix with biological activity. *Semin Cancer Biol*, 15 (5):378-386.

Knudson, C.B. and Knudson, W., (1993). Hyaluronan-binding proteins in development, tissue homeostasis, and disease. *FASEB J*, 7 (13):1233-1241.

Laflamme, M.A., Chen, K.Y., Naumova, A.V., Muskheli, V., Fugate, J.A., Dupras, S.K., Reinecke, H., Xu, C., Hassanipour, M., Police, S., O'Sullivan, C., Collins, L., Chen,

Y., Minami, E., Gill, E.A., Ueno, S., Yuan, C., Gold, J. and Murry, C.E., (2007). Cardiomyocytes derived from human embryonic stem cells in pro-survival factors enhance function of infarcted rat hearts. *Nat Biotechnol*, 25 (9):1015-1024.

Lam, B.S. and Adams, G.B., (2010). Hematopoietic stem cell lodgment in the adult bone marrow stem cell niche. *Int J Lab Hematol*, 32 (6 Pt 2):551-558.

Laurent, T.C. and Fraser, J.R., (1992). Hyaluronan. *FASEB J*, 6 (7):2397-2404.

Lennon, D.P. and Caplan, A.I., (2006). Isolation of human marrow-derived mesenchymal stem cells. *Exp Hematol*, 34 (11):1604-1605.

Liao, H., Munoz-Pinto, D., Qu, X., Hou, Y., Grunlan, M.A. and Hahn, M.S., (2008). 'Influence of hydrogel mechanical properties and mesh size on vocal fold fibroblast extracellular matrix production and phenotype'. *Acta Biomater*, 4 (5):1161-1171.

Lin, C.C. and Anseth, K.S., (2009). Glucagon-like peptide-1 functionalized PEG hydrogels promote survival and function of encapsulated pancreatic beta-cells. *Biomacromolecules*, 10 (9):2460-2467.

Liu, Y., Shu, X.Z. and Prestwich, G.D., (2006). Osteochondral defect repair with autologous bone marrow-derived mesenchymal stem cells in an injectable, in situ, cross-linked synthetic extracellular matrix. *Tissue Eng*, 12 (12):3405-3416.

Liu, Y., Shu, X.Z. and Prestwich, G.D., (2007). Tumor engineering: orthotopic cancer models in mice using cell-loaded, injectable, cross-linked hyaluronan-derived hydrogels. *Tissue Eng*, 13 (5):1091-1101.

Luca, G., Calvitti, M., Nastruzzi, C., Bilancetti, L., Becchetti, E., Angeletti, G., Mancuso, F. and Calafiore, R., (2007). Encapsulation, in vitro characterization, and in vivo biocompatibility of Sertoli cells in alginate-based microcapsules. *Tissue Eng*, 13 (3):641-648.

Lutolf, M.P. and Hubbell, J.A., (2005). Synthetic biomaterials as instructive extracellular microenvironments for morphogenesis in tissue engineering. *Nat Biotechnol*, 23 (1):47-55.

Lysaght, M.J., Jaklenec, A. and Deweerd, E., (2008). Great expectations: private sector activity in tissue engineering, regenerative medicine, and stem cell therapeutics. *Tissue Eng Part A*, 14 (2):305-315.

Mammoto, T. and Ingber, D.E., (2010). Mechanical control of tissue and organ development. *Development*, 137 (9):1407-1420.

Martens, T.P., Godier, A.F., Parks, J.J., Wan, L.Q., Koeckert, M.S., Eng, G.M., Hudson, B.I., Sherman, W. and Vunjak-Novakovic, G., (2009). Percutaneous cell delivery into the heart using hydrogels polymerizing in situ. *Cell Transplant*, 18 (3):297-304.

Martin, I., Baldomero, H., Tyndall, A., Niederwieser, D. and Gratwohl, A., (2010). A survey on cellular and engineered tissue therapies in europe in 2008. *Tissue Eng Part A*, 16 (8):2419-2427.

Mavroudis, D., Read, E., Cottler-Fox, M., Couriel, D., Molldrem, J., Carter, C., Yu, M., Dunbar, C. and Barrett, J., (1996). CD34+ cell dose predicts survival, posttransplant morbidity, and rate of hematologic recovery after allogeneic marrow transplants for hematologic malignancies. *Blood*, 88 (8):3223-3229.

Miles, E.J., Dunn, E., Howard, D. and Mangram, A., (2004). The role of laparoscopy in penetrating abdominal trauma. *JSLS*, 8 (4):304-309.

Montes, J.M., Wong, J.H., Fayad, P.B. and Awad, I.A., (2000). Stereotactic computed tomographic-guided aspiration and thrombolysis of intracerebral hematoma : protocol and preliminary experience. *Stroke*, 31 (4):834-840.

Moon, J.J., Hahn, M.S., Kim, I., Nsiah, B.A. and West, J.L., (2009). Micropatterning of poly(ethylene glycol) diacrylate hydrogels with biomolecules to regulate and guide endothelial morphogenesis. *Tissue Eng Part A*, 15 (3):579-585.

Mooney, D.J. and Vandenburgh, H., (2008). Cell delivery mechanisms for tissue repair. *Cell Stem Cell*, 2 (3):205-213.

Nagaoka, M., Si-Tayeb, K., Akaike, T. and Duncan, S.A., (2010). Culture of human pluripotent stem cells using completely defined conditions on a recombinant E-cadherin substratum. *BMC Dev Biol*, 10:60.

Newman, J., (2009). Review of soft tissue augmentation in the face. *Clin Cosmet Investig Dermatol*, 2:141-150.

Ochi, M., Adachi, N., Nobuto, H., Yanada, S., Ito, Y. and Agung, M., (2004). Articular cartilage repair using tissue engineering technique--novel approach with minimally invasive procedure. *Artif Organs*, 28 (1):28-32.

Patel, P.N., Smith, C.K. and Patrick, C.W., Jr., (2005). Rheological and recovery properties of poly(ethylene glycol) diacrylate hydrogels and human adipose tissue. *J Biomed Mater Res A*, 73 (3):313-319.

Penolazzi, L., Tavanti, E., Vecchiatini, R., Lambertini, E., Vesce, F., Gambari, R., Mazzitelli, S., Mancuso, F., Luca, G., Nastruzzi, C. and Piva, R., (2010). Encapsulation of mesenchymal stem cells from Wharton's jelly in alginate microbeads. *Tissue Eng Part C Methods*, 16 (1):141-155.

Phelps, E.A., Landazuri, N., Thule, P.M., Taylor, W.R. and Garcia, A.J., (2010). Bioartificial matrices for therapeutic vascularization. *Proc Natl Acad Sci U S A*, 107 (8):3323-3328.

Phillips, J.E., Burns, K.L., Le Doux, J.M., Guldberg, R.E. and Garcia, A.J., (2008). Engineering graded tissue interfaces. *Proc Natl Acad Sci U S A*, 105 (34):12170-12175.

Poltorak, A. and Lemer, P., (2004). *Essentials of Licensing Intellectual Property*. Hoboken, NJ: John Wiley and Sons, Inc.

Prestwich, G.D., (2008). Engineering a clinically-useful matrix for cell therapy. *Organogenesis*, 4 (1):42-47.

Raeber, G.P., Lutolf, M.P. and Hubbell, J.A., (2005). Molecularly engineered PEG hydrogels: a novel model system for proteolytically mediated cell migration. *Biophys J*, 89 (2):1374-1388.

Rajasingh, J., Lambers, E., Hamada, H., Bord, E., Thorne, T., Goukassian, I., Krishnamurthy, P., Rosen, K.M., Ahluwalia, D., Zhu, Y., Qin, G., Losordo, D.W. and Kishore, R., (2008). Cell-free embryonic stem cell extract-mediated derivation of multipotent stem cells from NIH3T3 fibroblasts for functional and anatomical ischemic tissue repair. *Circ Res*, 102 (11):e107-117.

Read, E. and Khuu, H. (eds), (2009). *Use of a facility master file to facilitate regulartory submissions for cell therapy products*. New York, NY: Springer.

Reilly, G.C. and Engler, A.J., (2009). Intrinsic extracellular matrix properties regulate stem cell differentiation. *J Biomech*, 43 (1):55-62.

Reiner, B.I., (2009). The challenges, opportunities, and imperative of structured reporting in medical imaging. *J Digit Imaging*, 22 (6):562-568.

Roth, J.C., Curiel, D.T. and Pereboeva, L., (2008). Cell vehicle targeting strategies. *Gene Ther*, 15 (10):716-729.

Ruoslahti, E., (1996). Brain extracellular matrix. *Glycobiology*, 6 (5):489-492.

Sands, R.W. and Mooney, D.J., (2007). Polymers to direct cell fate by controlling the microenvironment. *Curr Opin Biotechnol*, 18 (5):448-453.

Sasportas, L.S., Kasmieh, R., Wakimoto, H., Hingtgen, S., van de Water, J.A., Mohapatra, G., Figueiredo, J.L., Martuza, R.L., Weissleder, R. and Shah, K., (2009). Assessment of therapeutic efficacy and fate of engineered human mesenchymal stem cells for cancer therapy. *Proc Natl Acad Sci U S A*, 106 (12):4822-4827.

Scadden, D.T., (2006). The stem-cell niche as an entity of action. *Nature*, 441 (7097):1075-1079.

Schmitz I., Ariyoshi W., Takahashi N., Knudson C.B., and Knudson W. (2010). Hyaluronan oligosaccharide treatment of chondrocytes stimulates expression of both HAS-2 and MMP-3, but by different signaling pathways. *Osteoarthitis Cartilage*, 18 (3):447-454.

Semino, C.E., (2008). Self-assembling peptides: from bio-inspired materials to bone regeneration. *J Dent Res*, 87 (7):606-616.

Serban, M.A. and Prestwich, G.D., (2008). Modular extracellular matrices: solutions for the puzzle. *Methods*, 45 (1):93-98.

Sheyn, D., Pelled, G., Zilberman, Y., Talasazan, F., Frank, J.M., Gazit, D. and Gazit, Z., (2008). Nonvirally engineered porcine adipose tissue-derived stem cells: use in posterior spinal fusion. *Stem Cells*, 26 (4):1056-1064.

Shikanov, A., Xu, M., Woodruff, T.K. and Shea, L.D., (2009). Interpenetrating fibrin-alginate matrices for in vitro ovarian follicle development. *Biomaterials*, 30 (29):5476-5485.

Slayton, W.B., Li, X.M., Butler, J., Guthrie, S.M., Jorgensen, M.L., Wingard, J.R. and Scott, E.W., (2007). The role of the donor in the repair of the marrow vascular niche following hematopoietic stem cell transplant. *Stem Cells*, 25 (11):2945-2955.

Snyder, B.R., Chiu, A.M., Prockop, D.J. and Chan, A.W., (2010). Human multipotent stromal cells (MSCs) increase neurogenesis and decrease atrophy of the striatum in a transgenic mouse model for Huntington's disease. *PLoS One*, 5 (2):e9347.

Soen, Y., Mori, A., Palmer, T.D. and Brown, P.O., (2006). Exploring the regulation of human neural precursor cell differentiation using arrays of signaling microenvironments. *Mol Syst Biol*, 2:37.

Stern, E., Jay, S.M., Demento, S.L., Murelli, R.P., Reed, M.A., Malinski, T., Spiegel, D.A., Mooney, D.J. and Fahmy, T.M., (2009). Spatiotemporal control over molecular delivery and cellular encapsulation from electropolymerized micro- and nanopatterned surfaces. *Adv Funct Mater*, 19 (18):2888-2895.

Suzuki, K., Murtuza, B., Beauchamp, J.R., Smolenski, R.T., Varela-Carver, A., Fukushima, S., Coppen, S.R., Partridge, T.A. and Yacoub, M.H., (2004). Dynamics and mediators of acute graft attrition after myoblast transplantation to the heart. *FASEB J*, 18 (10):1153-1155.

Taichman, R.S., (2005). Blood and bone: two tissues whose fates are intertwined to create the hematopoietic stem-cell niche. *Blood*, 105 (7):2631-2639.

Takahashi, K., Tanabe, K., Ohnuki, M., Narita, M., Ichisaka, T., Tomoda, K. and Yamanaka, S., (2007). Induction of pluripotent stem cells from adult human fibroblasts by defined factors. *Cell*, 131 (5):861-872.

Tate, C.C., Shear, D.A., Tate, M.C., Archer, D.R., Stein, D.G. and LaPlaca, M.C., (2009). Laminin and fibronectin scaffolds enhance neural stem cell transplantation into the injured brain. *J Tissue Eng Regen Med*, 3 (3):208-217.

Taylor, T.N., Davis, P.H., Torner, J.C., Holmes, J., Meyer, J.W. and Jacobson, M.F., (1996). Lifetime cost of stroke in the United States. *Stroke*, 27 (9):1459-1466.

Terrovitis, J., Kwok, K.F., Lautamaki, R., Engles, J.M., Barth, A.S., Kizana, E., Miake, J., Leppo, M.K., Fox, J., Seidel, J., Pomper, M., Wahl, R.L., Tsui, B., Bengel, F., Marban, E. and Abraham, M.R., (2008). Ectopic expression of the sodium-iodide symporter enables imaging of transplanted cardiac stem cells in vivo by single-photon emission computed tomography or positron emission tomography. *J Am Coll Cardiol*, 52 (20):1652-1660.

Thompson, D. (ed), (2006). *Cyclodextrins-enabling excipients: a case study of the development of a new excipient- sulfobutylether b-cyclodextrin (Captisol)*. New York, NY: Informa Healthcare USA, Inc.

Thomson, J.A., Itskovitz-Eldor, J., Shapiro, S.S., Waknitz, M.A., Swiergiel, J.J., Marshall, V.S. and Jones, J.M., (1998). Embryonic stem cell lines derived from human blastocysts. *Science*, 282 (5391):1145-1147.

Thornton, A.J., Alsberg, E., Albertelli, M. and Mooney, D.J., (2004). 'Shape-defining scaffolds for minimally invasive tissue engineering'. *Transplantation*, 77 (12):1798-1803.

Tibbitt, M.W. and Anseth, K.S., (2009). Hydrogels as extracellular matrix mimics for 3D cell culture. *Biotechnol Bioeng*, 103 (4):655-663.

Toh, W.S., Lee, E.H., Guo, X.M., Chan, J.K., Yeow, C.H., Choo, A.B. and Cao, T., (2010). Cartilage repair using hyaluronan hydrogel-encapsulated human embryonic stem cell-derived chondrogenic cells. *Biomaterials*, 31 (27):6968-6980.

Vanderhooft, J.L., Alcoutlabi, M., Magda, J.J. and Prestwich, G.D., (2009). Rheological properties of cross-linked hyaluronan-gelatin hydrogels for tissue engineering. *Macromol Biosci*, 9 (1):20-28.

Ventura, C., Maioli, M., Asara, Y., Santoni, D., Scarlata, I., Cantoni, S. and Perbellini, A., (2004). Butyric and retinoic mixed ester of hyaluronan. A novel differentiating glycoconjugate affording a high throughput of cardiogenesis in embryonic stem cells. *J Biol Chem*, 279 (22):23574-23579.

Wei, Y.T., Tian, W.M., Yu, X., Cui, F.Z., Hou, S.P., Xu, Q.Y. and Lee, I.S., (2007). Hyaluronic acid hydrogels with IKVAV peptides for tissue repair and axonal regeneration in an injured rat brain. *Biomed Mater*, 2 (3):S142-146.

Weisel, J.W., (2004). The mechanical properties of fibrin for basic scientists and clinicians. *Biophys Chem*, 112 (2-3):267-276.

West, E.R., Xu, M., Woodruff, T.K. and Shea, L.D., (2007). Physical properties of alginate hydrogels and their effects on in vitro follicle development. *Biomaterials*, 28 (30):4439-4448.

Wilson, J.T. and Chaikof, E.L., (2008). 'Challenges and emerging technologies in the immunoisolation of cells and tissues'. *Adv Drug Deliv Rev*, 60 (2):124-145.

Wosnick, J. and Shoichet, M., (2008). Three-dimensional Chemical Patterning of Transparent Hydrogels. *Chemistry of Materials*, 20:55-60.

Wu, Y., Wu, J., Lee, D.Y., Yee, A., Cao, L., Zhang, Y., Kiani, C. and Yang, B.B., (2005). Versican protects cells from oxidative stress-induced apoptosis. *Matrix Biol*, 24 (1):3-13.

Xu, M., West, E., Shea, L.D. and Woodruff, T.K., (2006). Identification of a stage-specific permissive in vitro culture environment for follicle growth and oocyte development. *Biol Reprod*, 75 (6):916-923.

Yamaguchi, Y., (2000). Lecticans: organizers of the brain extracellular matrix. *Cell Mol Life Sci*, 57 (2):276-289.

Young, J.L. and Engler, A.J., (2010). Hydrogels with time-dependent material properties enhance cardiomyocyte differentiation in vitro. *Biomaterials*, 32 (4):1002-1009.

Yu, J., Vodyanik, M.A., Smuga-Otto, K., Antosiewicz-Bourget, J., Frane, J.L., Tian, S., Nie, J., Jonsdottir, G.A., Ruotti, V., Stewart, R., Slukvin, II and Thomson, J.A., (2007). Induced pluripotent stem cell lines derived from human somatic cells. *Science*, 318 (5858):1917-1920.

Zheng Shu, X., Liu, Y., Palumbo, F.S., Luo, Y. and Prestwich, G.D., (2004). In situ crosslinkable hyaluronan hydrogels for tissue engineering. *Biomaterials*, 25 (7-8):1339-1348.

Zhong, J., Chan, A., Morad, L., Kornblum, H.I., Fan, G. and Carmichael, S.T., (2010). Hydrogel matrix to support stem cell survival after brain transplantation in stroke. *Neurorehabil Neural Repair*, 24 (7):636-644.

Zuk, P.A., Zhu, M., Mizuno, H., Huang, J., Futrell, J.W., Katz, A.J., Benhaim, P., Lorenz, H.P. and Hedrick, M.H., (2001). Multilineage cells from human adipose tissue: implications for cell-based therapies. *Tissue Eng*, 7 (2):211-228.

17

Cockle Shell-Based Biocomposite Scaffold for Bone Tissue Engineering

Zuki Abu Bakar, Bahaa F. Hussein and Noordin Mohamed Mustapha
Faculty of Veterinary Medicine, Universiti Putra Malaysia
Malaysia

1. Introduction

Bone transplantation is a rapidly growing and expanding field, and has a significant impact on improving the quality of life of patients suffering from bone tissue damage and disease. Bone is the second most frequently transplanted tissue in humans after the blood. Secondary bone tumour, trauma or deformity often presents a significant problem for bone surgery (Boyne *et al.*, 2002). Grafting is regularly used in medical procedures to replace damaged tissue. Bone grafting involves replacing damaged bone with harvested bone from donors or from another location within the body. Traditionally, bone graft treatments, such as autografting, allografting, and xenografting were used to replace or repair damaged bone tissue. These processes can be long, painful, and have the possibility of being rejected by the body. Autografts are osteoinductive, osteoconductive and have osteogenic properties (Cypher and Grossman, 1996). Although autografts are considered the gold standard of bone transplantation, they also have certain limitations due to possible donor site morbidity, low tissue availability and may introduce additional medical complications (Younger and Chapman, 1989; Moore *et al.*, 2001). Allografts are grafts made of tissue from a human donor, usually during post-mortem. This method rules out the limitations of auotgrafts technique, but have their own limitations including donor shortages and risks of infections. Xenografts are bone grafts from different species such as bovine, porcine or coralline bone that can be implanted into human graft site. Xenografts offer the advantage of availability in a variety of shapes and sizes, but they are also subjected to problems of immunogenicity and have the tendency to denature or decompose in room temperature (Vaccaro *et al.*, 2002). Distraction osteogenesis is another technique for bone treatment used to promote bone growth using the body's innate bone-healing mechanisms. The process is lengthy and painful, in which the two sides of a bone fracture are separated by a short distance every few weeks until the desired length of bone has been regrown (Chang *et al.*, 2004). Other treatment options, the man-made devices, such as bone cement fillers and prosthetics made of metals, ceramics and polymers are also used for bone defect repair or replacing damaged bone tissue. All the conventional methods for bone repair and replacement can be long, painful and have the possibility of being rejected by the body (Ducheyne *et al.*, 1992). Alternative approaches have been heavily researched and investigated based on a tissue engineering strategy, in an effort to overcome the inherent limitations of the currently available solutions to bone defects. In this approach, a tissue engineered bone is produced by seeding cells that can become osteoblasts on highly porous biomaterials (Hardin-Young *et al.*, 2000).

Tissue engineering is one of the most exciting and rapidly growing areas of biotechnology. According to Williams (1987), tissue engineering is defined as an interdisciplinary field that applies the principles of engineering and the life sciences toward the development of biological substitutes that restore, maintain or improve tissue function. These substitutes are generally known as "scaffolds". The concepts of tissue engineering is to engineer autografts, by expanding autologous cells *in vitro* guided by a scaffold (artifical matrix), or by implanting an acellular scaffold *in vivo* and allow the patient's cells to repair the tissue guided by the scaffold. In both ways *in vitro* or *in vivo*, tissue engineering utilize a scaffold (supporting structural device), osteogenic cells and bioactive factors either alone or in various combinations (Meyer *et al.*, 2004b). The scaffold should be made of a biocompatible and bioresorbable material that degrades and disappears over time with the regeneration of tissues, so that once the tissue has developed the scaffold no longer exists as such and the newly formed tissue can perform the function of the damaged tissue (Ross, 1998). Therefore, tissue engineering offers a potential approach to create tissues, organs and synthetic graft products under laboratory conditions to overcome the problems of implant rejection, transmission of diseases associated with xenografts and allografts, and shortage in organ donation.

The main target for bone tissue engineering is bone repair, and it can be used for healing or repairing wide range of bone defects. There are many different strategies which may be used to develop bone tissue engineering. One such strategy involves seeding autologous osteogenic cells *in vitro* along a biodegradable scaffold to create a scaffold–cell hybrid that may be called a tissue-engineered construct. Osteoblasts, chondrocytes and mesenchymal stem cells obtained from hard and soft tissues of the patient may be expanded in culture and seeded onto a scaffold that should in some way fade away to allow an entirely natural bone tissue replacement (Langer and Vacanti, 1993). Millions of patients with bone problems worldwide each year could benefit from this new therapeutic strategy (Buckwalter, 2004; Holtorf *et al.*, 2005).

Scaffolds are the key to the development of tissue engineered bone construct for implantation into the bone critical defect. Successful tissue generation *in vitro* and/or *in vivo* requires highly specialized artificial bone substitutes, since they serve as a model in three dimensions for the interposition of tissue mimicking the extracellular matrix for adhesion and cell proliferation, differentiation and must also provide support to protect the tissue in the early stages of healing (Freed *et al.*, 1994b). Many artificial bone substitutes that introduced to maintain the function of bone containing metals, ceramics and polymers. Each material has specific disadvantages, and none of them can fully substitute for autografts in current clinical practice. More importantly, the bone graft substitute must be biocompatible, nonimmunogenic and noncarcinogenic to minimize host response. The bone graft substitute must also possess mechanical properties similar to that of the bone to prevent stress shielding (Hench and Wilson, 1993). Many of artificial bone substitutes, with its high biocompatibility and good bioaffinity, stimulate osteoconduction and are slowly replaced by host bone after implantation. An osteoconductive material allows a series of events of protein adsorption resulting from the anchor of osteoblasts to the surface and subsequent deposition of extracellular matrix and bone (Dziedic *et al.*, 1996). This allows for intimate contact to be made between the implant and surrounding tissue, thus accelerating the natural healing response. Finally, a potential bone graft material should be consistent in use and readily available in unlimited quantity in order to be considered suitable for surgical implantation (Hatcher, 2004).

The success of a tissue engineering scaffold will come into play to determine whether it can support cell attachment, growth and eventually cell differentiation into the appropriate tissue. For these reasons, the bioresorbable scaffold must be biocompatible and having porous interconnected network to facilitate vascularization and rapid growth of newly formed tissue (Laurencin *et al.*, 1999; Mooney and Mikos, 1999). Hence, several needs are identified as crucial for the production of scaffolds in tissue engineering. The scaffold should possess a) interconnecting pores of a scale appropriate to promote integration and vascularization of tissues, b) controlled biodegradability or bioresorbability, so that the host tissue will eventually replace the scaffold, c) appropriate surface chemistry to promote cell attachment, differentiation and proliferation, d) mechanical properties sufficient to match the space provided for the implementation and handling, e) not induce a negative response, and f) easily manufactured in a variety of shapes and sizes (Li and Li, 2005). Given these requirements in mind, several materials have been adopted or synthesized and fabricated into scaffolds. The basic principle of our research work was to apply the tissue engineering strategy that combines osteoprogenitor cells with a new bioceramic scaffold for regeneration of critical size bone tissue defect. The cockle (*Anadara granosa*) is by far, the most abundant species cultured in Malaysia. A possible advantage of using cockle shell as a biomineral is that they may act as analogs of calcium carbonate present *in vivo*.

2. Morphological and biomechanical properties of cockle shell-based biocomposite scaffolds

The information from the experimental study showed that the physical and mechanical properties including stiffness, degradability, porosity and mechanical integrity of this scaffold were appropriate for use as extracellular matrix materials for bone tissue engineering. All the scaffolds prepared have pore size between 20-420 µm. The ESEM examinations of the prepared scaffolds as shown in figure 1 revealed that the fabricated scaffolds possessing different sizes of porosity. The physical characteristic of porosity formed in the scaffolds are mainly attributed to the voids containing trapped air that make way to provide spaces for the swelling effect when the scaffolds get wet, and these voids formed between the granules or particles in the bulk material. Our observation on the formation of porosity is in accord with Lam *et al.* (2002). The macro porosities in the scaffolds were created as the result of homogeneously dispersed air bubbles throughout the mixture by stirrer magnetic machines at 800-900 rpm. The mixture was observed to contain a fine dispersed of air bubbles, after the formation of a dough-like solution. Different wt% of scaffolds solution had a suitable interconnected pores size between 20-420 µm. After evaporation of water by heating the mixture becomes more concentrated and thick, which could be sufficient to block the air bubbles to release and this will lead to an increase in the pores after the mixture dried. Similar finding was reported by Kang *et al.* (2006). In this study, we used dextran, gelatin, and dextrin in addition to the cockle shell powder as scaffold materials for tissue engineering applications because we thought that we could take advantage of the powders in microsphere processing for scaffold design. The dextran was particularly chosen since it has been shown to be resistant to both protein adsorption and cell adhesion, and allowed designing a scaffold with specific sites for cell recognition (Le´vesque *et al.*, 2005). The used of dextran for scaffold development not only due to high number of pores created but also their morphology which was more rounded, and the degree of interconnectivity observed were highly influenced by the amount of dextran. In fact, the natural dissolution of dextrin, dextran and gelatin in water was responsible for the porosity formation. Furthermore, it could be attributed to the fact that an increase in gelatin concentration increased the capacity for entrapment of air bubbles and particle-induced pores in the matrix (Askarzadeh *et al.*, 2005; Le´vesque *et al.*, 2005). Therefore, an inversely proportional relationship between the compressive stiffness and porosity was established due to the increased in pores and the degree of interconnection between pores. Also, there is a proportional relationship between the amount of pores in the scaffold and the degradation rate due to an increase in the surface area. The scaffolds prepared by freeze-drying had dual pore structures caused by air bubbles and ice crystals, for freezing the dispersion or solution results in the formation of ice crystals that force and aggregate the collagen molecules into the interstitial spaces. The ice crystals were then removed by freeze-drying. Also, the conventional freeze dried scaffold formed only small pores as a result of a fast freezing rate, and the ice crystal agglomeration can controlled the pore size as showed in figure 2. These results are consistent with the results of Le´vesque *et al.* (2005) and Sachlos and Czernuszka (2003).

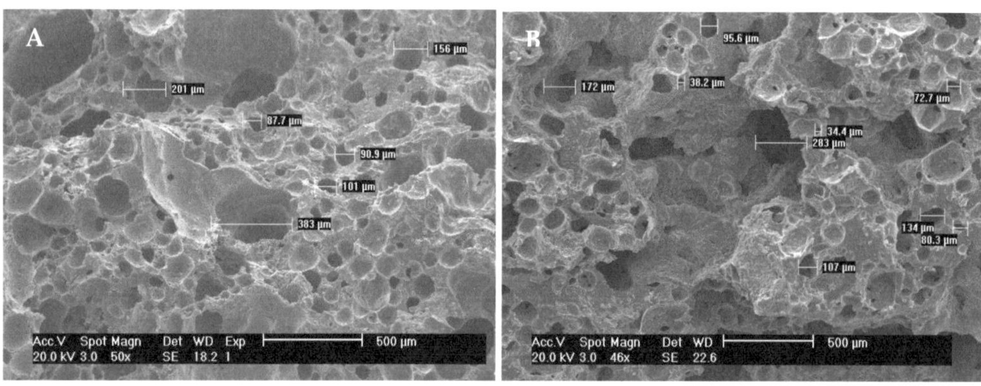

Fig. 1. ESEM microphotographs of scaffolds prepared by heated method showing porous structure. A) scaffold 244 and B) scaffold 334

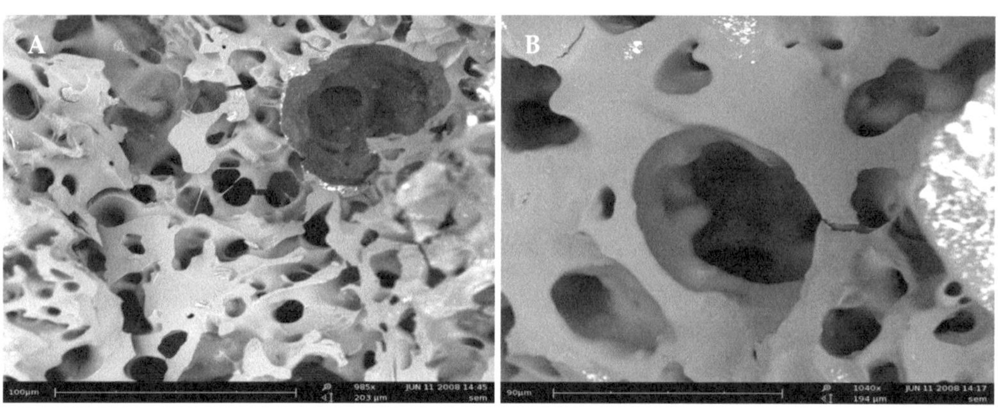

Fig. 2. Phenom SEM of the scaffolds prepared by freeze drying method show high porosity with different pore sizes and high interconnectivity. A) scaffold 244 and B) scaffold 352

Micro porosities may increase the degradation rate of the scaffolds due to the increased surface area available for attack. Scaffolds prepared with the freeze dryer method exhibit higher porosity. Accordingly, different scaffolds porosity and outer solid layer thickness was achieved by these two methods, which have direct influence on their degradation behaviour. In our study the scaffolds cross linked produced by heating method show a decrease in the acceleration and degradation rate. This result is in contrast with Oliverira and Reis (2004), who mention that most of the scaffolds lose half of their strength almost after the first days of immersion in the liquid, and the continuous soaking in the liquid will displayed only 5% of their initial strength and had lost almost half of their starting mass. In contrast, our study showed that the scaffolds dimensions did not change significantly during the degradation evaluation for up to 10 and 30 days for scaffolds prepared by freeze drying and heating methods, respectively. Thus, the decline in mechanical properties for these scaffolds is due to excessive mass loss, which was caused by the increased in porosity of the scaffolds. Scaffolds prepared by freeze drying method characterized by a rapid degradation than those fabricated by heating method, which was related to the short chains of amorphous product, consequently, this leads to increase in the number of sites susceptibility to chain scission (Porter *et al.*, 2001). The scaffolds prepared by freeze drying method exhibit significantly ($P<0.05$) higher water uptakes, compared to the samples prepared by heating method which have higher porosity. Nevertheless, the scaffolds produced in our study by both methods might be very useful in situations where a high mechanical property is necessary.

There is a conflict between a high porosity and the mechanical compression properties of the scaffolds. Dry infiltrated scaffolds 262 and 226 demonstrated lower mechanical properties (Table 2), which probably due to the higher porosity and interconnection between pores of the scaffolds prepared with this method. Although these porous structures are most appropriate for the final application of these scaffolds, still these scaffolds provide a good mechanical property when compared with other scaffolds that obtained from other biodegradable polymers which presented a compression module below 12 MPa, and proposed for use in tissue engineering of bone, for example, PLLA/hydroxyapatite composite foam (Zhang and Ma, 1999). Maria (2004) method for preparing more than one type of scaffolds was consistent with our technique. She used different processing methods to produced different scaffolds of different geometries, and the samples tested have also different geometries and dimensions.

The compression tests of scaffolds were carried out to obtain the stress-strain relations from which the yield strength and modulus were evaluated. The significant increase in modulus and yield strength for the scaffold can be attributed to the strong interactions between the four materials that form the scaffolds, and one of the major challenges in the fabrication of porous scaffolds is the tradeoff between adequate material porosity and mechanical strength (Thomson *et al.*, 1995b; Porter *et al.*, 2000; Zhensheng *et al.*, 2005). However, some of the scaffolds posses the similar yield strengths, which suggested that polymer infiltration, did not affect the yield strength of the dry sample. It was noted that for the porous designs, there was a significant increase in the yield strength in some because of different porosity scaffold after polymer infiltration. This reinforced the theory by infiltration with polymer to improve the mechanical properties of the scaffolds (Lam *et al.*, 2002). The role of gelatin in compression mechanical test was to provide support of the mechanical compression, and these were formed by the changes of chain of amino acids by cross linked. Thus the initial modulus of scaffolds gradually increased with increasing concentrations of gelatin in the mixture solution for both freeze-dried and heating stirrer-processed (heat method). This varying in mechanical property may be due to the increase and decrease of the scaffolds porosity, as showed in table 1. The decrease in the modulus MPa of scaffolds was due to the increase in scaffold porosity. Our finding agreed with what was reported by Kang *et al.* (2006). The variation in the ratio of gelatin to other materials, the degree of porosity composite and the degree of crosslinking may influence on modulus and ultimate compression test of the specimens.

No.	Sample	Yield strength MPa	Modulus MPa
244	Dry uninfiltrated	11.43	144.5
244	Dry infiltrated	13.95	182.3
244	Wet infiltrated	1.946	0.696
262	Dry uninfiltrated	3.628	74.57
262	Dry infiltrated	3.429	20.74
262	Wet infiltrated	0.132	0.428
352	Dry uninfiltrated	13.19	187.5
352	Dry infiltrated	9.676	71.41
352	Wet infiltrated	0.394	0.751
334	Dry uninfiltrated	7.271	123.1
334	Dry infiltrated	5.801	33.24
334	Wet infiltrated	0.007	4.077
226	Dry uninfiltrated	4.894	48.05
226	Dry infiltrated	5.250	148.8
226	Wet infiltrated	2.779	0.639

Table 1. Compressive stiffness (dry uninfiltrated, dry infiltrated and wet infiltrated) measured at 25°C for scaffolds prepared by heat method

The DSC analysis can be used to determine the physical transformations or phase transitions of the bioceramic scaffolds. A melting or glass transition temperature of the material can be determined, which gives an overview of the physical nature of these scaffolds. The melting transition temperature (T_m) is influential when the materials used are crystalline or semi-crystalline. When a glassy region and the stage of glass transition is followed a significant decrease in the graph without other peaks present, an average glass transition temperature (T_g) can be determined. During its melting, the material blend shows two peaks next to each other; the peak found at 97.78-112.97°C is attributed to gelatin, dextran and dextrin melting (Figure 3). These temperatures are compatible with those used to extrude the blends. Also, there are peaks less intense, approximately at 204-245°C which probably related to glass transition temperature (T_g). As the temperature increases the sample eventually reaches its melting temperature (T_m). The melting process results in an endothermic peak in the DSC curve. The ability to determine transition temperatures and enthalpies (thermodynamics and chemistry) makes DSC an invaluable tool in producing phase diagrams for various chemical systems. These results were in accord to that of Marcos *et al.* (2005). Glass transitions (T_g) may occur as the temperature of an amorphous solid is increased. These transitions appear as a step in the baseline of the recorded DSC signal. This is due to the sample undergoing a change in heat capacity; no formal phase change occurred as showed in figure 3. As the temperature increases, an amorphous solid will become less viscous. The main application of DSC is in studying phase transitions, such as melting, glass transitions, or exothermic decompositions. These transitions involve energy changes or heat capacity changes that can be detected by DSC with great sensitivity. The melting points and glass transition temperature for these materials are available from standard compilations and the method can show up possible materials degradation by the lowering of the expected melting point. Depends on the molecular weight of these materials, thus lower grades will have lower melting points than expected. This is possible because the temperature range over a mixture of compounds melts is dependent on their relative amount da Silva (2009).

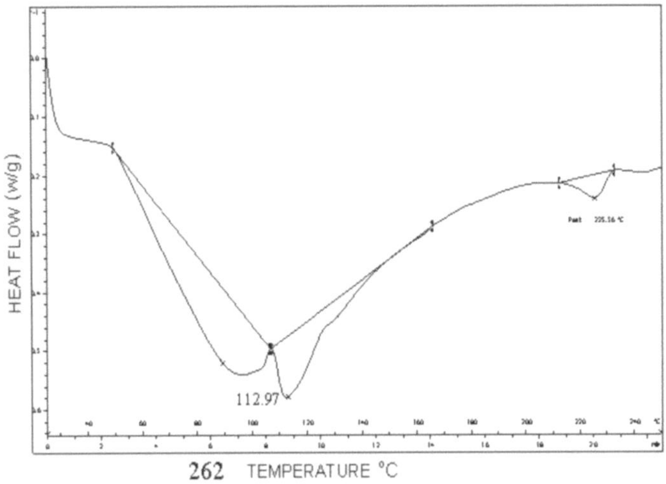

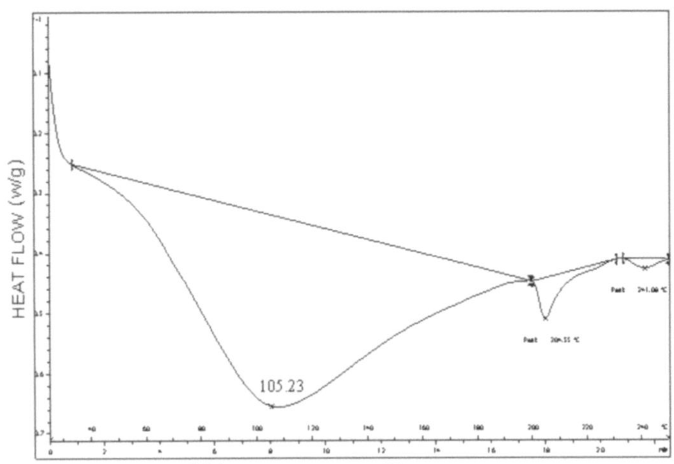

Fig. 3. The graphs show the endothermic DSC peak (denaturation) of the scaffold materials (gelatin, dextran, dextrin and $CaCO_3$) dispersion of the three samples with a maximum at 97.78-112.97°C. The first and second minimum melting peaks of the three samples observed to be semi symmetrical. The thermal signature was that of the scaffolds with the first and second peak of three samples (262, 334, 352)

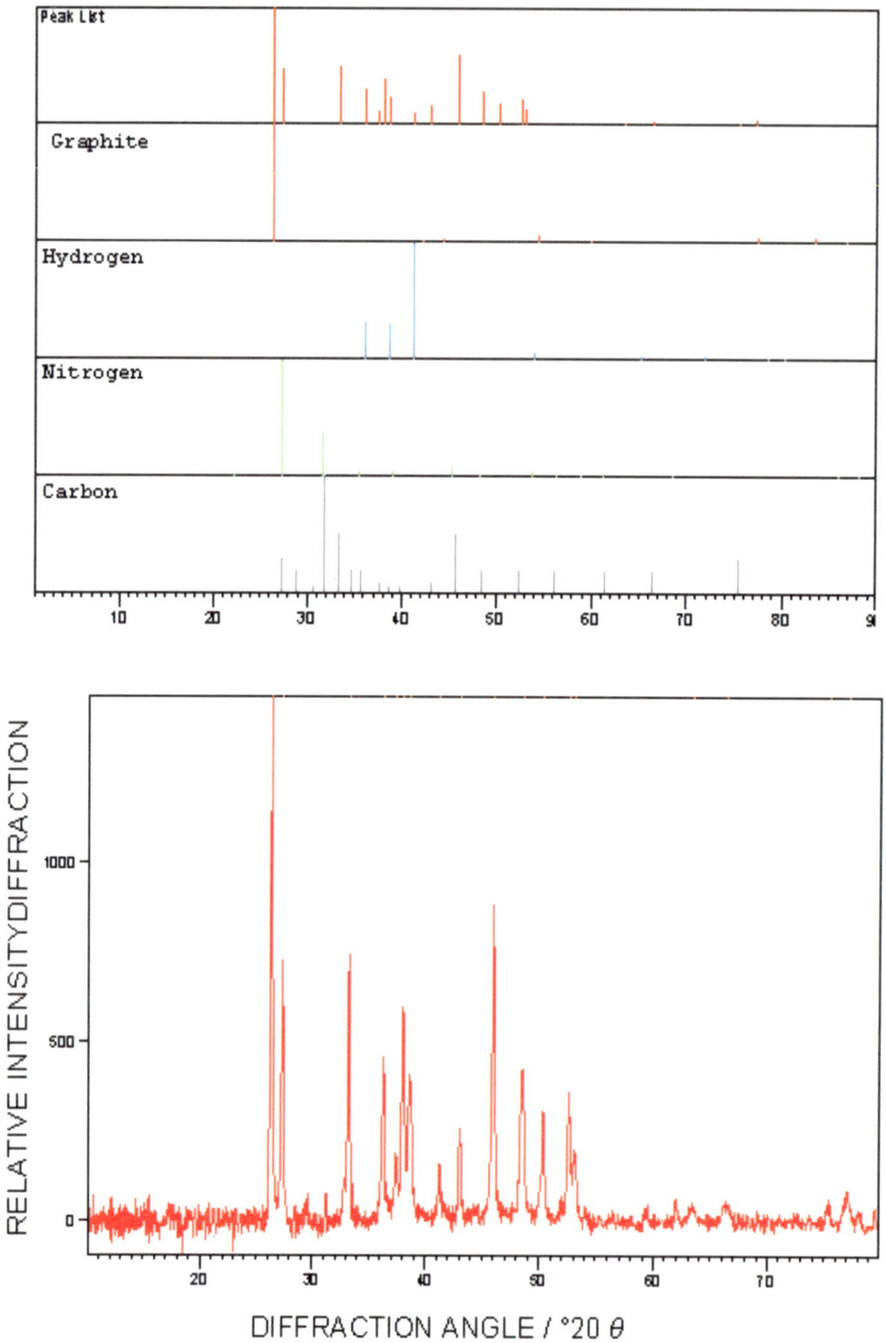

Fig. 4. The graphs show the XRD pattern of the scaffold products. The peak at 26°θ and 28°θ are characteristic of calcium like graphite, and other peaks are characterized of Calcium. This corresponds well to the specification (Graphite, Ca)

No.	Sample	Yield strength MPa	Modulus MPa
244	Dry uninfiltrated	10.10	2037
244	Dry infiltrated	12.62	2075
262	Dry uninfiltrated	23.99	1952
262	Dry infiltrated	23.70	1898
352	Dry uninfiltrated	7.141	268.5
352	Dry infiltrated	4.621	152.4
334	Dry uninfiltrated	10.80	825.3
334	Dry infiltrated	9.33	735.4
226	Dry uninfiltrated	4.871	550.2
226	Dry infiltrated	5.231	649.4

Table 2. Compressive stiffness (dry uninfiltrated and dry infiltrated) measured at 25°C for scaffolds prepared by freeze drying method

The XRD data show an increase in the crystal size for all the composites that do not exhibit an increase in compressive strength. This increase in crystal size, however, is not apparent in the micrographs, as the size of the crystals reduced from the XRD data is near the limit of resolution of the ESEM. Thus, the X-ray diffraction pattern of the scaffolds at different samples, no changes in peak pattern of crystallinity could be detected as shown in figure 4. On the other hand, the XRD data seem to indicate a similar mechanism at work in all the composites, namely improved mechanical properties due to smaller crystallites (Mickiewicz, 2001).

3. *In vitro* evaluation of the osteoblast seeded cockle shell-based biocomposite scaffolds

The ability of bone cells to colonize artificial bone graft substitute material is of clinical importance. Alkaline phophatase is an early differentiation marker. The ALP activity is a transient early marker of osteodifferentiation in MSCs and is generally a good indicator of differentiation. It was reported that ALP is associated with calcification and an enhanced expression of this enzyme is apparently needed just before the onset of matrix mineralization, providing localized enrichment of inorganic phosphate, one of the components of apatite, the mineral phase of bone (Genge *et al.*, 1988). In the current research work, the ALP results as showed in figure 5 agree with the results reported by many investigators, namely, Toquet *et al.* (1999), Kose *et al.* (2003), Park *et al.* (2004), Datta *et al.* (2005), and Wang *et al.* (2007). During the early period (from 5 to 14 days) the extracellular matrix undergoes a series of modifications in composition and organization that renders it competent for mineralization. During this matrix maturation phase, every cell has become alkaline phosphatase positive. Similar results were observed by Lian and Stein (1992). In the current study we observed that the peak activity occurred within 5-7 days, this was in agreement with Park *et al.* (2004) who reported that a high expression of ALP was prominent at 3 days of the culture. However, this finding disagreed with Kose, *et al.* (2003) who reported that ALP activity increased over time for all their samples and especially, after 14 days of incubation. The level of the ALP in sample 226 increased again after 14 days and this agrees with Kose *et al.* (2003) who reported the continuous activity of the cells despite the calcium deposition in the culture. Thus, in this study an initial rise in activity was expected, followed by a decrease corresponding to further differentiation of the cells, when alkaline phosphatase production slowed. The reduced activity of ALP could be attributed to the fact that more cells cultured on the scaffolds stepped into the next differentiation stage. This was in addition to the intracellular calcium increase which could determine an inhibitory effect. ALP decrease could represent a return to osteoprogenitor cells or maturation to osteocytes, which normally express small quantities of this enzyme. These results were in accord with those of Genge *et al.* (1988) and Wang *et al.* (2007). There was a major decline in ALP activity during matrix vesicle mineralization. The decline in ALP activity is very closely coupled with the rapid accumulation of Ca^{2+} by matrix vesicles. It is a highly reproducible and consistent finding and occurs in every instance of vesicle mineralization. ALP is known to be a transient marker of osteoblastic differentiation, being up-regulated initially and down-regulated as differentiation progresses. Figure 5 depicted the ALP activity during the period of osteoblast culture of different samples and this explains why it was possible to see a statistical difference in end-stage markers, such as calcium deposition, when none was seen for the earlier stage marker ALP.

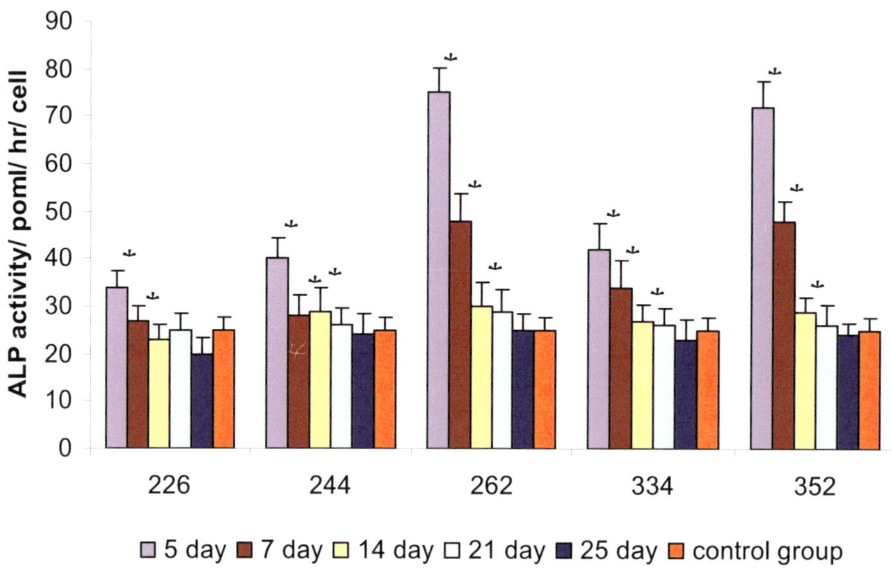

Fig. 5. The ALP activity of osteoblast cells at 5, 7, 14, 21 and 25 days post seeding on different scaffolds. Each bar denotes 1 standard deviation. * Significant difference ($p<0.05$) (one-sample t-test) within groups

All studies regarding calcium deposition in cultures of osteoblastic cells is a marker of full maturation. There is an important relation between the calcium deposition accumulation cultures and ALP activity. The studies reported here explore this phenomenon and revealed that a consistent marked loss of ALP activity does occur during the matrix vesicles mineralization. The time of onset and the extent of decline in ALP activity were found to mirror almost exactly the time of onset and the extent of Ca^{2+} accumulation by the matrix vesicles of all the scaffolds as shown in Figure 6. Such observations were also reported by

Genge, *et al.* (1988). The mineral accumulation is a consequence of the progression of pre-osteoblastic cells through the proliferation and matrix maturation stages of differentiation and it is an essential step for the further up-regulation or expression of genes responsible for the mineralization of the extracellular matrix. The findings here agreed with those reported by Lian and Stein (1992) and Maria (2004). After 15 days, a dramatic increase in calcium deposition was observed on all types of scaffolds that were cultured under normal flask conditions, clearly suggesting that scaffold 244 was responsible for the enhanced mineralization of marrow stromal cells more than other scaffolds. Possibly the dextrin and cockle shell ($CaCO_3$) concentrations were responsible for the observed, enhanced mineralization including the stimulation of the seeded cells on scaffolds. Additionally, the mitigation of potential nutrient transport limitations experienced by the cells cultured under these conditions enhanced the mineralization. The results clearly suggest that the new product composition is responsible for the enhanced mineralization of osteoblast cells. The higher calcium deposition reported for the scaffolds of some samples may be explained by the higher number of cells and cell density registered in these scaffolds. These observations were in agreement with the results of investigations carried out by Maria (2004). Therefore, calcium content of the scaffolds is indicative of the late stage differentiation of osteoblasts and a continual increase in calcium content over the culture period is expected. All constructs that were seeded with osteoblast cells showed a consistent increase in calcium content, with the highest value occurring at day-25 as seen in this study. Such results were in good agreement with the observations made by Datta *et al.* (2005) and Sikavitsas *et al.* (2005). It is well known to many researchers that the expression of calcium, a marker for mature osteoblasts, is strongly correlated with bone mineralization in osteoblast cultures. In the absence of mineralization, high levels of calcium cannot be expressed, and levels of ALP cannot decline. Such findings suggest that mineralization itself is closely associated with calcium expression, which may support our results in which the increased level of calcium (Ca^{2+}) occur during matrix vesicles mineralization formation. Such findings are in accord with that reported by Stein and Lian (1993) and Maeno *et al.* (2005).

From a cellular viewpoint, cell attachment and spreading are critical phenomena in formation of successful tissue-graft material interactions. In conclusion, this study has shown that there is no significant difference ($p > 0.05$) between groups, and this was an indication that all scaffold compositions have the same potential and even quality and can be used *in vivo*.

In the ESEM analysis, all the figures showed that all the osteoblasts on the cultured scaffold were covered by a dense matrix coating on the bottom and the top surfaces together. The cells were able to migrate throughout the scaffolds and filled the entire construct. The top surface of the scaffolds exhibited a thin layer of extracellular matrix, after 3 to 7 days in culture. ESEM examination of the surface of the seeded scaffolds revealed that cells had attached and begun to spread. However, the cells did not adhere closely to the irregular surface of the substrate but instead made numerous contacts via cytoplasmic extensions as seen in Figures 7. These results agreed with that of Sautier *et al.* (1990), who reported that the topographic surface of a substrate can influence the manner in which the cells attach to it. The ESEM images also confirm the adaptation of the cells to the 3D environment in the scaffolds; bridging the pores and attaching to the pore walls. Figures 8 showed a significant amount of calcium deposition in the bone matrix. Similar observations were also reported by Yoshikawa *et al.* (1997) and Yoshikawa *et al.* (2000). In addition, the attached cells exhibited numerous processes and surfaces studded with microvilli and ruffles. It is possible that fixation and dehydration caused cellular collapse and that the lack of close contact was an artifact. At the interface between the bone matrix and scaffold is an interfacial layer of collagenous unmineralized matrix, several micrometers thick. This layer is reportedly formed from secretion of extruded osteoblast processes and was seen in figures 9. These results agreed with those obtained by Yoshikawa *et al.* (1998). However, Sautier *et al.* (1990) reported that more protein synthesis occurred when cultured osteoblasts did not adhere closely to the substrate, but adopted stand-off morphology with considerable surface activity. Thus, the results confirmed the conclusion that the cultured scaffolds enhanced cell distribution *in vitro*.

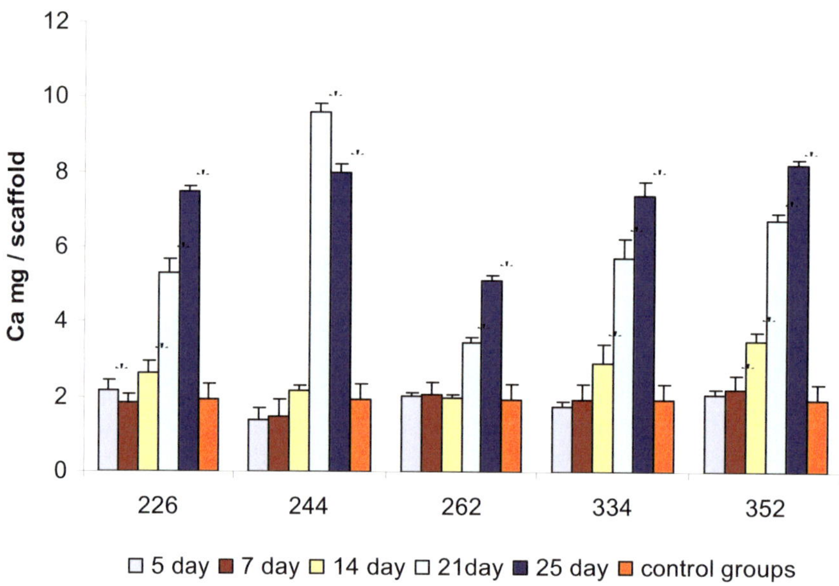

Fig. 6. Calcium depositions of osteoblast cells at 5, 7, 14, 21 and 25 days post seeding on different scaffolds. Each bar denotes 1 standard deviation. * Significant difference (p<0.05) (one-sample t-test) within groups

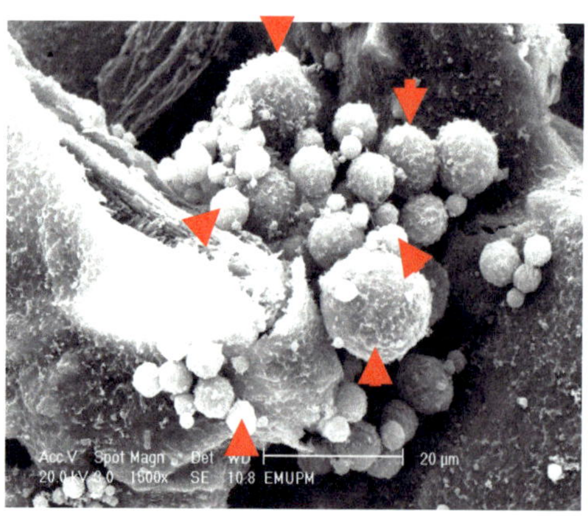

Fig. 7. ESEM microphotographs of rabbit osteoblasts cultured on different scaffolds for 3 days (arrowhead) and 7 days (arrow) show the structure of the osteoblast cells on scaffold 262 (bars = 20µm)

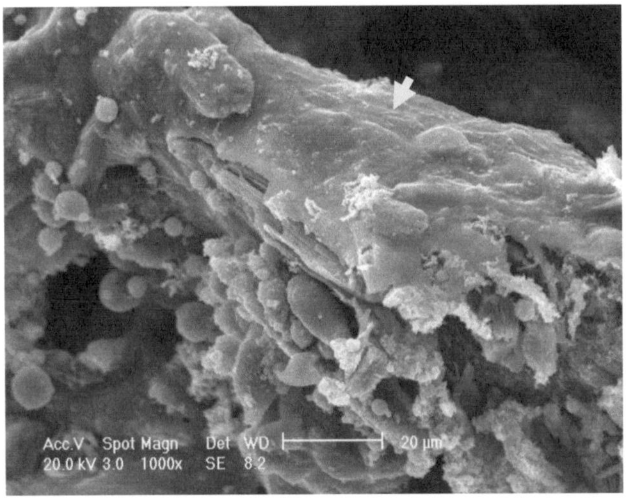

Fig. 8. ESEM microphotographs of scaffold section at 19 days post-seeding. Subculture composite *in vitro* shows the structure of the cells and their processes, and also the activity and creation of extracellular matrix (arrow) (bar = 20µm)

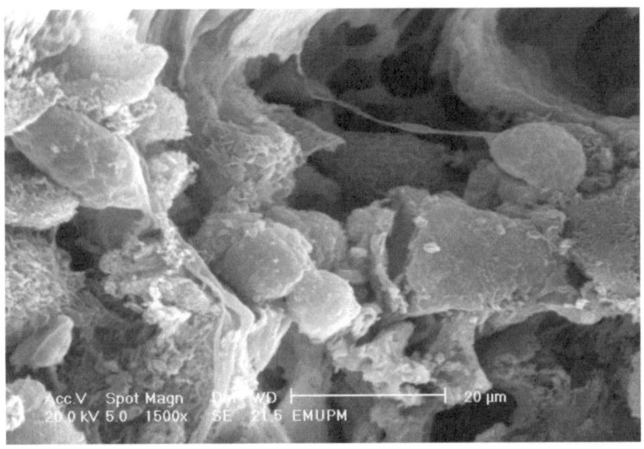

Fig. 9. ESEM microphotograph of rabbit osteoblasts cells on the cultured scaffold at 21 days post-seeding shows that some cells are well developed flat in shape and others are spindle or rounded during this period (bar = 20µm)

The EDX analysis showed that there are many nodules consisting of calcium carbonate. These nodules are calcium-rich minerals, as identified by energy-dispersive X-ray analysis. The formation of calcium carbonate, or mineral deposition, is a primary function of osteoblast cells. The energy dispersive analysis in Figure 10, were taken from the cell surfaces as well as the scaffold matrices (marked with a blue rectangle) and shown on the accompanying SEM images. Energy dispersive analysis of an osteoblast on these scaffolds showed signs of calcium after 19 days culture as well as the extracellular matrix, and their amounts increased from day 14 to day 25. These results suggest that both the individual cells and cell clusters on these new scaffolds contributed to the production of calcium. These results conform with Lu *et al.* (2003) and Li *et al.* (2005).

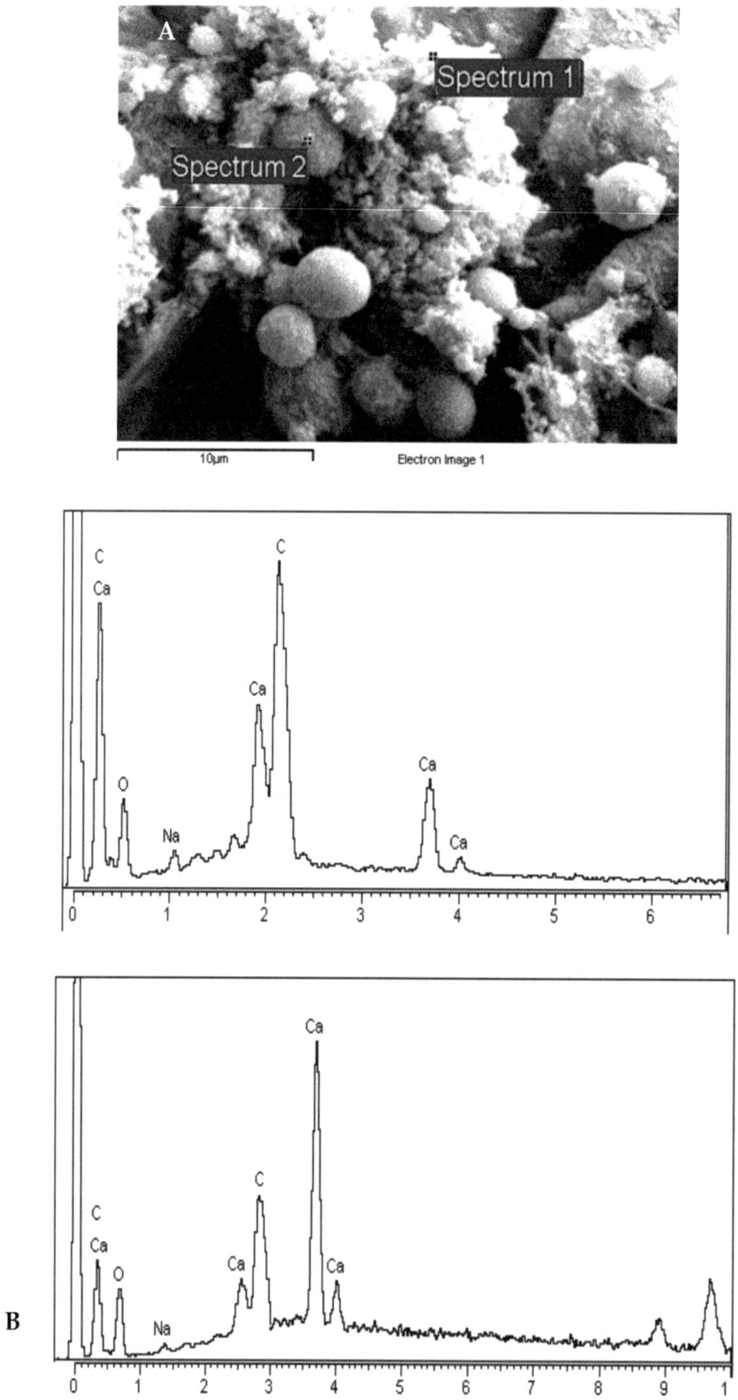

Fig. 10. SEM microphotograph (A) shows the osteoblasts grown on bioceramic scaffolds at 19 days of dynamic proliferation, (B) spectrum 1 depicts the extra cellular matrix and spectrum 2 the calcium carbonate as determined by EDX analysis

Histological examination showed that the scaffolds were covered and filled with new bone or osteoid tissue. Histological examination confirmed that the bone formation commenced after seeding of the scaffolds with the osteoblast cells. Thus, the osteoblast cells secreted both the collagen and the ground substance that constitutes the initial unmineralized bone or osteoid. The osteoblast is also responsible for the calcification of the matrix. The calcification process appeared to be initiated by the osteoblast through the secretion into the matrix of small, membrane-limited matrix vesicles. The vesicles are rich in alkaline phosphatase and are actively secreted only during the period in which the cell is producing the bone matrix. These vesicles are typically located at some distance from the cells where mineralization is to occur. These observations were in good accord with what was observed by Ross and Pawlina (2006). As the process continued, the newly organized tissue, at the presumptive bone site became more visualized, and the aggregated mesenchymal cells became larger and rounded as shown in Figures 11. Because of the collagen content, the bone matrix appeared denser than the surrounding mesenchyme in which the intercellular spaces revealed only delicate connective tissue fibers. The maturation and organization of the bone like extracellular matrix demonstrates the biological significance of the onset of mineralization as a second transition point in the osteoblast developmental sequence. This is further supported by studies that directly demonstrate a relationship between mineralization and the sequential expression of genes during the progressive development of the osteoblast phenotype (Chantal *et al.*, 2000). Furthermore, it is known that all cellular events depend on interactions between the cells and the extracellular matrix (ECMs) and that ECM protein can modify the surface chemistry of tissue-engineered substrates to enhance cell adhesion and promote growth (Jäger *et al.*, 2005).

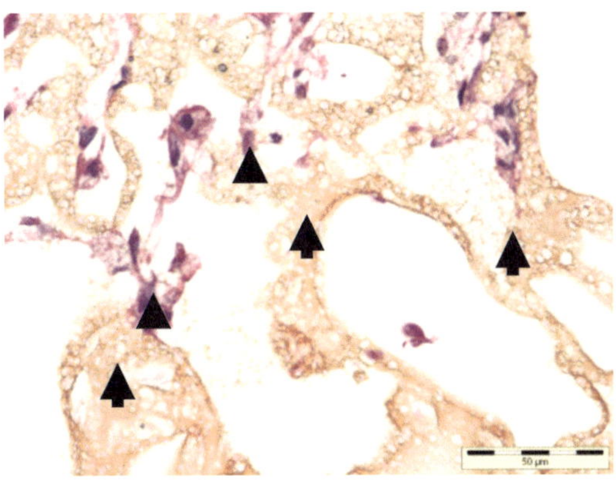

Fig. 11. Microphotograph of the scaffold stained with counterstained van Gison at 60 days post seeded show the osteoblasts (arrowheads) and the osteoid tissue presented in the most area (arrow) with (x 400)

4. *In vivo* evaluation of the osteoblast seeded cockle shell-based biocomposite scaffolds

Rabbit was the animal model used in this experiment because it correlates well with the clinical situations, and the model is well documented and reproducible. We used the segmental bone defect model of rabbit radius to evaluate the bone healing capacity for this new product. The radius is easily accessible; no fixation is needed because of the ulna support, and the short duration of fracture healing in rabbit model. Only male rabbits were chosen to standardize the sexes and to avoid the presence of hormonal fluctuation in female rabbits, which could give some influence on bone healing. The animal model, it is recommended to use adult rabbits to prevent epiphyseal slipping. For this we used 4-6 month-old rabbits. Such observations also reported by Herold, (1971) who was the first to described this model. The rabbit also is a well-established animal model in the field of BMP research (Zegzula *et al.*, 1997; Texeira and Urist, 1998; Yuehuei and Friedman, 1999). The critical-size bone defects (CSD) normally do not heal when left untreated, thus we used this size of defect to proof that the new developed scaffolds can be used to treat large bone defect. It is well accepted that the length of the CSD in long bone is twice the bone diameter, and bone defects of 15 millimeters were used in this study.

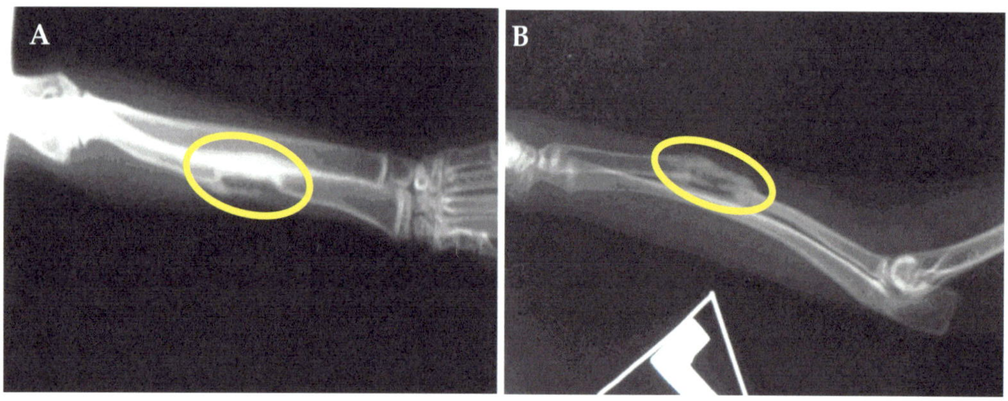

Fig. 12. Radiographs of the radial bone defect in group B at 4 weeks post-implantation show the extensive bone formation is seen over the defect implanted with a freeze dried scaffold which filled up the defect space (yellow circles). Radiograph of the radial bone defect in group B at 8 weeks post-implantation show the radio density of the implant is increased, and the borders appear smooth indicating bone remodeling

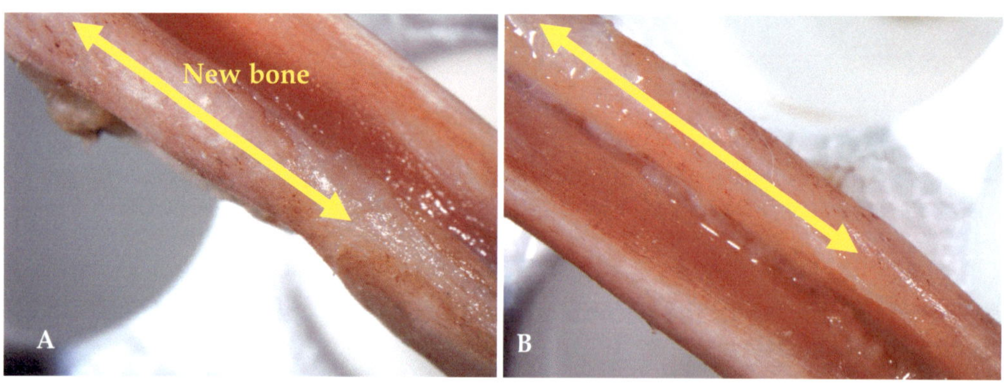

Fig. 13. Photographs of the radial bone implanted with seeded scaffold in group A at 8 weeks post-implantation show the new bone formation completely filled the defect area and restored the continuity of the radial bone (arrows). A, lateral view, B, medial view

The assessment of the rabbit radial bone was done at the end of 8th week. The subsequent radiographs at 4 and 8 weeks post-implantation showed the smooth unambiguous union evidence (Figure 12). Thus we decided that the 8th week group would be timely and adequate in giving the information of this new scaffold which produced as a bone substitute for repairing bone process. The duration of complete radial bone union and healing in large defect of 1.5 cm was evident at 8th week post-implantation (Figure 13).

Radiographic findings of the new bone formation of the radial bone implant presented at the margin and centre of the defect implanted with seeded scaffolds was diffuse at the centre and margin region of the implanted scaffold indicating the osteogenesis. The newly formed bone image in the defect implanted with seeded scaffold was visible completely bridging the radial bone defect, and the bone formation was faster than the previous studies for the tissue engineering of bone repair (Li and Li, 2005). The results of this study indicate that the bioactivity, stability and osteoconductivity of the scaffolds are present in repairing the radial bone defect.

The ALP and Ca^{2+} was evaluated in this study from the blood sample taken during the period of experiment. It seems that ALP concentration for both scaffolds of group A and B give indication that the secretion of ALP comes from the active osteoblast cells. ALP is an enzyme membrane of osteoblasts and its activity increases with time as the osteoblast differentiation and maturation proceed. Previous study has showed that when osteoblast cells were implanted, the ALP activity appeared after 2 weeks and Ca^{2+} was detected after 3 weeks (Kessel, 1989). Elevated serum ALP may be due to the rapid growth of bones, since it is produced by bone forming cells called osteoblasts. The level of ALP of group A during the 4-8 weeks post-implantation was maintained high as compared to the level of ALP in control. The high Ca^{2+} concentration stimulates the osteoblasts to secrete ALP, which increases the local concentration of PO_4^- ions. The high PO_4^- concentration stimulates further increases in Ca^{2+} concentration where mineralization initiated. At this stage of high extracellular Ca^{2+} and PO_4^- concentration, the osteoblasts release small matrix vesicles contain ALP and pyrophosphatase that cleave PO_4^- ions from other molecules of the matrix (Kessel, 1989; Ross and Pawlina, 2006). The ALP analysis also demonstrated a high osteogenic capacity from 1 week post-implantation. The results obtained at weeks 4 and 8 post-implantation indicated that heated and freeze dryer scaffolds was able to promote osteoblastic activity by seeding the osteoblast cells into the porous scaffolds, so the achievement of early bone formation should enable the rapid repair of bone defect, thereby allowing for early recovery and return to daily activities (Ross and Pawlina, 2006). This experiment shows that there was no significant difference ($P > 0.05$) between the seeded scaffolds and non seeded scaffold (control) in group B. However, there was a significant difference ($p > 0.05$) between the seeded scaffolds and non seeded scaffold (control) in group A. This indicated that the scaffold of group A is better than those in group B, since the concentration of the ALP was generally higher in group A.

The Ca^{2+} concentration as showed in Figures 14 can be mobilized to enter an adjacent capillary or it can be removed from the blood and deposited in the new bone matrix as needed. Ca^{2+} accumulation by matrix vesicles (MV) was produced by osteoblasts that increase in correlation with the level of bone regeneration. The matrix vesicles that accumulate Ca^{2+} and cleave PO_4^- ions cause the local isoelectric point to increase, which result in crystallization of $CaPO_4$ in the surrounding matrix vesicles. Thus, the $CaPO_4$

crystals initiate matrix mineralization by formation and deposition of $[Ca_{10}(PO_4)_6(OH)_2]$ (hydroxyapatite) crystals in the matrix surrounding the osteoblasts that was showing in period of bone formation during the level of Ca^{2+} was elevated (Kessel, 1989; Ross and Pawlina, 2006).

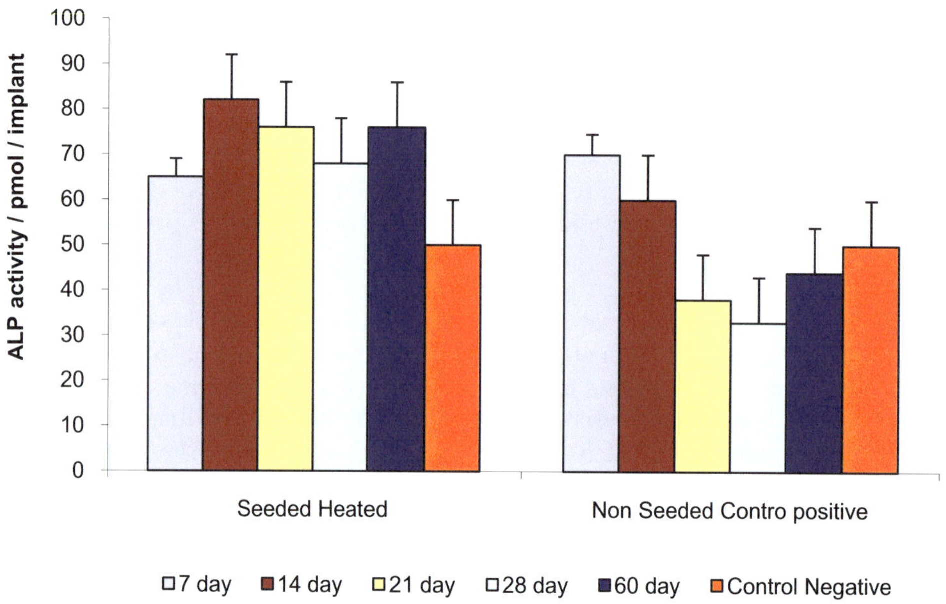

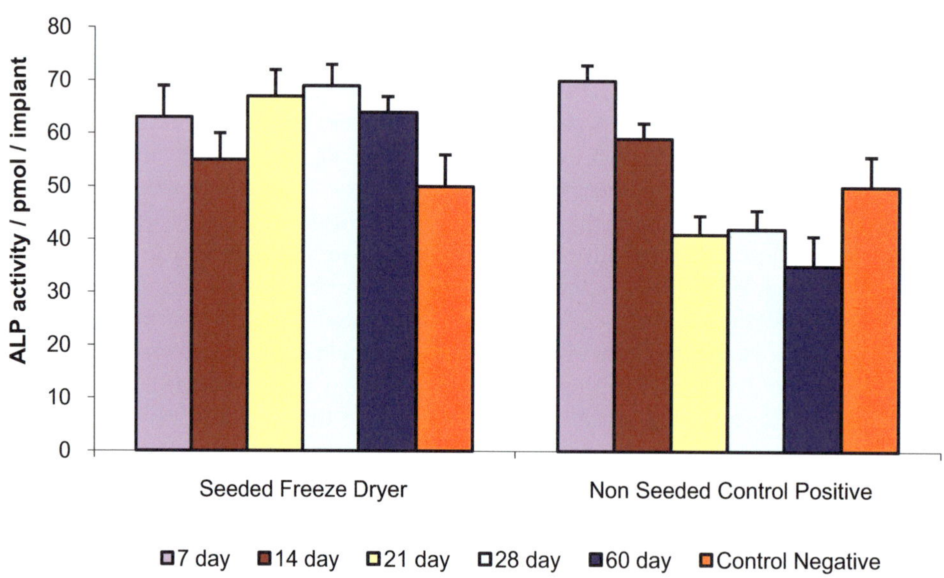

Fig. 14. The graphs show the value of total alkaline phosphatase (ALP) in both groups A and B at different time intervals post-implantation

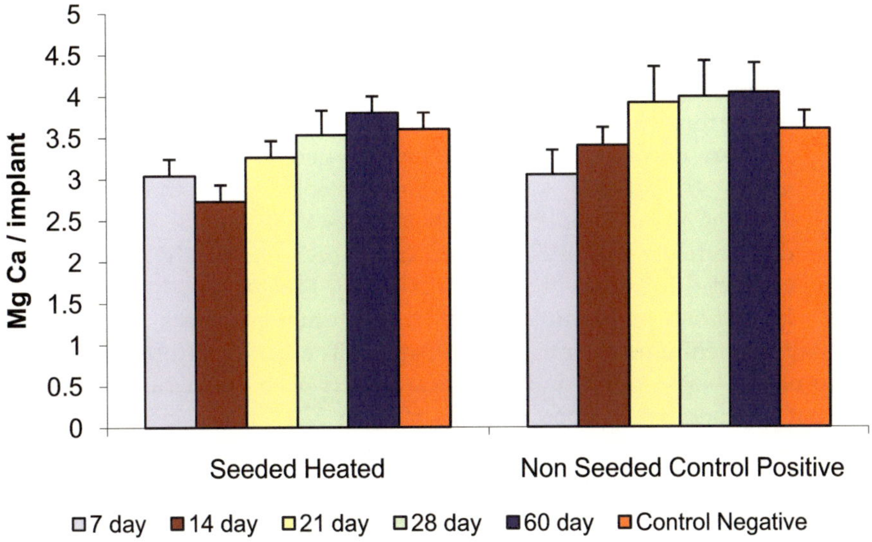

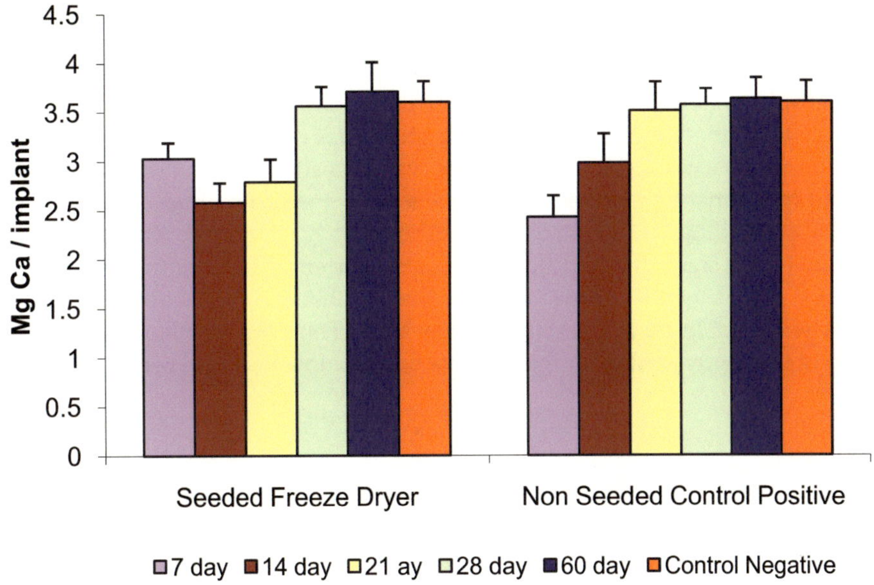

Fig. 15. The grafts show the value of total calcium in both group A and B at different time intervals post-implantation

Histological sections were assessed qualitatively by evaluating the bone formation at the defect area implanted with composite scaffolds. The histological analysis is widely used for examination of bone formation qualitatively. In this study, histological examination revealed complete new bone formation at the defect area replacing the seeded implant scaffolds in both groups A and B. However, in the control animals, it was observed that the bone formation was not complete and left empty defect space at the middle of the defect area, and there was no osteogenesis in the centre of implant. In the animals implanted with seeded scaffolds, the new bone formation present in the centre of implant. The seeded scaffold were completely resorbed by 8 weeks post-implantation and there were complete new bone formation replacing the implant, and mineralization of new bone and remodeling process was observed at this stage as demonstrated by the presence of osteoclast cells. From the observation, we thus infer that the rapid new bone formation in animals implanted with seeded scaffolds was due to the present of marrow-derived osteoblasts in the scaffolds. These observations showed that osteoconduction, osteoinduction and osteogenesis occur simultaneously in the animals implanted with seeded scaffolds, while in animals implanted with non seeded scaffolds, only osteoconduction was present, thus the new bone formation was slow. These results agree with Li and Li (2005). Previous studies on the new bone formation on scaffolds post-implantation using different types of scaffolds have demonstrated different time of new bone formation. The study performed by Martin *et al.* (1993) who placed blocks of HA in the cortical defects of the humerus and radius in dog revealed that interposition of bone in pores is from 52% at 16 weeks to 74% at 1 year (amount of bone relative to the pore space) post-implantation. The spongy bone regrowth was 38% after 4 weeks, and increased another 17% at 1 year, yet the pore spaces of HA rarely completely filled with bone during this period (Rosen *et al.* 1990). Vuola *et al.* (1995) who used the coral implant reported that after 6 weeks of implantation, the bone does not actually invade the pores of the sample of coral, but replaces the matrix. Yoshikawa (2000) founds that HA modified implants required a large amount of fresh cells from bone marrow and the method would be difficult to apply clinically. In addition, the bone forming capacity of these transplants have not been solved after discontinuation of immunosuppressive. The surface of chitosan modified PDLLA scaffolds study was conducted by Cai *et al.* (2007) after implantation for 12 weeks. A lot of new bone was formed within the scaffolds at this stage with a mild inflammatory reaction. A little new bone even connected them to form circles around the scaffolding, which had not yet completely adsorbed. The previous results demonstrated quite significantly different from the current results on the time taken for the defect to heal completely. The current results showed that at 8 weeks post-implantation, the defect area was completely bridged by the new bone. Thus, the new biocomposite scaffolds developed in this study seeded with osteoblast before implantation is better.

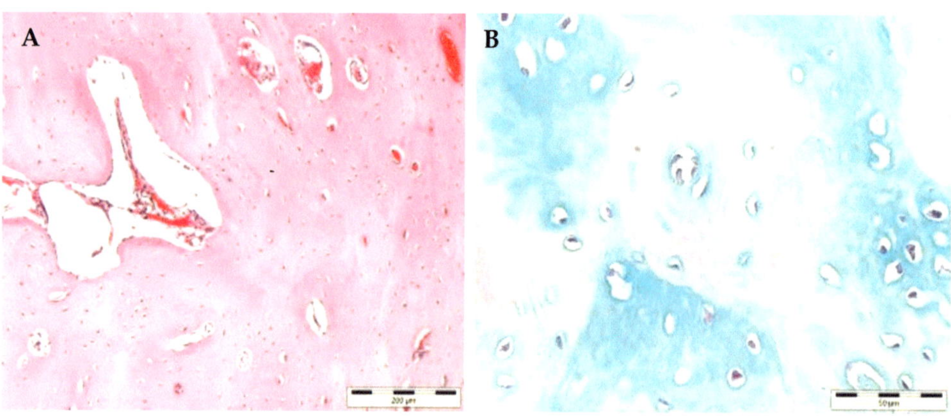

Fig. 16. Microphotographs of the decalcified specimen stained with H&E (A, x100) and Masson's trichrome (B, x400) show the mature bone. Note the osteon with concentric lamellae

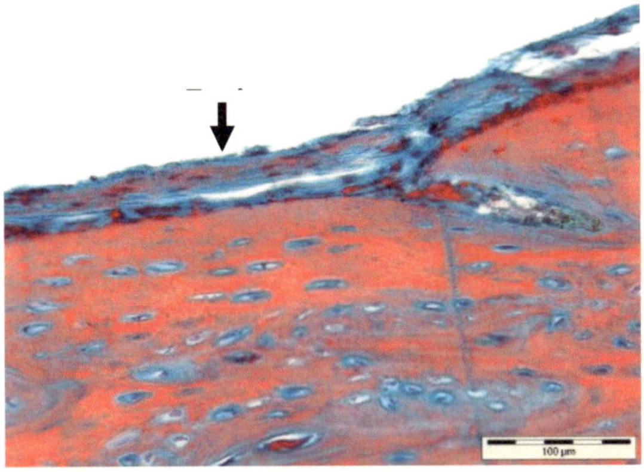

Fig. 17. Microphotograph of the cross section of the bone stained with Masson's trichrome (x200) show the complete bone formation with the present of periosteum (arrow)

5. Conclusion

Our observations indicate that the tissue engineering bone implant constructed by autogenous bone marrow derived osteoblasts and our scaffolds has the capability of osteogenesis, osteoconduction, and osteoinduction with a better osteogenetic effect and quality than scaffolds implant alone. Due to the availability of bone marrow and this scaffold, along with the bone regeneration potential, the tissue engineering bone might be an ideal scaffold for bone defect repair. The osteoinductive scaffold which is needed for the treatment of large bone defects are to be resorbed in a suitable time frame in order to minimize the amount of fibrous tissue and not to interfere with the remodelling of the new bone. In general, for observations in this study, the tissue engineering bone resulted in a higher quality and quantity of new bone formation when compared with the implanting of different types of scaffolds in previous studies, specially the HA, $CaPO_4$ bioceramic, TGF-ß, auto, allograft bone and scaffold alone at the same period within 8 weeks. These results cannot absolutely exclude the possibility that the scaffolds alone give the same result by 12 or 16 weeks. Without doubt, if the scaffolds alone give the same result by 12 or 16 weeks, the cost effectiveness of osteoblast seeding may not be worthwhile. This is an interesting question requiring further investigation due to its importance in the practical application of this tissue engineering bone graft.

This study has revealed a new approach to design and fabrication of scaffolds for tissue engineering by using new materials in combination with natural polymer and water. Scaffolds were characterized for their physical and chemical properties. The analysis and tests used in this study demonstrated that the porous 3D scaffolds created by a new blend of materials through the described procedures are achievable.

This study also provides the fundamental information for researchers and engineers working in advanced composite industry, to open a new approach to the development of bioengineered composites. Cross-disciplinary research efforts are definitely needed in bridging expertise from bio-, nano and advanced composite areas, to work closely along this new scientific and engineering direction.

The work described in this thesis is heavily focused on developing and using biomaterials as tissue engineering matrices. Tissue engineering has been described as "the principles and methods of engineering, materials science and cell and molecular biology to the development of viable substitutes that restore, maintain or improve the function of humans' bone". Advances in new biocompatible materials, both osteoconductive and osteoinductive, will also aid in the treatment of bone diseases, while new diagnostic methods will help identify those at risk of disease before symptoms are obvious. A better understanding of how bone perceives and responds to mechanical signals will certainly help accelerate the healing of fractures, increasing the osseointegration of implants, and help to ensure that diseases such as osteoporosis can be treated properly. These are undoubtedly exciting times for everyone working in the bone field, and the interdisciplinary nature that has brought us tremendous insight into bone's complexity on a structural and functional level hence serve us well in the future.

6. Acknowledgement

The study was supported by e-Science research grant (project no. 06-01-04-SF202) provided by Malaysia Government.

7. References

Askarzadeh, K.; Orang, F. & Moztarzadeh, F. (2005). Fabrication and Characterization of a Porous Composite Scaffold Based on Gelatin and Hydroxyapatite for Bone Tissue Engineering. *Iranian Polymer Journal*, vol. 14, pp.511-520.

Boyne, N. P. ; Lambrianides, A. L. & Pollard, C. (2002). Incisional hernia through iliac crest bone graft donor sites. *ANZ Journal of Surgery*, vol. 72, pp. 156-157.

Buckwalter, J. A. (2004). Can tissue engineering help orthopaedic patients. Clinical needs and criteria for success. In: Tissue Engineering in Musculoskeletal Clinical Practice, edited by L. J. Sandell and A. J. Grodzinsky. *American Academy of Orthopaedic Surgeons*, pp.3-16

Cai, K.; Yao, K.; Yang, Z.; Qu, Y. & Li, X. (2007). Histological study of surface modified three dimensional poly (D, L-lactic acid) scaffolds with chitosan *in vivo*. *Journal of Materials Science: Materials in Medicine*, vol. 18, pp. 2017-2024.

Chang, S. C.; Chuang, H.; Chen, Y. R.; Yang, L. C.; Chen, J. K.; Mardini, S.; Chung, H. Y.; Lu, Y. L.; Ma, W. C. & Lou, J. (2004). Cranial Repair Using BMP-2 Gene Engineered Bone Marrow Stromal Cells. *Journal of Surgical Research*, vol. 119, pp. 85-91.

Chantal, E. H.; Molly S. S. & John E. D. (2000). Engineering three dimensional bone tissue *in vitro* using biodegradable scaffolds: Investigating initial cell-seeding density and culture period. *Journal Biomedical Material Research*, vol. 51, pp. 376-382.

Cypher, T. J. & Grossman, J. P. (1996). Biological principles of bone graft healing. *Journal of Foot Ankle Surgery*. Vol. 35, pp. 413–417.

da Silva, V.M. (2009). Development of New Chitosan Based Biodegradable Blends for Bone and Cartilage Tissue Engineering. Thesis

Datta, N.; Holterf, H. L.; Sikuritsas, V. I.; Jansen, J. A. & Mikos, A. G. (2005). Effect of bone extracellular matrix synthesized *in vitro* on the osteoblastic differentiation of marrow stromal cells. *Biomaterials*, vol. 26, pp. 971-977.

Ducheyne, P. ; Bianco, P. D. & Kim, C. (1992). Bone tissue growth enhancement by calcium phosphate coatings on porous titanium alloys: The effect of shielding metal dissolution product. *Biomaterials*, vol. 13, pp. 617-624.

Dziedic, D. M., Savva, I. H., Wilkinson, D. S. and Davies, J.E. (1996). Osteoconduction on, and bonding to, calcium phosphate ceramic implants, *Materials Research* Society.*Symposium*, vol. 14, pp. 147-156.

Freed, L. E.; Vunjak-Novakovic, G.; Marquis, J. C. & Langer, R. (1994b). Kinetics of chondrocytes growth in cell-polymer implants. *Biotechnology and Bioengineering*. Vol. 43, pp. 597-604.

Genge, B. R.; Sauer, G. R.; Wu L. N.; Mc Lean, F. M. & Wuthier, R. E. (1988). Correlation between loss of alkaline phosphatase activity and accumulation of calcium during matrix vesicle-mediated mineralization. *Journal Biological Chemistry*, vol. 263, pp. 18513-18519.

Hardin-Young, J.; Teumer, J.; Ross, R. & Parenteau, N. (2000). Approaches to transplanting engineered cells and tissues in Principles of Tissue Engineering, Lanza, R, Langer, R, and Vacanti, J, (Eds). Academic Press. New York. Pp. 281

Hatcher, B. M. (2004). Bioactive Organic-Inorganic Hybrids for Tissue Engineerin Applications. PhD Thesis. University of Florida

Hench, L. L. and Wilson, J. (1993). An Zntro&ction to Bioceramics, Ed: Hench, L-L, and Wilson, J. World Scientific Publishing Co., New Jersey, pp. 1-198.

Herold, G. H.; Hurvitz, A. & Tadmor, A. (1971). The effect of growth hormone on the healing of experimental bone defects. *Acta Orthopaedica Scand.*, vol. 42, pp. 377-384.

Holtorf, H. L.; Sheffield, T. L.; Ambrose, C. G.; Jansen, J. A. & Mikos, A. G. (2005). Flow Perfusion Culture of Marrow Stromal Cells Seeded on Porous Biphasic Calcium Phosphate Ceramics. *Annals of Biomedical Engineering*, vol. 33, pp. 1238-1248.

Jäger, M.; Feser, T.; Denck, H. & Krauspe, R. (2005). Proliferation and Osteogenic Differentiation of Mesenchymal Stem Cells Cultured Onto Three Different Polymers *In Vitro*. *Annals of Biomedical Engineering*, vol. 33, pp. 1319-1332.

Kang, H. G.; Kim, S. Y. & Lee, Y. M. (2006). Novel Porous Gelatin Scaffolds by Overrun/Particle Leaching Process for Tissue Engineering Applications. *J Biomed Mater Res Part B: Applied Biomaterials*, vol. 79B, pp. 388–397.

Kessel, R. G. (1989). Basic Medical Histology: *The biology of Cells, Tissue, and Organs*. pp. 144-145.

Kose, G. T.; Korkusuz, F.; Korkusuz, P.; Purali, N.; Özkul, A. & Hasirci,V. (2003). Bone generation on PHBV matrices: an *in vitro* study. *Biomaterials*, vol.24, pp. 4999-5007.

Lam, C. X. F.; Mo, X. M.; Teoh, S. H. & Hutmacher, D. W. (2002). Scaffold development using 3D printing with a starch-based polymer. *Materials Science and Engineering*, vol. 20, pp. 49–56.

Langer, R. & Vacanti, J.P. (1993). Tissue Engineering, *Science*, vol. 260, pp. 920-926.

Laurencin, C. T.; Ambrosio, A. M. A.; Borden, M.D. & Cooper, J. A. (1999). Tissue engineering: Orthopedic applications, *Annual Review of Biomedical Engineering*, vol. 1, pp. 19-24.

Le´vesque, S. G. ; Lim, R. M. & Shoichet, M. S. (2005). Macroporous interconnected dextran scaffolds of controlled porosity for tissue-engineering applications. *Biomaterials,* vol. 26, pp. 7436–7446.

Lian, J. B. & Stein, G. S. (1992). Concepts of osteoblast growth and differentiation: basis for modulation of bone cell development and tissue formation. *Crticial Reviews in Oral Biology and Medicine,* vol. 3, pp. 269-305.

Li, Z. & Li, Z. B. (2005). Repair of Mandible Defect with Tissue Engineering Bone in Rabbits. *ANZ Journal of Surgery,* vol. 75, pp. 1017–1021.

Li, Z.; Ramay, H. R.; Hauch, K. D.; Xia, D. & Zhang, M. (2005). Chitosan–alginate hybrid scaffolds for bone tissue engineering. *Biomaterials,* vol. 26, pp. 3919-3928.

Lu, H. H.; El-Amin, S. F.; Scott, K. D. & Laurencin, C. T. (2003). Three-dimensional, bioactive, biodegradable, polymer-bioactive glass composite scaffolds with improved mechanical properties support collagen synthesis and mineralization of human osteoblast-like cells *in vitro. Journal Biomedical Materials,* vol. 64A, pp. 465-474.

Maeno, S.; Niki,Y.; Matsumoto, H.; Morioka, H.; Yatabe, T.; Funayama, A.; Toyama, Y.; Taguchi, T. & Tanaka, J. (2005). The effect of calcium ion concentration on osteoblast viability, proliferation and ifferentiation in monolayer and 3D culture. *Biomaterials,* vol. 26, pp. 4847–4855.

Marcos, M.; da S. Paula.; Fabiana, B. B.; Medeiros, R.; Adriano, M. B.; M´arcio, A. F. & El´ıdio, A. (2005). Characterization of aluminized polyethylene blends via mechanical recycling. *Materials Science and Engineering* A, vol. 403, pp. 37–41.

Maria, M. E-G. A. (2004). Bone Tissue Engineering Strategy basal on Starch Scaffolds and Bone Marrow Cells Cultured in a Flow Perfusion Bioreactor. PhD Thesis. Universidade Do Minho Escola De Engenharia

Martin, R. B.; Chapman, M. W.; Sharkey, N. A.; Zissimos, S. L.; Bay, B. & Shors, E. C. (1993). Bone ingrowth and mechanical properties of coralline hydroxyapatite1 yr after implantation. *Biomaterials,* vol. 14, pp. 341-348.

Meyer, U.; Joos, U. & Wiesmann, H. P. (2004b). Biological and biophysical principles in extracorporeal bone tissue engineering. Part I. *International Journal of Oral and Maxillofacial Surgery,* vol. 33, pp. 325-332.

Mickiewicz, R. A. (2001). Polymer-calcium phosphate composites for use as an injectable bone substitute. Department of Materials Science and Engineering MS Thesis. Massachusetts Institute of Technology, University of Toronto.

Mooney, D. J. & Mikos, A. G. (1999). Growing new organs, *Scientific American,* 280:38-43

Moore, W. R., Graves, S. E. and Bain, G. I. (2001). Synthetic bone graft substitutes. *ANZ Journal of Surgery,* vol. 71, pp. 354-361.

Oliverira, A.L. & Reis, R. L. (2004). Pre-mineralisation of starch/ polycrapolactone bone tissue engineering scaffolds by a calcium silicate-based process. *Journal of Materials Science: Materials in Medicine,* vol. 15, pp. 533-540.

Park, E. K.; Lee, Y. E.; Je-Yong, C.; Sun-Ho, O.; Hong-In, S.; Kyo-Han, K.; Kim, Shin-Yoon & K. S. (2004). Cellular biocompatibility and stimulatory effects of calcium metaphosphate on osteoblastic differentiation of human bone marrow-derived stromal cells. *Journal Biomaterials,* vol. 25, pp. 3403-3411.

Porter, B. D.; Oldham, J. B.; He, S. L.; Zobitz, M. E.; Payne, R. G. & An, K. N. (2000). Eng Mechanical Properties of A Biodegradable Bone Regeneration Scaffold. *Journal Biomechanics*, vol. 122, pp. 286–288.

Porter, N. L.; Pilliar, R. M. & Grynpas, M. D. (2001). Fabrication of porous calcium polyphosphate implants by solid freeform fabrication: A study of processing parameters and *in vitro* degradation characteristics. *Journal of Biomedical Material*, vol. 4, pp. 504-515.

Rosen, H. M. & McFarland, M. M. (1990). The biologic behavior of hydroxyapatite implanted into the maxillofacial skeleton. *Plastic Reconstructive Surgery*, vol. 85, pp. 718-723.

Ross, J. M. (1998). Cell-Extracellular Matrix Interactions. In: Patrick CW, Mikos AG, Mc.Intre L, editors. *Frontiers in Tissue Engineering*. Oxford: Elsevier Science Ltd., pp. 15-27.

Ross, M. H. & Pawlina, W. (2006). A Text and Atlas with Correlated Cell and Molecular Biology. Fifth Edition, pp. 224-225.

Sachlos, E. & Czernuszka, J. T. (2003). Making Tissue Engineering Scaffolds Work Review on the Application of Solid Freeform Fabrication Technology to the Production of Tissue Engineering Scaffolds. *European Cell and Materials*, vol. 5, pp. 29-40.

Sautier, J. M.; Nefussi, J. R.; Boulekbache, H. & Forest, N. (1990). *In Vitro* Bone Formation on Coral Granules. *In Vitro Cellular & Developmental Biology*, vol. 26, pp. 1079-1085.

Sikavitsas, V. I.; Bancroft, G. N.; Lemoine, J. J.; Liebschner, M. A. K.; Dauner, M. & Mikos, A. G. (2005). Flow Perfusion Enhances the Calcified Matrix Deposition of Marrow Stromal Cells in Biodegradable Nonwoven Fiber Mesh Scaffolds. *Annals of Biomedical Engineering*, vol. 33, pp. 63-70.

Stein, G. S. & Lian, J. B. (1993). Molecular mechanisms mediating proliferation/ differentiation interrelationships during progressive development of the osteoblastic phenotype. *Endocrine* Reviews, vol. 14, pp. 424-442.

Texeira, J. O. & Urist, M. R. (1998). Bone morphogenetic protein induced repair of compartmentalized segmental diaphyseal defects. *Archives of Orthopaedic and Trauma Surgery*, vol. 117, pp. 27-34.

Thomson, R. C.; Yaszemski, M. J.; Powers, J. M. & Mikos, A. G. (1995b). Fabrication of biodegradable polymer scaffolds to engineer trabecular bone. *Journal of Biomaterials Science, Polymer Edition*, vol. 7, pp. 23.

Toquet, J.; Rohanizadeh, R.; Guicheux, J.; Couillaud, S.; Passuti, N.; Daculsi, G. & Heymann, D. (1999). Osteogenic potential *in vitro* of human bone marrow cells cultured on macroporous biphasic calcium phosphate ceramic. *Journal of Biomedical Materials Research*, vol. 44, pp. 98-108.

Vaccaro, A. R.; Chiba, K.; Heller, J. G.; Patel, T. C.; Thalgott, J. S. & Truumees, E. (2002). Bone grafting alternatives in spinal surgery. *The Spine Journal*, vol. 2, pp. 206-215.

Vuola, J.; Göransson, H.; Böhling, T. & Seljavaara, A. (1995). Bone marrows induce osteogenesis in hydroxyapatite and calcium carbonate implants. *Journal of Biomaterials*, vol. 17, pp. 1761-1766.

Wang, Y.; Zhang, S.; Zeng, X.; Ma, L. L.; Wang, W.; Yan, W. & Qian, M. (2007). Osteoblastic cell response on fluoridated hydroxyapatite coatings. *Acta Biomaterialia*, vol. 3, pp. 191–197.

Williams, D. F. (1987). Definitions in biomaterials. *Proceedings of a consensus conference of the european society for biomaterials,* Vol. 4. Chester, England, March 3-5.New York: Elsevier.

Yoshikawa, T. (2000). Bone reconstruction by cultured bone graft. *Materials Science and Engineering C,* vol. 13, pp. 29-37.

Yoshikawa, T.; Davies, J. E.; Ichijima, K.; Ohgushi, H. & Tamai, S. (1998). Morphological study on mineralization of collagen fibers in *in vitro* bone formation, *Journal of Japanese Society Biomaterals,* vol. 16, pp. 72-77.

Yoshikawa, T.; Nakajima, H.; Yamada, E.; Akahane, M.; Dohi, Y.; Ohgushi, H.; Tamai, S. & Ichijima, K. (2000). *In Vivo* Osteogenic Capability of Cultured Allogeneic Bone in Porous Hydroxyapatite: Immunosuppressive and Osteogenic Potential of FK506 In *Vivo. Journal of Bone Miner Research,* vol. 15, pp. 1147-1157.

Yoshikawa, T.; Peel, S. A. F.; Gladstone, J. R. & Davies, J. E. (1997). Biochemical analysis of the response in rat bone marrow cell cultures to mechanical stimulation, *Biomedical Materials and Engineering,* vol. 7, pp. 369-377.

Younger, E. M. & Chapman, M.W. (1989). Morbidity at bone graft donor sites, *Journal of Orthopaedic Trauma,* vol. 3, pp. 192-195.

Yuehuei & Friedman. (1999). *Animal Models in Orthopedic Research.* CRC Press, Inc., Boca Raton, Florida, USA. Chapter 29, pp: 659-674.

Zegzula, H. D.; Buck, D. C.; Brekke, J.; Wozney, J. M. & Hollinger, J. O. (1997). Bone formation with use of rhBMP-2 (recombinant human bone morphogenetic protein-2). *Journal of Bone and Joint Surgery,* vol. 79, pp. 1778-1790.

Zhang, R. Y. & Ma, P. X. (1999). Poly(alpha-hydroxyl acids) hydroxyapatite porous composites for bonetissue engineering. I. Preparation and morphology. *Journal* of *Biomedical Materials Research,* vol. 44, pp. 446-455.

Zhensheng L.; Ramay, H. R.; Hauch, K. D.; Xiao, D. & Zhang, M. (2005). Chitosan–alginate hybrid scaffolds for bone tissue engineering. *Biomaterials,* vol. 26, pp. 3919-3928.

Part 4

Advanced Strategies

18

Bioactive Scaffolds for the Controlled Formation of Complex Skeletal Tissues

Sandra Hofmann[1] and Marcos Garcia-Fuentes[2]
[1]Institute for Biomechanics, ETH Zurich, Zurich,
[2]NANOBIOFAR Group, Dep. Pharmacy and Pharmaceutical Technology, Campus Vida, University of Santiago de Compostela, Santiago de Compostela
[1]Switzerland
[2]Spain

1. Introduction

Most skeletal tissues have the innate capacity to regenerate, at least to a certain extent, without therapeutic intervention. Bone for example has the capacity to repair fractures up to a critical size with the help of local or recruited cells of the patient's body, and the defects heal by forming new bone that is indistinguishable from uninjured tissue (Deschaseaux et al., 2009; Place et al., 2009). However, traumatic fractures of the lower extremities are one group of orthopedic injuries that often require prolonged rehabilitation, or multiple procedures to achieve maximal functional recovery, and comprise the bulk of nonunion or delayed-union fractures (Deschaseaux et al., 2009). In contrast, articular cartilage, ligaments and tendons are known to have a limited capacity for self-repair and methods to augment their natural healing response are still highly investigated.

Biomaterials used to restore the structure and function of lost tissues and organs have evolved greatly due to the large knowledge accumulated on material-protein and material-cell interactions. Current commercially available biomaterials are engineered to be non-toxic, non-immunogenic, and hemocompatible. In contrast to elements of the extracellular matrix of living systems, however, these scaffolds are very poor in information, which makes them suboptimal for many tissue engineering applications. These passive biomaterials are unlikely to guide cell migration and differentiation or controlled matrix deposition, a problem that becomes even more evident in complex tissues with more than one cell type. Furthermore, they also cannot induce tissue neoformation while preventing other undesirable tissue repair processes such as scarring, and they are unable -with the notable exception of some materials in bone regeneration- to promote functional tissue integration in the host, including vascular and/or nervous connectivity. Finally, these passive scaffolds largely lack the capacity to induce cell differentiation, which becomes a major limitation for their use together with current stem cell-based therapies (Huebsch & Mooney, 2009). A promising strategy to overcome these limitations is to design bioactive scaffolds capable of recruiting tissue-forming cells from the body in such a way that they promote cell adhesion and tissue formation within their matrix. Moreover, materials that can recruit endogenous cells of the patient into scaffolds, avoid the expense and difficulties

associated with culture, storage and distribution of cells, not to mention other immunological considerations (Discher et al., 2009; Place, ES et al., 2009).
Bioactivity is defined as having the capacity to interact with a living tissue or system. In terms of scaffold design, we define a bioactive scaffold as one in which biological functionality has been integrated to provide an information-rich support material for tissue engineering. Bioactive scaffolds are designed to control cell- and tissue responses, and to provide a more efficient integration with the host. In this manuscript we will cover the most important scaffolds presenting a biomimetic composition (Section 2), the main strategies to introduce biomimetic features in scaffolds (Section 3), their expected applications in skeletal tissue engineering and the main challenges remaining for the successful application of this concept (Section 4). We will highlight methods of how scaffold materials can be modified to be bioactive, how a scaffold can promote the body's innate capabilities to heal and recruit cells to the site of injury. Bioactive scaffolds have already found their way into clinics and hold great promise to serve as a manageable alternative to tissue engineered implants that face a vast array of regulatory and logistical issues. In the end, products on the market need to show convincing clinical performance as well as cost-effectiveness. We will therefore not cover the traditional tissue engineering approach to seed and culture scaffolds ex vivo with cells before implantation. However, many of the methods presented here have been discovered from in vitro cell cultures and might as well be of high importance in these strategies.

2. Scaffolds with a biomimetic composition

Three dimensional biopolymer networks are major players in physiological wound healing. It is assumed that wound healing progresses through three stages: (i) blood clotting and inflammation; (ii) granulation tissue; (iii) tissue remodeling. Each of these stages can be associated to the prevalence of a particular biopolymer network, i.e. fibrin (stage i), unstructured collagen/glycosaminoglycan (GAG, for an acronyms list refer to Section 6) extracellular matrix (stage ii), structured collagen/GAG extracellular matrix (stage iii) (Braiman-Wiksman et al., 2007). The delicate choreography of deposition and remodeling of these biopolymer networks is coupled to the activation of relevant signaling cascades, directing immune cell invasion, inflammatory processes, progenitor cell attraction, differentiation, extracellular matrix (ECM) deposition, ECM remodeling and revascularization of the newly formed tissue graft (Gillitzer & Goebeler, 2001). These critical processes in wound healing underline the importance of fibrin, collagens and GAGs in tissue formation and fundament their recognition as bioactive polymers for tissue engineering. Besides, this biomimetic approach has also suggested the potential of reprocessed tissues for tissue/organ engineering. Reprocessed tissues are complex, structured scaffolds prepared by performing different extraction treatments (i.e. demineralization, decellularization) to biological tissues.

2.1 Fibrin

Fibrin is a biopolymer derived from the crosslinking of fibrinogen, the major protein involved in blood coagulation. As a medical product, fibrin gel is a blood derivative, typically presented as a two syringe-system with a fibrinogen solution and a thrombin solution having a common port. Once casted, fibrinogen is cleaved by thrombin, resulting in the formation of a gel in a few seconds. In Europe, fibrin glue has been available as tissue adhesive, sealant and hemostat from the 1980's. Its approval by the FDA was delayed until 1998 due to security standards regarding blood derivative processing. Fibrin is now regularly used in surgery around the world for a broad number of indications (Spotnitz, 2010).

Fostered by its success in surgery, fibrin has gained increased attention in tissue engineering strategies. Fibrin has the great advantage that it can be delivered by minimally-invasive methods to the tissue defect, a property that is only shared with a few materials capable of gelling in situ (Kretlow et al., 2007). Due to this controlled gelling capacity, fibrin glue can be casted in defined shapes and be applied to fill wound beds. For cell delivery, fibrin glue excels in many ways, as cells can be injected with fibrinogen solution and are then retained at the wound site as the fibrin gels. As fibrin has been designed by nature as the emergency "physiological scaffold" used in wound healing, fibrin gels are well structured to accommodate cells (T.A.E. Ahmed et al., 2008). As a cell delivery matrix, fibrin glue has been used with mesenchymal stem cells (MSCs) for bone (Dozza et al., 2011; Liao et al., 2011), cartilage (T.A. Ahmed et al., 2011; J.S. Park et al., 2011), ligament and tendon regeneration (Lim et al., 2004; Soon et al., 2007). It has also been used with chondrocytes in cartilage regeneration (Eyrich et al., 2007; S.J. Lee et al., 2011; Peretti et al., 2006; Singh et al., 2011), with osteoblasts for bone regeneration (S.J Kim et al., 2007), and fibroblasts in tendon regeneration (Chun et al., 2003).

Fibrin glue can also be used as a drug delivery device for growth factors (GFs), either in combination with cell transplantation or not. Several studies have shown that fibrin glue can provide sustained release of GFs during a few days to one week (Spicer & Mikos, 2010). Longer GF retention times can be achieved by modifications in the GF (Schmoekel et al., 2004), by integrating high GF-affinity moieties in the fibrin chains (Merritt et al., 2010), or by covalent linking of the GFs (Drinnan et al., 2010; Schmoekel et al., 2005). Examples of GFs that have been integrated in fibrin gels in skeletal tissue engineering applications include BMP-2 (Kang et al., 2011; Schmoekel et al., 2004; Schmoekel et al., 2005), TGF-β1 and TGF-β3 (Drinnan et al., 2010; J.S. Park et al., 2011; W. Wang et al., 2010b). A similar concept to GF-loaded fibrin is platelet-rich fibrin, a new blood-derivative formed by platelet concentration. Platelet-rich fibrin also comprises fibrinogen for its polymerization in situ, although handling is different as compared to fibrin glue (Dohan et al., 2006). Due to this different processing, platelet-rich fibrin contains a concentrate of GFs that are present in platelets (i.e. PDGF, TGF, IGF, VEGF, EGF), transforming this product in a truly blood-clot biomimetic (Eppley et al., 2004). Platelet-rich fibrin, however, presents some disadvantages compared to fibrin glue encapsulating GFs. These are mainly related to potential variability in GF-content between batches and undefined GF doses.

Besides those indicated for GF delivery, fibrin glue has some important shortcomings. Firstly, fibrin is known to shrink as gelation occurs, and overall, fibrin lacks adequate mechanical properties (Sierra, 1993). Those limitations are particularly critical to skeletal tissue engineering, as many tissues to be regenerated are load bearing. Moreover, fibrin gels are known to degrade in a few days in vivo, due to the activation of plasminogen by the cells, which leads to formation of the serine protease plasmin and to subsequent fibrinolysis (Ye et al., 2000). Again, this short biodegradation time is probably inadequate for most applications in skeletal tissue engineering, as regeneration of these tissues usually takes several weeks or months. To address these limitations, fibrin glue scaffolds can be optimized by using a solid porous scaffold for reinforcement (Schagemann et al., 2010; Z.H. Wang et al., 2010a; W. Wang et al., 2010b), or by the use of enzymatic inhibitors (Fussenegger et al., 2003). Eyrich et al. reported some optimized conditions that lead to fibrin gels that were stable for up to three weeks in vitro (Eyrich et al., 2007).

Despite being a material with a long history of application, fibrin has been studied recently in some advanced tissue engineering strategies for bone regeneration. Dozza et al. showed that MSCs seeded in fibrin scaffolds resulted in better bone formation around an uncemented hip prosthesis and better bone-prosthesis contact compared to the control (Dozza et al., 2011). Liao et al. showed that MSCs seeded in either platelet-rich fibrin or a Medpor® sheet filled with platelet-rich fibrin resulted in approximately 4-fold higher bone formation as compared to empty defects in a mandibular reconstruction model (Liao et al., 2011). Leong et al. confirmed the osteogenic differentiation of MSCs in fibrin glue filled polycaprolactone/tricalcium phosphate scaffolds (Leong et al., 2008).

Fibrin-based devices have also been recently studied for cartilage and ligament/tendon regeneration. MSCs seeded in fibrin glue containing TGF-β3 showed the capacity to regenerate articular cartilage (J.S. Park et al., 2011). Eyrich et al. demonstrated the capacity to form cartilage with chondrocyte-seeded stable fibrin gels (Eyrich et al., 2007). Wegener et al. used polyester scaffolds filled with a fibrin gel (with or without chondrocytes) to repair cartilage defects in the femoral condyle in sheep (Wegener et al., 2010). This study confirmed cartilage formation and improved O'Driscoll scores in defects filled with these scaffolds. Lee et al. formed a cartilage layer by seeding chondrocytes in fibrin gels on top of auricular implants as a strategy to minimize morbidity associated with implant dislodgement (S.J. Lee et al., 2011). Wang et al. used polyester/fibrin gel constructs loaded with MSCs and TGF-β1 to promote restoration of full-thickness cartilage defects (W. Wang et al., 2010b). Also recently, Scotti et al. used fibrin gels to promote cell adhesion to devitalized spongiosa cylinders used for engineering osteochondral grafts (Scotti et al., 2010). In two studies by the laboratory of Lee, MSCs embedded in fibrin glue were able to promote better osseointegration of tendon grafts (Lim et al., 2004; Soon et al., 2007).

A clinical study on mandibular degree II furcation defects has shown that patients treated with platelet-rich fibrin glue and an open flap debridement presented statistical improvements in all clinical parameters as compared to patients treated with just open flap debridement (Pradeep & Sharma, 2011). In another clinical study, MSCs in platelet-rich fibrin glue were implanted in patients with condyle defects for cartilage regeneration (Haleem et al., 2010). The study showed improvement of patient's symptoms in all subjects over the follow-up period of 12 months, achieving nearly normal arthroscopic scores after the follow-up and partial to total cartilage surface regeneration.

2.2 Collagen

Collagen is a family of proteins that are the main components of connective tissue, and the most abundant proteins in mammals (>90% of the ECM) (Van der Rest & Garrone, 1991). Collagen type-I is a major component of scar tissue, while this and other types of collagens are present in physiological tissue (and regenerated tissue if correctly remodeled) (Fratzl 2008). Specifically in skeletal tissues, collagen type-I is preeminent in bone, tendons and muscle. Collagens in cartilage change with specific tissue types: collagen type-II is preeminent in hyaline cartilage, collagens type-I and –II are present in fibrocartilage and collagen type-X is a marker of hyperthrophic cartilage. Collagen type-I is a long, stiff, triple-stranded helical structure comprising two identical chains (i.e. one α1(I)-chain and one α2(I)-chain). This α2(I)-chain contains the sequence repeat (G-X-Y)n, X being frequently proline and Y hydroxyproline, that allow the formation of a triple helix (Ricard-Blum & Ruggiero, 2005). In physiological tissues, collagens show very defined structures. For instance, collagen type-I forms parallel, longitudinal fibers in tendons and ligaments and long bones. In lamellar bone, collagen type-I fibrils run in parallel to other fibrils in the same layer, but are aligned in the opposite direction in alternating layers (Fratzl 2008). In cartilage, collagen type-II is crosslinked to proteoglycans by the action of collagen type-IX that acts as a biological "spacer" (Fratzl 2008; Lodish, H et al., 2002).

As materials for tissue scaffolds, collagens have been thoroughly investigated due to their biomimetic nature, their biocompatibility and abundance. Indeed, collagen type-I has shown to be degradable, and highly biocompatible, particularly after proteolytic removal of telopeptides (Glowacki & S. Mizuno, 2008). Currently, collagen is FDA approved for implantation. Collagen scaffolds can be prepared in two forms: as hydrogels or as fibers in a mesh like structure (also named as "collagen sponges" or "collagen foams"). Collagen fibers can be reinforced by crosslinking, a modification that can be performed with either physical or chemical methods (Glowacki & S. Mizuno, 2008). Some chemical crosslinkers such as glutaraldehyde have been related to cytotoxicity (Middelkoop et al., 1995), although others have shown that this limitation can be overcome by protocol optimization (Nimni et al., 1987). Today, other more biocompatible crosslinkers can be selected (Kubow et al., 2009; Mekhail et al., 2010). Collagen fibers manufactured by optimized protocols present an adequate porous microstructure and promote cell adhesion (Glowacki & S. Mizuno, 2008). In skeletal tissue engineering, collagen scaffolds have been used for seeding osteoblasts (Laflamme & Rouabhia, 2008), periosteal cells (Ryu et al., 2010), MSCs (Brady et al., 2010; Dozza et al., 2011; Jäger et al., 2008), chondrocytes (Oliveira et al., 2010; Zhou et al., 2011) and fibroblasts (George et al., 2008; Mekhail et al., 2010). Collagen type-I has shown the capacity to specifically upregulate osteoblastic function and bone formation (Lynch et al., 1995), and this property has been linked to the interaction of this collagen with integrin $\alpha2\beta1$ (M. Mizuno et al., 2000). On the other hand, collagen type-II scaffolds have shown favorable interactions for cartilage tissue engineering and cartilage regeneration, particularly in combination with GAGs (Chang et al., 2007; Wu et al., 2010). These results highlight, once again, the importance of providing the cells with scaffold compositions that mimic their natural environment.

To enhance their regenerative properties even further, collagen scaffolds can be loaded with GFs such as BMPs (Laflamme & Rouabhia, 2008), TGF (Pabbruwe et al., 2010) and FGF (Pang et al., 2010). Collagen-based sponge constructs be impregnated in BMP-2 are already available as a FDA-approved commercial product for spinal reconstruction (Infuse™, Medtronic, Inc.). Nevertheless, collagen has rather limited controlled release capacity, with most occurring in the first days (Maeda et al., 1999). Release properties can be improved to a limited extent by crosslinking (Fujioka et al., 1998) or drastically, by embedding in the scaffold a suitable drug delivery device (Fujioka et al., 1998; J.E. Lee et al., 2004).

Apart from its limited capacity to control drug release, the main limitations of collagen scaffolds are their poor mechanical properties. Indeed, collagen scaffolds show no rigidity and shrink due to the cohesive forces that appear as ECM is deposited. As with fibrin, this shrinking is observed already in vitro, and clearly show the intrinsic limitations of this material to reconstruct tissue constructs with defined shapes: ultimately, one of the main objectives of scaffold implantation. Besides, this lack of mechanical competence might compromise the integrity of the scaffold upon implantation resulting in implant failure, non-union or even ectopic tissue formation. Again, it needs to be considered that for most skeletal tissue engineering applications, scaffolds will be exposed to considerable mechanical stress. The mechanical properties of collagen scaffolds can be improved by the integration of a reinforcing substructure (Zhou et al., 2011) or using composites of collagen with mechanically competent materials (Lee et al., 2004). Apart from this limitation, collagen scaffolds might lead to some minor immunological problems upon implantation if the telopeptides are not correctly removed from the materials by proteolysis (S. Mizuno & Glowacki, 1996).

Due to their many intrinsic advantages, collagen scaffolds have been applied to the regeneration of all skeletal tissues. Zhang et al. studied chitosan/collagen scaffolds bioactivated with an adenovirus carrying a BMP-7 transgene and seeded with periodontal ligament cells in dental implant defects (Y. Zhang et al., 2007). Histomorphological and qRT-PCR analysis were consistent with more bone formation and higher expression of bone-matrix markers (alkaline phosphatase, osteopontin, bone sialoprotein). Xu et al. studied collagen type-I scaffolds integrating bone sialoprotein for their capacity to induce osteoblast differentiation, and found that this treatment resulted in early mineral deposition by proliferating repair-cells in calvarial defects. Pabburke et al. implanted collagen-based constructs seeded with MSCs in sheep meniscal disks, and confirmed by histomorphometric analysis that they improve integration compared to the scaffolds alone and the control (Pabbruwe et al., 2009; Pabbruwe et al., 2010). Zhou et al. showed that substructure-reinforced collagen matrixes seeded with chondrocytes supported neocartilage formation (Zhou et al., 2011). Chen et al. tested the potential of genipin-crosslinked collagen type-II scaffolds seeded with MSCs for cartilage repair in an osteochondral defect and observed that after 2 months, there were clear signs of cartilage formation in the defect (W.C. Chen et al., 2011). After 24 weeks, cartilage in the defect presented the same structure as native cartilage. A recent study has confirmed the efficacy of knitted silk-collagen sponge scaffolds as support for embryonic stem cell-derived MSCs in tendon regeneration (J.L. Chen et al., 2010).

Being one of the first and most widely used materials studied for tissue engineering, collagen has already been investigated in several clinical studies (Rodkey et al., 1999; Stone et al., 1997) that have ultimately resulted in collagen regulatory approval. Intense clinical research continues, as new indications for collagen-based scaffold technologies are sought and long-term clinical outcomes are reevaluated. In a recent study, Crawford et al. studied the safety and healing potential of commercial cartilage matrix seeded with autologous chondrocytes (NeoCart) in the human knee (Crawford et al., 2009). The study revealed significantly reduced pain in the treatment group, with neocartilage formation that was well integrated. Welsch et al. compared cartilage repair in the femoral condyle after autologous chondrocyte transplantation in a hyaluronan-based scaffold or a collagen-based scaffold (Welsch et al., 2010). Results based on morphological scores and T2-mapping showed that most parameters were similar for both scaffolds, however, collagen showed significantly better constitution of the surface and higher T2-relaxation times.

2.3 Glycosaminoglycans

Glycosaminoglycans (GAGs) are linear polysaccharides consisting of repeating hexuronic acid bonded to a hexosamine (Iozzo, 2000). GAGs are usually sulfated and linked to a protein domain (a structure that is named proteoglycans). Hyaluronan is exceptional in this regard as it is the only kind of GAG that is neither modified by sulfonation nor attached covalently to proteins. Nevertheless, hyaluronan does attach by non-covalent interactions to other molecules of the ECM and to cells. Other important GAGs are dermatan sulfate, chondroitin sulfate, heparin, heparan sulfate and keratan sulfate. GAGs have important physiological functions as "space fillers", in shock absorption, in preventing coagulation, and in GF binding. Studies have also shown the importance of GAGs in embryogenesis and tissue regeneration (Fedarko et al., 1992; Toole, 2001). However, the complete functionality of GAGs is still a topic requiring further investigation.

Unmodified GAGs tend to form polymer solutions or gels that dissolve slowly due to the high molecular weight of the components. This dissolution of the matrices together with their poor mechanical properties are unsuitable for tissue engineering. Therefore, GAGs are most often not used by themselves to form scaffolds, but rather as composites (Garcia-Fuentes et al., 2008; Min et al., 2010; Moss et al., 2010; S.J. Park et al., 2009; Schagemann et al., 2010). This role is even more prominent for heparin and heparan sulfate, which are typically added in very small proportions as GF-complexing agents in drug delivery formulations (Biondi et al., 2008; N.X. Wang & von Recum, 2011). Alternatively, GAGs can be chemically modified as a more insoluble compound (e.g. Hyaff®, Fidia Farmaceutici s.p.a) or as crosslinked networks (Crescenzi et al., 2003; Zheng Shu et al., 2004). In skeletal tissue engineering studies, GAG-containing scaffolds have been seeded with osteoblast-like cells (Wagner et al., 2007), chondrocytes (Nuernberger et al., 2011; Schagemann et al., 2010), MSCs (J.L Chen et al., 2010; Min et al., 2010; Schwartz et al., 2011), and fibroblasts (Irie et al., 2011; Sawaguchi et al., 2010). Hyaluronan interacts specifically with cells through CD44, RHAMM and ICAM-1 (Tammi et al., 2002; Toole, 2004). These interactions trigger signaling cascades that could result in the observed capacity of hyaluronan to instruct tissue neo-formation and regeneration. For illustration, hyaluronan has shown pro-proliferative properties (Zou et al., 2004), capacity to enhance unspecific ECM deposition (Garcia-Fuentes et al., 2009), to induce cartilage (Allemann et al., 2001; Williams et al., 2003; Yamane et al., 2005) and ligament (Cristino et al., 2005) formation, capacity to maintain embryonic stem cells in undifferentiated state (Gerecht et al., 2007), and capacity to modulate inflammatory and catabolic markers (Grigolo et al., 2005; Homandberg et al., 2004). Hyaluronan biological properties, though, seem to be very dependent on its molecular weight, and the presence of hyaluronan oligosaccharides have been connected to chondrocyte-driven chondrolysis (Knudson et al., 2000).

GAG-based or GAG-modified scaffolds have also been applied with GFs for sustained delivery. Again, this mimics a physiological process, as sulfated GAGs are GF reservoirs in the ECM. For most GFs, GAG binding occurs through specific non-covalent interactions between a so-called heparin binding domain in the polypeptide and the sulfated GAG (Iozzo, 2000). Modification of drug delivery systems with sulfated GAGs for specific binding of GFs is a usual technological strategy, commonly practiced as a surface modification (i.e. "surface heparinization") (Biondi et al., 2008; N.X. Wang & von Recum, 2011). An alternative but still rather unexplored strategy is the design of delivery systems based on crosslinked sulfated GAGs for GF-delivery applications. Drug delivery devices modified with sulfated GAGs have been applied to the delivery of TGF-β1 (Chou et al., 2006; Lee et al., 2004) and FGF-2 (Mi et al., 2006) among others. Heparin and other sulfated GAGs known to interact with GFs can enhance markedly the affinity of these molecules for the drug delivery system and change the release profile of the device, as reviewed before (N.X. Wang & von Recum, 2011). In general terms, when sulfated GAGs are integrated in the composition of a hydrogel, they show high affinity for the GF and result in sustained release profiles (Benoit & Anseth, 2005b). In the work of Cai et al. (Cai et al., 2005), FGF-2 was released from several crosslinked GAG-based hydrogels in a sustained way for over a month. Moreover, it was found that hydrogels presenting immobilized heparin further enhanced the retention of FGF-2. Interestingly, the bioactivity of the released GF was confirmed both in the study of Benoit et al. (Benoit & Anseth, 2005b) and in that from Cai et al. (Cai et al., 2005).

GAG-based materials, particularly hyaluronan, have been extensively investigated for skeletal tissue regeneration. Patterson et al. prepared hyaluronan hydrogels with different degradation rates by photo-crosslinking to control the release of BMP-2 in a rat calvarial defect model (Patterson et al., 2010). They showed that tissue regeneration could be achieved with any of the BMP-2 loaded hydrogels tested. However, the fastest and the slowest degrading hydrogels resulted in more organized bone structures. Kang et al. investigated the possibility of coating porous polyester scaffolds with a fibrin/hyaluronan matrix loaded with BMP-2, and the effect of such constructs in MSCs (Kang et al., 2011). The results showed that MSCs were differentiated more efficiently and resulted in better bone formation when BMP-2 was sustainably released from the fibrin/hyaluronan constructs as compared to BMP-2 in solution. Chen et al. showed that polyester scaffolds coated with hyaluronan/methylated collagen were able to enhance bone formation by MSCs as compared to non-coated polyester systems (M. Chen, et al., 2010). In the study published by Irie et al., chitosan/hyaluronan fibers seeded with fibroblasts were tested for rabbit medial collateral ligament reconstruction (Irie et al., 2011). The in vivo test indicated that ligaments treated with cell-seeded scaffolds showed enhanced collagen type-I deposition and mechanical strength as compared to non cell-seeded ones.

Cartilage and osteochondral structures have been the main focus of attention for GAG-based scaffolds, as GAGs are particularly important components of cartilage ECM. Tan et al. showed the potential of new oxidized hyaluronan/N-succinyl chitosan systems that are able to gel in situ for the delivery of chondrocytes (Tan et al., 2009). Eggelet et al. tested a cell-free polyester/hyaluronan scaffold to cover microfractured full-thickness articular cartilage defects in sheep, and achieved significant improvements in histological structure, and in collagen type-II content as detected by immunohistochemistry (Erggelet et al., 2009). Im et al. designed a hyaluronan-atelocollagen/hydroxyapatite-tricalcium phosphate composite scaffold for testing in osteochondral defects in minipigs (Im et al., 2010). Clinical scoring, histological and mechanical results showed comparable results for this composite scaffold with or without seeded chondrocytes and for defects refilled with the removed osteochondral pieces. Moreover, all these groups showed significantly higher performance than the negative control (empty defects), but comparable in many aspects to native cartilage. In the studies of Moss et al. and Huang et al., GAG-based scaffolds were used for reconstitution of the nucleus pulposus of the intervertebral disc (Huang et al., 2011; Moss et al., 2010). The results of these studies sustained the possibility to maintain disc height and induce intervertebral disc regeneration upon implantation of these scaffolds. Interestingly, the study of Moss et al. compared hyaluronan scaffolds with a composite of hyaluronan and an elastin-based peptide (Moss et al., 2010). Although the composites presented better mechanical properties, this improvement did not translate in enhanced regeneration.

Nehrer et al. and Welsch et al. have published clinical evaluations of hyaluronan scaffolds seeded with chonodrocytes for cartilage regeneration (Nehrer et al., 2009; Welsch et al., 2010). The study by Nehrer supported the interest of Hyalograft C for cartilage regeneration in healthy young patients with isolated cartilage defects, but not for salvage procedures

(Nehrer et al., 2009). Similarly positive results were obtained by Welsch et al. for condyle regeneration, although this study showed some slight advantages of collagen-based scaffolds compared to the hyaluronan-based ones.

2.4 Reprocessed tissue scaffolds

From the point of view of biomimetism, reprocessed tissue scaffolds (e.g. demineralized bone matrix and decellularized tissues) represent probably the most advanced scaffolds available. These natural scaffolds recapitulate not only the complex composition, but also the structure of the native tissues (Danti et al., 2007; Traphagen & Yelick, 2009). This cannot be achieved with the scaffolds prepared from extracted ECM compounds such as those presented before, because the processing steps used to extract the biopolymers completely erases the morphology, histological topology, the precise interactions between different components, the molecular alignment, and other subtle cues from the materials. While both demineralized bone matrix (DBM) and decellularized tissue can be considered similar concepts ("reprocessed tissue scaffolds"), their processing methods and characteristics differ and are discussed separately herein.

Demineralized bone matrix is prepared by acid extraction of allograft bone (Urist, 1965). It results in the elimination of the major part of the mineral phase and the immunogenic components of bone, but retains collagen that provides a structured osteoconductive scaffold and a soluble protein fraction comprising several GFs, BMPs among them (Dinopoulos & Giannoudis, 2006; Reddi, 1998). It is also usual to retain a very low fraction of mineral bone phase after the demineralization step (2% approx.). The exact composition of DBM -including the GF content- will depend heavily on the exact protocol followed for mineral extraction, and thus, DBM from different commercial sources might show considerable variability. Extensive studies have shown that DBM is osteoinductive and osteoconductive (Dahners & Jacobs, 1985; Martin Jr et al., 1999). Osteoinduction mechanism by DBM seems to be very dependent on the environment. DBM implanted in submuscular/subcutaneous tissue undergoes ossification through a process recapitulating several steps from the endochondral bone formation process. DBM implanted in calvarial defects direct MSC differentiation to osteoblasts, similarly to the process followed in intramembranous ossification (J. Wang & Glimcher, 1999a, 1999b; J. Wang et al., 2000). DBM by itself is presented in the form of a powder, but is usually mixed with other suitable materials (e.g. hyaluronan, poloxamer) to form a paste that can be easily casted at the bone defect site (Dinopoulos & Giannoudis, 2006; Juang Ming Yee et al., 2003). Mixing DBM with other materials, however, changes the product from a regulatory perspective.

DBM can be supplemented with suitable cell types as a strategy to promote tissue regeneration. For instance, some articles report on seeding chondroblasts in DBM as a potential alternative in cartilage tissue engineering (Jin et al., 2006; Z.H. Wang et al., 2010a). A large number of studies have also investigated MSC-seeded DBM (e.g. see Gurevitch et al., 2003; Liu et al., 2010a). DBM can also be supplemented with additional GFs in an attempt to boost their regenerative capacity. DBM has been mainly loaded with BMPs (Lammens et al., 2009) -and particularly, with BMP-2- for bone tissue engineering (Lammens et al., 2009; H. Lin et al., 2008). DBM has also been loaded with other GFs such as TGF-β1 (Moxham et al., 2009). Overall, DBM presents very little capacity to bind additional GFs, and therefore is not an ideal choice as a sustained delivery system. Modification, for instance by heparin addition, can increase GF binding affinity. First DBMs were introduced to the clinics during the 1990's, and there are currently more than 20 commercial products available based on this technology. Therefore, there is currently large clinical evidence supporting their efficacy for several orthopedic conditions (Irinakis, 2011; Y.K. Kim et al., 2010; Topuz et al., 2009).

Decellularized tissue scaffolds are prepared by cell lysis, induced by physical and/or chemical treatments, followed by cell component removal treatments. Mostly, chemical methods are used for this last step (Gilbert et al., 2006). Similarly to DBM, decellularized tissues present the biomimetic topology, tissue-conductive capacity and regeneration-inductive properties arising from the presence of physiological amounts of GFs in the ECM. However, these beneficial characteristics are heavily dependent on the processing protocol followed for decellularization (Gilbert et al., 2006; Ingram et al., 2007; Woods & Gratzer, 2005). Moreover, this protocol might also influence other critical parameters such as the decellularization efficacy and the complete removal of chemical agents from the natural scaffold. Although decellularized tissues and organs are being investigated in several tissue engineering applications, they have not been so popular for skeletal tissue engineering, as scientist have focused on these technologies for engineering high-complexity, localized organs such as the heart (Ott et al., 2008). That is except for DBM, which can be considered also a decellularized tissue scaffold. Some articles have supported the concept of decellularized cartilage as a chondroconductive matrix in cartilage tissue engineering (Elder et al., 2010; Gong et al., 2011; Hou et al., 2011; Secretan et al., 2010; Stabile et al., 2010). Indeed, in vitro tests have shown the possibility to form cartilage tissue in vitro by chondrocyte seeding on a decellularized tissue scaffold (Gong et al., 2011). This ability to support cartilage formation translates to advanced in vivo models such as a for larynx reconstruction in a rabbit model (Hou et al., 2011), or for meniscus reconstruction in sheep (Stabile et al., 2010). First proof-of-concept of the therapeutic interest of this concept in humans is available through the successful transplantation of an MSC-seeded decellularized airway in a 30-years old patient with bronchomalacia (Macchiarini et al., 2008). Decellularized tissues have also been applied to ligament and tendon tissue engineering in vitro, with successful results (Abousleiman et al., 2009; Deeken et al., 2011; Ingram et al., 2007; MacLean & Gratzer, 2011; Woods & Gratzer, 2005). Further animal experiments and clinical studies will be necessary to validate the potential of this strategy. Similarly, other studies have indicated the possibility to use decellularized tissue scaffolds also for skeletal muscle repair and adipose-tissue engineering (L. Flynn et al., 2007; L.E. Flynn, 2010; Gillies et al., 2010; Merritt et al., 2010).

3. Strategies for scaffold bioactivation

As compared to nature's derived biopolymers, synthetic materials for tissue engineering offer important advantages such as more reproducible manufacture, tailored mechanical and biodegradation properties, and diverse processing options. With a few exceptions (e.g. titanium) synthetic materials do not interact favorably with the body, rendering them incapable of actively promoting tissue regeneration. Despite this limitation, synthetic biomaterials can be engineered to promote their capacity to cross-talk with relevant cell populations. Indeed, bioactive scaffolds can also be prepared from synthetic materials by physical adsorption or chemical immobilization of biomolecules or oligopeptides on the scaffold surface, or by physical entrapment of bioactive molecules alone or incorporated in a drug delivery system into the scaffold. These strategies can also be applied to enhance the bioactivity of scaffolds made from ECM-native materials such as those covered in the previous section.

3.1 Scaffolds as drug delivery systems

Engineered tissues need not only to fill a defect and to integrate into a host tissue, but they also need to meet the demands of a constantly changing - through growth or adaptation - tissue. It was hypothesized that those tissues capable of growing with time could be engineered by supplying growth stimulus signals to cells from the biomaterial used for cell transplantation (Alsberg et al., 2002). Smart drug delivery systems might not only be able to transmit one single signal to the cells, but multiple signals in concert in a timely controlled release pattern. This release may be controlled through properties of the drug delivery system itself: examples of those are biodegradation-controlled release devices or stimuli-sensitive systems (polymeric networks that react to changes in pH, temperature or to the presence of enzymes). In a further degree of sophistication, drug release can also be triggered through external physical stimuli such as ultrasounds or electrical fields (Deckers et al., 2008; Sirivisoot et al., 2011). Another method of remote drug delivery relies on magnetic nanoparticles and thermally sensitive vesicles (de Cogan et al., 2011; Mart et al., 2009a; Mart et al., 2009b). In these devices, heat originating from magnetic stimulation of nanoparticles induces swelling of the themosensitive polymer and the subsequent drug release.

In tissue engineering, polymeric matrices used as scaffolds can double as drug delivery systems, a strategy that has been mainly used for soluble signalling molecules such as GFs (e.g. BMPs or TGF-β). GFs and cytokines have shown to be potent inducers of cell migration, proliferation, differentiation and new tissue formation. Many of them are also important morphogens in embryogenesis and natural wound healing through their action on stem cells. Therefore, there are large expectations set on GF therapeutic efficacy, particularly in combination with stem-cell-based therapies. Besides, many researchers believe that through the chemoattractant properties of these molecules, drug delivery devices will be able to mobilize the resident stem cell pool from the injured host, making reiterative any strategy based on stem cell supplementation. Cell recruitment and migration to the site of injury may be promoted through various signaling molecules. Many of these factors, e.g. TGF-βs, BMPs and IGF-1, are not only involved in cell attraction but also affect cell proliferation and differentiation (Lieberman et al., 2002; Reddi, 2001a; Reddi, 2001b; Sundelacruz & Kaplan, 2009).

Growth factors and cytokines - although very potent - usually present short half-lives, especially in physiological media. Also, their potential is highly dependent on their spatial and temporal action pattern, and a change in local concentration may have a huge effect (Uebersax et al., 2009). It follows that systemic administration of GFs is inherently inefficient, since it results in fast GF elimination and indiscriminate biodistribution. Ultimately, the therapeutic concentration of the GF may never be reached at the site of injury. To overcome this, drug delivery strategies are designed to (i) provide a platform for the localized delivery of the GF at the site of implantation creating a favorable concentration gradient, (ii) to protect the bioactivity of the molecule and (iii) to provide a controlled release pattern of the drug over a desired time frame. To add a further layer complexity, in developmental pathways, different factors become active at different times, and GF release profiles that recapitulate these dynamics are likely to provide more leverage over cell behavior than those that apply these signals indiscriminately (Place et al., 2009). This is a central dilemma in tissue regeneration: on one hand, biomimetic platforms integrating complex release behavior are required for an optimal therapeutic outcome, but on the other hand, engineering needs to be kept to a reasonable level to provide technologically-feasible medical solutions (see section 4.2).

Two approaches have been mainly used for scaffold bioactivation: GFs can be encapsulated in a selected drug delivery system such as a microsphere or nanoparticle formulation, and these can be incorporated into the scaffolds. Alternatively, GFs might be incorporated directly into the scaffold itself (Holland & Mikos, 2006; Holland et al., 2007; Liu et al., 2010b; Luginbuehl et al., 2004; Shi et al., 2011; Uebersax et al., 2009). For example, IGF-1 has been incorporated into biodegradable poly(lactide-co-glycolide) microspheres and used as a treatment for 10-mm segmental tibial defects in sheep (Meinel et al., 2003). The application of 100 μg IGF-1 per defect resulted in the downregulation of inflammatory marker genes at the site of injury and the induction of new bone formation that bridged the defect within 8 weeks. Uebersax et al. explored the alternative approach, where IGF-1 was directly incorporated into porous 3D silk fibroin scaffolds (Uebersax et al., 2008). Silk scaffolds incorporating IGF-1 were able to preserve GF bioactivity, and prompted chondrogenic stimuli to seeded MSCs in vitro. By definition, implantation of GF-loaded scaffolds results in the localized delivery of the signaling molecule. Still, a certain fraction of the incorporated drug can reach the lymphatics or the circulation and distribute to non-target tissues. Therefore, even for these localized therapies, potential adverse effects of GF need to be carefully monitored. For example IGF-1 has been shown to be involved in diabetes or in the development of colorectal and breast cancers (Dunger et al., 2005; Putney & Burke, 1998; Sandhu et al., 2002).

Controlled or sustained release patterns may also be achieved through non-covalent association with matrix components, for example, with GAGs (section 2.3) (L. Zhang et al., 2006). In vivo, GAGs have critical roles in the regulation of GF activity. This includes sequestering GFs from the tissue liquid phase, acting as GF reservoir, preventing their degradation and presenting them to cell-surface receptors (Place et al., 2009). Regiospecific sulfation patterns that enable specific interactions between GFs and GAGs have been introduced in other biomaterials that are not natively from the ECM. This modification resulted in specific GF-scaffold bindings. For example, this has been done in alginate hydrogels and in silk fibroin scaffolds that have been decorated with sulfonated moieties capable of influencing the release of GFs such as FGF-2 (Freeman et al., 2008; Wenk et al., 2010).

Release upon cell demand is mainly based on protease-sensitive polymer modifications. For this strategy, peptide sequences that can be cleaved by proteases are introduced to hydrogel crosslinking groups, and the GF is loaded in this hydrogel. In the absence of proteases, the small mesh size of the hydrogels prevents the GF from being released. In their presence, proteolytic activity cleaves the crosslinks, increases the mesh size of the hydrogels resulting in the GF diffusion out of the polymeric mesh. This is again a biomimetic concept, since physiologically, the ECM not only provides structural and biochemical cues for cells in contact, but can also promote cell invasion and remodeling upon cell-triggered proteolysis, a process that ultimately results in tissue regeneration. Lutolf et al. used synthetic poly(ethylene glycol) (PEG) hydrogels containing crosslinking matrix metalloproteinase substrates that can undergo cell-mediated proteolytic degradation (Lutolf et al., 2003a). Upon degradation of the matrix, entrapped BMP-2 was released and bone regeneration was achieved within 5 weeks in critical size cranial defects in rats (Lutolf et al., 2003b). This effect was highly reduced when BMP-2 was entrapped into matrix metalloproteinase-insensitive hydrogels.

All drug release strategies mentioned above may also be combined resulting in even more complex strategies. For example Haberstroh et al. showed that human intervertebral disc-derived nucleus pulposus cell migration can be promoted through a serum fraction gradient in a concentration dependent manner and that both TGF-β3 and hyaluronan were able to promote cell differentiation and matrix generation (Haberstroh et al., 2009). They concluded that a bioactive scaffold containing serum and TGF-β3 or hyaluronan might be an excellent candidate for cell-free biological treatment strategies in preventive and curative approaches for degenerative disc disease.

There are other complex drug delivery strategies for tissue engineering that have not been covered in this manuscript. For example cells can be used as drug delivery systems. Especially immunocytes that are mobile and can migrate across impermeable barriers can be exploited as trojan horses for drug delivery (Batrakova et al., 2011). Applied cells may also be genetically modified in order to support immunomodulating, homing or paracrine activities at the implant (Myers et al., 2010). Besides, scaffolds can also be used for gene delivery, together with gene nanocarriers or by themselves. This last concept is usually referred as "gene activated matrices" and has attracted some attention for their use for skeletal tissue regeneration (Geiger et al., 2005; Guo et al., 2006). Besides this alternative methodologies based on genetically engineered cells and gene therapy, it is important to bear in mind that not only physiological factors can influence cell differentiation and tissue regeneration. Synthetic small molecular drugs such as dexamethasone, β-glycerophosphate or ions such as strontium released from bioactive glasses have also been shown to influence cell proliferation and differentiation (Chung et al., 1992; Isaac et al., 2011). Indeed, the potential of small molecular weight drugs for tissue engineering has been dramatically demonstrated by their capacity to markedly enhance cell-reprogramming efficiency, and even substitute many of the transgenes used in these processes (Huangfu et al., 2008).

3.2 Scaffolds bound to biological functionalities

In section 2, we highlighted the capacity of purified ECM compound scaffolds to provide sites for cell attachment and cell-substrate crosstalk (Fig.1). These scaffolds exert bioactivity because they bear complex information, coded in their physical and chemical structures. Unfortunately, scaffolds fully made out of natural ECM molecules present several limitations related to their difficult purification and processing and their suboptimal mechanical properties (Place et al., 2009). These limitations have broadened the range of scaffold materials investigated towards synthetic scaffolds that are functionalized with bioactive moieties.

Most cell receptors do not require interacting with the whole ligand for their activation. Often, a relatively short peptide fragment of about 40 amino acids is sufficient to provide the desired signal to the cells. Indeed, it has been shown that small peptides can activate specific cellular pathways: binding their respective receptors, triggering signal transduction and leading to appropriate cells responses (Chung & T.G. Park, 2007; Place et al., 2009). The required peptide sequences can either be obtained from biological sources or can be chemically synthesized (Sreejalekshmi & Nair, 2011). As these small fragments are easier to produce synthetically and usually show quick refolding under physiological conditions, they might represent a more cost-effective, easier to manipulate, alternative to GFs or other signaling proteins. Moreover, they have been shown to have a higher stability against conformational change, easy controllability of surface density and orientation, and more favorable ligand-receptor interaction (Chung & T.G. Park, 2007; Hersel et al., 2003; Lutolf et al., 2003b; Ruoslahti, 1996). Sreejalekshmi et al. has recently reviewed peptide-modified scaffolds, a manuscript that includes a decision-tree-type flow chart indicating probable cellular outcomes resulting from a given modification (Sreejalekshmi & Nair, 2011).

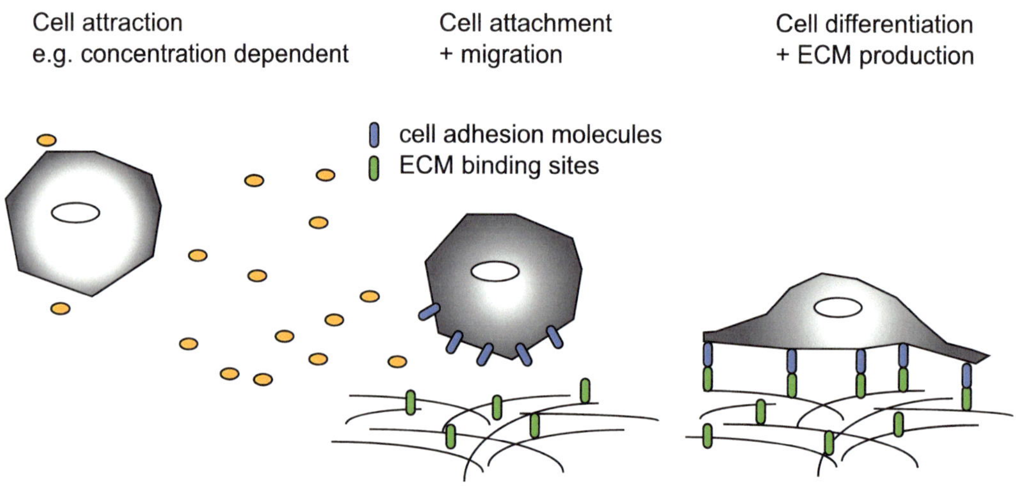

Fig. 1. Scaffolds as devices capable to activate the host stem cell pool. Growth factors activate the stem cell pool and induce their migration to the scaffold. Stem cells adhere to the scaffold by integrin binding and undergo GF/scaffold-directed differentiation

3.2.1 Fibronectin and small cell adhesion motifs

Cell adhesion is a prerequisite to the success of scaffold-based tissue engineering strategies. In native tissues, fibronectin is one essential component of the ECM that mediates cell-matrix interaction. Cells can bind to fibronectin through transmembrane receptor proteins of the integrin family, which mechanically interlink the actin cytoskeleton to the ECM through an elaborate adhesion complex. The binding process between ECM and the dimeric integrin receptors on the cells surface is then followed by a cascade of signaling events leading to the up- or downregulation of the expression of several genes. It is important to note that the binding of integrins happens through dimerization of different alpha and beta subunits, and therefore, association with diverse ligands can lead to different effects in the cell. Some specific functions of biopolymers can be attributed to small functional domains, and these may be incorporated in synthetic analogs in place of the full protein (Dunehoo et al., 2006; Schense et al., 2000; Silva et al., 2004). Probably the best-known sequence binding to integrins is arginine-glycine-aspartic acid (RGD). It is found in many ECM proteins, including fibronectin, laminin, collagen type-IV, tenascin and thrombospondin (Benoit & Anseth, 2005a; Comisar et al., 2007; Underwood et al., 1995). However, the RGD motif and its derivatives are not the only integrin-binding sequence used for scaffold modification. For example, polyethylene terephthalate surfaces modified with the cell adhesion motif

GRGDSPC showed improved differentiation of pre-osteoblastic cells as compared to non-grafted surfaces (Zouani et al., 2010). Alginate hydrogels modified with G_4RGDY were able to support the formation of growth-plate-like structures in a co-culture of mouse osteoblasts and chondrocytes (Alsberg et al., 2002). In order to promote cell attachment, an adhesion motif can be simply mixed with the scaffold material or coated on the scaffold surface. Syndecan-binding peptides such as AG73 (RKRLQVQLSIRT) have shown enhanced integrin-mediated biological activities in scaffolds made out of collagen, laminin-11 and fibronectin (Yamada et al., 2011). Morphological analysis indicated that these effects were, at least partially, mediated by enhanced cellular attachment and cell spreading. Attia et al. have shown that polyurethane scaffolds containing an anionic dihydroxy oligomer coated with fibronectin promoted not just cell attachment, but also cell alignment parallel to the scaffold fibers (Attia et al., 2010). This featured scaffold topology was able to increase collagen production after seeding with annulus fibrosus cells. Genetically engineered biopolymers provide another method for the integration of bioactive sequences. Girotti et al. engineered an ECM-analogue of a recombinant multi(bio)functional elastin-like protein polymer with integrated REDV cell adhesion sequences for tissue engineering purposes (Girotti et al., 2004). Cells are known to be very sensitive to the mechanical properties of the substrate (Engler et al., 2006), but also to directional mechanical forces transduced through the ECM (Kurpinski et al., 2006). In skeletal tissues, where mechanical stresses are so important and even necessary for cell stimulation, the integration of adhesion motifs capable of transducing mechanical forces into the right cell signals is of critical importance (Fig. 2).

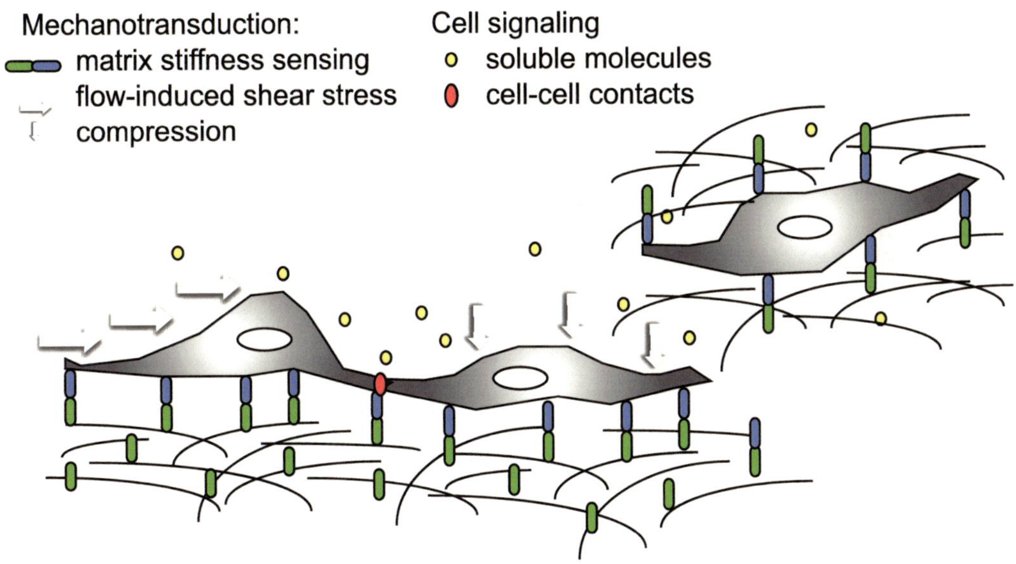

Fig. 2. Cell differentiation and extracellular matrix deposition can be affected by many different signals, e.g. mechanotransduction or cell signaling

3.2.2 Growth factors

Scaffolds may not only be modified with bioactive sequences, whole GFs have been shown to retain their bioactivity upon conjugation to scaffold materials. Through covalent immobilization on the scaffold surface in either random or specific orientation many disadvantages of controlled GF delivery can be circumvented. However, in order to transmit the desired signal to the cells, the moiety must be accessible. This requires the GF to be exposed on the scaffold surface and not incorporated in the polymer core. The application of polymeric spacers such as PEG can provide higher biological activity due to reduced interaction between the GF and the scaffold. For example, collagen matrices were covalently modified with horseradish peroxidase as a model protein with or without PEG as a linker. Introduction of PEG resulted in higher long-term stability of the enzyme (Chen et al., 2002). Recombinant TGF-β2 was either covalently bound through a PEG spacer or admixed to injectable bovine dermal fibrillar collagen (Bentz et al., 1998). PEG-TGF-β2 showed a significantly higher in vitro and in vivo bioactivity than admixed TGF-β2. The authors concluded that covalent binding of TGF-β2 by their method allowed to preserve the GF's biological activity in vitro, and even potentiated their response in vivo. Karageorgiou et al. have shown that immobilization of BMP-2 on a silk fibroin surface was more efficient in terms of osteogenic differentiation of MSCs in vitro than delivery of the GFs in solution (Karageorgiou et al., 2004). In a follow-up study, they could show that with the same protocol, porous 3D scaffolds modified with BMP-2 implanted into rat critical size femoral defects improved bone formation similarly to the same scaffolds seeded with either undifferentiated MSCs or with MSCs predifferentiated for 4 weeks in vitro before implantation (Kirker-Head et al., 2007). Clearly, the protocol used for biofunctionalization is a critical step for these technologies. De Mel et al. have recently published a review on biofunctionalization of biomaterials for accelerated in situ endothelialization (de Mel et al., 2008).

Not only complete GFs can be attached to scaffold surfaces, but also oligopeptides such as those derived from BMP-2 or BMP-7 that have been immobilized on alginate hydrogels and glass, respectively (Kirkwood et al., 2003; Saito et al., 2003; Suzuki et al., 2000). These in vitro studies, performed with a murine multipotent MSC culture or with primary rat calvarial osteoblastic cell populations showed that immobilized oligopeptides may have the capacity to induce osteoblastic differentiation and mineralization in a more predictable manner than the entire GF. This might be attributed to the lower complexity of the molecule and therefore increased stability. Implantation of a BMP-2 derived oligopeptide covalently coupled to alginate into the calf muscle of rats resulted in ectopic bone formation (Suzuki et al., 2000). Synthetic peptide mimics have also been applied for the promotion of angiogenic responses (X. Lin et al., 2006): F2A4-K-NS is a synthetic mimic of FGF-2 that has been shown to trigger signal transduction as monitored by the stimulation of ERK1/2 phosphorylation in human umbilical cord endothelial cells. In cell-based assays, it increased cell migration, cell proliferation and gelatinase secretion. When used in similar quantities, F2A4-K-NS achieved comparable response endpoints to those associated with FGF-2 stimulation.

As mentioned before, wound healing and tissue repair might be stimulated by GF release, but optimal effects are only achieved when several chemical signals are provided in the necessary order and at precise times during the regeneration process. This is probably one of the main disadvantages of GF conjugation to scaffolds, as there is no easy way to provide a time-dependent switch to activate/deactivate the provided signaling.

3.2.3 Other binding moieties: cleavage sequences and oligosaccharide domains

Incorporation of cleavage sequences is an interesting concept to provide on-demand GF delivery (see 3.1), and also to control cell invasion (Place et al., 2009). In this concept, cleavage sequences engineered in multidomain peptides are integrated in the polymeric network of the scaffold. This function mimics the multilayered bioactivity of the ECM, where enzymatic remodeling can liberate "cryptic sites" contained within the ECM proteins. A nice example was provided by Lutolf et al., who rendered a synthetic PEG hydrogel network as amenable to both proteolytic degradation and cell invasion through the integration of integrin binding sites and substrates for matrix metalloproteases (MMP) (Lutolf et al., 2003a). Primary human fibroblasts were demonstrated to proteolytically invade these networks, a process that depended on MMP substrate activity, adhesion ligand concentration and network crosslinking density.

While most ligands that have been tested for tissue engineering so far were proteins, many scientists are turning to other biomolecules of high biological functionality: polysaccharides. Indeed, small oligosaccharide domains are involved in many cell-cell and cell-substrate interactions (Brown et al., 2008; Scaglione et al., 2010). Recent developments in understanding the role of natural polysaccharides and improved chemical synthesis of defined oligosaccharides will potentially result in the discovery several new targets of interest in skeletal tissue engineering.

4. Key challenges

4.1 Engineering large tissue constructs – mechanistic issues

Since the first works showing the possibility to instruct cells to aggregate and form 3D-tissue constructs, tissue engineers have debated between forming tissues and organs de novo or just boosting the body's intrinsic capacity for self-repair. Although in the long run the first of these concepts will probably result in technologies and therapeutic approaches of high medical interests, nowadays the second approach has gathered all the big success stories from a clinical perspective (Place et al., 2009). Devices that enhance tissue regeneration take advantage of environmental cues present at the wound site (i.e. progenitor cells, GFs and tissue ECM present at the wound borders) and accelerate the repair process by establishing a conductive matrix and by providing supplementary simulative factors. These factors can coordinate cell chemokinesis, cell differentiation and induction of tissue-specific ECM formation by committed cells. Many of these technologies in the form of bioactivated scaffolds and drug delivery devices are already a reality (Infuse™, Medtronic, Inc.) and others will become part of the orthopaedist's arsenal in the next few years. A second generation of these technologies could integrate stem cells as a potential opportunity to promote further regeneration. However, it is still unclear whether additional stem cell transplantation results in enhanced regeneration in biomimetic scaffolds (Im et al., 2010). Devices driven by this concept of enhanced regeneration will most likely not be applicable to wounds above a certain size. In those cases, it is easy to argue that regeneration will be limited by nutrient and oxygen diffusion, as well as by the diffusion/migration of pro-regenerative factors (i.e. GFs, cells) (Muschler et al., 2004).

Tissue engineers study in vitro tissue formation as a potential way to generate tissues that could be later implanted in larger defects. While this concept presents several practical disadvantages compared to the most direct approach described above, it is possible that it could result in well-structured tissues that could ultimately be combined in the form of larger constructs. Still, this larger constructs will be deprived of oxygen once formed and implanted, making it critical to induce the quick vascularization of the implanted tissue. In this approach, enhanced tissue construct survival could be achieved by co-implantation of drug delivery devices capable of releasing proangiogenic GFs (Richardson et al., 2001) or by transplanting cells genetically modified to enhance their survival under hypoxic conditions (R.P. Ahmed et al., 2010). Still, conventional in vitro tissue engineering strategies will be unsuitable for very large tissue construct or for regenerating complex tissues/organs.

Some very recent work indicated that through the use of relevant progenitor cells and sophisticated bioreactors, complex and even vascularized tissues might be formed in vitro (Tsigkou et al., 2010). Additionally, other studies have shown that even small organs can be engineered in vitro by culturing cells in reprocessed tissue scaffolds (i.e. decellularized tissues) (Ott et al., 2008). While these advances are really impressive, we must bear in mind that the mechanisms used for tissue/organ neogenesis in vitro are far closer to those involved in tissue regeneration than in physiological organogenesis. In vitro tissue formation occurs by the penetration of committed cell populations into a scaffold, the formation of cell-cell and cell-scaffold interactions and the formation of new tissue. That would also be, simplified, the sequence of events leading to tissue remodeling after cell colonization of bloods clots during wound regeneration. On the other hand, organ morphogenesis is a highly orchestrated process occurring during embryogenesis where the tissues are formed by a sequence of very defined events in a growing construct. In morphogenesis, cells undergo sequential differentiation steps under a very defined spatial and temporal regulation. This regulation is usually defined by gradients of morphogens, but also by cell polarity occurring from integrin-mediated cell-cell interactions (Krasnow, 1997). Further studies will define if in vitro strategies based on tissue regeneration mechanisms such as those tested by Ott et al. and Tsigkou et al. can result in functional, clinically useful tissues and organs.

In any case, knowing which tissue formation mechanism you are trying to recapitulate is important when designing a tissue-engineering strategy. For example, this is pertinent to bone engineering, where most strategies try to induce MSC differentiation to osteoblasts despite of the fact that this process is only physiological in intramembranous bone regeneration. For long bones, the physiological growth and regeneration process is through endochondral ossification. Endochondral ossification is a multistep process where cartilage is initially formed at the defect site, the cartilage becomes hyperthropic, and then it is remodeled into bone by osteoclasts and osteoblasts. Some steps of endochondral bone formation are recapitulated in DBM implanted subcutaneously (J. Wang & Glimcher, 1999a), and this process has also been studied as an advanced tissue engineering strategy in some recent works (Oliveira et al., 2010).

4.2 Engineering a complex environment

Tissue regeneration and tissue neoformation are very complex phenomena. Indeed, the cascade of events after a lesion is formed comprises blood clot formation, inflammation, formation of scar tissue and tissue remodeling. Every one of these processes requires a finely orchestrated cascade of cellular and molecular events. As an illustration, Gerstenfeld et al. have investigated the levels of GFs, other pro-inflammatory cytokines and MMPs occurring during bone regeneration after a traumatic injury (Gerstenfeld et al., 2003), and their studies show the involvement of a myriad of different molecules, present only at specific stages in the process. This complexity is not as surprising: GFs are known to have synergistic effect for many applications as they trigger coordinated effects in their effector cells. Moreover, it is known that all GF effects are very context-specific and concentration-dependent, and therefore, changes in the target cells induced by one GF could drastically change the effect of a second GF (Massagué, 2000). This results in the necessity to design a very complex environment if we want to recreate all this cues in biomimetic scaffolds.

But is this complexity compatible with realistic engineering solutions for tissue regeneration? If something we have learnt from the decline and rise of the tissue engineering field it is the necessity to look for realistic objectives and to avoid overengineered technologies (Place et al., 2009). In short, technologies that are now a success are those that have kept concepts simple. Dealing with the problem of engineering acceptably simple technologies to manipulate a complex process requires a practical approach. First, it is necessary to study how sophisticated your system needs to be for each specific application. For many applications, just the addition of a good tissue-conductive material might suffice. For other applications, this material could be bioactivated with an inducer or seeded with relevant cell populations. Optimization of these strategies could probably rely on high-throughput analysis and systems biology to identify the main signals required for inducing tissue regeneration (Anderson et al., 2004; Anderson et al., 2005; Langer & Tirrell, 2004). For more complex challenges, engineers could turn to reprocessed tissue scaffolds that might be capable of producing some complex tissue structures. Addressing the challenges with technologies that integrate the level of complexity needed to meet the demands of the medical problem addresses complexity from a management perspective. From a technical perspective, many research groups are working in technical solutions that allow implementing complex signaling in the scaffolds with relatively simple chemistries. Examples highlighted in this chapter on the use of small synthetic peptides (Sreejalekshmi & Nair, 2011) or molecules (Ding et al., 2003) and on the formation of material composites including biomimetic compounds might be just some of these simple technologies.

4.3 Controlling stem cell fate and individual variability

New tissue engineering strategies can benefit enormously from the concomitant use of stem cells. However, to fulfill the potential of these technologies, improved control over stem cell fate needs to be exerted, particularly in some applications. For instance, some scaffolds have been designed to maintain stem cells in an undifferentiated state for several passages, thus enhancing their capacity for renovation and cell population growth. Differentiation of stem cells can be achieved by their exposition to defined media. This approach is very efficient for some skeletal tissues (e.g. bone), but leads to suboptimal results for others: hyaline cartilage, skeletal muscles and tendon (Brent et al., 2005; Buckwalter & Brown, 2004). When designing tissue engineering strategies for these applications it is important to consider whether environmental cues from the host tissue might help to improve the differentiation process. Indeed, some experiments have shown tissue repair upon implantation of stem cells despite the questionable ability of these cells to differentiate to these tissues. In cases with modest stem cell contribution to tissue regeneration, bioactive scaffold materials can help to direct stem cell differentiation by introducing important environmental cues: nanotopology (Yim & Leong, 2005), mechanical properties of the supporting scaffold (Engler et al., 2006), or chemical moieties (Anderson et al., 2004). In extremely difficult situations, directed cell differentiation might be addressed by gene therapy approaches (Selvaraj et al., 2010). Controlling stem cell differentiation is critical not only to engineer tissues matching the structure, composition and properties characteristics of native ones, but also to prevent undesirable processes such as orthotopic tissue formation, scar formation and even, potential tumor induction (Amariglio et al., 2009).

Reliance on stem cell therapies raises also the topic of individual variability (Odorico et al., 2001). Stem cells collected from different donors are known to respond very differently to inductive stimuli, to present different capacities for GF secretion and ultimately, to have extremely different regenerative capacities. When designing tissue engineering devices based on stem cells, it would be critical to have quick tests ensuring a minimum of activity, similarly to the approach taken with conventional drugs. However, this concern does not only apply to tissue engineering products containing stem cells, but also to scaffolds and drug delivery devices intended for conduction and induction of a stem cell-mediated responses. A case in point would be the elder or diseased population, where resident stem cell pools might be partially depleted and their full pro-regenerative capacity compromised. The efficacy of tissue engineering devices will need to be tested in these populations, and specific recommendations be issued to take into account this potential variability.
Tissue engineered products based on in vitro cultured stem cells face another important limitation related to their necessary culture time. In vitro, tissues need to be cultured for several weeks and in many cases, this time is not an option for a therapeutic intervention. Even in those cases where this time is available, the postponed surgery will result in an uncomfortable situation for the patient, in increased medical and social costs and in potential worsening of the medical condition. Medical technologies that can be applied immediately out-of-the-box are therefore preferred, and bioengineers should consider this important limitation and the competing therapeutic options when deciding for these technologies.

4.4 Technical challenges

Although encapsulation and controlled release of macromolecules has now been studied for several decades, they are difficult technologies to get right due to the easy denaturation of the drug, to the difficulties to predict release in vivo, and even to potential patient-specific issues that could change the release kinetics of the system. This problem translates to GF-release devices, where potential denaturation of the drug combines with the added concern of their very high manufacturing costs. Growth factor denaturation can occur at three steps during their integration in a drug delivery device: (i) during encapsulation, (ii) upon storage and (iii) upon implantation in the release phase. In practical terms degradation during storage (ii) is infrequent and only applies to formulations in aqueous phases, or with materials that could cross-react with the GFs. For most solid formulations where the matrix is stable for long periods of time, this process should not be a concern. GF degradation during encapsulation (i) is usually a concern, and in most cases, can only be minimized and not completely avoided (Sah, 1999). To prevent GF degradation, exposure to high temperatures should be avoided, and high shearing forces and solvent/water interfaces minimized as possible. In any case, it is possible to enhance GF stability by adding other proteins and protecting groups that block the solvent/water interface. This works particularly well with GFs because of the low loading required that could lead to a high disproportion between the protecting protein/compound and the GF. Growth factor denaturation during the release phase (iii) is a concern for many polymeric systems whose degradation byproducts might react with the proteins. This is best illustrated by polyesters, probably the most used materials for sustained release of GFs. When implanted in the body, polyesters are known to hydrate and start their decomposition, which can occur at the surface but also at the inner matrix (Rezwan et al., 2006; Zhu et al., 2000). As the polyester discompose in the core of the matrix forming acidic oligomers, these remain trapped in the polymeric network and induce an acidic microclimate that can degrade the GF. This general mechanism applies similarly to polymers where the degradation products can react with or just degrade the GFs. For GF delivery, it is probably best to directly select materials that do not result in reactive degradation products, or that undergo only surface-erosion (e.g. polyanhydrides) (Tabata et al., 1993). However, further protection of GFs in polyesters can be achieved by incorporating buffering agents (Zhu et al., 2000). Interestingly, some polyester composites have shown capacity to protect macromolecules both during encapsulation and release (Csaba et al., 2005), and these technologies have already been applied for GF-delivery (d'Angelo et al., 2010). In any case, designing devices capable of providing GF release for several months and in a bioactive form continues to be a challenging objective.

Another strategy to achieve sustained exposure of cells to GFs or other signaling molecules is covalent tethering to the scaffold. Covalent tethering of proteins presents the difficulty of unselective binding of the protein to the polymer. Indeed, although some more selective reactions have been designed (i.e. cysteine based linking) (Backer et al., 2006), most protocols use reactions where protein can be bound through several aminoacids of the peptidic chain. This lack of specificity could result in a potential bioactivity loss if a critical aminoacid region is modified with the polymer. However, in general terms good bioactivity has been observed with immobilized GFs (Bentz et al., 1998). On the other hand, lack of specificity when linking biomolecules to synthetic materials raises another concern, linked to a difficult regulatory pathway with materials where no clear molecular structure can be defined.

5. Concluding remarks

We have gone a long way in the field of tissue engineering from the first discoveries pointing to the conductive properties of tissue scaffolds. The experience from reprocessed tissues (DBM, for example) and other ECM-containing products soon pointed to the importance of a biomimetic approach in scaffold design. In the last years, biopolymer-derived and bioactivated synthetic scaffolds have been thoroughly studied and GF delivery integrated in many devices. With a good share of hype and bumps on the way, we have now several tissue engineering products available on the market, most for therapeutic indications related to skeletal tissue engineering, and some having been transformed in great commercial successes (e.g. Infuse™ from Medtronic, Inc.). Tissue engineering potential, however, does not need to be limited to regeneration in relatively simple injury models. Recent studies showed the possibility to engineer body parts integrating several tissues (e.g. ostochondral grafts, vascularized bone) or even small organs. Whether these studies could translate into products of clinical interest is still uncertain, but worth exploring. Probably further optimization of the biomimetic scaffolds together with improved bioreactors will be necessary to realize these challenging goals. Another technology to be expected commercially in the next few years are biomaterials integrating stem cells. For those applications, scaffolds integrating functionalities capable of instructing stem cell proliferation, differentiation and tissue formation will be required, and this concept should be a top priority in tissue engineering, bioregenerative medicine and in the government's scientific agendas.

6. Acronyms used

BMP: bone morphogenic protein
DBM: demineralized bone matrix
ECM: extracellular matrix
EGF: epidermal growth factor
FGF: fibroblast growth factor
GF: growth factor
IGF: insulin-like growth factor
MMP: matrix metalloprotease
MSC: mesenchymal stem cell
PDGF: platelet-derived growth factor
PEG: poly(ethylene glycol)
qRT-PCR: quantitative reverse transcription polymerase chain reaction
TGF: transforming growth factor
VEGF: vascular endothelial growth factor

7. Acknoledgments

MGF acknowledges an Isidro Parga Pondal contract from Xunta de Galicia. This work was financed by Xunta de Galicia (PGIDIT07PXIB203067PR) and the Spanish Ministry of Science and Innovation (Programa de Investigación en Salud, PS09/01786).

8. References

Abousleiman, R.I., Reyes, Y., McFetridge, P., & Sikavitsas, V. (2009). Tendon Tissue Engineering Using Cell-Seeded Umbilical Veins Cultured in a Mechanical Stimulator. *Tissue Engineering. Part A*, Vol.15, No.4, (April 2009), pp. (787-795), ISSN 1937-3341

Ahmed, T.A.E., Dare, E.V., & Hincke, M. (2008). Fibrin: A Versatile Scaffold for Tissue Engineering Applications. *Tissue Engineering Part B: Reviews*, Vol.14, No.2, (2008), pp. (199-215), ISSN 1937-3368

Ahmed, T.A., Giulivi, A., Griffith, M., & Hincke, M. (2011). Fibrin Glues in Combination with Mesenchymal Stem Cells to Develop a Tissue-Engineered Cartilage Substitute. *Tissue Engineering. Part A*, Vol.17, No.3-4, (February 2011), pp. (323-335), ISSN 1937-335X

Ahmed, R.P., Haider, K.H., Shujia, J., Afzal, M.R., & Ashraf, M. (2010). Sonic Hedgehog Gene Delivery to the Rodent Heart Promotes Angiogenesis Via Inos/Netrin-1/Pkc Pathway. *Plos One*, Vol.5, No.1, (2010), p. (e8576), ISSN 1932-6203

Allemann, F., Mizuno, S., Eid, K., Yates, K.E., Zaleske, D., & Glowacki, J. (2001). Effects of Hyaluronan on Engineered Articular Cartilage Extracellular Matrix Gene Expression in 3-Dimensional Collagen Scaffolds. *Journal of Biomedical Materials Research. Part A*, Vol.55, No.1, (2001), pp. (13-19), ISSN 1552-4965

Alsberg, E., Anderson, K.W., Albeiruti, A., Rowley, J.A., & Mooney, D.J. (2002). Engineering Growing Tissues. *Proceedings of the National Academy of Sciences of the United States of America*, Vol.99, No.19, (2002), p. (12025), ISSN 1091-6490

Amariglio, N., Hirshberg, A., Scheithauer, B.W., Cohen, Y., Loewenthal, R., Trakhtenbrot, L., Paz, N., Koren-Michowitz, M., Waldman, D., Leider-Trejo, L., Toren, A., Constantini, S., & Rechavi, G. (2009). Donor-Derived Brain Tumor Following Neural Stem Cell Transplantation in An Ataxia Telangiectasia Patient. *Plos Med*, Vol.6, No.2, (July 2009), p. (e1000029)

Anderson, D.G., Levenberg, S., & Langer, R. (2004). Nanoliter-Scale Synthesis of Arrayed Biomaterials and Application to Human Embryonic Stem Cells. *Nature Biotechnology*, Vol.22, No.7, (2004), pp. (863-866), ISSN 1087-0156

Anderson, D.G., Putnam, D., Lavik, E.B., Mahmood, T.A., & Langer, R. (2005). Biomaterial Microarrays: Rapid, Microscale Screening of Polymer-Cell Interaction. *Biomaterials*, Vol.26, No.23, (2005), pp. (4892-4897), ISSN 0142-9612

Attia, M., Santerre, J.P., & Kandel, R.A. (2010). The Response of Annulus Fibrosus Cell to Fibronectin-Coated Nanofibrous Polyurethane-Anionic Dihydroxyoligomer Scaffolds. *Biomaterials*, Vol.32, No.2, (January 2011), pp. (450-602010), ISSN 0142-9612

Backer, M.V., Patel, V., Jehning, B.T., Claffey, K.P., & Backer, J.M. (2006). Surface Immobilization of Active Vascular Endothelial Growth Factor Via a Cysteine-Containing Tag. *Biomaterials*, Vol.27, No.31, (2006), pp. (5452-5458), ISSN 0142-9612

Batrakova, E.V., Gendelman, H.E., & Kabanov, A.V. (2011). Cell-Mediated Drug Delivery. *Expert Opinion on Drug Delivery*, Vol. 8, No.4, (April 2011), pp. (415-433), ISSN 1742-5247

Benoit, D.S.W. & Anseth, K.S. (2005a). The Effect on Osteoblast Function of Colocalized RGD and PHSRN Epitopes on PEG Surfaces. *Biomaterials*, Vol.26, No.25, (2005), pp. (5209-5220), ISSN 0142-9612

Benoit, D.S.W. & Anseth, K.S. (2005b). Heparin Functionalized PEG Gels That Modulate Protein Adsorption for hMSC Adhesion and Differentiation. *Acta Biomaterialia*, Vol.1, No.4, (2005), pp. (461-470), ISSN 1742-7061

Bentz, H., Schroeder, J.A., & Estridge, T.D. (1998). Improved Local Delivery of Tgf-β2 by Binding to Injectable Fibrillar Collagen Via Difunctional Polyethylene Glycol. *Journal of Biomedical Materials Research*, Vol.39, No.4, (1998), pp. (539-548), ISSN 1097-4636

Biondi, M., Ungaro, F., Quaglia, F., & Netti, P.A. (2008). Controlled Drug Delivery in Tissue Engineering. *Advanced Drug Delivery Reviews*, Vol.60, No.2, (2008), pp. (229-242), ISSN 0169-409X

Brady, M.A., Sivananthan, S., Mudera, V., Liu, Q., Wiltfang, J., & Warnke, P.H. (2010). The Primordium of a Biological Joint Replacement: Coupling of Two Stem Cell Pathways in Biphasic Ultrarapid Compressed Gel Niches. *Journal of Cranio-Maxillo-Facial Surgery*, Epub ahead of print, (August 2010), ISSN 1878-4119

Braiman-Wiksman, L., Solomonik, I., Spira, R., & Tennenbaum, T. (2007). Novel Insights Into Wound Healing Sequence of Events. *Toxicologic Pathology*, Vol.35, No.6, (2007), pp. (767-779), ISSN 0192-6233

Brent, A.E., Braun, T., & Tabin, C.J. (2005). Genetic Analysis of Interactions Between the Somitic Muscle, Cartilage and Tendon Cell Lineages During Mouse Development. *Development*, Vol.132, No.3, (2005), p. (515), ISSN 0950-1991

Brown, A.J., Alicknavitch, M., D'Souza, S.S., Daikoku, T., Kirn-Safran, C.B., Marchetti, D., Carson, D.D., & Farach-Carson, M.C. (2008). Heparanase Expression and Activity

Influences Chondrogenic and Osteogenic Processes During Endochondral Bone Formation. *Bone*, Vol.43, No.4, (2008), pp. (689-699), ISSN 8756-3282

Buckwalter, J.A. & Brown, T.D. (2004). Joint Injury, Repair, and Remodeling: Roles in Post-Traumatic Osteoarthritis. *Clinical Orthopaedics and Related Research*, Vol.423, (2004), p. (7), ISSN 0009-921X

Cai, S., Liu, Y., Zheng Shu, X., & Prestwich, G.D. (2005). Injectable Glycosaminoglycan Hydrogels for Controlled Release of Human Basic Fibroblast Growth Factor. *Biomaterials*, Vol.26, No.30, (2005), pp. (6054-6067), ISSN 0142-9612

Chang, C.F., Lee, M.W., Kuo, P.Y., Wang, Y.J., Tu, Y.H., & Hung, S.C. (2007). Three-Dimensional Collagen Fiber Remodeling by Mesenchymal Stem Cells Requires the Integrin--Matrix Interaction. *Journal of Biomedical Materials Research. Part A*, Vol.80, No.2, (2007), pp. (466-474), ISSN 1552-4965

Chen, M., Le, D.Q., Baatrup, A., Nygaard, J.V., Hein, S., Bjerre, L., Kassem, M., Zou, X., & Bünger, C. (2010). Self-Assembled Composite Matrix in a Hierarchical 3-D Scaffold for Bone Tissue Engineering. *Acta Biomaterialia,* Epub ahead of print, (December 2010), ISSN 1878-7568

Chen, J.S., Noah, E.M., Pallua, N., & Steffens, G.C.M. (2002). The Use of Bifunctional Polyethyleneglycol Derivatives for Coupling of Proteins to and Cross-Linking of Collagen Matrices. *Journal of Materials Science. Materials in Medicine*, Vol.13, No.11, (2002), pp. (1029-1035), ISSN 0957-4530

Chen, W.C., Yao, C.L., Wei, Y.H., & Chu, I.M. (2011). Evaluating Osteochondral Defect Repair Potential of Autologous Rabbit Bone Marrow Cells on Type II Collagen Scaffold. *Cytotechnology*, Vol.63, No.1, (January 2011), pp. (13-23), ISSN 0920-9069

Chen, J.L., Yin, Z., Shen, W.L., Chen, X., Heng, B.C., Zou, X.H., & Ouyang, H.W. (2010). Efficacy of hESC-MSCs in Knitted Silk-Collagen Scaffold for Tendon Tissue Engineering and Their Roles. *Biomaterials*, Vol.31, No.36, (December 2010), pp. (9438-9451), ISSN 1878-5905

Chou, C.H., Cheng, W.T., Lin, C.C., Chang, C.H., Tsai, C.C., & Lin, F.H. (2006). Tgf-β1 Immobilized Tri-Co-Polymer for Articular Cartilage Tissue Engineering. *Journal of Biomedical Materials Research. Part B, Applied Biomaterials*, Vol.77, No.2, (May 2006), pp. (338-348), ISSN 1552-4973

Chun, J., Tuan, T.L., Han, B., Vangsness, C.T., & Nimni, M.E. (2003). Cultures of Ligament Fibroblasts in Fibrin Matrix Gel. *Connective Tissue Research*, Vol.44, No.2, (2003), pp. (81-87), ISSN 0300-8207

Chung, C.H., Golub, E.E., Forbes, E., Tokuoka, T., & Shapiro, I.M. (1992). Mechanism of Action of B-Glycerophosphate on Bone Cell Mineralization. *Calcified Tissue International*, Vol.51, No.4, (1992), pp. (305-311), ISSN 0171-967X

Chung, H.J. & Park, T.G. (2007). Surface Engineered and Drug Releasing Pre-Fabricated Scaffolds for Tissue Engineering. *Advanced Drug Delivery Reviews*, Vol.59, No.4-5, (2007), pp. (249-262), ISSN 0169-409X

de Cogan, F., Gough, J.E., & Webb, S.J. (2011). Adhesive Interactions Between Cells and Biotinylated Phospholipid Vesicles in Alginate: Towards New Responsive Biomaterials. *Journal of Materials Science. Materials in Medicine,* Epub ahead of print, (March 2011), pp. (1-7), ISSN 0957-4530

Comisar, W.A., Kazmers, N.H., Mooney, D.J., & Linderman, J.J. (2007). Engineering RGD Nanopatterned Hydrogels to Control Preosteoblast Behavior: A Combined

Computational and Experimental Approach. *Biomaterials*, Vol.28, No.30, (2007), pp. (4409-4417), ISSN 0142-9612

Crawford, D.C., Heveran, C.M., Cannon, W.D., Foo, L.F., & Potter, H.G. (2009). An Autologous Cartilage Tissue Implant Neocart for Treatment of Grade III Chondral Injury to the Distal Femur: Prospective Clinical Safety Trial at 2 Years. *The American Journal of Sports Medicine*, Vol.37, No.7, (July 2009), pp. (1334-1343), ISSN 1552-3365

Crescenzi, V., Francescangeli, A., Capitani, D., Mannina, L., Renier, D., & Bellini, D. (2003). Hyaluronan Networking Via Ugi's Condensation Using Lysine As Cross-Linker Diamine. *Carbohydrate Polymers*, Vol.53, No.3, (2003), pp. (311-316), ISSN 0144-8617

Cristino, S., Grassi, F., Toneguzzi, S., Piacentini, A., Grigolo, B., Santi, S., Riccio, M., Tognana, E., Facchini, A., & Lisignoli, G. (2005). Analysis of Mesenchymal Stem Cells Grown on a Three-Dimensional HYAFF 11®-Based Prototype Ligament Scaffold. *Journal of Biomedical Materials Research. Part A*, Vol.73, No.3, (2005), pp. (275-283), ISSN 1552-4965

Csaba, N., Caamaño, P., Sánchez, A., Domínguez, F., & Alonso, M.J. (2005). Plga:Poloxamer and Plga:Poloxamine Blend Nanoparticles: New Carriers for Gene Delivery. *Biomacromolecules*, Vol.6, No.1, (2005), pp. (271-278), ISSN 1525-7797

Dahners, L.E. & Jacobs, R.R.. (1985). Long Bone Defects Treated with Demineralized Bone. *Southern Medical Journal*, Vol.78, No.8, (August 1985), p. (933-934), ISSN 0038-4348

d'Angelo, I., Garcia-Fuentes, M., Parajó, Y., Welle, A., Vántus, T., Horváth, A., Bőkőnyi, G., Kéri, G., & Alonso, M.J. (2010). Nanoparticles Based on Plga:Poloxamer Blends for the Delivery of Proangiogenic Growth Factors. *Molecular Pharmaceutics*, Vol.7(5), (August 2010), pp. (1724-1733), ISSN 1543-8392

Danti, S., Lisanti, M., Berrettini, S., Petrini, M., & Pietrabissa, A. (2007). Bone Tissue Engineering: Natural Origination or Synthetic Polymeric Scaffolds? *Advanced Materials Research*, Vol.15, (2007), pp. (65-70), ISSN 1662-8985

Deckers, R., Rome, C., & Moonen, C.T.W. (2008). The Role of Ultrasound and Magnetic Resonance in Local Drug Delivery. *Journal of Magnetic Resonance Imaging*, Vol.27, No.2, (2008), pp. (400-409), ISSN 1522-2586

Deeken, C.R., White, A.K., Bachman, S.L., Ramshaw, B.J., Cleveland, D.S., Loy, T.S., & Grant, S.A. (2011). Method of Preparing a Decellularized Porcine Tendon Using Tributyl Phosphate. *Journal of Biomedical Materials Research. Part B, Applied Biomaterials*, Vol.96, No.2, (February 2011), pp. (199-206), ISSN 1552-4981

Deschaseaux, F., Sensébé, L., & Heymann, D. (2009). Mechanisms of Bone Repair and Regeneration. *Trends in Molecular Medicine*, Vol.15, No.9, (2009), pp. (417-429), ISSN 1471-4914

Ding, S., Wu, T.Y.H., Brinker, A., Peters, E.C., Hur, W., Gray, N.S., & Schultz, P.G. (2003). Synthetic Small Molecules That Control Stem Cell Fate. *Proceedings of the National Academy of Sciences of the United States of America*, Vol.100, No.13, (2003), p. (7632), ISSN 1091-6490

Dinopoulos, H.T. & Giannoudis, P.V. (2006). Safety and Efficacy of Use of Demineralised Bone Matrix in Orthopaedic and Trauma Surgery. *Expert Opinion in Drug Safety*, Vol.5, No.6, (November 2006), pp. (847-866), ISSN 1744-764X

Discher, D.E., Mooney, D.J., & Zandstra, P.W. (2009). Growth Factors, Matrices, and Forces Combine and Control Stem Cells. *Science*, Vol.324, No.5935, (June 2009), pp. (1673-1677), ISSN 0036-8075

Dohan, D.M., Choukroun, J., Diss, A., Dohan, S.L., Dohan, A.J., Mouhyi, J., & Gogly, B. (2006). Platelet-Rich Fibrin (PRF): A Second-Generation Platelet Concentrate. Part I: Technological Concepts and Evolution. *Oral Surgery, Oral Medicine, Oral Pathology, Oral Radiology, and Endodontology*, Vol.101, No.3, (2006), pp. (e37-e37)

Dozza, B., Di Bella, C., Lucarelli, E., Giavaresi, G., Fini, M., Tazzari, P.L., Giannini, S., & Donati, D. (2011). Mesenchymal Stem Cells and Platelet Lysate in Fibrin or Collagen Scaffold Promote Non-Cemented Hip Prosthesis Integration. *Journal of Orthopaedic Research,* Epub ahead of print, (January 2011), ISSN 1554-527X

Drinnan, C.T., Zhang, G., Alexander, M.A., Pulido, A.S., & Suggs, L.J. (2010). Multimodal Release of Transforming Growth Factor-β1 and the BB Isoform of Platelet Derived Growth Factor From Pegylated Fibrin Gels. *Journal of Controlled Release,* Vol.147, No.2, (October 2010), pp. (180-186), ISSN 1873-4995

Dunehoo, A.L., Anderson, M., Majumdar, S., Kobayashi, N., Berkland, C., & Siahaan, T.J. (2006). Cell Adhesion Molecules for Targeted Drug Delivery. *Journal of Pharmaceutical Sciences,* Vol.95, No.9, (2006), pp. (1856-1872), ISSN 1520-6017

Dunger, D.B., Regan, A.C.L., & Acerini, C.L. (2005). Childhood and Adolescent Diabetes. *Endocr Dev,* Vol.9, (2005), p. (107)

Elder, B.D., Kim, D.H., & Athanasiou, K.A. (2010). Developing An Articular Cartilage Decellularization Process Toward Facet Joint Cartilage Replacement. *Neurosurgery,* Vol.66, No.4, (April 2010), pp. (722-7), ISSN 1524-4040

Engler, A.J., Sen, S., Sweeney, H.L., & Discher, D.E. (2006). Matrix Elasticity Directs Stem Cell Lineage Specification. *Cell,* Vol.126, No.4, (2006), pp. (677-689), ISSN 0092-8674

Eppley, B.L., Woodell, J.E., & Higgins, J. (2004). Platelet Quantification and Growth Factor Analysis From Platelet-Rich Plasma: Implications for Wound Healing. *Plastic and Reconstructive Surgery,* Vol.114, No.6, (2004), p. (1502), ISSN 0032-1052

Erggelet, C., Endres, M., Neumann, K., Morawietz, L., Ringe, J., Haberstroh, K., Sittinger, M., & Kaps, C. (2009). Formation of Cartilage Repair Tissue in Articular Cartilage Defects Pretreated with Microfracture and Covered with Cell-Free Polymer-Based Implants. *Journal of Orthopaedic Research,* Vol.27, No.10, (October 2009), pp. (1353-1360), ISSN 1554-527X

Eyrich, D., Brandl, F., Appel, B., Wiese, H., Maier, G., Wenzel, M., Staudenmaier, R., Goepferich, A., & Blunk, T. (2007). Long-Term Stable Fibrin Gels for Cartilage Engineering. *Biomaterials,* Vol.28, No.1, (January 2007), pp. (55-65), ISSN 0142-9612

Fedarko, N.S., Vetter, U.K., Weinstein, S., & Robey, P.G. (1992). Age-Related Changes in Hyaluronan, Proteoglycan, Collagen, and Osteonectin Synthesis by Human Bone Cells. *Journal of Cellular Physiology,* Vol.151, No.2, (1992), pp. (215-227), ISSN 1097-4652

Flynn, L.E., (2010). The Use of Decellularized Adipose Tissue to Provide An Inductive Microenvironment for the Adipogenic Differentiation of Human Adipose-Derived Stem Cells. *Biomaterials,* Vol.31, No.17, (June 2010), pp. (4715-4724), ISSN 1878-5905

Flynn, L., Prestwich, G.D., Semple, J.L., & Woodhouse, K.A. (2007). Adipose Tissue Engineering with Naturally Derived Scaffolds and Adipose-Derived Stem Cells. *Biomaterials,* Vol.28, No.26, (September 2007), pp. (3834-3842), ISSN 0142-9612

Fratzl, P. (Ed(s).). (2008). *Collagen: structure and mechanics,* Springer, ISBN 038773905X, New York

Freeman, I., Kedem, A., & Cohen, S. (2008). The Effect of Sulfation of Alginate Hydrogels on the Specific Binding and Controlled Release of Heparin-Binding Proteins. *Biomaterials*, Vol.29, No.22, (2008), pp. (3260-3268), ISSN 0142-9612

Fujioka, K., Maeda, M., Hojo, T., & Sano, A. (1998). Protein Release From Collagen Matrices. *Advanced Drug Delivery Reviews*, Vol.31, No.3, (1998), pp. (247-266), ISSN 0169-409X

Fussenegger, M., Meinhart, J., Höbling, W., Kullich, W., Funk, S., & Bernatzky, G. (2003). Stabilized Autologous Fibrin-Chondrocyte Constructs for Cartilage Repair in Vivo. *Annals of Plastic Surgery*, Vol.51, No.5, (2003), p. (493-498), ISSN 0148-7043

Garcia-Fuentes, M., Giger, E., Meinel, L., & Merkle, H.P. (2008). The Effect of Hyaluronic Acid on Silk Fibroin Conformation. *Biomaterials*, Vol.29, No.6, (February 2008), pp. (633-642), ISSN 0142-9612

Garcia-Fuentes, M., Meinel, A.J., Hilbe, M., Meinel, L., & Merkle, H.P. (2009). Silk Fibroin/Hyaluronan Scaffolds for Human Mesenchymal Stem Cell Culture in Tissue Engineering. *Biomaterials*, Vol.30, No.28, (October 2009), pp. (5068-5076), ISSN 1878-5905

Geiger, F., Bertram, H., Berger, I., Lorenz, H., Wall, O., Eckhardt, C., Simank, H.G., & Richter, W. (2005). Vascular Endothelial Growth Factor Gene-Activated Matrix (VEGF165-GAM) Enhances Osteogenesis and Angiogenesis in Large Segmental Bone Defects. *Journal of Bone and Mineral Research*, Vol.20, No.11, (2005), pp. (2028-2035), ISSN 1523-4681

George, J., Onodera, J., & Miyata, T. (2008). Biodegradable Honeycomb Collagen Scaffold for Dermal Tissue Engineering. *Journal of Biomedical Materials Research. Part A*, Vol.87, No.4, (December 2008), pp. (1103-1111), ISSN 1552-4965

Gerecht, S., Burdick, J.A., Ferreira, L.S., Townsend, S.A., Langer, R., & Vunjak-Novakovic, G. (2007). Hyaluronic Acid Hydrogel for Controlled Self-Renewal and Differentiation of Human Embryonic Stem Cells. *Proceedings of the National Academy of Sciences*, Vol.104, No.27, (2007), p. (11298–11303)

Gerstenfeld, L.C., Cullinane, D.M., Barnes, G.L., Graves, D.T., & Einhorn, T.A. (2003). Fracture Healing As a Post-Natal Developmental Process: Molecular, Spatial, and Temporal Aspects of Its Regulation. *Journal of Cellular Biochemistry*, Vol.88, No.5, (2003), pp. (873-884)

Gilbert, T.W., Sellaro, T.L., & Badylak, S.F. (2006). Decellularization of Tissues and Organs. *Biomaterials*, Vol.27, No.19, (July 2006), pp. (3675-3683), ISSN 0142-9612

Gillies, A.R., Smith, L.R., Lieber, R.L., & Varghese, S. (2010). Method for Decellularizing Skeletal Muscle Without Detergents or Proteolytic Enzymes. *Tissue Eng Part C Methods*(December 2010), ISSN 1937-3392

Gillitzer, R. & Goebeler, M. (2001). Chemokines in Cutaneous Wound Healing. *Journal of Leukocyte Biology*, Vol.69, No.4, (April 2001), p. (513-521), ISSN 0741-5400

Girotti, A., Reguera, J., Rodriguez-Cabello, J.C., Arias, F.J., Alonso, M., & Testera, A.M. (2004). Design and Bioproduction of a Recombinant Multi (Bio) Functional Elastin-Like Protein Polymer Containing Cell Adhesion Sequences for Tissue Engineering Purposes. *Journal of Materials Science. Materials in Medicine*, Vol.15, No.4, (2004), pp. (479-484), ISSN 0957-4530

Glowacki, J. & Mizuno, S. (2008). Collagen Scaffolds for Tissue Engineering. *Biopolymers*, Vol.89, No.5, (2008), pp. (338-344), ISSN 1097-0282

Gong, Y.Y., Xue, J.X., Zhang, W.J., Zhou, G.D., Liu, W., & Cao, Y. (2011). A Sandwich Model for Engineering Cartilage with Acellular Cartilage Sheets and Chondrocytes. *Biomaterials*, Vol.32, No.9, (March 2011), pp. (2265-2273), ISSN 1878-5905

Grigolo, B., De Franceschi, L., Roseti, L., Cattini, L., & Facchini, A. (2005). Down Regulation of Degenerative Cartilage Molecules in Chondrocytes Grown on a Hyaluronan-Based Scaffold. *Biomaterials*, Vol.26, No.28, (2005), pp. (5668-5676), ISSN 0142-9612

Guo, T., Zeng, X., Hong, H., Diao, H., Wangrui, R., Zhao, J., Zhang, J., & Li, J. (2006). Gene-Activated Matrices for Cartilage Defect Reparation. *The International Journal of Artificial Organs*, Vol.29, No.6, (June 2006), p. (612-621), ISSN 0391-3988

Gurevitch, O., Kurkalli, B.G., Prigozhina, T., Kasir, J., Gaft, A., & Slavin, S. (2003). Reconstruction of Cartilage, Bone, and Hematopoietic Microenvironment with Demineralized Bone Matrix and Bone Marrow Cells. *Stem Cells (Dayton, Ohio)*, Vol.21, No.5, (2003), pp. (588-597), ISSN 1066-5099

Haberstroh, K., Enz, A., Zenclussen, M.L., Hegewald, A.A., Neumann, K., Abbushi, A., Thomé, C., Sittinger, M., Endres, M., & Kaps, C. (2009). Human Intervertebral Disc-Derived Cells Are Recruited by Human Serum and Form Nucleus Pulposus-Like Tissue Upon Stimulation with TFG-β3 or Hyaluronan in Vitro. *Tissue & Cell*, Vol.41, No.6, (2009), p. (414), ISSN 1532-3072

Haleem, A.M., Singergy, A.A., Sabry, D., Atta, H.M., Rashed, L.A., Chu, C.R., El Shewy, M.T., Azzam, A., & Abdel Aziz, M.T. (2010). The Clinical Use of Human Culture-Expanded Autologous Bone Marrow Mesenchymal Stem Cells Transplanted on Platelet-Rich Fibrin Glue in the Treatment of Articular Cartilage Defects: A Pilot Study and Preliminary Results. *Cartilage*, Vol.1, No.4, (October 2010), pp. (253-261), ISSN 1947-6043

Hersel, U., Dahmen, C., & Kessler, H. (2003). RGD Modified Polymers: Biomaterials for Stimulated Cell Adhesion and Beyond. *Biomaterials*, Vol.24, No.24, (2003), pp. (4385-4415), ISSN 0142-9612

Holland, T.A., Bodde, E.W.H., Cuijpers, V., Baggett, L.S., Tabata, Y., Mikos, A.G., & Jansen, J.A. (2007). Degradable Hydrogel Scaffolds for in Vivo Delivery of Single and Dual Growth Factors in Cartilage Repair. *Osteoarthritis and Cartilage*, Vol.15, No.2, (2007), pp. (187-197), ISSN 1063-4584

Holland, T.A. & Mikos, A.G. (2006). Biodegradable Polymeric Scaffolds. Improvements in Bone Tissue Engineering Through Controlled Drug Delivery. *Advances in Biochemical Engineering/Biotechnology*, Vol.102, (2006), p. (161), ISSN 0724-6145

Homandberg, G.A., Ummadi, V., & Kang, H. (2004). Hyaluronan Enhances Cartilage Repair Through Low Grade Tissue Remodeling Involving Cytokines and Matrix Metalloproteinases. *Inflammation Research*, Vol.53, No.10, (2004), pp. (534-543), ISSN 1023-3830

Hou, N., Cui, P., Luo, J., Ma, R., & Zhu, L. (2011). Tissue-Engineered Larynx Using Perfusion-Decellularized Technique and Mesenchymal Stem Cells in a Rabbit Model. *Acta Oto-laryngologica,* Epub ahead of print, (February 2011), ISSN 1651-2251

Huang, B., Zhuang, Y., Li, C.Q., Liu, L.T., & Zhou, Y. (2011). Regeneration of the Intervertebral Disc with Nucleus Pulposus Cell-Seeded Collagen-α/Hyaluronan/Chondroitin-6-Sulfate Tri-Copolymer Constructs in a Rabbit Disc Degeneration Model. *Spine,* Epub ahead of print, (February 2011), ISSN 1528-1159

Huangfu, D., Maehr, R., Guo, W., Eijkelenboom, A., Snitow, M., Chen, A.E., & Melton, D.A. (2008). Induction of Pluripotent Stem Cells by Defined Factors Is Greatly Improved by Small-Molecule Compounds. *Nature Biotechnology*, Vol.26, No.7, (2008), pp. (795-797), ISSN 1087-0156

Huebsch, N. & Mooney, D.J. (2009). Inspiration and Application in the Evolution of Biomaterials. *Nature*, Vol.462, No.7272, (November 2009), p. (426-432), ISSN 1476-4687

Im, G.I., Ahn, J.H., Kim, S.Y., Choi, B.S., & Lee, S.W. (2010). A Hyaluronate-Atelocollagen/Beta-Tricalcium Phosphate-Hydroxyapatite Biphasic Scaffold for the Repair of Osteochondral Defects: A Porcine Study. *Tissue Engineering. Part A*, Vol.16, No.4, (April 2010), pp. (1189-1200), ISSN 1937-335X

Ingram, J.H., Korossis, S., Howling, G., Fisher, J., & Ingham, E. (2007). The Use of Ultrasonication to Aid Recellularization of Acellular Natural Tissue Scaffolds for Use in Anterior Cruciate Ligament Reconstruction. *Tissue Engineering*, Vol.13, No.7, (July 2007), pp. (1561-1572), ISSN 1076-3279

Iozzo, R.V. (2000). *Proteoglycans: Structure, Biology, and Molecular Interactions*, Marcel Dekker, ISBN 0824703340, New York

Irie, T., Majima, T., Sawaguchi, N., Funakoshi, T., Nishimura, S.I., & Minami, A. (2011). Biomechanical and Histologic Evaluation of Tissue Engineered Ligaments Using Chitosan and Hyaluronan Hybrid Polymer Fibers: A Rabbit Medial Collateral Ligament Reconstruction Model. *Journal of Biomedical Materials Research. Part A*, Epub ahead of print, (March 2011), ISSN 1552-4965

Irinakis, T., (2011). Efficacy of Injectable Demineralized Bone Matrix As Graft Material During Sinus Elevation Surgery with Simultaneous Implant Placement in the Posterior Maxilla: Clinical Evaluation of 49 Sinuses. *J Oral Maxillofac Surg*, Vol.69, No.1, (January 2011), pp. (134-141), ISSN 1531-5053

Isaac, J., Nohra, J., Lao, J., Jallot, E., Nedelec, J.M., Berdal, A., & Sautier, J.M. (2011). Effects of Strontium-Doped Bioactive Glass on the Differentiation of Cultured Osteogenic Cells. *European Cells & Materials*, Vol.21, (February 2011), p. (130-143), ISSN 1473-2262

Jäger, M., Fischer, J., Dohrn, W., Li, X., Ayers, D.C., Czibere, A., Prall, W.C., Lensing-Höhn, S., & Krauspe, R. (2008). Dexamethasone Modulates BMP-2 Effects on Mesenchymal Stem Cells in Vitro. *Journal of Orthopaedic Research*, Vol.26, No.11, (November 2008), pp. (1440-1448), ISSN 1554-527X

Jin, X.B., Luo, Z.J., & Wang, J. (2006). Treatment of Rabbit Growth Plate Injuries with An Autologous Tissue-Engineered Composite. An Experimental Study. *Cells Tissues Organs*, Vol.183, No.2, (2006), pp. (62-67), ISSN 1422-6405

Juang Ming Yee, A., Bae, H.W., Friess, D., Robbin, M., Johnstone, B., & Yoo, J.U. (2003). Augmentation of Rabbit Posterolateral Spondylodesis Using a Novel Demineralized Bone Matrix-Hyaluronan Putty. *Spine*, Vol.28, No.21, (November 2003), p. (2435-2440), ISSN 0362-2436

Kang, S.W., Kim, J.S., Park, K.S., Cha, B.H., Shim, J.H., Kim, J.Y., Cho, D.W., Rhie, J.W., & Lee, S.H. (2011). Surface Modification with Fibrin/Hyaluronic Acid Hydrogel on Solid-Free Form-Based Scaffolds Followed by BMP-2 Loading to Enhance Bone Regeneration. *Bone*, Vol.48, No.2, (February 2011), pp. (298-306), ISSN 1873-2763

Karageorgiou, V., Meinel, L., Hofmann, S., Malhotra, A., Volloch, V., & Kaplan, D. (2004). Bone Morphogenetic Protein-2 Decorated Silk Fibroin Films Induce Osteogenic Differentiation of Human Bone Marrow Stromal Cells. *Journal of Biomedical Materials Research. Part A*, Vol.71, No.3, (2004), pp. (528-537), ISSN 1552-4965

Kim, S.J., Jang, J.D., & Lee, S.K. (2007). Treatment of Long Tubular Bone Defect of Rabbit Using Autologous Cultured Osteoblasts Mixed with Fibrin. *Cytotechnology*, Vol.54, No.2, (June 2007), pp. (115-120), ISSN 0920-9069

Kim, Y.K., Kim, S.G., Lim, S.C., Lee, H.J., & Yun, P.Y. (2010). A Clinical Study on Bone Formation Using a Demineralized Bone Matrix and Resorbable Membrane. *Oral Surg Oral Med Oral Pathol Oral Radiol Endod*, Vol.109, No.6, (June 2010), pp. (e6-11), ISSN 1528-395X

Kirker-Head, C., Karageorgiou, V., Hofmann, S., Fajardo, R., Betz, O., Merkle, H.P., Hilbe, M., Von Rechenberg, B., McCool, J., & Abrahamsen, L. (2007). BMP-Silk Composite Matrices Heal Critically Sized Femoral Defects. *Bone*, Vol.41, No.2, (2007), pp. (247-255), ISSN 8756-3282

Kirkwood, K., Rheude, B., Kim, Y.J., White, K., & Dee, K.C. (2003). In Vitro Mineralization Studies with Substrate-Immobilized Bone Morphogenetic Protein Peptides. *Journal of Oral Implantology*, Vol.29, No.2, (2003), pp. (57-65), ISSN 1548-1336

Knudson, W., Casey, B., Nishida, Y., Eger, W., Kuettner, K.E., & Knudson, C.B. (2000). Hyaluronan Oligosaccharides Perturb Cartilage Matrix Homeostasis and Induce Chondrocytic Chondrolysis. *Arthritis & Rheumatism*, Vol.43, No.5, (2000), pp. (1165-1174), ISSN 1529-0131

Krasnow, M.A., (1997). Genes That Control Organ Form: Lessons From Bone and Branching Morphogenesis. *Cold Spring Harbor Symposia on Quantitative Biology*, Vol.62, (1997), p. (235), ISSN 0091-7451

Kretlow, J.D., Klouda, L., & Mikos, A.G. (2007). Injectable Matrices and Scaffolds for Drug Delivery in Tissue Engineering. *Advanced Drug Delivery Reviews*, Vol.59, No.4-5, (2007), pp. (263 - 273), ISSN 0169-409X

Kubow, K.E., Klotzsch, E., Smith, M.L., Gourdon, D., Little, W.C., & Vogel, V. (2009). Crosslinking of Cell-Derived 3D Scaffolds Up-Regulates the Stretching and Unfolding of New Extracellular Matrix Assembled by Reseeded Cells. *Integr. Biol.*, Vol.1, No.11-12, (2009), pp. (635-648)

Kurpinski, K., Chu, J., Hashi, C., & Li, S. (2006). Anisotropic Mechanosensing by Mesenchymal Stem Cells. *Proceedings of the National Academy of Sciences of the United States of America*, Vol.103, No.44, (2006), pp. (16095-16100), ISSN 0027-8424

Laflamme, C. & Rouabhia, M. (2008). Effect of BMP-2 and BMP-7 Homodimers and a Mixture of BMP-2/BMP-7 Homodimers on Osteoblast Adhesion and Growth Following Culture on a Collagen Scaffold. *Biomedical Materials*, Vol.3, No.1, (March 2008), p. (015008), ISSN 1748-605X

Lammens, J., Nijs, J., Schepers, E., Ectors, N., Lismont, D., & Verduyckt, B. (2009). The Effect of Bone Morphogenetic Protein-7 (OP-1) and Demineralized Bone Matrix (DBM) in the Rabbit Tibial Distraction Model. *Acta Orthop Belg*, Vol.75, No.1, (February 2009), pp. (103-109), ISSN 0001-6462

Langer, R. & Tirrell, D.A. (2004). Designing Materials for Biology and Medicine. *Nature*, Vol.428, No.6982, (2004), pp. (487-492), ISSN 0028-0836

Lee, S.J., Broda, C., Atala, A., & Yoo, J.J. (2011). Engineered Cartilage Covered Ear Implants for Auricular Cartilage Reconstruction. *Biomacromolecules*, Vol.12, No.2, (February 2011), pp. (306-313), ISSN 1526-4602

Lee, J.E., Kim, K.E., Kwon, I.C., Ahn, H.J., Lee, S.H., Cho, H., Kim, H.J., Seong, S.C., & Lee, M.C. (2004). Effects of the Controlled-Released Tgf-β1 From Chitosan Microspheres on Chondrocytes Cultured in a Collagen/Chitosan/Glycosaminoglycan Scaffold. *Biomaterials*, Vol.25, No.18, (August 2004), pp. (4163-4173), ISSN 0142-9612

Leong, D.T., Nah, W.K., Gupta, A., Hutmacher, D.W., & Woodruff, M.A. (2008). The Osteogenic Differentiation of Adipose Tissue-Derived Precursor Cells in a 3D Scaffold/Matrix Environment. *Current Drug Discovery Technologies*, Vol.5, No.4, (December 2008), pp. (319-327), ISSN 1570-1638

Liao, H.T., Chen, C.T., Chen, C.H., Chen, J.P., & Tsai, J.C. (2011). Combination of Guided Osteogenesis with Autologous Platelet-Rich Fibrin Glue and Mesenchymal Stem Cell for Mandibular Reconstruction. *The Journal of Trauma*, Vol.70, No.1, (January 2011), pp. (228-237), ISSN 1529-8809

Lieberman, J.R., Daluiski, A., & Einhorn, T.A. (2002). The Role of Growth Factors in the Repair of Bone: Biology and Clinical Applications. *The Journal of Bone and Joint Surgery*, Vol.84, No.6, (June 2002), p. (1032-1044)

Lim, J.K., Hui, J., Li, L., Thambyah, A., Goh, J., & Lee, E.H. (2004). Enhancement of Tendon Graft Osteointegration Using Mesenchymal Stem Cells in a Rabbit Model of Anterior Cruciate Ligament Reconstruction. *Arthroscopy: The Journal of Arthroscopic & Related Surgery*, Vol.20, No.9, (November 2004), pp. (899-910), ISSN 1526-3231

Lin, X., Takahashi, K., Campion, S.L., Liu, Y., Gustavsen, G.G., Pena, L.A., & Zamora, P.O. (2006). Synthetic Peptide F2A4-K-NS Mimics Fibroblast Growth Factor-2 in Vitro and Is Angiogenic in Vivo. *International Journal of Molecular Medicine*, Vol.17, No.5, (2006), p. (833), ISSN 1107-3756

Lin, H., Zhao, Y., Sun, W., Chen, B., Zhang, J., Zhao, W., Xiao, Z., & Dai, J. (2008). The Effect of Crosslinking Heparin to Demineralized Bone Matrix on Mechanical Strength and Specific Binding to Human Bone Morphogenetic Protein-2. *Biomaterials*, Vol.29, No.9, (March 2008), pp. (1189-1197), ISSN 0142-9612

Liu, G., Li, Y., Sun, J., Zhou, H., Zhang, W., Cui, L., & Cao, Y. (2010a). In Vitro and in Vivo Evaluation of Osteogenesis of Human Umbilical Cord Blood-Derived Mesenchymal Stem Cells on Partially Demineralized Bone Matrix. *Tissue Engineering. Part A*, Vol.16, No.3, (March 2010), pp. (971-982), ISSN 1937-335X

Liu, H., Zhang, L., Shi, P., Zou, Q., Zuo, Y., & Li, Y. (2010b). Hydroxyapatite/Polyurethane Scaffold Incorporated with Drug-Loaded Ethyl Cellulose Microspheres for Bone Regeneration. *Journal of Biomedical Materials Research Part B: Applied Biomaterials*, Vol.95, No.1, (2010), pp. (36-46), ISSN 1552-4981

Lodish, H, Berk, A, & Zipursky, SL (2002). *Molecular Cell Biology* (4th), W H Freeman & Co, New York

Luginbuehl, V., Meinel, L., Merkle, H.P., & Gander, B. (2004). Localized Delivery of Growth Factors for Bone Repair. *European Journal of Pharmaceutics and Biopharmaceutics*, Vol.58, No.2, (2004), pp. (197-208), ISSN 0939-6411

Lutolf, M.P., Lauer-Fields, J.L., Schmoekel, H.G., Metters, A.T., Weber, F.E., Fields, G.B., & Hubbell, J.A. (2003a). Synthetic Matrix Metalloproteinase-Sensitive Hydrogels for the Conduction of Tissue Regeneration: Engineering Cell-Invasion Characteristics.

Proceedings of the National Academy of Sciences of the United States of America, Vol.100, No.9, (2003), p. (5413), ISSN 1091-6490

Lutolf, M.P., Weber, F.E., Schmoekel, H.G., Schense, J.C., Kohler, T., Müller, R., & Hubbell, J.A. (2003b). Repair of Bone Defects Using Synthetic Mimetics of Collagenous Extracellular Matrices. *Nature Biotechnology*, Vol.21, No.5, (2003), pp. (513-518), ISSN 1087-0156

Lynch, M.P., Stein, J.L., Stein, G.S., & Lian, J.B. (1995). The Influence of Type I Collagen on the Development and Maintenance of the Osteoblast Phenotype in Primary and Passaged Rat Calvarial Osteoblasts: Modification of Expression of Genes Supporting Cell Growth, Adhesion, and Extracellular Matrix Mineralization. *Experimental Cell Research*, Vol.216, No.1, (1995), pp. (35-45), ISSN 0014-4827

Macchiarini, P., Jungebluth, P., Go, T., Asnaghi, M., Rees, L.E., Cogan, T.A., Dodson, A., Martorell, J., Bellini, S., & Parnigotto, P.P. (2008). Clinical Transplantation of a Tissue-Engineered Airway. *The Lancet*, Vol.372, No.9655, (2008), pp. (2023-2030), ISSN 0140-6736

MacLean, S.B. & Gratzer, P.F. (2011). Effect of Basic Fibroblast Growth Factor on the Cellular Repopulation of Decellularized Anterior Cruciate Ligament Allografts. *J Tissue Eng Regen Med*, Vol.5, No.3, (March 2011), pp. (201-209), ISSN 1932-7005

Maeda, M., Tani, S., Sano, A., & Fujioka, K. (1999). Microstructure and Release Characteristics of the Minipellet, a Collagen-Based Drug Delivery System for Controlled Release of Protein Drugs. *Journal of Controlled Release*, Vol.62, No.3, (1999), pp. (313-324), ISSN 0168-3659

Mart, R.J., Liem, K.P., & Webb, S.J. (2009a). Creating Functional Vesicle Assemblies From Vesicles and Nanoparticles. *Pharmaceutical Research*, Vol.26, No.7, (2009), pp. (1701-1710), ISSN 0724-8741

Mart, R.J., Liem, K.P., & Webb, S.J. (2009b). Magnetically-Controlled Release From Hydrogel-Supported Vesicle Assemblies. *Chemical Communications*, Vol.7, No.17, (2009), pp. (2287-2289)

Martin Jr, G.J., Boden, S.D., Titus, L., & Scarborough, N.L. (1999). New Formulations of Demineralized Bone Matrix As a More Effective Graft Alternative in Experimental Posterolateral Lumbar Spine Arthrodesis. *Spine*, Vol.24, No.7, (1999), p. (637), ISSN 0362-2436

Massagué, J., (2000). How Cells Read Tgf-B Signals. *Nature Reviews Molecular Cell Biology*, Vol.1, No.3, (2000), pp. (169-178), ISSN 1471-0072

Meinel, L., Zoidis, E., Zapf, J., Hassa, P., Hottiger, M.O., Auer, J.A., Schneider, R., Gander, B., Luginbuehl, V., & Bettschart-Wolfisberger, R. (2003). Localized Insulin-Like Growth Factor I Delivery to Enhance New Bone Formation. *Bone*, Vol.33, No.4, (2003), pp. (660-672), ISSN 8756-3282

Mekhail, M., Wong, K.K., Padavan, D.T., Wu, Y., O'Gorman, D.B., & Wan, W. (2010). Genipin-Cross-Linked Electrospun Collagen Fibers. *Journal of Biomaterials Science. Polymer Edition*, Epub ahead of print, (November 2010), ISSN 1568-5624

de Mel, A., Jell, G., Stevens, M.M., & Seifalian, A.M. (2008). Biofunctionalization of Biomaterials for Accelerated in Situ Endothelialization: A Review. *Biomacromolecules*, Vol.9, No.11, (2008), pp. (2969-2979), ISSN 1525-7797

Merritt, E.K., Hammers, D.W., Tierney, M., Suggs, L.J., Walters, T.J., & Farrar, R.P. (2010). Functional Assessment of Skeletal Muscle Regeneration Utilizing Homologous

Extracellular Matrix As Scaffolding. *Tissue Engineering. Part A*, Vol.16, No.4, (April 2010), pp. (1395-1405), ISSN 1937-335X

Mi, F.L., Shyu, S.S., Peng, C.K., Wu, Y.B., Sung, H.W., Wang, P.S., & Huang, C.C. (2006). Fabrication of Chondroitin Sulfate-Chitosan Composite Artificial Extracellular Matrix for Stabilization of Fibroblast Growth Factor. *Journal of Biomedical Materials Research. Part A*, Vol.76, No.1, (January 2006), pp. (1-15), ISSN 1549-3296

Middelkoop, E., Vries, H.J.C., Ruuls, L., Everts, V., Wildevuur, C.H.R., & Westerhof, W. (1995). Adherence, Proliferation and Collagen Turnover by Human Fibroblasts Seeded Into Different Types of Collagen Sponges. *Cell and Tissue Research*, Vol.280, No.2, (1995), pp. (447-453), ISSN 0302-766X

Min, B.H., Park, S., Choi, B.H., & Park, S.R. (2010). Chondrogenesis of Rabbit Mesenchymal Stem Cells in Fibrin/Hyaluronan Composite Scaffold in Vitro. *Tissue Engineering. Part A*, Epub ahead of print, (December 2010), ISSN 1937-335X

Mizuno, M., Fujisawa, R., & Kuboki, Y. (2000). Type I Collagen-Induced Osteoblastic Differentiation of Bone-Marrow Cells Mediated by Collagen-α2β1 Integrin Interaction. *Journal of Cellular Physiology*, Vol.184, No.2, (2000), pp. (207-213), ISSN 1097-4652

Mizuno, S. & Glowacki, J. (1996). Three-Dimensional Composite of Demineralized Bone Powder and Collagen for in Vitro Analysis of Chondroinduction of Human Dermal Fibroblasts. *Biomaterials*, Vol.17, No.18, (1996), pp. (1819-1825), ISSN 0142-9612

Moss, I.L., Gordon, L., Woodhouse, K.A., Whyne, C.M., & Yee, A.J. (2010). A Novel Thiol-Modified-Hyaluronan and Elastin-Like Polypetide Composite Material for Tissue Engineering of the Nucleus Pulposus of the Intervertebral Disc. *Spine*, Epub ahead of print, (December 2010), ISSN 1528-1159

Moxham, J.P., Wong, K.K., & Kibblewhite, D.J. (2009). Transforming Growth Factor-β1 Shows An Incremental Osteoinductive Dose-Response Relationship. *Laryngoscope*, Vol.119, No.1, (January 2009), pp. (126-130), ISSN 1531-4995

Muschler, G.F., Nakamoto, C., & Griffith, L.G. (2004). Engineering Principles of Clinical Cell-Based Tissue Engineering. *The Journal of Bone and Joint Surgery*, Vol.86, No.7, (2004), p. (1541)

Myers, T.J., Granero-Molto, F., Longobardi, L., Li, T., Yan, Y., & Spagnoli, A. (2010). Mesenchymal Stem Cells at the Intersection of Cell and Gene Therapy. *Expert Opinion on Biological Therapy*, Vol.10, No.12, (2010), pp. (1663-1679), ISSN 1471-2598

Nehrer, S., Dorotka, R., Domayer, S., Stelzeneder, D., & Kotz, R. (2009). Treatment of Full-Thickness Chondral Defects with Hyalograft C in the Knee: A Prospective Clinical Case Series with 2 to 7 Years' Follow-Up. *The American Journal of Sports Medicine*, Vol.37 Suppl 1, (November 2009), pp. (81S-87S), ISSN 1552-3365

Nimni, M.E., Cheung, D., Strates, B., Kodama, M., & Sheikh, K. (1987). Chemically Modified Collagen: A Natural Biomaterial for Tissue Replacement. *Journal of Biomedical Materials Research*, Vol.21, No.6, (June 1987), pp. (741-771), ISSN 0021-9304

Nuernberger, S., Cyran, N., Albrecht, C., Redl, H., Vécsei, V., & Marlovits, S. (2011). The Influence of Scaffold Architecture on Chondrocyte Distribution and Behavior in Matrix-Associated Chondrocyte Transplantation Grafts. *Biomaterials*, Vol.32, No.4, (February 2011), pp. (1032-1040), ISSN 1878-5905

Odorico, J.S., Kaufman, D.S., & Thomson, J.A. (2001). Multilineage Differentiation From Human Embryonic Stem Cell Lines. *Stem Cells*, Vol.19, No.3, (2001), pp. (193-204), ISSN 1549-4918

Oliveira, S.M., Ringshia, R.A., Legeros, R.Z., Clark, E., Yost, M.J., Terracio, L., & Teixeira, C.C. (2010). An Improved Collagen Scaffold for Skeletal Regeneration. *Journal of Biomedical Materials Research. Part A*, Vol.94, No.2, (August 2010), pp. (371-379), ISSN 1552-4965

Ott, H.C., Matthiesen, T.S., Goh, S.-K., Black, L.D., Kren, S.M., Netoff, T.I., Taylor, D.A., Ott, H.C., Matthiesen, T.S., Goh, S.-K., Black, L.D., Kren, S.M., Netoff, T.I., & Taylor, D.A. (2008). Perfusion-Decellularized Matrix: Using Nature's Platform to Engineer a Bioartificial Heart. *Nature Medicine*, Vol.14, No.2, (January 2008), p. (213-221), ISSN 1078-8956

Pabbruwe, M.B., Esfandiari, E., Kafienah, W., Tarlton, J.F., & Hollander, A.P. (2009). Induction of Cartilage Integration by a Chondrocyte/Collagen-Scaffold Implant. *Biomaterials*, Vol.30, No.26, (September 2009), pp. (4277-4286), ISSN 1878-5905

Pabbruwe, M.B., Kafienah, W., Tarlton, J.F., Mistry, S., Fox, D.J., & Hollander, A.P. (2010). Repair of Meniscal Cartilage White Zone Tears Using a Stem Cell/Collagen-Scaffold Implant. *Biomaterials*, Vol.31, No.9, (March 2010), pp. (2583-2591), ISSN 1878-5905

Pang, Y., Wang, X., Ucuzian, A.A., Brey, E.M., Burgess, W.H., Jones, K.J., Alexander, T.D., & Greisler, H.P. (2010). Local Delivery of a Collagen-Binding FGF-1 Chimera to Smooth Muscle Cells in Collagen Scaffolds for Vascular Tissue Engineering. *Biomaterials*, Vol.31, No.5, (February 2010), pp. (878-885), ISSN 1878-5905

Park, S.H., Cui, J.H., Park, S.R., & Min, B.H. (2009). Potential of Fortified Fibrin/Hyaluronic Acid Composite Gel As a Cell Delivery Vehicle for Chondrocytes. *Artificial Organs*, Vol.33, No.6, (June 2009), pp. (439-447), ISSN 1525-1594

Park, J.S., Yang, H.N., Woo, D.G., Jeon, S.Y., & Park, K.H. (2011). Chondrogenesis of Human Mesenchymal Stem Cells in Fibrin Constructs Evaluated in Vitro and in Nude Mouse and Rabbit Defects Models. *Biomaterials*, Vol.32, No.6, (February 2011), pp. (1495-1507), ISSN 1878-5905

Patterson, J., Siew, R., Herring, S.W., Lin, A.S., Guldberg, R., & Stayton, P.S. (2010). Hyaluronic Acid Hydrogels with Controlled Degradation Properties for Oriented Bone Regeneration. *Biomaterials*, Vol.31, No.26, (September 2010), pp. (6772-6781), ISSN 1878-5905

Peretti, G.M., Xu, J.W., Bonassar, L.J., Kirchhoff, C.H., Yaremchuk, M.J., & Randolph, M.A. (2006). Review of Injectable Cartilage Engineering Using Fibrin Gel in Mice and Swine Models. *Tissue Engineering*, Vol.12, No.5, (May 2006), pp. (1151-1168), ISSN 1076-3279

Place, E.S., Evans, N.D., & Stevens, M.M. (2009). Complexity in Biomaterials for Tissue Engineering. *Nature Materials*, Vol.8, No.6, (2009), pp. (457-470), ISSN 1476-1122

Pradeep, A.R. & Sharma, A. (2011). Autologous Platelet Rich Fibrin in the Treatment of Mandibular Degree Ii Furcation Defects: A Randomized Clinical Trial. *Journal of Periodontology*, Epub ahead of print, (February 2011), ISSN 1943-3670

Putney, S.D. & Burke, P.A. (1998). Improving Protein Therapeutics with Sustained-Release Formulations. *Nature Biotechnology*, Vol.16, No.2, (1998), pp. (153-157), ISSN 1087-0156

Reddi, A.H., (1998). Role of Morphogenetic Proteins in Skeletal Tissue Engineering and Regeneration. *Nature Biotechnology*, Vol.16, No.3, (1998), pp. (247-252), ISSN 1087-0156

Reddi, A.H., (2001a). Bone Morphogenetic Proteins: From Basic Science to Clinical Applications. *The Journal of Bone and Joint Surgery*, Vol.83, No.S1, Part 1, (2001), p. (S1-S6)

Reddi, A.H., (2001b). Interplay Between Bone Morphogenetic Proteins and Cognate Binding Proteins in Bone and Cartilage Development: Noggin, Chordin and DAN. *Arthritis Research*, Vol.3, No.1, (2001), pp. (1-5), ISSN 1465-9905

Rezwan, K., Chen, Q.Z., Blaker, J.J., & Boccaccini, A.R. (2006). Biodegradable and Bioactive Porous Polymer/Inorganic Composite Scaffolds for Bone Tissue Engineering. *Biomaterials*, Vol.27, No.18, (2006), pp. (3413-3431), ISSN 0142-9612

Ricard-Blum, S. & Ruggiero, F. (2005). The Collagen Superfamily: From the Extracellular Matrix to the Cell Membrane. *Pathologie Biologie*, Vol.53, No.7, (2005), pp. (430-442), ISSN 0369-8114

Richardson, T.P., Peters, M.C., Ennett, A.B., & Mooney, D.J. (2001). Polymeric System for Dual Growth Factor Delivery. *Nature Biotechnology*, Vol.19, No.11, (2001), pp. (1029-1034), ISSN 1087-0156

Rodkey, W.G., Steadman, J.R., & Li, S.T. (1999). A Clinical Study of Collagen Meniscus Implants to Restore the Injured Meniscus. *Clinical Orthopaedics and Related Research*, Vol. 367, No.Suppl, (October 1999), pp. (S281-S292), ISSN 0009-921X

Ruoslahti, E., (1996). RGD and Other Recognition Sequences for Integrins. *Annual Review of Cell and Developmental Biology*, Vol.12, No.1, (1996), pp. (697-715), ISSN 1081-0706

Ryu, Y.M., Hah, Y.S., Park, B.W., Kim, D.R., Roh, G.S., Kim, J.R., Kim, U.K., Rho, G.J., Maeng, G.H., & Byun, J.H. (2010). Osteogenic Differentiation of Human Periosteal-Derived Cells in a Three-Dimensional Collagen Scaffold. *Molecular Biology Reports*, Epub ahead of print, (January 2010), ISSN 1573-4978

Sah, H., (1999). Protein Instability Toward Organic Solvent/Water Emulsification: Implications for Protein Microencapsulation Into Microspheres. *PDA Journal of Pharmaceutical Science and Technology*, Vol.53, No.1, (1999), p. (3), ISSN 1079-7440

Saito, A., Suzuki, Y., Ogata, S., Ohtsuki, C., & Tanihara, M. (2003). Activation of Osteo-Progenitor Cells by a Novel Synthetic Peptide Derived From the Bone Morphogenetic Protein-2 Knuckle Epitope. *Biochimica Et Biophysica Acta (BBA)-Proteins & Proteomics*, Vol.1651, No.1-2, (2003), pp. (60-67), ISSN 1570-9639

Sandhu, M.S., Dunger, D.B., & Giovannucci, E.L. (2002). Insulin, Insulin-Like Growth Factor-I (IGF-I), IGF Binding Proteins, Their Biologic Interactions, and Colorectal Cancer. *Journal of the National Cancer Institute*, Vol.94, No.13, (2002), p. (972), ISSN 0027-8874

Sawaguchi, N., Majima, T., Funakoshi, T., Shimode, K., Harada, K., Minami, A., & Nishimura, S. (2010). Effect of Cyclic Three-Dimensional Strain on Cell Proliferation and Collagen Synthesis of Fibroblast-Seeded Chitosan-Hyaluronan Hybrid Polymer Fiber. *Journal of Orthopaedic Science*, Vol.15, No.4, (July 2010), pp. (569-577), ISSN 1436-2023

Scaglione, S., Lazzarini, E., Ilengo, C., & Quarto, R. (2010). A Composite Material Model for Improved Bone Formation. *Journal of Tissue Engineering and Regenerative Medicine*, Vol.4, No.7, (2010), pp. (505-513), ISSN 1932-7005

Schagemann, J.C., Kurz, H., Casper, M.E., Stone, J.S., Dadsetan, M., Yu-Long, S., Mrosek, E.H., Fitzsimmons, J.S., O'Driscoll, S.W., & Reinholz, G.G. (2010). The Effect of Scaffold Composition on the Early Structural Characteristics of Chondrocytes and Expression of Adhesion Molecules. *Biomaterials*, Vol.31, No.10, (April 2010), pp. (2798-2805), ISSN 1878-5905

Schense, J.C., Bloch, J., Aebischer, P., & Hubbell, J.A. (2000). Enzymatic Incorporation of Bioactive Peptides Into Fibrin Matrices Enhances Neurite Extension. *Nature Biotechnology*, Vol.18, No.4, (2000), pp. (415-419), ISSN 1087-0156

Schmoekel, H., Schense, J.C., Weber, F.E., Grätz, K.W., Gnägi, D., Müller, R., & Hubbell, J.A. (2004). Bone Healing in the Rat and Dog with Nonglycosylated BMP-2 Demonstrating Low Solubility in Fibrin Matrices. *Journal of Orthopaedic Research*, Vol.22, No.2, (2004), pp. (376-381), ISSN 1554-527X

Schmoekel, H.G., Weber, F.E., Schense, J.C., Grätz, K.W., Schawalder, P., & Hubbell, J.A. (2005). Bone Repair with a Form of BMP-2 Engineered for Incorporation Into Fibrin Cell Ingrowth Matrices. *Biotechnology and Bioengineering*, Vol.89, No.3, (2005), pp. (253-262), ISSN 1097-0290

Schwartz, Z., Griffon, D.J., Fredericks, L.P., Lee, H.B., & Weng, H.Y. (2011). Hyaluronic Acid and Chondrogenesis of Murine Bone Marrow Mesenchymal Stem Cells in Chitosan Sponges. *American Journal of Veterinary Research*, Vol.72, No.1, (January 2011), pp. (42-50), ISSN 0002-9645

Scotti, C., Wirz, D., Wolf, F., Schaefer, D.J., Bürgin, V., Daniels, A.U., Valderrabano, V., Candrian, C., Jakob, M., Martin, I., & Barbero, A. (2010). Engineering Human Cell-Based, Functionally Integrated Osteochondral Grafts by Biological Bonding of Engineered Cartilage Tissues to Bony Scaffolds. *Biomaterials*, Vol.31, No.8, (March 2010), pp. (2252-2259), ISSN 1878-5905

Secretan, C., Bagnall, K.M., & Jomha, N.M. (2010). Effects of Introducing Cultured Human Chondrocytes Into a Human Articular Cartilage Explant Model. *Cell Tissue Res*, Vol.339, No.2, (February 2010), pp. (421-427), ISSN 1432-0878

Selvaraj, V., Plane, J.M., Williams, A.J., & Deng, W. (2010). Switching Cell Fate: The Remarkable Rise of Induced Pluripotent Stem Cells and Lineage Reprogramming Technologies. *Trends in Biotechnology*, Vol.28, No.4, (April 2010), pp. (214-223)

Shi, M., Kretlow, J.D., Spicer, P.P., Tabata, Y., Demian, N., Wong, M.E., Kasper, F.K., & Mikos, A.G. (2011). Antibiotic-Releasing Porous Polymethylmethacrylate/Gelatin/Antibiotic Constructs for Craniofacial Tissue Engineering. *Journal of Controlled Release*, Epub ahead of print, (2011), ISSN 0168-3659

Sierra, D.H., (1993). Fibrin Sealant Adhesive Systems: A Review of Their Chemistry, Material Properties and Clinical Applications. *Journal of Biomaterials Applications*, Vol.7, No.4, (April 1993), pp. (309-352), ISSN 0885-3282

Silva, G.A., Czeisler, C., Niece, K.L., Beniash, E., Harrington, D.A., Kessler, J.A., & Stupp, S.I. (2004). Selective Differentiation of Neural Progenitor Cells by High-Epitope Density Nanofibers. *Science*, Vol.303, No.5662, (2004), p. (1352), ISSN 0036-8075

Singh, K., Moyer, H., Williams, J.K., Schwartz, Z., & Boyan, B.D. (2011). Fibrin Glue: A Scaffold for Cellular-Based Therapy in a Critical-Sized Defect. *Annals of Plastic Surgery*, Vol.66, No.3, (March 2011), pp. (301-305), ISSN 1536-3708

Sirivisoot, S., Pareta, R., & Webster, T.J. (2011). Electrically Controlled Drug Release From Nanostructured Polypyrrole Coated on Titanium. *Nanotechnology*, Vol.22, (2011), p. (085101)

Soon, M.Y., Hassan, A., Hui, J.H., Goh, J.C., & Lee, E.H. (2007). An Analysis of Soft Tissue Allograft Anterior Cruciate Ligament Reconstruction in a Rabbit Model: A Short-Term Study of the Use of Mesenchymal Stem Cells to Enhance Tendon Osteointegration. *The American Journal of Sports Medicine*, Vol.35, No.6, (June 2007), pp. (962-971), ISSN 0363-5465

Spicer, P.P. & Mikos, A.G. (2010). Fibrin Glue As a Drug Delivery System. *Journal of Controlled Release*, Epub ahead of print (July 2010), ISSN 1873-4995

Spotnitz, W.D., (2010). Fibrin Sealant: Past, Present, and Future: A Brief Review. *World Journal of Surgery*, Vol.34, No.4, (April 2010), pp. (632-634), ISSN 1432-2323

Sreejalekshmi, K.G. & Nair, P.D. (2011). Biomimeticity in Tissue Engineering Scaffolds Through Synthetic Peptide Modifications-Altering Chemistry for Enhanced Biological Response. *Journal of Biomedical Materials Research. Part A*, Vol.96, No.2, (February 2011), pp. (477-491), ISSN 1552-4965

Stabile, K.J., Odom, D., Smith, T.L., Northam, C., Whitlock, P.W., Smith, B.P., Van Dyke, M.E., & Ferguson, C.M. (2010). An Acellular, Allograft-Derived Meniscus Scaffold in An Ovine Model. *Arthroscopy: The Journal of Arthroscopic & Related Surgery*, Vol.26, No.7, (July 2010), pp. (936-948), ISSN 1526-3231

Stone, K.R., Steadman, J.R., Rodkey, W.G., & Li, S.T. (1997). Regeneration of Meniscal Cartilage with Use of a Collagen Scaffold. Analysis of Preliminary Data. *The Journal of Bone and Joint Surgery. American Volume*, Vol.79, No.12, (December 1997), pp. (1770-1777), ISSN 0021-9355

Sundelacruz, S. & Kaplan, D.L. (2009). Stem Cell-And Scaffold-Based Tissue Engineering Approaches to Osteochondral Regenerative Medicine. *Seminars in Cell & Developmental Biology*, Vol.20, No.6, (August 2009), pp. (646-655), ISSN 1096-3634

Suzuki, Y., Tanihara, M., Suzuki, K., Saitou, A., Sufan, W., & Nishimura, Y. (2000). Alginate Hydrogel Linked with Synthetic Oligopeptide Derived From BMP-2 Allows Ectopic Osteoinduction in Vivo. *Journal of Biomedical Materials Research*, Vol.50, No.3, (2000), pp. (405-409), ISSN 1097-4636

Tabata, Y., Gutta, S., & Langer, R. (1993). Controlled Delivery Systems for Proteins Using Polyanhydride Microspheres. *Pharmaceutical Research*, Vol.10, No.4, (1993), pp. (487-496)

Tammi, M.I., Day, A.J., & Turley, E.A. (2002). Hyaluronan and Homeostasis: A Balancing Act. *The Journal of Biological Chemistry*, Vol.277, No.7, (February 2002), p. (4581-4584), ISSN 0021-9258

Tan, H., Chu, C.R., Payne, K.A., & Marra, K.G. (2009). Injectable in Situ Forming Biodegradable Chitosan-Hyaluronic Acid Based Hydrogels for Cartilage Tissue Engineering. *Biomaterials*, Vol.30, No.13, (May 2009), pp. (2499-2506), ISSN 1878-5905

Toole, B.P., (2001). Hyaluronan in Morphogenesis. *Seminars in Cell & Developmental Biology*, Vol.12, No.2, (April 2001), pp. (79-87), ISSN 1084-9521

Toole, B.P., (2004). Hyaluronan: From Extracellular Glue to Pericellular Cue. *Nature Reviews Cancer*, Vol.4, No.7, (2004), pp. (528-539), ISSN 1474-175X

Topuz, K., Colak, A., Kaya, S., Simsek, H., Kutlay, M., Demircan, M.N., & Velioglu, M. (2009). Two-Level Contiguous Cervical Disc Disease Treated with Peek Cages Packed with Demineralized Bone Matrix: Results of 3-Year Follow-Up. *Eur Spine J*, Vol.18, No.2, (February 2009), pp. (238-243), ISSN 1432-0932

Traphagen, S. & Yelick, P.C. (2009). Reclaiming a Natural Beauty: Whole-Organ Engineering with Natural Extracellular Materials. *Regenerative Medicine*, Vol.4, No.5, (2009), pp. (747-758)

Tsigkou, O., Pomerantseva, I., Spencer, J.A., Redondo, P.A., Hart, A.R., O'Doherty, E., Lin, Y., Friedrich, C.C., Daheron, L., Lin, C.P., & others (2010). Engineered Vascularized Bone Grafts. *Proceedings of the National Academy of Sciences*, Vol.107, No.8, (February 2010), p. (3311-3316)

Uebersax, L., Merkle, H.P., & Meinel, L. (2008). Insulin-Like Growth Factor I Releasing Silk Fibroin Scaffolds Induce Chondrogenic Differentiation of Human Mesenchymal Stem Cells. *Journal of Controlled Release*, Vol.127, No.1, (2008), pp. (12-21), ISSN 0168-3659

Uebersax, L., Merkle, H.P., & Meinel, L. (2009). Biopolymer-Based Growth Factor Delivery for Tissue Repair: From Natural Concepts to Engineered Systems. *Tissue Engineering Part B: Reviews*, Vol.15, No.3, (2009), pp. (263-289), ISSN 1937-3368

Underwood, P.A., Bennett, F.A., Kirkpatrick, A., Bean, P.A., & Moss, B.A. (1995). Evidence for the Location of a Binding Sequence for the Alpha 2 Beta 1 Integrin of Endothelial Cells, in the Beta 1 Subunit of Laminin. *Biochemical Journal*, Vol.309, No.Pt 3, (1995), p. (765)

Urist, M.R., (1965). Bone: Formation by Autoinduction. *Science*, Vol.150, No.3698, (1965), p. (893), ISSN 0036-8075

Van der Rest, M. & Garrone, R. (1991). Collagen Family of Proteins. *The FASEB Journal*, Vol.5, No.13, (1991), p. (2814-2823)

Wagner, M., Kiapur, N., Wiedmann-Al-Ahmad, M., Hübner, U., Al-Ahmad, A., Schön, R., Schmelzeisen, R., Mülhaupt, R., & Gellrich, N.C. (2007). Comparative in Vitro Study of the Cell Proliferation of Ovine and Human Osteoblast-Like Cells on Conventionally and Rapid Prototyping Produced Scaffolds Tailored for Application As Potential Bone Replacement Material. *Journal of Biomedical Materials Research. Part A*, Vol.83, No.4, (December 2007), pp. (1154-1164), ISSN 1549-3296

Wang, J. & Glimcher, M.J. (1999a). Characterization of Matrix-Induced Osteogenesis in Rat Calvarial Bone Defects: I. Differences in the Cellular Response to Demineralized Bone Matrix Implanted in Calvarial Defects and in Subcutaneous Sites. *Calcified Tissue International*, Vol.65, No.2, (1999), pp. (156-165), ISSN 0171-967X

Wang, J. & Glimcher, M.J. (1999b). Characterization of Matrix-Induced Osteogenesis in Rat Calvarial Bone Defects: II. Origins of Bone-Forming Cells. *Calcified Tissue International*, Vol.65, No.6, (1999), pp. (486-493), ISSN 0171-967X

Wang, Z.H., He, X.J., Yang, Z.Q., & Tu, J.B. (2010a). Cartilage Tissue Engineering with Demineralized Bone Matrix Gelatin and Fibrin Glue Hybrid Scaffold: An in Vitro Study. *Artificial Organs*, Vol.34, No.2, (February 2010), pp. (161-166), ISSN 1525-1594

Wang, W., Li, B., Yang, J., Xin, L., Li, Y., Yin, H., Qi, Y., Jiang, Y., Ouyang, H., & Gao, C. (2010b). The Restoration of Full-Thickness Cartilage Defects with BMSCs and TGF-β1 Loaded PLGA/Fibrin Gel Constructs. *Biomaterials*, Vol.31, No.34, (December 2010), pp. (8964-8973), ISSN 1878-5905

Wang, N.X. & von Recum, H.A. (2011). Affinity-Based Drug Delivery. *Macromolecular Bioscience*, Vol.11, No.3, (March 2011), pp. (321-332), ISSN 1616-5195

Wang, J., Yang, R., Gerstenfeld, L.C., & Glimcher, M.J. (2000). Characterization of Demineralized Bone Matrix-Induced Osteogenesis in Rat Calvarial Bone Defects: III. Gene and Protein Expression. *Calcified Tissue International*, Vol.67, No.4, (2000), pp. (314-320), ISSN 0171-967X

Wegener, B., Schrimpf, F.M., Bergschmidt, P., Pietschmann, M.F., Utzschneider, S., Milz, S., Jansson, V., & Müller, P.E. (2010). Cartilage Regeneration by Bone Marrow Cells-Seeded Scaffolds. *Journal of Biomedical Materials Research. Part A*, Vol.95, No.3, (December 2010), pp. (735-740), ISSN 1552-4965

Welsch, G.H., Mamisch, T.C., Zak, L., Blanke, M., Olk, A., Marlovits, S., & Trattnig, S. (2010). Evaluation of Cartilage Repair Tissue After Matrix-Associated Autologous Chondrocyte Transplantation Using a Hyaluronic-Based or a Collagen-Based Scaffold with Morphological MOCART Scoring and Biochemical T2 Mapping: Preliminary Results. *The American Journal of Sports Medicine*, Vol.38, No.5, (May 2010), pp. (934-942), ISSN 1552-3365

Wenk, E., Murphy, A.R., Kaplan, D.L., Meinel, L., Merkle, H.P., & Uebersax, L. (2010). The Use of Sulfonated Silk Fibroin Derivatives to Control Binding, Delivery and Potency of FGF-2 in Tissue Regeneration. *Biomaterials*, Vol.31, No.6, (2010), pp. (1403-1413), ISSN 0142-9612

Williams, C.G., Kim, T.K., Taboas, A., Malik, A., Manson, P., & Elisseeff, J. (2003). In Vitro Chondrogenesis of Bone Marrow-Derived Mesenchymal Stem Cells in a Photopolymerizing Hydrogel. *Tissue Engineering*, Vol.9, No.4, (2003), pp. (679-688), ISSN 1076-3279

Woods, T. & Gratzer, P.F. (2005). Effectiveness of Three Extraction Techniques in the Development of a Decellularized Bone-Anterior Cruciate Ligament-Bone Graft. *Biomaterials*, Vol.26, No.35, (December 2005), pp. (7339-7349), ISSN 0142-9612

Wu, C.H., Ko, C.S., Huang, J.W., Huang, H.J., & Chu, I.M. (2010). Effects of Exogenous Glycosaminoglycans on Human Chondrocytes Cultivated on Type II Collagen Scaffolds. *Journal of Materials Science. Materials in Medicine*, Vol.21, No.2, (2010), pp. (725-729), ISSN 0957-4530

Yamada, Y., Katagiri, F., Hozumi, K., Kikkawa, Y., & Nomizu, M. (2011). Cell Behavior on Protein Matrices Containing Laminin [Alpha] 1 Peptide AG73. *Biomaterials*, Article in press, (2011), ISSN 0142-9612

Yamane, S., Iwasaki, N., Majima, T., Funakoshi, T., Masuko, T., Harada, K., Minami, A., Monde, K., & Nishimura, S. (2005). Feasibility of Chitosan-Based Hyaluronic Acid Hybrid Biomaterial for a Novel Scaffold in Cartilage Tissue Engineering* 1. *Biomaterials*, Vol.26, No.6, (2005), pp. (611-619), ISSN 0142-9612

Ye, Q., Zünd, G., Jockenhoevel, S., Hoerstrup, S.P., Schoeberlein, A., Grunenfelder, J., & Turina, M. (2000). Tissue Engineering in Cardiovascular Surgery: New Approach to Develop Completely Human Autologous Tissue* 1. *European Journal of Cardio-Thoracic Surgery*, Vol.17, No.4, (2000), pp. (449-454), ISSN 1010-7940

Yim, E.K.F. & Leong, K.W. (2005). Significance of Synthetic Nanostructures in Dictating Cellular Response. *Nanomedicine: Nanotechnology, Biology, and Medicine*, Vol.1, No.1, (2005), pp. (10-21)

Zhang, L., Furst, E.M., & Kiick, K.L. (2006). Manipulation of Hydrogel Assembly and Growth Factor Delivery Via the Use of Peptide-Polysaccharide Interactions. *Journal of Controlled Release : Official Journal of the Controlled Release Society*, Vol.114, No.2, (2006), pp. (130-142), ISSN 0168-3659

Zhang, Y., Song, J., Shi, B., Wang, Y., Chen, X., Huang, C., Yang, X., Xu, D., Cheng, X., & Chen, X. (2007). Combination of Scaffold and Adenovirus Vectors Expressing Bone Morphogenetic Protein-7 for Alveolar Bone Regeneration at Dental Implant Defects. *Biomaterials*, Vol.28, No.31, (November 2007), pp. (4635-4642), ISSN 0142-9612

Zheng Shu, X., Liu, Y., Palumbo, F.S., Luo, Y., & Prestwich, G.D. (2004). In Situ Crosslinkable Hyaluronan Hydrogels for Tissue Engineering. *Biomaterials*, Vol.25, No.7-8, (2004), pp. (1339-1348), ISSN 0142-9612

Zhou, L., Pomerantseva, I., Bassett, E.K., Bowley, C.M., Zhao, X., Bichara, D.A., Kulig, K., Vacanti, J., Randolph, M., & Sundback, C. (2011). Engineering Ear Constructs with a Composite Scaffold to Maintain Dimensions. *Tissue Engineering. Part A*, Epub ahead of print, (February 2011), ISSN 1937-335X

Zhu, G., Mallery, S.R., & Schwendeman, S.P. (2000). Stabilization of Proteins Encapsulated in Injectable Poly (Lactide- Co-Glycolide). *Nature Biotechnology*, Vol.18, No.1, (January 2000), pp. (52-57), ISSN 1087-0156

Zou, X., Li, H., Chen, L., Baatrup, A., Bünger, C., & Lind, M. (2004). Stimulation of Porcine Bone Marrow Stromal Cells by Hyaluronan, Dexamethasone and rhBMP-2. *Biomaterials*, Vol.25, No.23, (2004), pp. (5375-5385), ISSN 0142-9612

Zouani, O.F., Chollet, C., Guillotin, B., & Durrieu, M.C. (2010). Differentiation of Pre-Osteoblast Cells on Poly (Ethylene Terephthalate) Grafted with RGD and/or BMPs Mimetic Peptides. *Biomaterials, Vol.31, No.32, (November* 2010), pp. (8245-8253), ISSN 0142-9612

19

Angiogenesis and Vascularity for Tissue Engineering Applications

Remo A. Largo[1], Venkat M. Ramakrishnan[1], Martin Ehrbar[2], Algirdas Ziogas[2], Jan A. Plock[3] and Daniel Eberli[1]
[1]*Division of Urology, University Hospital Zurich*
[2]*Division of Obstetrics, University Hospital Zurich*
[3]*Division of Plastic and Hand Surgery, University Hospital Zurich*
Switzerland

1. Introduction

Tissue engineering is a field of medicine that has experienced significant growth in prominence over the past three decades. Though traditional interventions exist for many medical maladies, tissue engineering aims to combat such disorders through the synthesis of body tissues and organs, resulting in functional implants. Tissue engineering takes an innovative approach, often utilizing autologous stem cells for tissue construction or materials that are biocompatible while avoiding immune rejection (Nomi *et al.*, 2002).

Despite its immense successes, a major hurdle still faces tissue engineering. Large volumes of implanted tissue are unable to stimulate the formation of necessary blood vessels required for their survival. In the body, naturally occurring, equivalent vascular networks serve vital functions in gas and nutrient exchange, metabolic processes, and waste expulsion. Though individual, large vessels have been successfully engineered for implant, it is still exceptionally difficult to fashion a stable and sustainable network of vessels for large volumes of tissue (Nomi *et al.*, 2002). Neovascularization after tissue damage requires a level of positive and negative control that has not been successfully replicated in a laboratory environment to date. As such, rapid *de novo* synthesis of a controlled, established vascular network remains a challenge today.

Angiogenesis is the morphogenic process of forming new blood vessels from pre-existing ones (Laschke *et al.*, 2006; Dai and Rabie, 2007; Li and Rabie, 2007). This event plays an important, normal physiological role in wound healing, tissue repair, pregnancy, and exercise (Ferrara and Davis-Smyth, 1997), and exists in contrast to vasculogenesis (the formation of the every first blood vessels in the body, and especially predominant in embryological development). Yet, the abuse of angiogenesis, leading to an uncontrolled vascular formation as a consequence of epigenetic influence, nucleotide polymorphisms, or endocrine irregularities can also result in tumor formation (Verbridge *et al.*, 2010). However, angiogenesis is clearly an activity that is central to development and tissue maintenance. Successful modulation of angiogenesis can have profound therapeutic outcomes for organs and tissues deprived of an adequate, stable vasculature. Studies from the last two decades have shown that the manipulation of various factors directly influences angiogenic outcome.

Angiogenesis requires the activity of soluble factors such as Vascular Endothelial Growth Factor (VEGF; outlined in the next section and in Tables 2 & 3), basic Fibroblast Growth Factor (bFGF), Platelet-derived Growth Factor (PDGF), Transforming Growth Factor-β (TGF- β) (Nomi *et al.*, 2002; Dai and Rabie, 2007; Li and Rabie, 2007; Kanczler and Oreffo, 2008; Bates, 2010), Keratinocyte Growth Factor (KGF) (Elia *et al.*, 2010), Hepatocyte Growth Factor (HGF) (Hoot *et al.*, 2010), Ephrin-B2 (Herbert *et al.*, 2009), and Angiopoietin(Han *et al.*, 2010) (Table 1). Morphologically, angiogenesis can be attributed to endothelial cell migration and proliferation as well as pericyte recruitment, migration and differentiation (Egginton, 2010).

Molecule	Known Properties	Citations
Basic Fibroblast Growth Factor (bFGF)	Stimulates activity of fibroblasts, neurons, smooth muscle cells, and endothelial cells; acts via tyrosine kinase receptors; induces production of VEGF during angiogenic stimulation.	(Pepper *et al.*, 1992; Lee *et al.*, 2003; Arkudas *et al.*, 2007; Jung *et al.*, 2010; Wu *et al.*, 2010)
Platelet-derived Growth Factor (PDGF)	Certain isoforms possess survival and mitogenic functions; implicated in tumor angiogenesis; compared to VEGF, PDGF: (a) has comparable angiogenic activity, (b) produces blood vessels with decreased permeability and leakage, and (c) is thought to produce functionally different blood vessels.	(Li *et al.*, 2010; Wu *et al.*, 2010)
Transforming Growth Factor β (TGF-β)	Promotes and inhibits angiogenesis and tumor invasion via stimulation of Hepatocyte Growth Factor (HGF) expression.	(Hoot *et al.*, 2010)
Keratinocyte Growth Factor (KGF)	Member of FGF family; certain isoforms have been implicated in wound healing as well as the inhibition of neovascularization.	(Wang *et al.*, 2010)
Hepatocyte Growth Factor (HGF)	Stimulates endothelial cell growth, migration, scatter, and elongation independently of VEGF.	(Hoot *et al.*, 2010)
Ephrin-B2	Implicated in arterial/venous differentiation (see section: Modulation of Notch Signaling).	(Herbert *et al.*, 2009)
Angiopoietin	Signals through Tie receptors; certain isoforms are angiogenic inhibitors, while others promote inhibition of apoptosis; Stabilize blood vessels and reduce leakage.	(Han *et al.*, 2010)

Table 1. Notable Ancillary Growth Factors and Their Properties

2. VEGF and its receptors: a brief overview

In studying the development of cells, tissues, and organs, Vascular Endothelial Growth Factor (VEGF) has been identified as a key, though not sole, proponent of angiogenesis (Takahashi and Shibuya, 2005; Arkudas *et al.*, 2007; Dai and Rabie, 2007; Yla-Herttuala, 2009;

Bates, 2010; Elia *et al.*, 2010). VEGF exists in six classes (Table 2), lettered A through F, though the vast amount of literature concerning VEGF-A indicates that it is by far the most widely understood of all. VEGF-A exists in at least eight homodimeric isoforms, all conceived by alternative mRNA splicing(Ferrara, 2009) and differing by their amino acid number. Of these, $VEGF_{121}$ and $VEGF_{165}$ are predominant and show great promise in tissue engineering applications.

VEGF Class	Known Properties	Citations
A	Induces proliferation of arterial, venous, and lymphatic vascular endothelial cells; stimulates monocyte chemotaxis; hematopoetic effects; eight isoforms in humans.	(Ferrara, 2009)
B	May play role in atrial conduction, but not required for cardiovascular development.	(Olofsson *et al.*, 1996a; Olofsson *et al.*, 1996b; Aase *et al.*, 2001)
C	Involved in embryonic angiogenesis, lymphangiogenesis, & lymphatic vessel maintenance; mitogenic for cultured endothelial cells.	(Kukk *et al.*, 1996; Orlandini *et al.*, 1996; Jeltsch *et al.*, 1997; Yamada *et al.*, 1997; Dumont *et al.*, 1998)
D	Thought to be involved in pulmonary development, endothelial cell mitogen; Function(s) still generally unclear.	(Orlandini *et al.*, 1996; Yamada *et al.*, 1997; Achen *et al.*, 1998; Farnebo *et al.*, 1999)
E	Proteins are encoded by Orf-viruses and predominantly expressed in sheep, goats, and rarely in humans.	(Lyttle *et al.*, 1994; Meyer *et al.*, 1999)
F	Derived from snake venom; enhances the formation of vascular fenestrations in guinea pigs.	(Klein and Catargi, 2007; Matsunaga *et al.*, 2009; Yamazaki *et al.*, 2009)
Placental Growth Factor (PlGF)	Induces vascular permeability; supplements VEGF activity during wound-healing; may enhance VEGF-driven angiogenesis.	(Carmeliet *et al.*, 2001; Adini *et al.*, 2002; Hattori *et al.*, 2002; Luttun *et al.*, 2002; Odorisio *et al.*, 2002)

Table 2. Six General VEGF Classes

The binding of VEGF to any of its many receptors, such as VEGF-R1 (Flt-1) (Shibuya *et al.*, 1990; de Vries *et al.*, 1992), VEGF-R2 (KDR) (Terman *et al.*, 1991), VEGF-R3 (Flt-4) (Fitz *et al.*, 1997), and Neuropilin (NRP-1 and -2) (Soker *et al.*, 1996; Soker *et al.*, 1998; Dallas *et al.*, 2008), has been shown to trigger a signaling cascade that results in the activation of angiogenesis (Stefanini *et al.*, 2009). Not surprisingly, VEGF receptor and ligand placement and density have been implicated in the successful transmission of the signaling cascades required for angiogenesis (Stefanini *et al.*, 2009). Modulating receptors and their densities appear to be a logical avenue for further efforts in modulating neovascularization. Further, Heparin Sulfate Proteoglycan (HSPG) has been implicated in the binding of VEGF to their receptors, but its exact function remains unclear (Ferrara *et al.*, 2003; Lee *et al.*, 2010). Preliminary findings suggest that HSPGs inhibit VEGF binding to its receptors, and consequently, angiogenic activity (Lee *et al.*, 2010).

3. Modulations to VEGF and its delivery

Though VEGF has a significant influence on cell migration, proliferation, and vasodilation, the uncontrolled or sole use of VEGF *in vivo* has been shown to result in the disordered growth of blood vessels into a dense mass (hemangioma), malignant tumor angiogenesis, and the assembly of leaky vessels (Takahashi and Shibuya, 2005; Bates, 2010). As such, the fundamental function of VEGF in physiological development and maintenance, contrasted by its role in tumorigenesis and vessel instability, is paradoxical. In instances where VEGF is heterogeneously present in a microenvironment, areas of high VEGF expression resulted in abnormal angiogenesis. However, implanting VEGF-transfected myoblasts, with each cell equally producing VEGF over time, led to the formation of a stable, normal vascular network (Misteli *et al.*, 2010). The latter provides a compelling case for carefully-controlled VEGF release, distribution, and kinetics in tissue engineering applications. As discussed below, several current tissue-engineering efforts are positioned to resolve this dilemma.

It should not be assumed that blind administration of VEGF is solely responsible for a successful angiogenic effort. Many studies have illustrated that the promotion of successful angiogenesis depends on a prolonged exposure to a low dose of VEGF (Wernike *et al.*, 2010), while other studies claim that micro-environmental conditions must be taken into account (Ferrara and Davis-Smyth, 1997; Misteli *et al.*, 2010). With regards to the latter, techniques like FACS purification (Misteli *et al.*, 2010) and microdialysis (Hoier *et al.*, 2010; Marcus *et al.*, 2010) are becoming increasingly prevalent in closely monitoring VEGF expression in the microenvironment. Further, many attempts at eliciting controlled angiogenesis also focus on coupling the properties of VEGF with certain other growth factors – most notably bFGF (Arkudas *et al.*, 2007) or KGF (Elia *et al.*, 2010) (see Table 1); stable vessels were formed when VEGF was combined with either of these two factors.

There exist contradictions regarding the parameters of tissue exposure to VEGF. In one comparison, a study highlighting growth factor implementation in orthopedic applications suggested that VEGF delivery for over 14 days may have interfered in the vascularization during bone healing and restoration (Wernike *et al.*, 2010). This suggestion challenges an argument that a longer VEGF exposure (of approximately one month) was necessary for the production of stable, but leaky, vessels as brief exposure to VEGF (less than 15 days) resulted in the formation of unstable vessels. Further, these vessels actually degenerated after VEGF delivery cessation (Tafuro *et al.*, 2009). Disparities such as this are common in the literature and demonstrate the sheer complexity of VEGF activity in various tissue environments.

The view that successful VEGF delivery and angiogenic response are strictly dose-dependent is challenged by factors such as delivery kinetics, which appear to be critical for proper vasculature (Borselli *et al.*, 2010; Wernike *et al.*, 2010). For instance, a characteristic problem with the bolus delivery of VEGF is the outcome of variable systemic effects (Matoka and Cheng, 2009) such as haemorrhage, hypotension, or flu-like symptoms (Benjamin *et al.*, 1999). This method of delivery also fails to achieve the prolonged supply of physiological low dose of VEGF necessary to produce a mature, lasting vascular network. The lack of lasting vessel formation observed in many therapeutic trials is probably due to difficulties in the delivery of VEGF, a growth factor with an apparently narrow therapeutic window (Hariawala *et al.*, 1996; Lee *et al.*, 2000; Dor *et al.*, 2003; Ozawa *et al.*, 2004), in the optimal time and dose for maintaining sufficient vascularity.

A need exists for biomaterials or stable scaffolds that enable slow, sustained VEGF release(Rocha *et al.*, 2008) for a predictable and functional outcome. Studies in different laboratories have indicated that incorporation of growth factor into slow-release polymer formulations could present a means for better control of dose, location, and duration of active signals in tissue (Edelman *et al.*, 1991; Lee *et al.*, 2000; Sheridan *et al.*, 2000; Ehrbar *et al.*, 2004). The kinetics of VEGF delivery today depends on the physical properties (such as cross-linking and porosity, for instance) of biocompatible conduits such as fibrin-gels (Ehrbar *et al.*, 2004; Arkudas *et al.*, 2007), gelatin microparticles (Patel *et al.*, 2008), collagen / fibronectin hydrogels (Glotzbach *et al.*, 2010), and PLG(A) scaffolds (Murphy *et al.*, 2000)(Rocha *et al.*, 2008; Matoka and Cheng, 2009; Borselli *et al.*, 2010; Golub *et al.*, 2010). Moreover, transfecting or transducing developing cells (muscle precursor cells or myoblasts, for instance) with VEGF and subsequently injecting them into a site can also provide a steady, longer-term delivery of the growth factor (Misteli *et al.*, 2010).

Once VEGF is coupled with a delivery conduit, a more intricate approach to controlling VEGF delivery involves modifying the factor itself. Extensive delivery-mechanics research was conducted with the isoform $VEGF_{121}$. Though $VEGF_{121}$ is initially confined within a biomaterial (such as a fibrin-gel matrix) upon implant, plasmin and metalloproteinase degradation of the implant over time allows for the rapid, free diffusion of $VEGF_{121}$ into the whole body environment (Ehrbar *et al.*, 2004), potentially resulting in angiogenesis that may not be localized. Additionally, the short biological half-life of VEGF impedes its use in long-term applications. A synthetic variant of $VEGF_{121}$, known as TG-$VEGF_{121}$, cross-links to fibrinogen by the transglutaminating activity of factor XIII during fibrin-gel polymerization. This covalently tethered TG-$VEGF_{121}$ is protected from rapid diffusion. Gradual degradation of the fibrin-gel by local fibrinolytic activities results in a local liberation of low levels of TG-$VEGF_{121}$ into tissue. Experimental animal models have shown that fibrin-conjugated TG-$VEGF_{121}$ produced more structurally stable vessels than $VEGF_{121}$ while avoiding vascular leakage (Ehrbar *et al.*, 2004; Ehrbar *et al.*, 2008).

4. Hypoxia-mediated control and modulation of notch signaling

Hypoxia presents another means of employing direct control on VEGF and ancillary angiogenic factors. In the low-oxygen environments of normal muscle, VEGF mRNA experiences decreased degradation and increased expression (Ikeda *et al.*, 1995; Levy *et al.*, 1995; Levy *et al.*, 1996; Tang *et al.*, 2004), though severely hypoxic settings will actually impede VEGF up-regulation (Milkiewicz *et al.*, 2004). VEGF mRNA stability is made possible by its interaction with HuR, a complex that binds and stabilizes RNAs, consequently regulating gene expression (Levy *et al.*, 1998). The increase in VEGF production can be attributed to the Internal Ribosome Entry Site (IRES), which is accountable for efficient factor synthesis under hypoxic conditions (Stein *et al.*, 1998). Other proteins can also impact VEGF efficiency and activity. Control of expressed VEGF lies with ORP150 (oxygen regulated protein), a chaperone that transports VEGF from the cell's endoplasmic reticulum to the Golgi apparatus (Kuwabara *et al.*, 1996; Ozawa *et al.*, 2001). An increase in ORP150 levels correlates with an increased production of VEGF during hypoxia (Ozawa *et al.*, 2001). Further, Hypoxia-inducing factor (HIF) is responsible for activating the transcription of genes associated with neovascularization (Covello and Simon, 2004; Ramirez-Bergeron *et al.*, 2006). These proteins represent points of control for VEGF modulation.

Because angiogenesis is the synthesis of vasculature from existing vessels, constant remodeling and modification of vessels takes place *in vivo*. Existing endothelial cells, such as stalk and tip cells, have receptors that respond to environmental conditions like hypoxia, in which Notch signaling (driven by VEGF-A presence) is most prevalent. The presence of VEGF-A leads to an increase in the presence of Delta-like Ligand 4 (DLL4), a major Notch ligand. It is believed that Notch signaling modulates the ratio of VEGF to its receptors through the inhibition of VEGF-R2.

Manipulation of the Notch signaling pathway presents another means of controlling angiogenesis. Cao, et al., discussed the role of Notch signaling in modulating VEGF activity, resulting in effective pruning and branching of vascular vessels (Cao *et al.*, 2009; Cao *et al.*, 2010). Activation of the Notch signaling pathway inhibits VEGF signaling by down-regulating the VEGF receptor synthesis. With fewer available receptors for free VEGF to bind, endothelial cell proliferation is effectually curbed. This process occurs at the cellular level of the lining endothelium, a major component of vascular vessels, and vascular growth in any particular direction is controlled.

5. Applications in regenerative medicine: notable case studies

Recent advances in understanding the angiogenic process and isolating potent and specific angiogenic growth factors prompted their therapeutic usage. Evidence that VEGF is a specific endothelial cell growth factor suggested its potential in therapeutic angiogenesis. Injection of the $VEGF_{165}$ protein enhanced revascularization of rabbit ischemic hindlimbs (Takeshita *et al.*, 1994). VEGF treatment induced collateral vessel formation, endothelium-dependent blood flow and tissue perfusion. VEGF has been tested for potential beneficial effects on wound repair in diabetic animal models (Greenhalgh *et al.*, 1990; Tsuboi and Rifkin, 1990; Frank *et al.*, 1995). The growth factor treatment regimens accelerated granulation, tissue formation, and wound closure. However, due to the high clearance and/or degradation of the proteins from the administration site, topically administered growth factors would require high dosages and frequent delivery.

An alternative strategy for therapeutic angiogenesis is gene therapy using recombinant angiogenic growth factors. The first study using $VEGF_{165}$ cDNA was performed by gene transfer into the iliac artery of an ischemic hindlimb of a rabbit (Bauters *et al.*, 1994; Takeshita *et al.*, 1996). VEGF protein was expressed at the site of injection, augmenting the formation of collateral vessels. Subsequently, intramuscular gene transfer of VEGF cDNA was used in a similar model with similar results (Asahara *et al.*, 1996). This technique was further employed in patients with peripheral vascular disease and critical limb ischemia (Isner *et al.*, 1996). Clinical trials showed significant improvement in collateral blood flow, healed ischemic ulcers, and most importantly, salvage of limbs in patients in whom amputation was imminent (Isner, 1998; Isner and Takayuki, 1998). VEGF protein and cDNA have been used for coronary revascularization, resulting in improved myocardial perfusion and increased collateral density (Isner and Losordo, 1999).

As previously discussed, a successful angiogenic outcome is not solely attributed to VEGF implementation alone (discussed in Section 3 and Table 1). Wilcke, et al., showcased the coupling of VEGF with bFGF in a fibrin dermal substitute (Wilcke *et al.*, 2007). This led to a marked improvement in factor delivery, with a notable, prolonged release and resulted in a higher density of newly developed vessels in *in vivo* murine models. Concurrently, an *in vivo* experiment performed by Zacchigna, et al., studied the effect of a VEGF and Angiopoietin 1 (Ang 1) combination (delivered via an adeno-associated viral vector, for long-term protein production and release) on the muscle blood flow (MBF) and vascular permeability of rat skeletal muscle (Zacchigna *et al.*, 2007). The grouping elicited a marked increase in both resting MBF and perfusion post exercise stimulation. To contrast, VEGF expression alone did not enhance the resting MBF and actually reduced tissue perfusion after exercise.

Elcin, et al., researched *in vitro* release kinetics and *in vivo* angiogenic effects of human VEGF-loaded PLGA sponges in rats (Elcin and Elcin, 2006). When compared with control sponges (containing no factor) and bolus injections of VEGF, the use of VEGF-loaded PLGA sponges led to the establishment of neovascularized sites suitable for tissue engineering purposes. Patel, et al., who utilized VEGF-loaded gelatin microparticles infused in biodegradable composite scaffolds, outlined another notable attempt in VEGF delivery modulation in the field of orthopedics (Patel *et al.*, 2008). Findings suggested that modulating the degree of gelatin cross-linking could affect VEGF release into the microenvironment and, consequently, angiogenic outcome. Ennett, et al., studied the temporally regulated delivery of VEGF *in vivo*, hypothesizing that the means of VEGF-loading into a delivery scaffold would have an impact on the factor release kinetics (Ennett *et al.*, 2006). They compared (a) VEGF loaded directly into a PLG scaffold with (b) VEGF pre-encapsulated in PLG microspheres that were later used to fabricate a PLG scaffold. Though pre-encapsulated VEGF microspheres further delayed the factor's release into the surrounding environment, this approach produced a desirable angiogenic outcome, with significant local angiogenesis and negligible systemic effects.

Pre-encapsulation of VEGF by nanoparticles or microspheres has also been investigated in conjunction with Matrigel hydrogels, PLGA-, and collagen-scaffolds. While it is clear that encapsulation protects the factor(s) within the nanoparticles and microspheres, it also offers a mechanism of controlled release - especially ideal for a potent angiogenic factor like VEGF. Even though pre-encapsulated VEGF-loaded PLGA scaffolds performed better than similarly loaded Matrigel hydrogels (with respect to release profiles in a saline solution; also in comparison with free VEGF) *in vitro*, both of the loaded delivery conduits improved angiogenesis *in vivo*. An increase in both endothelial cell counts and red blood cells was noted at the sites of implantation (des Rieux *et al.*, 2010). Further, VEGF-loaded PLGA microspheres combined with collagen elicited strong enhancements to vascular sprouting and activation of endothelial cells *in vivo* and *in vitro*, respectively (des Rieux *et al.*, 2010).

The idea of optimizing delivery agents combined with the notion that co-factor usage is ideal for neovascularization resulted in an experiment that analyzed the co-implementation of VEGF with FGF-2 via an acellular collagen scaffold implant (Nillesen *et al.*, 2007). An implant containing both factors resulted in the highest vessel density and the most mature blood vessels (characteristics of an enhanced, stable vasculature) in a rat model when compared with scaffolds containing either of the factors alone or no factor at all.

6. Conclusion

A significant challenge facing tissue engineering lies in eliciting controlled neovascularization. Controlled neovascularization can have a profound impact in vessel and organ synthesis, as well as in the treatment of damaged tissues. Many current attempts for such control entail the use of VEGF. Attempts at fabricating an ordered vascular network include changes to VEGF through structural modification, the employment of strategies controlling its release, and coupling VEGF to other growth factors.

Yet, despite the promises shown by VEGF, significant obstacles remain. Many researchers propose that use of growth factors alone will not ensure stable angiogenesis. Rather, a combination of growth factors (VEGF and bFGF, for example), delivery methods, and modulation of inflammatory responses (via fibroblast, macrophage, cell-adhesion molecule and cytokine manipulation, for instance) and pathways (such as hypoxia and Notch) are thought to facilitate adequate vessel fabrication and stability.

Additionally, investigation into the mechanisms of cellular crosstalk is necessary to better understand angiogenesis in general. Moreover, the refinement and implementation of microenvironment monitoring technologies are vital in ensuring proper vascular development. Such technologies would enable researchers to closely study the impact of gross- or modulated-release of growth factors on the delicate balance required for vascular formation and branching.

7. Acknowledgements

Remo A. Largo and Venkat M. Ramakrishnan contributed equally to the production of this chapter.

8. References

Aase, K., von Euler, G., Li, X., Ponten, A., Thoren, P., Cao, R., Cao, Y., Olofsson, B., Gebre-Medhin, S., Pekny, M., Alitalo, K., Betsholtz, C., and Eriksson, U. (2001). Vascular endothelial growth factor-B-deficient mice display an atrial conduction defect. *Circulation* 104, 358-364.

Achen, M.G., Jeltsch, M., Kukk, E., Makinen, T., Vitali, A., Wilks, A.F., Alitalo, K., and Stacker, S.A. (1998). Vascular endothelial growth factor D (VEGF-D) is a ligand for the tyrosine kinases VEGF receptor 2 (Flk1) and VEGF receptor 3 (Flt4). *Proc Natl Acad Sci U S A* 95, 548-553.

Adini, A., Kornaga, T., Firoozbakht, F., and Benjamin, L.E. (2002). Placental growth factor is a survival factor for tumor endothelial cells and macrophages. *Cancer Res* 62, 2749-2752.

Arkudas, A., Tjiawi, J., Bleiziffer, O., Grabinger, L., Polykandriotis, E., Beier, J.P., Sturzl, M., Horch, R.E., and Kneser, U. (2007). Fibrin gel-immobilized VEGF and bFGF efficiently stimulate angiogenesis in the AV loop model. *Mol Med* 13, 480-487.

Asahara, T., Tsurumi, Y., Takeshita, S., and Isner, J.M. (1996). Naked cDNA encoding secreted proteins for intra-arterial and intramuscular gene transfer. *Seminars in interventional cardiology : SIIC 1*, 225-232.

Bates, D.O. (2010). Vascular endothelial growth factors and vascular permeability. *Cardiovasc Res* 87, 262-271.

Bauters, C., Asahara, T., Zheng, L.P., Takeshita, S., Bunting, S., Ferrara, N., Symes, J.F., and Isner, J.M. (1994). Physiological assessment of augmented vascularity induced by VEGF in ischemic rabbit hindlimb. *The American journal of physiology* 267, H1263-1271.

Benjamin, L.E., Golijanin, D., Itin, A., Pode, D., and Keshet, E. (1999). Selective ablation of immature blood vessels in established human tumors follows vascular endothelial growth factor withdrawal. *J Clin Invest* 103, 159-165.

Borselli, C., Ungaro, F., Oliviero, O., d'Angelo, I., Quaglia, F., La Rotonda, M.I., and Netti, P.A. (2010). Bioactivation of collagen matrices through sustained VEGF release from PLGA microspheres. *J Biomed Mater Res* A 92, 94-102.

Cao, L., Arany, P.R., Kim, J., Rivera-Feliciano, J., Wang, Y.S., He, Z., Rask-Madsen, C., King, G.L., and Mooney, D.J. (2010). Modulating Notch signaling to enhance neovascularization and reperfusion in diabetic mice. *Biomaterials* 31, 9048-9056.

Cao, L., Arany, P.R., Wang, Y.S., and Mooney, D.J. (2009). Promoting angiogenesis via manipulation of VEGF responsiveness with notch signaling. *Biomaterials* 30, 4085-4093.

Carmeliet, P., Moons, L., Luttun, A., Vincenti, V., Compernolle, V., De Mol, M., Wu, Y., Bono, F., Devy, L., Beck, H., Scholz, D., Acker, T., DiPalma, T., Dewerchin, M., Noel, A., Stalmans, I., Barra, A., Blacher, S., Vandendriessche, T., Ponten, A., Eriksson, U., Plate, K.H., Foidart, J.M., Schaper, W., Charnock-Jones, D.S., Hicklin, D.J., Herbert, J.M., Collen, D., and Persico, M.G. (2001). Synergism between vascular endothelial growth factor and placental growth factor contributes to angiogenesis and plasma extravasation in pathological conditions. *Nat Med* 7, 575-583.

Covello, K.L., and Simon, M.C. (2004). HIFs, hypoxia, and vascular development. *Current topics in developmental biology* 62, 37-54.

Dai, J., and Rabie, A.B. (2007). VEGF: an essential mediator of both angiogenesis and endochondral ossification. *Journal of dental research* 86, 937-950.

Dallas, N.A., Gray, M.J., Xia, L., Fan, F., van Buren, G., 2nd, Gaur, P., Samuel, S., Lim, S.J., Arumugam, T., Ramachandran, V., Wang, H., and Ellis, L.M. (2008). Neuropilin-2-mediated tumor growth and angiogenesis in pancreatic adenocarcinoma. *Clin Cancer Res* 14, 8052-8060.

de Vries, C., Escobedo, J.A., Ueno, H., Houck, K., Ferrara, N., and Williams, L.T. (1992). The fms-like tyrosine kinase, a receptor for vascular endothelial growth factor. *Science* 255, 989-991.

des Rieux, A., Ucakar, B., Mupendwa, B.P., Colau, D., Feron, O., Carmeliet, P., and Preat, V. (2010). 3D systems delivering VEGF to promote angiogenesis for tissue engineering. *J Control Release*.

Dor, Y., Djonov, V., and Keshet, E. (2003). Induction of vascular networks in adult organs: implications to proangiogenic therapy. *Annals of the New York Academy of Sciences* 995, 208-216.

Dumont, D.J., Jussila, L., Taipale, J., Lymboussaki, A., Mustonen, T., Pajusola, K., Breitman, M., and Alitalo, K. (1998). Cardiovascular failure in mouse embryos deficient in VEGF receptor-3. *Science* 282, 946-949.

Edelman, E.R., Mathiowitz, E., Langer, R., and Klagsbrun, M. (1991). Controlled and modulated release of basic fibroblast growth factor. *Biomaterials* 12, 619-626.

Egginton, S. (2010). Physiological factors influencing capillary growth. *Acta physiologica.*

Ehrbar, M., Djonov, V.G., Schnell, C., Tschanz, S.A., Martiny-Baron, G., Schenk, U., Wood, J., Burri, P.H., Hubbell, J.A., and Zisch, A.H. (2004). Cell-demanded liberation of VEGF121 from fibrin implants induces local and controlled blood vessel growth. *Circulation research* 94, 1124-1132.

Ehrbar, M., Zeisberger, S.M., Raeber, G.P., Hubbell, J.A., Schnell, C., and Zisch, A.H. (2008). The role of actively released fibrin-conjugated VEGF for VEGF receptor 2 gene activation and the enhancement of angiogenesis. *Biomaterials* 29, 1720-1729.

Elcin, A.E., and Elcin, Y.M. (2006). Localized angiogenesis induced by human vascular endothelial growth factor-activated PLGA sponge. *Tissue engineering* 12, 959-968.

Elia, R., Fuegy, P.W., VanDelden, A., Firpo, M.A., Prestwich, G.D., and Peattie, R.A. (2010). Stimulation of in vivo angiogenesis by in situ crosslinked, dual growth factor-loaded, glycosaminoglycan hydrogels. *Biomaterials* 31, 4630-4638.

Ennett, A.B., Kaigler, D., and Mooney, D.J. (2006). Temporally regulated delivery of VEGF in vitro and in vivo. *J Biomed Mater Res* A 79, 176-184.

Farnebo, F., Piehl, F., and Lagercrantz, J. (1999). Restricted expression pattern of vegf-d in the adult and fetal mouse: high expression in the embryonic lung. *Biochem Biophys Res Commun* 257, 891-894.

Ferrara, N. (2009). VEGF-A: a critical regulator of blood vessel growth. *Eur Cytokine Netw* 20, 158-163.

Ferrara, N., and Davis-Smyth, T. (1997). The biology of vascular endothelial growth factor. *Endocr Rev* 18, 4-25.

Ferrara, N., Gerber, H.P., and LeCouter, J. (2003). The biology of VEGF and its receptors. *Nat Med* 9, 669-676.

Fitz, L.J., Morris, J.C., Towler, P., Long, A., Burgess, P., Greco, R., Wang, J., Gassaway, R., Nickbarg, E., Kovacic, S., Ciarletta, A., Giannotti, J., Finnerty, H., Zollner, R., Beier, D.R., Leak, L.V., Turner, K.J., and Wood, C.R. (1997). *Characterization of murine Flt4 ligand/VEGF-C. Oncogene* 15, 613-618.

Frank, S., Hubner, G., Breier, G., Longaker, M.T., Greenhalgh, D.G., and Werner, S. (1995). Regulation of vascular endothelial growth factor expression in cultured keratinocytes. Implications for normal and impaired wound healing. *J Biol Chem* 270, 12607-12613.

Glotzbach, J.P., Levi, B., Wong, V.W., Longaker, M.T., and Gurtner, G.C. (2010). The basic science of vascular biology: implications for the practicing surgeon. *Plast Reconstr Surg* 126, 1528-1538.

Golub, J.S., Kim, Y.T., Duvall, C.L., Bellamkonda, R.V., Gupta, D., Lin, A.S., Weiss, D., Robert Taylor, W., and Guldberg, R.E. (2010). Sustained VEGF delivery via PLGA nanoparticles promotes vascular growth. *Am J Physiol Heart Circ Physiol* 298, H1959-1965.

Greenhalgh, D.G., Sprugel, K.H., Murray, M.J., and Ross, R. (1990). PDGF and FGF stimulate wound healing in the genetically diabetic mouse. *Am J Pathol* 136, 1235-1246.

Han, S., Arnold, S.A., Sithu, S.D., Mahoney, E.T., Geralds, J.T., Tran, P., Benton, R.L., Maddie, M.A., D'Souza, S.E., Whittemore, S.R., and Hagg, T. (2010). Rescuing vasculature with intravenous angiopoietin-1 and alpha v beta 3 integrin peptide is protective after spinal cord injury. *Brain : a journal of neurology* 133, 1026-1042.

Hariawala, M.D., Horowitz, J.R., Esakof, D., Sheriff, D.D., Walter, D.H., Keyt, B., Isner, J.M., and Symes, J.F. (1996). VEGF improves myocardial blood flow but produces EDRF-mediated hypotension in porcine hearts. *J Surg Res* 63, 77-82.

Hattori, K., Heissig, B., Wu, Y., Dias, S., Tejada, R., Ferris, B., Hicklin, D.J., Zhu, Z., Bohlen, P., Witte, L., Hendrikx, J., Hackett, N.R., Crystal, R.G., Moore, M.A., Werb, Z., Lyden, D., and Rafii, S. (2002). Placental growth factor reconstitutes hematopoiesis by recruiting VEGFR1(+) stem cells from bone-marrow microenvironment. *Nat Med* 8, 841-849.

Herbert, S.P., Huisken, J., Kim, T.N., Feldman, M.E., Houseman, B.T., Wang, R.A., Shokat, K.M., and Stainier, D.Y. (2009). Arterial-venous segregation by selective cell sprouting: an alternative mode of blood vessel formation. *Science* 326, 294-298.

Hoier, B., Rufener, N., Bojsen-Moller, J., Bangsbo, J., and Hellsten, Y. (2010). The effect of passive movement training on angiogenic factors and capillary growth in human skeletal muscle. *The Journal of physiology* 588, 3833-3845.

Hoot, K.E., Oka, M., Han, G., Bottinger, E., Zhang, Q., and Wang, X.J. (2010). HGF upregulation contributes to angiogenesis in mice with keratinocyte-specific Smad2 deletion. *J Clin Invest* 120, 3606-3616.

Ikeda, E., Achen, M.G., Breier, G., and Risau, W. (1995). Hypoxia-induced transcriptional activation and increased mRNA stability of vascular endothelial growth factor in C6 glioma cells. *J Biol Chem* 270, 19761-19766.

Isner, J.M. (1998). Vascular endothelial growth factor: gene therapy and therapeutic angiogenesis. *The American journal of cardiology* 82, 63S-64S.

Isner, J.M., and Losordo, D.W. (1999). Therapeutic angiogenesis for heart failure. *Nat Med* 5, 491-492.

Isner, J.M., and Takayuki, A. (1998). Therapeutic angiogenesis. *Frontiers in bioscience : a journal and virtual library* 3, e49-69.

Isner, J.M., Walsh, K., Symes, J., Pieczek, A., Takeshita, S., Lowry, J., Rosenfield, K., Weir, L., Brogi, E., and Jurayj, D. (1996). Arterial gene transfer for therapeutic angiogenesis in patients with peripheral artery disease. *Human gene therapy* 7, 959-988.

Jeltsch, M., Kaipainen, A., Joukov, V., Meng, X., Lakso, M., Rauvala, H., Swartz, M., Fukumura, D., Jain, R.K., and Alitalo, K. (1997). Hyperplasia of lymphatic vessels in VEGF-C transgenic mice. *Science* 276, 1423-1425.

Jung, S., Wermker, K., Poetschik, H., Ziebura, T., and Kleinheinz, J. (2010). The impact of hyperbaric oxygen therapy on serological values of vascular endothelial growth factor (VEGF) and basic fibroblast growth factor (bFGF). *Head & face medicine* 6, 29.

Kanczler, J.M., and Oreffo, R.O. (2008). Osteogenesis and angiogenesis: the potential for engineering bone. *Eur Cell Mater* 15, 100-114.

Klein, M., and Catargi, B. (2007). VEGF in physiological process and thyroid disease. *Annales d'endocrinologie* 68, 438-448.

Kukk, E., Lymboussaki, A., Taira, S., Kaipainen, A., Jeltsch, M., Joukov, V., and Alitalo, K. (1996). VEGF-C receptor binding and pattern of expression with VEGFR-3 suggests a role in lymphatic vascular development. *Development* 122, 3829-3837.

Kuwabara, K., Matsumoto, M., Ikeda, J., Hori, O., Ogawa, S., Maeda, Y., Kitagawa, K., Imuta, N., Kinoshita, T., Stern, D.M., Yanagi, H., and Kamada, T. (1996). Purification and characterization of a novel stress protein, the 150-kDa oxygen-regulated protein (ORP150), from cultured rat astrocytes and its expression in ischemic mouse brain. *J Biol Chem* 271, 5025-5032.

Laschke, M.W., Harder, Y., Amon, M., Martin, I., Farhadi, J., Ring, A., Torio-Padron, N., Schramm, R., Rucker, M., Junker, D., Haufel, J.M., Carvalho, C., Heberer, M., Germann, G., Vollmar, B., and Menger, M.D. (2006). Angiogenesis in tissue engineering: breathing life into constructed tissue substitutes. *Tissue engineering* 12, 2093-2104.

Lee, K.Y., Peters, M.C., Anderson, K.W., and Mooney, D.J. (2000). Controlled growth factor release from synthetic extracellular matrices. *Nature* 408, 998-1000.

Lee, K.Y., Peters, M.C., and Mooney, D.J. (2003). Comparison of vascular endothelial growth factor and basic fibroblast growth factor on angiogenesis in SCID mice. *J Control Release* 87, 49-56.

Lee, T.Y., Folkman, J., and Javaherian, K. (2010). HSPG-binding peptide corresponding to the exon 6a-encoded domain of VEGF inhibits tumor growth by blocking angiogenesis in murine model. *PLoS One* 5, e9945.

Levy, A.P., Levy, N.S., and Goldberg, M.A. (1996). Post-transcriptional regulation of vascular endothelial growth factor by hypoxia. *J Biol Chem* 271, 2746-2753.

Levy, A.P., Levy, N.S., Wegner, S., and Goldberg, M.A. (1995). Transcriptional regulation of the rat vascular endothelial growth factor gene by hypoxia. *J Biol Chem* 270, 13333-13340.

Levy, N.S., Chung, S., Furneaux, H., and Levy, A.P. (1998). Hypoxic stabilization of vascular endothelial growth factor mRNA by the RNA-binding protein HuR. *J Biol Chem* 273, 6417-6423.

Li, Q.F., and Rabie, A.B. (2007). A new approach to control condylar growth by regulating angiogenesis. *Archives of oral biology* 52, 1009-1017.

Li, X., Kumar, A., Zhang, F., Lee, C., Li, Y., Tang, Z., and Arjuna, P. (2010). VEGF-independent angiogenic pathways induced by PDGF-C. *Oncotarget* 1, 309-314.

Luttun, A., Tjwa, M., Moons, L., Wu, Y., Angelillo-Scherrer, A., Liao, F., Nagy, J.A., Hooper, A., Priller, J., De Klerck, B., Compernolle, V., Daci, E., Bohlen, P., Dewerchin, M., Herbert, J.M., Fava, R., Matthys, P., Carmeliet, G., Collen, D., Dvorak, H.F., Hicklin, D.J., and Carmeliet, P. (2002). Revascularization of ischemic tissues by PlGF treatment, and inhibition of tumor angiogenesis, arthritis and atherosclerosis by anti-Flt1. *Nat Med* 8, 831-840.

Lyttle, D.J., Fraser, K.M., Fleming, S.B., Mercer, A.A., and Robinson, A.J. (1994). Homologs of vascular endothelial growth factor are encoded by the poxvirus orf virus. *Journal of virology* 68, 84-92.

Marcus, H.J., Carpenter, K.L., Price, S.J., and Hutchinson, P.J. (2010). In vivo assessment of high-grade glioma biochemistry using microdialysis: a study of energy-related molecules, growth factors and cytokines. *J Neurooncol* 97, 11-23.

Matoka, D.J., and Cheng, E.Y. (2009). Tissue engineering in urology. *Canadian Urological Association journal = Journal de l'Association des urologues du Canada* 3, 403-408.

Matsunaga, Y., Yamazaki, Y., Suzuki, H., and Morita, T. (2009). VEGF-A and VEGF-F evoke distinct changes in vascular ultrastructure. *Biochem Biophys Res Commun* 379, 872-875.

Meyer, M., Clauss, M., Lepple-Wienhues, A., Waltenberger, J., Augustin, H.G., Ziche, M., Lanz, C., Buttner, M., Rziha, H.J., and Dehio, C. (1999). A novel vascular endothelial growth factor encoded by Orf virus, VEGF-E, mediates angiogenesis via signalling through VEGFR-2 (KDR) but not VEGFR-1 (Flt-1) receptor tyrosine kinases. *Embo J* 18, 363-374.

Milkiewicz, M., Pugh, C.W., and Egginton, S. (2004). Inhibition of endogenous HIF inactivation induces angiogenesis in ischaemic skeletal muscles of mice. *The Journal of physiology* 560, 21-26.

Misteli, H., Wolff, T., Fuglistaler, P., Gianni-Barrera, R., Gurke, L., Heberer, M., and Banfi, A. (2010). High-throughput flow cytometry purification of transduced progenitors expressing defined levels of vascular endothelial growth factor induces controlled angiogenesis in vivo. *Stem Cells* 28, 611-619.

Murphy, W.L., Peters, M.C., Kohn, D.H., and Mooney, D.J. (2000). Sustained release of vascular endothelial growth factor from mineralized poly(lactide-co-glycolide) scaffolds for tissue engineering. *Biomaterials* 21, 2521-2527.

Nillesen, S.T., Geutjes, P.J., Wismans, R., Schalkwijk, J., Daamen, W.F., and van Kuppevelt, T.H. (2007). Increased angiogenesis and blood vessel maturation in acellular collagen-heparin scaffolds containing both FGF2 and VEGF. *Biomaterials* 28, 1123-1131.

Nomi, M., Atala, A., Coppi, P.D., and Soker, S. (2002). Principals of neovascularization for tissue engineering. *Molecular aspects of medicine* 23, 463-483.

Odorisio, T., Schietroma, C., Zaccaria, M.L., Cianfarani, F., Tiveron, C., Tatangelo, L., Failla, C.M., and Zambruno, G. (2002). Mice overexpressing placenta growth factor exhibit increased vascularization and vessel permeability. *J Cell Sci* 115, 2559-2567.

Olofsson, B., Pajusola, K., Kaipainen, A., von Euler, G., Joukov, V., Saksela, O., Orpana, A., Pettersson, R.F., Alitalo, K., and Eriksson, U. (1996a). Vascular endothelial growth factor B, a novel growth factor for endothelial cells. *Proc Natl Acad Sci U S A* 93, 2576-2581.

Olofsson, B., Pajusola, K., von Euler, G., Chilov, D., Alitalo, K., and Eriksson, U. (1996b). Genomic organization of the mouse and human genes for vascular endothelial growth factor B (VEGF-B) and characterization of a second splice isoform. *J Biol Chem* 271, 19310-19317.

Orlandini, M., Marconcini, L., Ferruzzi, R., and Oliviero, S. (1996). Identification of a c-fos-induced gene that is related to the platelet-derived growth factor/vascular endothelial growth factor family. *Proc Natl Acad Sci U S A* 93, 11675-11680.

Ozawa, C.R., Banfi, A., Glazer, N.L., Thurston, G., Springer, M.L., Kraft, P.E., McDonald, D.M., and Blau, H.M. (2004). Microenvironmental VEGF concentration, not total dose, determines a threshold between normal and aberrant angiogenesis. *J Clin Invest* 113, 516-527.

Ozawa, K., Kondo, T., Hori, O., Kitao, Y., Stern, D.M., Eisenmenger, W., Ogawa, S., and Ohshima, T. (2001). Expression of the oxygen-regulated protein ORP150 accelerates wound healing by modulating intracellular VEGF transport. *J Clin Invest* 108, 41-50.

Patel, Z.S., Ueda, H., Yamamoto, M., Tabata, Y., and Mikos, A.G. (2008). In vitro and in vivo release of vascular endothelial growth factor from gelatin microparticles and biodegradable composite scaffolds. *Pharm Res* 25, 2370-2378.

Pepper, M.S., Ferrara, N., Orci, L., and Montesano, R. (1992). Potent synergism between vascular endothelial growth factor and basic fibroblast growth factor in the induction of angiogenesis in vitro. *Biochem Biophys Res Commun* 189, 824-831.

Ramirez-Bergeron, D.L., Runge, A., Adelman, D.M., Gohil, M., and Simon, M.C. (2006). HIF-dependent hematopoietic factors regulate the development of the embryonic vasculature. *Developmental cell* 11, 81-92.

Rocha, F.G., Sundback, C.A., Krebs, N.J., Leach, J.K., Mooney, D.J., Ashley, S.W., Vacanti, J.P., and Whang, E.E. (2008). The effect of sustained delivery of vascular endothelial growth factor on angiogenesis in tissue-engineered intestine. *Biomaterials* 29, 2884-2890.

Sheridan, M.H., Shea, L.D., Peters, M.C., and Mooney, D.J. (2000). Bioabsorbable polymer scaffolds for tissue engineering capable of sustained growth factor delivery. *J Control Release* 64, 91-102.

Shibuya, M., Yamaguchi, S., Yamane, A., Ikeda, T., Tojo, A., Matsushime, H., and Sato, M. (1990). Nucleotide sequence and expression of a novel human receptor-type tyrosine kinase gene (flt) closely related to the fms family. *Oncogene* 5, 519-524.

Soker, S., Fidder, H., Neufeld, G., and Klagsbrun, M. (1996). Characterization of novel vascular endothelial growth factor (VEGF) receptors on tumor cells that bind VEGF165 via its exon 7-encoded domain. *J Biol Chem* 271, 5761-5767.

Soker, S., Takashima, S., Miao, H.Q., Neufeld, G., and Klagsbrun, M. (1998). Neuropilin-1 is expressed by endothelial and tumor cells as an isoform-specific receptor for vascular endothelial growth factor. *Cell* 92, 735-745.

Stefanini, M.O., Wu, F.T., Mac Gabhann, F., and Popel, A.S. (2009). The presence of VEGF receptors on the luminal surface of endothelial cells affects VEGF distribution and VEGF signaling. *PLoS computational biology* 5, e1000622.

Stein, I., Itin, A., Einat, P., Skaliter, R., Grossman, Z., and Keshet, E. (1998). Translation of vascular endothelial growth factor mRNA by internal ribosome entry: implications for translation under hypoxia. *Mol Cell Biol* 18, 3112-3119.

Tafuro, S., Ayuso, E., Zacchigna, S., Zentilin, L., Moimas, S., Dore, F., and Giacca, M. (2009). Inducible adeno-associated virus vectors promote functional angiogenesis in adult organisms via regulated vascular endothelial growth factor expression. *Cardiovasc Res* 83, 663-671.

Takahashi, H., and Shibuya, M. (2005). The vascular endothelial growth factor (VEGF)/VEGF receptor system and its role under physiological and pathological conditions. *Clinical science* 109, 227-241.

Takeshita, S., Weir, L., Chen, D., Zheng, L.P., Riessen, R., Bauters, C., Symes, J.F., Ferrara, N., and Isner, J.M. (1996). Therapeutic angiogenesis following arterial gene transfer of vascular endothelial growth factor in a rabbit model of hindlimb ischemia. *Biochem Biophys Res Commun* 227, 628-635.

Takeshita, S., Zheng, L.P., Brogi, E., Kearney, M., Pu, L.Q., Bunting, S., Ferrara, N., Symes, J.F., and Isner, J.M. (1994). Therapeutic angiogenesis. A single intraarterial bolus of vascular endothelial growth factor augments revascularization in a rabbit ischemic hind limb model. *J Clin Invest* 93, 662-670.

Tang, N., Wang, L., Esko, J., Giordano, F.J., Huang, Y., Gerber, H.P., Ferrara, N., and Johnson, R.S. (2004). Loss of HIF-1alpha in endothelial cells disrupts a hypoxia-driven VEGF autocrine loop necessary for tumorigenesis. *Cancer Cell* 6, 485-495.

Terman, B.I., Carrion, M.E., Kovacs, E., Rasmussen, B.A., Eddy, R.L., and Shows, T.B. (1991). Identification of a new endothelial cell growth factor receptor tyrosine kinase. *Oncogene* 6, 1677-1683.

Tsuboi, R., and Rifkin, D.B. (1990). Recombinant basic fibroblast growth factor stimulates wound healing in healing-impaired db/db mice. *The Journal of experimental medicine* 172, 245-251.

Verbridge, S.S., Choi, N.W., Zheng, Y., Brooks, D.J., Stroock, A.D., and Fischbach, C. (2010). Oxygen-controlled three-dimensional cultures to analyze tumor angiogenesis. *Tissue Eng Part* A 16, 2133-2141.

Wang, X., Zhou, X., Ma, J., Tian, H., Jiao, Y., Zhang, R., Huang, Z., Xiao, J., Zhao, B., Qian, H., and Li, X. (2010). Effects of keratinocyte growth factor-2 on corneal epithelial wound healing in a rabbit model of carbon dioxide laser injury. *Biological & pharmaceutical bulletin* 33, 971-976.

Wernike, E., Montjovent, M.O., Liu, Y., Wismeijer, D., Hunziker, E.B., Siebenrock, K.A., Hofstetter, W., and Klenke, F.M. (2010). VEGF incorporated into calcium phosphate ceramics promotes vascularisation and bone formation in vivo. *Eur Cell Mater* 19, 30-40.

Wilcke, I., Lohmeyer, J.A., Liu, S., Condurache, A., Kruger, S., Mailander, P., and Machens, H.G. (2007). VEGF(165) and bFGF protein-based therapy in a slow release system to improve angiogenesis in a bioartificial dermal substitute in vitro and in vivo. *Langenbeck's archives of surgery / Deutsche Gesellschaft fur Chirurgie* 392, 305-314.

Wu, S., Wu, X., Zhu, W., Cai, W.J., Schaper, J., and Schaper, W. (2010). Immunohistochemical study of the growth factors, aFGF, bFGF, PDGF-AB, VEGF-A and its receptor (Flk-1) during arteriogenesis. *Mol Cell Biochem* 343, 223-229.

Yamada, Y., Nezu, J., Shimane, M., and Hirata, Y. (1997). Molecular cloning of a novel vascular endothelial growth factor, VEGF-D. *Genomics* 42, 483-488.

Yamazaki, Y., Matsunaga, Y., Tokunaga, Y., Obayashi, S., Saito, M., and Morita, T. (2009). Snake venom Vascular Endothelial Growth Factors (VEGF-Fs) exclusively vary their structures and functions among species. *J Biol Chem* 284, 9885-9891.

Yla-Herttuala, S. (2009). Gene therapy with vascular endothelial growth factors. *Biochem Soc Trans* 37, 1198-1200.

Zacchigna, S., Tasciotti, E., Kusmic, C., Arsic, N., Sorace, O., Marini, C., Marzullo, P., Pardini, S., Petroni, D., Pattarini, L., Moimas, S., Giacca, M., and Sambuceti, G. (2007). In vivo imaging shows abnormal function of vascular endothelial growth factor-induced vasculature. *Human gene therapy* 18, 515-524.

20

The Liver Vascular Bed for Hepatocytes Cell Therapy and Tissue Engineering

Yaacov Baruch
Liver Unit, Rambam human health care campus, Haifa,
The Bruce Rappaport Faculty of Medicine, Technion-Israel Institute of Technology, Haifa
Israel

1. Introduction

Liver replacement is necessary for acute and chronic end stage liver failure and for correction of metabolic disorders. Even a temporary liver support is relevant under specific clinical conditions, giving cell based therapy a great potential. The decision on how much liver tissue should be replaced, depends on the degree of liver dysfunction and the chance of the native liver to recover. To minimize complications it is believed that the maximal percentage of engraftment that can be achieved is in the range of 2-5% of the host hepatocytes, whereas 10% are needed for an adequate support in liver failure. Several questions would still rise such as What is the best location for liver cell therapy? What kind of cells should be used? What strategy to increase cell engraftment should be employed? In this review we will discuss the experience gained with intravascular cell transplantation to the liver and future solutions using new technology for biomaterials production. We will expand on our experience with injectable hydrogel scaffold constructs consisting of PEGylated fibrinogen backbone and diacrylate (PEG-DA). The experience with their use for intravascular injection is still limited.

Hepatocyte transplantation is a promising alternative method for the treatment of acute liver failure and also for the treatment of end stage liver disease. Hepatic cell transplantation can be used to temporarily gain time for acutely failing liver to recover, and to restore liver function in the chronic stage until a liver transplant becomes available. Transplanted hepatocytes can serve for the correction of congenital metabolic defects and for the transfer of genes by ex vivo gene therapy. The techniques used for cell transplantation may be implemented for other kinds of cells for therapy, including stem cells and immortalized-hepatocyte- cell lines. Other advantages of cell therapy over whole organ transplantation include the option to perform multiple hepatocyte transplantations in the same patient, and several patients could potentially be treated from one donor at a cost believed to be one tenth that of orthotopic liver transplantation. Studies have been undertaken to improve the methods related to hepatocyte transplantation, which include hepatocyte isolation from donor livers, cell culture propagation of the hepatocytes, hepatocyte preservation, and genetic modification of the hepatocytes to provide liver-specific functions and longevity. In humans, parenchymal hepatocytes isolated from donor livers remain the best cell source for transplantation, and other cell sources, including stem cells, are at the preclinical and early clinical stages (Puppi et al. 2011). Major problems do not allow this method to be practical and these include the shortage of organs for cell recruitment and limitations in cell preservation. Moreover technical issues still exist. In rodents old reports suggested that transplanted cells into a normal host with normal liver architecture can integrate into the host liver and function for the long run but this was not proven to be the case (Ponder et al. 1991). New problems appeared when human studies were initiated, such as poor initial and long term engraftment and function. It was agreed in a recent consensus, that to obtain sufficient levels of repopulation of liver with donor cells in patients with metabolic liver disease, some form of liver preconditioning would be required to enhance the engraftment proliferation of donor cells (Puppi et al. 2011). Therefore the switch from the animal model to human studies in the set up of acute hepatic failure is not direct and the amount of cells to be transplanted based on animal studies is probably irrelevant to humans. Moreover, the time needed to reverse a failing human liver which is in the range of 3-6 months would be only a few days in the rodent model making issues of cell engraftment and long term function again irrelevant. What seems now to be the right volume of cells (2-5% of the hepatocytes), to be transplanted into the portal vein in the situation of acute hepatic failure in humans, can be later proven to be insufficient (Fisher and Strom. 2006).

2. Which cell to transplant?

The technique of cell transplantation in contrast to whole organ transplantation allows selecting the proper cells for transplantation as discussed in a recent review (Piscaglia et al. 2010). For the purpose of treating the acutely failing organ adult parenchymal cells will be most suitable. These can be harvested from livers denied for transplantation or from non heart beating livers (Hughes et al. 2006; Mitry et al. 2004). Fetal parenchymal cells may be preferred due to their growth potential (Weber et al. 2010). Both sources of cells are of course limited. Harvesting cryopreserved cells may allow using cells when no immediate donor is available (Terry et al. 2007). The use of stem cells from embryonic source or from induced pluripotent stem cells as a source for hepatocytes is still remote from applicability. It may take long periods to expand and differentiate in culture, and the risk of tumor development always exists (Piscaglia et al. 2010). Xenotransplantation is also not applicable even though some experience was gained using porcine hepatocytes on a bioartificial liver device (Soto-Gutierrez et al. 2006). The fear of transferring infection always exists even though the immune response to transplanted cells seems lesser than to a whole organ (Rhim et al. 1994).

2.1 Where to transplant

The liver naturally is the ideal target organ for cell transplantation, in terms of the unique hepatic organization, interactions with non-parenchymal liver cells and biliary drainage. The spleen is also a viable target tissue for transplantation of hepatocytes since it offers the ability to form differentiated chord structures and reform nearly normal hepatic architectures (Mito et al. 1979). Hepatocytes transplanted to the spleen are responsive to liver regeneration stimuli after partial hepatectomy (Aoki et al. 2005). However, the infusion of cells into the portal system by the intra splenic route may be associated with portal vein thrombosis, liver necrosis, hemorrhage and portal and pulmonary hypertension. For transplanted hepatocytes to engraft into the host liver and remain functionally viable over the long term, the most important outcome for these cells is to translocate from the portal pedicle into the liver microenvironment, as described by several groups (Ponder et al. 1991; Gupta et al. 1999). Another important limiting factor in utilizing either the intraportal or intrasplenic approach is the number of viable cells that can be engrafted without causing complications. It is believed that the maximal percentage of engraftment that can be achieved is in the range of 2-5% of the host hepatocytes whereas 10% are needed for an adequate support in liver failure (Sohlenius-Sternbeck. 2006; Asonuma et al. 1992). The first report on the intrasplenic transplantation via the intrasplenic arterial route was reported on patients with liver failure awaiting liver transplantation (Strom et al. 1999). Despite these complications and limitations, clinical trials of hepatocyte transplantation have been performed without causing any reported fatalities. The peritoneal cavity offers a large space for hepatocyte engraftment, and contact with portal flow but the peritoneal surface does not support long-term attachment and survival of liver cells. Attachment to collagen-coated beads or microencapsulation allowed for only minimal improvement (Selden et al. 2003; Baldini et al. 2008). Co transplantation with non parenchymal cells showed some promise yet, hepatocytes transplanted to the peritoneum or dorsal fat pad do not express genes as those transplanted in the liver (Gupta et al. 1994; Selden et al. 1995). To overcome the problems of intraportal transplantation in the advanced cirrhotic liver with portal hypertension, transplantation into an extrahepatic site is necessary. A strategy that allow repeated extrahepatic infusion of hepatocytes was tried, using a non-immunogenic self-assembling peptide nanofiber. Developed as a three-dimensional scaffold and combined with growth factors, injected into the muscle of small animals, it resulted in improved hepatocytes function, but applicability to humans seems remote (Navarro-Alvarez et al. 2010).

2.2 The microenvironment

Hepatocytes are the major cellular component of a bioartificial liver support system. Although they can elicit multiple functions by themselves, they depend on the cellular network of the non-parenchymal cells. The latter are major regulators of hepatocyte intermediary metabolism, growth and response to injury. Kupffer cells interact with hepatocytes to produce the acute-phase response , i.e., synthesis of C-reactive protein and alpha2-macroglobulin by IL-6. In -vitro, when Kupffer cells and hepatocytes are co-cultured, the production of Nitric Oxide is enhanced (Billiar et al. 1990). Hepatocytes interact with hepatic stellate cells (fat storing cells, Ito cells, lipocytes) for the production of cellular matrix *in vitro*. Hepatic stellate cells proliferate, transform into myofibroblast like cells ("activated" hepatic stellate cells) and synthesize large amounts of extracellular matrix components (Collagens, fibronectin, tenascin, undulin, laminin and proteoglycans). Interaction with hepatocytes is mostly important for tissue remodeling. Myofibroblast like cells participate in the recruitment, activation and migration of inflammatory cells at sites of liver injury and play a role in regulating hepatic microcirculation. The endothelial cells that line the blood vessels and the space of Disse, have an important function in cytokine production and the synthesis of other molecules that support the microvasculature and microcirculation. The vascular growth and remodeling is very much dependent on growth factors and extracellular matrix components. A beneficial effect of co-transplantation with non parenchymal cell has been shown before (Selden et al. 1995). When hepatocytes formed spheroids they kept their function within the matrices for a longer period of time (Lin and Bissell. 1993; Sakai et al. 2010). A recent paper described the beneficial effect for co-transplantation in the peritoneum and the importance of spheroids structure is also stressed (Selden et al. 2003; Hamazaki et al. 2002; Torok et al. 2011). Bioartificial liver support systems incorporating nonparenchymal cells obtained better results in terms of both survival and function of hepatocytes. A method to prepare a biomatrix by a novel, four-step perfusion decellularization protocol using conditions designed to keep all collagen types insoluble, for repopulation, was reported recently (Wang et al. 2011). The various growth factors attached to the extra cellular matrix allowed for better differentiation of stem cells that repopulated the natural bioscaffold.

3. Major growth factors involved in liver growth, regeneration and angiogenesis

The knowledge obtained to date on liver cell growth and development during regeneration is very applicable to the development of liver tissue bioengineering for therapeutic strategies. Liver regeneration after partial hepatectomy is one of the most studied models for tissue regeneration medicine (Michalopoulos. 2010). We have learned that hepatocytes growth factor (HGF), epidermal growth factor (EGF), transforming growth factor α (TGFα) and TNFα/IL6 system are the most potent mitogens *in vivo* (Nakamura et al. 1987; Michalopoulos. 1990). Mainly nonparenchymal liver cells, such as hepatic stellate cells , and endothelial cells produce HGF (Schirmacher et al. 1992). During liver regeneration TGFα is secreted by both parenchymal and non-parenchymal liver cells. TNFα and IL-6 family of cytokines were shown to have an important role in the regenerating liver as shown in knock out mice (Cressman et al. 1996). The proliferative and anti-apoptotic effect of these cytokines takes place only under special conditions such as those existing after partial hepatectomy (Streetz et al. 2000). Major inhibitors such as TGF beta, activin, p21 and p53, help regulate the regeneration process.

Endothelial GFs: fibroblast GF (FGF) and vascular endothelial GF (VEGF) participate in liver regeneration (Assy et al. 1999). FGF is a potent endothelial GF and a primary liver mitogen. In its acidic form, aFGF makes hepatocytes more responsive to EGF and TGFa. The basic form is mitogenic to rat hepatocytes *in vivo* (Baruch et al. 1991; Baruch et al. 1995). bFGF, attached to macroaggregates, enhanced the engraftment of transplanted hepatocytes, by improving vascularization (Perets et al. 2003). FGF is mostly important also for the differentiation of stem cells into definitive endoderm (Kubo et al. 2004).VEGF has a wide range of activities; it increases the endothelial cells' permeability, facilitates monocytes penetration, and enhances synthesis of plasminogen activator and plasminogen activator inhibitor in endothelial cells. VEGF has in addition a vasodilatatory effect mediated by nitric oxide. In the liver, receptors are found on endothelial cells and VEGF can be found in a small quantity within liver cells. It was reported that VEGF is released by both host and transplanted hepatocytes during cell entry into the liver plate (Slehria and Gupta. 1998). VEGF is secreted by replicating hepatocytes induce sinusoidal endothelial cell proliferation during regeneration after partial hepatectomy in rats. vWF translocates to sinusoidal endothelial cells during liver regeneration, a change that may have an important role in tissue remodeling during liver regeneration (Baruch et al. 2002).

4. Intravsacular cell based therapy and implantable constructs

The most frequently applied hepatocyte transplantation method in humans is the injection of hepatocyte solution into the portal vein. This method, aimed to allow engraftment of hepatocytes within a defected liver, were used by Fox for the first time to treat a girl with Crigler Najar syndrome (Fox et al. 1998; Chowdhury et al. 1991). This method has the problems of insufficient cell survival and engraftment (Chowdhury et al. 1991; Grossman et al. 1995; Raper et al. 1996). Studies concerning cell injection into the spleen showed rapid translocation of transplanted cell through the spleen into the liver (Gupta et al. 1997). By using radio labeled cells it was demonstrated that cells were largely entrapped within the liver vascular bed and not entered the systemic circulation in normal animals (Gupta et al. 1994; Ott and et al. 2002). Cell occlusion is due to mechanical mechanism, because the diameter of hepatocytes is 20-40 mm whereas that of the hepatic sinusoids is 6-9mm. The sinusoids can probably dilate up to the size of 40mm. The total volume of extracellular space is believed to be 15% of total liver volume. Due to this size difference, transplanted cells are deposited in the most proximal sinusoids, adjacent to the portal areas of the liver lobule (Gupta. 2002).While the number of cells passing through the sinusoids to the pulmonary circulation is less than 1% in the normal animal, in animals with portal hypertension and portosystemic collaterals even intra splenic injection results in 50% of heptocytes reaching the pulmonary circulation (Gupta et al. 1993). The vast majority of cells are destroyed instantaneously in the pulmonary capillaries possibly due to shear stress and some other mechanisms. These may include apoptosis and destruction by Kupffer cells secondary to possible ischemia reperfusion injury in this system, needing the use of vasodilators (Slehria et al. 2002). The transplanted cells occlude the hepatic sinusoids and the portal vein radicles, and portal hypertension is observed immediately. Interestingly the increased pressure is transient as was studied with albumin macroaggregates (Gupta et al. 1994). It is believed that the most common location for transplanted cells to engraft is the liver sinusoidal bed and manipulation to vasodilate the sinusoidal bed to improve engraftment was tried with some success (Slehria et al. 2002). We believe that once hepatocytes are in the sinusoids they will go all the way through, and those that remain and have the potential to engraft are those stacked in the portal radicles. This explains the very low transplantation efficiency on the one hand but on the other hand target the portal vein radicles as the site needs to be manipulated, rather than the sinusoidal bed. It is this site that probably allow for redistribution of blood after cell transplantation to reduce portal hypertension and to allow for transplanted cell aggregates to be included in the liver tissue. Therefore there is a rational for using angiogenic growth factors to accelerate this process and to allow better cellular engraftment (Shani-Peretz et al. 2005). We have used for this purpose VEGF and later Heparanase that on top of its angiogenic stimulation is capabale to release other growth factors from the injured liver that favor hepatocytes replication (Carmel et al. 2010; Tsiperson et al. 2008).

It is quite clear that isolated cell transplantation by itself would not be enough and cell based strategies for temporary support systems for both extracorporeal and intra corporeal devices need special constructs (Navarro-Alvarez et al. 2010; Allen and Bhatia. 2002b). These devices include hepatocytes or hepatocytes derived cell lines in order to process patient plasma in the case of extracorporeal systems. Implantable tissue engineered constructs have also been developed for the purpose of long-term liver support (Allen and Bhatia. 2002a).

Construction of thicker tissues, was not possible due to limited diffusion of nutrients and oxygen within the engineered cell mass. It is known that cells can only survive within an area close enough to a source of nutrients and oxygen (Folkman and Hochberg. 1973; Rouwkema et al. 2008). Various techniques in polymer biochemistry and scaffold design to mimic the structure of this vascular network in engineered tissues were used. Strategies have included the co seeding of endothelial cells that spontaneously form capillary-like networks, and the engineering of branching channels to mimic the vascular tree.(Wang et al. 2011; Kaihara et al. 2000). In addition, angiogenesis was induced in within engineered tissues by incorporating angiogenic peptides and growth factors into scaffolds and by engineering cells used in the organ constructs, to express these factors. However, these efforts have fallen short of producing scaffolds that contain a vascular tree with centralized inlet and outlet vessels that are suitable for transplantation and capable of nutrient and gas exchange. (Borselli et al. 2010). Scaffolds were fabricated from synthetic material that allowed micro-patterning of vascular tree-like structures. When growth factors are used alone, they tend to create only a microvasculature consisting of small and fragile capillaries, in a time consuming process, and therefore this technique is only applicable for the engineering of smaller size tissues (Perets et al. 2003).

Development of implantable constructs with vasculature, remains a complicated challenge in terms of finding solutions for, the high metabolic rate of the liver cells, the need for unimpeded transport for large macromolecules and the need for cell-cell interaction (Dixit and Gitnick. 1995). In this respect, adult hepatocytes must interact with the various biomaterials, a process which is poorly understood and one of the key challenges to overcome in this field. Various hepatic tissue engineering approaches have been explored including the attachment of hepatocytes to microcarriers and encapsulated hepatocytes spheroids allowing for immunological protection and cell interaction (Demetriou et al. 1986). Co transplantation is easier in within microcapsules that also have the advantage of immunological protection. The preferred site so far is the peritoneum that can accommodate large volume (Perets et al. 2003; Ambrosino et al. 2003; Teng et al. 2010; Shi et al. 2009; Mooney et al. 1997). Absorbing cells on biodegradable synthetic scaffolds to allow control over scaffold chemistry, and space for cell interaction and vascular growth was used (Perets et al. 2003; Mooney et al. 1997; Lee et al. 2009). A significant advancement in the field of bioscaffold design has been the utilization of decellularized liver tissue creating a three-dimensional scaffold for tissue engineering strategies (Baptista et al. 2011). This study demonstrated that human liver cells can be seeded through the portal vein of a liver bioscaffold, and can be maintained in a specialized bioreactor with constant culture medium perfusion up to one week. Progressive human liver tissue formation was documented, as well as liver-associated functions. Widespread cell proliferation inside the bioengineered liver tissue with low cell apoptosis was also observed. These studies showed the possibility of seeding these bioscaffolds with liver cells from animals, but the possibility of generating functional human hepatic tissue is still in question.

Cell encapsulation techniques their advantages and disadvantages are reviewed elsewhere (Hernandez et al. 2010). Recently, a biomaterial system developed that takes advantage of the biological properties of natural extracellular matrix proteins combined with precisely controllable properties of synthetic polymers for making scaffolds for tissue engineering (Almany and Seliktar. 2005; Seliktar. 2005). The biosynthetic hybrid biomaterials are ideally suitable for hepatic tissue engineering in that both the structure and function of the construct can be exactly regulated by the compositional alterations to the biological and synthetic constituents of the material (Dikovsky et al. 2006; Underhill et al. 2007). These hybrids are made from extra cellular matrix proteins such as fibrinogen, collagen and albumin, while the synthetic polymer is made from polyethylene glycol (PEG) (Gonen-Wadmany et al. 2007). The PEG constituent is biocompatible with liver cells and already received FDA approval. The PEGylated protein hydrogels have already been applied successfully in orthopedic applications and their biocompatibility has been established with 5-week implantation studies in rats. An additional advantage of the PEGylated protein approach is that the biomaterials are polymerizable using photo-initiation, which provide the ability to directly inject the biomaterial precursors solution together with hepatic cell suspension into the body and polymerize the construct in situ (Shapira-Schweitzer and Seliktar. 2007). The combination of PEG and extra cellular matrix proteins in a PEGylated protein hydrogel system represents an advanced approach in scaffold design.

These polymers have the advantage of size, shape and cell number control. Taking this advantage we have recently formed microcapsules at the size of 200-600 mm (Fig. 1). This size is injectable through a 25G syringe through the portal vain of rats. The injected microcapsules can protect groups of hepatocytes from shear stress and from passing through the sinusoids. These particles are stacked in the portal radicle allowing eventually for a group of cells to engraft into the host liver (Nayshool et al.).

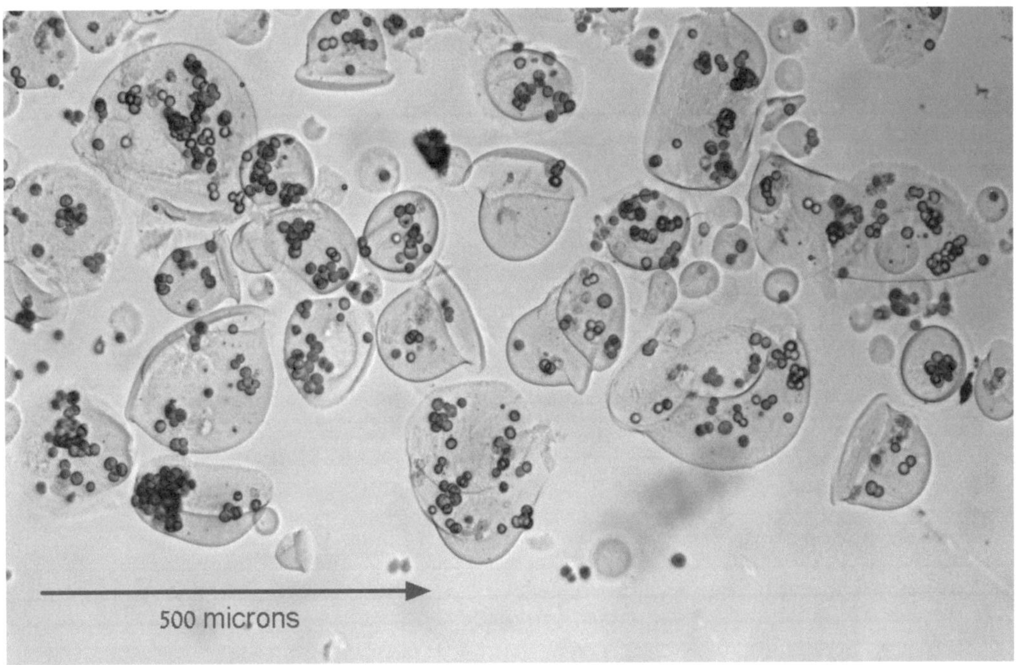

Fig. 1. HUH7 cells encapsulated in a PEG-Fibrinogen hydrogel scaffold. Radius of microcapsules 100-200 micrometers

In the long struggle against liver failure no one method would be most favorable. The strategy may be to employing different methods in the same individual and this would need the optimization of each one of them. The intravascular route for encapsulated cells is a promising method, solving the problem of shear stress protection, oxygen and nutrient supply, and allow for planning ahead the cell type mixture for transplantation. Still, portal hypertension remain a problem that would need to be resolved.

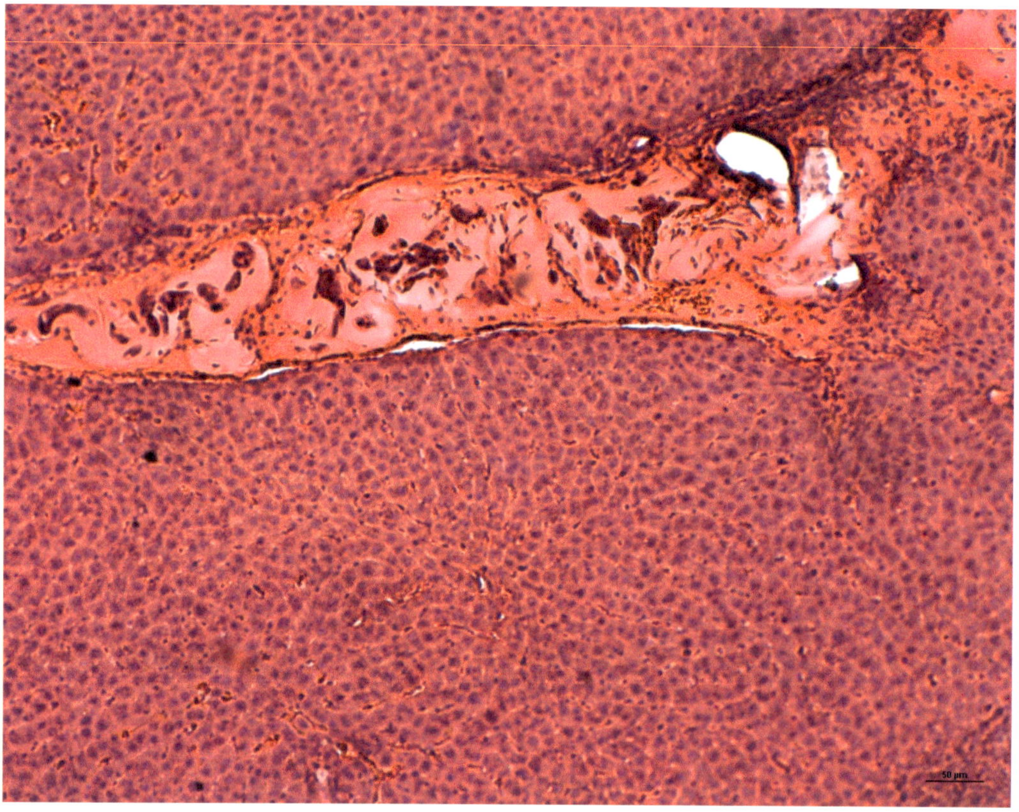

Fig. 2. HUH7 cells encapsulated in a PEG-Fibrinogen hydrogel scaffold in a portal radicle (H&E X40)

Disadvantages	**Advantages**
Portal hypertension	Natural microenvironment
Portal vein thrombosis	Linkage to bile system
Distorted tissue in acute failure and cirrhosis	Relative large vascular space
Cells bypass to the lungs	Favorable regenerative tissue
Insufficient space	
Coagulation defects	

Table 1. The liver as the target

1.	Use of Vasodilators.
2.	Combined use of angiogenic and proliferative growth factors.
3.	Control for the size of particles.
4.	Co-transplantation with non parenchymal cells.
5.	Liver preconditioning (partial hepatectomy, portal vein thrombosis, irradiation).
6.	Control the size of cell constructs

Table 2. Strategies to improve cell engraftment within the liver

5. References

[1] Allen JW and Bhatia SN. Engineering liver therapies for the future. *Tissue Eng.* 8: 5: 725-737, 2002a.

[2] Allen JW and Bhatia SN. Improving the next generation of bioartificial liver devices. *Semin.Cell Dev.Biol.* 13: 6: 447-454, 2002b.

[3] Almany L and Seliktar D. Biosynthetic hydrogel scaffolds made from fibrinogen and polyethylene glycol for 3D cell cultures. *Biomaterials* 26: 15: 2467-2477, 2005.

[4] Ambrosino G, Varotto S, Basso SM, Cecchetto A, Carraro P, Naso A, De Silvestro G, Plebani M, Abatangelo G, Donato D, Cestrone A, Giron G and D'Amico DF. Hepatocyte transplantation in the treatment of acute liver failure: microencapsulated hepatocytes versus hepatocytes attached to an autologous biomatrix. *Cell Transplant.* 12: 1: 43-49, 2003.

[5] Aoki T, Hui H, Umehara Y, LiCalzi S, Demetriou AA, Rozga J and Perfettit R. Intrasplenic transplantation of encapsulated genetically engineered mouse insulinoma cells reverses streptozotocin-induced diabetes in rats. *Cell Transplant.* 14: 6: 411-421, 2005.

[6] Asonuma K, Gilbert JC, Stein JE, Takeda T and Vacanti JP. Quantitation of transplanted hepatic mass necessary to cure the Gunn rat model of hyperbilirubinemia. *J.Pediatr.Surg.* 27: 3: 298-301, 1992.

[7] Assy N, Spira G, Paizi M, Shenkar L, Kraizer Y, Cohen T, Neufeld G, Dabbah B, Enat R and Baruch Y. Effect of vascular endothelial growth factor on hepatic regenerative activity following partial hepatectomy in rats. *J.Hepatol.* 30: 5: 911-915, 1999.

[8] Baldini E, Cursio R, De Sousa G, Margara A, Honiger J, Saint-Paul MC, Bayer P, Raimondi V, Rahmani R, Mouiel J and Gugenheim J. Peritoneal implantation of cryopreserved encapsulated porcine hepatocytes in rats without immunosuppression: viability and function. *Transplant.Proc.* 40: 6: 2049-2052, 2008.

[9] Baptista PM, Siddiqui MM, Lozier G, Rodriguez SR, Atala A and Soker S. The use of whole organ decellularization for the generation of a vascularized liver organoid. *Hepatology* 53: 2: 604-617, 2011.

[10] Baruch Y, Frith LO and Kirsch RE. Twenty-one percent partial hepatectomy. In vivo rat model for the study of liver regeneration. *J.Hepatol.* 12: 3: 316-320, 1991.

[11] Baruch Y, Neubauer K, Shenkar L, Sabo E, Ritzel A, Wilfling T and Ramadori G. Von Willebrand factor in plasma and in liver tissue after partial hepatectomy in the rat. *J.Hepatol.* 37: 4: 471-477, 2002.

[12] Baruch Y, Shoshany G, Neufeld G and Enat R. Basic fibroblast growth factor is hepatotropic for rat liver in regeneration. *J.Hepatol.* 23: 3: 328-332, 1995.

[13] Billiar TR, Curran RD, Ferrari FK, Williams DL and Simmons RL. Kupffer cell:hepatocyte cocultures release nitric oxide in response to bacterial endotoxin. *J.Surg.Res.* 48: 4: 349-353, 1990.

[14] Borselli C, Storrie H, Benesch-Lee F, Shvartsman D, Cezar C, Lichtman JW, Vandenburgh HH and Mooney DJ. Functional muscle regeneration with combined delivery of angiogenesis and myogenesis factors. *Proc.Natl.Acad.Sci.U.S.A.* 107: 8: 3287-3292, 2010.

[15] Carmel J, Arish A, Ilan N, Shoshany G, Vlodavsky I and Baruch Y. Heparanase accelerates proliferation of both hepatocyytes and sinusoidal endothelial cells after partial hepatectomy despite of IL-6 and TNF-a downregulation. Poster presented at: EASL 2003. *Proceedings of the 45-th annual meeting of EASL; 2010 Apr 14-18; Vienna, Austria. J.Hepatol.* 52: S383, 2010.

[16] Chowdhury JR, Grossman M, Gupta S, Chowdhury NR, Baker JR,Jr and Wilson JM. Long-term improvement of hypercholesterolemia after ex vivo gene therapy in LDLR-deficient rabbits. *Science* 254: 5039: 1802-1805, 1991.

[17] Cressman DE, Greenbaum LE, DeAngelis RA, Ciliberto G, Furth EE, Poli V and Taub R. Liver failure and defective hepatocyte regeneration in interleukin-6-deficient mice. *Science* 274: 5291: 1379-1383, 1996.

[18] Demetriou AA, Whiting JF, Feldman D, Levenson SM, Chowdhury NR, Moscioni AD, Kram M and Chowdhury JR. Replacement of liver function in rats by transplantation of microcarrier-attached hepatocytes. *Science* 233: 4769: 1190-1192, 1986.

[19] Dikovsky D, Bianco-Peled H and Seliktar D. The effect of structural alterations of PEG-fibrinogen hydrogel scaffolds on 3-D cellular morphology and cellular migration. *Biomaterials* 27: 8: 1496-1506, 2006.

[20] Dixit V and Gitnick G. Transplantation of microencapsulated hepatocytes for liver function replacement. *J.Biomater.Sci.Polym.Ed.* 7: 4: 343-357, 1995.

[21] Fisher RA and Strom SC. Human hepatocyte transplantation: worldwide results. *Transplantation* 82: 4: 441-449, 2006.

[22] Folkman J and Hochberg M. Self-regulation of growth in three dimensions. *J.Exp.Med.* 138: 4: 745-753, 1973.

[23] Fox IJ, Chowdhury JR, Kaufman SS, Goertzen TC, Chowdhury NR, Warkentin PI, Dorko K, Sauter BV and Strom SC. Treatment of the Crigler-Najjar syndrome type I with hepatocyte transplantation. *N.Engl.J.Med.* 338: 20: 1422-1426, 1998.

[24] Gonen-Wadmany M, Oss-Ronen L and Seliktar D. Protein-polymer conjugates for forming photopolymerizable biomimetic hydrogels for tissue engineering. *Biomaterials* 28: 26: 3876-3886, 2007.

[25] Grossman M, Rader DJ, Muller DW, Kolansky DM, Kozarsky K, Clark BJ,3rd, Stein EA, Lupien PJ, Brewer HB,Jr and Raper SE. A pilot study of ex vivo gene therapy for homozygous familial hypercholesterolaemia. *Nat.Med.* 1: 11: 1148-1154, 1995.

[26] Gupta S. Regulation of cell engraftment in the liver. In: *Hepatocyte transplantation. Proceedings of Falk symposium 126, 2002,* edited by Gupta S, Jansen P, Klempnauer J and Manns MP2002.

[27] Gupta S, Gorla GR and Irani AN. Hepatocyte transplantation: emerging insights into mechanisms of liver repopulation and their relevance to potential therapies. *J.Hepatol.* 30: 1: 162-170, 1999.

[28] Gupta S, Lee CD, Vemuru RP and Bhargava KK. 111Indium labeling of hepatocytes for analysis of short-term biodistribution of transplanted cells. *Hepatology* 19: 3: 750-757, 1994.
[29] Gupta S, Vasa SR, Rajvanshi P, Zuckier LS, Palestro CJ and Bhargava KK. Analysis of hepatocyte distribution and survival in vascular beds with cells marked by 99mTC or endogenous dipeptidyl peptidase IV activity. *Cell Transplant.* 6: 4: 377-386, 1997.
[30] Gupta S, Yerneni PR, Vemuru RP, Lee CD, Yellin EL and Bhargava KK. Studies on the safety of intrasplenic hepatocyte transplantation: relevance to ex vivo gene therapy and liver repopulation in acute hepatic failure. *Hum.Gene Ther.* 4: 3: 249-257, 1993.
[31] Hamazaki K, Doi Y and Koide N. Microencapsulated multicellular spheroid of rat hepatocytes transplanted intraperitoneally after 90% hepatectomy. *Hepatogastroenterology* 49: 48: 1514-1516, 2002.
[32] Hernandez RM, Orive G, Murua A and Pedraz JL. Microcapsules and microcarriers for in situ cell delivery. *Adv.Drug Deliv.Rev.* 62: 7-8: 711-730, 2010.
[33] Hughes RD, Mitry RR, Dhawan A, Lehec SC, Girlanda R, Rela M, Heaton ND and Muiesan P. Isolation of hepatocytes from livers from non-heart-beating donors for cell transplantation. *Liver Transpl.* 12: 5: 713-717, 2006.
[34] Kaihara S, Borenstein J, Koka R, Lalan S, Ochoa ER, Ravens M, Pien H, Cunningham B and Vacanti JP. Silicon micromachining to tissue engineer branched vascular channels for liver fabrication. *Tissue Eng.* 6: 2: 105-117, 2000.
[35] Kubo A, Shinozaki K, Shannon JM, Kouskoff V, Kennedy M, Woo S, Fehling HJ and Keller G. Development of definitive endoderm from embryonic stem cells in culture. *Development* 131: 7: 1651-1662, 2004.
[36] Lee J, Cuddihy MJ, Cater GM and Kotov NA. Engineering liver tissue spheroids with inverted colloidal crystal scaffolds. *Biomaterials* 30: 27: 4687-4694, 2009.
[37] Lin CQ and Bissell MJ. Multi-faceted regulation of cell differentiation by extracellular matrix. *FASEB J.* 7: 9: 737-743, 1993.
[38] Michalopoulos GK. Liver regeneration after partial hepatectomy: critical analysis of mechanistic dilemmas. *Am.J.Pathol.* 176: 1: 2-13, 2010.
[39] Michalopoulos GK. Liver regeneration: molecular mechanisms of growth control. *FASEB J.* 4: 2: 176-187, 1990.
[40] Mito M, Ebata H, Kusano M, Onishi T, Saito T and Sakamoto S. Morphology and function of isolated hepatocytes transplanted into rat spleen. *Transplantation* 28: 6: 499-505, 1979.
[41] Mitry RR, Dhawan A, Hughes RD, Bansal S, Lehec S, Terry C, Heaton ND, Karani JB, Mieli-Vergani G and Rela M. One liver, three recipients: segment IV from split-liver procedures as a source of hepatocytes for cell transplantation. *Transplantation* 77: 10: 1614-1616, 2004.
[42] Mooney DJ, Sano K, Kaufmann PM, Majahod K, Schloo B, Vacanti JP and Langer R. Long-term engraftment of hepatocytes transplanted on biodegradable polymer sponges. *J.Biomed.Mater.Res.* 37: 3: 413-420, 1997.
[43] Nakamura T, Nawa K, Ichihara A, Kaise N and Nishino T. Purification and subunit structure of hepatocyte growth factor from rat platelets. *FEBS Lett.* 224: 2: 311-316, 1987.
[44] Navarro-Alvarez N, Soto-Gutierrez A, Chen Y, Caballero-Corbalan J, Hassan W, Kobayashi S, Kondo Y, Iwamuro M, Yamamoto K, Kondo E, Tanaka N, Fox IJ and

Kobayashi N. Intramuscular transplantation of engineered hepatic tissue constructs corrects acute and chronic liver failure in mice. *J.Hepatol.* 52: 2: 211-219, 2010.

[45] Nayshool O, Carmel J, Saadi T, Arish A, Mironi-Harpaz I, Seliktar D and Baruch Y. In Vitro comparison between different biosynthetic hydrogel scaffolds, for liver tissue regeneration therapy. Poster presented at: Israeli Society for the Study of Liver Disease (IASLD) *on annual meeting ; 2011 Jan 26-29; Eilat, Israel.*

[46] Ott M and et al. Assessment of shunting, portal haemodynamic changes and liver damage after intraportal ifusion macro-aggregated albumin (MAA) particles and hepatocytes in rabbits. In: *Hepatocyte transplantation. Proceedings of Falk symposium 126, 2002*, edited by Gupta S, Jansen P, Klempnauer J and Manns MP2002.

[47] Perets A, Baruch Y, Weisbuch F, Shoshany G, Neufeld G and Cohen S. Enhancing the vascularization of three-dimensional porous alginate scaffolds by incorporating controlled release basic fibroblast growth factor microspheres. *J.Biomed.Mater.Res.A.* 65: 4: 489-497, 2003.

[48] Piscaglia AC, Campanale M, Gasbarrini A and Gasbarrini G. Stem cell-based therapies for liver diseases: state of the art and new perspectives. *Stem Cells Int.* 2010: 259461, 2010.

[49] Ponder KP, Gupta S, Leland F, Darlington G, Finegold M, DeMayo J, Ledley FD, Chowdhury JR and Woo SL. Mouse hepatocytes migrate to liver parenchyma and function indefinitely after intrasplenic transplantation. *Proc.Natl.Acad.Sci.U.S.A.* 88: 4: 1217-1221, 1991.

[50] Puppi J, Strom SC, Hughes RD, Bansal S, Castell JV, Dagher I, Ellis EC, Nowak G, Ericzon BG, Fox IJ, Gomez-Lechon MJ, Guha C, Gupta S, Mitry RR, Ohashi K, Ott M, Reid LM, Roy-Chowdhury J, Sokal E, Weber A and Dhawan A. Improving the Techniques for Human Hepatocyte Transplantation: Report from a Consensus Meeting in London. *Cell Transplant.* 2011.

[51] Raper SE, Grossman M, Rader DJ, Thoene JG, Clark BJ,3rd, Kolansky DM, Muller DW and Wilson JM. Safety and feasibility of liver-directed ex vivo gene therapy for homozygous familial hypercholesterolemia. *Ann.Surg.* 223: 2: 116-126, 1996.

[52] Rhim JA, Sandgren EP, Degen JL, Palmiter RD and Brinster RL. Replacement of diseased mouse liver by hepatic cell transplantation. *Science* 263: 5150: 1149-1152, 1994.

[53] Rouwkema J, Rivron NC and van Blitterswijk CA. Vascularization in tissue engineering. *Trends Biotechnol.* 26: 8: 434-441, 2008.

[54] Sakai Y, Yamagami S and Nakazawa K. Comparative analysis of gene expression in rat liver tissue and monolayer- and spheroid-cultured hepatocytes. *Cells Tissues Organs* 191: 4: 281-288, 2010.

[55] Schirmacher P, Geerts A, Pietrangelo A, Dienes HP and Rogler CE. Hepatocyte growth factor/hepatopoietin A is expressed in fat-storing cells from rat liver but not myofibroblast-like cells derived from fat-storing cells. *Hepatology* 15: 1: 5-11, 1992.

[56] Selden C, Calnan D, Morgan N, Wilcox H, Carr E and Hodgson HJ. Histidinemia in mice: a metabolic defect treated using a novel approach to hepatocellular transplantation. *Hepatology* 21: 5: 1405-1412, 1995.

[57] Selden C, Casbard A, Themis M and Hodgson HJ. Characterization of long-term survival of syngeneic hepatocytes in rat peritoneum. *Cell Transplant.* 12: 6: 569-578, 2003.

[58] Seliktar D. Extracellular stimulation in tissue engineering. *Ann.N.Y.Acad.Sci.* 1047: 386-394, 2005.

[59] Shani-Peretz H, Tsiperson V, Shoshani G, Veitzman E, Neufeld G and Baruch Y. HVEGF165 increases survival of transplanted hepatocytes within portal radicles: suggested mechanism for early cell engraftment. *Cell Transplant.* 14: 1: 49-57, 2005.

[60] Shapira-Schweitzer K and Seliktar D. Matrix stiffness affects spontaneous contraction of cardiomyocytes cultured within a PEGylated fibrinogen biomaterial. *Acta Biomater.* 3: 1: 33-41, 2007.

[61] Shi XL, Zhang Y, Gu JY and Ding YT. Coencapsulation of hepatocytes with bone marrow mesenchymal stem cells improves hepatocyte-specific functions. *Transplantation* 88: 10: 1178-1185, 2009.

[62] Slehria S and Gupta S. Vascular endothelial growth factor is released before sinusoidal endothelial disruption during liver repopulation with hepatocyte transplantation (abstract). *J.Invest.Med* 211A, 1998.

[63] Slehria S, Rajvanshi P, Ito Y, Sokhi RP, Bhargava KK, Palestro CJ, McCuskey RS and Gupta S. Hepatic sinusoidal vasodilators improve transplanted cell engraftment and ameliorate microcirculatory perturbations in the liver. *Hepatology* 35: 6: 1320-1328, 2002.

[64] Sohlenius-Sternbeck AK. Determination of the hepatocellularity number for human, dog, rabbit, rat and mouse livers from protein concentration measurements. *Toxicol.In.Vitro.* 20: 8: 1582-1586, 2006.

[65] Soto-Gutierrez A, Kobayashi N, Rivas-Carrillo JD, Navarro-Alvarez N, Zhao D, Okitsu T, Noguchi H, Basma H, Tabata Y, Chen Y, Tanaka K, Narushima M, Miki A, Ueda T, Jun HS, Yoon JW, Lebkowski J, Tanaka N and Fox IJ. Reversal of mouse hepatic failure using an implanted liver-assist device containing ES cell-derived hepatocytes. *Nat.Biotechnol.* 24: 11: 1412-1419, 2006.

[66] Streetz KL, Luedde T, Manns MP and Trautwein C. Interleukin 6 and liver regeneration. *Gut* 47: 2: 309-312, 2000.

[67] Strom SC, Chowdhury JR and Fox IJ. Hepatocyte transplantation for the treatment of human disease. *Semin.Liver Dis.* 19: 1: 39-48, 1999.

[68] Teng Y, Wang Y, Li S, Wang W, Gu R, Guo X, Nan X, Ma X and Pei X. Treatment of acute hepatic failure in mice by transplantation of mixed microencapsulation of rat hepatocytes and transgenic human fetal liver stromal cells. *Tissue Eng.Part C.Methods* 16: 5: 1125-1134, 2010.

[69] Terry C, Hughes RD, Mitry RR, Lehec SC and Dhawan A. Cryopreservation-induced nonattachment of human hepatocytes: role of adhesion molecules. *Cell Transplant.* 16: 6: 639-647, 2007.

[70] Torok E, Lutgehetmann M, Bierwolf J, Melbeck S, Dullmann J, Nashan B, Ma PX and Pollok JM. Primary human hepatocytes on biodegradable poly(l-lactic acid) matrices: a promising model for improving transplantation efficiency with tissue engineering. *Liver Transpl.* 17: 2: 104-114, 2011.

[71] Tsiperson V, Goldshmidt O, Ilan N, Shoshany G, Vlodavsky I, Veitsman E and Baruch Y. Heparanase enhances early hepatocyte inclusion in the recipient liver after transplantation in partially hepatectomized rats. *Tissue Eng.Part A.* 14: 3: 449-458, 2008.

[72] Underhill GH, Chen AA, Albrecht DR and Bhatia SN. Assessment of hepatocellular function within PEG hydrogels. *Biomaterials* 28: 2: 256-270, 2007.

[73] Wang Y, Cui CB, Yamauchi M, Miguez P, Roach M, Malavarca R, Costello MJ, Cardinale V, Wauthier E, Barbier C, Gerber DA, Alvaro D and Reid LM. Lineage restriction of human hepatic stem cells to mature fates is made efficient by tissue-specific biomatrix scaffolds. *Hepatology* 53: 1: 293-305, 2011.

[74] Weber A, Touboul T, Mainot S, Branger J and Mahieu-Caputo D. Human foetal hepatocytes: isolation, characterization, and transplantation. *Methods Mol.Biol.* 640: 41-55, 2010.

21

Nano- Coating with Titanium of Glutaraldehyde- Fixed Heart Valve Prostheses Enables a Reduced Immune Response and a Self-Seeding Within Circulation

Norbert W. Guldner[1], Hangörg Zimmermann[2] and Hans- H. Sievers[1]
[1]Clinic of Cardiac Surgery, University of Lübeck,
[2]Pfm Titan GmbH Nürnberg
Germany

1. Introduction

Glutaraldehyd-fixed heart valve prostheses with long durability and function are desirable as a treatment option of heart valve disease. They do not require chronic anticoagulation therapy, however break down quickly in young patients (<5 years) but last for > 15 years in patients older than 70 years. It is generally accepted that the break down of bioprosthetic heart valves is due to calcification occurring because of immunologic and chemical processes.

Thus, new nano- technological methods in combination with chemical and biochemical procedures concerning detoxification and titanization of glutardialdehyde- fixed collagen scaffolds are applied on glutardialdehyde fixed pericardial samples. Nano- technology with titanium using a plasma application under high alternating current, developed elsewhere (GfE Nürnberg, Germany, patent number EP 0 897 997 A1) might decrease the immune response caused by coating and improve endothelialization by detoxification and the biocompatibility of titanium. A plasma-activated surface of collagen allows a titanization under room temperature described later on. Immune response with secondary dystrophic calcification may be why glutaraldehyde fixed xenograft valves fail, especially in young patients. Hypothesis is, detoxification and titanization may reduce immunologic reactions and enable an endothelialization.

2. Detoxification and titanium coating of glutaraldehyde-fixed pericardium

Free aldehyde-groups of glutaraldehyde-fixed pericardium are responsible for its toxicity, which prevents an endothelialization.

2.1 Detoxification

Platelets of glutaraldehyde- fixed pericardium were treated differently for detoxification. Some were detoxified chemically by incubation with citric acid (CA) of 10%. Others were detoxified enzymatically by aldehyde dehydrogenase and received nano-technological methods using plasma activation either solely or combined with a treatment of organic bound titanium.

2.2 Determination of free aldehyde groups

A method for the determination of the amount of glutaraldehyde in specimens of glutaraldehyde fixed pericardium has been developed using aldehyde dehydrogenase (ALDH), (Guldner et.al. 2009). In presence of ALDH, aldehydes are oxidated to carbon acids while a reduction equivalent (here: NAD+) is reduced. Aldehyde dehydrogenases have broad substrate specificity. Thus, they process glutardialdehyde as well as other aldehydes

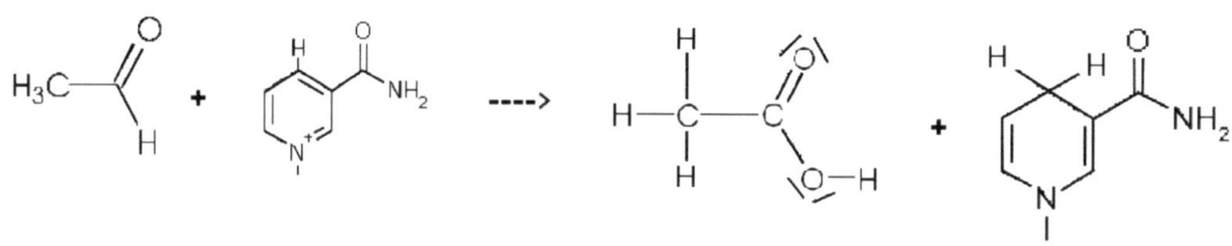

Acetaldehyde + NAD+ → Acetic acid + NADH

The determination of aldehydes with the ALDH method was performed photometrically. The increase of optical density (extension) by the reduced NADH could be measured at a wave length of 340nm. The amount of NADH being proportional to the original amount of aldehyde, statements on the remaining toxicity of the fixed pericardium could be made. Optimal duration for incubation was inquired.

Enzyme, NAD+ (tablets) and buffer solution from a commercially available kit, originally developed to determine acetaldehyde concentrations in a variety of materials, were used (Acetaldehyde, Enzymatic BioAnalysis; R-Biopharm, Darmstadt, Germany). Platelets of pericardium were incubated in a 24 well plate with 1ml enzyme solution each for 24 hours. Thereafter 100µl from each well were examined photometrically. Native pericardium served as control.

2.3 Chemical Vapor Deposition (CVD) by a titanium containing precursor

Before physical titanization of cardiovascular scaffolds by plasma activated chemical gas deposition (PACVD) it is necessary to extract its water completely, which is feasible in a vacuum by a slow drying process. PACVD is a coating technology (GfE Nürnberg, Germany, patent number EP 0 897 997 A1) where the so called precursor (Tetrakisdimethylamidotitan, Ti $[N(CH_3)_2]_4$) is transferred into the gas phase and brought into the reactor by a carrier gas (nitrogen). The precursor or parts of the precursor react with the substrate creating a resistant layer. Physical plasma however is able to supply the substrate with high energy while the temperature during deposition can kept low. Within this non isotherm plasma with solely high electron temperature and room temperature of the neutrons and ions the electrons can follow a quickly changing electrical field with typical values of a radio frequency (40kHz) low pressure plasma as described elsewhere (Grill,1994; Sivaram S.1995).Thickness of the titanium containing deposit is depend on sputtering time. After PACVD the dehydrated implants came to the same shape in physiological saline solution as before titanization.

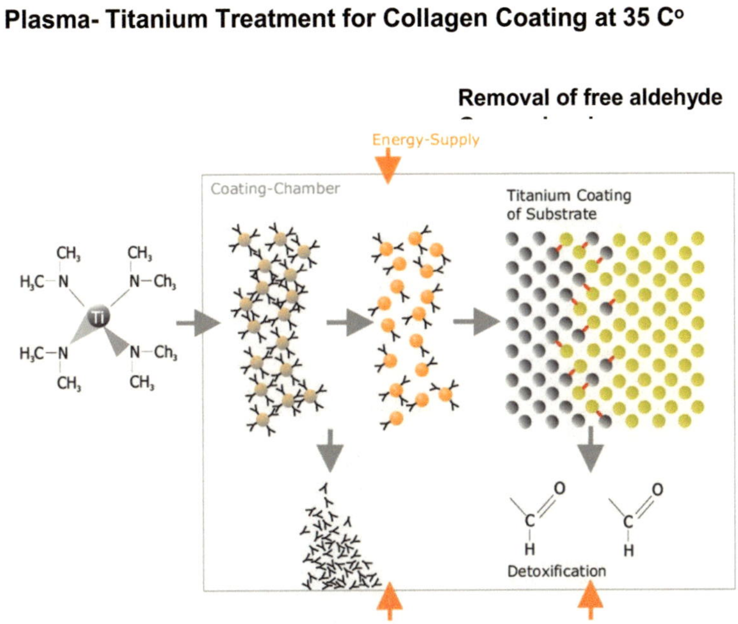

Fig. 1. Scheme of Chemical Vapor Deposition (CVD) with the gasiform titanium bounded organic molecule as precursor at 35 C^0. In a vacuum chamber with energy supply by high alternative current, the precursor is deconstructed and precipitated on the plasma activated collagen surface (Plasma Activated Chemical Vapor Deposition, PACVD) and electron pair bounded to collagen molecules

2.4 Examination of the titanium coat

Deposits composition and binding energy of titanium, carbon, oxygen and nitrogen atoms could become analyzed by X-ray Photoelectron Spectroscopy (XPS). X-ray activated atoms emit electrons from inner electron layers (photo effect) with a distinct kinetic energy being specific for each element. The kinetic energy of the emitted electron diagnoses the element and therefore the composition of elements within the titanic layer. The chemical binding in which an element is, influences the energy levels of its inner electrons. A more electronegative partner binds its electrons tighter on its inner electron layers. That influences the kinetic energy on the emitted photoelectrons during X-ray application. These values in kinetic energy of the emitted photoelectrons enable to determine the chemical binding energy between elements during XPS and so in titanic layers as for oxygen, titanium, nitrogen and carbon (Dag, 1997; Moulderm,1995). The limited intrinsic depth for XPS of about 1 nm requires a removal of substrate in layers using argon ions. The XPS analysis showed a maximal surface atom fraction of 21.1±3 8% titanium atoms within a removal time of about 300 seconds, which translates to a titanium thickness of 30 nm. The presence of oxygen indicated that titanium had been changed by subsequent oxidation into titanium dioxide, with an electron-binding energy for titanium of 463 eV. Two peaks for titanium represented titanium bound carbon and titanium bound with oxygen as titanium oxide. These analyses show an extremely tight electron binding of the titan precursor fragments to collagen. Titanium is non removable bounded to collagen in an extremely thin coating onto collagen with a diameter of about 30 nano-meters.

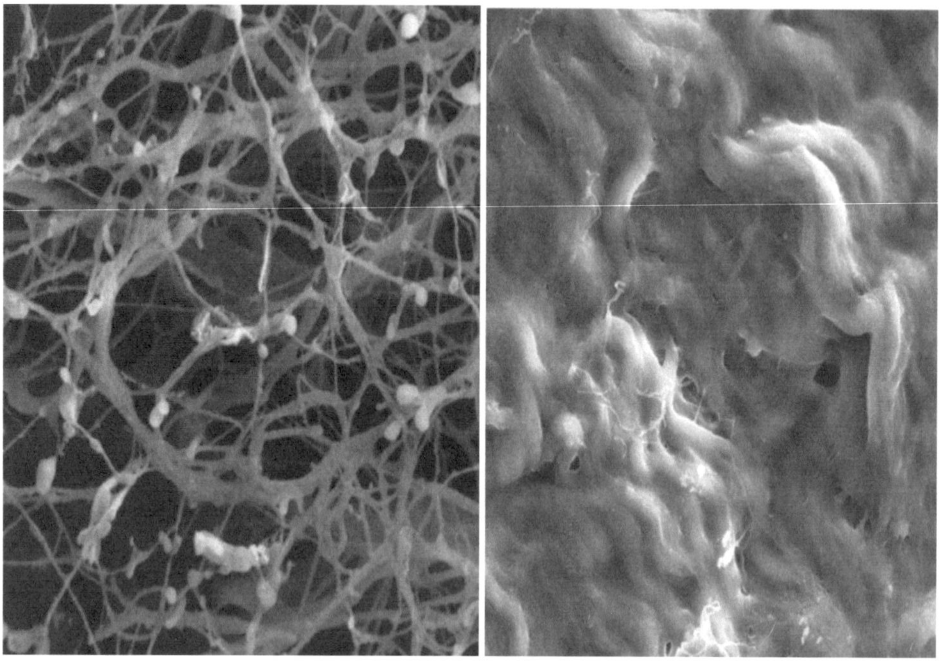

Fig. 2. REM of GA- fixed and dehydrated pericardium with a magnification of 5000X (left), and with a complete covering of the surface by titanium (right)

Scanning electron microskopic images of GA-fixed pericardium show a more filigree net like structure (Fig.2 left) which disappears after titanization (Fig.2 right). These morphologic changes however do not affect the mechanical stability of the pericardium as already investigated (unpublished data). After dehydration and a swelling of the soft collagen, the titanium coat may burst into grooves. This uncomplete surface covering may have important implications on iC3b deposits and PNM attraction as discussed in the following descriptions about the immune response. But nevertheless the decovering of collagene by these grooves is manly dependent on the chosen dehydration procedure.

3. Immunology of glutaraldehyde-fixed bovine pericardium

Glutaraldehyd-fixed heart valve prostheses calcify and fail in a time period lesser than 12 years (Williams et. al.,1982). Some investigators believe that the GA-process leads to calcification and failure of the valves due to chemical processes (Simionescu, 2004). Others are convinced the main break down of GA-fixed valves is caused by an immunologic response (Rocchini et. al.,1981; Williams et.al., 1982; Ueyama et.al, 2002; Rizwan et.al.,2006). They showed in a young animal model that GA-fixed xenograft valves undergo rejection and that inflammation is correlated to calcification (Rizwan et.al., 2006).

Young patients with a strong immune system show a valve break down very quickly (<5) years (Rocchini et. al.,1981; Williams et.al., 1982;). In patients > 70 years with a minor immune response bio-prosthetic valves function well > 10 years (Ueyama et.al, 2002). Xenogenic organ transplantation arises a very aggressive form of humoral and cellular immune rejection with tissue infiltrations by lymphocytes and monocytes (Fischbein et.al., 2000). Whereas the immunologic lymphocytes and monocytes response could become decreased by de-cellularization or GA-fixation the granulocyte recruitments in de-cellularized xenografts and GA-fixed heart valves were not influenced at all. The residual immuno-stimulatory activity of de-cellularized porcine vascular tissue towards polymorphonuclear (PNM) cells resp. granulocyte recruitments, observed in vitro (Fischbein et. al., 2000; Juthier et.al; 2006), correlates with in vivo findings with de-cellularized human and porcine heart valves (Friedhelm et.al; 2005; Rieder et.al.; 2006; Bastian et.al., 2008). For the evaluation of the immunologic activity of GA- fixed pericardium in vitro blood sampling, plasma preparation and isolation of PNMs are necessary.

3.1 Blood sampling, preparation of plasma, and isolation of PMN

Venous blood was collected from 3 healthy adult volunteers who were not using any medications and who gave informed consent (approved by the ethical committee, Medical University of Vienna). Whole blood was anti-coagulated with 5 IU/ml heparin and centrifuged at 2000 x *g* for 15 min to obtain plasma. Human PMN were retrieved from EDTA-anticoagulated venous blood by lysing 1ml blood with 5ml 0.9% (w/v) ammonium chloride for 15 min at 4°C, followed by centrifugation at 160 x *g* and 4°C for 10 min (Nilsson, 2001). The supernatant was discarded and the cell pellet washed three times with PBS. Cells were subsequently re-suspended in RPMI 1640 medium (BioWhittaker™, Verbiers, Belgium) to a final concentration of 5000/µl and immediately used for the experiments. The cell suspensions comprised 84.4±3.1% PMN, 11.6±2.7% lymphocytes, and 3.5±1.6% monocytes. The platelet contamination was less than 0.1 platelet per PMN. These procedures were necessary for the quantification of IC3b deposits on the blood contacting surface of the GA-fixed pericardium and the granulocyte-matrix adhesion experiments.

3.2 IC3b- deposits

IgG deposits(iC3b) were visualized by an immune staining as described elsewhere (Bastian et. al.; 2008) and shown in figure 4. They are documented as coloured pixels (red). Pixels are counted electronically.

IC3b- deposits within the control group with GA-fixed pericardium were regarded as 100%; in the titanized group IC3b deposits were evaluated as 34%. The authors opinion is, that these preliminary results can become further optimised (further decrease of immunologic response) by an advanced procedure of pericard drying and plasma deposition based on titanium.

3.3 Granulocyte/matrix adhesion experiments

Platelets of titanium coated, glutaraldehyde-fixed bovine pericardium were brought into an immunologic test setting inducing a complement-mediated granulocyte adhesion and activation as described elsewhere (Bastian et. al.; 2008).

Plasma used for pre-treating the tissue samples and granulocytes was obtained from the same donor during each experiment. Tissue specimens (0.5 x 0.5 cm) were either incubated with 20% of heparinized autologous plasma (diluted in PBS containing 0.15 mmol/l Ca^{2+} and 0.5 mmol/l Mg^{2+}) on a rocking platform for 5h each at 37°C, or left untreated. Cryostat sections were incubated with 5,000 PMN/µl for 15 minutes in a humidified incubation chamber at room temperature (RT). This incubation time was chosen since it proved to be optimal for demonstrating PMN adhesion in vitro.

PMN quantification was performed by counting the leucocytes (including 86% PMNs) near the blood contacting surface of the pericard within an areas of 3x3 mm. Five samples from each histologic slice were taken for the evaluation of PNMs.

Nano- coating with titanium remarkably reduced the early immunologic response to GA fixed bovine pericardium in vitro. Thus, the titanium coat is protective against iC3b and immune based PMN attraction in contact with human blood.

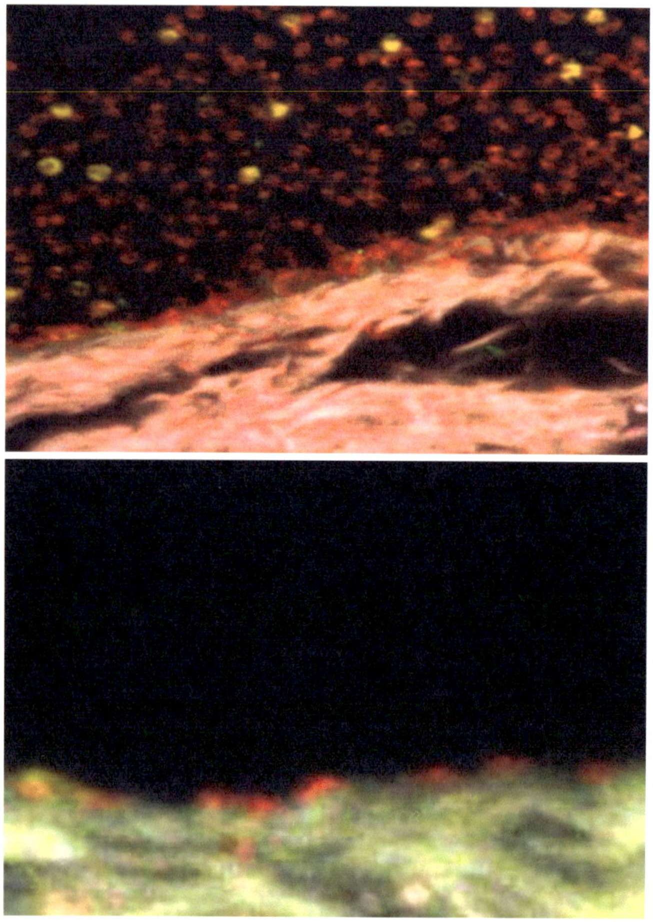

Fig. 3. PNM count in front of the GA-fixed pericard slices were regarded as 100% (top) GA- fixed and titanium coated pericardium however reduced the attraction of PNMs to 6.0% (bottom) as an expression for an extremely reduced immune response

4. Endothelialization

Glutardialdehyde's toxicity however prevents biological coating. Detoxification of glutaraldehyde-fixed scaffolds by amino acids (Fischlein et.al, 1994) and by citric acids (Lehner et.al., 1997; Gulbins et.al., 2003) in vitro were successfully applied for partial detoxification but endothelialization was only possible by pre-seeding with fibroblasts. Such pre-seeded layers on porcine heart valves were stable in a sheep model (Gulbins et.al., 2003,2006). Due to the fibroblasts however, the valves showed a thickening of the leaflets resulting in an impairment of function.

Recent investigations developed detoxification strategies and applied nano- technological methods such as ultrathin titanium surface coating on glutaraldehyde fixed bovine pericardium as the base material for prosthetic heart valves (Guldner et.al., 2009). Titanization by chemical vapor deposition was feasible at temperatures between 30-35°C (Grill et.al., 1994; Silvaram et.al., 1995; Dag,1995). For endothelial cell adhesion on a titanium surface endothelial cell cultures were performed.

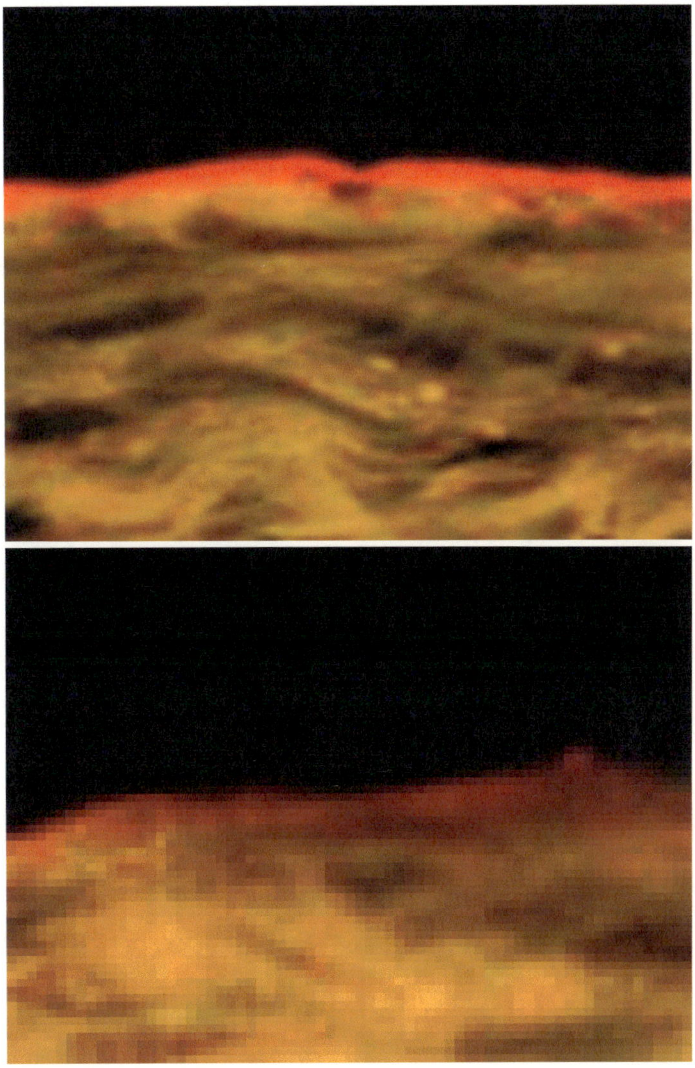

Fig. 4. Immune staining of iC3b deposits on GA-fixed pericardium (top) was control and defined as 100% pixels, titanium coated pericardium platelets showed a count of a decreased number of pixels to 34% (bottom) as an expression of an impressive lesser immune response as to the uncoated GA-fixed pericardium

4.1 Cell culture

Endothelial cells have been isolated from human saphena veins, gained from coronary bypass surgery. Ethical approval was obtained from the Ethical Commission on Research on Humans of the University of Schleswig Holstein, Campus Lübeck, Germany (registration

No.05-097). Patients had been informed before surgery that parts of their veins were to be used for research purposes. The vein pieces were cannulated, rinsed with cell medium then filled with 0.5% dispase and incubated for 20 minutes at 37°C and 5% CO_2. Then solution was then centrifugated at 1200 upm for 5 minutes, the cell pellet was re-suspended in endothelial growth medium (EGM-2; Cambrex, Walkersville, US) and plated on cell culture flasks. For passaging, endothelial cells were trypsinized after reaching confluence, centrifuged at 1200 upm for 5 minutes, then re-suspended in endothelial cell medium and plated out again. Cell counting has been performed by incubating 20µl of the cell suspension with an equal amount of trypan-blue and then counting using a Neubauer chamber.

Pre-treated platelets of glutardialdehyde fixed pericardium were examined by scanning electron microscopy (SEM) after 48 hours of incubation with human endothelial cells. SEM showed malformed non confluent endothelial cells on Citric Acid (10%) only treated pericardium (Figure 5a, n=5). In contrast to that, a completely confluent cell layer of regular human endothelial cells could be observed on pericardial surfaces treated with a combination of Citric Acid (10%) and plasma-titan method (n=5) (Figure 5).Toxicity of pericardium was reduced by combination of procedures using ALDH and plasma-titanium coating (n=12) to 17.3± 2.1% ($p \leq 0.01$).

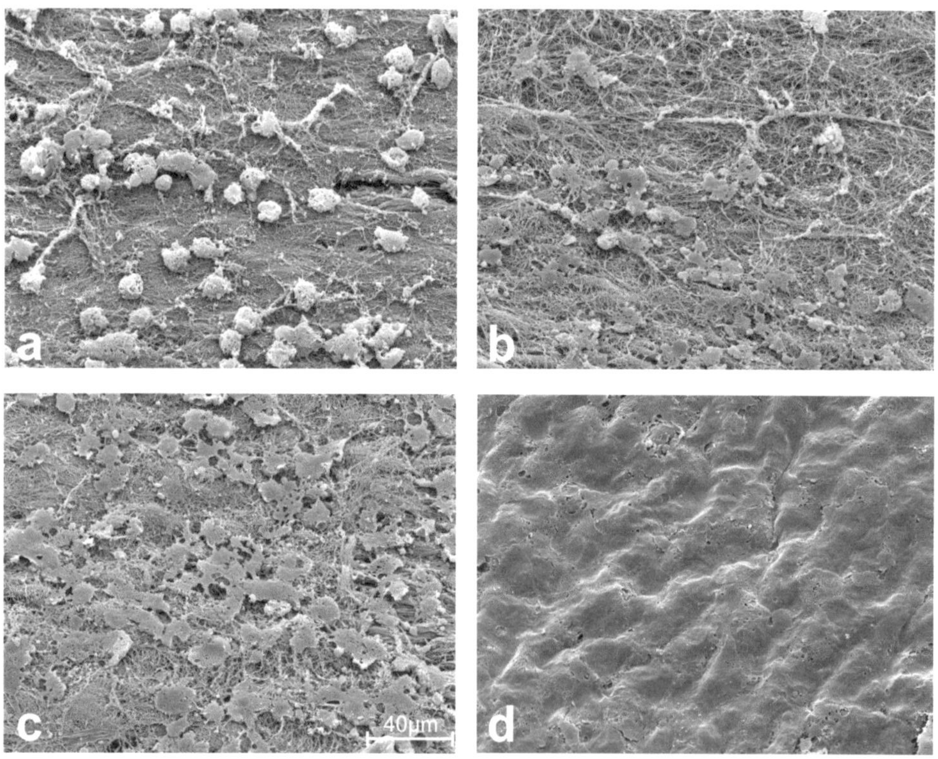

Fig. 5. Scanning electron microscopy of human endothelial cells on glutardialdehyde fixed bovine pericardium shows sporatic cell cadavers (a), treated with citric acid (10%) demonstrates malformed endothelial cells (b) detoxified with citric acid (10%) and ALDH results in non confluent endothelial layers (c) the nano-technological promoted procedure with a detoxification and a titan plasma administration enabled a confluent layer of endothelial cells (d)

Several platelets of pericardium differently detoxified were examined by scanning electron microscopy (SEM) after being seeded in vitro with human endothelial cells. Of each specimen, 10 visual fields were evaluated under 1000x magnification. Both cell morphology and confluence of the cell layer were assessed semi-quantitatively by different examiners.

4.2 Cell adhesion under flow conditions

A flow chamber made of V4A steel and glass was able to incorporate in its bottom six detoxified and endothelialized platelets within V4A steel rings.. It was constructed such, that no significant turbulences happened and that the laminar shear stress of 30 dyn/cm^2 and flow of 5 L/min was similar to the intra-aortic fluid dynamic.
Adhesion of human endothelial cells on glutaraldehyde fixed pericardial platelets treated by CA 10%, ALDH and a titanium coating was expressed by the vitality of the remaining cells after a flow of 5L/min for 24 hours within a flow chamber.
Vitality of the endothelial cells was measured by the extinction of the MTS-test of 72% (n=6) after a pre-seeding of 15 min and 81% (n=6) after pre-seeding of 60 min (Fig.6).

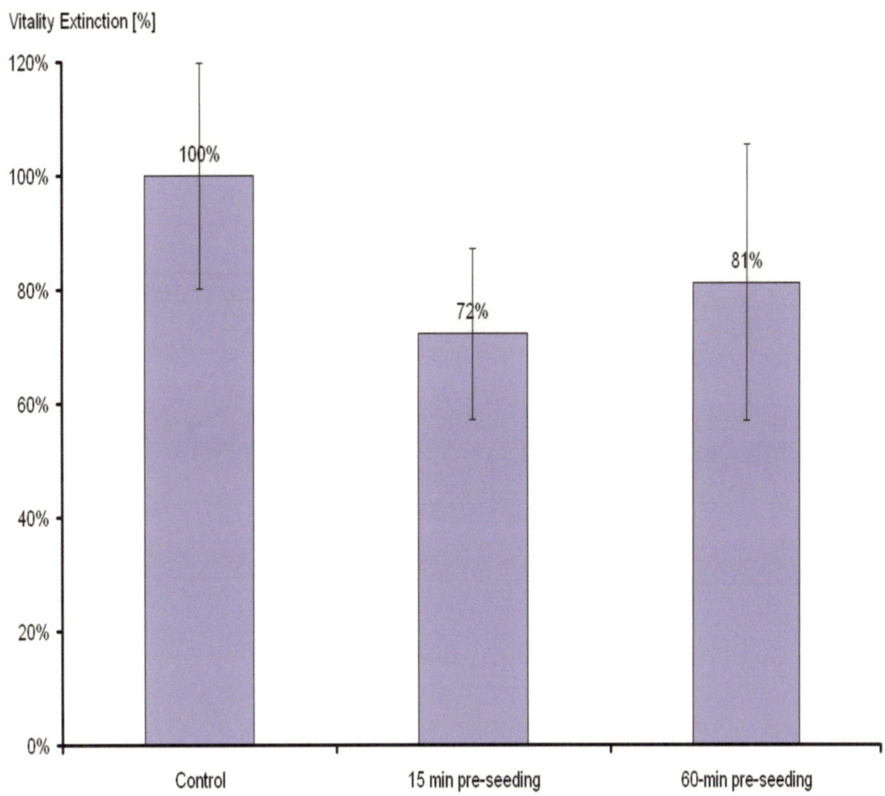

Fig. 6. Extinction of MTS-test for titanium treated (n=6) endothelialized glutaraldehyde-fixed bovine pericardium platelets after in a flow channel of 5L/min. Adhesion of human endothelial cells is 72% after a pre-seeding of 15 min and 81% after pre-seeding of 60 min

This study provides evidence that titanium coating in combination with a CA10% and ALDH treatment removed most of the free aldehyde ligands of glutardialdehyde fixed pericardium in-vitro and enabled a living and confluent cell layer of human endothelial cells with a high adhesion capacity. This study gave evidence that titanium coating applied to a porcine heart valve may enable a self-seeding with cells within circulation.

4.3 The first self-endothelialized titanium coated glutaraldehyde-fixed heart halve prosthesis within systemic circulation

Implantation of a pre-seeded aortic valve in the aorta descendens position in a sheep model without extra-corporal circulation was already described before (Gulbins et.al., 2006). We used a male Bore goat with a weight of 83 kilograms (ethical accreditation of the Ethics Committees for Animals in Kiel, V 362-72241.122-6) to create a muscular blood pump in aorta descendens position (Fig.7a). We integrated two glutaraldehyde-fixed, titanium coated and detoxified porcine heart valves into the inflow and outflow part of the pumping chamber of this muscular blood pump, the Biomechanical Heart (Guldner et.al., 2001). This skeletal muscle ventricle pumped with a frequency of 20 beats per minute. After 6 months of pumping the valves were removed from the pumping chamber for histological analysis.
Valve leaflets from the two valves were imbedded in paraffin, cut into sections of 4 μm and treated with a hematoxilin-eosin staining . Other sections were treated using polyclonal antibodies against factor VIII (Dako, Hamburg, Germany). After incubation with primary antibody, an anti-rabbit IgG antibody was applied which was conjugated with alkaline phosphatase (SIGMA,Steinheim, Germany). Photographic documentation was performed by using a magnification of 20X.

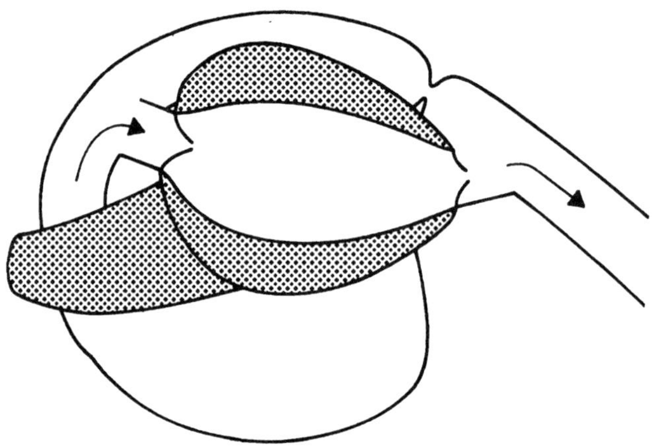

Fig. 7a. Valve's topography within the inflow and outflow part of the pumping chamber of a Biomechanical Heart, a muscular blood pump within aorto -aortic position

Porcine titanized and detoxified aortic valves, having been implanted heterotopic within the systemic circulation of a goat for 6 months, were macroscopically well and showed a normal opening and closure behaviour. Cross-section of a valve leaflet demonstrated a cell-seeding of its blood contacting surface (HE-staining magnification 20X). Immunhistochemic staining against factor VIII proved a complete endothelialization of the total blood contacting surface, labelled with alkaline phosphatase and visualized with immunefluorescense (7b, bottom).

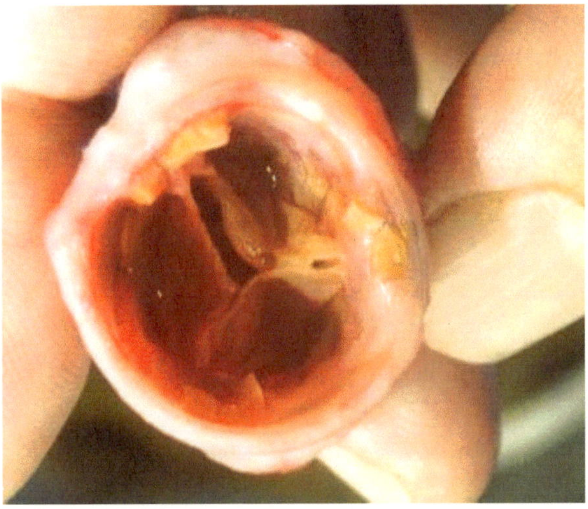

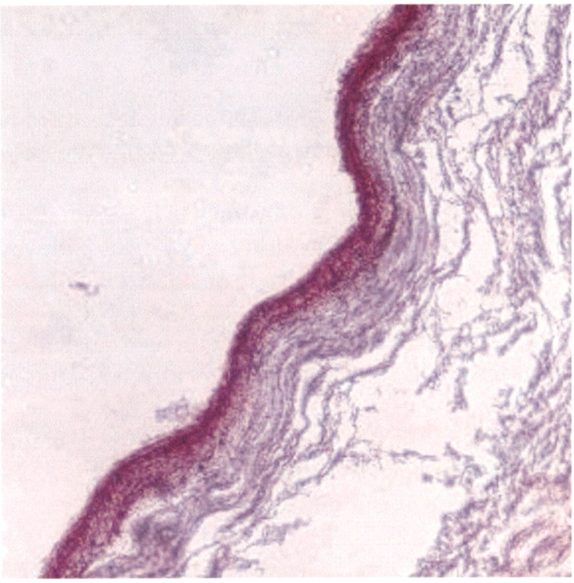

Fig. 7b. Porcine detoxified and titanized aortic valve, having been implanted heterotopic within the aorta of a goat for 6 months (top). Immunhistochemic staining against factor VIII (bottom) proves a complete endothelialization over the total blood contacting surface (bottom)

5. Conclusions

Tissue engineering of heart valves is considered to be a hopeful concept to generate improved substitute bioprostheses. However, because there is yet no realizable blueprint to construct semilunar valves (Sievers, 2007) our aim was to refine conventional, proven bioprostheses using chemical agents as citric acid and aldehyde dehydrogenase (ALDH) for detoxification and innovative nanotechnologies such as plasma deposition with the most biocompatible titanium coating. These procedures have shown to reduce the immune

response to the glutardialdehyde-based cross-linked collagen, probably the main reason why GA- fixed bioprostheses calcify and fail and enable the valve-protecting endothelialization in vitro under static and flow conditions..
Furthermore this new method combining biochemical methods of detoxification with nanotechnology including a titanium coating created the first self-seeded glutaraldehyde-fixed biologic heart valve within circulation under arterial pressure. This technology might open a new field of research for developing new heart valves with improved durability and function. In future big animal studies are necessary to approve this procedure mainly to increase durability and function in biologic glutaraldahyde fixed heart valves over several decades and avoid re-operations caused by valve degeneration

6. Acknowledgment

The authors thank for the support from M. Malaika, B. Girndt from Pfm Titan GmbH Nürnberg titanizing the GA-fixed pericardium and G.Weigl and F. Bastian of the Clinic of Cardiac Surgery University of Vienna (AKH) performing immune histologic investigations.

7. References

Bader A, Schlilling T, Teebken OE, et al. (1998); Tissue engineering of heart valves - human endothelial cell seeding of detergent acellularized porcine valves. *Eur J Cardiothorac Surg* 14:279-84.

Bastian F, Stelzmüller ME, Kratochwill K, Kasimir MT, Simon P, Weigel G(2008). IgG deposition and activation of the classical complement pathway involvement in the activation of human granulocytes by decellularized porcine heart valve tissue. *Biomaterials.*; 29(12):1824-32

Bengtsson L, Radegran K, Haegerstrand A(1993). In vitro endothelialization of commercially available heart valve bioprostheses with cultured adult human cells. *Eur J Cardiothorac Surg*; 7 :393-8.

Chen W, Schoen FJ and Levy RJ(2006). Glutaraldehyde-fixed bioprosthetic heart valve conduits calcify and fail from xenograft rejection *Circulation.;* 14: 318-32.

Dag B. (Ed.) (1997) Surface Characterization.Wiley-VCH, Weinheim..

Eberl T, Siedler S, Schumacher B, Zilla P, Schlaudraff K, Fasol R:(1992). Experimental in vitro endothelialization of cardiac valve leafets. *Ann Thorac Surg*; 53:487-92.

Fischbein MP, Abbas Ardehali, James Yun, Stephen Schoenberger, Hillel Laks, Yoshihito Irie, Paul Dempsey, Genhong Cheng, Michael C. Fishbein and Benjamin Bonavida (2000) CD40 Signaling Replaces CD4+ Lymphocytes and Its Blocking Prevents Chronic Rejection of Heart Transplants *The Journal of Immunology*, 165: 7316-7322.

Fischlein T, Lehner G, Lante W, Reichart B(1992). Endothelialization of aldehyde-fixed acrdiac valve bioprostheses. *Transplant Proc*; 24:2988.

Fischlein T, Lehner G, Lante W, et al. (1994); Endothelialization of cardiac valve bioprostheses: *Int J Artf Organs* 17:345-53.

Fischlein T, Fasol R(1996);. In vitro endothelialization of bioprosthetic heart valves. *J Heart Valve Dis* 5:58-65.

Fu P, Lan H, Wang D, Guan H. (1997);. Experimental study on modified treatment and endothelialization of bovine pericardial valves. *J Tongji Med Univ* 17:136-9.

Gemmell CH, Black JP, Yeo EL, Sefton MV (1996). Material-induced up-regulation of leukocyte CD11b during whole blood contact: material differences and a role for complement. *J Biomed Mater Res.* Sep; 32(1):29-35.

Gorbet B, M. V. Sefton(1974) Complement inhibition reduces material-induced leukocyte activation with PEG modified polystyrene beads (Tentagel™) but not polystyrene beads Volume A, Issue 4, pages 511–522.

Gudbjartsson TA, Cohn LH(2003);. Mechanical/bioprosthetic mitral valve replacement. In: Cardiac surgery in the adult. Cohn L.H., Edmunds L.H .Jr (Eds.), McGraw-Hill, New York USA. 51-986.

Gulbins H, Pritisanac A, Anderson I, et al(2003);. Myoblasts survive for 16 weeks after intracardiac transfer and start differentiation. *Thorac Cardiovasc Surg* 51:295-300

Gulbins H, Goldemund A, Anderson I, et al. (2003); Preseeding wih autologous fibroblasts improves endothelialization of glutaraldehyde-fixed porcine aortic valves. *J Thorac Cardiovasc Surg* 125:592-601.

Gulbins H, Pritisanac A, Piper K(2006). Successful endothelialization of porcine glutaraldehyde.fixed aortic valves in a heterotopic sheep model *Ann Thorac Surg.;* 81:1472-1479.

Guldner NW, Klapproth P, Großherr M., Rumpel E, Noel R, Sievers HH (2001); Biomechanical Hearts: Muscular Blood Pumps, Performed in a One-Step Operation, and Trained under Support of Clenbuterol, *Circulation* 104 717-22

Guldner N W, Jasmund I, Zimmermann H, Heinlein M, Girndt B, Meier V, Flüß F, Rohde D, Gebert A, Sievers H H(2009) Detoxification and Endothelialization of Glutaraldehyde-Fixed Bovine Pericardium With Titanium Coating; *Circulation;* 119:1653-1660

Guldner NW, I.Jasmund, H Zimmermann, M. Heinlein, B.Girndt, M.Großherr, A .Klinger, H.-H. Sievers(2009); The First Self-Endothelialized Titanium Coated Glutaraldehyde-Fixed Heart Valve Prosthesis within Systemic Circulation. *J Thorac Cardiovasc Surg,* 138: 248-250

Grill A.: (1994) Cold plasma in materials fabrication: from fundamentals to applications. *IEEE Press,* New York..

Hoerstrup SP, Kadner A, Melnitchouk S, et al. (2002) Tissue engineering of functional trileafletheart valves from human marrow stromal cells. *Circulation;*106:I143-50.

Jamieson WR, Burr LH, Munro AI, Miyagishima RT(1998): Carpentier-Edvards standard porcine bioprosthesis: a 21-year experience. *Ann.Thorac.Surg.;* 6 Suppl: 66-72.

Jansson K, Bengtssoon L, Swedenborg J, Haegerstrand A(2001). In vitro endothelialization of bioprosthetic heart valves provides a cell monolayer with proliferative capacities and resistance to pulsatile flow. *J Thorac Cardiovasc Surg;*121:108-15.

Juthier Francis, Andrė Vincentelli, Julien Gaudric, Delphine Corseaux, Olivier Fouquet, Christine Calet,Thierry Le Tourneau, Valérie Soenen, Christophe Zawadzki, Olivier Fabre, Sophie Susen, Alain Prat, Brigitte Jude(2006);. Decellularized heart valve as a scaffold for in vivo recellularization: Deleterious effects of granulocyte colony-stimulating factor. *J Thorac Cardiovasc Surg* 131:843-852

Köckerling F, Scheidbach P, Tannapfel A, Schmidt U, Lippert H.(2004). Influence of Titanium Coating on the Biocompatibility of a Heavyweight Polypropylene Mesh. *Eur Surg Res;* 36: 313-317.

Lehner G, Fischlein T, Baretton G, Murphy JG, Reicharr B(1997);. Endothelialized biological heart valve prostheses in the non-human primate model. *Eur J Cardiothorac Surg* 11:498-504.

Nilsson UR, (2001). Deposition of C3b/iC3b leads to the concealment of antigens, immunoglobulins and bound C1q in complement-activating immune complexes. . *Mol Immunol*. Aug;38(2-3):151-60.

Rieder E, Nigisch A, Dekan B, Kasimir MT, Mühlbacher F, Wolner E, Simon P, Weigel G (2006). Granulocyte-based immune response against decellularized or glutaraldehyde cross-linked vascular tissue. *Biomaterials*. Nov; 27(33):5634-42.

Rocchini AP, Weesner KM, Heidelberger K, Keren D, Behrendt D, Rosenthal (1981) A.Porcine xenograft valve failure in children: an immunologic response. *Circulation*. Aug;64(2 Pt 2):II162-71

Rizwan A. Manji, MD, PhD; Lin F. Zhu, MD; Nimrit K. Nijjar; David C. Rayner, MD; Greg S. Korbutt, PhD; Thomas A. Churchill, PhD; Ray V. Rajotte, PhD; Arvind Koshal, MD; David B. Ross, MD(2006); Glutaraldehyde-Fixed Bioprosthetic Heart Valve Conduits Calcify and Fail From Xenograft Rejection Circulation. 114:318-327.

Sayk Friedhelm, MDa, Inge Bos, MDa, Ute Schubert, BSb, Thilo Wedel, MDc, Hans-H. Sievers (2005); Histopathologic Findings in a Novel Decellularized Pulmonary Homograft: An Autopsy Study. *Ann Thorac Surg* 79:1755-1758.

Simionescu DT (2004); Prevention of calcification in bioprosthetic heart valves: challenges and perspectives. *Expert Opin Biol Ther*. 4:1971

Simon P, Kasimir MT, Seebacher G, Weigel G, Ullrich R, Salzer-Muhar U, Rieder E, Wolner (2003) E Early failure of the tissue engineered porcine heart valve SYNERGRAFT in pediatric patients. *Eur J Cardiothorac Surg*. Jun; 23(6):1002-6.

Sivaram S(1995). Chemical vapor deposition: thermal and plasma deposition of electronic materials. New York, NY: Van Nostrand Reinhold;.

Sievers H.H. (2007) In vivo tissue engineering an autologous semilunar biovalve:Can we get what we want? *Thorac Cardiovasc Surg*.; 134: 20-22.

Sodian R, Hoerstrup SP, Sperling JS, et al. (2000)Evaluation of biodegradable, three-dimensional matrices for tissue engineering of heart valves. *ASAIO J*;46:107-10.

Terramani TT, Eton D, Bui PA, Wang Y, Weaver FA, Yu H (2000). Human macrovascular endothelial cells: optimization of culture conditions. *In Vitro Cell Dev Biol Anim*; 36:125-31.

Ueyama K, Kamiya H, Kanamori T, Ohashi H, Yasushi T, Kawai T, Ohnaka M(2002). Long-term follow-up of cardiac valve replacement using bioprosthesis in patients 70 years old and older. *Artif Organs*. Dec;26(12):1059-62.

Vesely I (2003).The evolution of bioprosthetic heart valve design and its impact on durability. Cardiovasc . Pathol.; 12(5): 277-286.

Williams DB, GK Danielson, DC McGoon, FJ Puga, DD Mair and WD Edwards (1982); Porcine heterograft valve replacement in children. *Journal Thorac Cardiovasc Surg*. 84: 446-450.

22

Formation of Stable Vascular Networks in Engineered Tissues

Bin Jiang and Eric M. Brey
Illinois Institute of Technology
USA

1. Introduction

Proper vascular network structure is essential for normal tissue function. The vasculature provides oxygen, nutrients and immune cells, as well as removes tissue waste and by-products. (Jain 2003) In tissue engineering and regenerative medicine, the existence of functional blood vessels is critical to the survival of the newly generated tissue. Current strategies for promoting new blood vessel formation, or neovascularization, have mostly focused on angiogenesis, which is the formation of new capillaries from pre-existing vessels by sprouting of endothelial cells (ECs). (Carmeliet 2003) Another mechanism of neovascularization, vasculogenesis, which is the *in situ* assembly of endothelial progenitors into capillaries, has also been explored as a method for stimulating new vessel formation. Genes or proteins of angiogenic factors have been delivered systematically or directly to a target tissue to promote neovascularization via either angiogenesis or vasculogenesis. For example, proteins from the vascular endothelial growth factor (VEGF) and fibroblast growth factor (FGF) families delivered by various methods have been investigated extensively for enhancing neovascularization. (Brey *et al.*, 2005; Brey & McIntire, 2008)

While the delivery of angiogenic factors has been successful in promoting new capillary formation, the structure of the newly formed vessels is often non-ideal. The vessels can be immature, with small diameters and lack of some essential cellular and extracellular components required for proper function. Immature vascular structure may result in poor blood perfusion and possibly vessel regression upon a decrease in the vascular stimulus. (Benjamin *et al.*, 1999) In addition, the formation of small capillaries alone is not likely to be sufficient for vascularization of large, complex tissues, which are often needed for tissue engineering. Therefore, the goal of neovascularization in engineered tissue should be focused not only on the initial assembly of capillaries via angiogenesis or vasculogenesis, but also the expansion and stabilization of new vessels, which means the formation of mature, long-lasting vessels with dimensions that meet the requirement of high conductance blood flow. (Carmeliet & Conway, 2001) In this chapter, we will discuss current status in the generation of stable, long-lasting vascular networks in tissue engineering and regenerative medicine, identifying recent advances and limitations yet to be overcome.

2. Angiogenesis stabilization: basic mechanisms

The process of angiogenesis includes four stages: (a) initiation, in which a certain angiogenic stimulus increases vessel permeability and protein leakage; (b) progression, characterized by the production of proteolytic enzymes that degrade the basement membrane (BM) lining endothelial cells (ECs) and surrounding extracellular matrix (ECM) to allow EC invasion and migration; (c) differentiation, which involves the assembly of ECs into a network with a luminal structure; and (d) stabilization and maturation, where mural cells (pericytes and vascular smooth muscle cells) are recruited and ECs deposit new BM. (Bussolino *et al.*, 1997) It is important to note that the MCs may be recruited after initial network function or simultaneous with vessel assembly. (Brey *et al.*, 2004) Numerous cellular and molecular factors are involved in the regulation of this process. Failure in this regulation could result in excessive vessel generation or premature regression, which is seen in many pathological conditions. In the following sections we will describe the basic components of the vessel and their role in vessel stability.

2.1 Cells

The walls of stable blood vessels are composed of two distinct cell types: ECs and mural cells (MCs). (Figure 1) While ECs form the inner lining of vascular tubes, MCs associate with and coat the outside of the endothelial tube. (Gaengel *et al.*, 2008) In the case of capillaries the MCs, or pericytes, are intermittently present on the surface. For larger vessels the MCs, or vascular smooth muscle cells (VSMCs), coat the entire vessel surface often in multiple layers. Interactions between the ECs and MCs play a critical role in the regulation of vascular formation, stabilization, remodeling, and function. Abnormal interactions between the two cell types are implicated in a number of pathological conditions, including tumor angiogenesis, diabetic microangiopathy, ectopic tissue calcification, and stroke and dementia syndrome. (Armulik *et al.*, 2005; Carmeliet 2003)

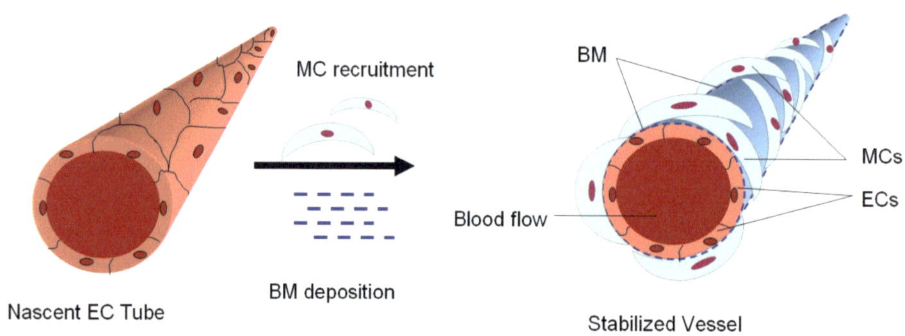

Fig. 1. Process of blood vessel stabilization. Endothelial cells (EC) in immature vessels secrete signals that stimulate the migration and proliferation of mural cells (MC). ECs also secrete a thin layer of matrix, known as the basement membrane (BM) that lines the vessels

Junctional structures, including adherent junctions and tight junctions, are present between ECs and provide the endothelium with integrity, mechanical strength and tightness, and a barrier to molecular transport. (Carmeliet 2003) During vascular sprouting, junctions are partially disorganized to allow cell migration and proliferation, and to increase vessel permeability. During vessel stabilization, junctional integrity is re-established and permeability is tightly controlled. (Dejana, 2004) As vessels sprout and migrate, they will form branches. These sprouts then inosculate with other vessels to form complex networks. Typically, a high vessel density is formed in response to the initial vascular stimuli. This vasculature must be remodeled for tissue-specific transport and function. Apoptosis of ECs are involved at this later stage to remove excessive vessels. (Darland & D'Amore, 2001) However, excessive apoptosis may result in premature vessel regression. Growth factors, which are essential for angiogenesis, not only stimulate EC proliferation and migration but also inhibit EC apoptosis. (Dimmeler & Zeiher, 2000) Thus, the prevention of EC apoptosis may improve the survival and durability of blood vessels formed in engineered tissues.

MCs are commonly subdivided into VSMCs, which are associated with arteries and veins forming multiple concentric layers; and pericytes, which are associated with small diameter capillaries. (Gaengel *et al.*, 2009) Intermediate-size vessels, arterioles and venules, have MCs with properties between those of typical VSMCs and pericytes. Since VSMCs and pericytes share some common cellular markers, such as α-smooth muscle actin (α-SMA) and desmin, they are believed to represent phenotypic variants of the same lineage, and may function as progenitors for each other depending on external stimuli. (Gerhardt & Betsholtz, 2003)

The presence of MCs on the surface of ECs plays a critical role in vessel stability. Intercellular adhesion between MCs and ECs is mediated by integrin $\alpha_4\beta_1$, which is expressed by proliferating ECs, and its ligand VCAM-1, which is expressed by proliferating MCs. (Garmy-Susini *et al.*, 2005) Studies have shown that the association of MCs to the newly formed endothelial plexus stabilizes the vasculature, while the disruption of EC-MC associations results in excessive regression of vasculature. (Benjamin *et al.*, 1998) Co-culture of EC with SMC in a 3D *in vitro* model results in a mature, quiescent EC phenotype with increased number junctional structures and increased resistance to apoptosis. (Korff *et al.*, 2001) However, the presence of MCs in the microvasculature does not always guarantee vessel stability. Vessels with extensive MC can still regress following loss of neovascular stimuli. (Brey *et al.*, 2004) Therefore, some additional signals, such as ECM and soluble growth factors, are also required for vessel stabilization.

2.2 Extracellular components

Extracellular matrices (ECM), including BM (primarily collagen IV and various laminin isoforms) and interstitial matrices (typically collagen I, elastin and fibronectin but depends on the particularly tissue), play important roles in angiogenesis. (Francis *et al.*, 2008) Firstly, the ECM scaffold provides mechanical support for ECs to form channels, preventing vessels from collapsing. In addition, EC adhesion to ECM regulates EC proliferation, migration, and survival, mediated by different signaling pathways. (Davis & Senger, 2005) And finally, the deposition of ECM, particularly BM, which occurs after MC recruitment stabilizes the vessel by decreasing permeability and preventing vessel regression. (Stratman *et al.*, 2009)

The ECM is dynamically regulated during angiogenesis. In the initiation of angiogenesis, ECM is degraded by proteinases, such as matrix metalloproteinases (MMPs). The degradation of ECM not only allows EC migration and proliferation, but also releases soluble growth factors trapped in the matrix. (Ghajar *et al.*, 2008) After lumen formation, MCs are recruited to EC tubes and induce the deposition of new BM as well as the secretion of tissue inhibitor of metalloproteinases (TIMPs), which inhibit MMPs proteolysis. (Stratman *et al.*, 2009) Dysregulation of ECM, such as glycation of collagen due to diabetes and aging, can slow ECM vessel assembly. (Francis-Sedlak *et al.*, 2010)

2.3 Signaling pathways

The formation of stable vasculature is tightly regulated by the spatial and temporal kinetics of numerous signaling pathways. A stable, operational vascular network results from a balance between signals that favor angiogenesis and vascular stabilization, and those that promote vascular regression. (Bussolino *et al.*, 1997) Signaling pathways affect vascular stabilization presumably by regulating proliferation and apoptosis of ECs and MCs, as well as the deposition and degradation of ECM. (Jain 2003) In this section, we will discuss some of the signaling pathways and their roles in vascular stabilization.

2.3.1 Platelet derived growth factor (PDGF) family

The PDGF family is composed of disulfide-bonded homodimers of four subunits, PDGF-A, PDGF-B, PDGF-C and PDGF-D, as well as the heterodimer PDGF-AB. PDGF isoforms exert their biological effects through the activation of two tyrosine kinase receptors, PDGFR-α and PDGF-β, which are present on MCs and messenchymal cells. (Hellberg *et al.*, 2010)

Specifically, PDGF-BB, secreted by proliferating ECs during angiogenic sprouting, stimulates the migration and proliferation of MCs, which express PDGFR-β. In addition, PDGF-BB induces the differentiation of messenchymal cells toward a MC lineage.(Hirschi *et al.*, 1999) Genetic knockout of *pdgf-b* or *pdgfr-β* in mice results in MC deficiency, vascular leakage and embryonic lethality (Hellstrom *et al.*, 1999; Lindahl *et al.*, 1997) The more recently described PDGF-C and PDGF-D isoforms are also involved in vessel maturation and stabilization. PDGF-C drives the recruitment and differentiation of MCs resulting in improved BM integrity and a more mature, stable vascular wall with a lower permeability. (di Tomaso *et al.*, 2009) PDGF-D induces macrophage recruitment and vessel stabilization by improving the SMC coating of angiogenic blood vessels and decreases their permeability. (Uutela *et al.*, 2004)

2.3.2 Angiopoietins

The angiopoietin (Ang) family consists of Ang-1, Ang-2, Ang-3 and Ang-4. The angiopoietins are ligands for the Ties, a family of receptors that are expressed within the vascular endothelium. (Yancopoulos *et al.*, 2000) Ang-1 is predominantly expressed by MCs, suggesting a paracrine mode of action; whereas Ang-2 is mostly expressed by ECs, suggesting an autocrine function, although expression in MCs has also been reported. (Gaengel *et al.*, 2009) Ang-3 and Ang-4 are less well understood at this time.

Ang-1 stabilizes nascent vessels and make them leak-resistant, presumably by facilitating interactions between ECs and MCs. Genetic deletion of Ang-1 or its receptor Tie2 in mice resulted in defects in vessel remodeling with poor association of ECs with BM, and failure to recruit MCs. (Sato *et al.*, 1995; Suri *et al.*, 1996) Transgenic overexpression of Ang-1 in mice lead to a significant increase in vessel diameter and increased resistance to vessel leakage. The resistance is presumably due to the ability of Ang-1 to maximize interactions between ECs, MCs and BM. (Thurston *et al.*, 1999, 2000) Ang-2, on the other hand, acts as an antagonist of Ang-1, destabilizes vessels and contributes to vessel regression. Transgenic overexpression of Ang-2 disrupts blood vessel formation in the mouse embryo, resulting in similar effects to Ang-1 or Tie 2 knockouts. (Maisonpierre *et al.*, 1997) Thus, angiopoietins positively or negatively regulate vessel stabilization, however, the detailed mechanism of the signaling pathways are not yet well-described.

2.3.3 Sphingosine-1-Phosphate (S1P)

S1P is a bioactive lipid mediator that is primarily secreted by platelets. It activates a family of G-protein coupled receptors ($S1P_{1\ to\ 5}$), formerly known as endothelial differentiation gene (Edg)-receptors. (Gaengel *et al.*, 2009) Mice deficient in S1P have severe vascular defects resulting in embryonic lethality, with incomplete coverage of MCs. Mice deficient in $S1P_1$, its receptor, also showed severe defect in vascular maturation, suggesting the role of $S1P/S1P_1$ signaling pathway in promoting vascular stabilization. (Mizugishi *et al.*, 2005) Moreover, knockout of the EC-specific $S1P_1$ in mice embryos resulted in a similar phenotype as obtained for the $S1P_1$ full knockout, suggesting that $S1P_1$ functions mainly through ECs instead of other cell types. (Allende *et al.*, 2003) It was found that $S1P/S1P_1$ signaling pathway promotes vascular stabilization by activation of N-Cadherin, which is a cell adhesion molecule connecting ECs and MCs. (Paik *et al.*, 2004)

2.3.4 Transforming growth factor-β (TGF-β)

TGF- β, a growth factor family with three different isoforms TGF-β1, TGF-β2 and TGF-β3, is expressed by a number of cell types, including ECs and MCs. Depending on its receptor TGF-β mediates distinct signaling pathways. TGF-β can be both pro- and anti-angiogenic. (Pepper 1997) The major receptors for TGF-β in regulating vascular stabilization include activin receptor-like kinase (Alk)-1, Alk-5, and endoglin. (Gaengel *et al.*, 2009)

Activation of Alk-1 triggers EC migration, proliferation, and inhibits vessel maturation and MC differentiation; whereas activation of Alk-5 instead mediates inhibition of EC migration, reduced proliferation, increased vessel maturation and MC differentiation, suggesting that TGF-β regulates vessel maturation and stabilization via a balance between Alk-1 and Alk-5. (Goumans *et al.*, 2002) The determination of which signaling pathway TGF-β activates appears to depend on its concentration. At low doses, TGF-β promotes EC proliferation by activating TGF-β /Alk-1 pathway; while at high doses, Alk-5 is activated instead. (van den Driesche *et al.*, 2003) Endoglin, a TGF-β co-receptor expressed specifically in ECs, modulates Alk-1 and Alk-5 mediated pathways, possibly by favoring the TGF-β/Alk-1 pathway. (Letamendia *et al.*, 1998)

2.3.5 Vascular endothelial growth factor (VEGF) family

VEGF is a family of angiogenic growth factors expressed by a variety of cells, including keratinocytes, macrophages, mast cells, vSMCS, and ECs. Members of the VEGF family include VEGF-A, VEGF-B, VEGF-C, VEGF-D and placental growth factor (PlGF). VEGF is a major mitogen for ECs, mediated via VEGF tyrosine kinase receptors (VEGFRs) expressed on ECs. (Bussolino *et al.*, 1997; Yancopoulos *et al.*, 2000) It has been shown that VEGF plays a central role in neovascularization by stimulating ECs proliferation, migration and increasing vascular permeability in a number of physiological and pathological processes. (Nor *et al.*, 1999)

However, the role of VEGF in regulating vascular stabilization can be both positive and negative, depending on its spatial and temporal kinetics. On the one hand, VEGF enhances

EC survival and protects EC from apoptosis by up-regulating the expression of Bcl-2 (B-cell lymphoma 2), an apoptosis regulator protein. (Nor *et al.*, 1999) Withdrawal of VEGF prior to stabilization and remodeling may result in excessive vessel regression by EC apoptosis. (Benjamin *et al.*, 1999; Jain, *et al.*, 1998) On the other hand, VEGF is also considered as a negative regulator for vessel maturation by disrupting VSMC function and pericyte coverage of nascent vascular sprouts, leading to vessel destabilization. (Greenberg *et al.*,2008). The overexpression of VEGF results in leaky, immature and unstable vessels, which contribute to a number of pathological conditions, such as tumors, diabetic retinopathy, and age-related macular degeneration (AMD). Therefore, the dosage of VEGF must be tightly regulated in a spatial, temporal and quantitative manner in therapeutic applications to avoid abnormal or aberrant vascular structure. (Yancopoulos *et al.*, 2000)

2.3.6 Ephrins and notch

Ephrins and Notch are two signals that are generally involved in the differentiation of many cell types. In angiogenesis, the two signaling pathways are involved in the determination of arterio-venous vessel fate. (Jain 2003) Thus, the ephrins and Notch signaling pathways are considered novel targets for therapeutic angiogenesis. (Sullivan & Bicknell, 2003; Cristofaro & Emanueli, 2009) The ephrin ligands and Eph receptors are both transmembrane molecules expressed on ECs and other cell types.(Yancopoulos *et al.*, 2000; Cristofaro & Emanueli, 2009) The ligand ephrin-B2 is expressed specifically in the arterial endothelium, while the receptor EphB4 expresses in the venous endothelium. The distribution suggest a role in defining boundaries between arterial and venous domains. (Adam *et al.*, 1999; Wang *et al.*, 1999). Notch is a family of transmembrane protein receptors that bind to ligands Delta/Serrate. The Notch pathway, in angiogenesis, is also required for arterial-venous differentiation. (Sullivan & Bicknell, 2003; Gaengel *et al.*, 2009) Activation of Notch signaling leads to repression of venous cell fate in EC, while ablation of Notch function resulted in defects in blood vessel formation similar to those associated with improper arterial-venous specification. (Lawson *et al.*, 2001; Liu *et al.*, 2003)

2.3.7 Others

There are a number of other signaling pathways involved in the formation of stable vasculature. For example, brain derived neurotrophic factor (BDNF) is a neurotrophin for neuron survival and differentiation. It is also an EC survival factor involved in intramyocardial vessel stabilization, by activation of trk B receptors expressed on cardiac microvascular ECs.(Donovan *et al.*, 2000) Monocyte chemotatic protein-1 (MCP-1) is a small cytokine best known for its ability to recruit monocytes after injury. However, recent studies have shown that MCP-1 improves vessel stabilization during angiogenesis by promoting MC recruitment and inducing VSMC proliferation. (Aplin *et al.*, 2010; Selzman *et al.*, 2002) Nitric oxide (NO), a multifunctional gaseous molecule that regulates various physiological functions is also involved in vascular stabilization. An *in vivo* tissue-engineered blood vessel model revealed that NO mediates EC-MC interaction and induces MC recruitment, vessel branching, and longitudinal extension and subsequent stabilization of the vessels. (Kashiwagi *et al.*, 2005) A summary of the of numerous signalling pathways is provided in Table 1, with ligand/receptor pairs, cells that express these molecules, and their roles in vessel maturation and stabilization.

Ligand (Cells)	Receptor (Cells)	Roles in vessel stabilization
Ang-1 (MCs)	Tie2 (ECs, MSCs)	Stabilizes vessels by mediating EC-MC and EC-ECM adhesion; Induces production of factors such as HB-EGF, HGF and MCP-1.
Ang-2 (ECs)	Tie2 (ECs; MSCs)	Destabilizes vessels by binding to Tie2 to attenuate the effect of Ang-1.
Bcl-2 (ECs)	ERKs (ECs)	Prevents EC apoptosis and enhances vessel durability.
BDNF (neurons; ECs)	Trk-B (neurons; ECs; MCs)	Promotes intramyocardial arteries and capillaries stabilization by enhancing EC survival during embryogenesis.
Delta/Serrate (ECs)	Notch (ECs)	Determines arterial-venous fate of ECs by repressing venous fate in arterial ECs.
Ephrin-B2 (Arterial ECs)	Eph-B4 (Venous ECs)	Determines arterial and venous EC specialization.
HB-EGF (ECs)	ErbB1, ErbB2 (MCs)	Promotes MCs proliferation and protects from apoptosis, involved in Ang(s) regulation.
HGF (ECs, MSCs)	C-met (ECs, SMCs)	Induces SMC migration, mediating angiopoietins regulated vessel stabilization.
MCP-1 (ECs, MCs)	CCR-2 (MCs)	Mediates spontaneous and Ang-1 induced MC recruitment.
PDGF-BB (ECs)	PDGFR-β (MCs; MSCs)	Promotes MC migration and proliferation; Induces MSCs differentiation to MCs.
S1P (Platelets)	S1P1 (MCs)	Activates adhesion molecule connecting ECs and MCs; directs MC recruitment and association to vessels.
TGF-β (ECs; MCs)	Alk1 (ECs; MSCs)	Promotes EC migration and proliferation; Inhibits vessel maturation and MC differentiation.
	Alk5 (ECs; MSCs)	Induces MC differentiation and ECM production; Inhibits EC migration and proliferation.
	Endoglin (ECs)	Modulates Alk-1 and Alk-5 mediated pathway by favoring Alk-1.
VEGF (Macrophages; MCs; fibroblasts)	VEGFRs (ECs)	Enhances EC survival by inducing Bcl-2 expression; Inhibits MC association and increases vessel permeability.

Table 1. Signaling Pathways Regulating Vessel Stabilization

Many of the signaling pathways mentioned above work in a coordinated fashion to regulate vessel stabilization. For example, MCP-1 was found to be one of the many cytokines released during angiopoietin induced MC recruitment.(Aplin *et al.*, 2010) Other cytokines involved in angiopoietin function include EC-derived heparin binding-epidermal growth factor (HB-EGF) (Iivanainen *et al.*, 2009; Stratman *et al.*, 2010) and hepatocyte growth factor (HGF) (Kobayashi *et al.*, 2006). PDGF-BB was found to promote MC proliferation partially by inducing the release of fibroblast growth factor-2 (FGF-2), which in turn transactivates FGFR-1. (Millette *et al.*, 2006) PDGF-B stimulation also induces an acute production of TGF-β, which in turn, negatively regulates Ang-1 expression induced by the PDGF-B stimulation. (Nishishita & Lin, 2004) Thus, a complex interaction of multiple signaling pathways is required to maintain vascular homeostasis and vascular stabilization.

3. Tissue engineering strategies

Extensive attention has been given to tissue engineering for its potential to restore, maintain or improve tissue functions. (Langer & Vacanti, 1993) However, the clinical application of engineered tissues requires the ability to form extensive, stable microvascular networks within engineered tissues. (Brey & McIntire, 2008) While strategies for neovascularization, including angiogenesis, vasculogenesis and arteriogenesis, have been extensively reviewed elsewhere (Brey *et al.*, 2005; Ennett & Mooney, 2002; Sneider *et al.*, 2009), increased attention is being given to the later stage of neovascularization, which is the maturation and stabilization of newly formed vessels. Different tissue engineering strategies, including protein delivery, gene therapy and cell therapy (Table 2) have been developed and tested in both *in vitro* and *in vivo* models.

	Advantages	**Disadvantages**
Protein Delivery	Better dose control than gene therapies; Temporal control over release can be achieved with certain delivery systems.	Expensive to produce and isolate proteins; Short protein half-lives *in vivo*;
Gene Therapy	Long-term active growth factor expression; Delivery restricted to area of modified cells	Safety issues based on delivery methods; Difficult to control the level and duration of gene expression; Limited transfection efficiency.
Cell Therapy	Introducing cell population that is sensitive to stimuli unlike many cells in the disease/damaged tissue Cells can stimulate neovascularization by incorporation into new vessels or release of soluble factors	Costly and time consuming for cell isolation and expansion; Limited cell sources for autologous transplantation; Immunosuppresive medications needed for allogeneic or exnogeneic transplantation.

Table 2. Advantages and Disadvantages of General Strategies for Promoting Stable Neovascularization in Engineered Tissues

3.1 Protein delivery

3.1.1 Single factor delivery

The formation of a stable vascular network during normal healing involves a complex spatial and temporal regulation of a large number of angiogenic stimulators and inhibitors. (Jain, 2003) While it may seem logical that all of these factors would be required in order to generate vessels in engineered tissues, researchers have attempted to initiate the entire process through the delivery of a single factor. Members of the VEGF and FGF families were some of the first identified angiogenic growth factors and have been used in clinical trials for the treatment of limb ischemia. (Cao *et al.*, 2005) However, the short term delivery of basic FGF (FGF-2) or VEGF may result in immature vessels, stable for only a short time prior to regression. (Dellian *et al.*, 1996) Based on these issues others have chosen to use factors with a known role in angiogenesis stabilization. For example, PDGF-BB, Ang-1, HGF, and ephrin-B2 are some of the factors currently under investigation in animal and *in vitro* models. (Brey & McIntire, 2008)

It is possible that the short lifetime and immaturity exhibited by vessels stimulated in these cases is not the result of using only a single factor, but instead due to the transient nature of their presence. The short lifetime of proteins and their rapid transport out of tissues can results in only a short-term stimulus. A single protein delivered in a controlled manner may be able to provide enough signal to prolong the vascularization response and, possibly, increase vessel maturation. A number of delivery systems, including natural and synthetic biomaterials, have been developed for growth factor delivery. Commonly used materials include naturally occurring materials, such as alginate, gelatin and fibrin, as well as synthetic materials, such as poly (lactic-co-glycolic acid) (PLGA), and modified polyethylene glycol (PEG). Sustained growth factor delivery can be achieved and controlled using those materials in the form of hydrogels, micro/nano particles, fibers, porous matrix, or their combinations. (Chen & Mooney, 2003; Zisch *et al.*, 2003a) The materials can be used to modulate delivery via diffusive resistance or through direct interactions between the materials and the growth factors. For example, the release of FGF-1 encapsulated in alginate microbeads can be modulated based on alginate composition. The release of FGF-1 from this relatively simple biomaterial can result in significantly prolonged increase in vascular density compared to a bolus injection of FGF-1. The enhancement in neovascularization can persist for greater than 6 weeks. In addition, release of FGF-1 from alginate can lead to a greater number of capillaries interacting with MCs but the fraction of MCs was not changed. (Moya *et al.*, 2009, 2010)

In addition to diffusion-controlled release, molecules can also be covalently immobilized to a polymeric surface or matrix for growth factor delivery and presentation. For example, VEGF has been covalently bound to fibrin gels resulting in cell demanded release. The immobilized VEGF cannot diffuse out and is only released as the fibrin is degraded. Compared to VEGF that diffuses from the fibrin gels, which formed leaky and chaotic vessels, the VEGF immobilized fibrin gels were able to stimulate the formation of nonleaky, organized vasculature with MC coverage and BM deposition. (Ehrbar *et al.*, 2004) Synthetic polymer matrixes have also been used for immobilized growth factor delivery. PEG hydrogels modified with MMP sensitive degradation sequences and cell adhesion sequences have been used in combination with a number of growth factors, including VEGF (Leslie-Barbick *et al.*, 2011; Moon *et al.*, 2010; Zisch *et al.*, 2003b), Ephrin-A1 (Moon *et al.*, 2007), and PDGF-BB (Saik *et al.*, 2011). Capillary-like structures were formed only in the contact area with the immobilized VEGF *in vitro*, while after *in vivo* implantation, functional vessels were formed with MC coverage. (Moon *et al.*, 2010) Compared to single dose delivery, the sustained delivery of growth factors allows a more precise spatial control over the formation of vascular networks and may be able to lead to a more persistent vascular response and mature vessel formation even in the absence of additional maturation factors.

3.1.2 Multiple factors delivery

Since the formation of a stable vascular network appears to require the collaboration of a number of growth factors, the delivery of a combination (cocktail) of angiogenic molecules has been investigated as a method for improved vascular maturation. For example, co-delivery of both FGF-2 and VEGF from an accellular collagen-heparin scaffold resulted in higher vessel density and more MC covered mature vessels, compared to single factor delivery. (Nillesen *et al.*, 2007) In addition, some of the molecules involved in the angiogenesis stabilization signaling pathways (Table 1) have been delivered to the target tissue together with stimuli involved in the initiation step. PDGF-BB, a signaling protein stimulating MC proliferation and recruitment, has been used in combined delivery with VEGF (Chen *et al.*, 2007; Richardson *et al.*, 2001) and FGF-2 (Cao *et al.*, 2003; Lu *et al.*, 2007). Tested in many animal models, the dual delivery of PDGF-BB, a vessel maturation factor, with VEGF or FGF-2 resulted in not only an increase in vessel quantity, but also improved vessel quality. In a mouse corneal micropocket model, the dual delivery of PDGF-BB with FGF-2 led to stable vessels lasting for more than a year, while vessels formed by single delivery of PDGF-BB or FGF-2 regressed within 70 days (Cao *et al.*, 2003). Other vessel maturation factors, such as Ang-1 (Peirce *et al.*, 2003) and S1P (Tengood *et al.*, 2010), have also been investigated in combined delivery systems with VEGF resulting in stable, long-lasting vessels compared to VEGF delivery alone.

Moreover, it appears that in dual factor delivery, the sequence of the molecules administrated matters. Angiogenic factors, such as VEGF and FGF, are involved in the initial stages of angiogenesis primarily acting on ECs, while vessel maturation factors, such as PDGF-BB, Ang-1 and S1P, are involved in the later stage of vessel stabilization, mostly targeting MCs. Therefore, a temporal separation for distinct kinetics is likely preferred with initial administration of angiogenic factor(s) followed by a vessel maturation factor(s). The sequential delivery of VEGF followed by S1P (Tengood *et al.*, 2010) and FGF-2 followed by PDGF (Tengood *et al.*, 2011) resulted in increased vessel density and stability compared to simultaneous dual delivery. A reverse in the sequential delivery results in decreased vessel density and stability, presumably due the factors working against each other. It was proposed that early delivery of vessel maturation factors restricts EC tube formation and expansion, resulting in decreased vessel density and vessel diameter. The late delivery of angiogenic factors was thought to then inhibit MC recruitment, resulting in excessive formation of immature vessels susceptible to regression. (Tengood *et al.*, 2010, 2011)

With certain modifications, sequential delivery of multiple growth factors can be achieved with delivery materials. For example, alginate can be made into multilayer microbeads with an internal region for slow protein release and external layer for faster protein release. (Khanna *et al.*, 2010) PLGA, a biodegradable copolymer, can be made into microspheres encapsulated porous matrix, in which the microspheres are loaded with factor required for later stage of angiogenesis (PDGF-BB) with slow release kinetics, while the porous matrix is loaded with factor required for initial stage of angiogenesis (VEGF) with fast release rate.

(Carmeliet & Conway 2001; Richardson *et al.*, 2001) Although the temporal and spatial kinetics of the growth factor(s) delivered to the target tissues can be controlled using certain delivery systems, there are still limitations existing with this strategy. The production and purification of massive amounts of active protein factors can be very expensive. And since most growth factors have relatively short half-lives, their bioactivity can be compromised due to storage, delivery routes, and metabolism *in vivo*. Moreover, the optimal combination and concentration of growth factors for different engineered tissues and the any side effect after long-term administration still need to be identified.

3.2 Gene delivery

3.2.1 Genes

While the use of protein delivery offers improved control of the dose for the growth factors, there are also some disadvantages for those strategies, including expensive mass production of the proteins, and short protein half lives *in vivo*. Those limitations can potentially be overcome by means of gene therapy. (Rissanen *et al.*, 2001) Gene therapy allows prolonged overexpression of proteins, theoretically leading to sustained therapeutic effects after a single application, with minor change in systemic growth factor concentrations. (Gosh *et al.*, 2008) Therefore, DNA sequences encoding for growth factors involved in angiogenesis can be used to improve neovascularization. In this section, we will only review genes used directly to the testing model or patients, either by systemic or local injection, or release from polymeric system. Genetic modification to cells for neovascularization will be introduced in the following section, cell therapies.

Transient *vegf* gene expression *in vivo* results in potent angiogenic sprouting, however, the resulting vasculature appears to be immature and leaky, similar to the results of VEGF protein delivery. (Rissanen *et al.*, 2003) When *vegf* gene is co-administrated with genes encoding angiogenic maturation factors, such as *pdgf-b* (Hao *et al.*, 2004a; Korpisalo *et al.*, 2008; Kupatt *et al.*, 2010) and *ang-1* (Chen *et al.*, 2007; Siddiqui *et al.*, 2003; Su *et al.*, 2009), the vasculature is longer-lasting and less permeable,with increased perfusion, which is also similar to the results of dual protein growth factor delivery. Other genes delivered for stable vascular networks formation includes *ang-1* (Shyu *et al.*, 1998), *pdgf-b* (Shea *et al.*, 1999), *fgf-2* with *pdgf-b* (de Paula *et al.*, 2009; Hao *et al.*, 2004b), *inf-β* (Dickson *et al.*, 2007), and *neutrophin-3* (Cristofaro *et al.*, 2010).

3.2.2 Vectors

Vectors are needed in order for the genetic material to be transferred into cells and expressed. Commonly used vectors for gene delivery are plasmids and viral vectors. (Rissanen *et al.*, 2001) Plasmids are DNA sequences that can automatically repeat and express in a host cells. They are easy to manufacture in a large quantity, and have low toxicity with a low immune response. However, naked plasmid DNA usually has very low transfection efficiency except in muscle tissue, therefore biomaterials have been used to increase plasmid-based gene transfer efficiency. (Yia-Herttuala & Alitalo, 2003) For example, PLGA, a degradable copolymer that has been widely used as a tissue engineering scaffold and drug delivery vehicle, has been applied as a carrier for plasmid. PLGA can be made into nanoparticles with the modification of cationic polymer chitosan to increase plasmid transfer efficiency. (Tahara *et al.*, 2008) It can also be prepared as a matrix for sustained plasmid delivery, which in turn, results in longer term growth factor expression. (Nie *et al.*,

2008) Viral vectors, on the other hand, have much higher transfection efficiency compared to naked plasmids, but have significant safety concerns. Adenoviruses (Levanon *et al.* 2006) and adeno-associated viruses (AAV) (Arsic *et al.*, 2003) are the most commonly used viral vectors for growth factor gene delivery. While adenoviral vectors may cause inflammatory reactions and lack sustained expression, AAV only causes a very mild immune response and has shown long-term transgene expression, which makes it more attractive in gene therapy. (Dickson *et al.*, 2007)

Despite the success of gene therapies for the formation of stable vasculature in some animal models, clinical efficacy has yet to be observed. The pharmacokinetics and pharmacodynamics of gene vectors and products remain largely unknown and difficult to control. In addition, data from long-term follow up of clinical studies is lacking. (Yia-Herttuala & Alitalo, 2003) Moreover, the method of gene delivery must be carefully selected to balance the need for maximum transfer efficiency with minimum risk of vector related safety issues. (Brey & McIntire, 2008)

3.3 Cell therapies

Although the delivery of growth factors or their genes has shown great potential to activate the host angiogenic response, certain pathological conditions may also require the transplantation of appropriate blood vessel forming cells, due to loss of, or damage to, the host vascular cells. In addition, these cells may rapidly assemble into vessels which can inosculate with the host, decreasing the time required for tissue perfusion relative to protein/gene approaches. In cell therapies, the cells would have to be isolated from donor or host sources, expanded and implanted into the target tissue, where the cells proliferate and incorporate into functional vessels. (Ennett & Mooney, 2002) ECs are of great interest in this strategy for their potentials to mimic vasculogenesis and assemble into capillary structures, as well as to release multiple angiogenic factors following implantation. (Brey *et al.*, 2005, 2008) Transplantation of mature ECs have been shown to increase neovascularization in engineered tissues. (Nor *et al.*, 2001) However, the limited availability and proliferation capability of mature ECs limtis this approach. (Kim & von Recum, 2008) Therefore, a number of other cell types and modifications may be needed in addition to the used of mature ECs for stable vasculature formation.

Recent preclinical studies have shown that stem and progenitor cells derived from embryos or adult bone marrows have the potential to restore tissue vascularization after ischemic events. (Rafi & Lyden, 2003) Unlike mature cells, stem and progenitor cells have the potential to self-renew and differentiate into multiple cell types. While mature ECs seeded into collagen-fibronectin gels and implanted into different animal models suffered from poor perfusion and fast regression (Au *et al.*, 2008b; Koike *et al.*, 2004), ECs derived from human embryonic stem (hES) cells formed durable and functional blood vessels *in vivo*, supporting perfusion for over 150 days. (Wang *et al.*, 2007) Co-delivery of ECs with mesenchymal stem cells (MSC) can also result in stable and functional vessels, lasting for more than 130 days, which was not achieved when either of the cell types were used alone. (Au *et al.*, 2008a)

Endothelial progenitor cells (EPCs) derived from circulating blood or bone marrow have also been used to introduce neovascularization *in vivo*. (Park *et al.*, 2004) However, in order to form stable and durable vascular network, EPCs are often co-implanted with other cell types, such as VSMCs (Melero-Martin *et al.*, 2007) or mesenchymal progenitor cells (MPCs)

(Melero-Martin *et al.*, 2008). EPCs are expected to form the lumen of new vessels and VSMCs/MPCs serve as perivascular cells adjacent to the lumen providing survival and stability signals. EPCs derived from adult peripheral blood (PB) as well as umbilical cord blood (CB) have been compared for their ability to form functional long-lasting vessels *in vivo*. When co-implanted with 10T1/2 cells, a type of mouse MPCs, PB-EPCs and CB-EPCs both formed new vessels. While PB-EPCs formed vessels were unstable and regressed in 3 weeks, CB-EPCs formed normal-functioning vessels that lasted for more than 4 months, possibly due to a greater proliferative capacity of these cells. (Au *et al.*, 2008b) Therefore, CB-EPCs hold great potential for their ability to form stable vasculature in engineered and regenerated tissues. Adipose-derived stromal cells (ASCs) have also been investigated as another potent candidate for neovascularization cell therapy, due to their ability to stimulate angiogenesis as well as vessel maturation *in vivo*. (Rubina *et al.*, 2009)

Cell therapies can also be used in combination with single or multiple growth factors or their genes to improve the function of the transplanted cells. Stromal cell derived factor-1 (SDF-1), a small chemokine involved in recruiting EPCs, was delivered in combination with EPCs to promote EPC engraftement in ischemic muscle. (Kuliszewski *et al.*, 2011; Yu *et al.*, 2009) Dual delivery of VEGF and MCP-1 from alginate microbeads has been shown to support EC transplantation, with VEGF improving survival of transplanted ECs and MCP-1 inducing MCs recruitment. (Jay *et al.*, 2010) A similar effect was observed when co-delivering FGF-2 and granulocyte-colony stimulating factor (G-CSF) with bone marrow cells transplanted in a rodent model of critical limb ischemia. While FGF-2 directs EC migration and proliferation, G-CSF promotes the homing of bone marrow stem cells (in particular EPCs) to the ischemic site. (Layman *et al.*, 2011) PLGA microparticles loaded with VEGF and hepatocyte growth factor (HGF), as well as Ang-1, were delivered in combination with cord blood derived vasculogenic progenitor cells. The delivery of triple growth factors significantly enhanced the effect of cell therapy in multiple animal models, with increased progenitor cell incorporation, improved vessel function, and stabilization. (Saif *et al.*, 2010)

In addition to combined gene or protein delivery with cells, ECs and EPCs can also be genetically modified *ex vivo* prior to implantation to enhance vascular stabilization. For example, Bcl-2 is an anti-apoptosis protein that is upregulated during angiogenesis. ECs transfected with the *bcl-2* gene to inhibit EC apoptosis have shown increased vascular density and stability. Vessels formed with *bcl-2* transduced ECs connect with host circulation and last for at least 2 months. (Enis *et al.*, 2005; Schechner *et al.*, 2000; Shepherd *et al.*, 2009) EPCs with *vegf* gene overexpression have been shown to increase EPC migration and reduce serum starvation-induced apoptosis *in vitro*. (Yu *et al.*, 2009) When implanted *in vivo*, the *vegf* transfected EPCs stimulated greater blood flow and angiogenesis in animal models of ischemia than EPCs alone. (Ikeda *et al.*, 2004)

In spite of the promising results seen with cell therapies, certain limitations still exist. First, the sources of cells for transplantation, either mature cells or progenitor cells, can be problematic. If autologous cells have to be used, there are usually limited number of cells available, not to mention that their viability and activity are compromised due to disease or aging of the patient. (Brey & McIntire, 2008) If allogeneic and xenogeneic cells can be used, they would initiate a host immune response, which requires immunosupressive medication, making the patient vulnerable to infection and other diseases. Secondly, the expansion of isolated cells *ex vivo* can be costly, time-consuming and technically challenging. If mature

ECs and VSMCs are used for transplantation, it often requires longer times to reach a sufficient number, due to their low self-renewal capability. Stem and progenitor cells, on the other hand, have much higher proliferation rate. However, the expansion of these cells while maintaining their multipotency may be a technical challenge. (Ennett & Mooney, 2002) In addition, determination of the optimal cell types and concentration, delivery route, as well as identifying growth factors or genes to couple with cell transplantation requires further research. And lastly, the exact roles and ultimate fates of the transplanted cells remain largely unknown. Whether or not the transplanted cells would cause some other adverse complications, such as tumor angiogenesis, atheroma formation and retinopathies, is still unknown. (Rafii & Lyden, 2003) Those questions need to be answered with preclinical studies before proceeding to extensive clinical applications.

3.4 Analytical methods

3.4.1 *In vitro* models

In order to investigate the efficacy of different tissue engineering strategies for neovascularization, a serial of preclinical tests should be done with different analytical models before jumping to any conclusion and proceeding with patients. *In vitro* models are usually used to allow the evaluation of a single or a combination of factors (protein, gene or cells) in a relatively simple, robust, and controlled system. Most early *in vitro* models of neovascularization only used ECs to spontaneous form tubes in a 2D or 3D environment to mimic angiogenesis or vasculogenesis. (Francis *et al.*, 2008) Those models have been used to investigate the effect of soluble or insoluble factors, as well as genetic manipulation to the formation of new vessels *in vitro*. However, since the stabilization and remodeling stage of angiogenesis involves the interaction between ECs and MCs, many later studies have incorporated SMCs co-cultured with ECs. ECs and SMCs with a defined number can be co-cultured as a spheroid and put into a 3D scaffold. By fluorescently labeling the two cell types with different dyes, the cells can be visualized and distinguished under fluorescence microscope. The coculture spheroid can spontaneously organize into a core of SMCs and a surface layer of ECs, which is a spherical mimic of cylindrical blood vessels where there is a surface of ECs and coating MCs. (Korff *et al.*, 2001) This model allows study of the interactions between ECs and SMCs and their roles in angiogenesis and vessel stabilization. (Brey *et al.*, 2005)

Another *in vitro* model commonly used in studying angiogenesis and vessel stabilization is called aortic ring assay. (Nicosia & Ottinetti, 1990) The model results from the isolation of a 1-2 mm cross-section of aorta from rats or mice which is then embedded in a 3D biomaterial. Under certain stimuli, new vessels form with an inner core of ECs and a outer layer of MCs, which can then be visualized by immunofluorescent staining for different cellular markers. By treating the culture with vessel maturation factors, such as Ang-1 and MCP-1, increased MC recruitment was observed in this model. (Aplin *et al.*, 2010; Iurlaro *et al.*, 2003) However, since the model uses an organ part with multiple cell types, many other undefined cell types can also contribute to the network formation. It is difficult to isolate the contribution of each cell type in this model. (Brey *et al.*, 2005)

3.4.2 *In vivo* models

While the *in vitro* models provide convenient evaluation methods for the study of vessel assembly, the controlled environment differs dramatically from the actual *in vivo* environment where other factors, such as blood flow, inflammatory response, endocrine regulation, etc. contribute to the process. Therefore, studies with *in vitro* models are often followed by *in vivo* studies. A number of *in vivo* models have been developed to isolate the process of neovascularization, with different aspects of focus in different organ systems. Commonly evaluated parameters in most *in vivo* models include: vessel density, SMC coverage rate, vessel lasting or regression time, vessel size distribution, blood perfusion rate, vessel permeability, etc.

Subcutaneous implantation of scaffold materials is a commonly used method to evaluate protein, gene or cell therapies. (Jay *et al.*, 2010; Shea *et al.*, 1999; Tengood *et al.*, 2010) Growth factors, genes or cells can be embedded a matrix material, which will then be implanted subcutaneously to animals as a "plug". The presence of therapeutic agents will affect local neovascularization and tissue invasion into the material. At different time points after the implantation, local tissues will be harvested for evaluation. Typically histologic analysis of vessel density, diameter, and SMC coverage are performed, but more sophisticated techniques such as confocal microscopy, vascular casts, and microCT following contrast enhancement can be used to provide a better understanding of vascular structure. This model is relatively easy to use, and can be applied to evaluate different therapeutic strategies. However, all animals need to be sacrificed in order for the neovascularization assessmment, and it only provides the evaluation of some basic parameters. Therefore, other animal models are often used to provide greater detail into the process.

The mouse corneal micropocket assay allows for monitoring neovascularization without sacrificing animals. The method has been used by Cao *et al.* for the evaluation of vascular stability in response to dual growth factor delivery. Briefly, growth factor-loaded micropellets are implanted into a mouse corneal micropocket. Vessel lengths and vascularization areas of the eyes can be measured at various time points. The method allows the observation of neovascularization frequently without invasive surgery, thus limiting the number of animals used. Moreover, it allows long-term observation of vessel persistence. This method has only been used for evaluating growth factors or gene delivery, either as soluble factors (Cao *et al.*, 2003; Lai *et al.*, 2001; Lu *et al.*, 2007) or immobilized in a hydrogel (Moon *et al.*, 2010; Saik *et al.*, 2010). Use of this model to evaluate cell therapies is still rare to our knowledge. On the other hand, cranial windows model of severe combined immunodeficient (SCID) mouse has been used to evaluate cell therapies. (Au *et al.*, 2008a,b; Wang *et al.*, 2007) Similar to the corneal model, the cranial windows model also allows observation of neovascularization for several months without sacrificing the animals.

Mice, rats or rabbits hindlimb ischemia models have been used to evaluate the efficacy of protein delivery, gene therapy and cell therapy in neovascularization. (Cao *et al.*, 2003; de Paula *et al.*, 2009; Laymen *et al.*, 2011) The model applies ligation to the femoral artery and its branches to create ischemia in one limb. This model attempts to mimic peripheral arterial occlusive disease (PAOD), however this model only recreates the ischemic event and not the chronic vascular disease that lead to the condition in humans. After a certain treatment strategy, a number of neovascularizaiton parameters can be assessed, including flow recovery rate. The intensity of blood perfusion in the created ischemic area can be measured with different imaging techniques include laser Doppler or after systemically injection of fluorescent- or radio-labeled particles. Another ischemic model is the myocardial infraction heart model. Instead of femoral artery, the left anterior descend artery (LAD) is ligated in the animal to create ischmia. (Hao *et al.*, 2004) Besides common neovascularation parameters, cardiac morphology and function can also be evaluated using this model.

There are also a number of other animal models used in evaluating angiogenesis and vascular stability. The chorioallantoic membrane (CAM) model with chicken or quail embryos is a simple method to evaluate vascularization in response to different therapies. (Saif *et al.*, 2010; Stratman *et al.*, 2010) After therapeutic agents injection or implantation, the angiogenic effect can be easily assessed by imaging vascular regions of the CAM. This model is cost efficient and easy to perform compared to other mammalian animal models. Suitable animal models should be selected based on the target function of the therapeutic strategy. In most cases, one single animal model may not be sufficient to testing all the aspects of interest in neovascularization, therefore multiple *in vivo* models may be required to provide extensive understanding of the potential strategy. In addition, if the technique is ultimately prepared for a specific tissue engineering application then the appropriate pre-clinical model of the diseased or damaged tissue should be used.

4. Conclusions

The formation of stable vascular networks is critical for the survival of regenerated tissue in tissue engineering applications. In many cases, current strategies can result in immature vessel formation, providing poor perfusion and eventually leading to vessel regression. Recent studies have focused more on the later stage of neovascularization, which is the stabilization and remodeling of new vasculature. The process is highly regulated by both soluble and insoluble factors. Studies in the cellular and molecular mechanisms involved in vascular maturation have provided insight towards the development of strategies for the formation of stable vascular networks in engineered tissues. Successes have seen in many preclinical studies with these strategies; however, future success requires collaboration from interdisciplinary fields, including molecular and cellular biology, engineering and medicine.

5. References

Adams RH, Wilkinson GA, Weiss C, Diella F, Gale NW, Deutsch U, Risau W & Klein R. (1999). Roles of ephrinB ligands and EphB receptors in cardiovascular development: demarcation of arterial/venous domains, vascular morphogenesis, and sprouting angiogenesis. *Genes Dev.* Vol.13, No. 3, (February 1999), pp.295-306, ISSN 0890-9369

Allende ML, Yamashita T & Proia RL. (2003) G-protein-coupled receptor S1P1 acts within endothelial cells to regulate vascular maturation. *Blood.* Vol.102, No.10, (November 2003), pp. 3665-3667, ISSN 0006-4971

Aplin AC, Fogel E, & Nicosia RF. (2010). MCP-1 promotes mural cell recruitment during angiogenesis in the aortic ring model. *Angiogenesis,* Vol. 13, No. 3, (September 2010), pp. 219-226. ISSN 0969-6970

Armulik A, Abramsson A & Betsholtz C. (2005) Endothelial/pericyte interactions.*Circ Res.* Vol.97, No.6, (September. 2005), pp.512-523. ISSN 0009-7300

Asahara T, Masuda H, Takahashi T, Kalka C, Pastore C, Silver M, Kearne M, Magner M & Isner JM. (1999) Bone marrow origin of endothelial progenitor cells responsible for postnatal vasculogenesis in physiological and pathological neovascularization. *Circ Res.* Vol.85, No.3, (August 1999), pp.221-228. ISSN 0009-7300

Arsic N, Zentilin L, Zacchigna S, Santoro D, Stanta G, Salvi A, Sinagra G & Giacca M. (2003) Induction of functional neovascularization by combined VEGF and angiopoietin-1 gene transfer using AAV vectors. *Mol Ther.* Vol.7, No.4, (April 2003), pp.450-459. ISSN 1525-0016

Au P, Tam J, Fukumura D & Jain RK. (2008a) Bone marrow-derived mesenchymal stem cells facilitate engineering of long-lasting functional vasculature. *Blood.* Vol.111, No.9, (May 2008), pp.4551-4558. ISSN 0006-4971

Au P, Daheron LM, Duda DG, Cohen KS, Tyrrell JA, Lanning RM, Fukumura D, Scadden DT & Jain RK. (2008b) Differential in vivo potential of endothelial progenitor cells from human umbilical cord blood and adult peripheral blood to form functional long-lasting vessels. *Blood.* Vol.111, No.3, (February 2008), pp.1302-1305. ISSN 0006-4971

Benjamin LE & Keshet E. (1997) Conditional switching of vascular endothelial growth factor (VEGF) expression in tumors: induction of endothelial cell shedding and regression of hemangioblastoma-like vessels by VEGF withdrawal. *Proc Natl Acad Sci U S A.* Vol.94, No.16, (August 1997), pp.8761-8766. ISSN 0027-8424

Benjamin LE, Hemo I, & Keshet E. (1998) A plasticity window for blood vessel remodelling is defined by pericyte coverage of the preformed endothelial network and is regulated by PDGF-B and VEGF. *Development.* Vol.125, No.9, (May 1998), pp.1591-1598. ISSN 1011-6370

Benjamin LE, Golijanin D, Itin A, Pode D, & Keshet E. (1999) Selective ablation of immature blood vessels in established human tumors follows vascular endothelial growth factor withdrawal.*J Clin Invest.* Vol.103, No.2, (January 1999), pp.159-165. ISSN 0021-9738

Brey EM, McIntire LV, Johnston CM, Reece GP & Patrick CW Jr. (2004) Three-dimensional, quantitative analysis of desmin and smooth muscle alpha actin expression during angiogenesis. *Ann Biomed Eng.* Vol.32, No.8, (August 2004), pp.1100-1107. ISSN 0090-6964

Brey EM, Uriel S, Greisler HP & McIntire LV. (2005) Therapeutic neovascularization: contributions from bioengineering. *Tissue Eng.* Vol.11, No,3-4, (March-April 2005), pp.567-584. ISSN 1937-3368

Brey, E.M.; & McIntire, L.V. (2008). Vascular assembly in engineered and nature tissues, In: *Principles of Regenerative Medicine*, A. Atala,(Ed.), pp.1020-1037, Elsiver, ISBN 978-0-12-369410-2, Burlington, MA, USA.

Bussolino F, Mantovani A & Persico G. (1997) Molecular mechanisms of blood vessel formation. *Trends Biochem Sci.* Vol.22, No.7, (July 1997), pp.251-256. ISSN 0968-0004

Cao R, Bråkenhielm E, Pawliuk R, Wariaro D, Post MJ, Wahlberg E, Leboulch P & Cao Y. (2003) Angiogenic synergism, vascular stability and improvement of hind-limb ischemia by a combination of PDGF-BB and FGF-2. *Nat Med.* Vol.9, No.5, (May 2003), pp.604-613. ISSN 1078-8956

Cao Y, Hong A, Schulten H & Post MJ. (2005) Update on therapeutic neovascularization. *Cardiovasc Res.* Vol.65, No.3, (February 2005), pp.639-648. ISSN 2151-920X

Carmeliet P. (2000) Mechanisms of angiogenesis and arteriogenesis. *Nat Med.* Vol.6, No.4, (April 2000), pp.389-395. ISSN 1078-8956

Carmeliet P & Conway EM. (2001) Growing better blood vessels. *Nat Biotechnol.* Vol.19, No.11, (November 2001), pp.1019-1020. ISSN 1087-0156

Carmeliet P. (2003) Angiogenesis in health and disease. *Nat Med.* Vol.9, No.6, (June 2003), pp.653-660. ISSN 1078-8956

Chen F, Tan Z, Dong CY, Chen X & Guo SF. (2007) Adeno-associated virus vectors simultaneously encoding VEGF and angiopoietin-1 enhances neovascularization in ischemic rabbit hind-limbs. *Acta Pharmacol Sin.* Vol.28, No.4, (April 2007), pp.493-502. ISSN 1671-4083

Chen RR & Mooney DJ. (2003) Polymeric growth factor delivery strategies for tissue engineering. *Pharm Res.* Vol.20, No.8, (August 2003), pp.1103-1112. ISSN 0724-8741

Chen RR, Silva EA, Yuen WW & Mooney DJ. (2007) Spatio-temporal VEGF and PDGF delivery patterns blood vessel formation and maturation. *Pharm Res.* Vol.24, No.2, (February 2007), pp.258-264. ISSN 0724-8741

Cristofaro B & Emanueli C. (2009) Possible novel targets for therapeutic angiogenesis. *Curr Opin Pharmacol.* ; Vol.9, No.2, (April 2009), pp.102-108. ISSN 1471-4892

Cristofaro B, Stone OA, Caporali A, Dawbarn D, Ieronimakis N, Reyes M, Madeddu P, Bates DO & Emanueli C. (2010) Neurotrophin-3 is a novel angiogenic factor capable of therapeutic neovascularization in a mouse model of limb ischemia. *Arterioscler Thromb Vasc Biol.* Vol.30, No.6, (June 2010), pp.1143-1150. ISSN 1079-5642

Darland DC & D'Amore PA. (1999) Blood vessel maturation: vascular development comes of age. *J Clin Invest.* Vol.103, No.2, (January 1999), pp.157-158. ISSN 0021-9738

Davis GE & Senger DR. (2005) Endothelial extracellular matrix: biosynthesis, remodeling, and functions during vascular morphogenesis and neovessel stabilization.*Circ Res.* Vol.97, No.11, (November 2005), pp.1093-1107. ISSN 0009-7300

Dejana E. (2004) Endothelial cell-cell junctions: happy together. *Nat Rev Mol Cell Biol.* Vol.5, No.4, (April 2004), pp.261-270. ISSN 1471-0072

Dellian M, Witwer BP, Salehi HA, Yuan F & Jain RK. (1996) Quantitation and physiological characterization of angiogenic vessels in mice: effect of basic fibroblast growth factor, vascular endothelial growth factor/vascular permeability factor, and host microenvironment. *Am J Pathol.*Vol.149, No.1, (July 1996), pp.59-71. ISSN 0002-9440

de Paula EV, Flores-Nascimento MC, Arruda VR, Garcia RA, Ramos CD, Guillaumon AT & Annichino-Bizzacchi JM. (2009) Dual gene transfer of fibroblast growth factor-2 and platelet derived growth factor-BB using plasmid deoxyribonucleic acid promotes effective angiogenesis and arteriogenesis in a rodent model of hindlimb ischemia. *Transl Res.* Vol.153, No.5, (May 2009), pp.232-239. ISSN 1943-8141

Dimmeler S & Zeiher AM. (2000) Endothelial cell apoptosis in angiogenesis and vessel regression. *Circ Res.* Vol.87, No.6, (September 2000), pp.434-439. ISSN 0009-7300

Donovan MJ, Lin MI, Wiegn P, Ringstedt T, Kraemer R, Hahn R, Wang S, Ibañez CF, Rafii S & Hempstead BL. (2000) Brain derived neurotrophic factor is an endothelial cell survival factor required for intramyocardial vessel stabilization. *Development.* Vol.127, No.21, (November 2000), pp.4531-4540. ISSN 1011-6370

Ehrbar M, Djonov VG, Schnell C, Tschanz SA, Martiny-Baron G, Schenk U, Wood J, Burri PH, Hubbell JA & Zisch AH. (2004) Cell-demanded liberation of VEGF121 from fibrin implants induces local and controlled blood vessel growth. *Circ Res.* Vol.94, No.8, (April 2004), pp.1124-1132. ISSN 0009-7300

Enis DR, Shepherd BR, Wang Y, Qasim A, Shanahan CM, Weissberg PL, Kashgarian M, Pober JS & Schechner JS. (2005) Induction, differentiation, and remodeling of blood vessels after transplantation of Bcl-2-transduced endothelial cells. *Proc Natl Acad Sci U S A.* Vol.102, No.2, (January 2005), pp.425-430. ISSN 0027-8424

Ennett AB & Mooney DJ. (2002) Tissue engineering strategies for in vivo neovascularisation. *Expert Opin Biol Ther.* Vol.2, No.8, (December 2002), pp.805-818. ISSN 1471-2598

Francis ME, Uriel S & Brey EM. (2008) Endothelial cell-matrix interactions in neovascularization. *Tissue Eng Part B Rev.* Vol.14, No.1, (March 2008), pp.19-32. ISSN 1937-3368

Francis-Sedlak ME, Moya ML, Huang JJ, Lucas SA, Chandrasekharan N, Larson JC, Cheng MH & Brey EM. (2010) Collagen glycation alters neovascularization in vitro and in vivo. *Microvasc Res.* Vol.80, No.1, (July 2010), pp.3-9. ISSN 0026-2862

Gaengel K, Genové G, Armulik A & Betsholtz C. (2009) Endothelial-mural cell signaling in vascular development and angiogenesis. *Arterioscler Thromb Vasc Biol.* Vol.29, No.5, (May 2009), pp.630-638. ISSN 1079-5642

Garmy-Susini B, Jin H, Zhu Y, Sung RJ, Hwang R & Varner J. (2005) Integrin alpha4beta1-VCAM-1-mediated adhesion between endothelial and mural cells is required for blood vessel maturation.*J Clin Invest.* Vol.115, No.6, (June 2005), pp.1542-1551. ISSN 0021-9738

Gerhardt H & Betsholtz C. (2003) Endothelial-pericyte interactions in angiogenesis. *Cell Tissue Res.* Vol.314, No.1, (October 2003), pp.15-23. ISSN 0302-766X

Ghajar CM, George SC & Putnam AJ. (2008) Matrix metalloproteinase control of capillary morphogenesis.*Crit Rev Eukaryot Gene Expr.*Vol.18,No.3,pp.251-278. ISSN 1045-4403

Ghosh R, Walsh SR, Tang TY, Noorani A & Hayes PD. (2008) Gene therapy as a novel therapeutic option in the treatment of peripheral vascular disease: systematic review and meta-analysis. *Int J Clin Pract.* Vol.62, No.9, (September 2008), pp.1383-1390. ISSN 1368-5031

Goumans MJ, Valdimarsdottir G, Itoh S, Rosendahl A, Sideras P & ten Dijke P. (2002) Balancing the activation state of the endothelium via two distinct TGF-beta type I receptors. *EMBO J.* Vol.21, No.7, (April 2002), pp.1743-1753. ISSN 0261-4189

Greenberg JI, Shields DJ, Barillas SG, Acevedo LM, Murphy E, Huang J, Scheppke L, Stockmann C, Johnson RS, Angle N & Cheresh DA. (2008) A role for VEGF as a negative regulator of pericyte function and vessel maturation. *Nature.* Vol.456, No.7223, (December 2008), pp.809-813. ISSN 0028-0836

Hao X, Månsson-Broberg A, Blomberg P, Dellgren G, Siddiqui AJ, Grinnemo KH, Wärdell E & Sylvén C. (2004a) Angiogenic and cardiac functional effects of dual gene transfer of VEGF-A165 and PDGF-BB after myocardial infarction.*Biochem Biophys Res Commun.* Vol.322, No.1, (September 2004), pp.292-296. ISSN 0006-291X

Hao X, Månsson-Broberg A, Gustafsson T, Grinnemo KH, Blomberg P, Siddiqui AJ, Wärdell E & Sylvén C. (2004b) Angiogenic effects of dual gene transfer of bFGF and PDGF-BB after myocardial infarction. *Biochem Biophys Res Commun.* Vol.315, No.4, (March 2004), pp.1058-1063. ISSN 0006-291X

Hellberg C, Ostman A & Heldin CH. (2010) PDGF and vessel maturation. *Recent Results Cancer Res.* Vol.180, pp.103-114. ISSN 0080-0015

Hellström M, Kalén M, Lindahl P, Abramsson A & Betsholtz C. (1999) Role of PDGF-B and PDGFR-beta in recruitment of vascular smooth muscle cells and pericytes during embryonic blood vessel formation in the mouse. *Development.* Vol.126, No.14, (June 1999), pp.3047-3055. ISSN 1011-6370

Iivanainen E, Nelimarkka L, Elenius V, Heikkinen SM, Junttila TT, Sihombing L, Sundvall M, Maatta JA, Laine VJ, Yla-Herttuala S, Higashiyama S, Alitalo K & Elenius K. (2003) Angiopoietin-regulated recruitment of vascular smooth muscle cells by endothelial-derived heparin binding EGF-like growth factor. *FASEB J.* Vol.17, No.12, (September 2003), pp.1609-1621. ISSN 0892-6638

Ikeda Y, Fukuda N, Wada M, Matsumoto T, Satomi A, Yokoyama S, Saito S, Matsumoto K, Kanmatsuse K & Mugishima H. (2004) Development of angiogenic cell and gene therapy by transplantation of umbilical cord blood with vascular endothelial growth factor gene. *Hypertens Res.* Vol. 27, No.2, (February 2004), pp.119-28. ISSN 0916-9636

Iurlaro M, Scatena M, Zhu WH, Fogel E, Wieting SL & Nicosia RF. (2003) Rat aorta-derived mural precursor cells express the Tie2 receptor and respond directly to stimulation by angiopoietins.*J Cell Sci.* Vol. 116, No.17, (September 2003), pp.3635-3643. ISSN 0021-9533

Jain RK, Safabakhsh N, Sckell A, Chen Y, Jiang P, Benjamin L, Yuan F & Keshet E. (1998) Endothelial cell death, angiogenesis, and microvascular function after castration in an androgen-dependent tumor: role of vascular endothelial growth factor. *Proc Natl Acad Sci U S A.* Vol.95, No.18, (September 1998), pp.10820-10825. ISSN 0027-8424

Jain RK. (2003) Molecular regulation of vessel maturation. *Nat Med.* Vol. 9, No.6, (June 2003), pp.685-93. ISSN 1078-8956

Jay SM, Shepherd BR, Andrejecsk JW, Kyriakides TR, Pober JS & Saltzman WM. (2010) Dual delivery of VEGF and MCP-1 to support endothelial cell transplantation for therapeutic vascularization. *Biomaterials.* Vol.31, No.11, (April 2010), pp.3054-3062. ISSN 0142-9612

Kashiwagi S, Izumi Y, Gohongi T, Demou ZN, Xu L, Huang PL, Buerk DG, Munn LL, Jain RK & Fukumura D. (2005) NO mediates mural cell recruitment and vessel morphogenesis in murine melanomas and tissue-engineered blood vessels. *J Clin Invest.* Vol.115, No.7, (July 2005), pp.1816-1827. ISSN 0021-9738

Khanna O, Moya ML, Greisler HP, Opara EC & Brey EM. (2010) Multilayered microcapsules for the sustained-release of angiogenic proteins from encapsulated cells. *Am J Surg.* Vol.200, No.5, (November 2010), pp.655-658. ISSN 0147-5185

Kobayashi H, DeBusk LM, Babichev YO, Dumont DJ & Lin PC. (2006) Hepatocyte growth factor mediates angiopoietin-induced smooth muscle cell recruitment. *Blood.* Vol.108, No.4, (August 2006), pp.1260-1266. ISSN 0006-4971

Koike N, Fukumura D, Gralla O, Au P, Schechner JS & Jain RK. (2004) Tissue engineering: creation of long-lasting blood vessels. *Nature.* Vol.428, No.6979, (March 2004), pp.138-139. ISSN 0028-0836

Korff T, Kimmina S, Martiny-Baron G & Augustin HG. (2001) Blood vessel maturation in a 3-dimensional spheroidal coculture model: direct contact with smooth muscle cells regulates endothelial cell quiescence and abrogates VEGF responsiveness. *FASEB J.* Vol.15, No.2, (February 2001), pp.447-457. ISSN 0892-6638

Korpisalo P, Karvinen H, Rissanen TT, Kilpijoki J, Marjomäki V, Baluk P, McDonald DM, Cao Y, Eriksson U, Alitalo K & Ylä-Herttuala S. (2008) Vascular endothelial growth factor-A and platelet-derived growth factor-B combination gene therapy prolongs angiogenic effects via recruitment of interstitial mononuclear cells and paracrine effects rather than improved pericyte coverage of angiogenic vessels. *Circ Res.* Vol.103, No.10, (November 2008), pp.1092-1099. ISSN 0009-7300

Kuliszewski MA, Kobulnik J, Lindner JR, Stewart DJ & Leong-Poi H. (2011) Vascular Gene Transfer of SDF-1 Promotes Endothelial Progenitor Cell Engraftment and Enhances Angiogenesis in Ischemic Muscle. *Mol Ther.* [ahead of print] ISSN 1525-0016

Kupatt C, Hinkel R, Pfosser A, El-Aouni C, Wuchrer A, Fritz A, Globisch F, Thormann M, Horstkotte J, Lebherz C, Thein E, Banfi A & Boekstegers P. (2010) Cotransfection of vascular endothelial growth factor-A and platelet-derived growth factor-B via recombinant adeno-associated virus resolves chronic ischemic malperfusion role of vessel maturation. *J Am Coll Cardiol.* Vol.56, No.5, (July 2010), pp.414-422. ISSN 1558-3597

Lai CM, Brankov M, Zaknich T, Lai YK, Shen WY, Constable IJ, Kovesdi I & Rakoczy PE. (2001) Inhibition of angiogenesis by adenovirus-mediated sFlt-1 expression in a rat model of corneal neovascularization. *Hum Gene Ther.* Vol.12, No.10, (July 2001), pp.1299-1310. ISSN 1043-0342

Langer R & Vacanti JP. (1993) Tissue engineering. *Science.* Vol.260, No.5110, (May 1993), pp.920-926. ISSN 0036-8075

Lawson ND, Scheer N, Pham VN, Kim CH, Chitnis AB, Campos-Ortega JA & Weinstein BM. (2001) Notch signaling is required for arterial-venous differentiation during embryonic vascular development. *Development.* Vol.128, No.19, (October 2001), pp.3675-3683. ISSN 1011-6370

Layman H, Rahnemai-Azar AA, Pham SM, Tsechpenakis G & Andreopoulos FM. (2011) Synergistic angiogenic effect of codelivering fibroblast growth factor 2 and granulocyte-colony stimulating factor from fibrin scaffolds and bone marrow transplantation in critical limb ischemia. *Tissue Eng Part A.* Vol.17, No.1-2, (January 2011), pp.243-254. ISSN 1937-3368

Leslie-Barbick JE, Shen C, Chen C & West JL. (2011) Micron-Scale Spatially Patterned, Covalently Immobilized Vascular Endothelial Growth Factor on Hydrogels Accelerates Endothelial Tubulogenesis and Increases Cellular Angiogenic Responses. *Tissue Eng Part A.*Vol.17, No.1-2, (January 2011), pp.221-229. ISSN 1937-3368

Letamendía A, Lastres P, Botella LM, Raab U, Langa C, Velasco B, Attisano L & Bernabeu C. (1998) Role of endoglin in cellular responses to transforming growth factor-beta. A comparative study with betaglycan. *J Biol Chem.*Vol.273, No.49, (December 1998), pp.33011-33019. ISSN 0021-9258

Levanon K, Varda-Bloom N, Greenberger S, Barshack I, Goldberg I, Orenstein A, Breitbart E, Shaish A & Harats D. (2006) Vascular wall maturation and prolonged angiogenic effect by endothelial-specific platelet-derived growth factor expression. *Pathobiology.* Vol.73, No.3, pp.149-158. ISSN 1015-2008

Lindahl P, Johansson BR, Levéen P & Betsholtz C. (1997) Pericyte loss and microaneurysm formation in PDGF-B-deficient mice. *Science.* Vol.277, No.5323, (July 1997), pp.242-245. ISSN 0036-8075

Liu ZJ, Shirakawa T, Li Y, Soma A, Oka M, Dotto GP, Fairman RM, Velazquez OC & Herlyn M. (2003) Regulation of Notch1 and Dll4 by vascular endothelial growth factor in arterial endothelial cells: implications for modulating arteriogenesis and angiogenesis. *Mol Cell Biol.* Vol.23, No.1, (January 2003), pp.14-25. ISSN 0270-7306

Lu H, Xu X, Zhang M, Cao R, Bråkenhielm E, Li C, Lin H, Yao G, Sun H, Qi L, Tang M, Dai H, Zhang Y, Su R, Bi Y, Zhang Y & Cao Y. (2007) Combinatorial protein therapy of angiogenic and arteriogenic factors remarkably improves collaterogenesis and cardiac function in pigs. *Proc Natl Acad Sci U S A.*Vol.104, No.29, (July 2007), pp.12140-12145. ISSN 0027-8424

Maisonpierre PC, Suri C, Jones PF, Bartunkova S, Wiegand SJ, Radziejewski C, Compton D, McClain J, Aldrich TH, Papadopoulos N, Daly TJ, Davis S, Sato TN & Yancopoulos GD. (1997) Angiopoietin-2, a natural antagonist for Tie2 that disrupts in vivo angiogenesis. *Science.* Vol. 77, No.5322, (July 1997), pp.55-60. ISSN 0036-8075

Melero-Martin JM, Khan ZA, Picard A, Wu X, Paruchuri S & Bischoff J. (2007) In vivo vasculogenic potential of human blood-derived endothelial progenitor cells. *Blood.* Vol.109, No.11, (June 2007), pp.4761-4768. ISSN 0006-4971

Melero-Martin JM, De Obaldia ME, Kang SY, Khan ZA, Yuan L, Oettgen P & Bischoff J. (2008) Engineering robust and functional vascular networks in vivo with human adult and cord blood-derived progenitor cells. *Circ Res.* Vol.103, No.2, (July 2008), pp.194-202. ISSN 0009-7300

Mizugishi K, Yamashita T, Olivera A, Miller GF, Spiegel S & Proia RL. (2005) Essential role for sphingosine kinases in neural and vascular development. *Mol Cell Biol.* Vol.25, No.24, (December 2005), pp.11113-11121. ISSN 0270-7306

Moon JJ, Lee SH & West JL. (2007) Synthetic biomimetic hydrogels incorporated with ephrin-A1 for therapeutic angiogenesis. *Biomacromolecules.* Vol.8, No.1, (January 2007), pp.42-49. ISSN 1525-7797

Moon JJ, Saik JE, Poché RA, Leslie-Barbick JE, Lee SH, Smith AA, Dickinson ME & West JL. (2010) Biomimetic hydrogels with pro-angiogenic properties. *Biomaterials.* Vol.31, No.14, (May 2010), pp.3840-3847. ISSN 0142-9612

Moya ML, Lucas S, Francis-Sedlak M, Liu X, Garfinkel MR, Huang JJ, Cheng MH, Opara EC, & Brey EM. (2009) Sustained delivery of FGF-1 increases vascular density in comparison to bolus administration. *Microvasc Res.* Vol. 78, No.2, (September 2009), pp.142-147. ISSN 0026-2862.

Moya ML, Garfinkel MR, Liu X, Lucas S, Opara EC, Greisler HP & Brey EM. (2010) Fibroblast growth factor-1 (FGF-1) loaded microbeads enhance local capillary neovascularization.*J Surg Res.*Vol.160, No.2, (May 2010), pp.208-212. ISSN0022-4804.

Nie H, Lee LY, Tong H & Wang CH. (2008) PLGA/chitosan composites from a combination of spray drying and supercritical fluid foaming techniques: new carriers for DNA delivery. *J Control Release.* Vol.129, No.3, (August 2008), pp.207-214. ISSN 0168-3659

Nicosia RF & Ottinetti A (1990) Growth of microvessels in serumfree matrix culture of rat aorta. A quantitative assay of angiogenesis in vitro. *Lab Invest.* Vol.63, pp.115–122 ISSN 0023-6837

Nishishita T & Lin PC. (2004) Angiopoietin 1, PDGF-B, and TGF-beta gene regulation in endothelial cell and smooth muscle cell interaction. *J Cell Biochem.* Vol.91, No.3, (February 2004), pp.584-593. ISSN 0730-2312

Nör JE, Christensen J, Mooney DJ & Polverini PJ. (1999) Vascular endothelial growth factor (VEGF)-mediated angiogenesis is associated with enhanced endothelial cell survival and induction of Bcl-2 expression. *Am J Pathol.*Vol.154, No.2, (February 1999), pp.375-384. ISSN 0002-9440

Nör JE, Peters MC, Christensen JB, Sutorik MM, Linn S, Khan MK, Addison CL, Mooney DJ & Polverini PJ. (2001) Engineering and characterization of functional human microvessels in immunodeficient mice. *Lab Invest.* Vol.81, No.4, (April 2001), pp.453-463. ISSN 0023-6837

Paik JH, Skoura A, Chae SS, Cowan AE, Han DK, Proia RL & Hla T. (2004) Sphingosine 1-phosphate receptor regulation of N-cadherin mediates vascular stabilization.*Genes Dev.* Vol.18, No.19, (October 2004), pp.2392-9403. ISSN 0890-9369

Peirce SM, Price RJ & Skalak TC. (2004) Spatial and temporal control of angiogenesis and arterialization using focal applications of VEGF164 and Ang-1. *Am J Physiol Heart Circ Physiol.* Vol.286, No.3, (March 2004), pp.918-925. ISSN 0009-7300

Pepper MS. (1997) Transforming growth factor-beta: vasculogenesis, angiogenesis, and vessel wall integrity.*Cytokine Growth Factor Rev.* Vol.8, No.1, (March 1997), pp.21-43. ISSN 1359-6101

Rafii S & Lyden D. (2003) Therapeutic stem and progenitor cell transplantation for organ vascularization and regeneration. *Nat Med.* Vol.9, No.6, (June 2003), pp.702-712. ISSN 1078-8956

Richardson TP, Peters MC, Ennett AB & Mooney DJ.(2001) Polymeric system for dual growth factor delivery. *Nat Biotechnol.* Vol.19, No.11, (November 2001), pp.1029-1034. ISSN 1087-0156

Rissanen TT, Vajanto I & Ylä-Herttuala S. (2001) Gene therapy for therapeutic angiogenesis in critically ischaemic lower limb - on the way to the clinic. *Eur J Clin Invest.* Vol.31, No.8, (August 2001), pp.651-666. ISSN 0014-2972

Rissanen TT, Markkanen JE, Gruchala M, Heikura T, Puranen A, Kettunen MI, Kholová I, Kauppinen RA, Achen MG, Stacker SA, Alitalo K & Ylä-Herttuala S. (2003) VEGF-D is the strongest angiogenic and lymphangiogenic effector among VEGFs delivered into skeletal muscle via adenoviruses. *Circ Res.* Vol.92, No.10, (May 2003), pp.1098-1106. ISSN 0009-7300

Rubina K, Kalinina N, Efimenko A, Lopatina T, Melikhova V, Tsokolaeva Z, Sysoeva V, Tkachuk V & Parfyonova Y. (2009) Adipose stromal cells stimulate angiogenesis via promoting progenitor cell differentiation, secretion of angiogenic factors, and enhancing vessel maturation.*Tissue Eng Part A.* Vol.15, No.8, (August 2009), pp.2039-2050. ISSN 1937-3368

Saif J, Schwarz TM, Chau DY, Henstock J, Sami P, Leicht SF, Hermann PC, Alcala S, Mulero F, Shakesheff KM, Heeschen C & Aicher A. (2010) Combination of injectable multiple growth factor-releasing scaffolds and cell therapy as an advanced modality to enhance tissue neovascularization. *Arterioscler Thromb Vasc Biol.* Vol.30, No.10, (October 2010), pp.1897-1904. ISSN 1079-5642

Saik JE, Gould DJ, Watkins EM, Dickinson ME & West JL.(2011) Covalently immobilized platelet-derived growth factor-BB promotes angiogenesis in biomimetic poly(ethylene glycol) hydrogels. *Acta Biomater.* Vol.7, No.1, (January 2011), pp.133-143. ISSN 1742-7061

Sato TN, Tozawa Y, Deutsch U, Wolburg-Buchholz K, Fujiwara Y, Gendron-Maguire M, Gridley T, Wolburg H, Risau W & Qin Y. (1995) Distinct roles of the receptor tyrosine kinases Tie-1 and Tie-2 in blood vessel formation. *Nature.* Vol.376, No.6535, (July 1995), pp.70-74. ISSN 0028-0836

Schechner JS, Nath AK, Zheng L, Kluger MS, Hughes CC, Sierra-Honigmann MR, Lorber MI, Tellides G, Kashgarian M, Bothwell AL & Pober JS. (2000) In vivo formation of complex microvessels lined by human endothelial cells in an immunodeficient mouse. *Proc Natl Acad Sci USA.*Vol.97, No.16, (August 2000), pp.9191-9196. ISSN 0027-8424

Shea LD, Smiley E, Bonadio J & Mooney DJ. (1999) DNA delivery from polymer matrices for tissue engineering. *Nat Biotechnol.* Vol.17, No.6, (June 1999), pp .551-554. ISSN1087-0156

Shepherd BR, Jay SM, Saltzman WM, Tellides G & Pober JS. (2009) Human aortic smooth muscle cells promote arteriole formation by coengrafted endothelial cells. *Tissue Eng Part A.* Vol.15, No.1, (January 2009), pp.165-173. ISSN 1937-3368

Shyu KG, Manor O, Magner M, Yancopoulos GD & Isner JM. (1998) Direct intramuscular injection of plasmid DNA encoding angiopoietin-1 but not angiopoietin-2 augments revascularization in the rabbit ischemic hindlimb. *Circulation.* Vol.98, No.19, (November 1998), pp.2081-2087. ISSN 0009-7322

Siddiqui AJ, Blomberg P, Wärdell E, Hellgren I, Eskandarpour M, Islam KB & Sylvén C. (2003) Combination of angiopoietin-1 and vascular endothelial growth factor gene therapy enhances arteriogenesis in the ischemic myocardium. *Biochem Biophys Res Commun.* Vol.310, No.3, (October 2003), pp.1002-1009. ISSN 0006-291X

Sneider EB, Nowicki PT & Messina LM. (2009) Regenerative medicine in the treatment of peripheral arterial disease. *J Cell Biochem.*Vol.108, No.4, (November 2009), pp.753-761. ISSN 0730-2312

Stratman AN, Malotte KM, Mahan RD, Davis MJ & Davis GE. (2009) Pericyte recruitment during vasculogenic tube assembly stimulates endothelial basement membrane matrix formation. *Blood.*Vol.114, No.24, (December 2009), pp.5091-5101. ISSN 0006-4971

Stratman AN, Schwindt AE, Malotte KM & Davis GE. (2010) Endothelial-derived PDGF-BB and HB-EGF coordinately regulate pericyte recruitment during vasculogenic tube assembly and stabilization. *Blood*.Vol.116, No.22, (November 2010), pp.4720-4730. ISSN 0006-4971

Su H, Takagawa J, Huang Y, Arakawa-Hoyt J, Pons J, Grossman W & Kan YW. (2009) Additive effect of AAV-mediated angiopoietin-1 and VEGF expression on the therapy of infarcted heart. *Int J Cardiol.* Vol.133, No.2, (April 2009), pp.191-197. ISSN 0167-5273

Sullivan DC & Bicknell R. (2003) New molecular pathways in angiogenesis. *Br J Cancer.* Vol.89, No.2, (July 2003), pp.228-231. ISSN 0007-0920

Suri C, Jones PF, Patan S, Bartunkova S, Maisonpierre PC, Davis S, Sato TN & Yancopoulos GD. (1996) Requisite role of angiopoietin-1, a ligand for the TIE2 receptor, during embryonic angiogenesis. *Cell.* Vol.87, No.7, (December 1996), pp.1171-1180.ISSN 0092-8674

Tahara K, Sakai T, Yamamoto H, Takeuchi H & Kawashima Y. (2008) Establishing chitosan coated PLGA nanosphere platform loaded with wide variety of nucleic acid by complexation with cationic compound for gene delivery. *Int J Pharm.* Vol.354, No.1-2, (April 2008), pp.210-216. ISSN 0378-5173

Tengood JE, Kovach KM, Vescovi PE, Russell AJ & Little SR. (2010) Sequential delivery of vascular endothelial growth factor and sphingosine 1-phosphate for angiogenesis. *Biomaterials.* Vol.31, No.30, (October 2010), pp.7805-7812. ISSN 0142-9612

Tengood JE, Ridenour R, Brodsky R, Russell AJ & Little SR. (2011) Sequential Delivery of Basic Fibroblast Growth Factor and Platelet-Derived Growth Factor for Angiogenesis. *Tissue Eng Part A.* [ahead of print] ISSN 1937-3368

Thurston G, Suri C, Smith K, McClain J, Sato TN, Yancopoulos GD & McDonald DM. (1999) Leakage-resistant blood vessels in mice transgenically overexpressing angiopoietin-1. *Science.* Vol.286, No.5449, (December 1999), pp.2511-2514. ISSN 0036-8075

Thurston G, Rudge JS, Ioffe E, Zhou H, Ross L, Croll SD, Glazer N, Holash J, McDonald DM & Yancopoulos GD. (2000) Angiopoietin-1 protects the adult vasculature against plasma leakage. *Nat Med.* Vol.6, No.4, (April 2000), pp.460-463. ISSN 1078-8956

Uutela M, Wirzenius M, Paavonen K, Rajantie I, He Y, Karpanen T, Lohela M, Wiig H, Salven P, Pajusola K, Eriksson U & Alitalo K. (2004) PDGF-D induces macrophage recruitment, increased interstitial pressure, and blood vessel maturation during angiogenesis. *Blood*.Vol.104, No.10, (November 2004), pp.3198-3204. ISSN 0006-4971

van den Driesche S, Mummery CL & Westermann CJ. (2003) Hereditary hemorrhagic telangiectasia: an update on transforming growth factor beta signaling in vasculogenesis and angiogenesis. *Cardiovasc Res.* Vol.58, No.1, (April 2003), pp.20-31. ISSN 2151-920X.

Wang HU, Chen ZF & Anderson DJ. (1998) Molecular distinction and angiogenic interaction between embryonic arteries and veins revealed by ephrin-B2 and its receptor Eph-B4. *Cell.* Vol.93, No.5, (May 1998), pp.741-753. ISSN 0092-8674

Wang ZZ, Au P, Chen T, Shao Y, Daheron LM, Bai H, Arzigian M, Fukumura D, Jain RK & Scadden DT. (2007) Endothelial cells derived from human embryonic stem cells form durable blood vessels in vivo. *Nat Biotechnol.* Vol.25, No.3, (March 2007), pp.317-318. ISSN 1087-0156

Ylä-Herttuala S & Alitalo K. (2003) Gene transfer as a tool to induce therapeutic vascular growth. *Nat Med.* Vol.9, No.6, (June 2003), pp.694-701. ISSN 1078-8956

Yu JX, Huang XF, Lv WM, Ye CS, Peng XZ, Zhang H, Xiao LB & Wang SM. (2009) Combination of stromal-derived factor-1alpha and vascular endothelial growth factor gene-modified endothelial progenitor cells is more effective for ischemic neovascularization. *J Vasc Surg.* Vol.50, No.3, (September 2009), pp.608-616. ISSN 1078-5884

Zisch AH, Lutolf MP & Hubbell JA. (2003a) Biopolymeric delivery matrices for angiogenic growth factors. *Cardiovasc Pathol.* Vol.12, No.6, (November-December 2003), pp.295-310. ISSN 1054-8807

Zisch AH, Lutolf MP, Ehrbar M, Raeber GP, Rizzi SC, Davies N, Schmökel H, Bezuidenhout D, Djonov V, Zilla P & Hubbell JA. (2003b) Cell-demanded release of VEGF from synthetic, biointeractive cell ingrowth matrices for vascularized tissue growth. *FASEB J.* Vol.17, No.15, (December 2003), pp.2260-2262. ISSN 0892-6638

23

Tunable Stimuli-Responsive Polymers for Cell Sheet Engineering

Nithya Joseph, Anil Kumar P R and TV Kumary
Division of Implant Biology, Biomedical Technology Wing
Sree Chitra Tirunal Institute for Medical Sciences and Technology
India

1. Introduction

Anchorage dependent cells require a suitable substrate or scaffold for attachment, adhesion and growth. Conventionally cells are detached from substrates by mechanical or enzyme treatment, which can affect cell re-adhesion, viability etc. Poly (N-isopropylacrylamide) [PNIPAAm] is a well known temperature sensitive polymer with LCST of 32 °C and serve as a good substrate facilitating cell detachment without the use of proteolytic enzymes. Polymer and copolymers with N-isopropylacrylamide [NIPAAm] been proved to be excellent for cell detachment by temperature treatment. The co-monomers provide functional groups that bind to biological molecules and form hybrid polymer for enhanced cell function. This chapter reviews and compares different smart intelligent surfaces, method of preparation, characterization techniques and applications of cell sheet engineering. PNIPAAm based surfaces are able support cell growth and then to detach without any chemical reagent step. These thermo responsive substrates avoid the problem associated with scaffold based tissue engineering.

2. Cell sheet engineering

Tissue engineering involves ex vivo expansion of anchorage dependent cells in three dimensions to provide functional tissues (Langer and Vacanti, 1993). Prof. Robert Langer, Dr. Joseph and Dr. Charles Vacanti proposed the concept of tissue engineering in order to overcome the limitations of organ transplantation. Over the past 20 years, using the technology of tissue engineering, many kinds of tissues such as bone, heart valves, cartilage, corneal epithelium, cardiac muscles and skin have been successfully reconstructed (Yang *et al*, 2005; Nishida *et al*, 2004). Most anchorage dependent mammalian cells must adhere to a solid substrate in order to proliferate and manifest their function. The substrates used for cell growth should possess many features like the ability to encourage cell adhesion, spreading and proliferation. Also it is essential that the cells are easily detached from the substrate after it has grown. Conventionally enzymatic proteolysis of extracellular matrix (ECM) with trypsin or other proteolytic enzymes used to detach cells from substrates which can affect cell re-adhesion, cell viability etc (Kumashiro *et al*, 2010). Thermoresponsive polymers are of great interest in tissue engineering applications specifically in cell sheet detachment. The effect of these stimuli responsive polymers was first studied by Heskins and Guillet in early 1960's (Heskins and Guillet, 1968). The smart behavior of these polymers is because of its change in molecular conformation in response to a temperature change in the surrounding environment. N-isopropylacrylamide [NIPAAm] is a well characterized thermo responsive polymer that shifts its phase properties above and below its lower critical solution temperature (LCST) of 32 °C. Above the LCST, polymer will be hydrophobic resulting in the attachment and growth of cells while below the LCST it turns hydrophilic enabling the detachment of cells as a single cell or as a uniform cell sheet. Since the cell sheet could keep the cell junctions and deposited extra cellular matrix along with it, physiological functions and bioactivity of the cell sheet are similar to the cells in tissue, which could provide potential application as an engineered tissue. Polymer and copolymers with NIPAAm has been proved to be excellent for cell detachment by temperature variation. The group of Okano and coworkers done a lot of studies in thermoresponsive, Poly(NIPPAm) based polymers and proposed a new approach in tissue engineering called cell sheet engineering (Yamada *et al*, 1990).

3. Mechanism of cell attachment and detachment from the thermoresponsive surfaces

Cells initially attach to a surface by a process called passive adhesion in which the adherence is due to the physicochemical interaction such as hydrophobic interactions, coulomb forces and van der Waal's forces. Then the cells bind to a specific site on adhesive protein by a process called active adhesion via a receptor protein called integrin. Conventionally enzymes are used to detach cells from substrates, resulting in the destruction of cell membrane proteins and ECM which can affect cell re-adhesion and viability. But cell detachment from thermoresponsive surfaces retain their cellular structure and functions in which cell detachment occurs in a two step mechanism via a passive step, followed by an active step. In the passive step cell detachment is induced by the hydration of PNIPAAm chains on the substrate whereas in active step change of cell shape and detachment from the surface is driven by cytoskeletal action and metabolic process (Okano *et al*, 1995).

4. Methods of preparation of temperature responsive intelligent surfaces

Thermoresponsive polymers can be grafted on different substrates such as silicon, glass, quartz, polyethyleneterephthalate sheets etc but commonly grafted on Tissue Culture grade Poly Styrene (TCPS). The choice of substrate, method of application of thermoresponsive polymer on the substrate and thickness of the grafted polymer has a major role on cell attachment and detachment phenomena (Elloumi-Hannachi *et al*, 2010). It is reported that a grafting thickness of 15 to 20 nm is a prerequisite for efficient cell attachment and detachment (Akiyama *et al*, 2004).

Many types of polymerization techniques can be used for grafting the thermoresponsive surfaces such as Electron beam irradiation, Gamma radiation, Plasma polymerization, UV irradiation and Atom transfer radical polymerization.

4.1 Electron beam induced polymerization

Electron beam polymerization (EB) is a popular method used for producing thermoresponsive surfaces for cell sheet engineering (Yamato *et alj*, 2010). NIPAAm monomer is covalently bonded onto the TCPS (tissue culture polystyrene) dishes by the irradiation of electron beam on to the monomer. Electron beam method facilitates even, thin grafting and the large scale production of temperature responsive culture dishes even though it is expensive. The thickness is controlled by monomer concentration and radiation energy (Nagase *et al*, 2009).

4.2 Gamma irradiation

Gamma radiation is another method for the grafting of thermoresponsive surfaces on TCPS (Akiyama *et al*, 2004; Anil Kumar *et al*, 2007). This study also revealed that grafting by gamma irradiation is comparable with conventional electron beam irradiation and facilitates batch processing of polymerization and grafting of PNIPAAm. Moreover it avoids most sophisticated equipment like electron beam accelerator. The efficacy of the grafted surface was confirmed by the successful growth of different cell lines such as L-929 (mouse subcutaneous connective tissue fibroblast), NRK-49F (normal rat kidney), SIRC (rabbit corneal epithelium), and HOS (human osteosarcoma) and detachment of the cell sheets while maintaining the cell-cell and cell-extra cellular matrix contact.

4.3 Plasma polymerisatiom

Plasma polymerization is a different single step method which facilitates the preparation of thermoresponsive coatings on a solid substrate. A plasma glow discharge of NIPAAm monomer vapour was used to deposit PNIPAAm onto solid surfaces such as silicon, glass or TCPS. The cell adhesion/detachment studies showed to be insensitive to the grafted layer thickness. But this method is not suitable for large scale production due to the difficulties related with continuous treatment and size (Canavan *et al*, 2005).

4.4 UV irradiation

UV irradiation is a yet another approach for the grafting of thermoresponsive coatings on to the solid substrate. Photopolymerisation and photografting of PNIPAAm is done on TCPS by UV irradiation (365 nm for 5 to 30 min) using photo initiators such as benzophenone (Nagase *et al*, 2009).

4.5 Atom transfer radical polymerization technique (ATRP)

ATRP facilitates the preparation of surfaces with dense polymer brushes from surface immobilized ATRP initiators (Mizutani *et al*, 2008). Preparation of PNIPAAm brushes on poly(4 vinyl benzyl chloride) coated TCPS surface using ATRP and application of these grafted surfaces to thermoresponsive cell culture substrates were studied by Mizutami etal. Attachment/ detachment process is favorable only in thinner surface less than 30 nm.

4.6 Reversible addition fragmentation chain transfer polymerization technique (RAFT)

RAFT polymerization technique allows the preparation of thermoresponsive PNIPAAm brushes with controlled chain length and graft density for cell sheet harvest. Chain length and density of PNIPAAm grafted on the surfaces have a significant effect on the thermoresponsive cellular behavior.

4.7 Oxygen plasma treated PIPAAm surface

Oxygen plasma treated PNIPAAm film is a promising material as a cell culture substrate in cell sheet engineering (Shimizu *et al*, 2010).A cast and dried film of thermoresponsive

PNIPAAm was fabricated and treated with high intensity oxygen plasma. This novel method of cell sheet has mechanical strength with high stability and low cost.

4.8 Solution casting method

Most of the above techniques are expensive and might not be reachable to all researchers. An auxiliary simple and cost effective approach is the coating of thermoreponsive polymer onto the TCPS by solution casting method. Thickness of coating obtained by this method is reported to be in micron level and astonishingly doesn't affect the cell attachment and proved to be efficient for cell sheet attachment and detachment process (Joseph *et al,* 2010; Varghese *et al*, 2010).

5. Different methods of characterization of thermoresponsive substrates

Different characterization techniques are employed for the physicochemical and biological evaluation of thermoresponsive substrates which is very important. Physicochemical characterization techniques such as Attenuated total reflectance Fourier Transform spectroscopy (ATR-FTIR), Nuclear magnetic Resonance spectroscopy (NMR) are used for the qualitative and quantitative detection, while Atomic Force Microscopy (AFM), Ellipsometry, surface Plasmon resonance, Profilometry are used for the determination of thickness of the grafting or coating.

The presence and quantity of PNIPAAm on the thermoresponsive surface can be determined by the technique ATR-FTIR (Akiyama *et al*, 2007). The presence of characteristic peak of amide carbonyl group around 1650 cm^{-1} showed the presence of NIPAAm. The amount of PNIPAAm on the surface of TCPS significantly affects cell adhesion behaviour. The quantity of grafted PNIPAAm could be detected by the ratio of peak intensities (I_{1650}/I_{1600}) in ATR-FTIR.

NMR technology also provides quantitative as well as qualitative assessment of thermoresponsive surface (Joseph *et al,* 2010).This techniques provides detailed structural information about the polymers and gives a better understanding about the polymerization mechanism.

Water contact angle measures the hydrophobicity and hydrophilicity of the surface and a contact angle of 70° proved to be optimal for cell sheet adhesion and detachment (da Silva *et al,* 2007). The thermoresponsive surface wettability changes of free end linear and multipoint attached PNIPAAm surfaces were investigated using the Wilhelmy plate technique. A large contact angle change was observed for an end grafted PNIPAAm surfaces whereas a small contact angle change for multipoint attached surfaces due to the restricted chain conformation of the multipoint attached grafted surfaces.

Differential scanning calorimetry (DSC) is widely used for the detection of LCST. The LCST of homopolymer PNIPAAm lies around 32 °C and copolymers of PNIPAAm centered below or above 32°C depending upon the hydrophobicity and hydrophilicity of the comonomer.

Xray Photoelectron Spectroscopy (XPS) is one of the efficient methods for the qualitative assessment of the presence of PNIPAAm on different substrates (Jun and Okano, 2010)The elemental composition of carbon, nitrogen and oxygen on the grafted thermoresponsive surface, could be determined and should be in good agreement with the values based on the stochiometry of the monomer, NIPAAm.

AFM can be used for the quantitative and qualitative characterization of thermoresponsive surfaces such as to observe the nano texture of the surface and to measure surface roughness and the thickness of the PNIPAAm coating (Jun and Okano, 2010). Profilometry is another method to visualize the surface morphology and the quantitative analysis of surface roughness and thickness from the surface profile measurement (Joseph *et al,* 2010). Determination of thickness from ellipsometry and surface plasmon resonance is very difficult because of the similarity in refractive index of both PNIPAAm and polymeric substrates (Nagase *et al,* 2009).

Biological evaluations of thermoresponsive surface is the most important characterization technique for cell sheet engineering and are done by cytotoxicity analysis, cell activity measurement, thermoresponsive efficacy, viability analysis and cell sheet characterization etc. The cytocompatibility, specific cytocompatibility, thermoresponsive efficacy studies are usually done using L929 cells, SIRC cells, HOS, NRK 49 F,Human fetal lung fibroblasts (TIG-1), primary cells from rabbit cornea, rat primary hepatocytes etc (Anil Kumar *et al,* 2007; Joseph *et al,* 2010; Varghese *et al,* 2010). Cells were maintained in MEM (Minimum essential medium) supplemented with 10% FBS (fetal bovine serum), 100 IU/mL pencilin and 100 μg/ml streptomycin at 37° C in a 95 % humidified atmosphere with 5 % CO_2. The cytotoxicity analysis can be done by using direct contact method, indirect contact method;.High density polyethylene and zinc diethyl dithiocarbamate stabilized polyvinylchloride discs are usually used as negative and positive controls respectively. The cytotoxic evaluation is done by the comparison of the morphology of cells with positive and negative controls. MTT {(3-(4,5-dimethyl thiazol -2-yl) -2,5-diphenyl tetrazolium bromide} cell culture assay is usually used to evaluate the cell proliferation. Absorbance is measured using a multiwell plate reader at 540 nm. Cell viability is usually determined using neutral red staining. The thermoresponsive efficacy of the substrate is evaluated by retrieving the cell sheets by lowering the temperature below the LCST of the polymer and the viability of the retrieved cell sheet is determined using FDA (fluorescein diacetate) staining and observed under a fluorescence microscope. The cell sheet characterization is done by immuno staining technique using antibodies against proteins characteristic to tissue of interest.

6. Various thermoresponsive substrates for cell sheet engineering

Thermoresponsive substrates designed for cell sheet engineering have mainly used PIPAAm and its copolymers for cell adhesion and detachment. The use of thermoresponsive substrate to detach the confluent cell sheets without the use of conventional enzymatic treatments was first reported by Takezawa et al in 1990 (Takezawa *et al, 1990*). In this study they have used PIPAAm as a substratum by conjugating it with collagen for the culture of human dermal fibroblasts. These fibroblasts monolayer were harvested at reduced temperature by the dissolution of the dish coating. By the same time Yamado *et al* reported the successful culture of bovine hepatocytes on PNIPAAm grafted TCPS (Yamada *et al, 1990*). Homopolymer of NIPAAm has been reported to be grafted by various methods such as Electron beam polymerization, UV irradiation, Gamma irradiation etc and proved to be an excellent substrate for cell sheet engineering.

Copolymerization of NIPAAm with hydrophobic or hydrophilic monomer can modulate LCST for the systematic regulation of cell attachment and detachment. Thermoresponsive copolymers based on N-isopropylacrylamide-Co-glycidylmethacrylate(NGMA), N-isopropylacrylamide-methylmethacrylate (NMMA), (Varghese *et al,* 2010) N-isopropylacrylamide-butylmethacrylate (Tsuda *et al,* 2004) and NIPAAm-MMA-phosphorylated HEMA (Thomas *et al,* 2010) were successfully used as a substrate for cell sheet harvesting. Thermoresponsive copolymer NGMA was clearly identified as potential substrate for cell culture harvesting system for generating 3D synthetic tissue. NGMA was synthesized by solution polymerization followed by the coating on TCPS surfaces by solution casting method. Copolymerisation of PIPAAm with the hydrophobic monomer glycidylmethacrylate results in a decrease of LCST. Thermoresponsive copolymer NMMA was also reported to be synthesized and coated by the same methodology mentioned above and proved to be an ideal thermoresponsive substrate with good cytocompatibility for cell culture and tissue reconstruction. Hydrophobic monomer n butylmethacrylate (BMA) has been copolymerized with PNIPAAm to reduce the LCST of NBMA inorder to modulate the cell adhesion/detachment from culture dishes and revealed to be an efficient substrate for cell sheet engineering. Poly (N-vinylcaprolactum-co- N- isopropylacrylamide) PNVCL-co-PNIPAAm grafted on TCPS dishes by EB irradiation were efficiently applied to detach viable cell sheets from culture surfaces (Lim et al, 2007).

The spontaneous recovery of cell sheet is important for maintaining the viability of retrieved cell sheets. PNIPAAm grafted onto porous membranes facilitate the acceleration of cell sheet detachment by providing rapid water movement between the interface of cell sheets and membrane surfaces (Kwon *et al,* 2000a). Another method for the acceleration of cell sheet detachment is by cografting of a hydrophilic polymer Polyethylene glycol (PEG) with the thermoresponsive polymer PNIPAAm (Kwon *et al,* 2000b). Copolymerisation with 2-carboxyisopropylacrylamide (CIPAAm), {P(IPAAm-co-CIPAAm)} grafted surface accelerates the cell detachment process because of the presence of hydrophilic carboxyl group (Ebara *et al,* 2003).

7. Biomolecule immobilizing thermoresponsive substrates

Biomolecule immobilizing thermoresponsive cell culture dishes are known to be the next generation thermoresponsive substrates in cell sheet engineering (Jun and Okano, 2010). Regulation of cellular metabolisms and functions in culture are important aspects in cell sheet engineering. The simple approach to enhance the cellular activities are by immobilization or coating of bioactive compounds such as collagen, fibrin, elastin and arginine–glycine–aspartic acid (RGD) containing peptides on the cell culture substrate (Ebara *et al,* 2004). Anchoring of these bioactive compounds to this thermoresponsive substrate through chemical bonds is expected to increase the stability of the modified surface to a prolonged period. The interactions of these biomolecule ligands and cell membrane receptors have an important role in cell spreading, proliferation, differentiation, and signal transduction. This can be achieved either through coating or by the use of coupling agents or else by copolymerizing with an analogue of Isopropylacrylamide.

Cell adherence and growth on the thermoresponsive substrate is improved by coating cell adhesion promoters such as collagen, laminin and fibrin (Moran *et al,* 2007). But repeated washing will remove the coating of biomolecules. Immobilization is the process which helps to increase the stability and make possible their repeated or continued use.

Copolymerizing with an analogue of isopropylacrylamide, 2-carboxyisopropylacrylamide (CIPAAm) which has a carboxylate side chain that favours the immobilization of bioactive molecules onto thermoresponsive substrates. The synthetic cell adhesive peptide Arg-Gly-Asp (RGD) found in extracellular matrix proteins has been successfully immobilized onto the grafted polymer chains via the above mentioned method using the coupling agents 1-Ethyl-3-[3-dimethylaminopropyl]carbodiimide hydrochloride [EDC]. The cells adhered and spread well on these surface even in the absence of serum. Another method of immobilization is by the coupling of biotinylated biomolecules onto streptavidin immobilized thermoresponsive substrate by affinity binding without any coupling agents. The co-immobilisation of cell adhesive peptide RGDs and cell growth factor insulin facilitated the cell adhesion and cell proliferation respectively.

The epoxy ring opening of GMA moiety is well known for incorporating diverse type of biomolecules. The uniqueness of the thermoresponsive substrate such as N-isopropylacrylamide-co-glycidylmethacrylate is that, it facilitates the immobilization of biomolecules without any coupling agents by the presence of unreacted epoxy rings.

The biomolecule immobilized surfaces facilitates cell adhesion, culture and harvest of cell sheets which enables serum free cell culture and enzyme free cell harvesting.

8. Supports for the transfer of cell sheet

A cell sheet carrier is a support which facilitates the transfer of retrieved cell sheets. An ideal material as a cell sheet carrier should possess the features like biocompatibility, surface pliability, mechanical strength, hydrophilicity, presence of surface reactive groups which enable easy modification etc. Polyvinylidene fluoride membranes in the form of annular ring, fibrin gel coated membranes are some of the systems which have been used till today. Polyion complex gels are further developed as cell sheet carriers (Tang *et al*, 2007) but possess limited mechanical strength and have application difficulties. A further innovative development in this field is the cell culture substrate with combined properties of thermoresponsiveness and cell sheet carrier tool. Modified over head projection transparency sheet based on polyethyleneterephthalate coated with thermoresponsive N-isopropylacrylamide-Co-glycidylmethacrylate is reported to be an efficient cell culture substrate as well as a cell sheet carrier tool (Joseph *et al*, 2010). The study demonstrated that a simple modification of commercially available PET sheets by hydrolysis followed by coating of thermoresponsive NGMA forms a biocompatible substrate for scaffold free cell sheets as well as a cell sheet carrier tool.

9. Clinical applications of cell sheet engineering

Cell sheet engineering favours the formation of scaffold free transplantable three dimensional tissues from thermoresponsive cell culture substrate. This technology has been used in corneal surface reconstruction, myocardial tissue reconstruction, regeneration of damaged periodontal tissue, treatment of oesophageal ulceration, treatment for type 1 diabetes etc (Elloumi-Hannachi *et al*, 2010). Some of the clinical applications reported till dates are reported below. This technology in corneal surface reconstruction has been successfully applied to the patients suffering from unilateral or bilateral corneal stem cell deficiencies due to alkali burns or Stevens Johnson syndrome. These reconstructed corneal surfaces are able to restore the vision. Cell sheet technology for the treatment of oesophageal ulceration enhances the wound healing and reduces the host inflammatory responses. This is reported to be the first clinical case in the world which is applied to endoscopy with successful wound healing and improved post operative quality of life. Transplantation using cell sheets derived from periodontal ligament is reported to be useful for the regeneration of damaged periodontal tissue. Cell sheet technology using pancreatic islets has been used for the treatment of type1 diabetes. Another clinical application of cell sheet engineering is its use in the treatment for dilated cardiomyopathy and cardiovascular tissue repair by myocardial tissue reconstruction. This smart culture surfaces have already been used in clinical trials and thus it offers a new frontier in the biomedical field.

10. References

Abraham, T. N., V. Raj, T. Prasad, P. R. Anil Kumar, K. Sreenivasan and T. V. Kumary (2010). A novel thermoresponsive graft copolymer containing phosphorylated HEMA for generating detachable cell layers. *Journal of Applied Polymer Science* 115(1): 52-62.

Akiyama, Y., A. Kikuchi, M. Yamato and T. Okano (2004). Ultrathin poly(N-isopropylacrylamide) grafted layer on polystyrene surfaces for cell adhesion/detachment control. *Langmuir* 20(13): 5506-5511.

Akiyama, Y., A. Kushida, M. Yamato, A. Kikuchi and T. Okano (2007). Surface characterization of poly(N-isopropylacrylamide) grafted tissue culture polystyrene by electron beam irradiation, using atomic force microscopy, and X-ray photoelectron spectroscopy. *J Nanosci Nanotechnol* 7(3): 796-802.

Anil Kumar, P. R., K. Sreenivasan and T. V. Kumary (2007). Alternate method for grafting thermoresponsive polymer for transferring in vitro cell sheet structures. *Journal of Applied Polymer Science* 105(4): 2245-2251.

Canavan, H. E., X. Cheng, D. J. Graham, B. D. Ratner and D. G. Castner (2005). Cell sheet detachment affects the extracellular matrix: a surface science study comparing thermal liftoff, enzymatic, and mechanical methods. *J Biomed Mater Res* A 75(1): 1-13.

da Silva, R. M., J. F. Mano and R. L. Reis (2007). Smart thermoresponsive coatings and surfaces for tissue engineering: switching cell-material boundaries. *Trends Biotechnol* 25(12): 577-583.

Ebara, M., M. Yamato, M. Hirose, T. Aoyagi, A. Kikuchi, K. Sakai and T. Okano (2003). Copolymerization of 2-carboxyisopropylacrylamide with N-isopropylacrylamide accelerates cell detachment from grafted surfaces by reducing temperature. *Biomacromolecules* 4(2): 344-349.

Ebara, M., M. Yamato, T. Aoyagi, A. Kikuchi, K. Sakai and T. Okano (2004). Immobilization of cell-adhesive peptides to temperature-responsive surfaces facilitates both serum-free cell adhesion and noninvasive cell harvest. *Tissue Eng* 10(7-8): 1125-1135.

Elloumi-Hannachi, I., M. Yamato and T. Okano (2010). Cell sheet engineering: a unique nanotechnology for scaffold-free tissue reconstruction with clinical applications in regenerative medicine. *J Intern Med* 267(1): 54-70.

Heskins, M. and J. E. Guillet (1968). Solution Properties of Poly(N-isopropylacrylamide). *J. Macromol. Sci. Chem.* 2(8): 1441-1455.

Hyeong Kwon, O., A. Kikuchi, M. Yamato and T. Okano (2003). Accelerated cell sheet recovery by co-grafting of PEG with PIPAAm onto porous cell culture membranes. *Biomaterials* 24(7): 1223-1232.

Joseph, N., Anil Kumar P.R., T. Prasad, L. Joseph, S. K and K. T.V. (2010). Intelligent thermoresponsive substrate from modified overhead projection sheet as a tool for construction and support of cell sheets in vitro. *Tissue Engineering:* Part C 17(2): 181-191.

Joseph, N., T. Prasad, V. Raj, P. R. Anil Kumar, K. Sreenivasan and T. V. Kumary (2010). A Cytocompatible Poly(N-isopropylacrylamide-co-glycidylmethacrylate) Coated Surface as New Substrate for Corneal Tissue Engineering. *Journal of Bioactive and Compatible Polymers* 25(1): 58-74.

Jun, K. and O. Teruo (2010). Fabrication of a thermoresponsive cell culture dish: a key technology for cell sheet tissue engineering. *Science and Technology of Advanced Materials* 11(1): 014111.

Kumashiro, Y., M. Yamato and T. Okano (2010). Cell attachment-detachment control on temperature-responsive thin surfaces for novel tissue engineering. *Ann Biomed Eng* 38(6): 1977-1988.

Kwon, O. H., A. Kikuchi, M. Yamato, Y. Sakurai and T. Okano (2000). Rapid cell sheet detachment from poly(N-isopropylacrylamide)-grafted porous cell culture membranes. *J Biomed Mater Res* 50(1): 82-89.

Langer, R. and J. P. Vacanti (1993). Tissue engineering. *Science* 260(5110): 920-926.

Lim, Y. M., J. P. Jeun, J. H. Lee, Y. M. Lee and Y. C. Nho (2007). Cell Sheet Detachment from Poly(N-vinylcaprolactam-co-N-isopropylacrylamide) Grafted onto Tissue Culture Polystyrene Dishes. *Journal of Industrial and Engineering Chemistry* 13(1): 21-26.

Mizutani, A., A. Kikuchi, M. Yamato, H. Kanazawa and T. Okano (2008). Preparation of thermoresponsive polymer brush surfaces and their interaction with cells. *Biomaterials* 29(13): 2073-2081.

Moran, M. T., W. M. Carroll, I. Selezneva, A. Gorelov and Y. Rochev (2007). Cell growth and detachment from protein-coated PNIPAAm-based copolymers. *J Biomed Mater Res* A 81(4): 870-876.

Nagase, K., J. Kobayashi and T. Okano (2009). Temperature-responsive intelligent interfaces for biomolecular separation and cell sheet engineering. *J R Soc Interface* 6 Suppl 3: S293-309.

Nishida, K., M. Yamato, Y. Hayashida, K. Watanabe, K. Yamamoto, E. Adachi, S. Nagai, A. Kikuchi, N. Maeda, H. Watanabe, T. Okano and Y. Tano (2004). Corneal reconstruction with tissue-engineered cell sheets composed of autologous oral mucosal epithelium. *N Engl J Med* 351(12): 1187-1196.

Okano, T., N. Yamada, M. Okuhara, H. Sakai and Y. Sakurai (1995). Mechanism of cell detachment from temperature-modulated, hydrophilic-hydrophobic polymer surfaces. *Biomaterials* 16(4): 297-303.

Shimizu, K., H. Fujita and E. Nagamori (2010). Oxygen plasma-treated thermoresponsive polymer surfaces for cell sheet engineering. *Biotechnol Bioeng* 106(2): 303-310.

Takezawa, T., Y. Mori and K. Yoshizato (1990). Cell Culture on a Thermo-Responsive Polymer Surface. *Nat Biotech* 8(9): 854-856.

Tang, Z., A. Kikuchi, Y. Akiyama and T. Okano (2007). Novel cell sheet carriers using polyion complex gel modified membranes for tissue engineering technology for cell sheet manipulation and transplantation. *Reactive and Functional Polymers* 67(11): 1388-1397.

Tsuda, Y., A. Kikuchi, M. Yamato, Y. Sakurai, M. Umezu and T. Okano (2004). Control of cell adhesion and detachment using temperature and thermoresponsive copolymer grafted culture surfaces. *J Biomed Mater* Res A 69(1): 70-78.

Varghese, V., V. Raj, K. Sreenivasan and T. Kumary (2010). In vitro cytocompatibility evaluation of a thermoresponsive NIPAAm-MMA copolymeric surface using L929 cells. *Journal of Materials Science: Materials in Medicine* 21(5): 1631-1639.

Yamada, N., T. Okano, H. Sakai, F. Karikusa, Y. Sawasaki and Y. Sakurai (1990). Thermo-responsive polymeric surfaces; control of attachment and detachment of cultured cells. *Die Makromolekulare Chemie, Rapid Communications* 11(11): 571-576.

Yamato, M., O. H. Kwon, M. Hirose, A. Kikuchi and T. Okano (2001). Novel patterned cell coculture utilizing thermally responsive grafted polymer surfaces. *J Biomed Mater Res* 55(1): 137-140.

Yang, J., M. Yamato, C. Kohno, A. Nishimoto, H. Sekine, F. Fukai and T. Okano (2005). Cell sheet engineering: recreating tissues without biodegradable scaffolds. *Biomaterials* 26(33): 6415-6422.

24

Gene-Silencing for Treatment of Cardiovascular Diseases

A. Nolte, M. Schneider, T. Walker and H. P. Wendel
Clinical Research Laboratory, Department of Congenital and Paediatric Cardiac Surgery, University Children´s Hospital Tuebingen, Tuebingen
Germany

1. Introduction

In the last decade, great advances in RNA biology have been achieved. Micro- and short-interfering RNAs were once thought to be degradation products of larger RNA molecules. With the knowledge of today we know that they represent independent classes of small noncoding (~ 20-30 nt) RNAs regulating various cellular processes across the eukaryotes. The first miRNA lin4 was discovered in the Nematode *C. elegans* in the year 1993, when Lee et al. [1] demonstrated that this transcript works as an endogenous regulator of genes that control developmental timing. Five years later Fire et al. [2] showed the ability of exogenous double stranded RNA to targeted posttranscriptional gene silencing, which recognizes each target by Watson-Crick base pairing. This scientific breakthrough was awarded with the novel price 2007.

The RNA interference (RNAi) is a naturally happening catalytic process, offering the possibility to silence every pathological interesting gene which is defective expressed in a given disease. Thus it will give new approaches in the development and applications of siRNA-based therapeutics.

Since cardiovascular diseases (CVDs) are the main cause of mortality and morbidity in the Western world this chapter will discuss possible applications of si- and miRNA-based therapeutics and delivery systems with special regard to atherosclerosis, ischemic heart-disease and hypertension.

2. Molecular mechanism of RNAi

2.1 siRNA pathway

SiRNAs are ~ 20-25 nt regulatory molecules that play a primitive role in the defence against any foreign nucleic acid molecule derived from viruses or tranposons to preserve genome integrity. This hypothesis is supported by a study that demonstrates the passive uptake of siRNAs trough a membrane receptor protein called Systemic RNA Interference - Defective 1 (SID-1) [3].

RNAi is induced by linear, long perfectly complementary dsRNA which is directly introduced in the cytoplasm or is taken up by the environment [4]. First the dsRNA is processed by the RNase III-type endonuclease Dicer in 21-23 bp long siRNA duplexes that have 3´overhangs and 5`phosphates. In general Dicer posseses six domains including DExH

Helicase, DUF283, PAZ, RNAse IIIa, RNAse IIIb and RNA Binding Domain (RBD). The PAZ domain (Piwi, Argonaute, Zwille) binds to the 3'-nt overhangs of the cleaved RNA substrate while RBD recognizes duplex structures of RNA [5]. The next step is the incorporation of the duplex siRNAs into the nuclease-containing multiprotein complex RISC (RNA-induced silencing complex). The duplex is then unwound by the helicase activity (PAZ-domain) of the nuclease Ago2 which activates RISC. The unwinding process starts from the end of the siRNA with lower thermal energy. One strand called guide strand remains within the RISC, while the passenger strand is degraded by exonucleases [6]. The 5`end of the guide strand contains the seed region between the ribonucleotides at position 2-7 and are responsible for governing the RISC to its binding to target sequences [7]. Then the guide strand activates the RNase activity (PIWI domain) of Ago 2 which cleaves the complementary target mRNA. As result of these two unprotected mRNA ends are built, that are fast degraded by intracellular nucleases. This gene silencing by mRNA cleavage is an effective and catalytic process due to the fact that the activated RISC is freed after one round and is ready for further destruction of target mRNAs resulting in target repression [8] (See also Figure 1).

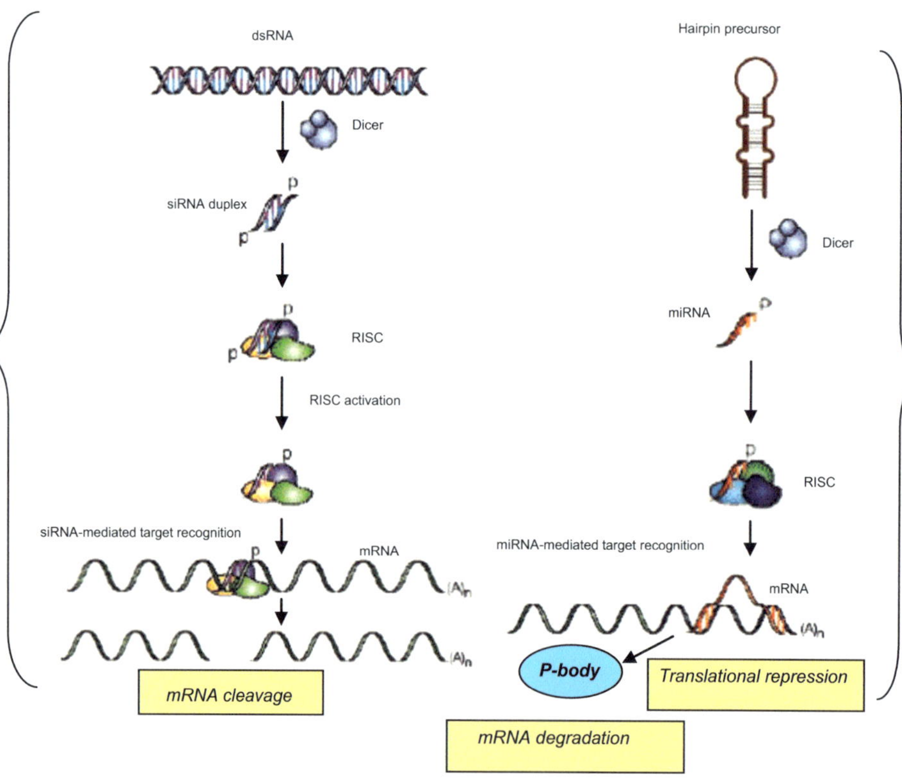

Fig. 1. Mechanism of RNA interference (RNAi) in mammalian systems

2.2 miRNA pathway

MiRNAs are ~22 nt long single-stranded RNAs with endogenous distinct that serve as regulators for proteins that repress gene expression [9]. Initially their precursors (pri-

miRNAs) are transcribed by the RNA polymerase II as single-stranded transcript that contain a local hairpin structure and also possess 5`cap and poly-A tail [10]. Secondly the pri-miRNAs are processed by the Drosha/Pasha complex (RNAse III) and they are excised as stem-loops (pre-miRNAs). Some pri-miRNAs are originated from introns and are called mitrons. They are generated by the pre-miRNA splicing machinery rather than by Drosha [11]. The pre-miRNA are then transported out of the nucleus in the cytoplasm and this transport is mediated by Exportin 5 and Ran-GTPase [12]. Subsequently the pre-miRNA is processed and shortened by Dicer to generate imperfectly-matched ds miRNA. Then the duplex miRNA is loaded on the Ago2-RISC complex and in case of imperfect sequence complementary the passenger strand is unwound by Ago2 producing a mature miRNA that is bound to active RISC.
The repression of the mature miRNA to genetic processes is made on two different levels: it recognizes and binds the 3`UTR of target mRNAs blocking translation or this binding results in mRNA degradation in special processing bodies, termed p-bodies [13, 14] (See also Figure 1).

3. Delivery systems and designs of siRNA- and miRNA-based therapeutics

According to its large molecular weight (~13 kDa) and high content of anionic charge (~40 negative phosphate charge), unmodified naked siRNA does not freely enter cell membranes. Therefore delivery systems are needed to give an access for the nucleic acids to its intracellular sites of actions since all other contributing parts of the RNAi based machinery are provided by the target cell / target tissue [39, 75]. Another critical hurdle to overcome is the low biological stability of nucleic acids under physiological conditions, which is the topic of the following passage.

3.1 Chemical modifications of siRNA: improving biological stability

Both single-stranded nucleic acids and double-stranded nucleic acids are relatively fast degraded trough nuclease attack if they are applied in unmodified and naked form. These finding indicates that they have short half-lives in blood and serum *in vivo* because of the activity of endo-and exonucleases [15]. To avoid the rapidly degradation there have been made efforts in the chemical modifications of siRNAs including strategies that improve the cellular uptake due to the conjugation with cholesterol, peptides, aptamers or antibodies.
These chemical modifications mainly focus on the backbone, base or sugar of the RNA. One strategy is to modify the 2`-position of the ribose enhancing duplex stability (Tm) and nuclease resistance of the phosphodiester bond between the nucleotides against hydrolysis. The 2`O-Methyl-modification (2`OMe) is a naturally found RNA which is untoxic but can implement the potency of the siRNA. To undergo this effect an alternation of 2`OMe with RNA or other 2`modifications can be used so that siRNA function and nuclease stability is preserved [16,17]. Another well studied 2`modification is the 2`fluoro (2`-F) modification at pyrimidine positions which show compliance with the siRNA potency and stabilisation of the duplex against nuclease degradation in vitro [18] and in vivo [19]. Morrissey et al [20] demonstrated that a combination of 2`OMe purines with 2`-F pyrimidines can generate RNA duplexes that are highly stable in serum and have good in vivo performance. The modification of internucleotide phosphate linkage through the replacement of the non-bridging oxygen with sulfur is also an efficient approach to

improve nuclease stability without reducing the potency of siRNA. This was revealed through several studies [21-23].
Furthermore the covalent conjugation of siRNA with various chemical groups can direct cell uptake and alter biodistribution. The attachment of hydrophobic ligands like chosterol is a prominent example to extend serum lifetime. Soutschek et al. demonstrated in a mice model that i.v. administration of anti-ApoB siRNA that contained two 2`OMe RNA residues, phosphorthioate 3`-end modification and a cholesterol group resulted in silencing of the apolipoprotein B gene and reduced total cholesterol and plasma levels of apoB protein [24]. These data suggest that chemical modifications are a useful tool to improve stability and cellular uptake of siRNAs in general.

3.2 Off-targets: activation of the innate immune system

In the design and selection of a siRNA sequence their effect on innate immune system has also to be considered. It have been shown that the introduction of too long ds RNA molecules (>30 nt) can initiate an antiviral IFN-response of the cell or can activate a cellular pathway that involves the serin/threonine protein kinase PKR [25, 26]. A suppression of these effects could be the use of smaller (21-23 nt) siRNAs and additional in vitro tests or BLAST search of desired siRNA sequences to avoid further immune responses. SiRNAs have also the potential to interfere with the innate immune system due to the activation of Toll-like receptors (TLRs). Especially TLR 3, 7 and 8 are expressed in endosomal compartments and exposure to these receptors is affected by strength of the entry of synthetic RNA enters the cell. TLR 3 recognizes motifs in ssRNA and TLR 7 and 8 notice motifs in dsRNA. TLR 7 is binding favourable to GU-rich sequences and selecting of sequences that are not recognized by this receptor will help to solve this problem [13]. Moreover it has been demonstrated that the use of chemical modification like 2`OMe can prevent the detection of siRNAs by the immune system. The study of Judge et al. showed that incorporation of two or three 2`OMe residues in a siRNA duplex can be enough to evade immune detection [27]. Another category of off-targets are the cross-reactions between the si- and mi-RNA-pathway. If one of the siRNA strands have partly complementary regions with the 3`UTR of not targeted mRNA it can result in imperfect base-pairing of guide strand and "target" strand. This will end in translational repression or exonucleolytic degradation in a way similar like miRNA silencing [28]. This mismatch effect would be avoided if the seed-region of the guide strand between the 2 and 8 nucleotid position (5`end) is fully complementary with the target mRNA sequence. Thus homology screenings of seed regions of siRNA in the 3`UTR of all interfering and target genes would be useful [29].

3.3 In vivo delivery systems

Beyond the importance of effective design of siRNA to achieve specific and potent target gene silencing, the therapeutic use of siRNA need reliable delivery into tissues and target cells. Due to the fact that small RNA molecules like si- and miRNA have a strong anionic charge of their backbone consisting of phosphodiesters and they have a considerable molecular weight it is not possible that they can simply enter the cell membrane by passive diffusion. This lead to the development of two main methods of delivery including viral and nonviral strategies, which have been already successfully applied *in vitro* and *in vivo*. Although viral transfection in patients is regarded very critically, the next paragraph should give a small overview of this technique.

Viral delivery sytems	Nonviral delivery systems
• Viral vectors: • Adenovirus, adenovirus-associated virus, retrovirus (e.g. lentivirus)	• Systemic delivery *in vivo*: cationic delivery systems (e.g. PEI, Acetocollagen, Chitosan) cationic liposomal delivery systems (e.g. *in vitro*: Lipofectamin, e.g. *in vivo*: SNALPs) • Local delivery strategies *in vivo*

Table 1. Overview of following discussed delivery systems

3.3.1 Viral delivery systems

Viral vectors commonly consists of nucleic acids (shRNAs or miRNA mimics) that are incorporated in the backbone of the virus genome and are induced through a Pol II or Pol III promoter. They are surrounded by a viral capsid of proteins that usually interact with ECM-molecules to enter the cell by membrane fusion or receptor binding [31]. The gene transfer with viral vectors is a highly effective method because only one application is able to silence the target gene for a long time *in vivo* [32]. But despite the great potency of viral vectors for delivery, one has to consider that threats like host immune responses, potential oncogen and mutagen properties can occur. The common used viral vectors are adenovirus, adenovirus-associated virus (AAV) and retrovirus [31]. Especially lentiviral vectors, a class of retroviruses show great potential of infecting a wide variety of non-dividing and dividing cells, stable integration in host genome, and results in long term expression of the transgene [32]. This kind of vector are of great interest for the injection in non-dividing cells like neurons and are suitable in the therapy of brain specific diseases or expression of shRNAs targeting disease associated genes [33, 34]. A possible target of shsiRNA could be in the case of a neurodegenerative disease like alzheimer, the γ-secretase BACE 1 which generates the toxic amyloid ß-peptid. This was demonstrated in a Alzheimer mouse model with reduction of endogenous levels of BACE1 by the use of a lentiviral vector expressing shRNA [35]. Adenoviral vectors have also been used for delivery of a shRNA in the brain directed against the mRNA expressing the polyQ-harbouring SCA-1 encoding transcript of spinocerebellar ataxia type 1 [36].

Another field of treatment would be chronic viral infections like chronic hepapitis b that need long-term treatment. Kim et al. demonstrated the *in vitro* lentivirus-mediated delivery of shRNA against HBx can effectively suppress the replication of HBV and reduce HBV covalently closed circular DNA [37]. Nevertheless, the safety issue remains a parameter in the decision in the use of a viral vector.

3.3.2 Nonviral delivery systems

There have been made many investigations for the use of nonviral delivery systems, including formulations like cationic polymers, cationic lipids, antibodies, conjugates or naked siRNA. The first consideration of the choice of a delivery strategy must be if a local or systemic administration is needed for the intended disease target.

Systemic delivery strategies in vivo

Cationic delivery systems

All cationic vectors share the ability to form complexes with the polyanionic nucleic acids through electrostatic interactions. Another possibility is the formation of nanoparticles in which the nucleic acid is entrapped within the particle through ionic interactions. The advantage of these formulations is that the siRNAs are not reachable for nucleases until it reaches its destination through vesicular transport (endocytosis and/ or macropinocytosis). Nanoparticles and complexes must have a limited size of 100 nm to be taken up by cells and to suppress renal excretion [38].

Cationic polymer delivery systems of siRNA

Cationic polymers that have been applied include both natural (e.g., acetocollagen, chitosan) and synthetic (e.g, polyethylenimine). Polyetylenimine (PEI) is one of the best investigated polymers for nucleic acid delivery [39].

The polymer is linear or branched and has high cationic charge density.

The PEI-siRNA complexes are endocytosed in the cell through electrostatic interactions where it escapes the endosome through the "proton-sponge-effect".

PEI has the ability to accept protons which results in a buffering-effect in the vesicular system and a subsequent influx of Cl- and water leading to a osmotic release of the complexes in the cytoplasm [40]. PEI is available in different weights, but low-molecular PEI is favourable because high molecular weight is highly toxic [41]. The linear 22kDa jetPEI was successfully used in a antiviral therapy of guinea pigs against a lethal Ebola virus [42]. PEI-siRNA complexes have also been effective applied as a antiviral agent in a murine model of influenza infection [43]. PEI-siRNA complexes have also been efficient targeted against many proteins like HER2 [44], VEGF [45] and many more [39].

Acetocollagen is a natural polymer which is produced through pepsin treatment of type I collagen of calf dermis [46]. Acetocollagen/siRNA-mediated targeting have been demonstrated in the treatment of prostata and pancreatic cancer as well as virus replication [47, 48, 49]. Another well-tolerated natural polymer is chitosan that builds cationic complexes with siRNAs. Howard et al. showed effective RNAi delivery in epithelial cells in the lung of mice through the intrasal application of transgenic enhanced green fluorescent protein (EGFP) [50].

Cationic liposomal delivery systems

Cationic liposomes and lipoplexes have been successfully applied for in vitro and in vivo delivery of siRNA. Liposomes are vesicels that can improve drug delivery through their fusion with cell membranes and they consist of a soluable compartment enclosed in a bilayer of phosholipids in which polar drugs can be entrapped. Lipoplexes are particles that are built through a complex formation of nucleic acids and lipids. The advantage of lipid formulations is the increase of siRNA stability in serum and the protection to overcome renal excretion, but unfortunately cationic liposomes can interfere with proteins like lipoproteins or serum proteins which can result in off effects as aggregation. For in vitro use there are many commercial available formulations like Oligofectamine, Lipofectamine (Invitrogen), RNAifect (Qiagen). An example of liposomes is SNALP (stable nucleic acid lipid particles) that have been efficient used in the *in vivo* silencing of the apoB gene in mice and primates [51]. The silencing effect of only one intravenous dose of 2.5 mg per kg SNALP-formulated siRNA lasted for 11 days and showed a reduction of 90% of the apoB mRNA.

Local delivery strategies in vivo

Especially the local application of siRNA at site-specific delivery targets offers the possibility to reduce the doses of use as well as preventing the threat of systemic off targets.

Intraocular delivery

Up to now it is possible to treat wet-age related macular degeneration (AMD) with si-RNA based therapeutics. This disease affects blood vessels behind the retina that show a overgrowth which can cause a loss of vision. A human clinical study with complete status show the potency of VEGF-targeted siRNA Bevasiranib (Acuity Pharmaceuticals) to reduce neovascularisation and is also thought for the treatment of diabetic macular oedema (clinical status: complete). A mouse model in which Bevasiranib (Cand 5) was used also showed downregulation of VEGF after ocular injection [52]. The Drug Ranibizumap (Sirna-027/AGN211745) which was developed by the Merck-Sirna Therapeutics is also already under complete status in the treatment of AMD. Another siRNA called RTP-801i from Silence Therapeutics is also under clinical investigation for AMD and renal injury use.

Intratumoral delivery

Calando Pharmaceuticals has induced a now active (not recruiting) status of a clinical trial in which the siRNA CALLAA-01 is encapsulated in the polymer cyclodextrin to built particles that are linked with transferrin. This siRNA is directed against a subunit of the ribonucleotide reductase to stop the synthesis of DNA required for growth within solid tumors. This is the first clinical attempt of connecting a siRNA particle with a receptor that is highly expressed on cancer cells, like transferrin to facilitate the uptake. Another example is the Atu-027 siRNA (Silence Therapeutics) which is thought to be effective in the treatment of gastrointestinal cancer that has achieved the status recruiting.

Intranasal delivery

Intranasal application of formulated siRNA is resulting in effective gene silencing in the lung. This has great benefit in the treatment of pulmonary diseases.

Alnylam Pharmaceuticals have developed the first antiviral siRNA ALN-RSV01 which is directed against the respiratory syntical virus through the silencing of the nucleocapsid N-gene of the virus [75].

Other clinical trials are summed up in the following table

siRNA/ miRNA	Disease	Clinical Status
TD101	Pachyonychia congenita	Completed
PRO-040201	Hypercholesterolemia	Recruiting
Anti-miRNA SPC3649	Chronic Hepatitis C	Recruiting

Table 2. Clinical trials (a selection)

4. Therapeutical applications of si- and miRNAs with focus on CVDs

The treatment of CVDs with si- and mi-RNA based therapeutics is a very novel strategy with high potential for clinical applications.

4.1 Possible targets of atherosclerosis therapy

Atherosclerosis is a chronic, multifactorial, smoldering inflammatory disease of medium sized and large conductive arteries characterized by lipid-fuelled lesions. The major players in the development of the disease are endothelial cells, monocytes, leukocytes and intimal smooth muscle cells [53]. Atherosclerotic lesions start to develop under an intact but activated, dysfunctional and partly leaky endothelium. Through the sites of defective endothelium especially lipoprotein particles can accumulate in the vessel wall where they can undergo modifications like oxidations. OxLDL is highly proatherogenic and is generated by myeloperoxidase, 15-lipoxygenase (or 12/15-LO), and/ or nitric oxide synthase (NOS) [53]. Especially the 12-15-LO is associated with cellular growth, migration, adhesion, and inflammatory gene expression in monocytes/macrophages, endothelial cells and vascular smooth muscle cells (VSMC). Li et al. demonstrated that shRNA was able to knockdown 12/15-LO in mouse macrophages and also in rat and mouse VSMCs. The knockdown of 12/15-LO had also functional effects which was showed through reduced monocyte-chemoattractant protein-1 (MCP-1) expression in a differentiated mouse monocyte line as well as reduced cellular adhesion and fibronectin expression in VSMCs [54]. Making 12/15-LO an interesting target of atherosclerotic RNAi based prevention. Endothelium is activated through atherogenic stimuli like oxLDL and cytokines, that induces expression of adhesion molecules like VCAM-1, ICAM-1 and selectins like P- and E-selectin which have proven to be important in atherosclerotic lesion development [55, 56, 57]. These adhesion molecules can contribute to the recruitment of monocytes as well as leukocytes and VSMCS. During the development of atherosclerosis and neointimal hyperplasia a transformation of VSMCs in the media from a contractile into a synthetic phenotype is ongoing and after the arrival of VSMCs in the intima of the arterial wall they begin to generate extracellular matrix, resulting in the formation of intimal lesions. Petersen et al. isolated VSMCs from the aorta of C57BL/6 mice and transfected them with siRNA targeting VCAM-1. In a migration assay they revealed that with the treatment of VCAM-1 siRNA the number of migrated VSMCs was significant reduced [58]. These findings indicate that VCAM-1 is necessary for the migration of VSMCs in the intima and VCAM-1 siRNA expression is an eventually potent approach to prevent and treat atherosclerosis and restenosis. Another possible target of atherosclerotic therapy is the inhibition of leukocyte entry to lesion areas to stop inflammatory progression. Pluvinet et al demonstrated that efficient blockage of the CD40-CD154 signaling by RNAi-mediated silencing of human CD40 expression on vascular endothelial cells leads to inhibition of VCAM-1, ICAM-1 and E-selectin expression and to a concomitant reduction of leukocyte adherence on these cells [59]. This signalling way is very interesting because the CD40-CD154 interaction also triggers matrix metalloproteinase (MMP) expression and these enzymes destabilize lipid-rich core of plaques which can cause thrombosis. SMCs are also responsible for the fibrin cap production which contributes also to neointimal thickening after arterial injury which can lead to restenosis. The discoidin domain receptor 2 (DDR2) plays potential roles in the regulation of collagen turnover mediated by VSMCs in atherosclerosis. DDR 1 and DDR 2 are nonintegrin receptors of collagen [60]. Shyu et al applied in a balloon injured rat model the DDR2 siRNA for attenuating the neointimal formation and decreasing the MMP2 protein labeling *in vitro* and *in vivo* [61]. Another study also showed successful *in vivo* transplantation of coated stents with a cationized pullulan-based hydrogel loaded with MMP2 siRNA. San Juan et al demonstrated in this study an uptake of siRNA into the arterial wall and a decrease of pro-MMP2 activity [62]. This stent coating technology could be a auspicious therapeutic approach for prevention of restenosis.

4.2 Possible targets of hypertension therapy

Hypertension also represents a severe riskfactor for many chronic CVDs. The use of cationic liposomal gene delivery vectors like DOTAP could be one possible strategy to reduce blood pressure. This finding was recently confirmed by Arnold et al., who showed that a single dose of 1mg/kg i.v. of siRNA-DOTAP (N-[1-(2,3-dioleoyloxy)]-N-N-N trimethyl ammonium propane) lipoplexes lead to the selective inhibition of β1-adrenoreceptor expression and reduced blood pressure lasting for 12 days [63]. The reduced blood pressure was also not greater if the β1-adrenoreceptor were blocked by β-blockers.

Sun et al. also used RNAi to inhibit α1D-adrenerig receptor gene in rat VSMCs and the therapeutic implications of this treatment are of great interest, for the reason that pharmacologic blockage of α1-ARs is a commonly used treatment for hypertension [64]. A other area of gene targeting in hypertension are the blockage of receptors that are affected by potent vasoconstrictors like Angiotensin II (Ang II). Ang II exerts its physiological effects by activating multiple subtypes of its receptor such as AT1a-, AT1-b and AT2-receptors. It regulates diverse functions of the cardiovascular system as increases in blood pressure, extracellular fluid volume, hormone secretion, stimulation of sympathetic nerve activity, damping of baroreflexes and vascular and cardiac remodelling [65, 66]. Mostly the effects are mediated through AT1R. Vázquez et al. used AT1147siRNA to silence the AT1a receptor for up to 7 days, with decrease of Ang II binding to cells that were transfected [67]. Thus this target will be an interesting field in CVD therapy.

4.3 Possible targets of ischemic heart-disease (IHD) therapy

The therapeutic opportunities for the treatment of IHD range from medical therapy achieving to decrease myocardial oxygen consumption and increase coronary flow to revascularization. Revascularization is done through coronary artery bypass grafting (CABG) or percutaneous coronary angioplasty (PCI) in most cases with stent positioning [68].

Through hypoxia the hypoxia inducible factor-1 transcriptional factor (HIF1-TF) can activate several angiogenic genes. HIF1-TF is naturally degraded by prolyl hydroxylase-2 (PHD2). Huang et al. imaged the biological role of shRNA therapy for improving cardiac function. Inhibition of PH2D by shRNA led to significant improvement in angiogenesis and contractility by in vivo and in vitro experiments and this process can be followed by molecular imaging [69]. Markkanen et al. reviews also that combined use of growth-factors or of factors that are capable of up-regulating other factors (e.g. HIF-1) will improve collateral vessel growth in case of IHD [70]. The potential use of miRNAs could also be a strategy for IHD treatment. MiRNAs contribute in the regulation of angiogenesis. Especially miR-126 can regulate angiogenesis and vascular integrity due to two recent studies [71, 72]. MiR-126 was demonstrated to be one of the most expressed miRNAs in cardiovascular tissue. Wang et al. generated miR-126 null mice that showed decreased sprouting potential of endothelial cells in vitro and also defective response to angiogenic factors in vivo [72]. These mice also showed decreased vascularization of infarcted myocardium, stronger fibrosis and loss of functional myocardium and were more prone to cardiac rupture. The search of mir-126 targets identified Spred-1, which is an negative regulator of the MAP kinase pathway that signals the pathway of angiogenic regulators like VEGF [68].

Thus an upregulation of miR-126 could be effective in the repression of antiangiogenesis. Another study by Fasanaro et al. also revealed a hypoxia induced upregulation of miR-210 in endothelial cells [73]. They used endothelial cells that were cultured under oxygen deprivation, which were derived from human, umbilical veins. It was shown that the overexpression of miR-210 stimulated angiogenesis in mormoxic endothelial cells, while miR-210 blockage inhibited it [73, 68]. A target of miR-210 is Ephrin-A3 which is a regulator of angiogenesis and VEGF signaling.

5. Conclusions and future perspectives

Since CVDs are the leading cause of morbidity and mortality in the world, siRNA and miRNA-based therapeutics represent a new approach to treat CVDs. Due to the fact, that RNAi technology has shown a fast development from the research level to human clinical trials as an effective gene-silencing method since its discovery in 1998, it shows great advantages for use in routine clinical practice as an adjuvance of existing therapies. The challenging hurdle remains the delivery of nucleic acids *in vivo* including the improvement of biological properties such as delivery efficacy, cellular uptake and well pharmacokinetics to achieve drug release to the wished target cell type. But there have been already made great efforts in the establishment of siRNA delivery methods like cationic polymer-mediated siRNA delivery, targeted siRNA delivery and conjugation or chemical modifications of siRNAs. Thus on going research and pratical use will improve the safety issue in general making RNAi-based therapeutics a new class of drugs with high potential, especially in the treatment of a wide field of diseases.

6. References

[1] Lee RC, Feinbaum RL, Ambros V. The C. elegans heterochronic gene lin-4 encodes small RNAs with antisense complementarity lin-14. *Cell.* 1993; 75(5): 843-54.

[2] Fire A, Xu S, Montgomery M, et al. Potent and specific genetic interference by double-stranded RNA in Caenorhabditis elegans. *Nature.* 1998; 391(6669): 806-11.

[3] Feinberg EH, Hunter CP. Transport of dsRNA into cells by the transmembran protein SID-1. *Science.* 2003; 301(5639): 1545-7.

[4] Mello CC, Conte D Jr, Revealing the world of RNA interference. *Nature* 2004; 431: 338-42.

[5] Yan KS, Yan S, Farooq A, et al. Structure and conserved RNA binding of the PAZ domain. *Nature.* 2003; 426(6965): 468-74.

[6] Matranga C, Tomari Y, Shin C, Bartel DP, Zamore PD. Passenger-strand cleavage facilitates assembly of siRNA into Ago2-containing RNAi enzyme complexes. *Cell.* 2005; 123(4): 607-20.

[7] Lewis B, Burge C, Bartel D. Conserved seed pairing, often flanked by adenosines, indicates that thousands of human genes are microRNA targets. *Cell.* 2005; 120(1): 15-20.

[8] Hutvanger G, Zamore P.D. A micro RNA in a multiple-turnover RNAi enzyme complex. *Science.* 2002; 297: 2056-2060.

[9] Matranga C, Zamore PD. Small silencing RNAs. *Current Biology.* 2007; 17:R789-93.

[10] Bartel DP. MicroRNAs: genomics, biogenesis, mechanism, and function. *Cell.* 2004; 116(2): 281-97.

[11] Barringhaus KG, Zamore PD. miRNAs: Regulating a change of heart. *Circulation.* 2009; 119(16): 2217-24.

[12] Lund E, Guttinger S, Calado A, Dahlberg JE, Kutay U. Nuclear export of microRNA precursors. *Science*. 2004; 303(5654): 95-8.

[13] De Fougerolles A, Vornlocher HP, Maraganore J, Liebermann J. Interfering with disease: a progress report on siRNA-based therapeutics. *Nature Reviews Drug Disvovery*. 2007; 6:443-52.

[14] Lui J et al. MicroRNA-dependent localization of targeted mRNAs to mammalian P-bodies. *Nature Cell Biol*. 2005; 7:719-23.

[15] Dowler T, Bergeron D, Tedeschi AL, Paquet L, Ferrari N, Damha MJ. Improvements in siRNA properties mediated by 2'-deoxy-2'-fluoro-beta-D-arabinonucleic acid (FANA). *Nucleic Acids Res*. 2006; 34: 1669-75.

[16] Czauderba F, Fechtner M, Dames S, Aygun H, Klippel A., Pronk GJ , Giese K, Kaufmann J. Structural variations and stabilising modifi cations ofsynthetic siRNAs in mammalian cells. *Nucleic Acids Res*. 2003; 31:2705–16.

[17] Choung S , Kim YJ, Kim S, Park HO, Choi, YC.Chemical modification of siRNAs to improve serum stability withoutloss of efficacy. *Biochem. Biophys. Res. Commun*. 2006; 342:919–27.

[18] Kraynack BA, Baker BF. (2006). Small interfering RNAs containing full 2′-O-methylribonucleotide-modified sense strands display Argonaute2/eIF2C2-dependent activity. *RNA*. 2006; 12:163–76.

[19] Layzer JM, MCCaffrey AP, Tanner AK, Huang Z, Kay MA, Sullenger BA. In vivo activity of nuclease-resistant siRNAs. *RNA*. 2004; 10:766–71.

[20] Morrissey DV et al. Potent and persistent in vivo anti-HBV activity of chemically modified siRNAs. *Nat. Biotechnol*. 2005; 23:1002–07.

[21] Amarzguioui M, Holen T, Babaie E, Prydz H. Tolerance for mutations and chemical modifi cations in a siRNA. *Nucleic Acids Res*. 2003; 31:589–95.

[22] Braasch DA, Jensen S, Liu Y, Kaur K, Arar K, White MA, Corey DR. RNA interference in mammalian cells by chemically-modified RNA. *Biochemistry*. 2003; 42:7967–75.

[23] Li ZY, Mao H, Kallick DA, Gorenstein DG. The effects of thiophosphate substitutions on native siRNA gene silencing. Biochem. *Biophys. Res. Commun*. 2005; 329:1026–30.

[24] Soutschek J, Akinc A, Bramlage B, Charisse K, Constien R, Donoghue M, et al. Therapeutic silencing of an endogenous gene by systemic administration of modified siRNAs. *Nature*. 2004; 432:173-8.

[25] Sledz CA, Holko M, de Veer MJ, Silverman RH, Williams BR. Activation of the interferon system by short-interfering RNAs. *Nat Cell Biol* 2003; 5: 834-9.

[26] Schlee M et al. SiRNA and isRNA: two edges of one sword. *Mol. Ther*. 2006; 14: 463-70

[27] Judge AD, Bola G, Lee AC, Machlachlan I.Design of noninflammatory synthetic siRNA mediating potent gene silencing in vivo. *Mol. Ther*. 2006; 13:494–505.

[28] Carthew RW, Sontheimer EJ. Origins and mechanisms of miRNAs and siRNAs. *Cell*. 2009; 136(4): 642-655.

[29] Anderson EM , Birmingham A, Baskerville S, Reynold A , Maksimova E, Laeke D, Federov Y, Karpilow J, Hvorova A. Experimental validation of the importance of seed complement frequency to siRNA specificity. *RNA*. 2008; 14: 853–86.

[30] Koostra NA, Verma IM. Gene therapy with viral vectors. *Ann. Rev. Pharmacol. Toxicol*. 2003; 43: 413-39.

[31] Loser P et al. Advances in the development of nonhuman viral DNA-vectors for gene delivery. *Curr. Gene Ther*. 2002; 2(2): 161-71;

[32] Singer O, Verma IM. Application of lentiviral vectors for shRNA delivery and transgenesis. *Curr. Gene Ther.* 2008; 8(6): 483-88.

[33] Naldini L, Blomer U, Gallay P et al. In vivo gene delivery and stable transduction of nondividing cells by a lentiviral vector. *Science.* 1996; 272: 263-67.

[34] Marr RA, Rockenstein E, Mukherjee A, et al. Neprilysin gene transfer reduces human amyloid pathology in transgenic mice. *J Neurosci.* 2003; 23: 1992-96.

[35] Singer O, Marr RA, Rockenstein E, et al. Targeting BACE1 with siRNA ameliorates Alzheimer disease neuropathology in a transgenic model. *Nat Neurosci.* 2005; 8: 1343-49.

[36] Xia H, et al. RNAi supress polyglutamine-induced neurodegeneration in a model of spinocerebellar ataxia. *Nature Med.* 2004; 10: 816-820.

[37] Kim JW, et al. Inhibition of in vitro hepatitis B virus replication by lentivirus-mediated short-hairpin RNA against HBx. *Korean J Hepatol.* 2009; 15(1): 15-24.

[38] Li W, Szoka FC Jr. Lipid-based nanoparticles for nucleic acid delivery. *Pharm. Res.* 2007; 24: 438-49.

[39] Aigner A. Cellular delivery in vivo of siRNA-based therapeutics. *Current Pharmaceutical Design.* 2008; 14:1-17.

[40] Boussif O, et al. A versatile vector for gene and oligonucleotide transfer into cells in culture and in vivo: polyethylenimine. *Proc. Natl Acad. Sci. USA.* 1995; 92:7297–7301.

[41] Kichler, A. Gene transfer with modified polyethylenimine. *J.Gene Med.* 2004; 6:3-10.

[42] Geisbert, T. W. et al. Postexposure protection of guinea pigs against lethal ebola virus challenge is conferred by RNA interference. *J. Infect. Dis.* 2006;193:1650–1657.

[43] Ge Q, et al. Inhibition of Influenza virus production in virus-infected mice by RNA interference. *Proc. Natl Acad. Sci. USA.* 2004; 101: 8676-81.

[44] Urban-Klein B, Werth S, Abuharbeid S, Czubayko F, Aigner A.RNAi-mediated gene-targeting through systemic application ofpolyethylenimine (PEI)-complexed siRNA in vivo. *Gene Ther.* 2005; 12: 461-6.

[45] Kim SH, Jeong JH, Lee SH, Kim SW, Park TG. Local and systemic delivery of VEGF siRNA using polyelectrolyte complex micelles for effective treatment of cancer. *J Control Release* 2008.

[46] Ochiya T, Nagahara S, Sano A, Itoh H, Terada M. Biomaterials for gene delivery: atelocollagen-mediated controlled release of molecularmedicines. *Curr Gene Ther* 2001; 1: 31-52.

[47] Takei Y, Kadomatsu K, Yuzawa Y, Matsuo S, Muramatsu T. A small interfering RNA targeting vascular endothelial growth factor as cancer therapeutics. *Cancer Res* 2004; 64: 3365-70.

[48] Iwaki K, Shibata K, Ohta M, Endo Y, Uchida H, Tominaga M, et al. A small interfering RNA targeting proteinase-activated receptor-2 is effective in suppression of tumor growth in a Panc1 xenograft model. *Int J Cancer* 2007.

[49] Matoba T, Orba Y, Suzuki T, Makino Y, Shichinohe H, Kuroda S, et al. An siRNA against JC virus (JCV) agnoprotein inhibits JCV infection in JCV-producing cells inoculated in nude mice. *Neuropathology.* 2008; 28: 286-94.

[50] Howard, K. A. et al. RNA interference in vitro and in vivo using a novel chitosan/siRNA nanoparticlesystem. *Mol. Ther.* 2006; 14: 476–84.

[51] Zimmermann TS et al. RNAi-mediated gene silencing in non-human primates. *Nature.* 2006; 441:111-14.

[52] Shen J, et al. Suppression of ocular neovascularisation with siRNA targeting VEGF receptor 1. *Gene Ther.* 2006; 13: 225-234.
[53] Falk E. Pathogenesis of atheroclerosis. *J. Am. Coll. Cardiol.* 2006; 47: C7-C12.
[54] Li S et al. Effects of silencing leucocyte-type 12/15-lipoxygenase using short interfering RNAs. *Journal of Lipid Research.* 2005; 46: 220-29.
[55] Dong ZM, et al. The combined role of P- and E-selectins in atherosclerosis. *J Clin Invest.* 1998; 102: 145-52.
[56] Shih PT, Brennan ML, Vora DK, Territo MC, Strahl D, Elices MJ, et al. Blocking very late antigen-4 integrin decreases leukocyte entry and fatty streak formation in mice fed an atherogenic diet. *Circ Res.* 1999;84: 345–51.
[57] Collins RG, Velji R, Guevara NV, Hicks MJ, Chan L, Beaudet AL. P-Selectin or intercellular adhesion molecule (ICAM)-1 deficiency substantially protects against atherosclerosis in apolipoprotein E deficient mice. *J Exp Med.* 2000; 191: 189–94.
[58] Petersen EJ, et al. siRNA silencing reveals role of vascular adhesion molecule-1 in vascular smooth muscle cell migration. *Atherosclerosis.* 2008; 198 (2): 301-306.
[59] Pluvinet R, et al. RNAi-mediated silencing of CD40 prevents leukocyte adhesion on CD154-activated endothelial cells. *Blood.* 104(12): 3642-46.
[60] Vogel WF, et al. Sensing extracellular matrix: an update on discoidin domain receptor function. *Cellular Signaling.* 2006; 18:1108-16.
[61] Shyu KG, et al. RNA interference for discoidin domain receptor 2 attenuates neointimal formation in balloon injured rat carotid artery. *Ateriosler Thromb Vasc Biol.* 2008; 28: 1447-53.
[62] San Juan A, et al. Development of a functionlized polymer for stent coating in aterial delivery of small interfering RNA. *Biomacromolecules.* 2009.
[63] Arnold AS, et al. Specific beta1-adrenergic receptor silencing with small interfering RNA lowers high blood pressure and improves cardiac function in myocardial ischemia. *J. Hypertens.* 2007; 25: 197-205.
[64] Sun B, et al. Inhibition of α1-adrenergic receptor gene by RNAi in rat vascular smooth muscle cells and its effects on other adrenergic receptors. *Vascul Pharmacol.* 2007; 46(5): 367-72.
[65] Saavedra JM. Brain and pituitary angiotensin. *Endocr Rev.* 1992; 13: 329-380.
[66] Veerasingham SJ, et al. Brain renin-angiotensin system dysfunction in hypertension: recent advances and perpectives. *Br J Pharmacol.* 2003; 139: 191-202.
[67] Vázquez J et al. Selective silencing angiotensin receptor subtype 1a (AT1aR) by RNA interference. *Hypertension.* 2005; 45: 115-19.
[68] Silvestri P et al. MicroRNAs and ischemic heart disease: towards a better comprehension of pathogenesis, new diagnostic tools and new therapeutic targets. Recent Patents of Cardiovascular *Drug Discovery.* 2009; 4: 109-118.
[69] Huang M, et al. Short hairpin RNA interference therapy for ischemic heart disease. *Circulation.* 2008; 118: 226-33.
[70] Markkanen JE, et al. growth factor-induced therapeutic angiogenesis and ateriogenesis in the heart-gene therapy. *Cardiovascular Research.* 2005; 65:656-64.
[71] Fish JE, et al. miR-126 regulates the angiogenic signaling and vascular integrity. *Dev Cell.* 2008; 15: 272-284.
[72] Wang S, et al. The endothelial-specific microRNA miR-126 governs vascular integrity and angiogenesis. *Dev Cell.* 2008; 15:261-71.

[73] Fasanaro P, et al. MicroRNA-219 modulates endothelial cell response of hypoxia and inhibits the receptor tyrosine-kinase ligand Ephrin-A3. *J Biol Chem*. 2008; 283: 15878-83.

[74] Akhtar S, Benter IF. Nonviral delivery of synthetic siRNAs in vivo. *The Journal of Clinical investigations*. 2007; 117(12): 3623-32.

[75] Castanotto D, Rossi JJ. The promises and pitfalls of RNA-interference-based therapeutics. 2009; *Nature*. 457(7228): 426-33.

25

Hydrodynamic 3D Culture for Bone Tissue Engineering

Dajiang Du, Takashi Ushida and Katsuko S Furukawa
Department of Bioengineering, Department of Mechanical Engineering,
School of Engineering, University of Tokyo
Japan

1. Introduction

Bone tissue engineering provided a promising approach for treatment of large bone defects resulting from maladies such as birth defects, trauma, or tumor resection. In vitro culture of a porous scaffold seeded with osteoprogenitor cells may enhance its bone regeneration potential. In this chapter, we describe the design of a novel perfusion bioreactor system with oscillatory flow for cultivating multiple 3D cellular constructs with clinical applicable size. It can perfuse culture media in a small volume repeatedly through out the scaffolds. The mixing effect of the oscillatory flow may enhance the uniform growth of cells in the scaffold in prolonged culture in vitro. The system was characterized. Mouse osteoblast-like cells, MC 3T3-E1, were dynamically seeded and cultured in large porous ceramic β-tricalcium phosphate (β-TCP) scaffolds by the oscillatory perfusion system, and then the cell growth within the scaffolds was evaluated by fluorescent study. The oscillatory perfusion system could be a simple and effective bioreactor for bone tissue engineering.

2. Bone regeneration

Although bone tissue is able to regenerate after injuries and can remodel in relation to local stresses, treatment of large bone defects resulting from maladies such as birth defects, trauma, or tumor resection has produced an extremely large clinical demand for bone graft well distributed across the population. Autologous bone grafts have been considered the gold standard for augmenting bone regeneration. However, there are limited sites where bone may be harvested without loss of function (Brown and Cruess, 1982; Enneking et al 1980). Autografts are less effective in irregularly shaped defects and may be resorbed prior to complete healing. Furthermore, autografts harvested from the iliac crest of the hip are associated with a 10% complication rate including infection, fracture, pain, paresthesia, nerve injury, and donor-site morbidity (Gitelise and Saiz, 2002; Younger and Chapman, 1989). Allografts derived from cadavers are another commonly used bone graft material. However, disease transmission and immunologic rejection are serious concerns with unprocessed allografts, and processed allografts, such as demineralized bone matrix, lack bone growth inducing factors necessary for efficacy. Xenografts, or bone grafts obtained from different species, are also a poor option due to the danger of disease transmission or immunological rejection (Erbe et al, 2001).

2.1 *In vitro* bone tissue engineering

Tissue engineering may provide functional substitutes of native tissues, to serve as grafts for implantation (Langer and Vacanti, 1993) and physiologically relevant models for controlled studies of cell function and tissue development. In one of the most typical approaches, 3D structures are generated by the association of osteoprogenitor cells (autologous or allogeneic) with porous scaffolds. This may enhance the bone regeneration potential of the graft, because the seeded cells are not only a cell source for regeneration, but can also coat the scaffold surface with their osteoinductive extracellular matrix even if they did not survive after implantation of the biomaterial (Holtorf et al, 2005; Goldstein et al, 2001; Sikavitsas et al, 2003).

But in conventional static culture, nutrients and oxygen transported through a diffusion process could favor cells to live within 1 millimeter from the outer and upper surface of the scaffold, with only dead cells in the center of the scaffolds after an extended period of time (Figure 1A).

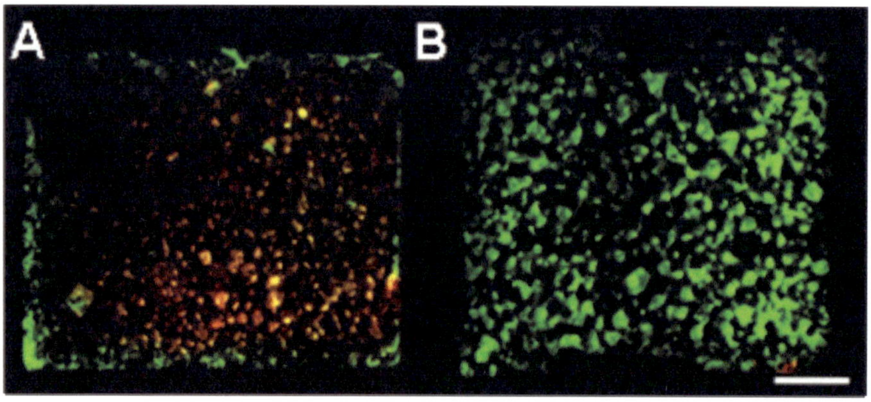

Fig. 1. Calcein-AM/PI double staining of mouse osteoblast-like cells after 6 days of culture on the porous ceramic scaffolds under fluorescent stereomicroscope (middle section view). A) Static culture: living shell of cells with a dead center; B) Perfusion culture by oscillatory fluid flow: living cells distribute uniformly throughout the scaffold. Dead cells were stained red and living cells green. Scale bar=2mm

2.1.1 Development and evaluation of an oscilatory perfusion bioreactor for bone regeneration in vitro

Thus various bioreactors have been designed and developed for in vitro culture of 3D cell-scaffold constructs under conditions that support better nutrition of cells, possibly combined with the application of mechanical forces to direct cellular activity and phenotype.

Spinner flasks and rotating wall vessel (RWV) bioreactors are two kinds of most basic bioreactors applied for bone tissue engineering, as shown in Figure 2 A-B. In the spinner flasks, scaffolds are attached to the needles hanging from the lid of the flask, and convective forces generated by a magnetic stirrer bar allow continuous mixing of the media surrounding the scaffolds[10]. Medium stirring enhances external mass-transfer but also generates turbulent eddies, which could be detrimental for the development of the tissue. In the rotating wall vessel reactors, the vessel walls are rotated at a rate that enables the drag force (Fd), centrifugal force (Fc) and net gravitational force (G) on the construct to be balanced; the construct thus

remains in a state of free-fall through the culture medium (Sikavitsas et al, 2002; Botchwey et al., 2001; Unsworth et al., 1998) also presenting a low fluid shear stress. The microgravity state may present some advantages because it avoids cell deposition, and at the same time promotes cellular interactions (Botchwey et al., 2001). However, it is also known that microgravity is deleterious for bone, leading often to losses in total bone mass (Sinha et al., 2002). A common aspect of both systems was that the cells were not homogeneously distributed throughout the scaffolds structure in the two systems (Skavitsas et al., 2002) probably due to the inefficient internal mass-transfer while the external limitation reduced.
Direct perfusion bioreactors have been applied to fabricate uniform artificial bone grafts *in vitro* (Figure 2C).

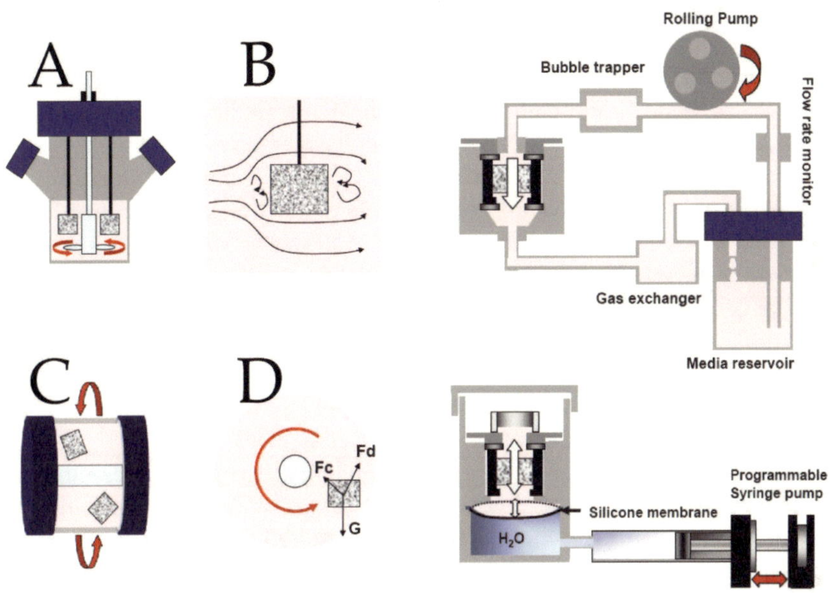

Fig. 2. Illustrations of existing bioreactors for bone tissue engineering. 1) Spinner flasks; 2) rotating wall vessel (RWV); 3) One-way perfusion system; 4) Oscillatory perfusion system

To reduce mass transfer limitations both at the construct periphery and within its internal pores, the media is directly conveyed throughout the interconnected pores to continuously introduce nutrients and remove wastes. Direct perfusion bioreactors have been shown to enhance early cell proliferation, osteogenic differentiation, and mineralized matrix production of bone marrow stromal osteoblasts seeded in three-dimensional scaffolds(Goldstein et al., 2001; Cartmell et al., 2003; Wang et al., 2003; Meinel et al., 2004; Dolder et al., 2003; Bancroft et al., 2002; Sikavitsas et al., 2003; Porter et al., 2005; Holtorf et al., 2005; Botchwey et al., 2003). In addition to enhancing chemotransport by shear flow over bone cells attached to the scaffold surface, a perfusion system may also simulate the fluidic mechanical environment bone cells experience *in vivo* and elicit a flow-induced osteogenic response of the bone cells, as evidenced by increased activity of alkaline phosphatase (ALP), an early osteogenic differentiation marker(Pavalko et al., 1998; Kapir et al., 2003; Jessop et al., 2002; McAllister et al., 1999), and the expression of osteocalcin (Lan et al., 2003), osteopontin (Kreke et al., 2004; Kreke et al., 2005), and bone sialoprotein (Kreke et al., 2005). So perfusion bioreactors are very important tools for research on bone tissue engineering.

Most existing perfusion bioreactors have still some problems to be addressed: 1) the culture media circuit, the typical structure of them, includes multiple tubes and junctions, making them complicated, difficult to handle. The bubbles are also easy trapped and alter the flow field. 2) Cell seeding of scaffolds - that is, the dissemination of isolated cells within a scaffold - is the first step in establishing a 3D culture, and might play a crucial role in determining the progression of tissue formation (Vunjak-Novakovic et al., 1998). But the large volume of medium necessary for circulation does not allow for a small seeding volume required for efficient inoculation; therefore, to improve seeding efficiency, researchers prefer to seed cells into scaffolds outside the perfusion system and then transfer them to the system, significantly increasing the risk of contamination. 3) Growth factors are essential for tissue formation and play an important role in tissue engineering. Bone morphogenetic proteins (BMPs), transforming growth factor beta (TGFb), fibroblast growth factors (FGFs), insulin growth factor I and II (IGF I/II), and platelet derived growth factor (PDGF) have been proposed for bone tissue engineering applications (Jadlowiec et al., 2003; Lind and Bunger, 2001; Yoon and Boden, 2002). Unfortunately, they are still very expensive now. So the large volume of medium required also made traditional perfusion cultures very costly when such kinds of expensive conditioning agents as well as valuable transfected cells or vectors are used. It also prevented the retention of newly synthesized extracellular matrix components or transfected gene products within the construct. In addition, it has been suggested that oscillatory fluid flow instead of unidirectional flow mimics the physiologic fluid flow profile observed in bone during mechanical loading and demonstrated to have beneficial effects on osteogenesis, such as increasing osteopontin and osteocalcin expression (Vance et al., 2002; Ponik et al., 2006; Wu et al., 2006; Li et al., 2004; You et al., 2000; You et al., 2001; Donahue et al., 2003; Batra et al., 2005; Qin et al., 2003).

We have designed and developed a compact oscillatory perfusion bioreactor recently. As shown in Figure 2D, the base of flow well was sealed with a 0.3-mm silicon film. The movement of the syringe back and forth was controlled by a continuous cycle syringe pump, consequently forcing the silicon film to move up and down, thus driving the media in the flow wells to perfuse up and down through the scaffolds in an oscillatory manner. Glass cloning rings were attached to the caps to enable the system to hold a limited volume (<500μm) of cell suspension during seeding for high seeding efficiency as well as enable conditioned culture in small volume. Because the syringe could be released from the pump easily, we were able to move only the syringe-connected chamber (without pump) onto a clean bench and change the medium just as one would do with a conventional 6-well plate. We have successfully fabricated a uniform tissue engineering bone with a clinical relevant size (10 mm in diameter, 8 mm in height) in only 1.5ml culture osteogenic media by this oscillatory perfusion bioreactor for prolonged period.

At first, the cellularity of the scaffold was evaluated by determining the double-standed DNA (dsDNA) content using the PicoGreen assay kit (Molecular Probes). The oscillatory perfusion group scaffold had significant higher number of cells than the static group 24 postseeding ($p<0.05$), as determined from measuring the DNA content (Fig. 3A). The scaffold DAN content after 5 days of differentiating culture was also higher in the oscillatory perfusion culture than in the static culture ($p<0.1$). During the 5 days of differentiation culture by oscillatory perfusion, the doubling time of the cells was significantly longer than that by the static culture method(Fig. 3B), indicating that the proliferation rate in the perfusion culure during this period ($p<0.05$).

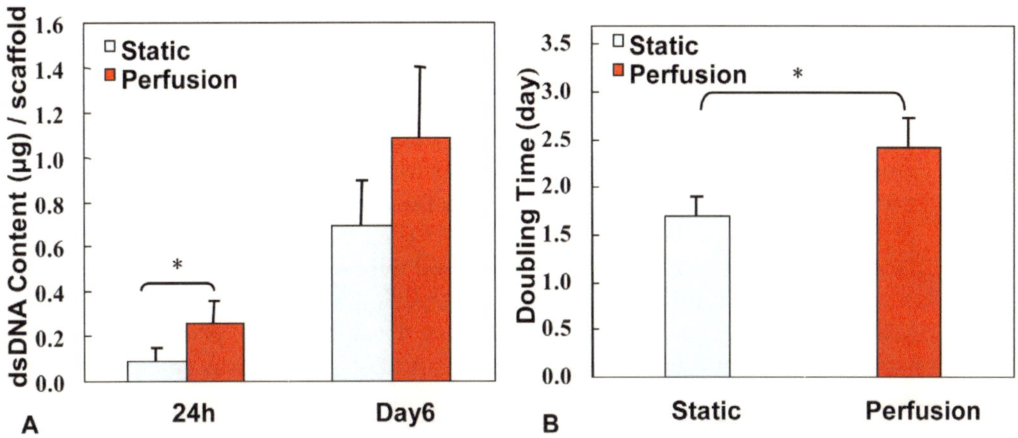

Fig. 3. A: DNA content of MC 3T3-E1 cells per scaffold after 24 h of seeding and 5 days of differentiation (day6). B: Doubling time of static culture and oscillatory perfusion culture during 5 days of differentiation culture, calculated from dsDNA content. All the experiments were conducted independently (static group: n=4, oscillatory perfusion group: n=3). Error bars represent SD. The asterisk (*) indicates a statisticallly significant difference between the static group and the perfusion group ($p<0.05$). (Du et al., 2007)

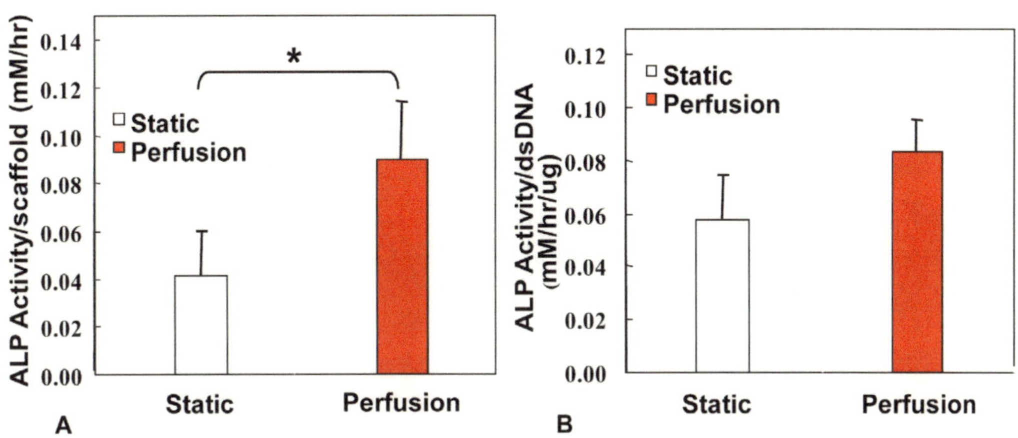

Fig. 4. ALP activity of seeded MC3T3-E1 cells after 5 days of differentiation culture. A: Total ALP activity per scaffold. B: ALP activity per dsDNA content per scaffold, as an estimate of cellular ALP activity. All the experiments were conducted independently (static group: n=4, oscillatory perfusion group: n=3). Error bars represent SD. The asterisk (*) indicates a statistically significant difference betweeen the static group and the perfusion group ($p<0.05$). (Du et al., 2007)

To evaluate the osteogenic function of the attached MC 3T3-E1 cells, activity of ALP, an early osteogenic differentiation marker, was assayed. The activity of ALP was measured by a clolorimetric endpoint assay. The same supernatant samples as those prepared for the DNA content assay were used. After 5 days of differentiating culture, the total ALP activity was significantly higher in the oscillatory perfusion culture than in the static culture ($p<0.05$) (Fig. 4A). As shown in Figure 4B, perfusion culture showed a higher average ALP activity per dsDNA content than that in the static culture ($p<0.1$), suggesting higher levels of ALP per cell in the oscillatory perfusion culture.

Stereomicroscopic observation of living cells darkly stained by MTT indicated that after 24 h of static seeding by top dropping, cells indeed penetrated into the center of the scaffolds (Fig. 5C) but remained at a much higher density on the top surface (Fig. 5A) and with cells seeded very sparsely in the lower regions (Fig. 5 D, E). After 5 days of differentiation culture, the density of cells increased significantly, and the cells appeared as a uniform layer on the top surface of the scaffolds (Fig. 5); far fewer living cells were found in the deeper pores (Fig. 5L-O). The cell density also increased along the side surfaces of the scaffolds with a top-to-bottom gradient of cells (data not shown), but MTT staining was pale in the center of the scaffold (Fig. 5M). On the contrary, the MTT staining of the oscillatory perfusion culture scaffolds showed a uniform seeeding density (Fi.g 5F-J) and subsequent homogeneous cell proliferation throught the scaffold (Fig. 5P-T).

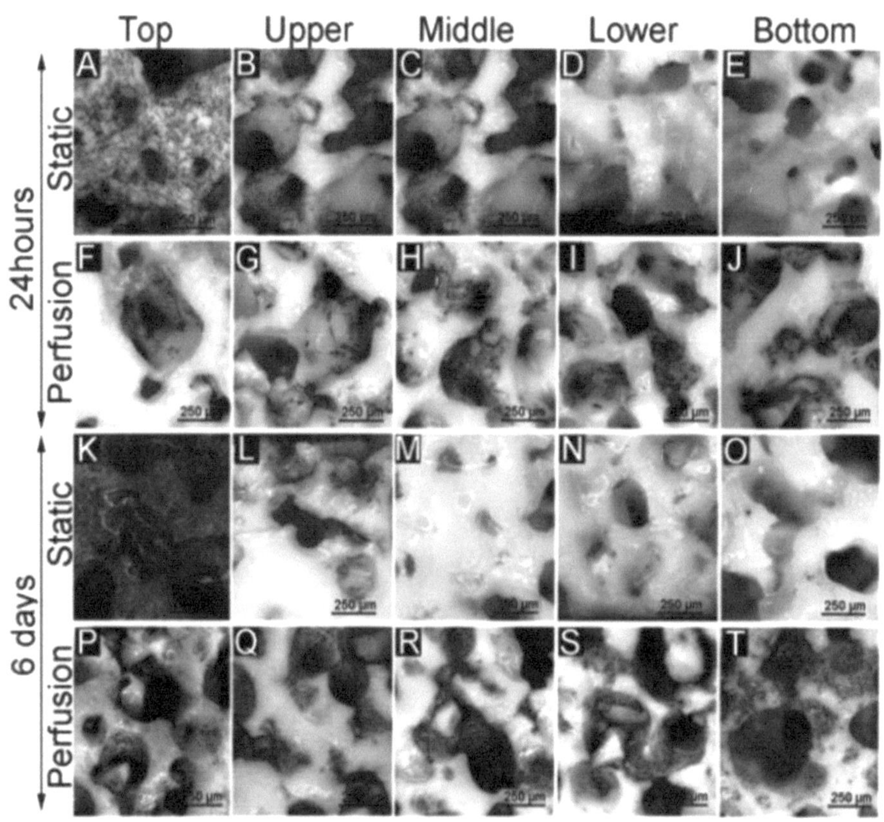

Fig. 5. MTT staining observed by fluorescence stereoscopy. A-J, 24 h after seeding; KT, 5 days of differentiation culture. A-E, and K-O are images from a statuc culture sample at the indicated region; F-J, and P-T are images from an oscillatory perfusion culture sample at the indicated regions of the scaffold. The dark crystals layering the pores represent the locations of living cells. Scale bar = 250 μm. (Du et al., 2007)

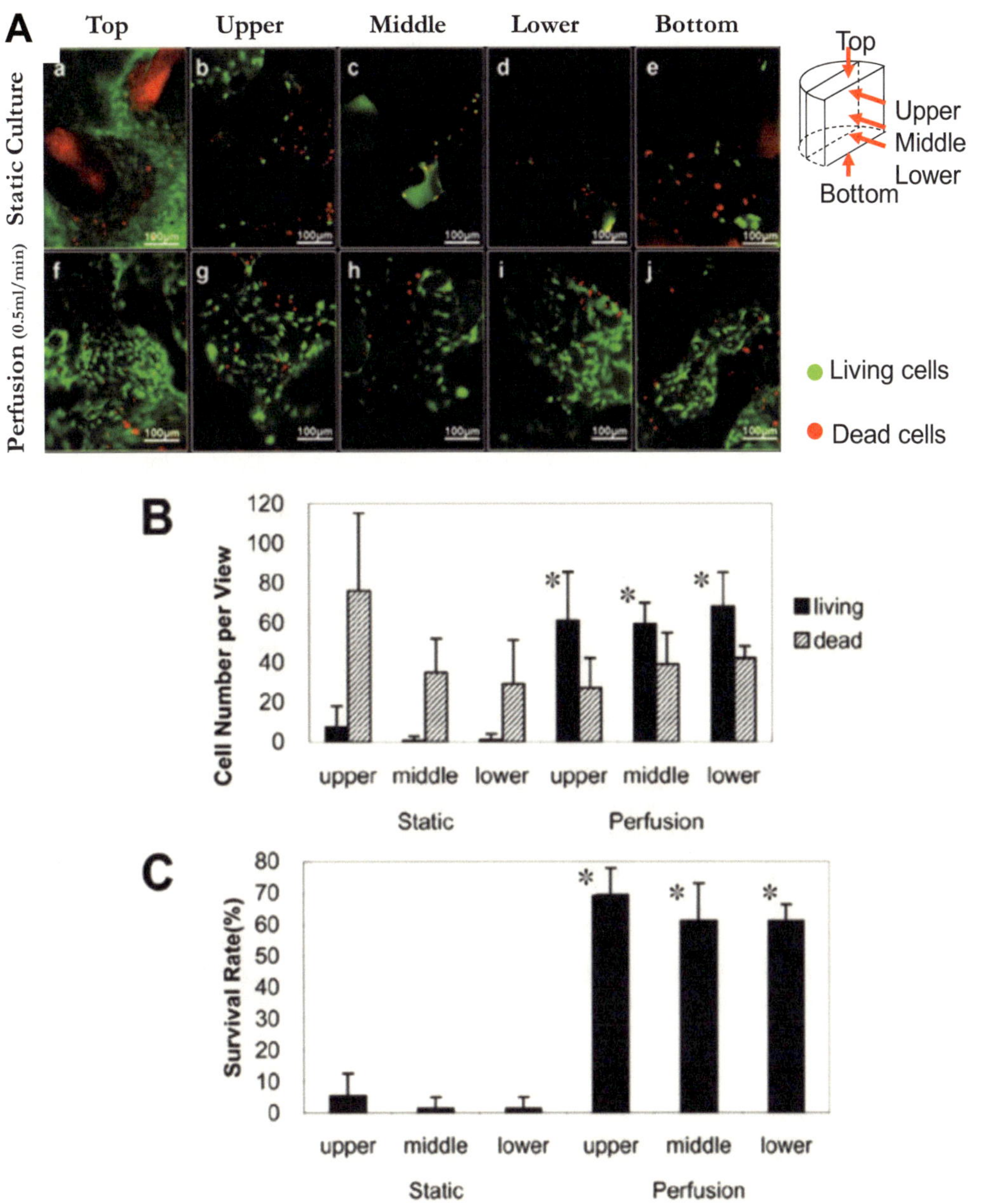

Fig. 6. A: Calcein-AM/PI double staining of the middle section of the scaffolds after 5 days of differentiation culture. a-e images from fluorescence stereomicroscopy of the indicated regions from a scaffold in static culture; f-j, images of the same regions froma scaffold in an oscillatory perfusion culture. Living cells are stained green and dead cells red. Scale bar = 100 mm. B: Number of living and dead cells per view at the same magnification of A. C: Cell viavility in the indicated regions. The results are derived from B. Error bars represent SD, n=6. Asterisks (*) indicate a statistically significant dirfference between the static group and the oscillatory perfusion group ($p<0.05$). (Du et al., 2007)

To elucidate the fate of cells seeded in the center of the scaffold in the static culture and whether the oscillatory shear flow led to severe damage to cells, Calcein-AM.PI double staining to visualize living and dead cells, respectively, and quantitative study of cell distribution and viability were conducted. The double staining showed that after 6 days of static culture, green (calcein) and red (PI) fluorescence were both strong on the top surface (Fig. 6Aa). However, only a few PI-stained cells-apparently undergoing apoptosis or necrosis-were found in the center of the scaffold (Fig. 6Ac), and only a sparse presence of shrunken red-stained nuclei of dead cells was noted in the lower region (Fig. 6Ad). On the other hand, green-stained living cells were distributed everly throughout the scaffold cultured in the oscillatory perfusion system with a similar ration of sparse dead cells to more abundant living cells throughout the scaffold (Fig. 6Af-j). The cell disctribution was quantified (Fi.g 6B). Living cells were disctibuted homogeneously throughout the section views in the oscillatory perfusion culture, wheareas mainly dead cells were distributed heterogeneously in the static culture scaffolds. Analysis of the viability of the cells by scaffold section revealed homogeneous cell viability in the oscillatory higher than the viability of the static cultured cells ($p<0.05$, $n=6$; Fig. 6C).
In summary, the present study demonstrates that a relatively large engineering bone construct (10 mm diameter x 8 mm hight) could be developed in vitro by culturing composites of osteoblast-like cells and porous ceramic blocks with an oscillatory perfusion system. The oscillatory flow condition not only allow a better seeding efficiency and homogeneity, but also facilitates uniform culture and early osteogenic differentiation,which suggests that the oscillatory perfusion culture system may be a valuable and convenient tool for bone tissue engineering. Another singnificant advantage of the perfusion system is that only a small volume of medium is needed to load oscillatory flow in this bioreactor, which makes it an especially efficient tool for conditioned culture with expensive growth factors or gene vectors with more economical feasibility. The produced cellular matrix might also be concentrated in such a small volume. In addition, similar studies by other kinds of cell types and scaffold materials would be helpful to investigate furtherly about the biological effects of oscillatory flow on 3D culture for tissue engineering.

2.1.2 Oscillatory perfusion culture of CaP-based tissue engineering bone with and without dexamethasone (Helvetica, 9 pt, bold)

It is important to develop not only a feasible 3D culture sytstem to maintain the tissue, but also an effective osteogenic system enhance and maintain the osteogenic potential of the graft. Dexamethasone is a key component in the classic osteogenic supplements (ascorbic acid, sodium b-glycerophosphate, and dexamethasone). It exerts a powerful effect on the osteogenic differentiation. Continual exposure to dexamethasone, beginning shortly after cell harvest, is required to drive and maintain the osteoblastic phenotype of marrow-derived progenitor cells. Although maintenance of the osteoblastic phenotype is desirable for long-term cultures, dexamethasone has been suggested to drive osteoprogenitors down the osteogenic pathway so quickly that they do not proliferate well, and even to induce apoptosis. Because previous studies have suggested synergistic osteogenic effects when both dexamethasone and unidirectional flow perfusion culture were used, (Holtorf et al., 2005) we supposed that beneficial synergistic effects could also be induced by dexamethasone and oscillatory fluid flow. Because the flow rates would have different biologocal effects on bone cells not only in two-dimensional (2D) culture on parallel plates (Bacabac et al., 2005; Bacabac et al., 2004; Bakker et al., 2001; Kreke et al., 2004; Thi et al., 2003), but alsol in 3D bone construction by a perfusion sytem (Cartmell et al., 2003), it would also be important to study the influence of different flow rates of the oscillatory flow on the administration of dexamethasone.

Moreover, CaP ceramic scaffolds were most widely used in clinical practice, but CaP materials had potential absorbing ability which might influence the performance of dexamethasone for osteogenesis. Therefore, it would be important for bone tissue engineering to confirm whether it would work the same way as 2D culture or not when CaP-based scaffolds were used for 3D culture. In addition, the 3D hydrodynamic environments in bioreactor would also make dexamethasone work differently from that of 2D static culture condition.

The mouse osteoblast-like cell line MC 3T3-E1, which has the potential to be fulther differentiated, was used in this work. The cells were seeded onto porous ceramic scaffolds using the oscillatory perfusion method to achieve a similar cellularity before culture, and eliminate any confounding effects arising from the use of different cell-seeding techniques. The seeded constructs were then cultured by either the static method or the oscillatory perfusion method continuously for 6 days either with or whithout dexamethasone. The flow rate of oscillatory perfusion was set at 0.05, 0.5, 1 mL/min per scaffold. The cell proliferation, early osteogenic effects and viability were evaluated.

From an MC3T3 sample of a known cell number, which was mixed with an equal volume of knifeminced scaffodl, the DNA content of MC3T3 per cell was determined to be 6.02 pg/cell, so the cell number could be evaluated by DNA content measurement. After seeding all cells by the oscillatory method, there were no significant differences among the groups, as shown in Fig. 7.

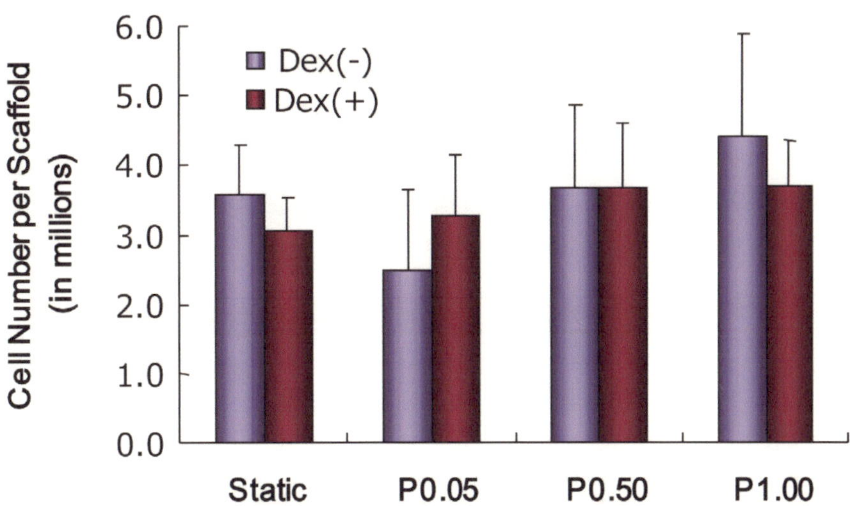

Fig. 7. The cell number per scaffold after 6 days of cluture. P0.05,P0.50, 1.00 referred to the perfusion groups with the flow rate of, respectively, 0.05m 0.50, and 1.00 mL/min. Error bars represent the means = SD. N=6. (Du et al., 2008)

ALP activity is an important early osteogenic marker. As shown in Fig. 8, there wrere no significant differences in ALP activity between the cultures with and those without dexamethasone in either the perfusion or static culture groups. Total ALP activity per scaffold was higher in all the perfusion culture groups than the static culture groups (p<0.05), except for the group perfused at 0.05 mL/min without dexamethasone. The total ALP activity was highest in the 0.5 mL/min perfusion groups, which were significantly higher than static culture and all perfusion groups at all other fow rates (p<0.05). The average ALP activity per cell in the 0.5 min/mL perfusion groups was significantly higher than that in the static culture group (p<0.05). The 1.0 mL/min perfusion group without dexamethasone had the lowest average ALP activity, and the activity in this group was significantly lopwer than that by other flow rates in the same medium.

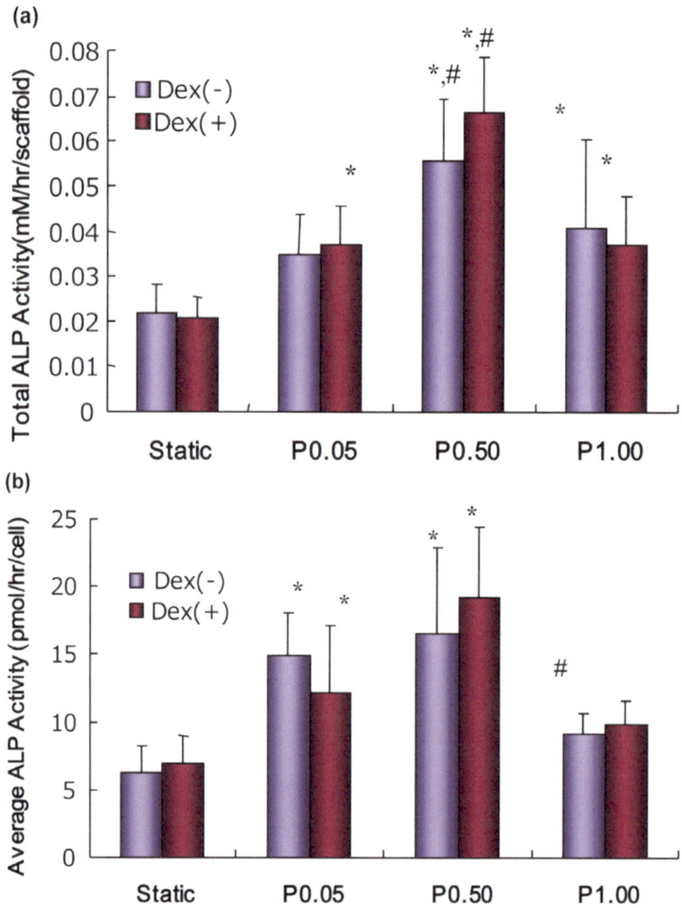

Fig. 8. ALP activity analysis. (a) The total ALP activity per scaffod after 6 days of culture and (b) the average ALP activity per scaffold after 6 days of culture. P0.05,P0.50,P1.00, referred to the perfusion groups with the flow rate of, respectively, 0.05, 0.50, and 1.00 mL/min. Error bars represent the means = SE. N=6. The asterisk (*) undicated a statisticallly significant difference between the static group and the perfusion group (p<0.05); (#) indicates a statistically significant difference from groups of other flow rates in the same medium (p<0.05). (Du et al., 2008)

The status of the cells in the center of the scaffolds under different culture conditions was observed under SEM, as shown in Fig. 9. No living cells were found in the center of the scaffolds subjected to static culture; higher magnification revealed dead cells that were shrunken and slack, with a preserved porous membrane and no surrounding extracellular matrix (Fig. 9a). However, numerous living cells were found in the center of the scaffold formed by perfusion culture; these cells were smooth, stretched out, and surrouned with a cellular matrix (Fig. 9b).

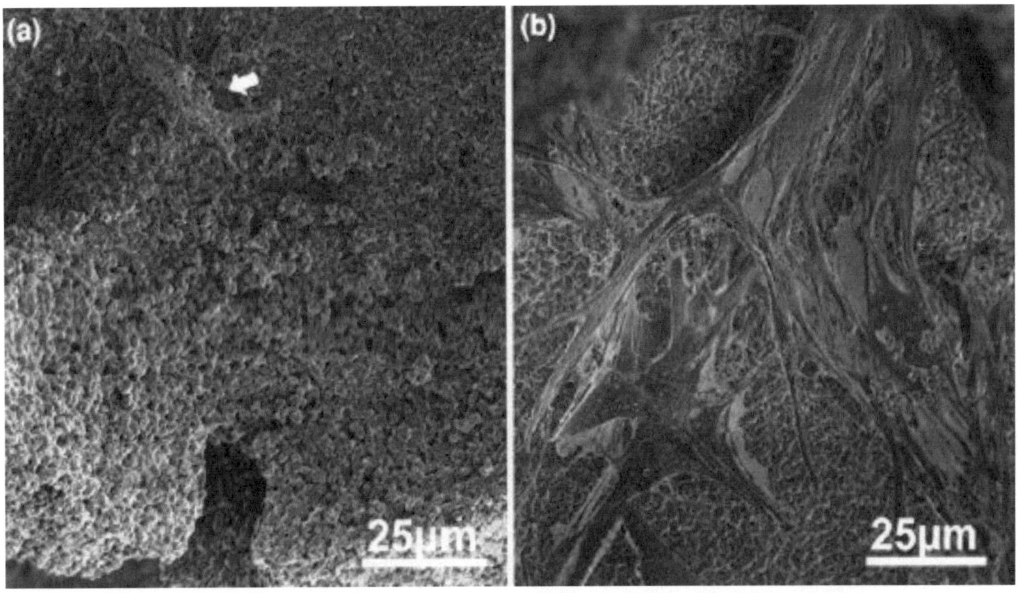

Fig. 9. The cells in the center of scaffolds after 3 days of culture observed under a scannning electron microscope. (a) Static culutre: no stretched cells were found except the poroous and shrunken dead cells marked with an arrow and (b) perfusion culutre (0.5 mL/min): living cells covered the surface of the pores in a stretched and intact shape. Scale bar =25 μm. (Du et al., 2008)

A gross view of the cell viability distribution throughout the various sections is shown in Fig. 10a. We can see that there is only a thin shell of living cells covering the surface of the scaffolds cutlrured by the staci method, with few living cells inside; in the 1 mL/min perfusion group; however, the living cells were distributed uniformly throughout the scaffold. The uniformity of distribution in the perfusion groups was dependent on the flow rate with the necrotic area decreasing as the flow rate increased. There was no clear increase in necrotic area by the administration of dexamethasone. As shown in Fig. 10b, 10x objective of the cells in the centerof the scaffolds revealed that there were almost no greem-stained living cells in the scaffolds subjected to static culture, but the cellular viability in the center of the scaffolds in the perfusion culture increased as the perfusion rates incerease, especially in the groups perfused at 0.5 and 1.0 mL/min, in which the green-stained living cells covered the pore surfaces uniformly with a streched shape. The groups treaed with dexamethasone did not show any groups treated with dexamethasone did not show any clear difference compared to those without dexamethasone.

The cell viability was further studied quantitatively under 10x objective. The uniformity of the viacilityw as the total viability decreased as the flow rate was decrased to 0.05 mL/min (Fig. 11). The perfusion culture had higher viability than static culture from the section view.

In conclusion, the biological effects of dexamethasone on oscillatory perfusion culture of tissue eigineering bone on a CaP scaffold was investigated. The results showed that the oscillatory flow could enhance early osteogenesis of osteoblast-like cells in 3D culture on ceramic scaffolds, whit a peak function at the flow rate of 0.5 mL/min. The cell viability was significantly higher and more uniform in the perfusion groups than in the static culture groups. The uniformity groups than in the static culture groups. The uniformity degreased as the perfusion rates decreased. However, dexamethasone was suggested to have no effect when CaP ceramics were used as scaffold, and thus other osteogenic growth factors should be used under this condition. Because most osteogenic growth factors are very expensive, the oscillatory perfusion culture using small medium-to-volume ratio perfusion culture using CaP scaffolds. Such a dynamic conditioned culture could also produce bioactive engineered bone tissue that would function as a DDS for bone regeneration.

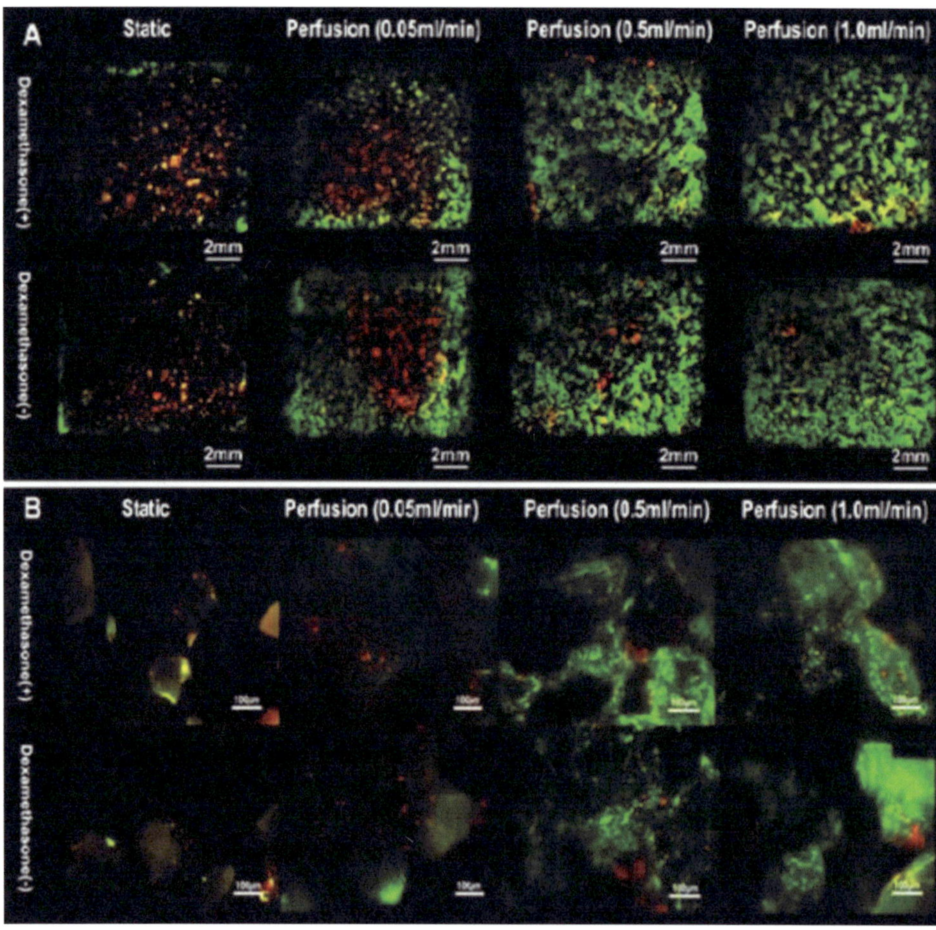

Fig. 10. Calcein-AM/PI double staining of the midline section of the scaffolds culuture with MC 3T3-E1 cells for 6 days by the static method or the perfusion method observed under a fluorescent stereomicroscope. The living cells were stained green and the dead cells red. (a) The gross views of the whole sections, scale bar=2 mm and (b) the center of scaffolds, scale bar = 100 μm. (Du et al., 2008)

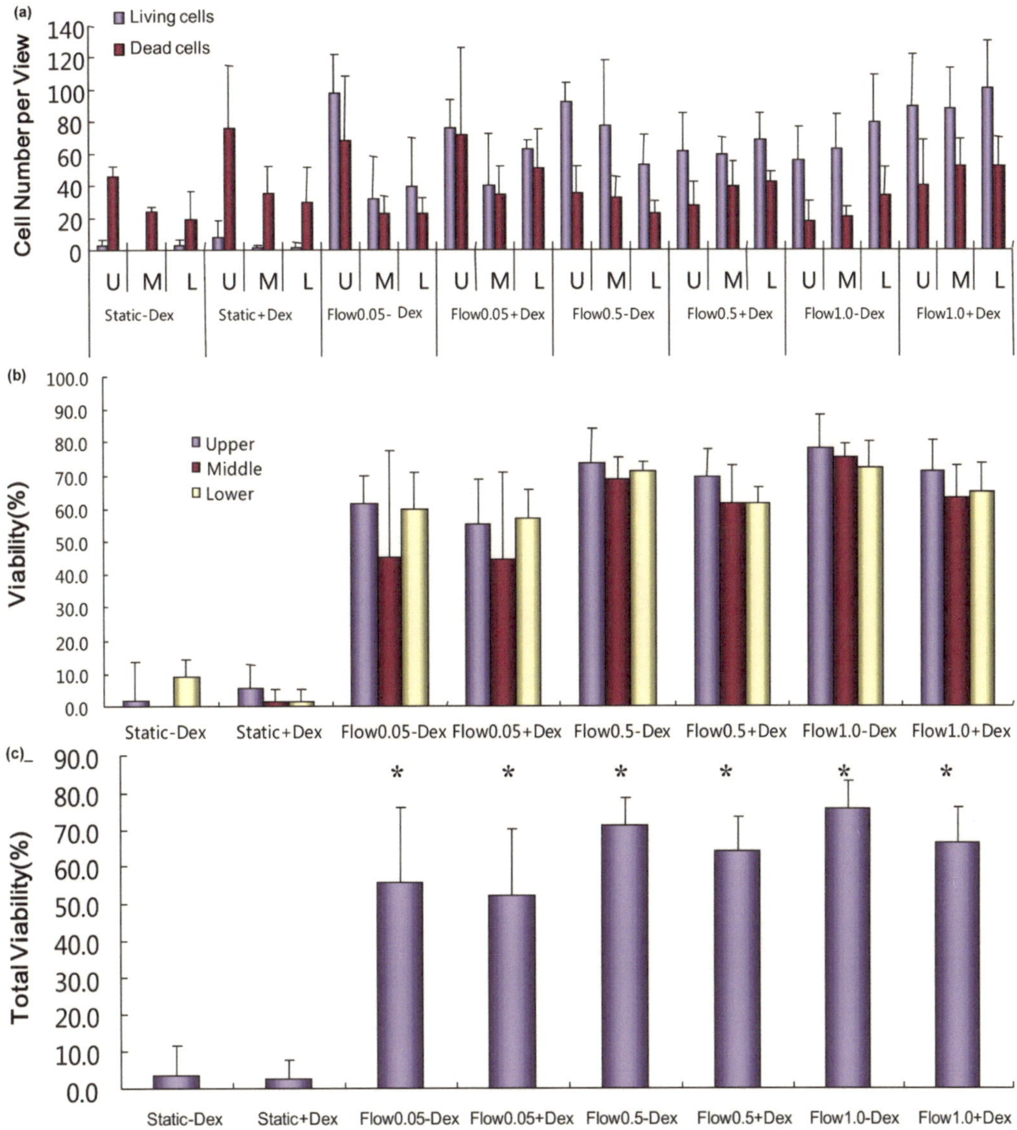

Fig. 11. Quantitative study of the cell vioability distribution according to Calcein-AM/PI double staining of the middle section after 6 days of culuture observed under a fluorescent stereomicroscope under 10x objective; (b) the viability of the corresponding regiions in (a); and (c) the total cell viability of each culture method. Error bars represent the means = SE, n=6. The asterisk (*) indicates a statistically significant difference between the static group and the perfusion group ($p<0.01$). (Du et al., 2008)

2.1.3 3D culture by unidirectional or oscillatory flow for bone tissue engineering

As we mentinoned in 2.1.2, the compact perfusion system with oscillatory flow appeared to enhance early osteogenesis and the uniformity of cultured bone constructs. Here, we compared the biological effects of a perfusion system with unifirectional flow on the 3D

construction of cell-seeded bone grafts against those of a perfusion system with oscillatory flow. Mouse osteoblast-like cells, MC 3T3-E1 were cultured in porous ceramic scaffolds by either static, unifirectional perfusion, or oscillatory perfusion culture for 6 days. Cell proliferation, early osteogenic effects, and viability were then evaluated.
A unifirectional perfusion system was designed as shown in Figure 12A. A media tank was set on each end of the perfusion camber as both media reservoirs and gas bubble trappers.

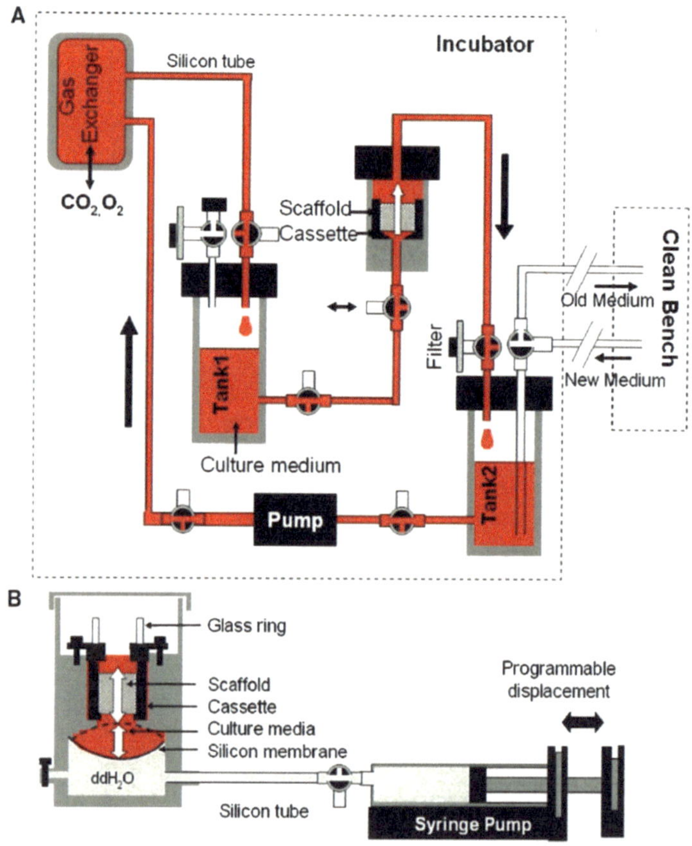

Fig. 12. Illustrations of the 3D perfusion culture systems for tissue engineering bone by either unifirectional flow or oscillatory flwo. A: The unidirectional perfusion system; (b) the oscillatory perfusion system. (Du et al., 2009)

A gas-permeable bag was set in the circuit loop as a gas exchanger. The unifirectional media flow was driven bya syringe ponp. A couple of silicon tubes led from Tnak 2 to the clean bench, so that the medium could be exchanged in the sterile environment without moving the whole complicated system. The perfusion sytem with the scaffolds set inside was sterilized by ethylene oxide gas.
As shown in Figure 13, the calcein-AM/PI double staining of the midline section of the samples after 6 days of culture demonstrated that the living cells grew only on the surfaces of the scaffolds cultured statically (Fig. 13A; the cells grew extremely inhomogeneously in unidirectional perfusion culture, where there was a clear inverted arch-shaped interface between living cells and the non-living area (Fig. 13B).

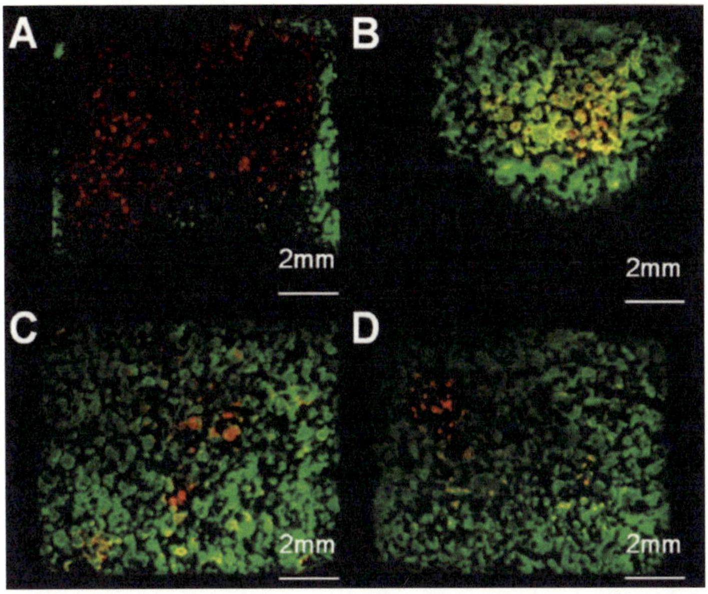

Fig. 13. Calcein-AM (green; living cells)/propium iodide (red; dead cells) double fluorescence staining of the scaffolds cultured with MC3T3-E1 cells for 6 days by static and perfusional methods with either unidirectional flow (1mL/min) or oscillatory flwo (0.5 mL/min): (A) static culture; (B) unidirectional perfusion culture at 1 mL/min; (C) oscillatory perfusion culuture at 0.5 mL/min; (D) oscillatory perfusion culture at 1 mL/min. (Du et al., 2009)

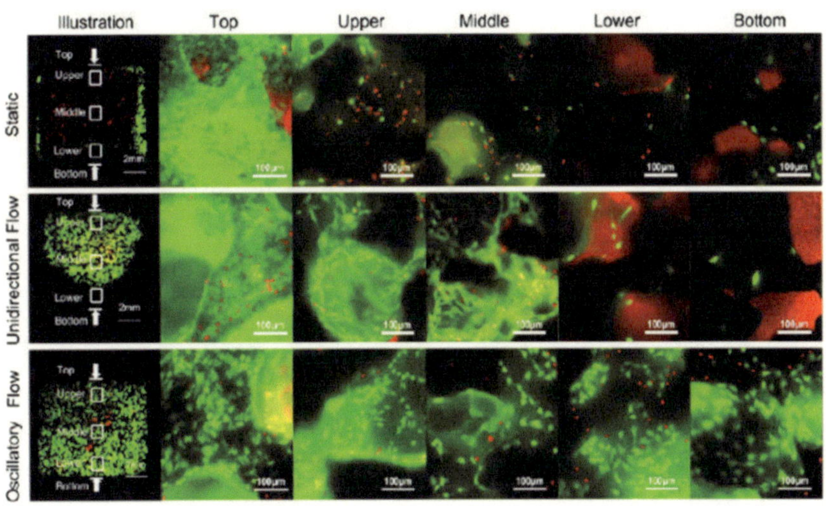

Fig. 14. Calcein-AM (green; living cells)/propidium iodide (red; dead cells) double fluoresce staining of the scaffolds cultured with MC3T3-E1 cells for 6 days by static and perfusional methods with either unidirectional flow (1mL/min) or oscillatory flow (0.5 mL/min). Images of scaffolds observed at 10x objective of a fluorescence stereomicroscope are shown afcording to the regions indicated by the broad views on the left. (Du et al., 2009)

In contrast, the scaffolds cultured by the oscillatory flow had a relatively uniform distribution of livnig cells throughout the scaffolds. The inhomogeneous living cells distribution of the unidirectional perfusion culture was further verified by higher magnificaion, as shown in Figure 14. In the unidirectional perfusion culture, the top surfaces of the scaffolds wre covered by a thick layer of cells and, above the arch-shaped interfac, there were dense living cells over the pore surface. However, very few cells were libing below the interface or at the bottom of the scafoolds.

As shown in Figure 15, the quatitative study of cell viability in the section further confirmed the findings presentaed above. In static culture, there were few libing cells in the section view, and these were accumulated only near the top surface. In the unidirectional perfusion culture, the living cells and dead cells were distributed extremely inhomogeneously, with living cells favoring the upper positions (Fig. 15A).

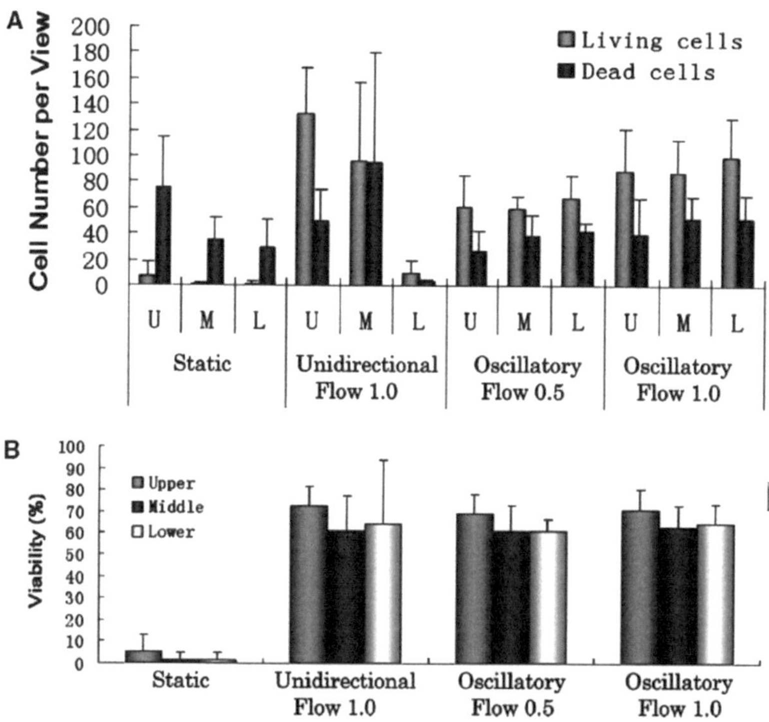

Fig. 15. Quantitative study of cell viability distribution according to the calcein-AM/PI double staining of the middle section after 6 days of culuture observed under fluorescent stereomicroscope under 10x objective. The number s 0.5 and 1.0 refer to the flow rates of these groups in mL/min. A: Living and dead cell numbers per view of respective regions wre counted at the same magnification as in Figure 3. U: upper; M: middle; L:lower, (B) viability of respective region derived from (A); Error bars represent means = SD. N=6. (Du et al., 2009)

However, although there were fewer cells in the lower part of the scaffold in the unifirectional perfusion culture, their viability was not low (Fig. 16B); in the oscillatory perfusion culture, the cells proliferated uniformly throughout the scaffolds.

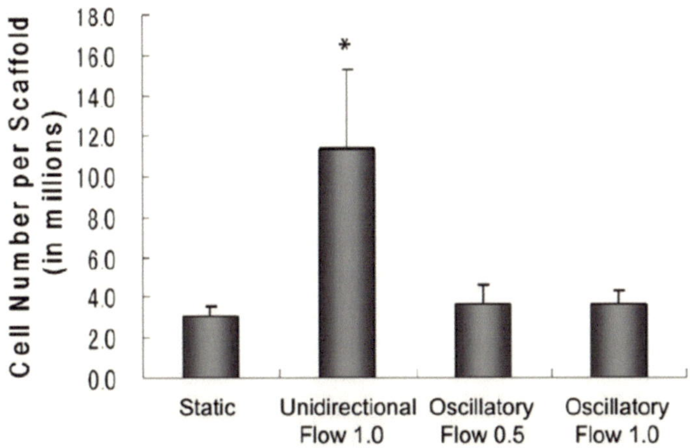

Fig. 16. The cell number per scaffold after 6 days of culture. The numbers 0.5 and 1.0 reger to the flow rates of the groups in mL/min. Error bars represent means = SD, n=6. The asterisk (*) indicates a statistically significant difference between the static group and all other groups. (Du et al., 2009)

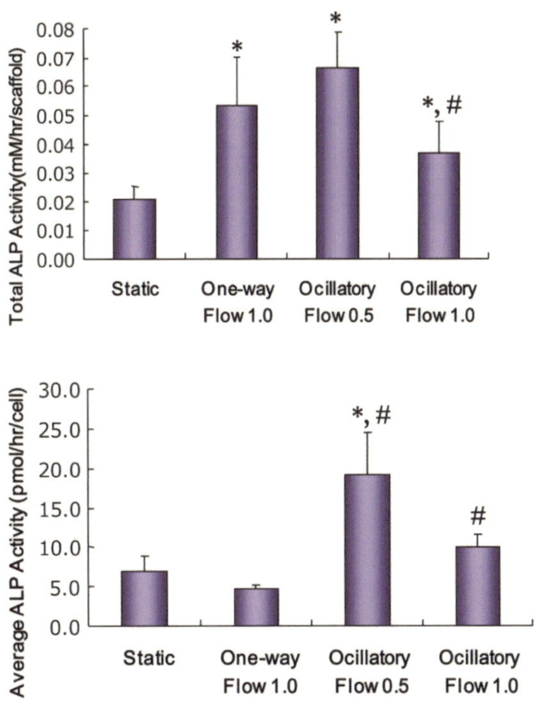

Fig. 17. The average ALP activity per cell after 6 days of culture. The numbers 0.5 and 1.0 refer to the flow rates of the groups in mL/min. Error bars represent neans = SD, n=6. The asterisk (*) indicates a statistically significant difference between the static group and the perfusion group (P<0.05). # indicates a statistically significant difference betweent unidirectional perfusion group and the oscillatory perfusion group (P<0.05). (Du et al., 2009)

The total cellularity of the scaffolds after 6 days of culture was evaluated by DNA content analysis, as show in Figure 17. The total cell number in unidirectional flow perfusion was significantly higher than in any of the other groups ($p<0.05$).
The average ALP activity did not differ significantly between the unidirectional perfusion culture and the static culture ($P>0.05$), although that of the oscillatory perfusion at 1 mL/min was also not significantly higher than that of the static group (Fig. 17).
Osteoblast-like cells were culture with porous ceramic scaffold three-dimensionally in vitro for 6 days under static and hydrodynamic conditions with either unifirectional or oscillatory flow. Although the unidirectional flow increased cell proliferation, the proliferation was extremely inhomogeneous, which rendered the engineered bone unsuitable for transplantation. On the other hand, the oscillatory flow enabled uniform proliferation of osteogenic cells and increased early osteogenesis. This suggested that the oscillatory fluid flow might be better than unidirectional flow for 3D culture of engineered bone in vitro. The oscillatory perfusion system could be a compact, safe, and efficient bioractor for bone tissue engineering.

3. Conclusion

As bone is a loading-dependent remodeling tissue, design of bioreactors for bone tissue engineering could not only culture a uniform "living" tissue, with suitable type and appropriate amount of physical stimuli, we may also "exercise" the bone graft in vitro and achieve a functional active engineering bone. As well as an efficient culture device for chemotransport and physical environment simulation, the design of future bioreactors would also be preferred to be safe, economy acceptable, customized-scaffold and nondestructive evaluation supported culture system.

4. Acknowledgment

This work was supported by a leading-edge research promotion fund from Japan Society for the Promotion of Science (JSPS), and a Grant-in-Aid for Scientific Research from the Japanese Ministry of Education, Culture, Sports, Science and Technology, GCOE Program. We gracefully acknowledge Olympus Terumo Biomaterials Corp. for the supply of scaffolds.

5. References

Bacabac, R.G., Smit, G.T., Mullender, M.G, Van Loon, J.J., & Klein-Nulend, J. (2005). Initial stress-kick is required for fluid shear stress-induced rate dependent activation of bone cells. *Annals Biomedical Engineering*, Vol. 33, No. 1, 104-110

Bacabac, R.G., Smit, T.H., Mullender, M.G., Dijcks, S.J., Van Loon, J.J., & Klein-Nulend, J. (2004). Nitric oxide production by bone cells is fluid shear stress rate dependent. *Biochemitry Biophysics Research Communications*, Vol. 315, No. 4, 823-829

Bakker, A.D., Soejima, K., Klein-Nulend, J., & Burger, E.H. (2001). The production of nitric oxide and prostaglandin E(2) by primary bone cells is shear stress dependent. *Journal of Biotechnology*, Vol. 34, No. 5, 671-677

Bancroft, G.N., Sikavitsas, V.I., van den Dolder, J., Sheffield, T.L., Ambrose, C.G., Jansen, J.A., & Mikos, A.G. (2002). Fluid flow increases mineralized matrix deposition in 3D perfusion culture of marrow stromal osteoblasts in a dose-dependent manner. *Proceedings of the National Academy of Sciences of the United States of America*, Vol. 99(20), 12600-12605, Epub 2002 Sep 19

Batra, N.N., Li, Y.J., Yellowley, C.E., You, L., Malone, A.M., Kim, C.H., & Jacobs, C.R. (2005). Effects of short-term recovery periods on fluid-induced signaling in osteoblastic cells. *Journal of Biomechanics*, Vol.38, No.9, 1909-1917

Botchwey, E.A., Pollack, S.R., Levine, E.M., Laurencin, C.T. (2001). Bone tissue engineering in a rotating bioreactor using a microcarrier matrix system. *Journal of Biomedical Materials Research*, May, Vol.55, No.2, 242-253

Botchwey, E.A., Pollack, S.R., El-Amin, S., Levine, E.M., Tuan, & R.S., Laurencin, C.T. (2003) Human osteoblast-like cells in three-dimensional culture with fluid flow. *Biorheology*, Vol.40, No.1-3, 299-306

Brown, K.L., & Cruess, R.L. (1982). Bone and cartilage transplantation in orthopaedic surgery. A review.*Journal of Bone Joint Surgery America*, Feb, Vol. 64, No. 2, 270-279

Cartmell, S..H., Porter, B.D., Garcia, A.J., & Guldberg, R.E. (2003). Effects of medium perfusion rate on cell-seeded three-dimensional bone constructs in vitro. *Tissue Engineering*, Vol. 9, 1197

Cartmell, S.H., Porter, B.D., Garcia, A.J., & Guldberg, R.E. (2003). Effects of medium perfusion rate on cell-seeded three-dimensional bone constructs in vitro. *Tissue Engineering*, Vol. 9, 1197-1203

Donahue, T.L., Haut, T.R., Yellowley, C.E., Donahue, H.J., & Jacobs, C.R. (2003). Mechanosensitivity of bone cells to oscillating fluid flow induced shear stress may be modulated by chemotransport. *Journal of Biomechanics*, Vol.36, No.9, 1363-1371

Du, D., Furukawa, K.S., Ushida, T. (2007). Oscillatory perfusion seeding and culturing of osteoblast-like cells on porous beta-tricalcium phophate scaffolds. *Journal of Biomedical Material Research*, Vo. 86A, 796-803

Du, D., Furukawa, K.S., Ushida, T. (2008). Oscillatory perfusion culture of CaP-based tissue engineering bone with and without dexamethasone. *Annals of Biomedical Engineering*, Vol. 37, No. 1, 146-155

Du, D., Furukawa, K.S., Ushida, T. (2009). 3D culture of osteoblast-like cells by unidirectional or oscillatory flow for bone tissue engineering, *Bioetechnology and Bioengineering*, Vol. 102, 1670-1678

Enneking, W.F., & Eady, J.L. (1980). Burchardt H. Autogenous cortical bone grafts in the reconstruction of segmental skeletal defects. *Journal of Bone Joint Surgery American*, Oct, Vol. 62, No. 7, 1039-1058

Erbe, E.M., Marx J.G., Clineff, T.D., & Bellincampi, L.D. (2001). Potential of an ultraporous beta-tricalcium phosphate synthetic cancellous bone void filler and bone marrow aspirate composite graft. *European Spine Journal*, Oct, Vol.10, Suppl. 2, S141-S146

Gitelis, S., Saiz, P. (2002). What's new in orthopaedic surgery. *Journal of the American College Surgeons*, Jun, Vol. 194, No. 6, 788-791

Goldstein, A.S., Juarez, T.M., Helmke, C.D., Gustin, M.C., & Mikos, A.G. (2001). Effect of convection on osteoblastic cell growth and function in biodegradable polymer foam scaffolds. *Biomaterials*, Vol.22, No.11, 1279-1288

Holtorf, H.L., Jansen, J.A., & Mikos, A.G. (2005). Flow perfusion culture induces the osteoblastic differentiation of marrow stroma cell-scaffold constructs in the absence of dexamethasone. *Journal of Biomedical Material Research* part A, Vol. 72, 326-334

Holtorf, H.L., Sheffield, T.L., Ambrose, C.G., Jansen, J.A., & Mikos, A.G. (2005). Flow perfusion culture of Marrow stromal cells seeded on porous biphasic calcium phosphate ceramics. *Annals of Biomedical Engineering*, Vol.33, No.9, 1238-1248

Jadlowiec, J.A., Celil, A.B., & Hollinger, J.O. (2003). Bone tissue engineering: recent advances and promising therapeutic agents. *Expert Opinion on Biological Therapy*, Jun, Vol.3, No.3, 409-423

Jessop, H.L., Rawlinson, S.C., Pitsillides, A.A., & Lanyon, L.E. (2002). Mechanical strain and fluid movement both activate extracellular regulated kinase (ERK) in osteoblast-like cells but via different signaling pathways. *Bone*, Vol.31, No.1, 186-94

Kapur, S., Baylink, D.J., & Lau, K.H. (2003). Fluid flow shear stress stimulates human osteoblast proliferation and differentiation through multiple interacting and competing signal transduction pathways. *Bone*, Vol.32, No.3, 241-51

Kreke, M.R., & Goldstein A.S. (2004). Hydrodynamic shear stimulates osteocalcin expression but not proliferation of bone marrow stromal cells. *Tissue Engineering*, Vol. 10, No. 5-6, 780-788

Kreke, M.R., & Goldstein, A.S. (2005). Hydrodynamic shear stimulates osteocalcin expression but not proliferation of bone marrow stromal cells. *Tissue Engineering*, Vol.10, No.5-6, 780-788

Kreke, M.R., Huckle W.R., & Goldstein A.S. (2005). Fluid flow stimulates expression of osteopontin and bone sialoprotein by bone marrow stromal cells in a temporally dependent manner. *Bone*, Vol.36, No.6, 1047-55

Lan, C.W., Wang, F.F., & Wang, Y.J. (2003). Osteogenic enrichment of bone-marrow stromal cells with the use of flow chamber and type I collagen-coated surface. *Journal of Biomedical Materials Research. Part A*, Vol.66, No.1, 38-46

Langer, R., & Vacanti, J.P. (1993). Tissue Engineering, Science, Vol.260, 920–926

Lind, M., & Bunger C. (2001). Factors stimulating bone formation. *European Spine Journal*, Oct, Vol.10, Suppl 2, S102-S109

Li, Y.J., Batra, N.N., You, L., Meier, S.C., Coe, I.A., Yellowley, C.E., & Jacobs, C.R. (2004). Oscillatory fluid flow affects human marrow stromal cell proliferation and differentiation. *Journal of Orthopaedic Research*, Vol.22, No.6, 1283-1289

McAllister, T.N., & Frangos, J.A. (1999). Steady and transient fluid shear stress stimulate NO release in osteoblasts through distinct biochemical pathways. *Journal of Bone and Mineral Research*, Vol.14, No.6, 930-936

Meinel, L., Karageorgiou, V., Fajardo, R., Snyder, B., Shinde-Patil, V., Zichner, L., Kaplan, D., Langer, R., & Vunjak-Novakovic, G. (2004). Bone tissue engineering using human mesenchymal stem cells: effects of scaffold material and medium flow. *Annals of Biomedical Engineering*, Vol.32, No.1, 112-122

Pavalko, F.M., Chen, N.X., Turner, C.H., Burr, D.B., Atkinson, S., Hsieh, Y.F., Qiu, J., & Duncan, R.L. (1998). Fluid shear-induced mechanical signaling in MC3T3-E1 osteoblasts requires cytoskeleton-integrin interactions. *American Journal of Physiology*, Vol.275, C1591-C1601

Ponik, S.M., Triplett, J.W., & Pavalko, F.M. (2006). Osteoblasts and osteocytes respond differently to oscillatory and unidirectional fluid flow profiles. *Journal of Cellular Biochemistry*, Vol.100, No.3, 794-807

Porter, B., Zauel, R., Stockman, H., Guldberg, & R., Fyhrie, D. (2005). 3-D computational modeling of media flow through scaffolds in a perfusion bioreactor. *Journal of Biomechanics*, Vol.38, No.3, 543-549

Qin, Y.X., Kaplan, T., Saldanha, A., & Rubin, C. (2003). Fluid pressure gradients, arising from oscillations in intramedullary pressure, is correlated with the formation of bone and inhibition of intracortical porosity. *Journal of Biomechanics*, Vol.36, No.10, 1427-1437

Sikavitsas, V.I., Bancroft, G.N., & Mikos, A.G. (2002). Formation of three-dimensional cell/polymer constructs for bone tissue engineering in a spinner flask and a rotating wall vessel bioreactor. *Journal of Biomedical Materials Research*, Oct, Vol.62, No.1, 136-148

Sikavitsas, V.I., Bancroft, G.N., Holtorf, H.L., Jansen, J.A., & Mikos, A.G. (2003). Mineralized matrix deposition by marrow stromal osteoblasts in 3D perfusion culture increases with increasing fluid shear forces. *Proceedings of the National Academy of Sciences of the United States of America*, Vol.100, No.25, 14683-14688

Sikavitsas, V.I., van den Dolder, J., Bancroft, G.N., Jansen, J.A., & Mikos, A.G. (2003). Influence of the in vitro culture period on the in vivo performance of cell/titanium bone tissue-engineered constructs using a rat cranial critical size defect model. *Journal of Biomedical Materials Research Part A*, Vol.67, No.3, 944-951

Sinha, R.K., Shah S.A., Hume E.L., & Tuan, R.S. (2002). The effect of a 5-day space flight on the immature rat spine. *Spine Journal*, Jul-Aug, Vol.2, No.4, 239-243

Thi, M.M., Kojima, T., Cowin S.C., Weinbaum, S., & Spray, D.C. (2003). Fluid shear stress remodels expression and function of junctional proteins in cultured bone cells. *Americal Journal of Physiology*, Vol. 284, No. 2, C389-C403

Unsworth, B.R., & Lelkes, P.I. (1998). Growing tissues in microgravity. *Nature Medecine*, Vol. 4, 901–907

Vance, J., Galley, S., Liu, D.F., & Donahue, S.W. (2005). Mechanical stimulation of MC3T3 osteoblastic cells in a bone tissue-engineering bioreactor enhances prostaglandin E2 release. *Tissue Engineering*, Vol.11(11-12), 1832-1839

van den Dolder, J., Bancroft, G.N., Sikavitsas, V.I., Spauwen, P.H., Jansen, J.A., & Mikos, A.G. (2003). Flow perfusion culture of marrow stromal osteoblasts in titanium fiber mesh. *Journal of Biomedical Materials Research Part A*, Vol.64, No.2, 235-241

Vunjak-Novakovic, G. et al. (1998). Dynamic cell seeding of polymer scaffolds for cartilage tissue engineering. *Biotechnology Progress*, Vol.14, 193–202

Wang, Y., Uemura T., Dong J., Kojima H., Tanaka J., & Tateishi T. (2003). Application of perfusion culture system improves in vitro and in vivo osteogenesis of bone marrow-derived osteoblastic cells in porous ceramic materials. *Tissue Engineering*, Vol.9, No.6, 1205-1214

Wu, C.C., Li, Y.S., Haga, J.H., Wang, N., Lian, I.Y., Su, F.C., Usami, S., & Chien S. (2006). Roles of MAP kinases in the regulation of bone matrix gene expressions in human osteoblasts by oscillatory fluid flow. *Journal of Cellular Biochemistry*, Jun 1, Vol.98, No.3, 632-641

Yoon, S.T., & Boden, S.D. (2002). Osteoinductive molecules in orthopaedics: basic science and preclinical studies. *Clinical Orthopaedics and Related Research*, Feb, Vol.395, 33-43

You, J., Yellowley, C.E., Donahue, H.J., Zhang, Y., Chen, Q., & Jacobs, C.R. (2000). Substrate deformation levels associated with routine physical activity are less stimulatory to bone cells relative to loading-induced oscillatory fluid flow. *Journal of Biomechanical Engineering* , Vol.122, No.4, 387-393

You, J., Reilly, G.C., Zhen, X., Yellowley, C.E., Chen, Q., Donahue, H.J., & Jacobs, C.R. (2001). Osteopontin gene regulation by oscillatory fluid flow via intracellular calcium mobilization and activation of mitogen-activated protein kinase in MC3T3-E1 osteoblasts. *Journal of Biological Chemistry*, Vol.276, No.16, 13365-13371

Younger, E.M., & Chapman, M.W. (1989). Morbidity at bone graft donor sites. *Journal of Orthopaedic Trauma*. Vol. 3, No. 3, 192-195

Part 5

Cell - Biomaterial Interaction

26

The Fibrotic Response to Implanted Biomaterials: Implications for Tissue Engineering

Barbara Rolfe et al.*
Australian Institute for Bioengineering and Nanotechnology,
University of Queensland, Brisbane
Australia

1. Introduction

The foreign body response describes the non-specific immune response to implanted foreign materials (Coleman et al., 1974; Anderson, 2001; Luttikhuizen et al., 2006). It is characterised by the infiltration of inflammatory cells to the area to destroy or remove this material, followed by the repair or regeneration of the injured tissue. However, if the foreign material cannot be phagocytosed and removed, the inflammatory response persists until the material becomes encapsulated in a dense layer of fibrotic connective tissue (Anderson, 2001) which shields it from the immune system and isolates it from the surrounding tissues.

The foreign body response has developed as a protective mechanism to limit exposure to toxic or allergenic materials, but also presents a problem for modern medicine. Biomedical devices now serve in a vast number of medical applications, including orthopedic, dental and breast implants, pacemakers, sutures, vascular grafts, heart valves, intraocular and contact lenses, controlled drug delivery devices and biosensors. This response is common to all medical devices or prostheses implanted into living tissue, and ultimately results in fibrosis or fibrous encapsulation which compromises the efficiency of the device and frequently leads to device failure (reviewed in Anderson et al., 2008). For example, the contraction of the myofibroblast-rich capsules around breast implants leads to 'implant shrinkage' (Abbondanzo et al., 1999) while encapsulating tissue prevents the diffusion of molecules to biosensors or from implanted drug delivery pumps (Anderson et al., 2008).

The response to implanted materials varies depending on their physicochemical properties (eg shape, size, surface chemistry, morphology and porosity; see Morais et al (2010) for review). Jones et al (2007; 2008) have shown that macrophage adhesion and fusion is higher on hydrophobic surfaces than hydrophilic/neutral surfaces while McBane and co-workers (McBane et al., 2011) found that compared with 2-dimensional films, 3-dimensional porous polyurethane scaffolds induced a low inflammatory, wound healing phenotype and may

---

* Jane Mooney, Bing Zhang, Sani Jahnke, Sarah-Jane Le, Yu-Qian Chau, Qiping Huang, Hao Wang, Gordon Campbell and Julie Campbell
Australian Institute of Bioengineering and Nanotechnology, University of Queensland, Brisbane, Australia

reduce the negative effects of the foreign body reaction. However despite attempts to identify non-immunogenic implant materials, or to mask surface properties of the implant material with biocompatible coatings (Quinn et al., 1995; Shive & Anderson, 1997; Draye et al., 1998; Paradossi et al., 2003), the inflammatory response cannot be completely avoided (Cao et al., 2008). This is thought to be due to the adsorption of proteins such as fibrinogen, complement and antibodies to the material immediately after implantation (Kao et al., 1999; Hu et al., 2001; Gretzer et al., 2006). Thus as outlined by Wisniewski et al (2001), the key to long-term functionality of implanted devices such as glucose sensors is modulation of the tissue response. In order to do this, it is important to first understand the mechanisms underlying the foreign body response to implanted biomaterials, the cells involved and their molecular mediators.

2. Application of the foreign body response to tissue engineering

While the foreign body response is an unwanted consequence of implantation of biomedical devices, the fibrotic response to implanted scaffold material has been investigated as a strategy for tissue engineering purposes. Sparks (1969, 1973) used the host inflammatory response to a foreign material to create living autologous tissue suitable for arterial bypass grafting. In this procedure a mandrel, composed of a smooth silicone rubber rod of desired diameter and length, covered with a large-mesh knitted dacron tube, was implanted subcutaneously near the location of the artery to be grafted. However despite intensive testing, the lack of compliance/strength of the resulting tissue, as well as the absence of an endothelial lining, resulted in unacceptably high rates of thrombosis, dilatation and aneurysm (Conte, 1998). More recently, our laboratory and others have used the peritoneal cavity as a 'bioreactor' to produce autologous tissue for replacement/repair of arteries (Campbell et al., 1999, 2000, 2008; Chue et al., 2004) and other hollow smooth muscle organs, specifically bladder, uterus, vas deferens (Campbell et al., 2008) and urethra (Gu et al., 2009). De Visscher and co-workers used a similar technique to pre-seed acellular matrix scaffolds from bovine pericardium for use as tissue engineered heart valves (De Visscher et al., 2007, 2008; Vranken et al., 2008) while Hayashida et al (2007) reported satisfactory function and mechanical properties for 'biovalves' prepared by embedding synthetic scaffolds subcutaneously for 4 weeks. This tissue has also been used as a source of growth factors to stimulate bone formation to repair a femoral bone defect in a sheep model (Lutton et al., 2009).

In the procedure described by our laboratory, sterile foreign objects of the appropriate shape are implanted into the peritoneal cavities of animals for 2-3 weeks, then the encapsulating tissue removed for grafting into the same animal. Our original studies showed that in the first 3 days after implantation of a foreign object, undifferentiated cells of bone marrow origin, either resident within the peritoneal fluid or recruited to it, encapsulated the object (Campbell et al., 2000). Most of these cells expressed the common leukocyte antigen (CD45) and had the morphological appearance of monocyte/macrophages (Campbell et al., 2000). By day 7 a distinct capsule of round cells and extracellular matrix (ECM) had formed, and by day 14 cells had elongated and organised into multilayered strata within a fibrillar matrix. Ultrastructurally, these elongated cells had the characteristics of myofibroblasts and contained large amounts of rough endoplasmic reticulum and bundles of peripherally distributed myofilaments (Campbell & Ryan, 1983; Campbell et al., 1999). A layer of mesothelial cells was also observed to cover the developing capsule. The tissue encapsulating free-floating foreign objects in the peritoneal cavity is avascular, in contrast to tissue surrounding foreign material at other anatomical sites which is highly vascularised (Campbell & Ryan, 1983).

The capacity of cells within this myofibroblast-rich tissue capsule to differentiate further, if subject to the appropriate environmental cues, has also been demonstrated. For example, when grafted into an autologous artery to replace excised segments, they gradually (over 1-2 months) developed the characteristics of mature vascular smooth muscle cells (SMC) and expressed the smooth muscle differentiation markers smoothelin and smooth muscle myosin heavy chain isoform SM-2 (Efendy et al., 2000; Chue et al., 2004). Tissue remodelling occurred such that the grafted tissue developed morphological characteristics of the native artery, with the luminal surface of the smooth muscle tube becoming lined by endothelial-like cells, an outer 'adventitial' layer comprising fibroblasts, collagen matrix and vasa vasora also developed. SMC differentiation could be similarly induced by subjecting graft tissue to active intermittent stretch in vitro (Efendy et al., 2000). Similarly, when myofibroblast-rich capsules were grafted into bladder, vasa deferens or uterine horn, the graft tissue gradually remodelled to resemble the host organ, both structurally and functionally (Campbell et al., 2008).

3. What are myofibroblasts?

Myofibroblasts are heterogeneous cells of diverse origin with a morphology intermediate between fibroblasts and smooth muscle (Gabbiani et al., 1971; Powell et al., 1999; Hinz et al., 2007; Hinz, 2010). They are characterised by expression of the smooth muscle actin isoform (α-SM actin), the fibronectin splice variant ectodomain (ED-A FN) and synthesis of ECM proteins such as collagen I (Serini et al., 1998). During normal wound repair, myofibroblasts are transiently present at the wound site where they play essential roles in wound contraction and restoration of tissue integrity. Once the wound has regained normal structure and function, myofibroblasts disappear as a result of apoptosis (Gabbiani, 1996). However the prolonged presence of these cells leads to excessive collagen production and tissue contraction, and ultimately reduced tissue function and fibrosis (Mutsaers et al., 1997). Thus the timely appearance, differentiation and removal of myofibroblasts are critical for appropriate wound healing. However, despite the important roles played by myofibroblasts, further research is required to clarify the regulatory mechanisms controlling their proliferation, differentiation and apoptosis, and the factors that turn a normal repair process into pathology.

4. What is the origin of foreign body-induced myofibroblasts?`

Myofibroblasts were originally believed to be derived from tissue fibroblasts (Serini and Gabbiani, 1999), but there is now mounting evidence for alternative origins, depending on the tissue location and surrounding microenvironment. These include epithelial cells (via epithelial-mesenchymal transition; EMT; Iwano et al., 2002; Zeisberg et al., 2007; Kim et al., 2009), smooth muscle cells (Humphreys et al., 2010) and fibrocytes (Bucala et al., 1994; Abe et al., 2001; Quan et al., 2006). As major cellular constituents of the healthy peritoneal membrane, mesothelial cells are also thought to be a source of myofibroblasts within the peritoneal cavity, via EMT (Pollock, 2005). Indeed intermediate cell types (co-expressing mesothelial and myofibroblast markers) have been reported in dialysis effluent and parietal peritoneum of peritoneal dialysis patients (Yanez-Mo et al., 2003; Jimenez-Heffernan et al., 2004). There is also evidence that myofibroblasts can derive from a bone marrow progenitor. Our early investigations used a chimaeric mouse model to demonstrate that cells of haematopoietic origin form the myofibroblast capsule in the peritoneal cavity (Campbell et al., 2000). These findings have been corroborated by numerous studies showing bone marrow-derived myofibroblasts in many organs including lung (Hashimoto et al., 2004; Brocker et al., 2006), stomach, oesophagus, skin and kidney (Direkze et al., 2003).

Given the prevalence of macrophage-like cells in the early tissue capsule, and the fact that myofibroblasts within this tissue can be derived from haemopoietic origin, we proposed that macrophages are a likely source of myofibroblasts in the peritoneal foreign body response. The capacity for peritoneal macrophages to transdifferentiate was first proposed by Kouri and Ancheta (1972) who demonstrated the presence of cells with intermediate morphologies between macrophages and fibroblasts within tissue capsules that formed around Epon lamina implants. Our laboratory also identified similar cells in the tissue capsule around foreign material (boiled blood clots) implanted in the peritoneal cavity (Campbell & Ryan, 1983; Mosse et al., 1985).

To further investigate this hypothesis, we used transgenic 'MacGreen' mice (in which the enhanced green fluorescent protein (EGFP) transgene is driven by the colony stimulating factor-1 receptor (*csf1r*) proximal promotor to direct myeloid-restricted expression; Sasmono et al., 2003, 2007) to show that the majority of cells recruited to encapsulate the foreign body were of myeloid (monocyte/macrophage) origin (Mooney et al., 2010). Although a small subset of $EGFP^-$ cells, comprising mainly lymphocytes and mast cells was also observed, these cells are unlikely to contribute directly to fibrotic tissue formation. Indeed Rodriguez et al (2009) showed that the foreign body response is similar in T-cell deficient mice, indicating that T lymphocytes do not play a significant role in this process.

Characterisation of the myeloid cell response by FACS analysis showed that in the early phase (day 2) there was a rapid recruitment of $EGFP^+$ $Gr1^+$ ($Ly6C^+$) subsets to the peritoneal cavity to encapsulate the foreign object; these cells are granulocytes and monocytes, similar to those described by Geissmann et al. (2003) and Sunderkotter et al. (2004) As the inflammatory response progressed, expression of Gr1 (Ly6C) was down-regulated, with concomitant up-regulation of F4/80 (indicative of mature macrophages) and the *csf1r*-EGFP transgene. Macrophages persisted throughout the period of study, such that by day 28, mature 'resident-like' macrophages ($Gr1^-$ $EGFP^{hi}$ $F4/80^{hi}$) were the predominant cell type within the tissue capsule. The $EGFP^+$ cells within the day 28 tissue capsule included many cells with spindle-shaped myofibroblastic morphology, although macrophages, multinucleated giant cells and a small number of neutrophils were also present. Foreign body giant cells are formed by fusion of macrophages (Anderson, 2000) and are considered a hallmark of the foreign body response (Jay et al., 2010). The presence of macrophages up to 28 days after foreign body implantation is in accord with a previous report by Gretzer et al. (2006) who showed increasing proportions of $ED2^+$ mature macrophages over time in exudates surrounding subcutaneous implants in rats. However in our experience, the peritoneal foreign body response is more rapid in rats than mice. In contrast to the mouse where only a small proportion of cells express α-SM actin at day 14 (Mooney et al., 2010), at this time-point in rats, the majority of cells within the tissue capsule no longer express haemopoietic markers (CD45 or CD68) and most express myofibroblast markers α-SM actin and SM22 (Le et al., 2010).

The essential role of macrophages in the peritoneal foreign body response was confirmed by experiments using MacGreen mice in which macrophage depletion with clodronate liposomes almost completely abrogated tissue capsule development (Mooney et al., 2010). We further showed that as the tissue capsule developed around foreign body implants in the peritoneal cavity, a sub-population of $EGFP^+$ cells appeared that co-expressed the myofibroblast marker α-SM actin. The proportion of $EGFP^+$ α-SM $actin^+$ cells increased with time, reaching 51±1% of total cells (approx 80% of total α-SM $actin^+$ cells) at later stages of tissue development. The morphology of $EGFP^+$ α-SM $actin^+$ cells also changed from a rounded macrophage-like appearance to a more spindle-shaped myofibroblastic phenotype, thus providing evidence that cells of myeloid origin can transdifferentiate to myofibroblasts (Mooney et al., 2010). These results are in agreement with those of Jabs et al (2005) who demonstrated that labelled peripheral blood mononuclear cells contributed to tissue capsule formation and that from day 14 onwards, a proportion of α-SM actin-expressing spindle-shaped cells co-expressed macrophage markers (ED1/ED2).

Macrophage transdifferentiation has been documented in a number of other settings. Cultured peritoneal macrophages from mice chronically infected with *Schistosoma manosi* exhibited fibroblast-like characteristics and co-expressed fibroblast (pro-collagen) and macrophage (mac-1/mac-2) markers (Godoy et al., 1989; Bertrand et al., 1992) while monocyte-derived macrophages in infarcted myocardium have been reported to differentiate to myofibroblasts (Fujita et al., 2007). In response to transforming growth factor (TGF)-β, cultured peritoneal-derived macrophages have also been shown to transdifferentiate into smooth muscle-like cells/myofibroblasts, expressing smooth muscle/myofibroblast markers such as calponin and α-SM actin and down-regulating expression of the macrophage marker CD11b (Ninomiya et al., 2006).

Although there is now convincing evidence for the direct involvement of macrophages as cellular progenitors of fibrotic tissue, other cellular sources of peritoneal myofibroblasts are also possible. Vranken et al (2008) identified stem/progenitor cells expressing Sca-1, c-kit, CD34 and CD271 as major contributors to the early foreign body response to bovine pericardium patches implanted in the peritoneal cavity. Importantly these latter cells were shown to have the potential to differentiate to a number of lineages, including myofibroblastic. More recently this same group has suggested that fibrocytes ($CD68^+CD34^+$), rather than macrophages ($CD68^+$ $CD34^-$) are able to differentiate to myofibroblasts (Mesure et al., 2010).

Macrophages also have indirect (paracrine) roles in the fibrotic response to foreign material, releasing cytokines, growth factors, other inflammatory mediators and matrix degrading enzymes to modulate the inflammatory response and regulate tissue repair (Xia & Triffitt, 2006). In serosal wound healing and fibrosis, macrophages have been implicated to play a supporting role via the release of cytokines/growth factors which stimulate mesothelial cell proliferation (Mutsaers et al., 2002) and fibrogenic processes such as ECM synthesis (Sakai et al., 2006). Macrophages also play a key role in angiogenesis and tissue repair, releasing matrix metalloproteinases and angiogenic growth factors (Murdoch et al., 2008) and co-operating with progenitor cells (Anghelina et al., 2006).

5. What regulates tissue capsule development and myofibroblast differentiation?

In order to identify key transcriptional events associated with development of the non-adhered, avascular myofibroblast-rich tissue encapsulating foreign objects implanted in the peritoneal cavity, our laboratory performed microarray expression profiling of tissue from different stages of capsule development in a rat model (Le et al., 2010). Consistent with changes in cellular composition, the data showed a change in gene expression over time from inflammatory, particularly myeloid cell-associated (including genes for CD14, CSF-1 and its receptor, CSF-1R) at the early stages of capsule formation, to myofibroblast-related (including SM-22 and fibulin) at later stages. The temporal changes in gene expression included the early up-regulation of genes for inflammatory mediators and chemokines (such as monocyte chemoattractant protein (MCP)-1, monocyte inflammatory protein (MIP-1α) and its receptor CCR1, and stromal-derived factor (SDF)-1) to attract inflammatory cells (mainly macrophages) to the foreign object, as well as altered expression of adhesion molecules associated with inflammatory responses and (later) tissue morphogenesis.

Also identified were growth factors and cytokines (including platelet-derived growth factor (PDGF) and TGF-β) known to be released by macrophages at the onset of the foreign body response (Luttikhuizen et al., 2006), as well as ECM proteins (collagens I and 3, biglycan, decorin, syndecans-1 and –2) and enzymes associated with fibrosis and tissue remodelling (matrix metalloproteinases MMP-2 and -9) and their inhibitors (plasminogen activator inhibitor (PAI)-1 and tissue inhibitor of metalloproteinase (TIMP)-1). In addition to its role as a potent mitogenic and chemotactic agent for myofibroblast progenitors (Lindahl & Betsholtz, 1998), PDGF-BB is associated with the early stages of myofibroblast differentiation from progenitor cells (Oh et al., 1998). TGF-β is the principal mediator of myofibroblast differentiation in wound healing, inducing fibroblasts (and possibly other cell types) to differentiate into α-SM actin-expressing myofibroblasts with the capacity for contraction and ECM synthesis (Leask & Abraham, 2004). TGF-β signalling was significant throughout tissue development, as evidenced by the continued expression of its receptor (TGFβRII) and downstream signalling molecules (SMAD-1,-2,-4 and latent TGF-β binding protein (LTBP)-2), as well as up-regulated expression of TGF-β inducible genes including connective tissue growth factor (CTGF), insulin-like growth factor binding protein (IGFBP)-3, TIMP-1, PAI-1, decorin and collagen I subunits (Verrecchia et al., 2001, 2006). The biological relevance of the data was confirmed by cell culture studies which showed that PDGF-BB stimulated the proliferation of tissue capsule cells, while TGF-β_1 inhibited the response to serum mitogens but induced expression of α-SM actin (Zhang et al., unpublished data). Myofibroblast differentiation was further enhanced by a cocktail of PDGF, interleukin (IL)-13 and TGF-β. Moreover inhibition of TGF-β signaling, by either chemical inhibition of TGFβRI (ALK5) (with LY-364947; Sawyer et al., 2003) or siRNA inhibition of TGFβRII, reduced α-SM actin expression by these cells *in vitro* and inhibited tissue development *in vivo*, demonstrating the critical role of this growth factor in the peritoneal foreign body response (Chau et al., unpublished data).

Functional analysis of the gene array data identified *Immune Response* and *Immune System Development and Function* as significant during the early stages and *Connective Tissue and Development and Function* at later stages of tissue development. The importance of fibrotic signalling was corroborated by pathway analysis identifying '*Hepatic Fibrosis/Hepatic Stellate*

Cell Activation' as significantly up-regulated at days 14 and 21. Although peritoneal tissue capsules show no evidence of vascularisation, the gene expression data provides additional evidence of the potential of capsular cells for angiogenesis, with genes for angiopoietins-1 and -2, vascular endothelial growth factor (VEGF) and its receptors Flt-1 and Flk-1 up-regulated. Moreover, the identification of *Cardiovascular System Development and Function* and *Skeletal and Muscular System Development and Function* as significant at day 14 indicates that, although the tissue capsule does not develop these functions, cells within it may have the potential to differentiate further along multiple pathways. This finding provides a molecular basis for our demonstration that when transplanted into smooth muscle organs within the same animal, cells within the tissue differentiate further towards a smooth muscle phenotype (Campbell et al., 1999, 2008; Efendy et al., 2000). The identification of genes associated with other mesenchymal lineages such as endothelial, cardiac and skeletal muscle suggests the capacity to differentiate to these cell types, given the appropriate environmental conditions.

While this study provided information regarding global changes in gene expression associated with the tissue response to foreign body implantation, information regarding gene expression by individual cell types is also required. To this end, a recent study by Mesure and co-workers (2010) showed up-regulated expression of cytokines and inflammatory response genes in CD68$^+$ cells isolated from tissue 3 days after foreign body implantation; pathways related to the innate immune response (IL-1α, IL-1β, IL-6, IL-10 and tumour necrosis factor (TNF)-α), cell adhesion (ICAM-1, VCAM-1) and matrix remodelling (MMP-13) were identified. In vitro culture of these cells with fibrinogen showed a switch over time from inflammatory to wound healing macrophages, evidenced by over-expression of genes such as IL-13Rα IL4Rα and arginase 1, and up-regulation of TGF-β signalling. These results are in line with an earlier study by Garrigues et al (2005) who identified a small number of genes which are highly regulated in macrophages exposed to wear debris from components of joint replacement prostheses. These included early changes (30 mins to 8 hours) in genes associated with ECM remodelling and angiogenesis, and also chemokines (IL-6), cytokines (RANTES, MIP-1α) and their receptors (TNFR1, 2 and IL2αR); other inflammatory cytokines (IL-1β, TNF-α, TGFβ3, oncostatin M) and IL6R-α subunit were up-regulated slightly later (at 24 hours).

Examination of cytokine production by biomaterial-adherent macrophages similarly showed expression of pro-inflammatory cytokines (IL-1β, IL-6, IL-8 and MIP-1β) highest at early stages (day 3) while IL-10 expression increased later, suggesting a phenotypic switch over time from classically activated to alternatively activated macrophages (Anderson & Jones, 2007). This has been corroborated by a recent study in our laboratory which identified two major phases: an inflammatory phase characterised by high levels of expression of Th1/M1 cytokines (IL-1β and TNF-α) at day 2, and a fibrotic phase regulated by TGF-β signalling pathways (Jahnke et al., unpublished data). Early expression of chemokines MCP-1, MIP-1α, RANTES and GRO/KC highlighted the importance of recruitment of inflammatory cells to the foreign object.

6. Modification of the foreign body response

The identification of strategies to regulate the foreign body response has important implications both for tissue engineering and for the proper functioning of implanted medical devices. *In vitro* studies have shown that although the foreign body response cannot be completely avoided, monocyte adhesion is influenced by surface chemistry of the biomaterial, the type and amount of adsorbed protein, and adhesion time (Shen & Horbett, 2001). Possibilities to modulate the response and the cellular content of resulting tissue also exist.

Attempts to reduce the fibrotic response to implanted medical devices have included the use of biocompatible coatings to mask the underlying material and reduce protein adsorption and cellular interaction. Coating materials tested include synthetic polymers such as poly(vinyl alcohol) (Galeska et al., 2005), poly(lactic acid) and poly(lactic co-glycolic) acid (Athanasiou et al., 1996). Naturally occurring materials such as chitosan (Borchard & Junginger, 2001), collagen (Geiger et al., 2003) and alginate (de Vos et al., 2002) have also been used, although these are frequently immunogenic and subject to natural variability in their macromolecular structure (Morais et al., 2010). They also allow better cell adhesion and therefore may be more suitable for tissue engineering applications requiring enhanced tissue production (Cheung et al., 2007). The physical attributes of the material may also be important, with both porous PLA (Koschwanez et al., 2008) and collagen (Ju et al., 2008) coatings shown to reduce fibrosis and/or promote blood vessel formation to enhance function and life-time of implantable glucose sensors.

Hydrogels composed of polar, uncharged, flexible materials such as poly (hydroxyethyl methacrylate) (PHEMA) or polyethylene glycol (PEG) form a hydrophilic interface between the underlying surface and the surrounding tissue, and allow analyte diffusion (Wisniewski & Reichert, 2000). PEG-based hydrogels have been shown to substantially reduce the immune response around biosensors implanted in rats (Quinn et al., 1997) while PHEMA coatings reduced clotting and protein adsorption to calcium monitors in dogs (McKinley et al., 1981). Phospholipid-containing materials designed to mimic the cell membrane have been shown to reduce adhesion of inflammatory cells and fibrous capsule formation around vascular devices (Goreish et al., 2004) while Abraham et al. (2005) showed that formulations incorporating PEG and phosphorylcholine into PHEMA-based hydrogels greatly reduced protein adsorption.

Alternatively, strategies may be directed towards augmentation of the foreign body response for tissue engineering. In our attempts to enhance the production of tissue for transplantation as autologous grafts for hollow smooth muscle organs, we tested the tissue response to peritoneal implantation of poly(lactic acid) tubular scaffolds with different layer-by-layer biomolecule coatings (as described by Croll et al., 2006). Immunohistochemical analysis of the resulting tissue showed that Matrigel-coated surfaces supported the strongest cellular response whereas multilayer coatings with elastin, collagen I, collagen III or chitosan outermost showed the lowest levels of cellular interaction. While differences in capsule thickness and growth characteristics were observed, all of the biomolecule coatings tested induced the peritoneal foreign body response, even in the presence of a non-adsorptive hyaluronic acid undercoat (Cao et al., 2008).

The variable ability of polymeric coatings to reduce acute and chronic inflammatory responses *in vivo* (Shen et al., 2002; Park & Bae, 2003) has led to investigation of anti-inflammatory drugs to inhibit the tissue response to biomaterials. The most commonly used drug has been dexamethasone which modulates macrophage behaviour and reduces the levels of pro-inflammatory cytokines such as tumor necrosis factor (TNF)-α (Joyce et al., 1997; Umland et al., 2002). Burgess and co-workers embedded dexamethasone-containing

PLGA microspheres in PVA hydrogels to create a 'smart' coating for glucose sensors which allowed rapid diffusion of analytes and slow release of dexamethasone (Hickey et al., 2002; Patil et al., 2004). Other anti-inflammatory strategies investigated include heparin-based coatings to reduce protein adsorption and leukocyte recruitment (Rele et al., 2005) and covalent conjugation of a superoxide dismutase mimetic to the surface of biomaterials, resulting in reduced neutrophil recruitment and inhibition of foreign body giant cell and fibrous tissue capsule formation (Udipi et al., 2000). However Jones (2007) has suggested that minimisation of the inflammatory response to implanted biomaterials may be counter-productive, and that a preferable strategy would be to design materials to direct the response towards reparative/wound healing. For example, it may be possible to tip the balance away from production of TGF-β1 (which promotes fibro-proliferation) towards TGFβ3 (which promotes tissue repair) (Ask et al., 2008). Indeed glucocorticocoid drugs have been shown to modulate the phenotype of infiltrating macrophages and lymphocytes (Peek et al., 2005; Mosser & Zhang, 2008) and could thus be used locally to regulate the cellular response.

By altering the cell populations recruited to the foreign body it may be possible to modify the inflammatory response to implanted foreign material, as well as the cellular nature of the subsequent tissue response. Given that chemokines regulate cell trafficking (Gerard & Rollins, 2001) and play a key role in the recruitment of inflammatory cells in the peritoneal foreign body response (Luttikhuizen et al., 2007), chemokines may be an important target for intervention. De Visscher et al. (2010) have shown that impregnation of bioprosthetic heart valves with SDF-1 and fibronectin modulated the cellular response to produce more biologically relevant tissue with properties very similar to native valves, whilst Thevenot et al. (2010) showed that delivery of SDF-1α to the site of biomaterial implantation increased the recruitment of host stem cells, and at the same time reduced the inflammatory response, such that the fibrotic response to scaffold implants was ameliorated. They suggested that enhanced recruitment of autologous stem cells can improve the tissue responses to biomaterial implants through modifying/bypassing inflammatory cell responses and stimulating stem cell participation in healing at the implant interface. Our preliminary experiments show that continuous infusion of AMD3100 (a specific antagonist of the SDF-1 receptor, CXCR4; Matthys et al., 2001) does not inhibit encapsulation of foreign material implanted in the peritoneal cavity, but alters the cellular composition of the encapsulating tissue (Le et al, unpublished data). Another important chemokine, MCP-1, is highly expressed at the early stages of the cellular response to foreign body implantation (Le et al., 2010), and has been implicated in the pathogenesis of progressive fibrosis in lungs (Moore et al., 2001) and kidney (Kitagawa et al., 2004). Via its receptor CCR2, MCP-1 plays an important role in the recruitment of inflammatory monocyte subsets from the bone marrow into injured tissues (Geissmann et al., 2003; Karlmark et al., 2009). However Kyriakides et al. (2004) demonstrated that the lack of MCP-1 resulted in reduced foreign body giant cell formation, but did not affect either the recruitment/migration of macrophages to the site of biomaterial implantation or capsule formation.

Biomaterials may also be used to deliver anti-fibrotic drugs/inhibitors. As mentioned above, we have shown that inhibition of TGF-β signalling either by siRNA knockdown of TGFβRII or chemical inhibition of TGFβRII (ALK5) inhibits myofibroblast differentiation in vitro and peritoneal tissue capsule formation in vivo (Chau et al, unpublished data). However in light of the pleiotropic roles of TGF-β, a more suitable target for selective intervention may be the downstream effector, connective tissue growth factor (CTGF) which is responsible for many of the pro-fibrogenic effects of TGF-β (Ward et al., 2008; Brigstock, 2009).

Conversely, chemokines/growth factors may be incorporated into biomaterials to promote tissue production for replacement/repair. For example, polylactic/glycolic acid scaffolds (as described by Cao et al., 2006) stabilise and prolong the half-life of growth factors, and provide a means for localised release of the growth factor at a controlled dose and rate of delivery over a prolonged period (Richardson et al., 2001). Chemokines and growth factors could be incorporated into different layers, then released sequentially to first recruit cells to the scaffold, then promote the proliferation of adherent cells, and finally induce their differentiation to produce mature tissue.

7. Conclusion

The ability to regulate the fibrotic response to implanted materials has important implications for bioengineering, both to control the deleterious response to implanted medical devices and to enhance the production of tissue for organ repair. Myofibroblasts are critical for appropriate wound healing and tissue repair, but are also responsible for fibrosis. Hence understanding the origins of cells involved in the development of myofibroblast-rich tissue, and identification of the mechanisms regulating their (trans)differentiation and biology, is the key to successful bioengineering strategies.

Our research into the peritoneal foreign body response questions the traditional notion of distinct terminally differentiated cell types with specific functions. The results demonstrate a developmental continuum from monocyte (or granulocyte) through macrophage to myofibroblast, and potentially smooth muscle and/or other cell types. Given that cellular plasticity is a hallmark of the myeloid lineage (Hume, 2008), these findings extend the prevailing concepts of adult cell fate. We propose that, at least for some cell lineages, cellular identity is more fluid than previously recognised. Thus within its life-time, a single cell has the capacity to adopt a range of phenotypes and functions according to physiological needs and local regulatory milieu. While research to date has focussed on strategies to minimise/inhibit the inflammatory response to biomaterials, a preferable strategy may be to direct the response towards immune tolerance and tissue regeneration.

8. References

Abbondanzo, S. L., V. L. Young, M. Q. Wei and F. W. Miller (1999). Silicone gel-filled breast and testicular implant capsules: a histologic and immunophenotypic study. *Mod Pathol* 12(7): 706-13.

Abe, R., S. C. Donnelly, T. Peng, R. Bucala and C. N. Metz (2001). Peripheral blood fibrocytes: differentiation pathway and migration to wound sites. *J Immunol* 166(12): 7556-62.

Abraham, S., S. Brahim, K. Ishihara and A. Guiseppi-Elie (2005). Molecularly engineered p(HEMA)-based hydrogels for implant biochip biocompatibility. *Biomaterials* 26(23): 4767-78.

Anderson, J. M. (2000). Multinucleated giant cells. *Curr Opin Hematol* 7(1): 40-7.

Anderson, J. M. (2001). Biological responses to materials. *Annu Rev Mater Res* 31: 81-110.

Anderson, J. M. and J. A. Jones (2007). Phenotypic dichotomies in the foreign body reaction. *Biomaterials* 28(34): 5114-20.

Anderson, J. M., A. Rodriguez and D. T. Chang (2008). Foreign body reaction to biomaterials. *Semin Immunol* 20(2): 86-100.

Anghelina, M., P. Krishnan, L. Moldovan and N. I. Moldovan (2006). Monocytes/macrophages cooperate with progenitor cells during neovascularization and tissue repair: conversion of cell columns into fibrovascular bundles. *Am J Pathol* 168(2): 529-41.

Ask, K., P. Bonniaud, K. Maass, O. Eickelberg, P. J. Margetts, D. Warburton, J. Groffen, J. Gauldie and M. Kolb (2008). Progressive pulmonary fibrosis is mediated by TGF-beta isoform 1 but not TGF-beta3. *Int J Biochem Cell Biol* 40(3): 484-95.

Athanasiou, K. A., G. G. Niederauer and C. M. Agrawal (1996). Sterilization, toxicity, biocompatibility and clinical applications of polylactic acid/polyglycolic acid copolymers. *Biomaterials* 17(2): 93-102.

Bertrand, S., M. Godoy, P. Semal and P. Van Gansen (1992). Transdifferentiation of macrophages into fibroblasts as a result of Schistosoma mansoni infection. *Int J Dev Biol* 36(1): 179-84.

Borchard, G. and H. E. Junginger (2001). Modern drug delivery applications of chitosan. *Adv Drug Deliv Rev* 52(2): 103.

Brigstock, D. R. (2009). Strategies for blocking the fibrogenic actions of connective tissue growth factor (CCN2): From pharmacological inhibition in vitro to targeted siRNA therapy in vivo. *J Cell Commun Signal* 3(1): 5-18.

Brocker, V., F. Langer, T. G. Fellous, M. Mengel, M. Brittan, M. Bredt, S. Milde, T. Welte, M. Eder, A. Haverich, M. R. Alison, H. Kreipe and U. Lehmann (2006). Fibroblasts of recipient origin contribute to bronchiolitis obliterans in human lung transplants. *Am J Respir Crit Care Med* 173(11): 1276-82.

Bucala, R., L. A. Spiegel, J. Chesney, M. Hogan and A. Cerami (1994). Circulating fibrocytes define a new leukocyte subpopulation that mediates tissue repair. *Mol Med* 1(1): 71-81.

Campbell, G. R. and G. B. Ryan (1983). Origin of myofibroblasts in the avascular capsule around free-floating intraperitoneal blood clots. *Pathology* 15(3): 253-64.

Campbell, G. R., G. Turnbull, L. Xiang, M. Haines, S. Armstrong, B. E. Rolfe and J. H. Campbell (2008). The peritoneal cavity as a bioreactor for tissue engineering visceral organs: bladder, uterus and vas deferens. *J Tissue Eng Regen Med* 2(1): 50-60.

Campbell, J. H., J. L. Efendy and G. R. Campbell (1999). Novel vascular graft grown within recipient's own peritoneal cavity. *Circ Res* 85(12): 1173-8.

Campbell, J. H., J. L. Efendy, C. Han, A. A. Girjes and G. R. Campbell (2000). Haemopoietic origin of myofibroblasts formed in the peritoneal cavity in response to a foreign body. *J Vasc Res* 37(5): 364-71.

Campbell, J. H., J. L. Efendy, C. L. Han and G. R. Campbell (2000). Blood vessels from bone marrow. *Ann N Y Acad Sci* 902: 224-9.

Cao, Y., Croll, T., O'Connor, A., Stevens, G., Cooper-White, J. (2006). Systematic selection of solvents for the fabrication of 3D combined micro- and macroporous poly (lactic-co-glycolic acid) scaffold for soft tissue engineering. *J Biomaterial Science* In press.

Cao, Y., B. Zhang, T. Croll, B. E. Rolfe, J. H. Campbell, G. R. Campbell, D. Martin and J. J. Cooper-White (2008). Engineering tissue tubes using novel multilayered scaffolds in the rat peritoneal cavity. *J Biomed Mater Res A* 87(3): 719-27.

Cheung, H.-Y., Lau, K.-T., Lu T.-P., Hui, D. (2007). A critical review on polymer-based bio-engineered materials for scaffold development . *Composites Part B: Engineering* 38(3): 291-300.

Chue, W. L., G. R. Campbell, N. Caplice, A. Muhammed, C. L. Berry, A. C. Thomas, M. B. Bennett and J. H. Campbell (2004). Dog peritoneal and pleural cavities as bioreactors to grow autologous vascular grafts. *J Vasc Surg* 39(4): 859-67.

Coleman, D. L., R. N. King and J. D. Andrade (1974). The foreign body reaction: a chronic inflammatory response. *J Biomed Mater Res* 8(5): 199-211.

Conte, M. S. (1998). The ideal small arterial substitute: a search for the Holy Grail? *Faseb J* 12(1): 43-5.

Croll, T. I., A. J. O'Connor, G. W. Stevens and J. J. Cooper-White (2006). A blank slate? Layer-by-layer deposition of hyaluronic acid and chitosan onto various surfaces. *Biomacromolecules* 7(5): 1610-22.

De Visscher, G., H. Blockx, B. Meuris, H. Van Oosterwyck, E. Verbeken, M. C. Herregods and W. Flameng (2008). Functional and biomechanical evaluation of a completely recellularized stentless pulmonary bioprosthesis in sheep. *J Thorac Cardiovasc Surg* 135(2): 395-404.

De Visscher, G., A. Lebacq, L. Mesure, H. Blockx, I. Vranken, R. Plusquin, B. Meuris, M. C. Herregods, H. Van Oosterwyck and W. Flameng (2010). The remodeling of cardiovascular bioprostheses under influence of stem cell homing signal pathways. *Biomaterials* 31(1): 20-8.

De Visscher, G., I. Vranken, A. Lebacq, C. Van Kerrebroeck, J. Ganame, E. Verbeken and W. Flameng (2007). In vivo cellularization of a cross-linked matrix by intraperitoneal implantation: a new tool in heart valve tissue engineering. *Eur Heart J* 28(11): 1389-96.

de Vos, P., C. G. Hoogmoed and H. J. Busscher (2002). Chemistry and biocompatibility of alginate-PLL capsules for immunoprotection of mammalian cells. *J Biomed Mater Res* 60(2): 252-9.

Direkze, N. C., S. J. Forbes, M. Brittan, T. Hunt, R. Jeffery, S. L. Preston, R. Poulsom, K. Hodivala-Dilke, M. R. Alison and N. A. Wright (2003). "Multiple organ engraftment by bone-marrow-derived myofibroblasts and fibroblasts in bone-marrow-transplanted mice. *Stem Cells* 21(5): 514-20.

Draye, J. P., B. Delaey, A. Van de Voorde, A. Van Den Bulcke, B. De Reu and E. Schacht (1998). In vitro and in vivo biocompatibility of dextran dialdehyde cross-linked gelatin hydrogel films. *Biomaterials* 19(18): 1677-87.

Efendy, J. L., G. R. Campbell and J. H. Campbell (2000). The effect of environmental cues on the differentiation of myofibroblasts in peritoneal granulation tissue. *J Pathol* 192(2): 257-62.

Fujita, J., M. Mori, H. Kawada, Y. Ieda, M. Tsuma, Y. Matsuzaki, H. Kawaguchi, T. Yagi, S. Yuasa, J. Endo, T. Hotta, S. Ogawa, H. Okano, R. Yozu, K. Ando and K. Fukuda (2007). Administration of granulocyte colony-stimulating factor after myocardial infarction enhances the recruitment of hematopoietic stem cell-derived myofibroblasts and contributes to cardiac repair. *Stem Cells* 25(11): 2750-9.

Gabbiani, G. (1996). The cellular derivation and the life span of the myofibroblast. *Pathol Res Pract* 192(7): 708-11.
Gabbiani, G., G. B. Ryan and G. Majne (1971). Presence of modified fibroblasts in granulation tissue and their possible role in wound contraction. *Experientia* 27(5): 549-50.
Galeska, I., T. K. Kim, S. D. Patil, U. Bhardwaj, D. Chatttopadhyay, F. Papadimitrakopoulos and D. J. Burgess (2005). Controlled release of dexamethasone from PLGA microspheres embedded within polyacid-containing PVA hydrogels. *AAPS J* 7(1): E231-40.
Garrigues, G. E., D. R. Cho, H. E. Rubash, S. R. Goldring, J. H. Herndon and A. S. Shanbhag (2005). Gene expression clustering using self-organizing maps: analysis of the macrophage response to particulate biomaterials. *Biomaterials* 26(16): 2933-45.
Geiger, M., R. H. Li and W. Friess (2003). Collagen sponges for bone regeneration with rhBMP-2. *Adv Drug Deliv Rev* 55(12): 1613-29.
Geissmann, F., S. Jung and D. R. Littman (2003). Blood monocytes consist of two principal subsets with distinct migratory properties. *Immunity* 19(1): 71-82.
Gerard, C. and B. J. Rollins (2001). Chemokines and disease. *Nat Immunol* 2(2): 108-15.
Godoy, M., M. Geuskens, E. A. Van Marck, R. Borojevic and P. Van Gansen (1989). Schistosomiasis and in vitro transdifferentiation of murine peritoneal macrophages into fibroblastic cells. *Parasitol Res* 76(2): 150-61.
Goreish, H. H., A. L. Lewis, S. Rose and A. W. Lloyd (2004). The effect of phosphorylcholine-coated materials on the inflammatory response and fibrous capsule formation: in vitro and in vivo observations. *J Biomed Mater Res A* 68(1): 1-9.
Gretzer, C., L. Emanuelsson, E. Liljensten and P. Thomsen (2006). The inflammatory cell influx and cytokines changes during transition from acute inflammation to fibrous repair around implanted materials. *J Biomater Sci Polym Ed* 17(6): 669-87.
Gu, G. L., Y. J. Zhu, S. J. Xia, J. Zhang, J. T. Jiang, Y. Hong and G. H. Liu (2008). Peritoneal cavity as bioreactor to grow autologous tubular urethral grafts in a rabbit model. *World J Urol* 28(2): 227-32.
Hashimoto, N., H. Jin, T. Liu, S. W. Chensue and S. H. Phan (2004). Bone marrow-derived progenitor cells in pulmonary fibrosis. *J Clin Invest* 113(2): 243-52.
Hayashida, K., K. Kanda, H. Yaku, J. Ando and Y. Nakayama (2007). Development of an in vivo tissue-engineered, autologous heart valve (the biovalve): preparation of a prototype model. *J Thorac Cardiovasc Surg* 134(1): 152-9.
Hickey, T., D. Kreutzer, D. J. Burgess and F. Moussy (2002). Dexamethasone/PLGA microspheres for continuous delivery of an anti-inflammatory drug for implantable medical devices. *Biomaterials* 23(7): 1649-56.
Hinz, B. (2010). The myofibroblast: paradigm for a mechanically active cell. *J Biomech* 43(1): 146-55.
Hinz, B., S. H. Phan, V. J. Thannickal, A. Galli, M. L. Bochaton-Piallat and G. Gabbiani (2007). The myofibroblast: one function, multiple origins. *Am J Pathol* 170(6): 1807-16.
Hu, W. J., J. W. Eaton, T. P. Ugarova and L. Tang (2001). Molecular basis of biomaterial-mediated foreign body reactions. *Blood* 98(4): 1231-8.
Hume, D. A. (2008). "Differentiation and heterogeneity in the mononuclear phagocyte system. *Mucosal Immunol* 1(6): 432-41.

Humphreys, B. D., S. L. Lin, A. Kobayashi, T. E. Hudson, B. T. Nowlin, J. V. Bonventre, M. T. Valerius, A. P. McMahon and J. S. Duffield (2010). Fate tracing reveals the pericyte and not epithelial origin of myofibroblasts in kidney fibrosis. *Am J Pathol* 176(1): 85-97.

Iwano, M., D. Plieth, T. M. Danoff, C. Xue, H. Okada and E. G. Neilson (2002). Evidence that fibroblasts derive from epithelium during tissue fibrosis. *J Clin Invest* 110(3): 341-50.

Jabs, A., G. A. Moncada, C. E. Nichols, E. K. Waller and J. N. Wilcox (2005). Peripheral blood mononuclear cells acquire myofibroblast characteristics in granulation tissue. *J Vasc Res* 42(2): 174-80.

Jay, S. M., E. A. Skokos, J. Zeng, K. Knox and T. R. Kyriakides (2010). Macrophage fusion leading to foreign body giant cell formation persists under phagocytic stimulation by microspheres in vitro and in vivo in mouse models. *J Biomed Mater Res A* 93(1): 189-99.

Jimenez-Heffernan, J. A., A. Aguilera, L. S. Aroeira, E. Lara-Pezzi, M. A. Bajo, G. del Peso, M. Ramirez, C. Gamallo, J. A. Sanchez-Tomero, V. Alvarez, M. Lopez-Cabrera and R. Selgas (2004). Immunohistochemical characterization of fibroblast subpopulations in normal peritoneal tissue and in peritoneal dialysis-induced fibrosis. *Virchows Arch* 444(3): 247-56.

Jones, J. A., D. T. Chang, H. Meyerson, E. Colton, I. K. Kwon, T. Matsuda and J. M. Anderson (2007). Proteomic analysis and quantification of cytokines and chemokines from biomaterial surface-adherent macrophages and foreign body giant cells. *J Biomed Mater Res A* 83(3): 585-96.

Jones, J. A., A. K. McNally, D. T. Chang, L. A. Qin, H. Meyerson, E. Colton, I. L. Kwon, T. Matsuda and J. M. Anderson (2008). Matrix metalloproteinases and their inhibitors in the foreign body reaction on biomaterials. *J Biomed Mater Res A* 84(1): 158-66.

Joyce, D. A., J. H. Steer and L. J. Abraham (1997). Glucocorticoid modulation of human monocyte/macrophage function: control of TNF-alpha secretion. *Inflamm Res* 46(11): 447-51.

Ju, Y. M., B. Yu, T. J. Koob, Y. Moussy and F. Moussy (2008). A novel porous collagen scaffold around an implantable biosensor for improving biocompatibility. I. In vitro/in vivo stability of the scaffold and in vitro sensitivity of the glucose sensor with scaffold. *J Biomed Mater Res A* 87(1): 136-46.

Kao, W. J., J. A. Hubbell and J. M. Anderson (1999). Protein-mediated macrophage adhesion and activation on biomaterials: a model for modulating cell behavior. *J Mater Sci Mater Med* 10(10/11): 601-5.

Karlmark, K. R., R. Weiskirchen, H. W. Zimmermann, N. Gassler, F. Ginhoux, C. Weber, M. Merad, T. Luedde, C. Trautwein and F. Tacke (2009). Hepatic recruitment of the inflammatory Gr1+ monocyte subset upon liver injury promotes hepatic fibrosis. *Hepatology* 50(1): 261-74.

Kim, K. K., Y. Wei, C. Szekeres, M. C. Kugler, P. J. Wolters, M. L. Hill, J. A. Frank, A. N. Brumwell, S. E. Wheeler, J. A. Kreidberg and H. A. Chapman (2009). Epithelial cell alpha3beta1 integrin links beta-catenin and Smad signaling to promote myofibroblast formation and pulmonary fibrosis. *J Clin Invest* 119(1): 213-24.

Kitagawa, K., T. Wada, K. Furuichi, H. Hashimoto, Y. Ishiwata, M. Asano, M. Takeya, W. A. Kuziel, K. Matsushima, N. Mukaida and H. Yokoyama (2004). Blockade of CCR2 ameliorates progressive fibrosis in kidney. *Am J Pathol* 165(1): 237-46.

Koschwanez, H. E., F. Y. Yap, B. Klitzman and W. M. Reichert (2008). In vitro and in vivo characterization of porous poly-L-lactic acid coatings for subcutaneously implanted glucose sensors. *J Biomed Mater Res A* 87(3): 792-807.

Kouri, J. and O. Ancheta (1972). Transformation of macrophages into fibroblasts. *Exp Cell Res* 71(1): 168-76.

Kyriakides, T. R., M. J. Foster, G. E. Keeney, A. Tsai, C. M. Giachelli, I. Clark-Lewis, B. J. Rollins and P. Bornstein (2004). The CC chemokine ligand, CCL2/MCP1, participates in macrophage fusion and foreign body giant cell formation. *Am J Pathol* 165(6): 2157-66.

Le, S. J., M. Gongora, B. Zhang, S. Grimmond, G. R. Campbell, J. H. Campbell and B. E. Rolfe (2010). Gene expression profile of the fibrotic response in the peritoneal cavity. *Differentiation* 79(4-5): 232-43.

Leask, A. and D. J. Abraham (2004). TGF-beta signaling and the fibrotic response. *Faseb J* 18(7): 816-27.

Lindahl, P. and C. Betsholtz (1998). Not all myofibroblasts are alike: revisiting the role of PDGF-A and PDGF-B using PDGF-targeted mice. *Curr Opin Nephrol Hypertens* 7(1): 21-6.

Luttikhuizen, D. T., M. C. Harmsen and M. J. Luyn (2006). Cellular and Molecular Dynamics in the Foreign Body Reaction. *Tissue Eng* 12: 1955-1970.

Luttikhuizen, D. T., M. C. Harmsen and M. J. van Luyn (2007). Cytokine and chemokine dynamics differ between rats and mice after collagen implantation. *J Tissue Eng Regen Med* 1(5): 398-405.

Lutton, C., S. Sugiyama, M. E. Wullschleger, R. Williams, J. H. Campbell, R. Crawford and B. Goss (2009). Transplanted abdominal granulation tissue induced bone formation--an in vivo study in sheep. *Connect Tissue Res* 50(4): 256-62.

Matthys, P., S. Hatse, K. Vermeire, A. Wuyts, G. Bridger, G. W. Henson, E. De Clercq, A. Billiau and D. Schols (2001). AMD3100, a potent and specific antagonist of the stromal cell-derived factor-1 chemokine receptor CXCR4, inhibits autoimmune joint inflammation in IFN-gamma receptor-deficient mice. *J Immunol* 167(8): 4686-92.

McBane, J. E., D. Ebadi, S. Sharifpoor, R. S. Labow and J. P. Santerre (2011). Differentiation of monocytes on a degradable, polar, hydrophobic, ionic polyurethane: Two-dimensional films vs. three-dimensional scaffolds. *Acta Biomater* 7(1): 115-22.

McKinley, B. A., K. C. Wong, J. Janata, W. S. Jordan and D. R. Westenskow (1981). In vivo continuous monitoring of ionized calcium in dogs using ion sensitive field effect transistors. *Crit Care Med* 9(4): 333-9.

Mesure, L., G. De Visscher, I. Vranken, A. Lebacq and W. Flameng (2010). Gene expression study of monocytes/macrophages during early foreign body reaction and identification of potential precursors of myofibroblasts. *PLoS ONE* 5(9): e12949.

Mooney, J. E., B. E. Rolfe, G. W. Osborne, D. P. Sester, N. van Rooijen, G. R. Campbell, D. A. Hume and J. H. Campbell (2010). Cellular plasticity of inflammatory myeloid cells in the peritoneal foreign body response. *Am J Pathol* 176(1): 369-80.

Moore, B. B., R. Paine, 3rd, P. J. Christensen, T. A. Moore, S. Sitterding, R. Ngan, C. A. Wilke, W. A. Kuziel and G. B. Toews (2001). Protection from pulmonary fibrosis in the absence of CCR2 signaling. *J Immunol* 167(8): 4368-77.

Morais, J. M., F. Papadimitrakopoulos and D. J. Burgess (2010). Biomaterials/tissue interactions: possible solutions to overcome foreign body response. *AAPS J* 12(2): 188-96.

Mosse, P. R., G. R. Campbell and G. B. Ryan (1985). A comparison of the avascular capsule surrounding free floating intraperitoneal blood clots in mice and rabbits. *Pathology* 17(3): 401-7.

Mosser, D. M. and X. Zhang (2008). Activation of murine macrophages. *Curr Protoc Immunol* Chapter 14: Unit 14 2.

Murdoch, C., M. Muthana, S. B. Coffelt and C. E. Lewis (2008). The role of myeloid cells in the promotion of tumour angiogenesis. *Nat Rev Cancer* 8(8): 618-31.

Mutsaers, S. E., J. E. Bishop, G. McGrouther and G. J. Laurent (1997). Mechanisms of tissue repair: from wound healing to fibrosis. *Int J Biochem Cell Biol* 29(1): 5-17.

Mutsaers, S. E., D. Whitaker and J. M. Papadimitriou (2002). Stimulation of mesothelial cell proliferation by exudate macrophages enhances serosal wound healing in a murine model. *Am J Pathol* 160(2): 681-92.

Ninomiya, K., A. Takahashi, Y. Fujioka, Y. Ishikawa and M. Yokoyama (2006). Transforming growth factor-beta signaling enhances transdifferentiation of macrophages into smooth muscle-like cells. *Hypertens Res* 29(4): 269-76.

Oh, S. J., H. Kurz, B. Christ and J. Wilting (1998). Platelet-derived growth factor-B induces transformation of fibrocytes into spindle-shaped myofibroblasts in vivo. *Histochem Cell Biol* 109(4): 349-57.

Paradossi, G., F. Cavalieri, E. Chiessi, C. Spagnoli and M. K. Cowman (2003). "Poly(vinyl alcohol) as versatile biomaterial for potential biomedical applications." *J Mater Sci Mater Med* 14(8): 687-91.

Park, H. J. and Y. H. Bae (2003). Hydrogels based on poly(ethylene oxide) and poly(tetramethylene oxide) or poly(dimethyl siloxane). III. In vivo biocompatibility and biostability. *J Biomed Mater Res A* 64(2): 309-19.

Patil, S. D., F. Papadimitrakopoulos and D. J. Burgess (2004). Dexamethasone-loaded poly(lactic-co-glycolic) acid microspheres/poly(vinyl alcohol) hydrogel composite coatings for inflammation control. *Diabetes Technol Ther* 6(6): 887-97.

Peek, E. J., D. F. Richards, A. Faith, P. Lavender, T. H. Lee, C. J. Corrigan and C. M. Hawrylowicz (2005). Interleukin-10-secreting "regulatory" T cells induced by glucocorticoids and beta2-agonists. *Am J Respir Cell Mol Biol* 33(1): 105-11.

Pollock, C. (2005). Pathogenesis of peritoneal sclerosis. *Int J Artif Organs* 28(2): 90-6.

Powell, D. W., R. C. Mifflin, J. D. Valentich, S. E. Crowe, J. I. Saada and A. B. West (1999). Myofibroblasts. I. Paracrine cells important in health and disease. *Am J Physiol* 277(1 Pt 1): C1-9.

Quan, T. E., S. E. Cowper and R. Bucala (2006). The role of circulating fibrocytes in fibrosis. *Curr Rheumatol Rep* 8(2): 145-50.

Quinn, C. A., R. E. Connor and A. Heller (1997). Biocompatible, glucose-permeable hydrogel for in situ coating of implantable biosensors. *Biomaterials* 18(24): 1665-70.

Quinn, C. P., C. P. Pathak, A. Heller and J. A. Hubbell (1995). Photo-crosslinked copolymers of 2-hydroxyethyl methacrylate, poly(ethylene glycol) tetra-acrylate and ethylene dimethacrylate for improving biocompatibility of biosensors. *Biomaterials* 16(5): 389-96.

Rele, S. M., W. Cui, L. Wang, S. Hou, G. Barr-Zarse, D. Tatton, Y. Gnanou, J. D. Esko and E. L. Chaikof (2005). Dendrimer-like PEO glycopolymers exhibit anti-inflammatory properties. *J Am Chem Soc* 127(29): 10132-3.

Richardson, T. P., M. C. Peters, A. B. Ennett and D. J. Mooney (2001). Polymeric system for dual growth factor delivery. *Nat Biotechnol* 19(11): 1029-34.

Rodriguez, A., S. R. Macewan, H. Meyerson, J. T. Kirk and J. M. Anderson (2009). The foreign body reaction in T-cell-deficient mice. *J Biomed Mater Res A* 90(1): 106-13.

Sakai, N., T. Wada, K. Furuichi, K. Shimizu, S. Kokubo, A. Hara, J. Yamahana, T. Okumura, K. Matsushima, H. Yokoyama and S. Kaneko (2006). MCP-1/CCR2-dependent loop for fibrogenesis in human peripheral CD14-positive monocytes. *J Leukoc Biol* 79(3): 555-63.

Sasmono, R. T., A. Ehrnsperger, S. L. Cronau, T. Ravasi, R. Kandane, M. J. Hickey, A. D. Cook, S. R. Himes, J. A. Hamilton and D. A. Hume (2007). Mouse neutrophilic granulocytes express mRNA encoding the macrophage colony-stimulating factor receptor (CSF-1R) as well as many other macrophage-specific transcripts and can transdifferentiate into macrophages in vitro in response to CSF-1. *J Leukoc Biol* 82(1): 111-23.

Sasmono, R. T., D. Oceandy, J. W. Pollard, W. Tong, P. Pavli, B. J. Wainwright, M. C. Ostrowski, S. R. Himes and D. A. Hume (2003). A macrophage colony-stimulating factor receptor-green fluorescent protein transgene is expressed throughout the mononuclear phagocyte system of the mouse. *Blood* 101(3): 1155-63.

Sawyer, J. S., B. D. Anderson, D. W. Beight, R. M. Campbell, M. L. Jones, D. K. Herron, J. W. Lampe, J. R. McCowan, W. T. McMillen, N. Mort, S. Parsons, E. C. Smith, M. Vieth, L. C. Weir, L. Yan, F. Zhang and J. M. Yingling (2003). Synthesis and activity of new aryl- and heteroaryl-substituted pyrazole inhibitors of the transforming growth factor-beta type I receptor kinase domain. *J Med Chem* 46(19): 3953-6.

Serini, G., M. L. Bochaton-Piallat, P. Ropraz, A. Geinoz, L. Borsi, L. Zardi and G. Gabbiani (1998). The fibronectin domain ED-A is crucial for myofibroblastic phenotype induction by transforming growth factor-beta1. *J Cell Biol* 142(3): 873-81.

Serini, G. and G. Gabbiani (1999). Mechanisms of myofibroblast activity and phenotypic modulation. *Exp Cell Res* 250(2): 273-83.

Shen, M. and T. A. Horbett (2001). The effects of surface chemistry and adsorbed proteins on monocyte/macrophage adhesion to chemically modified polystyrene surfaces. *J Biomed Mater Res* 57(3): 336-45.

Shen, M., L. Martinson, M. S. Wagner, D. G. Castner, B. D. Ratner and T. A. Horbett (2002). PEO-like plasma polymerized tetraglyme surface interactions with leukocytes and proteins: in vitro and in vivo studies. *J Biomater Sci Polym Ed* 13(4): 367-90.

Shive, M. S. and J. M. Anderson (1997). Biodegradation and biocompatibility of PLA and PLGA microspheres. *Adv Drug Deliv Rev* 28(1): 5-24.

Sparks, C. H. (1969). Autogenous grafts made to order. *Ann Thorac Surg* 8(2): 104-13.

Sparks, C. H. (1973)."Silicone mandril method for growing reinforced autogenous femoro-popliteal artery grafts in situ. *Ann Surg* 177(3): 293-300.

Sunderkotter, C., T. Nikolic, M. J. Dillon, N. Van Rooijen, M. Stehling, D. A. Drevets and P. J. Leenen (2004). Subpopulations of mouse blood monocytes differ in maturation stage and inflammatory response. *J Immunol* 172(7): 4410-7.

Thevenot, P. T., A. M. Nair, J. Shen, P. Lotfi, C. Y. Ko and L. Tang (2010). The effect of incorporation of SDF-1alpha into PLGA scaffolds on stem cell recruitment and the inflammatory response. *Biomaterials* 31(14): 3997-4008.

Udipi, K., R. L. Ornberg, K. B. Thurmond, 2nd, S. L. Settle, D. Forster and D. Riley (2000). Modification of inflammatory response to implanted biomedical materials in vivo by surface bound superoxide dismutase mimics. *J Biomed Mater Res* 51(4): 549-60.

Umland, S. P., R. P. Schleimer and S. L. Johnston (2002). Review of the molecular and cellular mechanisms of action of glucocorticoids for use in asthma. *Pulm Pharmacol Ther* 15(1): 35-50.

Verrecchia, F., M. L. Chu and A. Mauviel (2001). Identification of novel TGF-beta /Smad gene targets in dermal fibroblasts using a combined cDNA microarray/promoter transactivation approach. *J Biol Chem* 276(20): 17058-62.

Verrecchia, F., A. Mauviel and D. Farge (2006). Transforming growth factor-beta signaling through the Smad proteins: role in systemic sclerosis. *Autoimmun Rev* 5(8): 563-9.

Vranken, I., G. De Visscher, A. Lebacq, E. Verbeken and W. Flameng (2008). The recruitment of primitive Lin(-) Sca-1(+), CD34(+), c-kit(+) and CD271(+) cells during the early intraperitoneal foreign body reaction. *Biomaterials* 29(7): 797-808.

Ward, W. K., A. G. Li, Y. Siddiqui, I. F. Federiuk and X. J. Wang (2008). Increased expression of Interleukin-13 and connective tissue growth factor, and their potential roles during foreign body encapsulation of subcutaneous implants. *J Biomater Sci Polym Ed* 19(8): 1065-72.

Wisniewski, N., B. Klitzman, B. Miller and W. M. Reichert (2001). Decreased analyte transport through implanted membranes: differentiation of biofouling from tissue effects. *J Biomed Mater Res* 57(4): 513-21.

Wisniewski, N. and M. Reichert (2000). Methods for reducing biosensor membrane biofouling. *Colloids Surf B Biointerfaces* 18(3-4): 197-219.

Xia, Z. and J. T. Triffitt (2006). A review on macrophage responses to biomaterials. *Biomed Mater* 1(1): R1-9.

Yanez-Mo, M., E. Lara-Pezzi, R. Selgas, M. Ramirez-Huesca, C. Dominguez-Jimenez, J. A. Jimenez-Heffernan, A. Aguilera, J. A. Sanchez-Tomero, M. A. Bajo, V. Alvarez, M. A. Castro, G. del Peso, A. Cirujeda, C. Gamallo, F. Sanchez-Madrid and M. Lopez-Cabrera (2003). Peritoneal dialysis and epithelial-to-mesenchymal transition of mesothelial cells. *N Engl J Med* 348(5): 403-13.

Zeisberg, E. M., O. Tarnavski, M. Zeisberg, A. L. Dorfman, J. R. McMullen, E. Gustafsson, A. Chandraker, X. Yuan, W. T. Pu, A. B. Roberts, E. G. Neilson, M. H. Sayegh, S. Izumo and R. Kalluri (2007). Endothelial-to-mesenchymal transition contributes to cardiac fibrosis. *Nat Med* 13(8): 952-61.

27

Cell Responses to Surface and Architecture of Tissue Engineering Scaffolds

Hsin-I Chang[1] and Yiwei Wang[2]
[1]National Chia Yi University,
[2]ANZAC Research Institute,
[1]Taiwan (ROC)
[2]Australia

1. Introduction

Tissue engineering is an interdisciplinary field that combines the knowledge and technology of cells, engineering materials, and suitable biochemical factor to create artificial organs and tissues, or to regenerate damage tissues (Langer & Vacanti, 1993). It involves cell seeding on a scaffold followed by culturing in vitro prior to implantation in vivo. The ideal scaffolds provide a framework and initial support for the cells to attach, proliferate and differentiate, and form an extracellular matrix (ECM) (Agrawal & Ray, 2001). It should be noted that scaffold surface topography and chemistry (wettability, softness and stiffness, roughness); microstructure (porosity, pore size, pore shape, interconnectivity, specific surface area) (O'Brien & Gibson, 2005b) and mechanical properties (Engler & Discher, 2006; Peyton & Putnam, 2005) have been shown to significantly influence cell behaviors such as adhesion, growth and differentiation, and to affect the bioactivity of scaffolds used for in vivo regeneration applications of various tissues, such as cartilage, skin and peripheral nerves. For tissue engineering purposes, understanding cell behavior and responses on extracellular scaffolds within physiological relevant 3D construct can aid the design of optimal bioactive tissue engineering scaffolds. Controlling cell behavior and remodeling by modulating the local engineered extracellular environment process is also a critical step in the development of the next generation of bioactive tissue engineering scaffolds. The present chapter will discuss cell responses to surface chemistry and various architecture parameters; current approaches and technologies to optimize tissue engineering scaffolds and challenges in studying the cell interaction with scaffolds.

2. Cell responses on surface chemistry of tissue engineering scaffolds

When cells adhere to surface of a scaffold, a sequence of physico -chemical reactions will happen between cells and the scaffold. Immediately after a tissue engineering scaffold is implanted into an organism or comes into contact with cell culture environments, protein adsorption to its surface occurs and which mediates the cell adhesion, and also provides signals to the cell through the cell adhesion receptors, mainly integrins. Cells can adhere on the surfaces of tissue engineering scaffolds and release active compounds for signaling, extra-cellular matrix deposition, cell proliferation and differentiation. The interaction between cells and biomaterial scaffolds is called focal adhesion. To understand the factors that influence cell adhesive ability is a key in the development and application of new tissue engineering scaffold.

Cell attachment is a complex process, affected by numerous aspects, such as cell behavior, material surface properties, and environmental factors. Material surface properties comprise the hydrophobicity, charge, roughness, softness and chemical composition of the biomaterial surface itself.

2.1 Surface hydrophobicity

Biomaterial development has been focusing on surface modifications of biomaterials over years in order to promote a greater understanding and control of the material characteristics for regulating biocompatibility. The surface hydrophobicity is well known as a key factor to govern cell response. The surface hydrophobicity can be assessed by measuring contact angle through water spread of a droplet on a surface. The lower the contact angle, the more hydrophilic the surface is. Previous studies showed the more hydrophilic surface of material films is the much more cell adhesion on the surface (Goddard & Hotchkiss, 2007; Xu, 2007). For example, osteoblast adhesion was reported decrease when the contact angle of surface increased from 0° to 106°. Fibroblasts were found to have maximum adhesion when contact angles were between 60° and 80° (Tamada & Ikada, 1993; Wei et al., 2009). Interestingly, Vogler mentioned that the hydrophilic surface were suitable for the attachment of Madin-Darby Canine Kidney (MDCK) cells but more hydrophilic surfaces (contact angel $\theta < 65°$) did not yield progressively high level of attachment efficiency (Vogler, 1999). Furthermore, surface hydrophobicity is related to the rate of cell spreading and differentiation. On hydrophilic surfaces, cells generally showed good spreading, proliferation and differentiation. Mouse osteoblast-like cell line MC3T3-E1 showed more fractal morphology on hydrophilic surface (contact angel $\theta = 0°$) (Wei et al., 2009). 7F2 mouse osteoblasts on hydrophilic surface (contact angel $\theta = 24\text{-}31°$) demonstrated accelerated metabolic activity and osteodifferentiation compared to their unmodified counterparts (contact angel $\theta = 72°$) (Yildirim et al., 2010). The same phenomenon was observed in neuronal spreading and neurite outgrowth when the material surfaces reduced their hydrophobicity (Khorasani & Irani, 2008; Lee et al., 2003).

2.2 Protein adsorption

Since cell adhesion to material surface requires a series of cytoplasmic, transmembranal and extracellular proteins that assemble into stable contact sites (Geiger & Bendori, 1987), cell adhesion and behaviors is likely involved the adsorption onto the material surface of serum and ECM proteins (Brynda & Andrade, 1990; Hattori et al., 1985). Many proteins, including immunoglobulins, vitronectin, fibrinogen, and fibronectin (Fn), adsorb onto implant surfaces immediately upon contact with physiological fluids and modulate subsequent inflammatory responses. For example, adsorbed adhesive proteins mediate the attachment and activation of neutrophils, macrophages, and other inflammatory cells. Many literature studies mentioned that different cell behaviors, related to different hydrophobicities, may be mediated by protein absorption, because surface wettability modified the sort and the quantity of adsorbed cell adhesion molecules. Hydrophobic surfaces tend to adsorb more proteins, while hydrophilic surfaces tend to resist protein adsorption (Xu, 2007). Tamada et al. used bovine serum albumin (BSA), bovine γ-globulin and plasma Fn to study the protein absorption onto various polymer substrates and the maximal protein absorption was observed on surfaces with water contact angle ranging from 60° to 80° (Tamada & Ikada, 1993). Hence wetting has been discredited as an adequate predictor of protein adsorption. However, there is a growing concern that surface hydrophobicity does not guarantee the protein adsorption. Certain hydrogel-like materials (e.g. oligo(ethylene glycol)) are resistant to protein adsorption even though these surfaces are (apparently) only modestly wettable (Noh & Vogler, 2006). Tamada reported that preadsorption of serum albumin prevented cell adhesion of fibroblasts to all substrates, whereas preadsorbed Fn enhanced cell adhesion of fibroblasts to all the substrates, independent of their water wettability. With preadsorption of Fn and BSA, similar pattern of cell attachment was investigated on a series of *N*-isospropylacrylamide and *N*-tert-butylacrylamide based copolymer films (Allen et al., 2006). Therefore, the composition and conformation of the adsorbed protein layer is considered to be one of the major factors in determining the nature of cell interaction with the materials.

2.3 Surface charge

After surface hydrophobicity, surface charge has been recently described a lot in the cell attachment phenomenon. Firstly, the amount of surface charges can influence cell behavior (Ishikawa et al., 2007). As the degree of charge density of poly(styrene-*ran*-acrylic acid) increased, more cell adhesion and proliferation were observed (Jung et al., 2008). Fig. 1 presented similar effect on 2-hydroxyethyl methacrylate (HEMA) and 2-methacryloxyethyl trimethyl ammonium chloride (MAETAC) copolymer hydrogels (Kim & Kihm, 2009). Secondly, many researchers reported the improved-biocompatibility, cell affinity and cell differentiation on the implanted surfaces by using the positive ions and the negative ions (Bet et al., 2003). For instance, HEMA hydrogels incorporated with positive charges supported significantly more cell attachment and spreading of osteoblasts and fibroblasts as compared to negative or neutral charges (Schneider et al., 2004). Yaszemski's groups also investigated that negatively charged oligo(poly(ethylene glycol) fumarate) hydrogels increased the extent of chondrocyte differentiation, such as collagen and glycosaminoglycan expression, in comparison with that on the neutral or positively charged hydrogel scaffolds (Dadsetan et al., 2011). Similar pattern was also performed on neuronal growth and differentiation (Makohliso et al., 1993). Positively charged coating materials such as polylysine improve neuronal attachment in vitro. On positive fluorinated ethylenepropylene (FEP) films, neurite outgrowth was significantly higher comparing to negative and uncharged substrates. Finally, the surface charges can be used to modify cell behvaior through the chemical functionalities of the polymer materials (Table 1). Lee et al. prepared polyethylene (PE) surfaces with differently chargeable functional groups (-COOH, $-CH_2OH$, $-CONH_2$ and $–CH_2NH_2$ groups) by corona discharge treatment, graft copolymerization and substitution reaction to study the effect on cell behavior (J. H. Lee et al., 1994). Results indicated that Chinese hamster ovary (CHO) cells were more adhesive to the functional group-grafted surfaces than the control PE surface due to the increased wettability by grafting hydrophilic functional groups. The best cell adhesion, growth and spreading rate were recorded on polar and positively charged surfaces (amine group-grafted PE) while the negatively charged surface (carboxylic acid group-grafted PE) still had poor growth. Moreover, the surfaces grafted with neutral amide and hydroxyl groups showed a similar number of cell attachments; however; the morphology of cells attached on the surfaces was quite distinct. The cells were spread much more on the hydroxyl group-grafted surface than the amide group-grated one. On the other hand, surface charge may modulate protein adsorption to direct integrin binding and specificity, thereby controlling cell adhesion. Thevenot et al. mentioned that the incorporation of negative charges may facilitate adsorption of proteins which promote cell adhesion and responses (Thevenot et al., 2008). Keselowsky et al. reported that surfaces with differently chargeable functional groups ($-CH_3$, -OH, -COOH, and $-NH_2$ groups) modulated Fn adsorption and direct integrin binding and specificity to control cell adhesion of MC3T3 osteoblasts to Fn-coated surfaces followed the trend: OH >COOH =NH_2 > CH3 (Keselowsky et al., 2003). Same group also demonstrated that surfaces grafted with hydroxyl and amine groups up-regulated osteoblast-specific gene expression, alkaline phosphatase enzymatic activity, and matrix mineralization compared with surfaces presenting carboxyl and alkyl groups (Keselowsky et al., 2005). Although the molecular mechanisms in how to modulate surface charge-dependent cellular activities still remain poorly understood, these latest findings confirm that surface charge plays an important role in the application of cell biology and tissue engineering.

Functional group	Properties	Effect on cells
$-CH_3$	Neutral, hydrophobic	promotes increased leukocyte adhesion and phagocyte migration
-OH	Neutral, hydrophilic	increases osteoblast differentiation
-COOH	Negative, hydrophilic	Increase osteoblast attachment
$-NH_2$	Positive, hydrophilic	promotes myoblast and endothelial proliferation and osteoblast differentiation
$-CH_2NH_2$	Neutral, hydrophilic	enhance CHO attachment of Chinese hamster ovary cells

Table 1. The effect of material surface functional groups on proteins and cells (Schmidt et al., 2000)

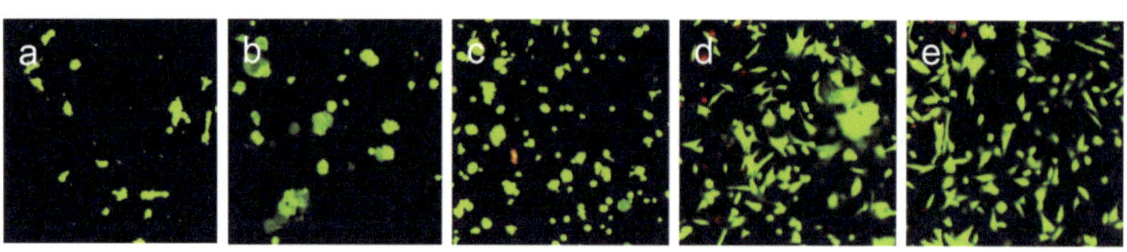

Fig. 1. Chondrocyte viability on hydrogels with different concentrations of MAETAC (positive charge functional group) in their formulation on days 1: (a) 0%; (b) 5%; (c) 10%; (d) 20%; (e) 30%. (S. Kim et al., 2009)

2.4 Surface roughness

Material surface roughness (or topography) is another important factor influencing cell adhesion and behavior. Indeed, roughness modulates the biological response of tissues in contact with the implant. Material surface roughness has a direct influence in vitro as well as in vivo on cellular morphology, proliferation, and phenotype expression. Literature papers have been reported that cells grown on microrough surfaces, were stimulated towards differentiation; as shown by their gene expression in comparison with cells growing on smooth surfaces. For instance, primary rat osteoblasts had higher proliferation and elevated alkaline phosphatase (ALP) activity and osteocalcin expression on the rough surface (0.81 μm) in comparison with smooth one (Hatano et al., 1999). In the case of human foetal osteoblastic cells (hFOB 1.19), a similar increase in cell spreading and proliferation on rough surfaces was reported (Lim, Hansen, Siedlecki, Runt, & Donahue, 2005). Depending on the scale of irregularities of the material surface, surface roughness can be divided to macroroughness (100 µm - milimeters), microroughness (100 nm - 100 µm), and nanoroughness (less than 100 nm), each with its specific influence (AGASKÁ et al., 2010). The response of cells to roughness is different depending on the cell type. For larger cells, such as osteoblasts and neurons, macroscopic descriptions of the surface roughness could be reasonable (Donoso et al, 2007). Lee et al. examined the behavior of MG63 osteoblast-like cells cultured on a polycarbonate (PC) membrane surfaces with different micropore sizes (200 nm–8.0 µm) (Lee et al., 2004). It seems that the cell adhesion and proliferation were progressively inhibited as the PC membranes had micropores with increasing size, probably due to surface discontinuities produced by track-etched pores (Fig. 2). On the other hand, increasing micropore size of the PC membrane resulted in improved cell differentiation such as higher osteocalcin expression and ALP specific activity in isolated cells. Bartolo et al. also investigated neuronal cell behavior in the surfaces with nanoscale (6.26 nm) to microscale (200 nm) roughness (Bartolo et al., 2008). The axonal length increased and the neuritis becomes highly branched on the nonoscale rough surfaces (6.26-49.38 nm). In the case of microscale rough membranes (87.2-200nm), the neurons were less developed as demonstrated by the round-shaped soma and poorly branched processes. Therefore, the nanoscale rough membranes seem to be more supportive of neurite outgrowth modulating the development process of the neurons. For smaller cells, such as human vein endothelial cells, increasing surface roughness of biomaterial surfaces at nanometer scale (10–102nm) could enhance cell adhesion and growth on roughness surfaces (Chung et al., 2003). Furthermore, Kim et al. used the dendrimer-immobilized surfaces to study nanoscale modifications and discovered that the human mammary epithelial cells (hTERT-HME1) cultured on the naked dendrimer surface (4.0 nm) were abundant in F-actin filaments of peripheral stress fibers and filopodia, compared with those cultured on the plain surface (Kim et al., 2007). However, when the surface roughness was larger than 4.0 nm, such cell stretching was inhibited, resulting in the predominant existence of round-shaped cells. Similar investigation was also reported by Dalby in the development of F-actin filaments in fibroblasts (Dalby, 2005). Interestingly, MC3T3-E1 osteoblastic cells showed that the rate of proliferation on the smooth regions (0.55nm) of the films is much greater than that on the rough regions (13nm) (Washburn et al., 2004). Therefore, the selectivity of cells on surface roughness could be highly advantage on the development of implanted devices.

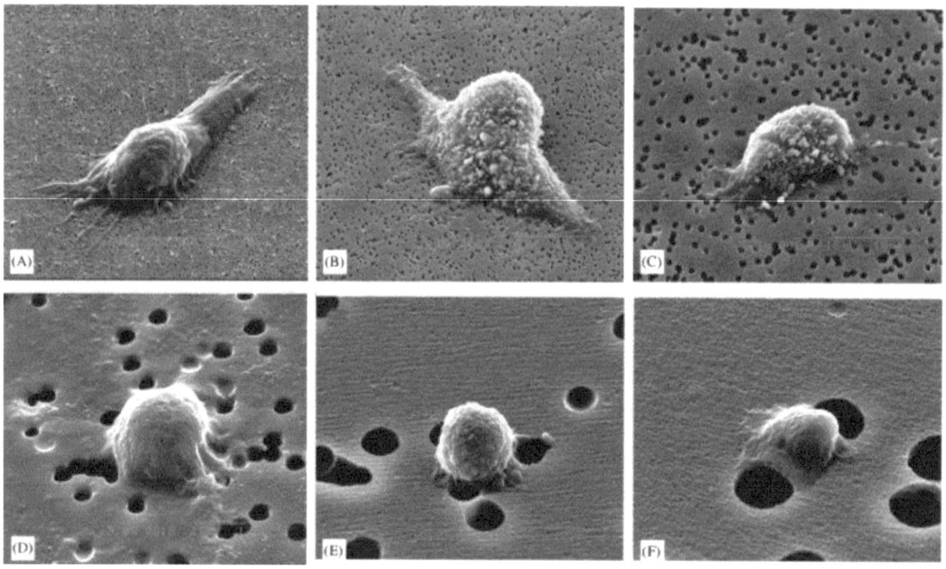

Fig. 2. SEM pictures of the MG63 cells attached on the PC membrane surfaces with different micropore sizes: (A) 0.2, (B) 0.4, (C) 1.0, (D) 3.0, (E) 5.0, and (F) 8.0 mm. (S. J. Lee et al., 2004)

2.5 Surface softness and stiffness

Surface stiffness is measure of how soft (as silk) or stiff (as rock) a materials surface is. Several studies have reported that cell attachment, proliferation and differentiation are all modulated by the substrate rigidity to a degree dependent upon the substrate stiffness in relation to the stiffness of the native tissunmee (Engler et al., 2006; Khatiwala et al., 2007). In addition, at the tissue-implant interface, cells can actively modify surfaces of the implants, altering the stiffness of microenviront of their own or other cells (Marquez et al., 2006). Tan and Teoh demonstrated that 3T3 fibroblasts preferred soft surface for proliferation. Su group used the rheometer to measure the cell adhesion force of MDCK cells on different substrate softness. The results showed that the adhesion force of MDCK cells increased with the decrease of substrate softness and which is correlated to the cell spreading area. Engler prepared polyacrylamide (PA) gels with different softness to study the correlation between cell spreading area of smooth muscle cells and elastic modulus of substrates. Cell spreading was found remarkably dependent on elastic modulus of PA gel substrates. Therefore, the softness and stiffness of substrates may regulate the mechanism between cell-ECM and seem to correlate broadly with cell adhesion response.

A limited number of attempts have yielded exciting findings (Lahann & Langer, 2005; Mrksich, 2005), in which dynamic changes of surface softness and stiffness were induced largely through application of environmental factors (e.g. temperature, pH and electric field). As a result, a study in neuronal regeneration (Jiang et al., 2008) indicted that initial softer substrates foster axonal elongation and stiffening of the substrate at a later stage could encourage outgrowth of more primary dendrites of neurons, thus promoting synaptogenesis. Moreover, cells response to mechanical alterations as demonstrated in the cell projection area and polarity was found vary depending on range of stiffness changes (Jiang et al., 2010).

3. Cell responses on architecture of tissue engineering scaffolds

On a macroscopic level, the overall shape of the scaffold provides boundaries for tissue regrowth. On a microscope level, the material provides a framework and capillary networks for local cell growth and tissue organization, permitting cell attachment, distribution and proliferation within a controllable microenvironment (Saltzman, 2002). Apart from tissue engineered skin and vascular grafts that have been progressed into clinical use, the most other tissue engineered human tissue or organs (e.g. liver and kidney) are still unsuccessful (Mikos et al., 2006). Simply produce a highly porous scaffold and cultivating it with the appropriate types of cells in most cases does not reproduce the desired feature of a normal tissue as tissue structure and function are known to be highly inter-related (Bhaia & Chen, 1999). Many tissues have a hierarchical structure that varies over length scales of 0.1-1mm (Griffith, 2002). The subcellular structures (1-10μm) control cell-cell inter-relationships and supracellular scale structures (100-1000μm) build the essential functional units of the tissue. In order to maintaining the activity of function cell, regulating cell behavior, and reconstructing 3-dimentional multicellular masses, scaffold must be optimized to satisfy cell and tissue growth including proper networks to provide fresh culture medium to all cells and remove metabolites from the cells and maintain the hierarchical cellular architectures to mimic the functional cells living environment. Altering the micro-architecture, such as the material crystallinity or the microporosity, and/or the macro-architecture of the scaffold can be achieved by changing the pores size, porosity, pore interconnectivity and tortuosity, to match the characteristics of the native tissue whilst retaining integrity (Hutmacher, 2001). A common problem encountered when using scaffolds in tissue engineering is the rapid cells attachment and proliferation on the outer edge of scaffold which restrict cell penetration to the scaffold center, resulting in a necrotic core (Freed et al., 1999). This can be addressed by altering the culture conditions use to growth tissue, for example using a flow perfusion culture system (Botchwey et al., 2001), but it is only relevant to tissue engineering in vitro. Another option or further method of addressing this is to design an optimized scaffold that will improve nutrient and cell transfer to the scaffold center, both in vitro and in vivo. As discussed in above, characterization of surface wettability, charges and softness; modification of surface chemistry by coating with adhesion molecules together with optimized the internal structure and architecture will all help to deal with this issue. Scaffold porosity in particular controls the key processes of nutrient supply to cells, metabolite dispersal, local pH stability and cell signaling. The size of the pores can affect how close the cells are at the initial stages of cultivation (allowing for cell-cell communication in three dimensions), but also influences the amount of space the cells have for 3-D organization in the later stages of tissue growth. In addition, a porous surface is known to improve mechanical interlocking between the implanted scaffolds and the surrounding natural tissue, providing greater mechanical stability at this critical interface (Karageorgiou & Kaplan, 2005). Cell seeding in the center of the scaffold and feeding the inner surfaces of the scaffolds are limited when the pores are too small whereas larger pores affect the stability of the scaffold and its ability to provide physical support for the seeded cells (Levenberg & Langer, 2004). To date cell seeding on 2-D scaffold surfaces has been shown to be easy to perform but the preparation of 3-D cell-scaffold constructs for regeneration of organs is far more complex. For example, pores of adequate size allow cells to migrate or adhere to the surface of a material, but interconnecting pores are necessary to permit cell growth into the scaffold interior.

3.1 Pore size of tissue engineering scaffold

Cell migration is modulated by a complex, spatiotemporally integrated set of biophysical mechanisms that are influenced not only by the biochemistry of extracellular and intracellular signaling, but also by the biophysics of the surrounding extracellular environment. Specific cells require different pore sizes for optimal attachment, growth and motility (Table 2) (Ranucci et al, 2000). A recent study (Yang et al., 2010)on variable pore size collagen gel found that cell migration is hindered by small pore size that invasive distance was not very sensitive in the pore size rang of 5-12μm. At small pore size, a variety of factors, including high ligand density in collagen gel that does not encourage the cell polarity and release seen in mesenchymal migration likely contributes to the limited invasion (Ulrich et al., 2010) whilst very large pore size in scaffolds have insufficient tethers on which to generate traction would also limit cell migration. As a result, many researches in tissue engineering are aimed at obtaining polymeric or bioceramic scaffolds with a very high porosity and simultaneous good control over pore size and morphology (Hou et al., 2000). The presence of pores smaller than 160μm in PLA and PLGA scaffolds, produced by salt leaching, has been reported to be optimal for attachment of human skin fibroblasts (Yang et al., 2002). Bony ingrowth was found to predominate in porous PMMA implanted in bone when the pore size was around 450μm (Ashman & Moss, 1977). Connective tissue formed when the pore size was below 100μm and extensive vascular infiltration was only observed with pores around 10000μm. In the case of polyurethane meniscal implants, structures comprising macropores (150-300 μm), highly interconnected by micropores (<50 μm) have been found to be conducive to ingrowth of fibrocartilaginous tissue (deGroot et al., 1996). The cell infiltration depth (120μm in 28 days) found in elastin scaffolds, for example, probably results from the material's high porosity and inter-connectivity (Lu, Ganesan et al., 2004). Osteoblasts was found to migrate faster inside the larger pore (100μm) of microcellular polyHIPE scaffolds; however, pore size did not affect cell penetration depth or mineralization extent (Akay et al., 2004). It has also been noticed in previous studies that cell-scaffold binding can block pores of inadequate size and geometry (Freed & Vunjak-Novakovic, 1998; Yannas, 2000). High inter-connectivity of pores is also essential to supply nutrients and allows oxygen exchange in the inner regions of a scaffold to maintain cell viability, especially for complex tissue engineering of organs.

Cell/tissue type	Optimal pore size (μm)	Scaffold material	Reference
Human skin fibroblasts	<160μm	PLA/PLG	(Yang.J et al., 2002)
Bone	450μm	PMMA	(Ashman & Moss, 1977)
Fibrocartilaginous tissue	150-300μm	Polyurethane	(deGroot et al., 1996)
Adult mammalian skin cells	20-125μm	Collagen-glycocaminoglycan	(Yannas, Lee, Orgill, SKrabut, & Murphy, 1989)
Osteogenic cells	100-150μm	Collagen-GAG	(O'Brien, Harley, Yannas, & Gibson, 2005a)
Smooth muscle cells	60-150μm	PLA	(Zeltinger, Sherwood, Graham, Mueller, & Griffith, 2001)
Endothelial cells	<80μm	Silicon nitride	(Salem et al., 2002)

Table 2. Optimal pore size for cell infiltration and host tissue ingrowth

3.2 Porosity of tissue engineering scaffold

The porosity, that is, the percentage of void volume in the materials, is also use as a means of quantifying the structure of a tissue engineering scaffolds. Researches have been focusing on the design of the scaffold to ensure appropriate porosity and porous structure for cells penetration and ingrowth. However, attempts to link scaffold porosity to cell performance have not been particularly successful. Toth et al (Toth et al., 1995) report that improvements in bone ingrowth occur with increasing porosity of macroporous biphasic calcium phosphate ceramic samples. However, they also report no discernible differences in bone union after six months implantation between scaffolds that have 30%, 50% and 70% porosity. This observation could be attributed to the complicated internal structure of scaffolds that consist of pores of different types (open, closed and blind-end pores), sizes and geometry, (Fig. 3). The presence of both random and anisotropic open porous architectures of PLA scaffolds were prepared using supercritical CO_2 aims to find the optimal channel diameter and geometry for osteosarcoma cell penetration. The results show that cells penetrate into scaffolds containing aligned channels (400μm) more extensively than those that did not.

3.3 Connectivity and tortuosity of tissue engineering scaffold

The pore structures typically consist of irregularly shaped voids and connecting channels (connects) that can be difficult to defined due to merging of adjacent cavities, resulting in the presence of fenestrations (windows) in the void walls. Beyond the fundamental requirements of adequate pore size and inter-connectivity, pore tortuosity also plays a key role in cells interaction with scaffolds. Tortuosity is defined as the ratio of the actual path length through connected pores to the Euclidean distance (shortest linear distance) (Fig.4). Tortuosity is another key factor in optimizing and designing tissue engineering scaffold which is known to influence molecules and oxygen diffusion and cell migration rate. Silva et al reported (Silva et al., 2006) that aligned channel in both hyroxyapatite (HA) and poly(D,L-lactic acid) ($P_{DL}LA$) scaffolds enhanced cell penetration and infiltration into the central region of the scaffold in comparison with tortuous channels. Analysis of human osteosarcoma cell penetration into the aligned channels revealed that cell coverage increased with increasing channel diameter from around 22% in the 170μm diameter channel to approximately 38% into the 420μm channel(Rose et al., 2004). In addition, cell penetration into 420μm channel was significantly greater than that observed within the 170μm channel. However, determination of tortuosity and cell responses on tortuous scaffold is still rarely quoted in the literature. A common method to calculate tortuosity is via the results from dissolution measurements. In this method, the tortuosity is calculated from several parameters related to the dissolution of a molecules from a matrix (Desai et al., 1966; Foster & Parrott, 1990). This approach can result in unrealistic values of more than one thousand (Papadokostaki et al., 1998) or below one (Foster & Parrott, 1990). Tortuosity can also be measured from the porosity and diffusion coefficients obtained from spin echo NMR measurements (Wu et al., 2006). Mercury intrusion porosimetry has also been suggested for determining tortuosity. Another example of tortuosity calculating is to use the inflection count metric (ICM). This approach adds the number inflections of a 3-D frame representation of a pore connecting two points and multiplies this number by the path length (Bullitt et al., 2003). Wu et al (Wu et al., 2006) described a method to find the shortest route through the pores in images of compacts using an algorithm called 'grey-weighted distance transform (GDT)′ which provide precise measurements of tortuosity. A recent study (Leber et al., 2010) showed that tortuous channels with 1 or 2 of 90°bends had faster osteoblasts growth than the control (non-bend). It could be hypothesize that the cell sidewall affinity could have contributed to this increase in cell quantity. These observations are in harmony with studies of osteoblast alignment with parallel grooves fabricated in various material surfaces, but expand the study to tortuous channels (Ber et al., 2005). No other reports have been noticed to show the relationship between cells attachment and migration with tortuous channels.

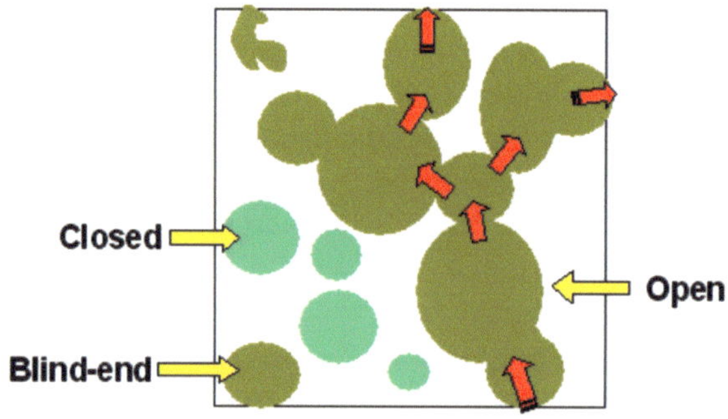

Fig. 3. Schematic of the different pore types found in tissue engineering scaffolds (Y. Wang et al., 2010)

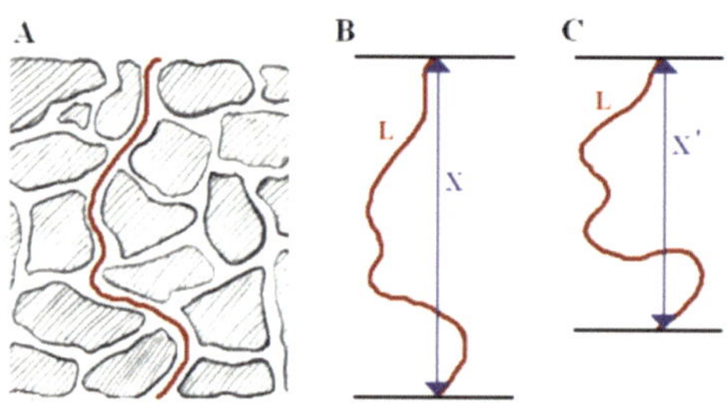

Fig. 4. Determination of tortuosity through a porous material using the arc-chord ratio (O'Connell et al., 2010)

3.4 Cell responses to dynamic scaffolds

For biodegradable polymer scaffolds, polymers slowly degrade and then dissolve following implantation. The dissolution rates represent an additional and important parameter in determining the properties of the scaffolds and can be turned to the need of specific cell and tissue. For example, cells that proliferate rapidly require scaffolds with higher degradation rates, whereas tissue structures that require stability and strength may benefit from longer-lasting material (e.g. bone, skin and tendon).

Furthermore, biodegradable polymers may provide an additional level of control over cell responses: during polymer degradation, the surface of the polymer is constantly renewed, providing a dynamic substrate for cell attachment and growth. Cells do not live in static surroundings in this situation; they exist in highly evolving dynamic environment. During cell adhesion and migration, cells adapt and communicate to their environment by numerous methods ranging from differentiation, gene expression, growth, and apoptosis. Dynamic substrates that can alter the presentation of ligands to an attached cell will generate immediate opportunities for studies of scaffolds-cell adhesion, signaling, migration and differentiation. Dynamic substrates may also be important for generating substrates that can control the spatial and temporal interactions between two or more different populations of attached cells in tissue engineering (Yousaf, 2009). However, much less effort has been invested in studying cell responses on dynamic substrates. Currently, there is no available method to generate dynamic gradient substrates for studying cell polarity and directed cell migration. One of recent studies found that programmed erasure of substrate topography cause a decrease in cell alignment as evidenced by an increase in angular dispersion with corresponding remodeling of the actin cytoskeleton. Cell viability remained greater than 95% before and after topography change (Davis et al., 2011).

Beyond the biodegradability, mechanical input on scaffolds also significant influence the reorientation of cell shape (J. Wang et al., 2003), actin cytoskeleton remodeling (J. Wang et al., 2000) and the synthesis of extraceullar matrix (Carver et al., 1991). Many studies showed that cells are capable of surveying the external mechanical properties of their surrounding environment, respond to changes in the balance of intra- and extra- cellular forces, and regulate many important physiological proves (Pelham & Wang, 1997). As most tissue engineering scaffolds are made of biomaterials which have certain elasticity, rigidity and stretching-tension, cell responses to such scaffolds relate mechanical stimulation is becoming more and more interesting. In case of fibroblasts differentiation, mechanical stretching of silicone dishes can induce differentiation of fibroblasts into myofibrolasts, which is known to form scar tissue in vivo. Tock et al (Tock et al., 2003) reported that mechanical loading impressed α-SMA expression, a marker of myofibroblasts hat mechanical tension in granulation tissues controls myofibroblast differentiation (Hinz et al., 2001).

4. Challenges in studying cell behaviour on biomaterials and complex tissue engineering scaffolds

Cell adhesion to a material surface is an important phenomenon that controls the behavior of cells, such as their morphology, migration, growth and differentiation. Counting the cells adhered to material's surface is the most common way to evaluate a material's affinity to the cells but this method could not quantify the cell adhesion force on the material's surface. Many studies used biological procedures to measure the characteristics of adhesion between cells and biomaterials. For example, cell spreading and migration are often used as indirect indicators of adhesion strength and this lack of quantitative understanding of adhesion strength limits the interpretation of functional studies of structural and signaling adhesive components. Studying cell adhesion molecules such as focal adhesion kinase involved the binding with biomaterial's surface can provide more direct information about cell adhesion strength. However, protein expression is working on a population of cells and not a single cell due to the sensitivity of western blot or protein binding assay. Recently, a number of techniques have been developed to study the cell adhesion behavior on materials from mechanical point of view. For instance, cell adhesion strength has been studied as centrifugation force by centrifuge, tensile force by micropipette manipulation, shear force by parallel flow chamber and chemical binding force by atomic force microscope (Thoumine &

Ott, 1997; Truskey & Proulx, 1993; Leonenko et al., 2007). McClay and Lotz groups determined the cell detachment force under different speed of centrifugation and Bouafsoun et al. used flow chamber and jet impingement techniques to study cell detachment forces. However, centrifugation force by centrifuge and shear force by parallel flow chamber quantify the cell-material adhesion strength for a population of cells but not for an individual cell. Thus, tensile force by micropipette manipulation and chemical binding force by atomic force microscope (AFM) are the more suitable techniques to study the cell adhesion strength of a single cell on biomaterial surface. In 2009, Hung et al. utilized the dielectrophoresis force acting on the human bladder epithelial (ECV) cells to induce spatial movement for studying the cell adhesion strength. Dielectrophoresis is the phenomenon in which a particle, such as a living cell, is polarized and moved by the electrical gravity in a non-uniform electric field (Jones, 2005). In our study, dielectrophoresis force was also used to determine cell adhesion strength of human bladder epithelial cells on Fn and collagen type 1 coated surfaces and the cell adhesion force was similar to the one measured by AFM but much smaller than the one measured by flow chamber techniques. We suggested that the cell adhesion strength between a single cell and biomaterial surface would be different from a population of cells. Therefore, new trends and possible long-term directions for determining both adhesion process and force are highlighted.

For characterizing internal structure of tissue engineering scaffold and their correlation with cell behviour, a variety of techniques have been used to evaluate scaffold porosity including theoretical assessment, scanning electron microscopy (Flynn et al., 2006), mercury porosimetry, gas pycnometry and adsorption. SEM analysis complements the theoretical calculations of porosity (Kellomake et al., 2000; Walsh et al., 2001; Zein et al., 2002) and allows direct measurements of pore size and wall thickness and cell morphology on the surface. However, SEM cannot examine the scaffold interior without sample sectioning which introduces uncertainty due to unwanted material compression and edge effects and cells damage. Mercury porosimetry is a well known and established method, but it neither measures small mesopores (2-50nm pores) due to lack of mercury penetration nor measures very large pores as the mercury penetrates the structure before measurements can be made. The gas adsorption method is relevant to the study of porosity in nano-featured and nano-modified scaffolds (Ma, 2004), and is based on the electrical forces of attraction that bind atoms in solids. To counter the net inward attractive forces, surface atoms bind surrounding gas molecules via Van der Waals and electrical forces. Researchers have used gas adsorption to assess scaffolds with pore sizes ranging from 0.35-400 nm or 3.5 to 2000μm but the analysis does not evaluate closed pore content and cell proliferation or migration.

For analyzing the 3-D construction of tissue scaffolds and cell-material interactions, new imagining techniques such as micro-computed tomographic (micro-CT) have been developed. Feldkamp et al (Feldkamp et al., 1989) pioneered micro-CT imaging technology to analyze trabecular bone samples at a spatial resolution of 50μm. Since then, micro-CT has been used extensively in the study of bone architecture and other tissue types. Micro-CT images the specimen through exposure to small quantities of ionizing radiation and corresponding measurements of absorption. The resulting grey-scale images form a series of 2-D sequential slices which build up into a density map of the sample. With relevant computerized reconstruction, micro-CT provides precise quantitative and qualitative information on the 3D morphology of specimens (Darling & Sun, 2004; Thurner et al., 2005; Thurner et al., 2004; Washburn et al., 2004; Williams et al., 2005) and the interior can be studied in great detail without resorting to physical sectioning or the use of toxic chemicals. Williams et al (Williams et al., 2005) recently used Micro-CT to visualize PCL scaffolds produced by selective laser sintering (SLS) and to assess the porosity and subsequent bone formation following cell seeding and implantation in mice. Micro-CT has also been used to quantify scaffold micro-architectural parameters related to compressive mechanical properties (Lin et al., 2003). Thurner et al (Thurner et al., 2005) explored X-ray Micro-CT for morphological characterization of cell cultures on filamentous 3-D scaffolds (Fig.5.) and Synchrotron Micro-CT has highlighted the subtlety of cell-scaffold interactions - fibroblasts tend to span between multi-filament yarns whereas osteoblast-like cells are confined to the filament surface (Thurner et al., 2004). Ongoing development of micro-CT techniques is improving qualitative and quantitative analysis of tissue engineering scaffolds. Jones et al (Jones et al., 2007) applied three algorithms to identify pores, interconnects and pore size distribution in bioceramic scaffolds to predict the permeability of the pore network and thus optimize bioreactor conditions for cell seeding.

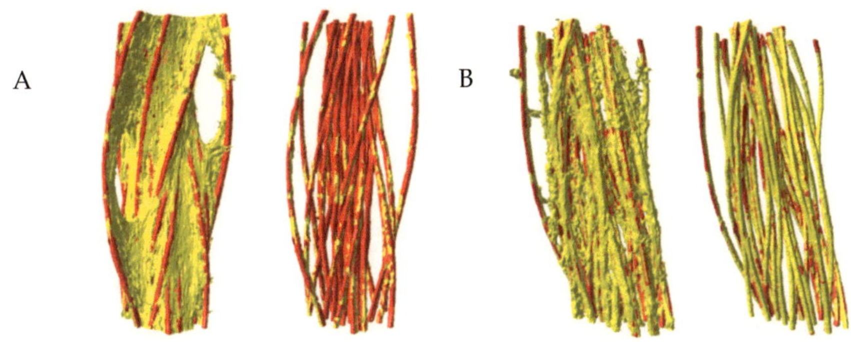

Fig. 5. 3D visualization of human foreskin fibroblasts (A) and mouse calvarial osteoblast-like cells (B). Adhesive surface (right) (both yellow) on the yarn (red) (Thurner et al., 2005)

5. Conclusion

In order to prove clinically use, tissue engineering scaffolds must consider many surface properties and 3-D structure design to maintain cell attachment, proliferation and phenotype expression. Surface and bulk mechanics, control of the scaffold interface and cell biology are essential for the development of tissue engineering. Clinical research has been demonstrating the value of tissue engineering approach on *in vivo* therapies and will likely continue to use cellular biology and signaling pathways to assess the corrections between tissue engineering and organ repair. Hence, testing cell response to tissue engineering scaffolds in vivo will provide better understanding host-implant response *in vivo* environment.

6. References

AGASKÁ, B., BAČÁKOVÁ, L., FILOVÁ, E., & BALÍK, K. (2010). Osteogenic Cells on Bio-Inspired Materials for Bone Tissue Engineering. *Physiological Research, 59*, 309-322.

Agrawal, C. M., & Ray, R. B. (2001). Biodegradable polymeric scaffolds for musculoskeletal tissue engineering. *Journal of Biomedical Materials Research, 55*, 141-150.

Akay, G., Birch, M. A., & Bokhari, M. A. (2004). Microcellular polyHIPE polymer supports osteoblast growth and bone formation in vitro. *Bone, 24*, 35-41.

Allen, L. T., Tosetto, M., Miller, I. S., O'Connor, D. P., Penney, S. C., Lynch, I., et al. (2006). Surface-induced changes in protein adsorption and implications for cellular phenotypic responses to surface interaction. *Biomaterials, 27*, 3096-3108.

Ashman, A., & Moss, M. L. (1977). Implantation of porous polymethlmethacrylate resin for tooth and bone replacement. *The Journal of Prosthetic Dentistry, 37*, 657-665.

Bartolo, L. D., Rende, M., Morelli, S., Salerno, S., A, P., Gordano, A., et al. (2008). Influence of membrane surface properties on the growth of neuronal cells isolated form hippocampus. *Journal of Membrane Science, 325*, 139-149.

Ber, S., Torun Koseb, G., & Hasircia, V. (2005). Bone tissue engineering on patterned collagen films: an in vitro study. *Biomaterials, 26*, 1977-1986.

Bet, M. R., Goissis, G., Vargas, S., & Selistre-de-Araujo, H. S. (2003). Cell adhesion and cytotoxicity studies over polyanionic collagen surfaces with variable negative charge and wettability. *Biomaterials, 24*, 131-137.

Bhaia, S. N., & Chen, C. S. (1999). Tissue engineering at the microscale. *Biomedical Microdevices, 2*, 131-144.

Botchwey, E. A., Dupree, M. A., Pollack, S. R., Levine, E. M., & Laurencin, C. T. (2001). Tissue engineered bone: measurement of nutrient transport in three-dimensional matrices. *Journal of Biomedical Materials Research, 67*, 357-367.

Brynda, E., Hlady, V., & Andrade, J. D. (1990). Protein packing in adsorbed layers studied by excitation energy transfer. *Journal of Colloid and Interface Science, 139*, 374-380.

Bullitt, E., Gerig, G., Aylward, S., Joshi, S., Smith, K., Ewendl, M., et al. (2003). Tortuosity. In *Attributes and Malignant Brain Tumors In MICCAI 2003 proceedings LNCS 2878* (pp. 671-679).

Carver, W., Nagpal, M. L., Nachigal, M., Borg, T. K., & Terrracio, L. (1991). Collagen expression in mechanically stimulated cardiac fibroblasts. *Circulation Research, 69*, 116-122.

Chung, T.-W., Liu, D.-Z., Wang, S.-Y., & Wang, S.-S. (2003). Enhancement of the growth of human endothelial cells by surface roughness at nanometer scale. *Biomaterials, 24*, 4655-4661.

Dadsetan, M., Pumberger, M., Casper, M. E., Shogren, K., Giuliani, M., Ruesink, T., et al. (2011). The effects of fixed electrical charge on chondrocyte behavior. *Acta Biomaterialia*.

Dalby, M. J. (2005). Topographically induced direct cell mechanotransduction. *Medical Engineering and Physics, 27*, 730-742.

Darling, A. L., & Sun, W. (2004). 3D microtomographic characterization of precision extruded poly-epsilon-caprolactone scaffolds. *Journal of Biomedical Materials Research. Part B, Applied Biomaterials. 70B, 2*, 311-317.

Davis, K. A., Burke, K. A., Mather, P. T., & Henderson, J. H. (2011). Dynamic cell behavior on shape memory polymer substrates. *Biomaterials, 32*, 2285-2293.

deGroot, J. H., de Vrijer, R., Pennings, A. J., Klompmaker, J., Veth, R. P. H., & Jansen, H. W. B. (1996). Use of porous polyurethanes for meniscal reconstruction and meniscal prostheses. *Biomaterials, 17*, 163-173.

Desai, S. J., Singh, P., Simonelli, A. P., & Higuchi, W. I. (1966). Investigation of factors influencing release of solid drug dispersed in inert matrices II. *Journal of Pharmaceutical Sciences, 55*, 1224-1229.

Donoso, M. G., Me´ndez-Vilas, A., Bruque, J. M., & Gonza´lez-Martin, M. L. (2007). On the relationship between common amplitude surface roughness parameters and surface area: Implications for the study of cell-material interactions. *International Biodeterioration and Biodegradation, 59*, 245-251.

Engler, A. J., Sen, S., Sweeney, H. L., & Discher, D. E. (2006). Matrix elasticity directs stem cell lineage specification. *Cell 126*, 677-689.

Feldkamp, L. A., Goldstein, S. A., Parfitt, A. M., Jesion, G., & Kleerekope, M. (1989). The direct examination of three-dimensional bone architecture in vitro by computed tomography. *Journal of Bone & Mineral Research, 4*, 3-11.

Flynn, L., Semple, J. L., & Woodhouse, K. A. (2006). Decellularized placental matrices for adipose tissue engineering. *Journal of Biomedical Materials Research A, 79*, 359-369.

Foster, T. P., & Parrott, E. L. (1990). Release of highly water-soluble medicinal compounds from inert, heterogeneous matrixes. *Journal of Pharmaceutical Sciences, 79*, 806-810.

Freed, L. E., Martin, I., & Vunjak-Novakovic, G. (1999). Frontiers in tissue engineering. In vitro modulation of chondrogenesis. *Clinical Orthopaedics & Related Research, 367*(58).

Freed, L. E., & Vunjak-Novakovic, G. (1998). Culture of organized cell communities. *Advanced Drug Delivery Reviews, 33*, 15-30.

Geiger, B., Volk, T., Volberg, T., & Bendori, R. (1987). Molecular ineractions in adherens-type contacts. *Journal of Cell Science.Supplement, 8*, 251-272.

Goddard, J. M., & Hotchkiss, J. H. (2007). Polymer surface modification for the attachment of bioactive compounds. *Progress in Polymer Science, 32*, 698-725.

Griffith, L. G. (2002). Emerging design principles in biomaterials and scaffold for tissue engineering. *Annals of the New York Academy of Sciences, 961*, 83-95.

Hatano, K., Inoue, H., Kojo, T., Tsujisawa, T., Uchiyama, C., & Uchida, Y. (1999). Effect of Surface Roughness on Proliferation and Alkaline Phosphatase Expression of Rat Calvarial Cells Cultured on Polystyrene. *Bone, 25*, 439-445.

Hattori, S., Andrade, J. D., Hibbs, J. B., Gregonis, D. E., & King, R. N. (1985). Fibroblasts cell proliferation of charged hydroxyethyl methacrylate copolymers. *Journal of Colloid and Interface Science, 104*, 73-78.

Hinz, B., Mastrangelo, D., Iselin, C. E., Chaponnier, C., & Gabbiani, G. (2001). Mechanical tension controls granulation tissue contractile activity and myofibroblast differentiation *American Journal of Pathology, 159*, 1009-1020.

Hou, Q., Grijpma, D. W., & Feijen, J. (2000). Porous polymeric structures for tissue engineering prepared by a coagulation, compression moulding and salt leaching technique. *Biomaterials, 24*(11), 1937-1947.

Hutmacher, D. W. (2001). Scaffold design and fabrication technologies for engineering tissues--state of the art and future perspectives. *Journal of biomaterials science, 12*, 107-124.

Ishikawa, J., Tsuji, H., Sato, H., & Gotoh, Y. (2007). Ion implantation of negative ions for cell growth manipulation and nervous system repair. *Surface and coatings technology, 201*, 8083-8090.

Jiang, F. X., Yurke, B., Firesein, B. L., & Langrana, N. A. (2008). Neurite ougrowth on a DNA crosslinked hyrogel with tunable stiffnesses. *Annals of Biomeical Engineering, 36*, 1565-1579.

Jiang, F. X., Yurke, B., Schloss, R. S., Firesein, B. L., & Langrana, N. A. (2010). The relationship between fibroblast growth and the dynamic stiffness of a DNA corsslinked hydrogel. *Biomaterials, 31*, 1199-1212.

Jones, J. R., Poologasundarampillai, G., Atwood, R. C., Bernard, D., & Lee, P. D. (2007). Non-destructive quantitative 3D analysis for the optimisation of tissue scaffolds. *Biomaterials, 28*, 1404-1413.

Jung, H., Kwak, B., Yang, H. S., Tae, G., Kim, J.-S., & Shin, K. (2008). Attachment of cells to poly(styrene-co-acrylic acid) thin films with various charge densities. *Colloids and Surfaces A: Physiochemical and Engineering Aspect, 313*, 562-566.

Karageorgiou, V., & Kaplan, D. (2005). Porosity of 3D biomaterial scaffolds and oseogenesis. *Biomaterials, 26*, 5474-5491.

Kellomake, M., Niiranen, H., Puumanen, K., Ashammakhi, N., Waris, T., & tormala, P. (2000). Bioabsorbable scaffolds for guided bone regeneration and generation. *Biomaterials, 21*, 2495-2505.

Keselowsky, B. G., Collard, D. M., & García, A. J. (2005). Integrin binding specificity regulates biomaterial surface chemistry effects on cell differentiation. *Proc Natl Acad Sci U S A, 102*, 5953-5957.

Keselowsky, B. G., Collard, D. M., & J., G. A. (2003). Surface chemistry modulates fibronectin conformation and directs integrin binding and specificity to control cell adhesion. *Jourmal of Biomedical Material Research A, 66*, 247-259.

Khatiwala, C. B., Peyton, S. R., Metzke, M., & Putnam, D. (2007). The regulation of osteogenesis by ECM rigidity in MC3T3-E1 cells requires MAPK activation. *Journal of Cell Phsiology, 211*, 661-672.

Khorasani, M. T., Mirzadeh, H., & Irani, S. (2008). Plasma surface modification of poly(l-lactic acid) and poly (lactic-co-glycolic acid) films for improvement of nerve cells adhesion. *Radiation Physics and Chemistry, 77*, 280-287.

Kim, M. H., Kino-Oka, M., Kawase, M., Yagi, K., & Taya, M. (2007). Response of human epithelial cells to culture surfaces with varied roughnesses prepared by immobilizing dendrimers with/without D-glucose display. *Journal of Bioscience and Bioengineering, 103*, 192-199.

Kim, S., E., E. A., & Kihm, K. D. (2009). Surface elasticity and charge concentration-dependent endothelial cell attachment to copolymer polyelectrolyte hydrogel. *Acta Biomaterialia, 5*, 144-151.

Lahann, J., & Langer, R. (2005). Smart materials with dynamically controllable surfaces. *MRS Bull, 30*, 185-188.

Langer, R., & Vacanti, J. P. (1993). Tissue engineering. *Science, 260*, 920-927.

Leber, C., Choi, H., Bose, S., & Bandyopadhyay, A. (2010). Micromachined Si channel width and tortuosity on human osteoblast cell attachment and proliferation. *Material science and engineering, 30*, 71-77.

Lee, J. H., Jung, H. W., Kang, I. K., & Lee, H. B. (1994). Cell behaviour on polymer surfaces with different functional groups. *Biomaterials, 15,* 705-711.

Lee, S. J., Choi, J. S., Park, K. S., Khang, G., Lee, Y. M., & Lee, H. B. (2004). Response of MG63 osteoblast-like cells onto polycarbonate membrane surfaces with different micropore sizes. *Biomaterials, 25,* 4699-4707.

Lee, S. J., Khang, G., Lee, Y. M., & Lee, H. B. (2003). The effect of surface wettability on induction and growth of neurites from the PC-12 cell on a polymer surface. *Journal of Collid and Interface Science, 259,* 228-235.

Levenberg.S., & Langer, R. (2004). Advances in tissue engineering. *Current Topics in Developmental Biology, 61,* 113-134.

Lim, J. Y., Hansen, J. C., Siedlecki, C. A., Runt, J., & Donahue, H. J. (2005). Human foetal osteoblastic cell response to polymer-demixed nanotopographic interfaces. *Journal of Royal Society Interface,* 2, 97-108.

Lin, A. S., Barrows, T. H., Cartmell, S. H., & Guldberg, R. E. (2003). Microarchitectural and mechanical characterization of oriented porous polymer scaffolds. *Biomaterials, 24*(3), 481-489.

Lu, Q., Ganesan, K., Simionescu, D. T., & Vyavahare, N. R. (2004). Novel porous aortic elastin and collagen scaffolds for tissue engineering. *Biomaterials, 25*(22), 5227-5237.

Ma, P. X. (2004). Scaffolds for tissue fabrication. *Material today, 7*(5), 30-40.

Makohliso, S. A., Valentini, R. F., & Aebischer, P. (1993). Magnitude and polarity of a fluoroethylene propylene electret substrate charge influences neurite outgrowth in vitro. *Journal of Biomedical Materials Research, 27,* 1075-1085.

Marquez, J. P., Genin, G. M., Pryse, K. M., & Elson, E. L. (2006). Celular and matrix contributions to tissue construct stiffness increase with cellular concentration. *Annals of Biomeical Engineering, 34,* 1475-1482.

Mikos, A. G., Herring, S. W., Ochareon, P., Elissenff, J., Lu, H. H., & Mooney, D. J. (2006). Engineering complex tissues. *Tissue Engineering, 12,* 3307-3340.

Mrksich, M. (2005). Dynamic substrates for cell biology. *MRS Bull, 30,* 180-184.

Noh, H., & Vogler, E. A. (2006). Volumetric interpretation of protein adsorption: mass and energy balance for albumin adsorption to particulate adsorbents with incrementally increasing hydrophilicity. *Biomaterials, 27,* 5801-5812.

O'Brien, F. J., Harley, B. A., Yannas, I. V., & Gibson, L. J. (2005a). The effect of pore size on cell-adhesion in collagen-GAG scaffolds. *Biomaterials, 26,* 433-441.

O'Brien, F. J., Harley, B. A., Yannas, I. V., & Gibson, L. J. (2005b). The effect of pore size on cell adhesion in collagen-GAG scaffolds. *Biomaterials, 26,* 433-441.

O'Connell, B., McGloughlin, T., & Walsh, M. T. (2010). Factors that affect mass transport from drug eluting stents into the artery wall. *BioMedical Engineering, 9,* 15.

Papadokostaki, K. G., Amarantos, S. G., & Petropoulos, J. H. (1998). Kinetics of release of particulate solutes incorporated in cellulosic polymer matrices as a function of solute solubility and polymer swellability. *Journal of applied polymer science, 67,* 227-287.

Pelham, R. J., & Wang, Y. L. (1997). Cell locomotion and focal adhesions are regulated by substrate flexibility. *Proc Natl Acad Sci U S A, 94,* 136661 -136665.

Peyton, S. R., & Putnam, D. (2005). Extracellular matrix rigidity governs smoth muscle cell motility in a biphasic fashion. *Journal of Cell Phsiology, 204*, 918-923.

Ranucci, C. S., Kumar, A., Batra, S. P., & Moghe, P. V. (2000). Control of hepatocyte function on collagen foames: sizing matrix pores toward selective induction of 2-D and 3-D cellular morphogenesis. *Biomaterials, 21*, 783-793.

Rose, F. R., Cyster, L. A., Grant, P., Scotchford, C. A., Howdle, S. M., & Shakesheff, K. M. (2004). In vitro assessment of cell penetration into porous hydroxyapatite scaffolds with a central aligned channel. *Biomaterials, 25*, 5507-5514.

Salem, A. K., Stevens, R., Pearson, R. G., Davies, M. C., Tendler, S. J., & Roberts, C. J. (2002). Interactions of 3T3 fibroblasts and endothelial cells with defined pore features. *Journal of Biomedical Materials Research, 61*(2), 212-217.

Saltzman, W. M. (2002). Delivery of Molecular Agents in Tissue Engineering. In *Tissue Engineering* (Vol. 17, pp. 23-30).

Schmidt, D. R., Waldeck, H., & Kao, W. J. (2000). Protein adsorption to Biomaterials.

Schneider, G. B., English, A., Abraham, M., Zaharias, R., Stanford, C., & Keller, J. (2004). The effect of hydrogel charge density on cell attachment. *Biomaterials, 25*, 3023-3028.

Silva, M. M. C. G., Cyster, L. A., Barry, J. J. A., Yang, X. B., & Rose, F. R. A. J. (2006). The effect of anisotropic architecture on cell and tissue infiltration into tissue engineering scaffolds. *Biomaterials, 27*, 5909-5917.

Tamada, Y., & Ikada, Q. (1993). Effect of preadsorbed proteins on cell adhesion to polymer surfaces. *Journal of Colloid and Interface Science, 155*, 334-339.

Thevenot, P., Hu, W. J., & Tang, L. P. (2008). Surface chemistry influences implant biocompatibility. *Current Topics in Medicinal Chemistry, 8*, 270-280.

Thurner, P., Muller, R., Raeber, G., Sennhauser, U., & Hubbell, J. A. (2005). 3D morphology of cell cultures: a quantitative approach using micrometer synchrotron light tomography. *Microscopy Research and Technique, 66*(6), 289-298.

Thurner, P., Muller, R., Sennhauser, U., Hubbell, J. A., & Muller, B. (2004). Tomography studies of biological cells on polymer scaffolds. *Journal of Physical Condensed Matter, 16*, S3499-3510.

Tock, J., Van Putten, V., Stenmark, K. R., & Nemenoff, R. A. (2003). Induction of SM-alpha-actin expression by mechanical strain in adult vascular smooth muscle cells in mediated through activation of JNK and p38 MAP kinase. *Biochemical & Biophysical Research Communications, 301*, 1116-1121.

Toth, J. M., An, H. S., Lim, T., Ran, Y., Weiss, N. G., Lundberg, W. R., et al. (1995). Evaluation of porous biphasic calcim phosphate ceramics for anterior cervial interbody fusion in a caprine model. *Spine, 20*, 2203-2210.

Ulrich, T. A., Jain, A., Tanner, K., MacKay, J. L., & Kumar, S. (2010). Probing cellular mechanobiology in three dimensional culture with collagen-agarose matrices. *Biomaterials, 31*, 1875-1884.

Vogler, E. A. (1999). Water and the acute biological response to surfaces. *Journal of Biomaterials Science Polymer Edition, 10*, 1015-1045.

Walsh, D., Furuzono, T., & Tanaka, J. (2001). Preparation of porous composite implant materials by in situ polymerization of prous apatite containing caprolactone or methyl methacrylate. *Biomaterials, 22*, 1205-1212.

Wang, J., Grood, E., Florer, J., & Wenstrup, R. (2000). Alignment and proliferation of MC3T3-E1 osteoblasts in microgrooved silicone substrata subjected to cyclic stretching. *Journal of Biomechanics, 33*, 729-735.

Wang, J., Jia, F., Yang, G., Campbell, B., Stone, D., & Woo, S. (2003). Cell orientation determines the alignment of cell-produced collagenous matrix. *Journal of Biomechanics, 36*, 97-102.

Wang, Y., Tomlins, P., Coombes, A., & Rides, M. (2010). On the determination of Darcy permeability coefficient for a porous tissue scaffold. *Tissue Engineering Part C Method, 16*(2), 281-289.

Washburn, N. R., Weir, M., Anderson, P., & Potter, K. (2004). Bone formation in polymeric scaffolds evaluated by proton magnetic resonance microscopy and X-ray microtomography. *Journal of Biomedical Materials Research. Part A, 69*(4), 738-747.

Washburn, N. R., Yamada, K. M., Simon, C. G., Jr., K., S. B. , & Amis, E. J. (2004). High-throughput investigation of osteoblast response to polymer crystallinity: influence of nanometer-scale roughness on proliferation. *Biomaterials, 25*, 1215-1224.

Wei, J., Igarashi, T., Okumori, N., Maetani, T., Liu, B. S., & Yoshinari, M. (2009). Influence of surface wettability on competitive protein adsorption and initial attachment of osteoblasts. *Biomedical Materials, 4*, 045002.

Williams, J. M., Adewunmi, A., Schek, R., Flanagan, C. L., Krebsbach, P. H., Feiberg, S. E., et al. (2005). Bone tissue engineering using polycaprolactone scaffolds fabricated via selective laser sintering. *Biomaterials, 26*, 4817-4827.

Wu, Y. S., Lucas, J., Frijlink, H. W., & Maarscalk, K. (2006). The determination of relative path length as a measure for tortuosity in compacts using image analysis. *European Journal of Pharmaceutical Sciences, 28*, 433-440.

Xu, L. C. (2007). Effect of surface wettability and contact time on protein adhesion to biomaterail surfaces. *Biomaterials, 28*, 3273-3283.

Yang, Y., Motte, S., & Kaufman, L. J. (2010). Pore size variable type I collagen gels and their interaction with glioma cells. *Biomaterials, 31*, 5678-5688.

Yang.J, Shi, G., Bei, J., Wang, S., Cao, Y., Shang, Q., et al. (2002). Fabrication and surface modification of macroporous poly(L-lactic acid) and poly(L-lactic co-glycolic acid) (70/30) cell scaffolds for human skin fibroblasts cell culture. *Journal of Biomedical Materials Research, 62*, 438-446.

Yannas, I. V. (2000). In vivo synthesis of tissues and organs. In R. p. Lanza, R. Langer & J. P. Vacanti (Eds.), *Principles of tissue engineering* (pp. 167-178).

Yannas, I. V., Lee, E., Orgill, D. P., SKrabut, E. M., & Murphy, G. F. (1989). Synthesis and characterization of a model extracellular matrix that induces partial regeneration of adult mammalian skin. *Pro Natl cad Sci USA, 86*(3), 933-937.

Yildirim, E. D., Besunder, R., Pappas, D., Allen, F., Guceri, S., & Sun, W. (2010). Accelerated differentiation of osteoblast cells on polycaprolactone scaffolds driven by a combined effect of protein coating and plasma modification. *Biofabrication, 2*, 014019.

Yousaf, M. N. (2009). Model substrates for studies of cell mobility. *Current Opinion in Chemical Biology, 13*, 697-704.

Zein, I., Hutmacher, D. W., Tan, K. C., & Teoh, S. H. (2002). Fused deposition modeling of novel scaffold architectures for tissue engineering applications. *Biomaterials*, 1169-1185.

Zeltinger, J., Sherwood, J. K., Graham, D. A., Mueller, R., & Griffith, L. G. (2001). Effect of pore size and void fraction on cellular adhesion, proliferation, and matrix depostion. *Tissue Engineering*, *7*(5), 557-572.